Motorbooks International

Authentic Restorati

MUSTANG '64½ - '70 RESTORATION GUIDE

Tom Corcoran & Earl Davis

First published in 1992 by Motorbooks International Publishers & Wholesalers, PO Box 2, 729 Prospect Avenue, Osceola, WI 54020 USA

© Tom Corcoran and Earl Davis, 1992

All rights reserved. With the exception of quoting brief passages for the purposes of review no part of this publication may be reproduced without prior written permission from the Publisher

Motorbooks International is a certified trademark, registered with the United States Patent Office

The information in this book is true and complete to the best of our knowledge. All recommendations are made without any guarantee on the part of the author or Publisher, who also disclaim any liability incurred in connection with the use of this data or specific details

We recognize that some words, model names and designations, for example, mentioned herein are the property of the trademark holder. We use them for identification purposes only. This is not an official publication

Motorbooks International books are also available at discounts in bulk quantity for industrial or sales-promotional use. For details write to Special Sales Manager at the Publisher's address

Library of Congress Cataloging-in-Publication Data
Corcoran, Tom.
Mustang '64½-'70 restoration guide / Tom Corcoran, Earl Davis.
p. cm. — (Motorbooks International authentic restoration guides)
Includes index.
ISBN 0-87938-598-7
1. Mustang automobile—Conservation and restoration—Handbooks, manuals, etc. I. Davis, Earl. 1951- . II. Title. III. Series.
TL215.M8C67 1992
629.222'2—dc20 92-38793
CIP

On the front cover: The 1968½ Mustang 428 Cobra Jet owned by Chris and Deborah Teeling of Enfield, Connecticut. The Ram Air with special striping, foglamps, C-stripes, styled steel wheels with F70 tires, and GT identification were all stock features. *Mike Mueller*

Printed and bound in the United States of America

Contents

	Acknowledgments	4
	Introduction	5
Chapter 1	**Data Plates and Decals**	6
Chapter 2	**Exterior**	24
Chapter 3	**Interior**	88
Chapter 4	**Engine**	129
Chapter 5	**Drivetrain**	212
Chapter 6	**Suspension, Brakes, Steering, Wheels, and Exhaust**	248
Chapter 7	**Wiring and Electrical Systems**	307
Chapter 8	**Options**	340
	Index	352

Acknowledgments

Any book of this size must be the work of many people. We have been fans of the 1964½-1970 Mustangs most of our lives, yet, through the help of others, found plenty to learn while assembling this volume. Donald Farr of Dobbs Publications, a previous editor of both *Mustang Monthly* and *Super Ford,* and currently editorial director of Dobbs Publications Group, gave us our first opportunities to get paid for learning our hobby. Larry Dobbs, who founded Dobbs Publications Group, has provided a work environment that encouraged our investigations into the details and fun of restoring Mustangs.

The Product Analysis and Publications Department, Parts and Services Division, Ford Motor Company, granted us permission to extract and reformat valuable vintage information from the *1965/72 Ford Car Master Parts and Accessories Catalog.* For this we are grateful. Much of the information in the huge catalog is difficult for the average hobbyist to decipher, but the 20in high volume is available from Ford Motor Company. Send inquiries regarding the catalog (#FPS 7635-A&B) to B.U.D. Co., PO Box 27146, Detroit, MI 48227. Its cost in late 1992 was $140.50 but, in case of changes, it would be wise to check with B.U.D. Co., before sending money.

Several people offered direct support beyond the call of duty: Barbara Fahrenholz and Debbie Miller (art director of *Mustang Monthly*) patiently and meticulously entered chart data on their computers; Rob Reaser and Tom Wilson, our associates at DPG, offered understanding and assistance; Jim Osborn of Jim Osborn Reproductions in Lawrenceville, Georgia (404-962-7556), provided the top-notch decals used in the first chapter; Harvey Hester, Rob Reaser, and Franette Ansel contributed expert darkroom work; Michael Dregni of Motorbooks International summoned patience and diplomacy undeserved by such rookie authors; Dinah George assisted greatly with computer and laser-printing knowledge; and our family members, Carissa Davis, Diane Davis, and Sebastian Corcoran, endured many months of strained schedules and late nights.

Many car owners, club members, automotive journalists, restorers, and parts vendors have helped to shape our ideas regarding the Mustang hobby. We have managed to learn from practically every hobbyist we've met over the past fifteen years. The world of Mustangs has become a significant hobby and a fertile atmosphere for the founding of hundreds of small related businesses. We owe thanks to everyone we've met over the years, especially Bob Aliberto, Pete Geisler, Melvin Little, Paul Newitt (author of the fine *GT/California Special Recognition Guide*), Don Walsh and Hank Dertian of Ford Motorsport/SVO, Al Sedita of Classic Auto Air, Marcia Klossner, Bob Perkins, Steve Statham, Tom Daniel, Gregg Cly, Paul Fix, Rick Byrnes, Frank Reynolds, Bruce Weiss, Lyle Johnston, and the many owners of the Mustangs photographed for this book. A special thanks must be offered the officials and members of the Mustang Club of America for setting standards of restoration and judging that improve the automobiles and encourage the hobby participation of thousands of individuals around the world.

Tom Corcoran
Earl Davis

Introduction

The *Mustang 1964½-1970 Restoration Guide* is intended to be a foundation for the disassembly, rebuilding, restoration, reassembly, research, authentication, and detailing of a top-notch collectible car. Throughout this volume can be found information—both major and trivial—to assist any hobbyist in locating, identifying, matching, and, if needed, swapping or substituting original Ford parts in the course of a restoration.

Authentication is always a challenge in the process of analyzing a twenty-five-year-old car, and the 175 charts in this book will serve as a helpful guideline for both beginner and veteran. In some cases the components listed are described by both number codes and dimensions, plus further descriptions and applications where needed.

The reader will notice that many parts, especially exterior trim and components, will be identified by both a photograph and an exploded line drawing, while most suspension, drivetrain, and "out-of-sight" parts will be identified by part number and line drawing only. This is because certain parts are difficult to photograph in a manner that communicates needed information; it is also the way the world of Mustangs works. If you are restoring a 1966 GT and need a gas cap, you should not need the gas cap part number in order to order one. Experienced Mustang parts vendors know what a 1966 GT gas cap looks like and, by referencing the photo in this book, you will too. If, however, you need to identify a part you already own, or need a particular attaching bracket for a power-steering hose, both you and the vendor can eliminate confusion by your having a part number handy. This book will provide such information.

Let us mention at the outset, that parts of this book are based upon the May 1975 (Final Issue) Edition of the 1965/72 Ford Car Master Parts and Accessories Catalog. During the years up to 1975, many components were deleted and consolidated with similar and adaptable parts in order to streamline Ford's procurement, storage, and distribution process. Some of the parts numbers listed may not represent the factory original pieces for vintage Mustangs and Shelby Mustangs. Nevertheless, the ability to identify any part included in the construction of a Mustang can simplify identification and communication in regard to parts purchases, swaps, trades, and shopping.

Please understand that the numbers shown in most of the exploded line drawings indicate unit base numbers only, and cannot be presumed to be complete part numbers. For example 9002 is the base number for all Ford gas tanks. The prefix and suffix of the actual part number would identify the correct model, year, body type, engine type, application, and so on. Additionally, while Ford still manages to stock parts that are listed for vintage Mustangs, there is no guarantee that any or all of the parts listed throughout this book will be available through the Ford dealer network.

As a final note, we must point out that the three assembly plants for Mustangs and Shelbys—San Jose, Dearborn, and Metuchen—were often inconsistent in their procedures and parts suppliers. Also, there was no guarantee that the "correct" part—any number of brackets, for instance—was placed on a car during its original assembly. For instance, if the supply of Mustang-correct air cleaners went dry, the plant did not shut down. Instead, compatible air cleaners from Fairlanes could have been installed over a period of days until the proper pieces were back in stock. Throughout this book, primarily in the exploded component views, you will see the word "typical." These drawings generally cover several models and styles, and should not be used as absolute guidelines to authenticity.

The hobby of restoration is, by and large, an enjoyable pastime. The main point is to have fun, and we trust that this book will make it easier for you to fulfill your desire to own and drive a vintage Mustang. So, have fun, and don't quit. There is a light at the end of the tunnel.

Chapter 1

Data Plates and Decals

VIN/Data Plate Decoding System

The coding system for Mustang identification involves four distinct areas, each using alphanumeric series and direct codes.

The VIN format remained consistent through early Mustang production. The data plate took different forms over the years, with coded items such as paint and interior colors changing from year to year (as well as the style of the plates and labels). The body buck tag, used on some but not all Mustangs, combined data and intra-plant guidelines; and the broadcast sheets, single pieces of paper, also provided assembly-line instructions for many Mustangs.

We have yet to completely decipher all body buck tag and broadcast sheet codes, but the VIN and data plate system shown here is straightforward.

The charts for applicable exterior body colors is included in Chapter 2 on Exteriors; the chart for interior trim code is included in Chapter 3 on Interiors.

Note that 1964-1969 warranty plates (or data plates), located on the lock face of the driver's door, could have been shaded with a black or a gray background. In 1964 and 1965, a black plate indicated a vehicle painted with M30J Non-Acrylic Enamel; a gray plate indicated a vehicle painted with M32J Acrylic Enamel. From 1966 to 1969, a black warranty plate identified a vehicle painted with M32J Acrylic Enamel, and a gray plate indicated a vehicle painted with M30J Non-Acrylic Enamel.

Vehicle Identification Numbers

By Federal law, every vehicle produced or sold in the United States must have a Vehicle Identification Number, VIN. The Mustang's coded number indicated the model year, production plant, body style, engine code, and consecutive assembly-line unit number. Each of the three plants that produced Mustangs had its own series of consecutive unit numbers; a 1966 Mustang from the Metuchen plant could have a VIN ending in 112233, and a completely different Mustang made in Dearborn could have those same final six digits. The makeup of the VIN remained constant through the first generation of Mustang manufacture.

The first position on the left of the VIN is the final digit of the model year; the number 7 would indicate a 1967 Mustang. (Note that the number 5 could indicate either the 1964½ or the 1965 model.)

The second position from the left indicated the assembly plant code. Dearborn cars received an F; Metuchen cars a T and San Jose cars an R.

The next two numbers, the third and fourth positions from the left, provided the body serial code. For 1965 and 1966 models, 07 meant hardtop, 08 meant convertible, and 09 indicated a fastback. For 1967 through 1970, the numbers were 01 for the hardtop, 02 for the fastback, and 03 for a convertible. Additionally, in 1970 (through 1973), 04 indicated a Grandé and 05 meant a Mach 1.

The next letter, in the fifth position from the left, identified the engine in the car (see charts in this chapter and the Engines chapter).

The final six digits were the consecutive assembly line unit number.

To clarify this decoding system, analyze VIN 6T08A211150. It is a 1966 convertible built in Metuchen, with an A-code, or 289-4V, engine, and consecutive unit number 211150. A Mustang with VIN 0F05M157894 would be a 1970 Mach built in Dearborn with a 351-4V engine.

Engine Compartment Decals

One of the biggest problem areas for the final detailing of any Mustang or Shelby restoration is the correct selection and placement of engine compartment decals. The easiest and most effective way to determine the individual needs of any car is to view similar cars at a concours exhibition such as a Mustang Club of America or Shelby American Automobile Club national event.

Experienced restorers are another good source for information, and the decals shown on the following pages represent a major portion of the Mustang needs, but there is always a chance that your car requires additional markings for correctness. Our main source for these decals and stickers was Jim Osborn Reproductions, Inc., 101 Ridgecrest Drive, Lawrenceville, GA 30245, (404) 962-7556.

We recommend diligent research and careful confirmation before buying and applying any decal or application label. Bear in mind that the three Mustang assembly plants did things in different ways, even from year to year in the same plant, and consistency from car to car, from model to model, and from week to week was not a Ford Motor Company hallmark.

Mustang Warranty Plate Date Codes

From 1965 to 1969 an alphanumeric code above or below the word "Date" identified the day and month a vehicle was built. A code of 17K meant the 17th day of October. Beginning in 1970, the date code was shown in "10/70" format at the upper left corner of the warranty label.

Code	Month, First Production Year
A	January
B	February
C	March
D	April
E	May
F	June
G	July
H	August
J	September
K	October
L	November
M	December

Code	Month, Second Production Year
N	January
P	February
Q	March
R	April
S	May
T	June
U	July
V	August
W	September
X	October
Y	November
Z	December

Note: Confusion often clouds the date codes of the 1964½ and 1965 Mustangs. At the outset of Mustang production, in March 1964, the codes began with C, D, E, and so on. In January and February 1965, the letters A and B were used, but were followed by Q for March and the remaining second-year codes through August (V).

1965-1970 Mustang Body Style Codes

Code	Years	Body Style	Interior Designation
63A	1965-1968	Fastback	Standard Interior, Bucket Seats
	1969-1970	SportsRoof	Standard Interior, Bucket Seats
63B	1965-1968	Fastback	Luxury Interior, Bucket Seats
	1969-1970	SportsRoof	Luxury Interior, Bucket Seats
63C	1968	Fastback	Standard Interior, Bench Seat
	1969-1970	Mach 1	Mach 1 Interior, Bucket Seats
63D	1968	Fastback	Luxury Interior, Bench Seat
65A	1965-1970	Hardtop	Standard Interior, Bucket Seats
65B	1965-1970	Hardtop	Luxury Interior, Bucket Seats
65C	1965-1969	Hardtop	Standard Interior, Bench Seat
65D	1968-1969	Hardtop	Luxury Interior, Bench Seat
65E	1969-1970	Grandé	Grandé Interior, Bucket Seats
76A	1965-1970	Convertible	Standard Interior, Bucket Seats
76B	1965-1970	Convertible	Luxury Interior, Bucket Seats
76C	1965-1967	Convertible	Standard Interior, Bench Seat

1964-1970 Ford Motor Company DSO Codes

These two-digit numbers indicated District Sales Offices. Certain DSOs began and ceased operation in the middle of model years. Often a Mustang will have a DSO code that was not officially in use until the next year.

Code	Years	Location
11	1965-1970	Boston
12	1965-1966	Buffalo
13	1965-1970	New York
14	1965-1966	Pittsburgh
15	1965-1970	Newark
16	1967-1970	Philadelphia
17	1967-1970	Washington DC
21	1965-1970	Atlanta
22	1965-1970	Charlotte
23	1965-1966	Philadelphia
24	1965-1970	Jacksonville

Code	Years	Location
25	1965-1970	Richmond
26	1965-1966	Washington DC
27	1967-1968	Cincinnati
28	1967-1970	Louisville
31	1965-1966	Cincinnati
32	1965-1970	Cleveland
33	1965-1970	Detroit
34	1965-1968	Indianapolis
35	1965-1970	Lansing
36	1965-1966	Louisville
37	1967-1970	Buffalo
38	1967-1970	Pittsburgh
41	1965-1970	Chicago
42	1965-1968	Fargo
43	1965-1966	Rockford
43	1967-1970	Milwaukee

Code	Years	Location
44	1965-1970	Twin Cities
45	1965-1968	Davenport
46	1969-1970	Indianapolis
47	1969-1970	Cincinnati
51	1965-1970	Denver
52	1965-1968	Des Moines
53	1965-1970	Kansas City
54	1965-1970	Omaha
55	1965-1970	St. Louis
56	1969-1970	Davenport
61	1965-1970	Dallas
62	1965-1970	Houston
63	1965-1970	Memphis
64	1965-1970	New Orleans
65	1965-1970	Oklahoma City
71	1965-1970	Los Angeles
72	1965-1970	San Jose
73	1965-1970	Salt Lake City
74	1965-1970	Seattle
75	1967-1970	Phoenix
81	1965-1968	Ford of Canada
83	1965-1970	Government
84	1965-1970	Home Office Reserve
85	1965-1970	American Red Cross
89	1965-1970	Transportation Services
90-99	1965-1970	Export

1964-1970 Mustang Axle Codes

Code	Years	Axle Ratio	Remarks
0	1967	2.79:1	
1	1965-1967	3.00:1	
	1968	2.75:1	
	1969	2.50:1	
2	1965-1967	2.83:1	
	1968	2.79:1	
	1969-1970	2.75:1	
3	1965-1967	3.20:1	
	1969-1970	2.79:1	
4	1967	3.25:1	
	1968	2.83:1	
	1969-1970	2.80:1	
5	1965-1967	3.50:1	
	1968	3.00:1	
	1969-1970	2.83:1	
6	1965-1967	2.80:1	
	1968	3.20:1	
	1969-1970	3.00:1	
7	1968	3.25:1	
	1969-1970	3.10:1	
8	1965-1966	3.89:1	
	1968	3.50:1	
	1969-1970	3.20:1	
9	1965-1966	4.11:1	
	1968	3.10:1	
	1969-1970	3.25:1	
A	1965-1967	3.00:1	Limited Slip
	1969-1970	3.50:1	
B	1965-1966	2.83:1	Limited Slip
	1969-1970	3.07:1	
C	1965-1967	3.20:1	Limited Slip
	1969-1970	3.08:1	
D	1967	3.25:1	Limited Slip
	1969	3.91:1	
E	1965-1967	3.50:1	Limited Slip
	1968	3.00:1	Limited Slip
	1969	4.30:1	
F	1965-1966	2.80:1	Limited Slip
	1968	3.20:1	Limited Slip
	1970	2.33:1	
G	1968	3.25:1	Limited Slip
H	1965	3.89:1	Limited Slip
	1968	3.50:1	Limited Slip
I	1965	4.11:1	Limited Slip
J	1969	2.50:1	Limited Slip
K	1969-1970	2.75:1	Limited Slip
L	1969	2.79:1	Limited Slip
M	1969-1970	2.80:1	Limited Slip
N	1969	2.83:1	Limited Slip
O	1969-1970	3.00:1	Limited Slip
P	1969	3.10:1	Limited Slip
Q	1969	3.20:1	Limited Slip
R	1969-1970	3.25:1	Limited Slip
S	1969-1970	3.50:1	Limited Slip
T	1969	3.07:1	Limited Slip
U	1969	3.08:1	Limited Slip
V	1969-1970	3.91:1	Limited Slip
W	1969-1970	4.30:1	Limited Slip
X	1970	2.33:1	Limited Slip

1965-1970 Mustang Data Plate Transmission Codes

Code	Years	Transmission Type	Remarks
1	1965-1967	Three-speed Manual	2.77:1
1	1968-1970	Three-speed Manual	
3	1966-1967	Three-speed Manual	3.03:1
5	1965-1967	Four-speed Manual	
5	1968-1970	Four-speed Manual	2.78:1

Code	Years	Transmission Type	Remarks
6	1965-1966	C-4 Automatic	
6	1967-1970	Four-speed Manual	2.32:1 Close Ratio
U	1967-1970	C-6 Automatic	
W	1967-1970	C-4 Automatic	
X	1969-1970	FMX Automatic	

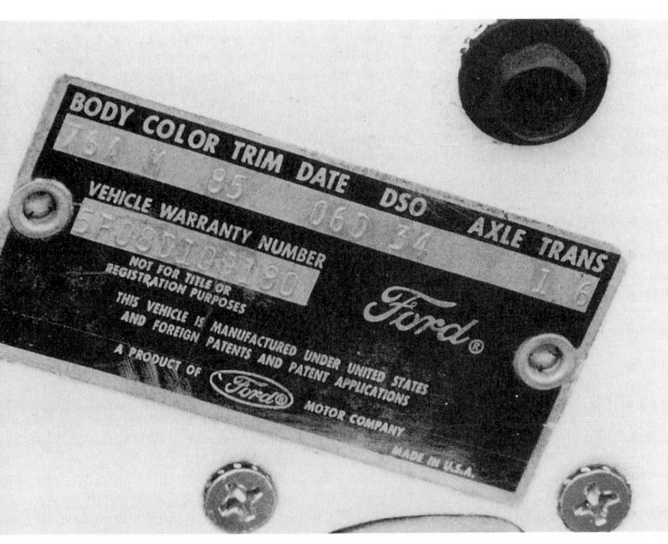

Typical 1964-1965 warranty plate. This data plate from a 1964½ Mustang tells us that the car is a standard interior convertible (76A), with Wimbledon White exterior paint (M); a red vinyl interior (85); a scheduled build date of April 6, 1964; destined for Indianapolis (DSO 34); with a 3.00:1 open rear end, C-4 automatic transmission (6), and a D-code 289ci four-barrel engine (the D in the 5th position of the eleven-character VIN).

This 1966 convertible's warranty plate was installed above the latch mechanism. This seeming inconsistency was not unusual on Ford production lines.

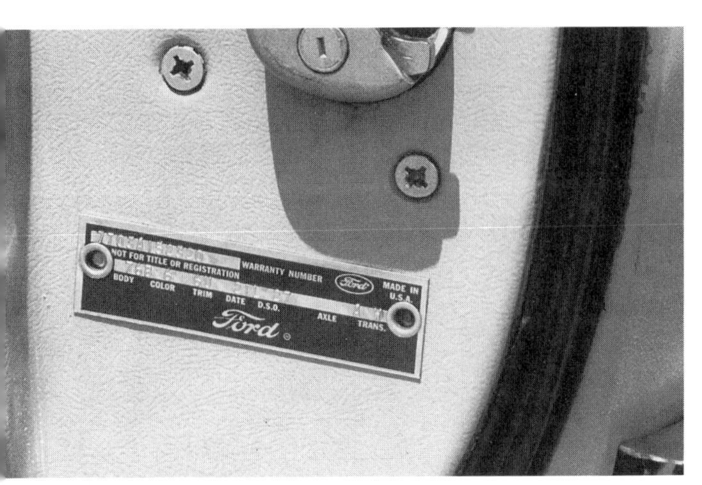

Typical 1966-1969 warranty plate. This 1967 convertible's driver's door data plate has engine code A in the fifth position of its eleven-character VIN. That indicates the 289ci four-barrel engine, and the car might be called an A-code car. Many Mustangs have had their driver's side door replaced, and are missing this plate. Reproductions can be obtained but for many such data is lost.

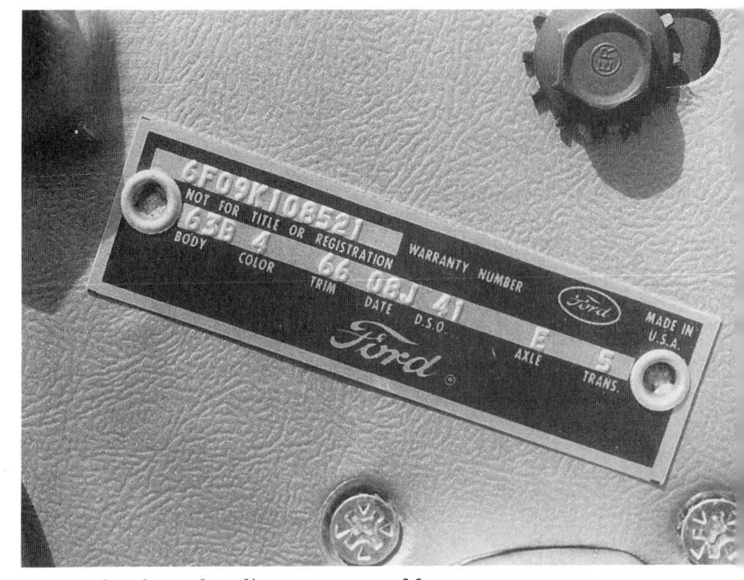

Using the data decoding system, a Mustang owner can determine that this K-code, Deluxe interior, Silver Frost 1966 fastback has a black vinyl interior, 3.50:1 limited slip rear end, and a four-speed manual transmission. Its scheduled build date was September 8, 1965, and its destination was Chicago.

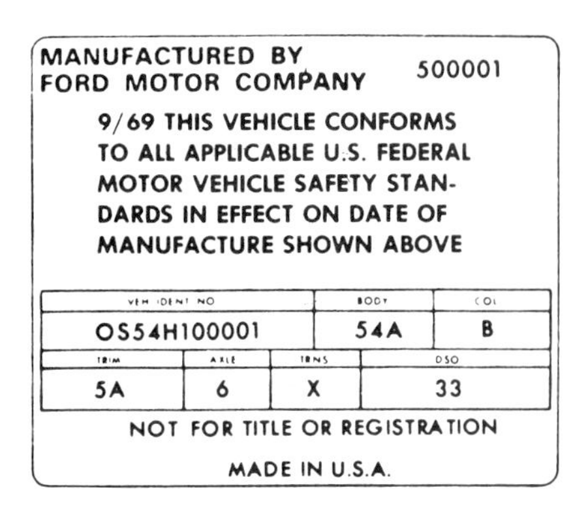
MANUFACTURED BY
FORD MOTOR COMPANY 500001

9/69 THIS VEHICLE CONFORMS TO ALL APPLICABLE U.S. FEDERAL MOTOR VEHICLE SAFETY STANDARDS IN EFFECT ON DATE OF MANUFACTURE SHOWN ABOVE

OS54H100001	54A	B

5A	6	X	33

NOT FOR TITLE OR REGISTRATION

MADE IN U.S.A.

In 1970 Ford replaced the metal warranty plate with a Safety Standards Certification Label that included the VIN and applicable vehicle data codes. This example is typical of the labels in use in 1970 and later.

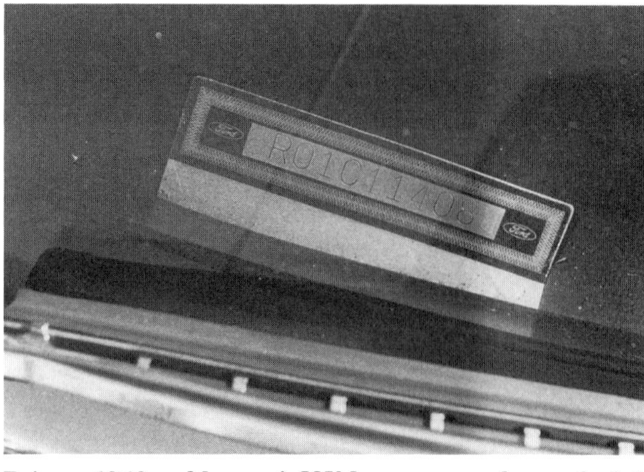

Prior to 1968, a Mustang's VIN was stamped atop the left front inner fender. Only in 1968, as shown here, did the official VIN appear on an aluminum plate riveted to the passenger-side upper dash adjacent to the windshield glass.

Mustang VIN Body Serial Codes, 1965-1970
These numerals are the third and fourth VIN positions.

Code	Years	Body Style
01	'67-'70	2-Door Hardtop
02	'67-'70	2-Door Fastback (SportsRoof, '69, '70)
03	'67-'70	2-Door Convertible
04	1970	2-Door Hardtop, Grande
05	1970	2-Door SportsRoof, Mach 1
06	n/a	Not used.
07	'65-'66	2-Door Hardtop
08	'65-'66	2-Door Convertible
09	'65-'66	2-Door Fastback

Beginning in 1969 the official stamped aluminum VIN plate was riveted inside the windshield in the driver's side of the Mustang.

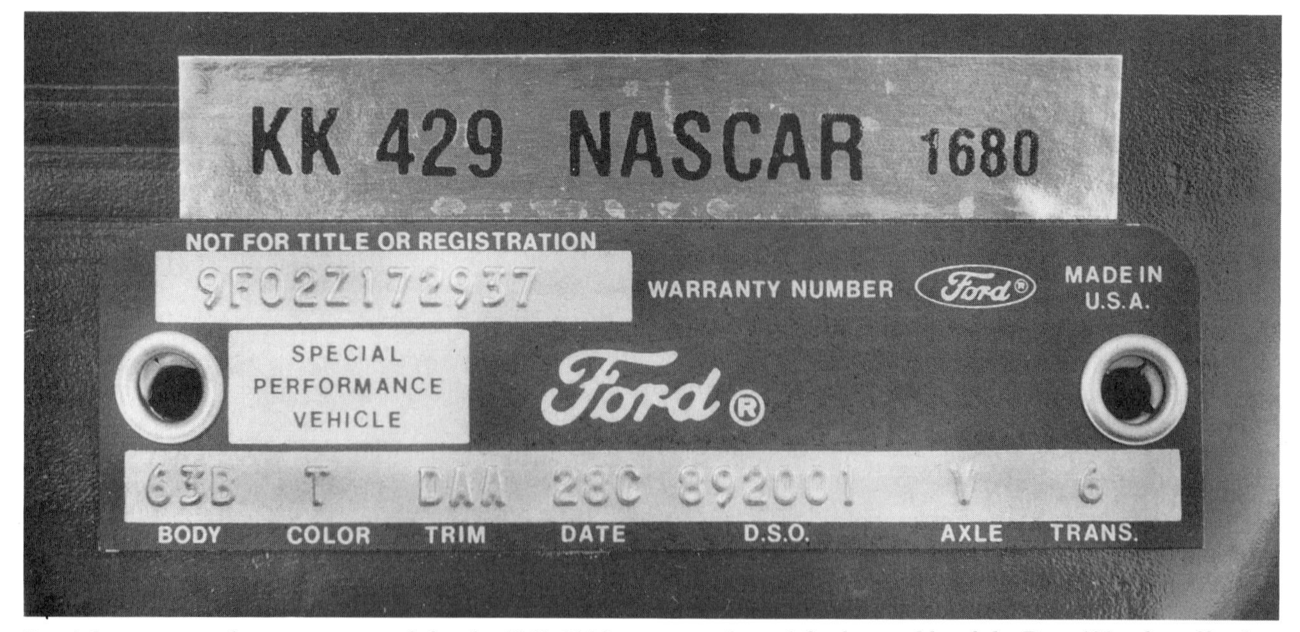

Special warranty plates were created for the 1969-1970 Boss 429 and Shelby Mustangs. Ford noted that these cars were Special Performance Vehicles, and gave the Shelbys identifying DSO codes. Kar Kraft, the company that performed final assembly of the Boss 429, also affixed an identifying plate above the Special Performance Vehicle data plate.

This VIN plate, on a 1970 Boss 302, is located on the driver's side.

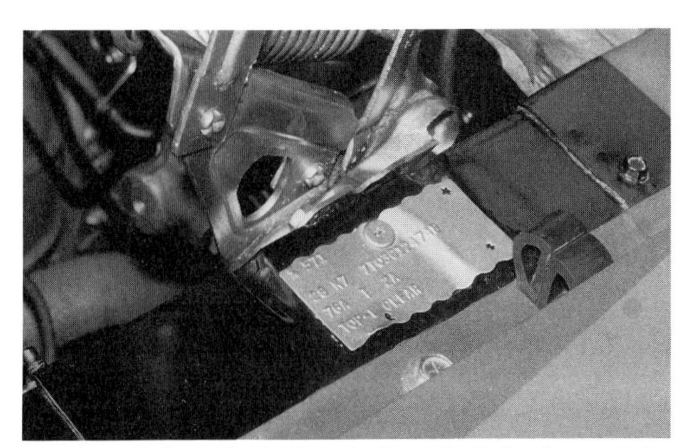

From 1964 through 1968, underhood body buck tags (also called build plates, build sheets, and broadcast sheets) were attached to all Metuchen-built cars, some Dearborn-built cars, and, it is thought, no San Jose-built cars. The tags provided instructions to assembly-line workers so that special options, special orders, and build exceptions could be noted on each vehicle. This body buck tag, on a 1967 convertible, indicates the assembly-line sequence number, beginning build date, VIN, exterior paint code, interior color code, convertible top color code, and a note to install clear rather than tinted glass. Metuchen tags, through 1967, generally had wavy edges; workers acknowledged their instructions by punching small holes of varying shapes in the tags.

In 1968 the body buck tags tended to have straight edges. According to best-guess decoding, this Metuchen-built fastback received power brakes, dual exhaust, and the GT package (including suspension). Because all 1968 Shelbys originated at Metuchen, most have this style body buck tag.

This Silver Frost 1966 convertible was built in Metuchen on February 18, 1966, with a black Deluxe interior, black top, A-code 289ci engine, remote mirror option, factory air conditioning, and an untinted windshield.

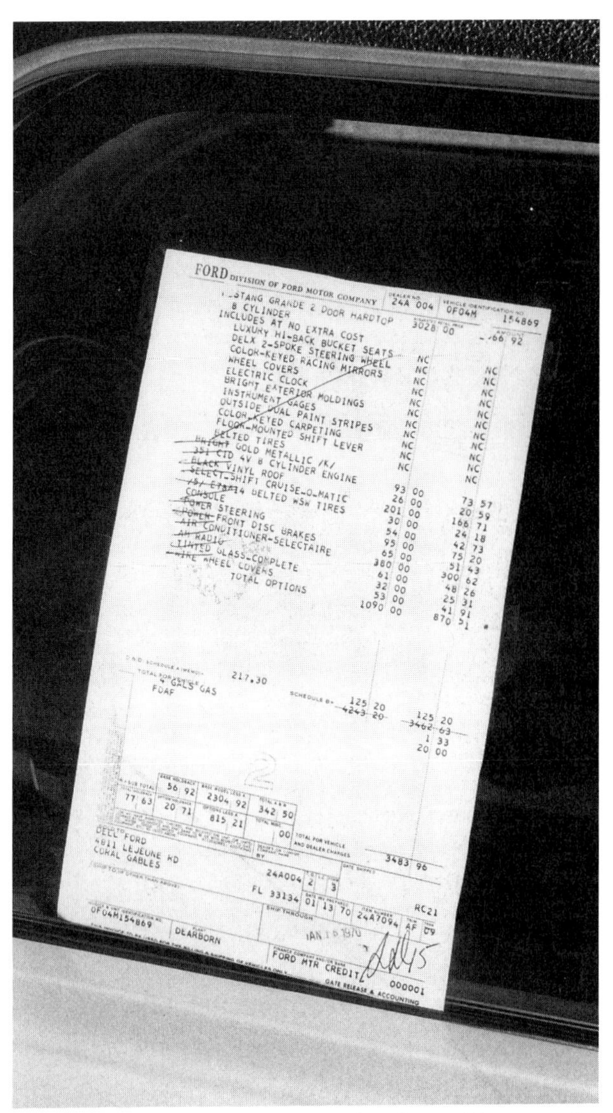

FORD DIVISION OF FORD MOTOR COMPANY 24A 004 0F04M 154869

STANG GRANDE 2 DOOR HARDTOP 3028 00
8 CYLINDER
INCLUDES AT NO EXTRA COST
LUXURY HI-BACK BUCKET SEATS
DELX 2-SPOKE STEERING WHEEL
COLOR-KEYED RACING MIRRORS
WHEEL COVERS
ELECTRIC CLOCK
BRIGHT EXTERIOR MOLDINGS
INSTRUMENT GAGES
OUTSIDE DUAL PAINT STRIPES
COLOR-KEYED CARPETING
FLOOR-MOUNTED SHIFT LEVER
BELTED TIRES
BRIGHT GOLD METALLIC /K/
351 CID 4V 8 CYLINDER ENGINE
BLACK VINYL ROOF
SELECT-SHIFT CRUISE-O-MATIC
/B/ E78X14 BELTED WSW TIRES
CONSOLE
POWER STEERING
POWER FRONT DISC BRAKES
AIR CONDITIONER-SELECTAIRE
AM RADIO
TINTED GLASS-COMPLETE
WIRE WHEEL COVERS
TOTAL OPTIONS

NC NC NC NC NC NC NC NC NC NC NC
93 00 / 26 00 / 201 00 / 30 00 / 54 00 / 95 00 / 65 00 / 380 00 / 61 00 / 32 00 / 53 00 / 1090 00

66 92
NC NC NC NC NC NC NC NC NC NC NC
73 57 / 20 59 / 166 71 / 24 18 / 42 73 / 75 20 / 51 43 / 300 62 / 48 26 / 25 31 / 41 91 / 870 51

217.30
4 GALS GAS
FOAF
SCHEDULE B 125 20 4243 20
125 20 3462 63
1 33
20 00

56 92 / 77 63 / 2304 92 / 20 71 / 342 50 / 815 21 / 00

TOTAL FOR VEHICLE AND DEALER CHARGES 3483 96

DELL FORD
4811 LEJEUNE RD
CORAL GABLES
FL 33134
24A004 2 3
01 13 70 24A7094 AF 09 RC21
0F04M154869 DEARBORN
JAN 1970
FORD MTR CREDIT
000001
GATE RELEASE & ACCOUNTING

The ultimate prize for a restorer is a bona fide copy of a Mustang's original sales invoice or window sticker. Both show the vehicle's VIN, as well as the standard and optional features on that particular car.

PERMANENT TYPE ANTI-FREEZE

(Do Not Over Fill Radiator)

PROTECTION ____________________ °F

DATE ____________________

NUMBER OF QUARTS ____________

Guaranteed All Winter Anti-Freeze Protection

The Ford Dealer listed on the reverse side warrants that this cooling system is guaranteed for all winter anti-freeze protection as noted on dealership repair order provided that the owner of the vehicle maintains the cooling system in a water-tight condition. This warranty is given in lieu of all other warranties expressed or implied. The obligation of the dealer is limited to the installation of such additional Geunine FoMoCo Anti-Freeze as may be necessary to maintain the guaranteed degree indicated on top portion of this tag and to referenced repair order.

The 1964-1970 Mustang FoMoCo Antifreeze tag.

NEAR THE RADIATOR FILLER NECK

YOU'RE FULLY PROTECTED FOR TWO YEARS AGAINST ANTI-FREEZE LOSS

Ford

Ford Permanent Anti-Freeze is a top quality, highly concentrated formula with a long lasting, rust inhibiting agent. A minimum quantity provides protection down to the required temperature level.

For this reason it is possible to offer an exceptionally liberal guarantee which protects your anti-freeze investment...

IF THE ANTI-FREEZE LOSES ITS DEGREE OF PROTECTION DUE TO ANY DEFECT IN THE ANTI-FREEZE...

(SEE COMPLETE DETAILS ON REVERSE SIDE)

FORD FULL STRENGTH
TWO YEAR PERMANENT ANTI-FREEZE
has been added to the cooling system of this vehicle as follows (see Guarantee on other side):

No. OF QUARTS ____________ PROTECTION ______ °F

OWNER'S NAME ____________

ADDRESS ____________

MODEL ____________

CAR MILEAGE ____________

The 1967-1972 Mustang Ford Antifreeze tag.

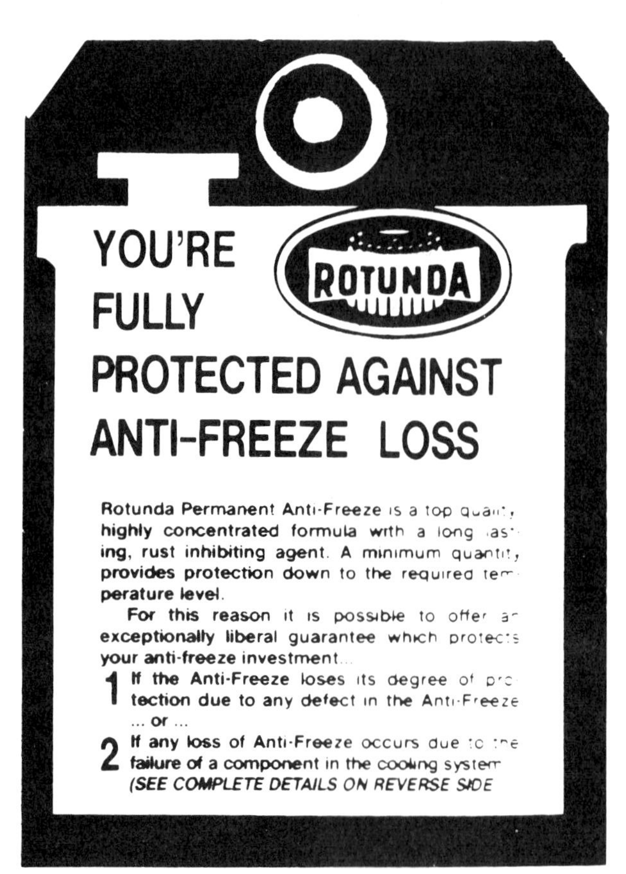

YOU'RE FULLY PROTECTED AGAINST ANTI-FREEZE LOSS

ROTUNDA

Rotunda Permanent Anti-Freeze is a top quality, highly concentrated formula with a long lasting, rust inhibiting agent. A minimum quantity provides protection down to the required temperature level.

For this reason it is possible to offer an exceptionally liberal guarantee which protects your anti-freeze investment...

1 If the Anti-Freeze loses its degree of protection due to any defect in the Anti-Freeze ... or ...

2 If any loss of Anti-Freeze occurs due to the failure of a component in the cooling system

(SEE COMPLETE DETAILS ON REVERSE SIDE

ROTUNDA FULL STRENGTH
ALL-WINTER PERMANENT ANTI-FREEZE
has been added to the cooling system of this vehicle as follows (see Guarantee on other side):

No. OF QUARTS ____________ PROTECTION ______

OWNER'S NAME ____________

ADDRESS ____________

MODEL ____________

CAR MILEAGE ____________ DATE ____________

The 1964-1966 Mustang Rotunda Antifreeze tag.

WHEN REFUELING:
- Check engine oil
  ONLY add same type oil as recommended under oil change.

ONCE A MONTH:
- Check battery fluid level.
  Check coolant-antifreeze level. DO NOT OVERFILL. Fill only to 1" below bottom of filler neck at operating temperature. ONLY add same type antifreeze as that used for coolant change.

EVERY 6000 MILES OR 6 MONTHS:
- Change oil and oil filter.
  ONLY use 10W-30 GRADE OIL (for temperatures above -10° F.) labeled. "Meets or exceeds car makers requirements sequence tested M.S."
  Use ROTUNDA Oil Filter for best service.

ANTI-FREEZE-COOLANT:
- Change every 36,000 miles or 2 years.
  Cooling system originally protected to −35° F.
- ROTUNDA permanent antifreeze mixed 50-50 with water gives -35° F. protection.

CRANKCASE VENTILATION SYSTEM:
- Service periodically in accordance with directions in the Owner's Manual.

ALWAYS USE GENUINE FORD MOTOR COMPANY REPLACEMENT PARTS THAT BEAR THESE INSIGNIA

ROTUNDA FoMoCo AUTOLITE

FOR COMPLETE SERVICE RECOMMENDATIONS SEE OWNER'S MANUAL

The 1965-1966 Mustang Service Specifications decal.

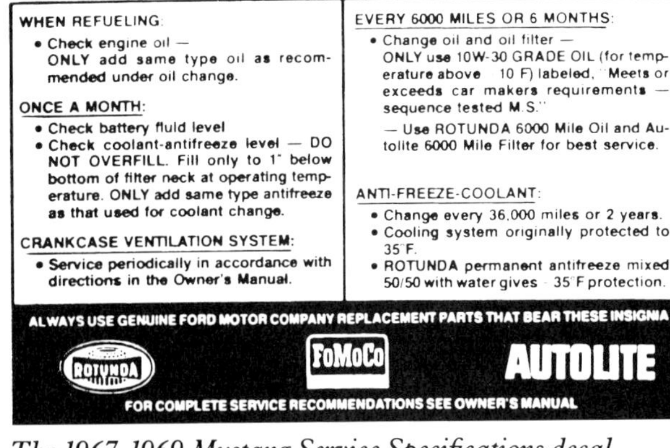

The 1967-1969 Mustang Service Specifications decal.

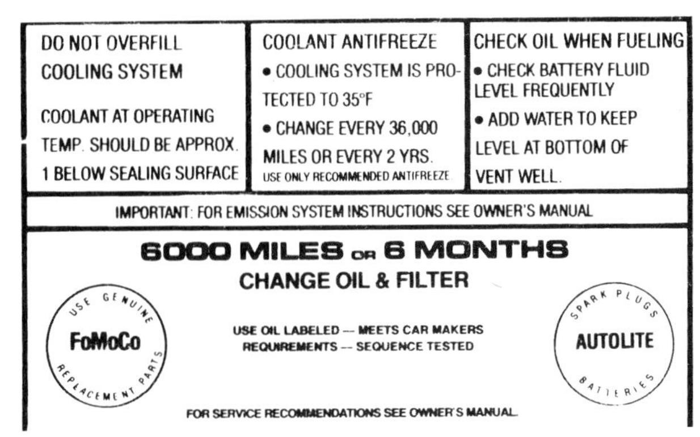

The 1964 Mustang Service Specifications decal (chrome).

The 1970 Mustang Service Specifications decal.

The 1969 Mustang Boss 302 and 428CJ non-Ram Air Autolite Parts Air Cleaner decal.

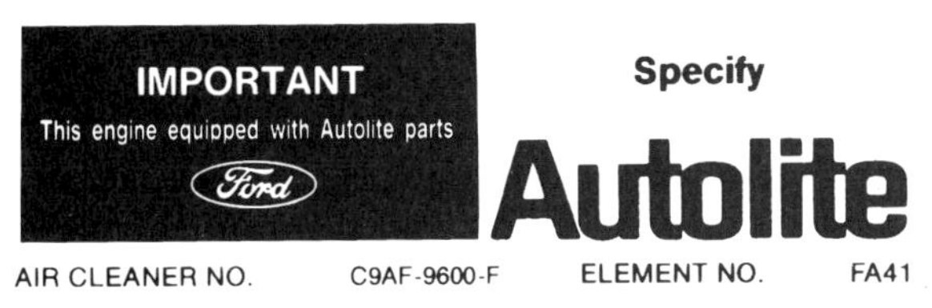

The 1969-1970 Mustang 428CJ Ram Air Autolite Parts Air Cleaner decal.

The 1970 Mustang Boss 302 Autolite Parts Air Cleaner decal.

The 1964-1970 Mustang Air Conditioner Compressor Aluminum tag.

The 1968-1975 Mustang Autolite Parts Air Cleaner decal.

The 1966 Mustang Disc Brake Master Cylinder decal.

CAUTION-FAN

The 1968-1969 Mustang Caution Fan decal.

CRANKCASE VENTILATION SYSTEM

THE CRANKCASE VENTILATION SYSTEM ON THIS ENGINE SHOULD BE SERVICED PERIODICALLY IN ACCORDANCE WITH INSTRUCTIONS IN OWNER'S MANUAL

SEE YOUR DEALER

The 1964-1966 Mustang Crankcase Vent Air Cleaner decal.

IMPORTANT: COOLING SYSTEM IS PROTECTED FOR -35°F. FOR PROTECTION BELOW -35°F ADD ROTUNDA PERMANENT TYPE ANTI-FREEZE. FOR YEAR AROUND PROTECTION USE A 50% MIXTURE OF ROTUNDA ANTI-FREEZE AND WATER FOR ADDITIONAL FILL.

The 1964-1966 Mustang Cooling System Info decal.

The 1964-1966 Mustang Oil Filler Cap decal.

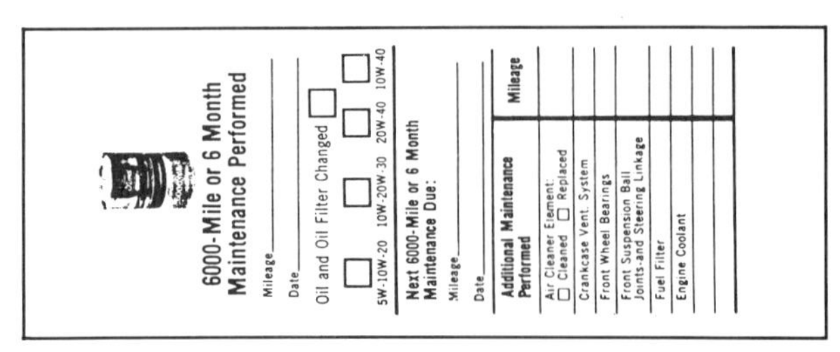

Ford 6000 Mile or 6 Month Lube sticker.

WARNING

YOUR NEW CAR IS EQUIPPED WITH FRONT DISC BRAKES.

RIDING WITH A FOOT RESTING ON THE BRAKE PEDAL CAN RESULT IN RAPID LINING WEAR, POSSIBLE DAMAGE TO THE BRAKES, AND REDUCED FUEL ECONOMY.

DEALER: LEAVE TAG IN PLACE AND CALL TO ATTENTION OF PURCHASER.

C5SA-2B182-A Printed in U.S.A.

The 1965 Mustang Disc Brake Warning tag (for interior of car).

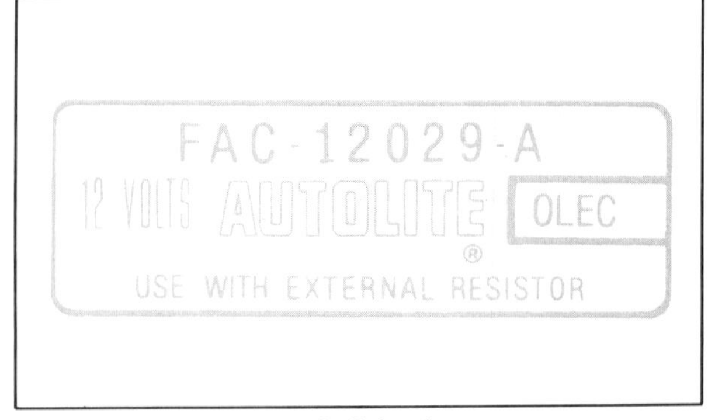

The 1965-1972 Mustang Coil decal.

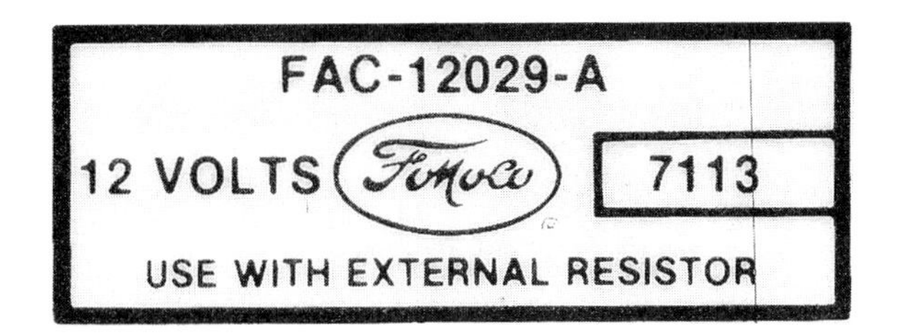

The 1964 Mustang Coil decal.

The 1965-1966 Mustang non-Air Conditioning Voltage Regulator decal.

The 1964 Mustang non-Air Conditioning Voltage Regulator decal.

The 1968-early 1970 Mustang non-Air Conditioning Voltage Regulator decal.

The 1965-1966 Mustang w/ Air Conditioning Voltage Regulator decal.

The 1964 Mustang w/ Air Conditioning Voltage Regulator decal.

The 1967 Voltage Regulator decal.

The 1968-1971 Mustang Hi-Performance Voltage Regulator decal.

1968-early 1970 Mustang w/ Air Conditioning Voltage Regulator decal.

The 1964-1968 Mustang Autolite Spark Plug Air Cleaner decal.

The 1964-1966 Mustang 6-Cylinder Autolite Spark Plug Air Cleaner decal.

The 1967 Mustang Disc Brake Master Cylinder decal (chrome).

The 1964-1972 Mustang Autolite Sta-Ful Battery tag.

The 1969-1970 Mustang Boss 429 Valve Cover decal.

The 1969-1970 Mustang Boss 302 Valve Cover decal.

The 1965-1970 Shelby Accessory Ventra-Flow Air Cleaner decal.

The 1967-1970 Mustang Air Conditioner Clutch decal.

The 1967 Mustang Air Cleaner Part Number decal (typical).

The 1964-1973 Mustang Air Cleaner front decal.

DATE

C4AA-19959-B

DO NOT EXPOSE TO SUSTAINED
TEMPERATURES IN EXCESS OF 250°F

The 1964-1966 Mustang Air Conditioner Dryer decal.

C7ZZ-19959-A

DO NOT EXPOSE TO SUSTAINED
TEMPERATURES IN EXCESS OF 250°F.

The 1967-1968 Mustang Air Conditioner Dryer decal.

JAN 10 1969 FLOW

C9AA-19959-C3

DO NOT EXPOSE TO SUSTAINED
TEMPERATURES IN EXCESS OF 250°F

The 1969 Mustang Air Conditioner Dryer decal.

SEP 10 1969 FLOW

DOAA-19959-C3

DO NOT EXPOSE TO SUSTAINED
TEMPERATURES IN EXCESS OF 250°F

The 1970 Mustang Air Conditioner Dryer decal.

BATTERY OK

CHARGE READING	DATE	BADGE

The 1964-1966 Battery Test OK decal (for inner fender).

The 1964½ Mustang 170ci Air Cleaner decal.

The 1965-1968 Mustang 200ci Air Cleaner decal.

The 1968-1969 Mustang 250ci Air Cleaner decal.

The 1966 Mustang Sprint 200 Air Cleaner decal.

The 1967-1968 Mustang Sports Sprint Air Cleaner decal.

The 1964½ Mustang 260ci Air Cleaner decal.

The 1967-1968 Mustang 289ci Air Cleaner decal (red/black checks).

The 1965-1966 Mustang 289ci Air Cleaner decal (white/black checks).

The 1967 Mustang 289-4V Air Cleaner decal (red/black checks).

The 1965-1966 Mustang 289-4V Air Cleaner decal (white/black checks).

The 1964-1967 Mustang 289 High Performance Air Cleaner decal.

The 1968-1969 Mustang 302 High Performance Air Cleaner decal.

The 1967-1968 Mustang 390 High Performance Air Cleaner decal.

The 1968-1969 Mustang 302-2V Air Cleaner decal.

The 1968-1969 Mustang 302-4V Premium Fuel Air Cleaner decal.

The 1969 Mustang 428-4V Premium Fuel Air Cleaner decal.

The 1969 Mustang 351-2V Air Cleaner decal.

The 1968 Mustang 428-4V Premium Fuel Air Cleaner decal.

The 1969 Mustang 351-4V Premium Fuel Air Cleaner decal.

The 1967 Shelby Cobra Powered by Ford Air Cleaner decal.

The 1967-1969 Mustang 390-4V Premium Fuel Air Cleaner decal.

The 1970 Mustang 250ci Air Cleaner decal.

The 1970 Mustang 302-2V Regular Fuel Air Cleaner decal.

The 1970 Mustang 302-4V Air Cleaner decal.

The 1970 Mustang Boss 302 Air Cleaner decal (green).

The 1969 Mustang Boss 302 Air Cleaner decal (orange).

The 1970 Mustang 351-4V Premium Fuel Air Cleaner decal.

The 1970 Mustang 351-2V Regular Fuel Air Cleaner decal.

The 1969 Shelby 351ci Ram Air Air Cleaner decal.

The 1969-1970 Shelby 428ci Cobra Jet Ram Air Air Cleaner decal.

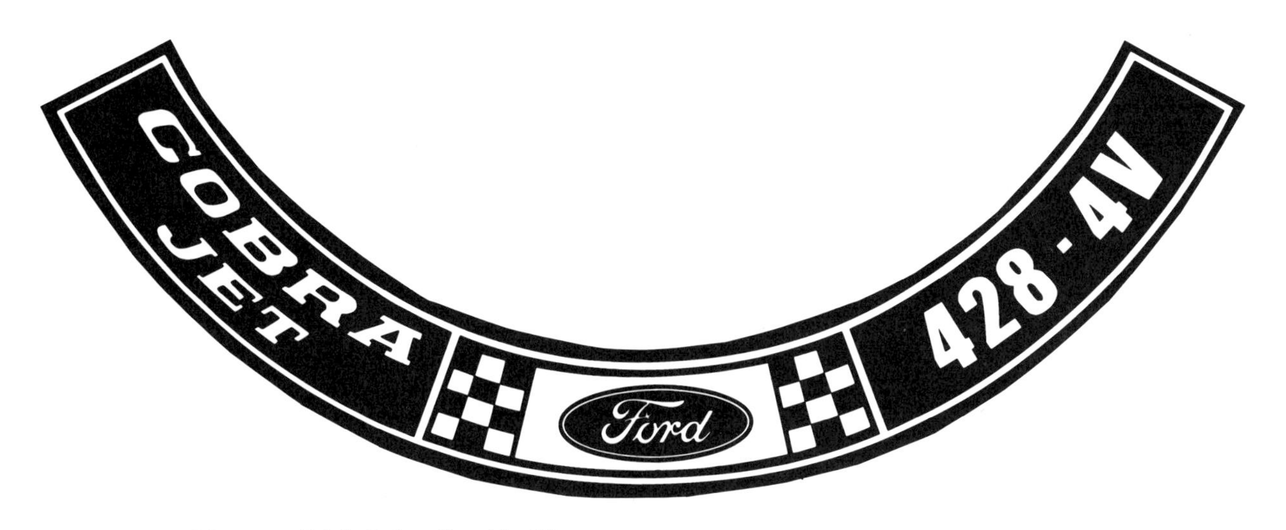

The 1969-1970 Mustang 428ci Cobra Jet Air Cleaner decal.

Chapter 2

Exterior

For many years experts and devotees have tried to analyze the appeal and success of the Ford Mustang. They also have tried to define the reasons for its long-term popularity. After all the discussion is over, one word applies: Style.

It is amazing that the unibody structure of the Mustang sprouted from the Ford Falcon, a somewhat staid design of the late 1950s. In the interest of reducing development costs for the Mustang, Ford engineers built upon the existing Falcon platform and drivetrain. Fortunately, the exterior styling did not suffer for those mechanical requirements. Even the interior used only a few Falcon items such as the door handles, instrument cluster, and heater box.

Mustang Style came from several years of development, effective judging of consumer demographics, a legendary intra-company design contest, the edicts of Lee Iacocca, and the known appeal of the long-hood, short-deck early Thunderbirds. The Mustang was to have a rear seat, sporty performance, and a lively appearance. Ford succeeded beyond its most optimistic sales forecasts, and the public took the Mustang to heart.

Looking back through the early years, Mustang design clusters into three groups: the 1964½ through 1966 models, the 1967 and 1968 models, and the 1969 and 1970 Mustangs. Those were the days when some form of annual design change came out of Detroit but, looking back to the end of World War II, such two- and three-year clusters were common at Ford.

Many of the exterior trim items depicted in this chapter will fall into those three groupings. Antenna bases, rearview mirrors, and bumpers will not differ in the 1967-1968 group, for instance, but most exterior badges and trim, the wheelcovers, and the grilles will allow a viewer to differentiate between each model year.

Introducing the Mustang

Even though the basic concept and design of the Mustang was well established prior to its April 1964 introduction, the 1964½ model was a hurried attempt to combine older components with fresh styling and to get a product out the door quickly. For that reason, many Falcon components are found on the earliest Mustangs. A number of those parts were phased out during the changeover to the 1965 Mustang (along with the introduction of the alternator and several engine changes), but many parts were used into the 1965 model year. These minor carryovers represented assembly-line economies: existing stock was used until supplies ran out, then newer components were phased into the production process. Many changes were due to the Mustang's ability to sell and to achieve its own image. Once its success seemed inevitable, designers felt compelled to make the Mustang even more distinctive.

Several exterior identifiers can help an observer spot a 1964½ Mustang. The top of the grille surround falls off at a 45-degree angle before angling another 45 degrees downward toward the sculptured louvers. Where the hood meets this angle, its edge also is beveled. An obvious carryover from the Falcon was the 13in wheel. A four-ply 6.50x13 tire came standard on all six-cylinder Mustangs plus all eight-cylinder models without the optional handling package (except for air-conditioned cars which received a four-ply 7.00x13 tire). Mustangs with the handling package received 6.50x14 tires during 1964½ production or, optionally, 5.90x15s. The 15in wheel was phased out in late August 1964. This standard wheelcover with its optional three-bladed spinner was offered for 13in wheels until the mid-summer 1964 changeover to the 1965 Mustang. A nearly identical wheelcover with spinner continued in the 14in size throughout the 1965 model year.

When the Mustang was introduced in April 1964, Ford felt the need to tie the Mustang name to the manufacturer. Hence the hood letters F-O-R-D, and a number of other smaller components which related both to Falcons and to full-size Fords. By 1967 the Mustang had attained its own (strong) identity, and there was no need for the word Ford trim on the hood.

The marketing of the Ford Mustang has been long regarded as a public relations masterpiece. America was teased about the product for weeks, then the print and broadcast media became saturated with "lifestyle" advertising and aggressive promotion. Dealership point-of-sale efforts and word of mouth added impetus to the craze. The week that Ford introduced the Mustang at the 1964 World's Fair in New York, both *Time* and *Newsweek* featured the new car in cover stories. The Mustang looked like no other American automobile. It was one of the first cars to show body color (on the valances) below the bumpers, and the effect made the car seem lower and more sporty. Ford's tremendous promotional campaign was boosted by the naming of the Mustang as the official 1964 Indianapolis 500 Pace Car.

Less than one month after the introduction of the Mustang, Ford offered their 289ci 271hp four-barrel V-8 engine. The Hi-Po option set the tone for the Mustang's performance image, and the marque was off and running.

1964½-1965 Mustang Exterior Paint Codes

Data Code	Color Name
A	Raven Black
B*	Pagoda Green
D	Dynasty Green
F*	Guardsman Blue
H	Caspian Blue
J	Rangoon Red
K	Silver Smoke Grey
M	Wimbledon White
P	Prairie Bronze
S*	Cascade Green
U	Sunlight Yellow
X	Vintage Burgundy
Y*	Skylight Blue
Z*	Chantilly Beige
3	Poppy Red
5	Twilight Turquoise
7	Phoenician Yellow

Colors marked with an asterisk were offered only on 1964½ Mustangs. All other colors carried through the 1965 model year.

1965 Mustang Exterior Paint Codes

Data Code	Name of Color
A	Raven Black
B	Midnight Turquoise
C	Honey Gold
D	Dynasty Green
H	Caspian Blue
I	Champagne Beige
J	Rangoon Red
K	Silver Smoke Grey
M	Wimbledon White
O	Tropical Turquoise
P	Prairie Bronze
R	Ivy Green
V	Sunlight Yellow
X	Vintage Burgundy
Y	Silver Blue
3	Poppy Red

1966 Mustang Exterior Paint Codes

Data Code	Name of Color
A	Raven Black
F	Arcadian Blue
H	Sahara Beige
K	Nightmist Blue
M	Wimbledon White
P	Antique Bronze
R	Ivy Green
T	Candyapple Red
U	Tahoe Turquoise
V	Emberglo
X	Vintage Burgundy
Y	Silver Blue
Z	Sauterne Gold
4	Silver Frost
5	Signalflare Red
8	Springtime Yellow

1966 Special Order Colors (No codes)
Medium Palomino Metallic
Medium Silver Metallic
Ivy Green Metallic
Tahoe Turquoise Metallic
Maroon Metallic
Silver Blue Metallic
Sauterne Gold Metallic
Light Beige

1967 Mustang Exterior Paint Codes

Data Code	Name of Color
A	Raven Black
B	Frost Turquoise
D	Acapulco Blue
E	Beige Mist
H	Arcadian Blue
I	Lime Gold
K	Nightmist Blue
M	Wimbledon White
N	Diamond Blue

Data Code	Name of Color
Q	Brittany Blue
S	Dusk Rose
T	Candyapple Red
V	Burnt Amber
W	Clearwater Aqua
X	Vintage Burgundy
Y	Dark Moss Green
Z	Sauterne Gold
4	Silver Frost
6	Pebble Beige
8	Springtime Yellow

1967 Special Order Colors (No codes)
Playboy Pink
Anniversary Gold
Columbine Blue
Aspen Gold
Blue Bonnet
Timberline Green
Lavender
Bright Red

1968 Mustang Exterior Paint Codes

Data Code	Name of Color
A	Raven Black
B	Royal Maroon
D	Acapulco Blue
F	Gulfstream Aqua
I	Lime Gold
M	Wimbledon White
N	Diamond Blue
O	Seafoam Green
Q	Brittany Blue
R	Highland Green
T	Candyapple Red
U	Tahoe Turquoise
W	Meadowlark Yellow
X	Presidential Blue
Y	Sunlit Gold
6	Pebble Beige

The 1964½ Mustang gas caps were not attached with a wire bail. From the introduction of the 1965 Mustangs, the bail was standard.

1969 Mustang Exterior Paint Codes

Data Code	Name of Color
A	Raven Black
B	Royal Maroon
C	Black Jade
D	Pastel Grey
D	Acapulco Blue*
E	Aztec Aqua
F	Gulfstream Aqua
I	Lime Gold
M	Wimbledon White
P	Winter Blue
S	Champagne Gold
T	Candyapple Red
W	Meadowlark Yellow
Y	Indian Fire
2	New Lime
4	Silver Jade
6	Acapulco Blue*

Note: All Mach 1 Color Codes ending in 5 denote flat black hood paint.

*Acapulco Blue is shown as both code D and code 6 on two conflicting vintage Ford reference charts.

1970 Mustang Exterior Paint Codes

Data Code	Name of Color
A	Black
C	Dark Ivy Green Metallic
D	Bright Yellow
G	Medium Lime Metallic
J	Grabber Blue
K	Bright Gold Metallic
M	White
N	Pastel Blue
Q	Medium Blue Metallic
S	Medium Gold Metallic
T	Red
U	Grabber Orange
Z	Grabber Green
1	Calypso Coral
2	Light Ivy Yellow
6	Bright Blue Metallic

The 1964½ Mustang nameplate that adorned the lower front fenders measured 4⅜in long. It was lengthened to 5in at the beginning of the 1965 model year.

Because it was a completely new model and had barely achieved its own identity, the 1964½ Mustang was adorned with Ford corporate insignias and logos. For example these early spinner wire wheelcovers bore the vintage Ford crest in their centers.

It was common for components and options such as this version of the spinner wire wheelcover to be shared by other Ford models.

Four out of the available six full disc wheelcovers installed on 1964½ Mustangs were manufactured in both 13in and 14in sizes, including this rare base cover with spinner.

A distinctive feature indicating a true 1964½ Mustang is this bevel on both sides of the grille opening. The bevel matches the contour of the unique 1964½ hood.

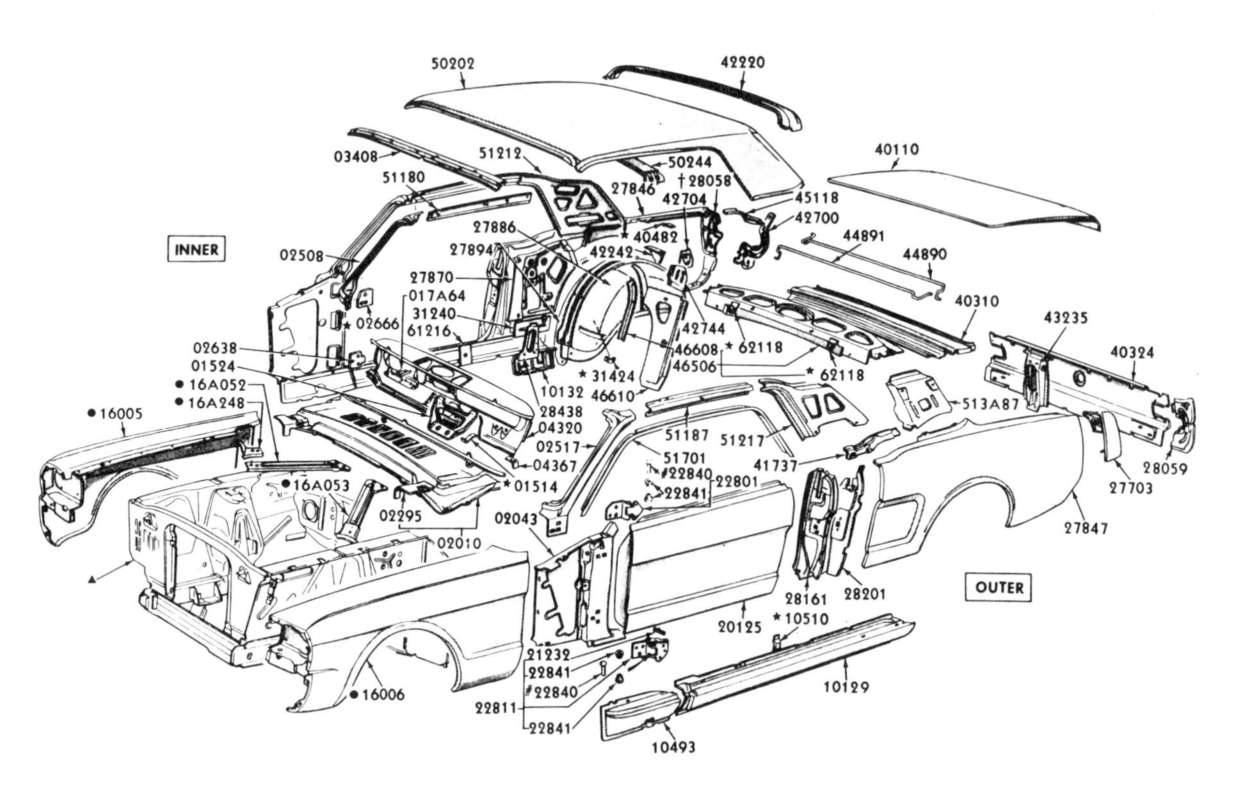

The 1965-1966 Mustang hardtop upper-body structural components and sheet-metal panels.

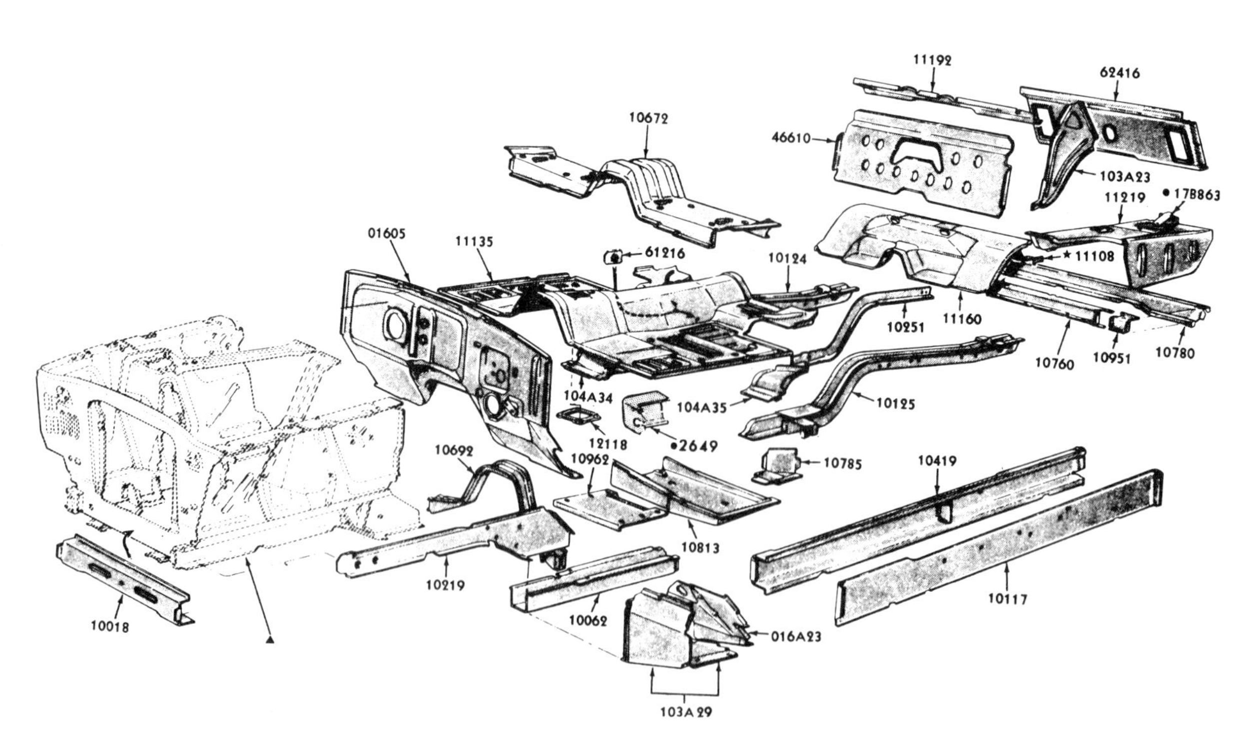

The 1965-1966 Mustang convertible underbody structural components and sheet-metal panels.

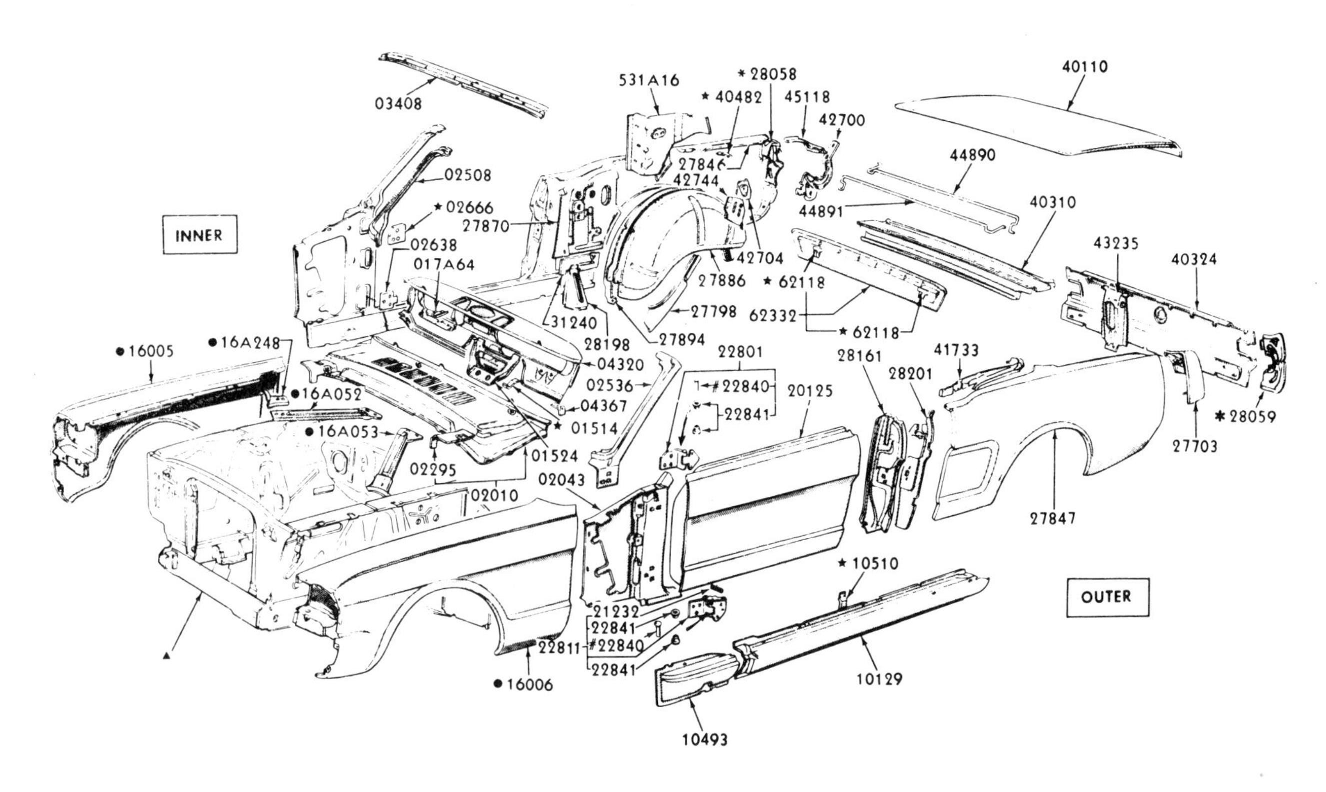

The 1965-1966 Mustang convertible upper sheet-metal panels and structural members.

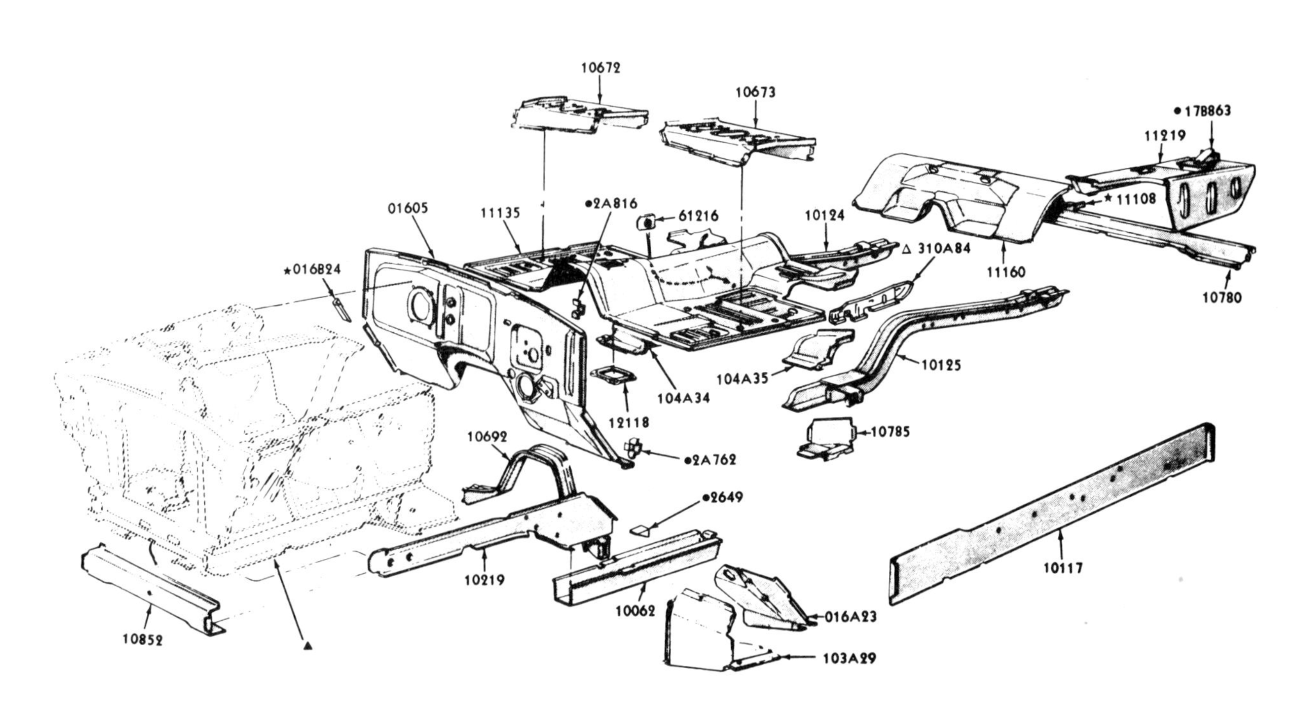

The 1967-1970 Mustang SportsRoof and 1967-1969 Mustang hardtop underbody sheet-metal panels and structural members.

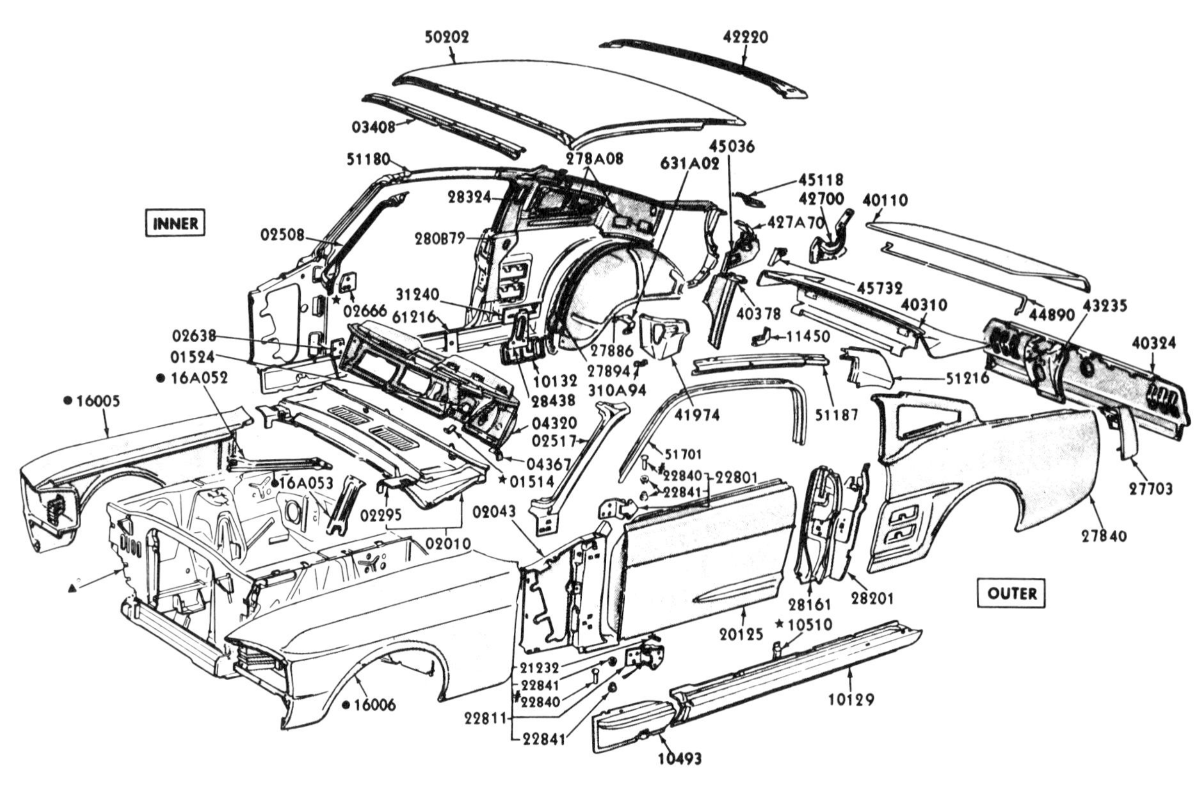

The 1967-1968 Mustang SportsRoof upper-sheet-metal panels and structural members.

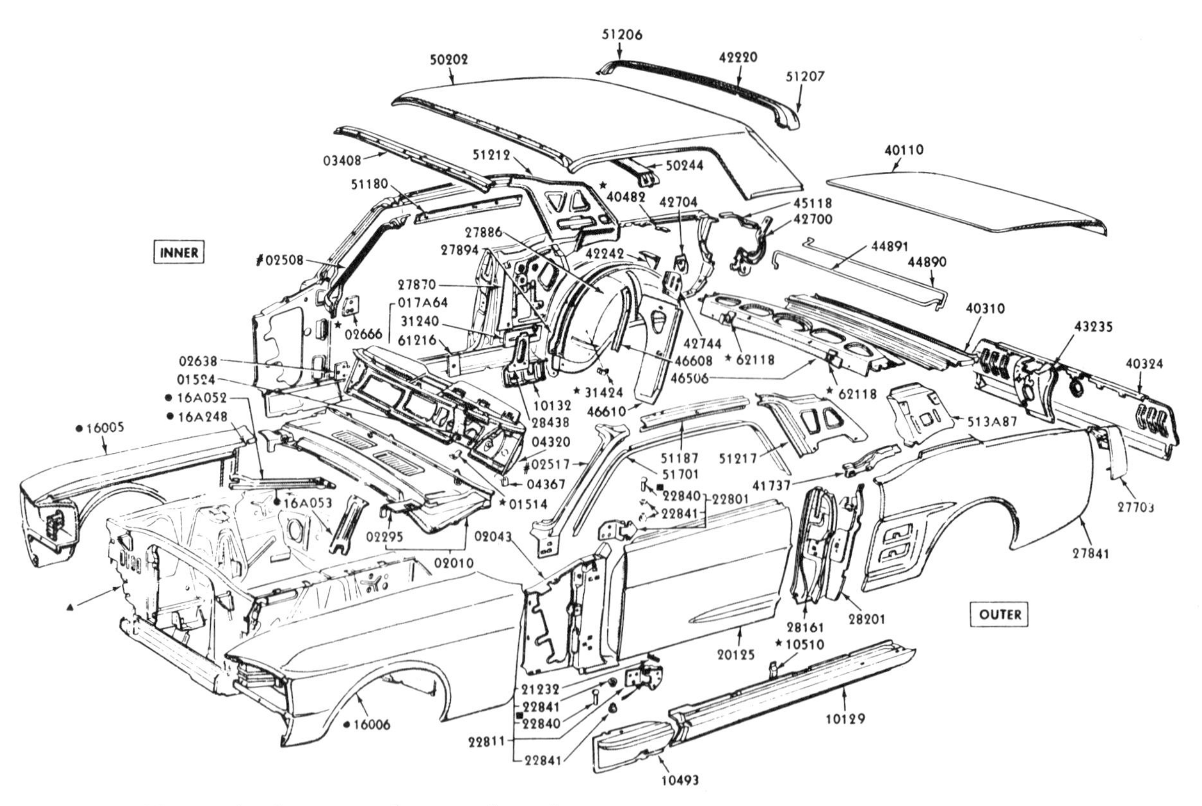

The 1967-1968 Mustang hardtop upper-sheet-metal panels and structural members.

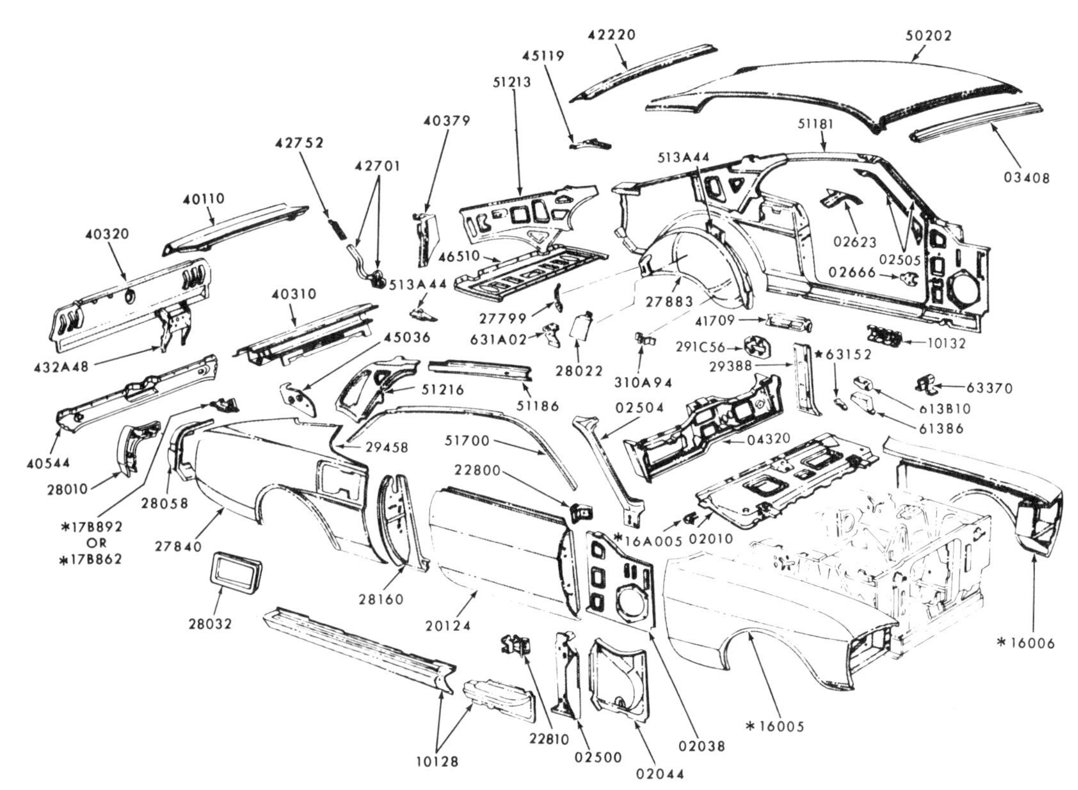

The 1969-1970 Mustang SportsRoof upper-body sheet-metal panels and structural members.

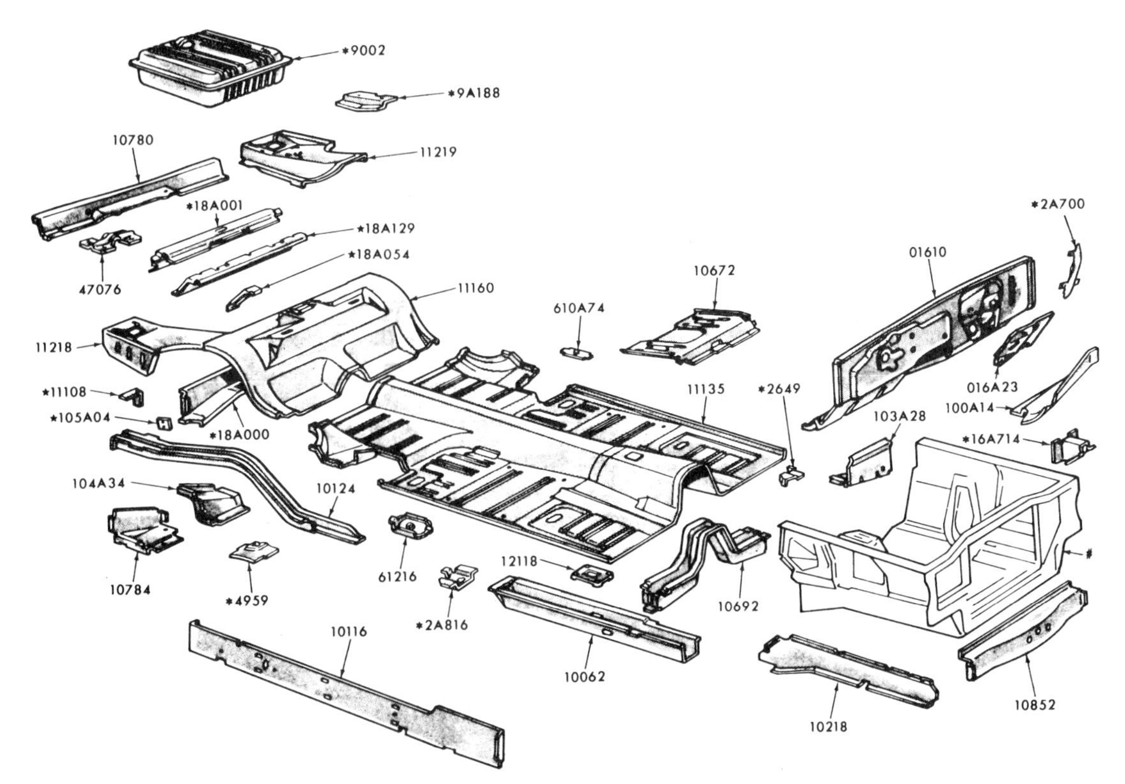

The 1969-1970 Mustang hardtop underbody sheet-metal panels and structural members.

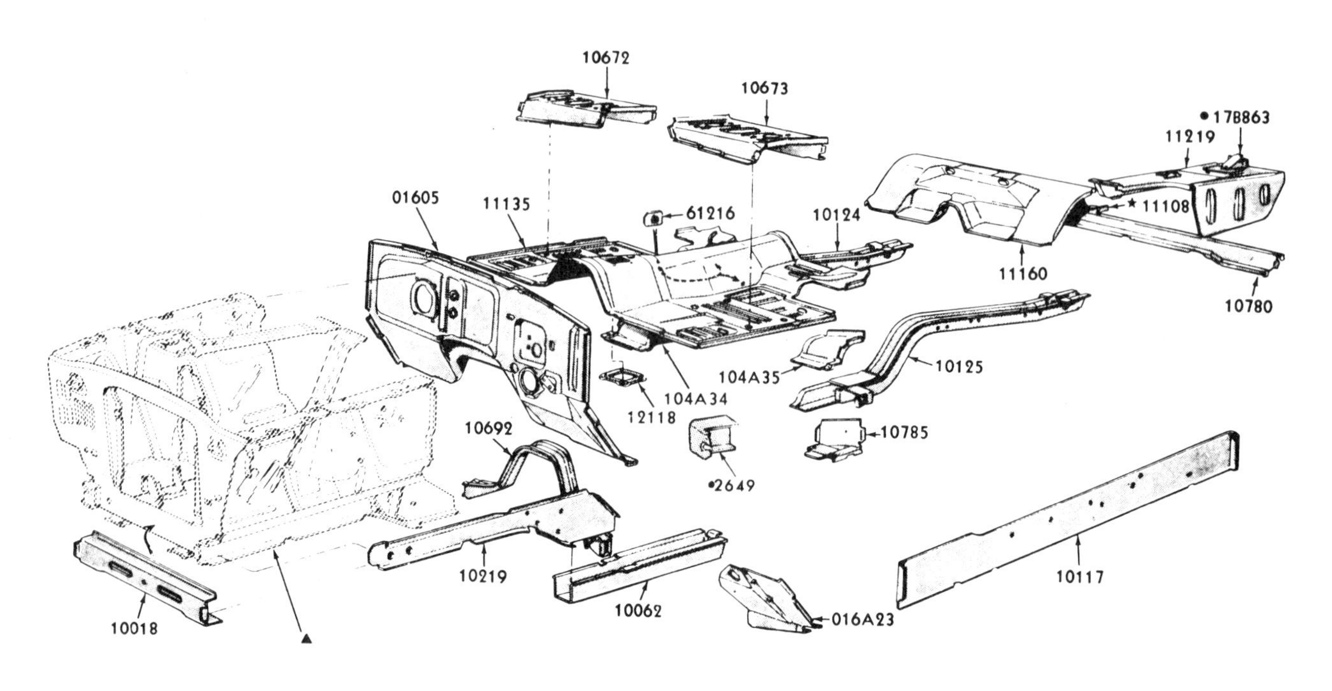

The 1965-1966 Mustang SportsRoof and hardtop underbody structural components and panels.

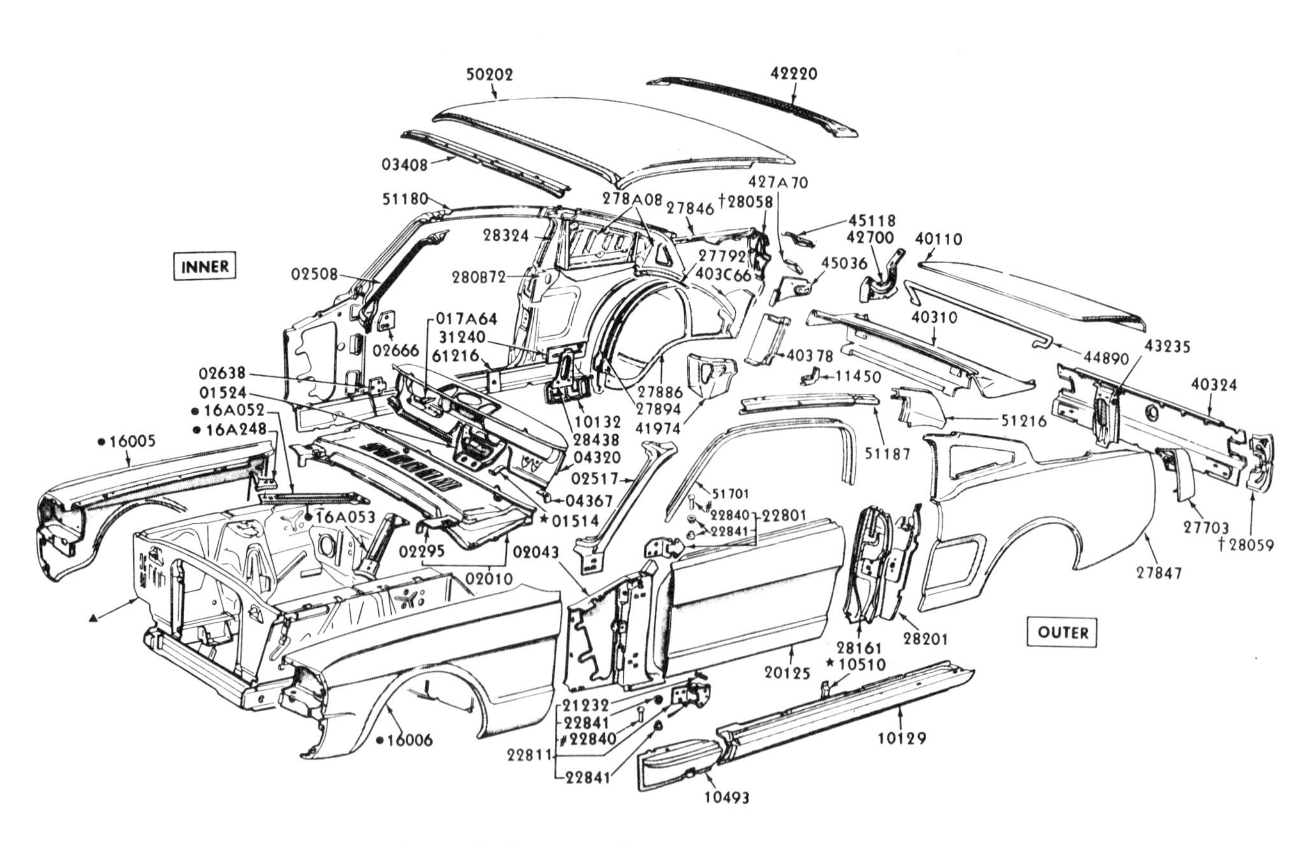

The 1965-1966 Mustang SportsRoof upper-body structural components and sheet-metal panels.

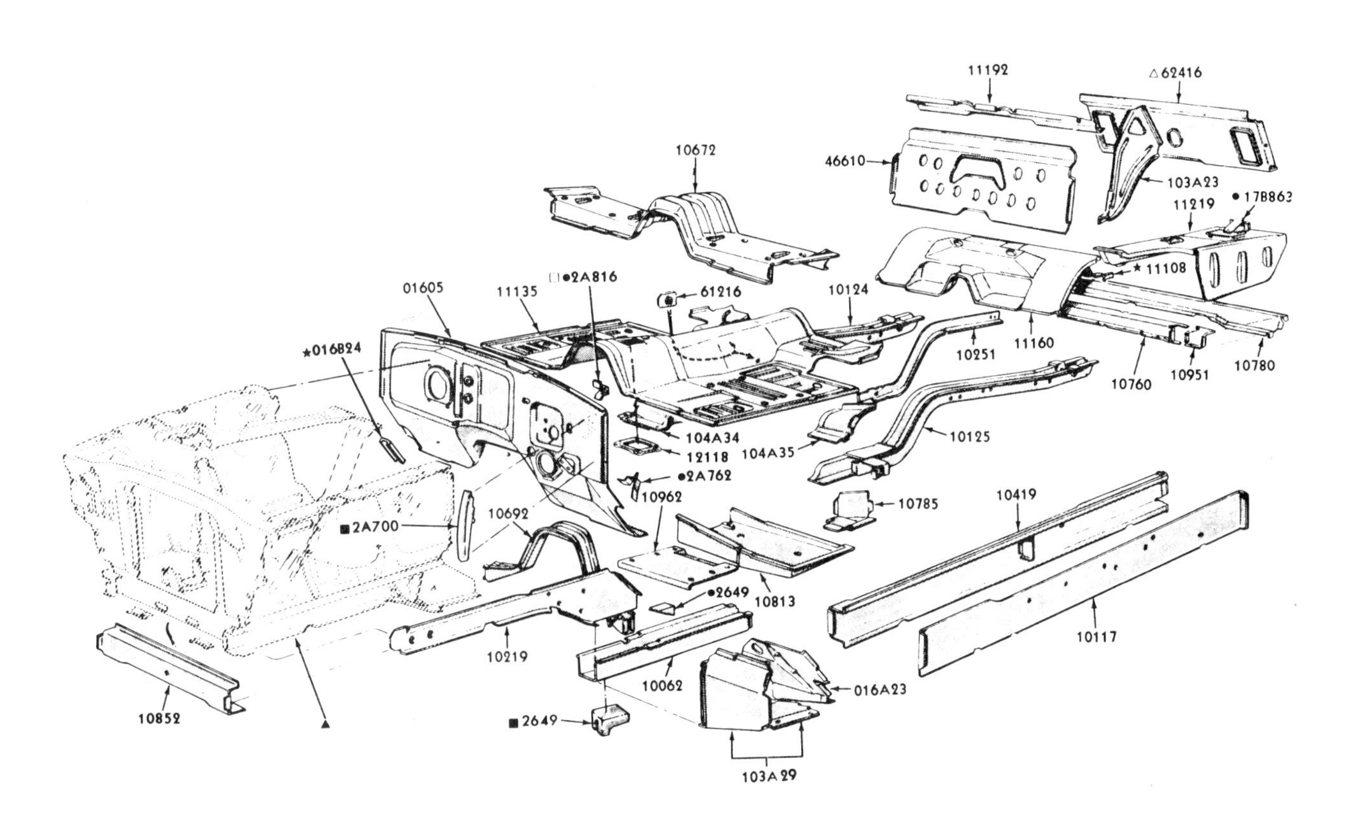

The 1967-1968 Mustang convertible underbody sheet-metal panels and structural members.

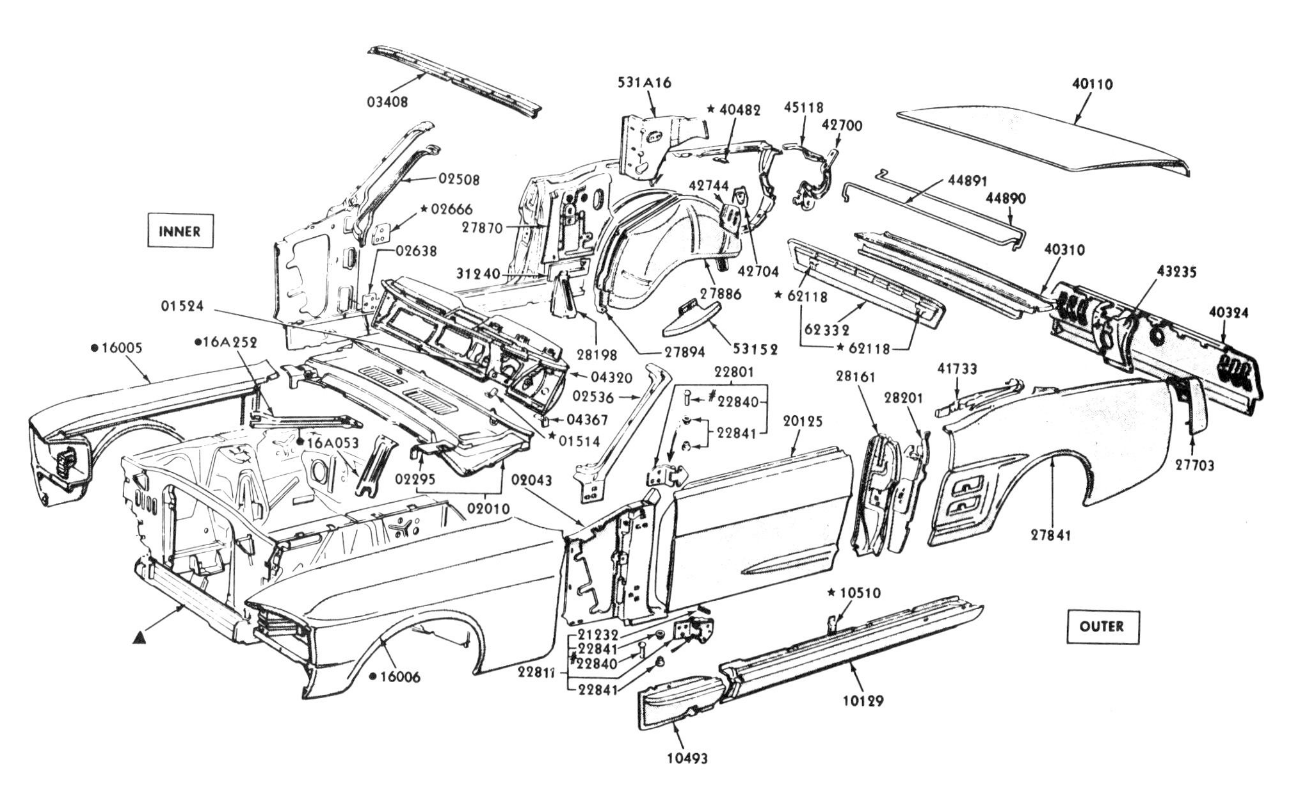

The 1967-1968 Mustang convertible upper-body sheet-metal panels and structural members.

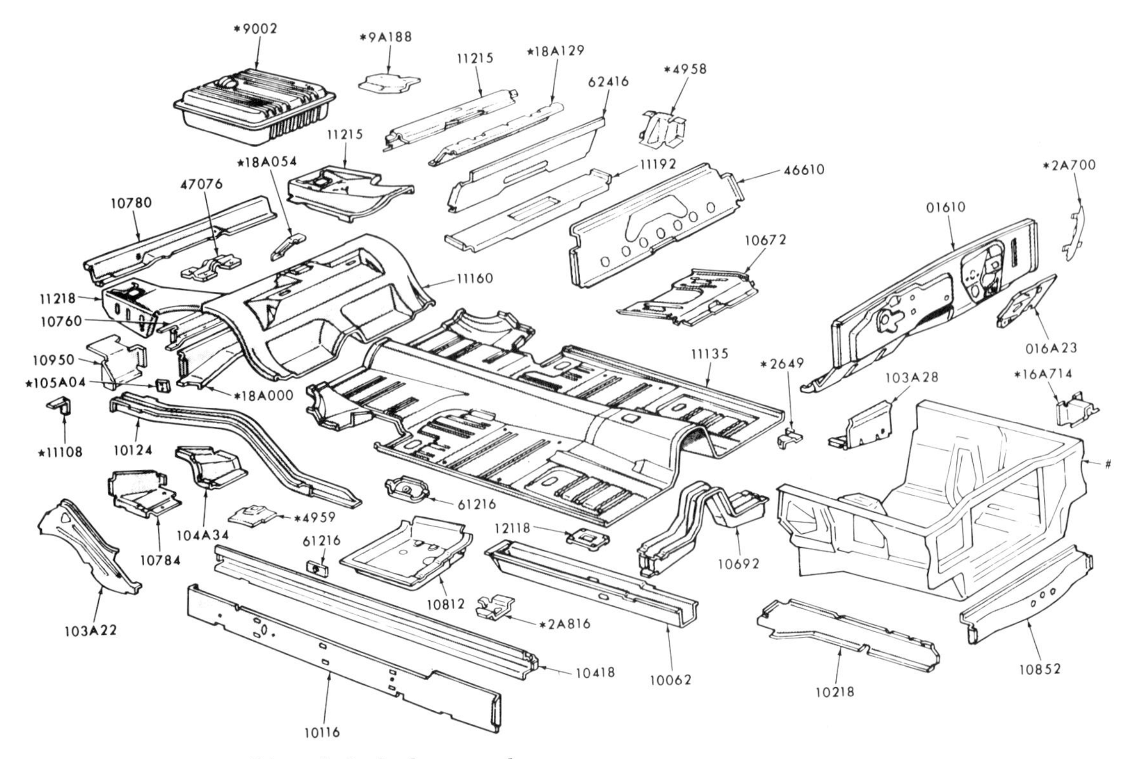

The 1969-1970 Mustang convertible underbody sheet-metal panels and structural members.

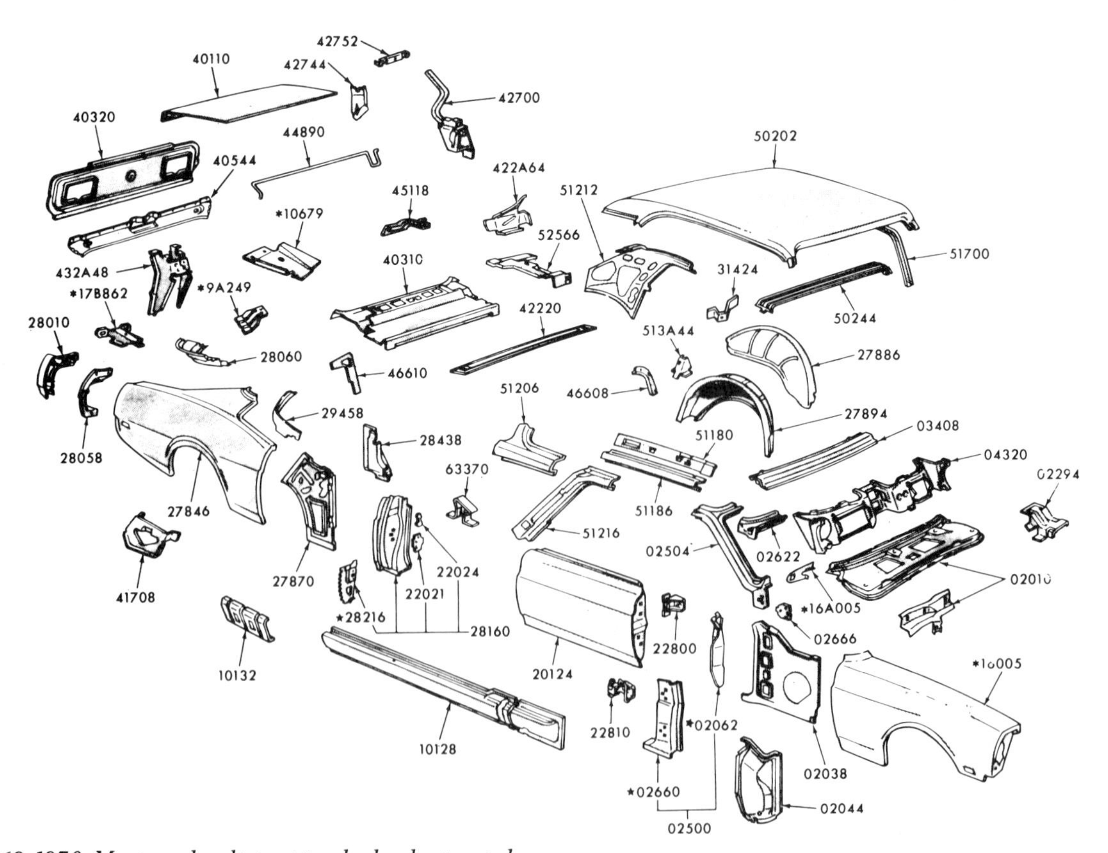

The 1969-1970 Mustang hardtop upper-body sheet-metal panels and structural members.

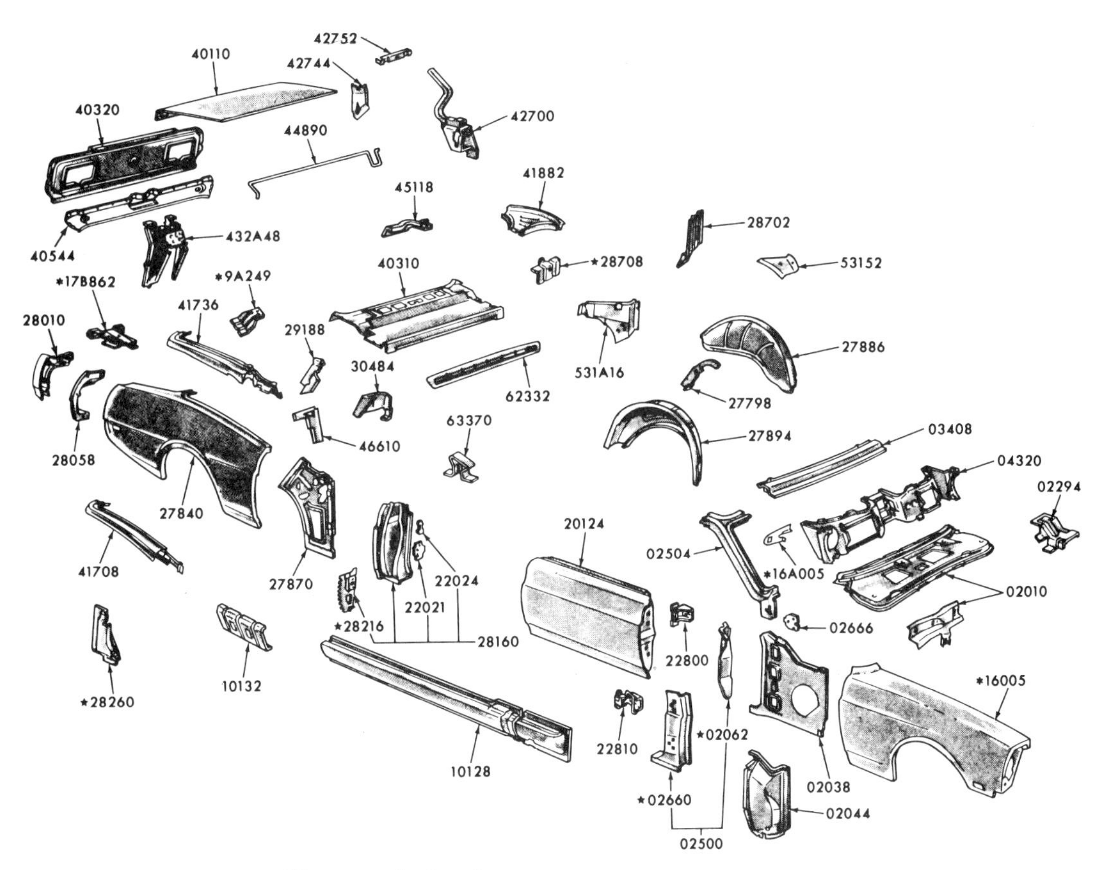

The 1969-1970 Mustang convertible upper-body sheet-metal panels and structural members.

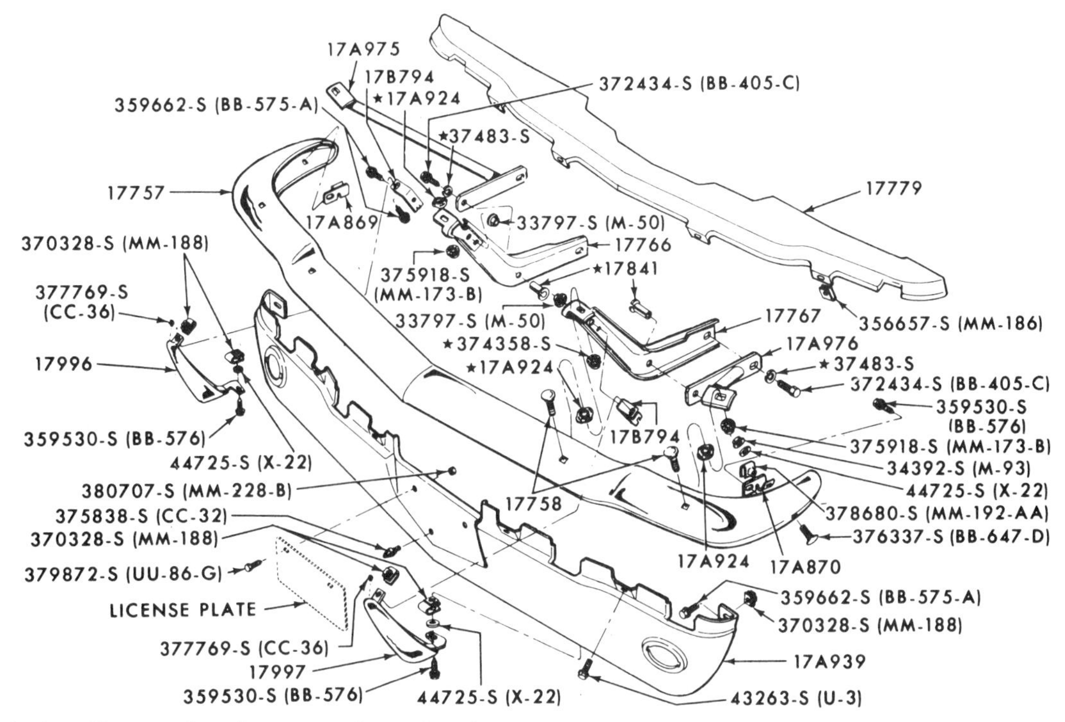

The 1965-1966 Mustang front bumper, valance, brackets, and related mounting hardware.

The standard 1965 grille with argent-colored honeycomb screen and conventional running horse and corral.

The 1965 GT grille had blacked-out screen and foglamps at the ends of the bars.

The 1966 GT grille uses the 1965 light bar on a blacked-out screen.

The 1966 grille loses the bars extending from the corral and replaces the screen with horizontal chrome strips.

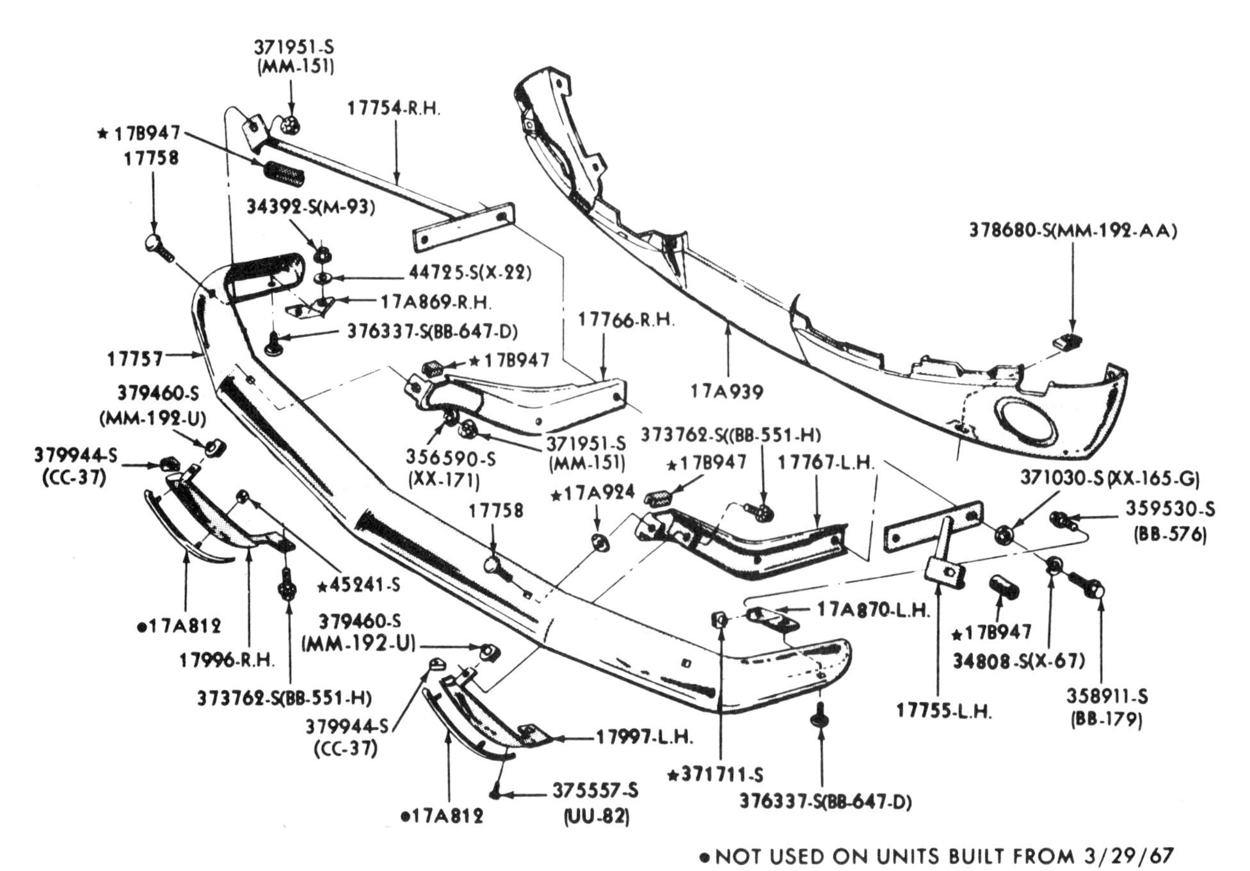

The 1967-1968 Mustang front bumper, valance, brackets, and related mounting hardware.

As in previous years, the 1967 Mustang GT grille includes a blacked-out screen and foglamps.

The standard 1967 Mustang grille reinstates the horizontal bars.

A 1968 GT California Special sets a pair of rectangular foglamps in front of a plain blacked-out screen.

The 1968 Mustang GT grille simply adds foglamps to a standard grille.

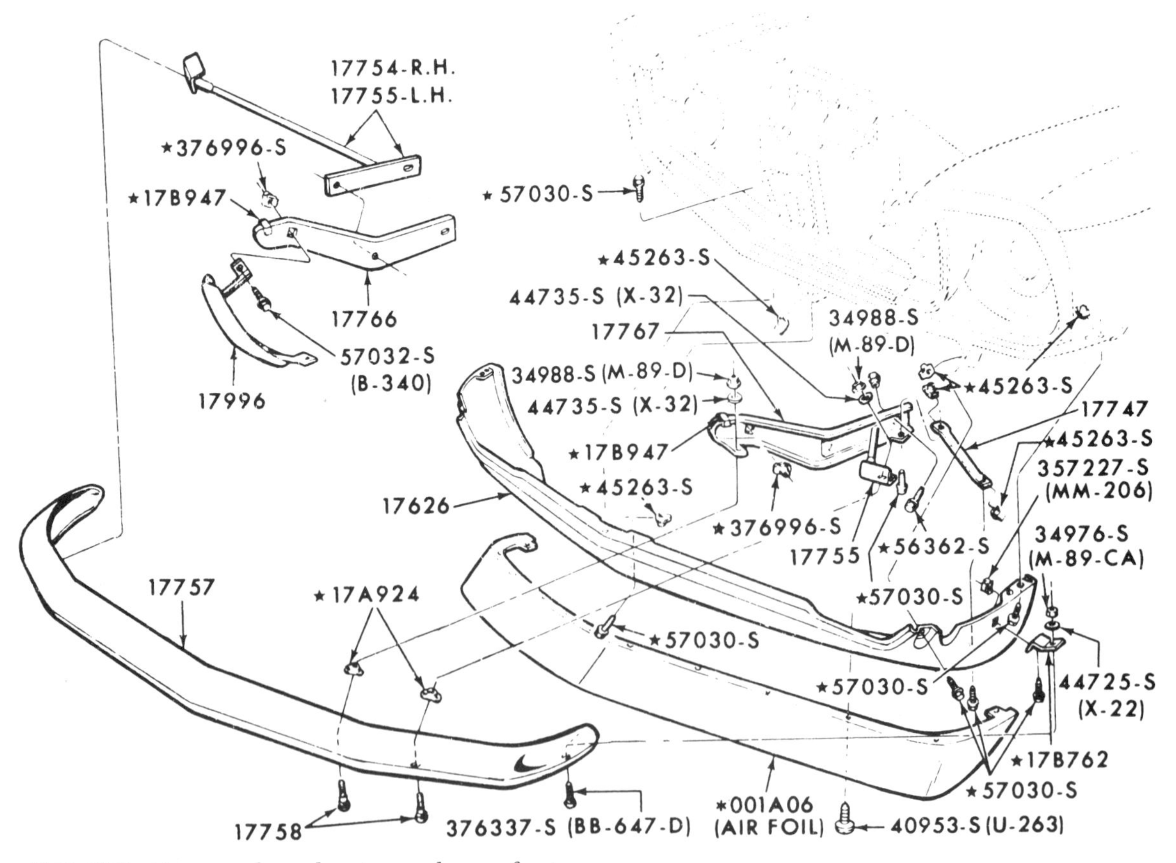

The 1969-1970 Mustang front bumper, valance, front spoiler, and attaching hardware.

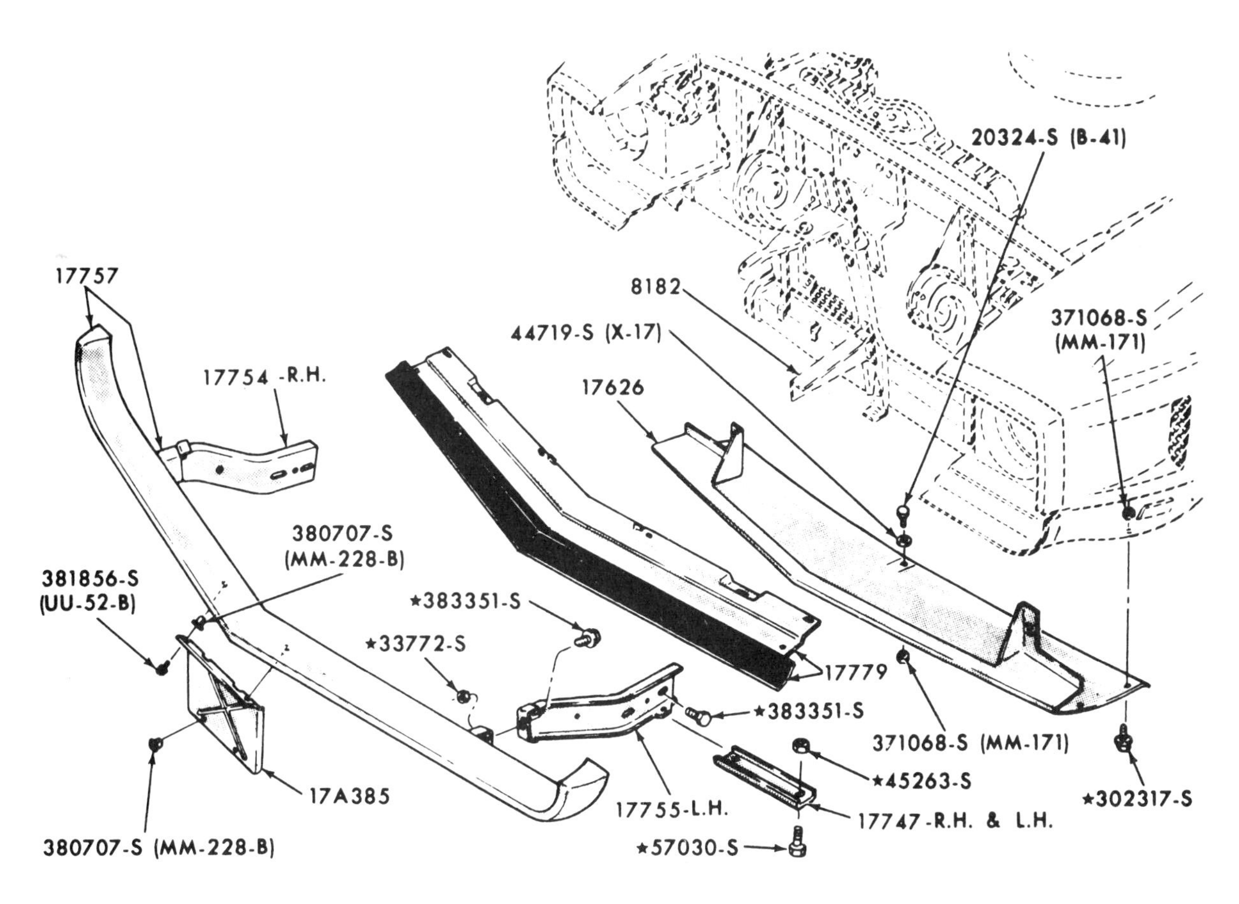

The 1969-1970 Shelby Mustang front bumper, valance, and attaching hardware.

The 1969 grille emblem, inner headlight, and chrome grille trim.

The 1970 Mustang headlight door with fake scoop.

The running horse and tri-colored bar emblem is located at the center of the grille in 1970.

The grille emblem was deleted from the 1970 Mach 1 grille, and replaced by two outboard sport lamps.

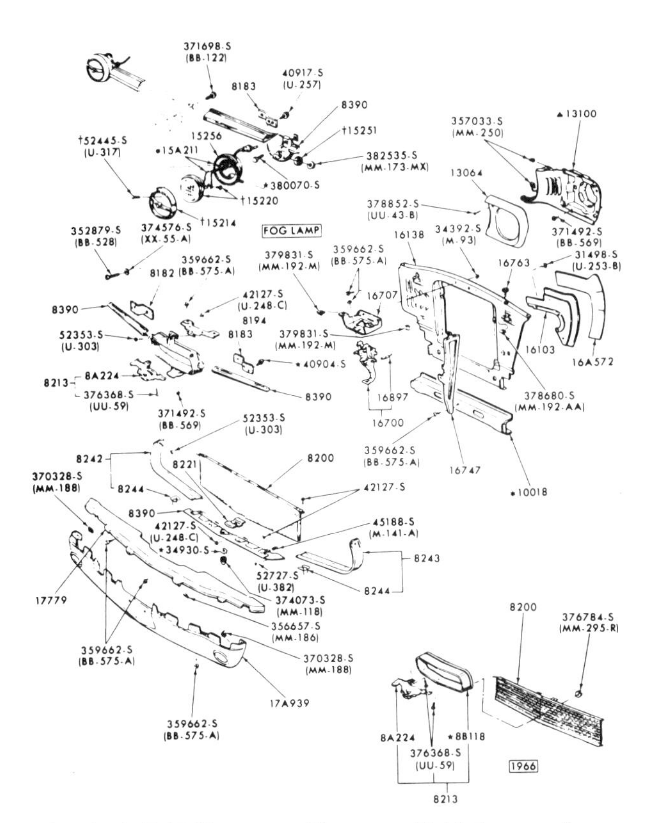

The 1965-1966 Mustang grille, stone shield, lower valance, headlight bucket, radiator-core support, GT light bar, and related hardware.

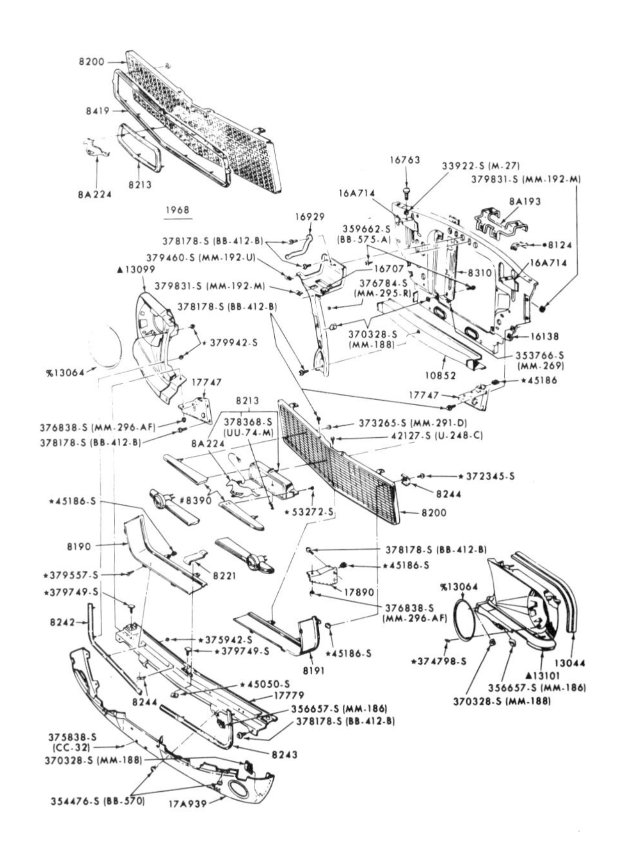

The 1967-1968 Mustang grille, fender extension, valance, trim, radiator-core support, and attaching hardware.

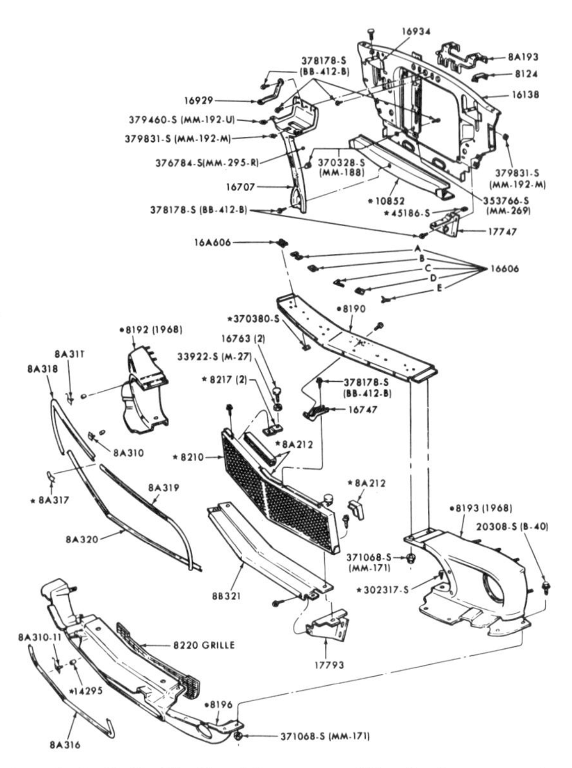

The 1967-1968 Shelby Mustang grille, fender extension, valance, header panel, radiator-core support, and attaching hardware.

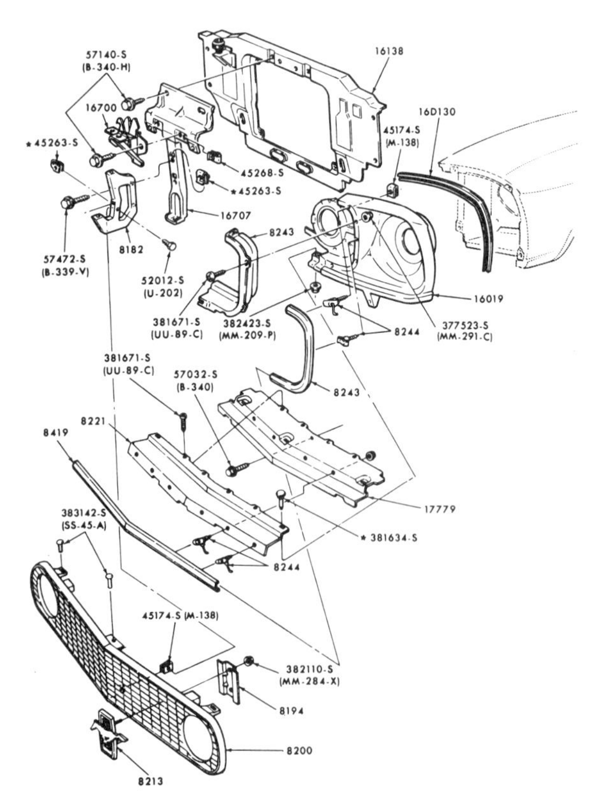

The 1969 Mustang grille, headlight bucket, stone shield, hood latch, radiator-core support, and related hardware.

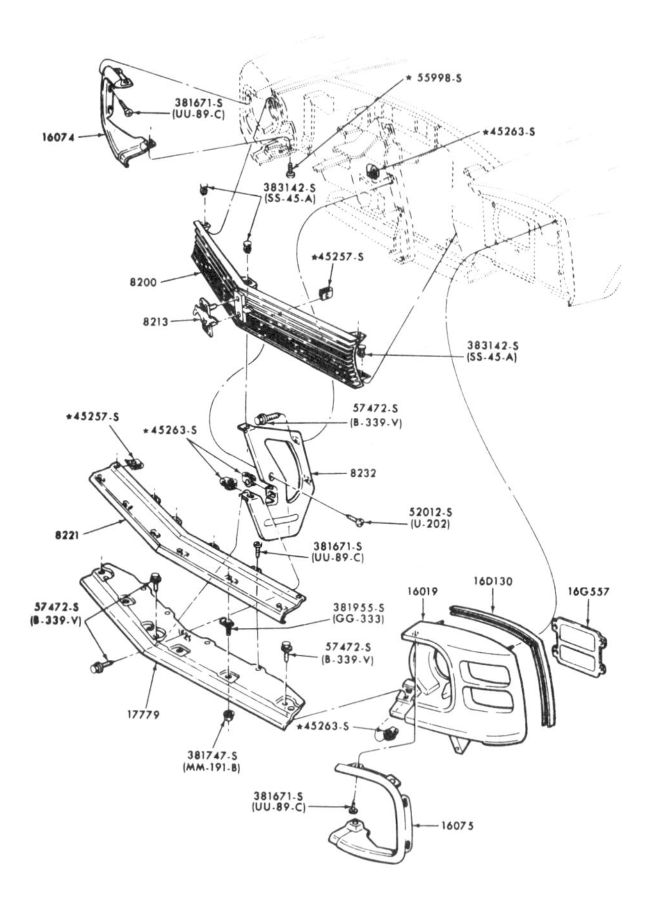

The 1970 Mustang grille, fender extension, stone shield, trim, and related hardware.

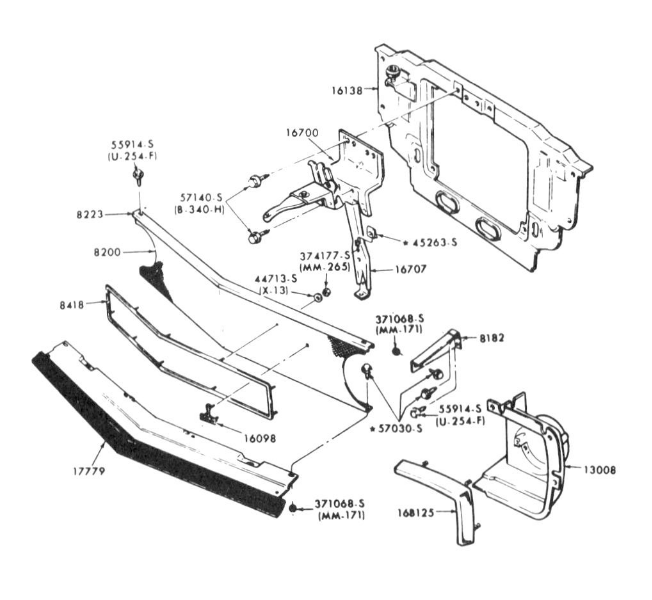

The 1969-1970 Shelby Mustang grille, stone shield, headlight bucket, hood latch, radiator-core support, and attaching hardware.

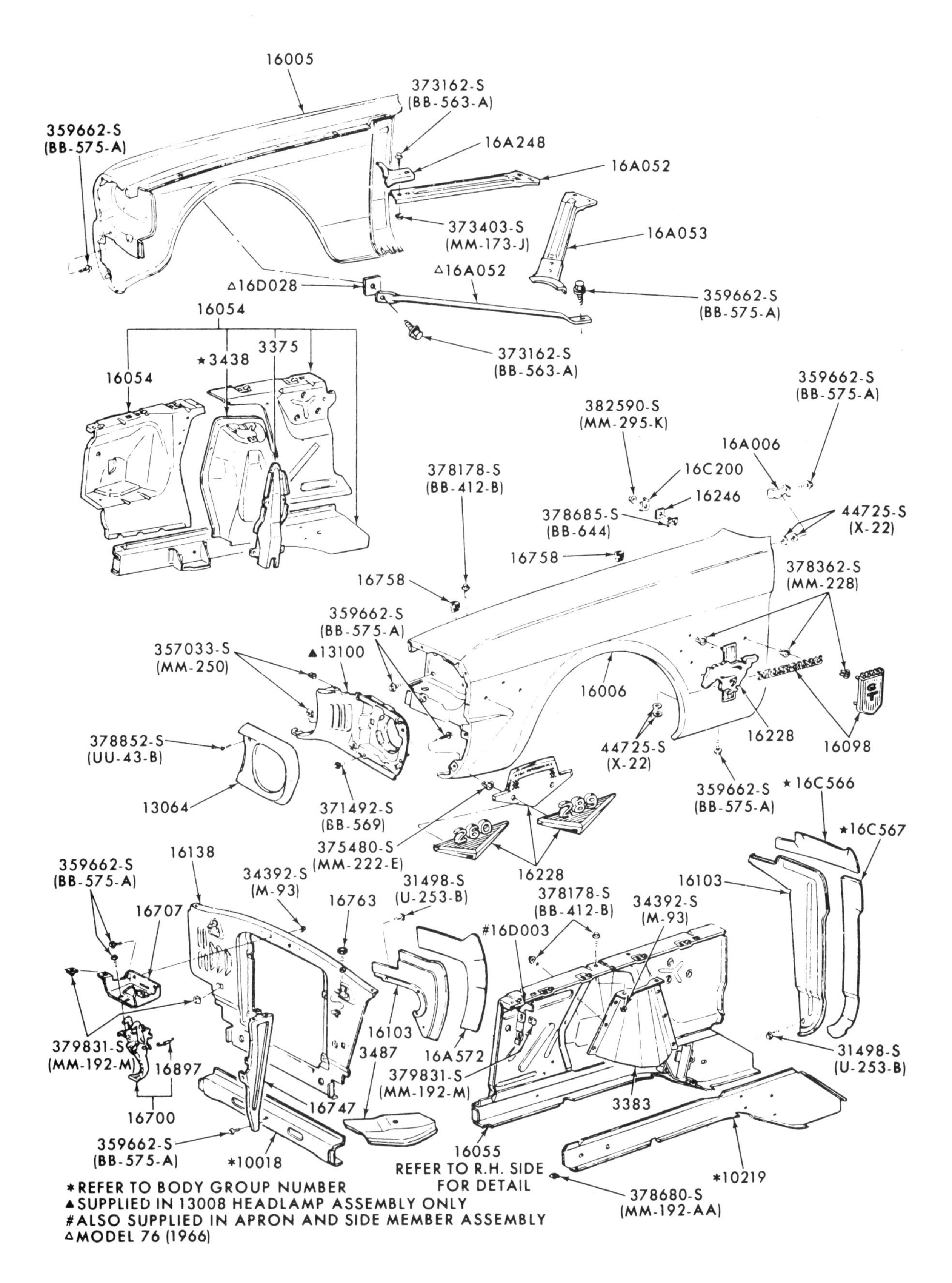

The 1965-1966 Mustang front fenders, aprons, and related parts.

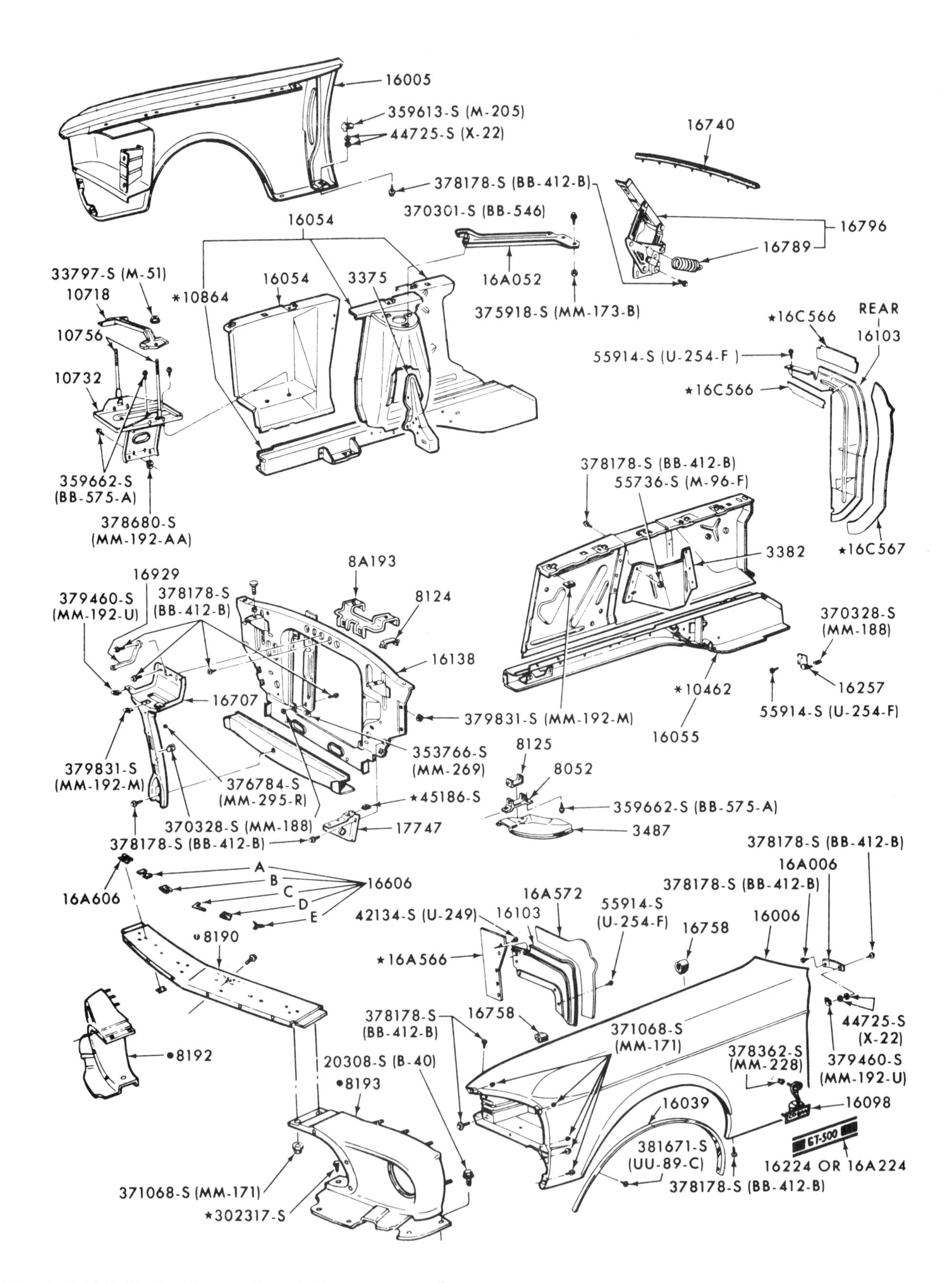

The 1967-1968 Shelby Mustang front fenders, aprons, and related parts.

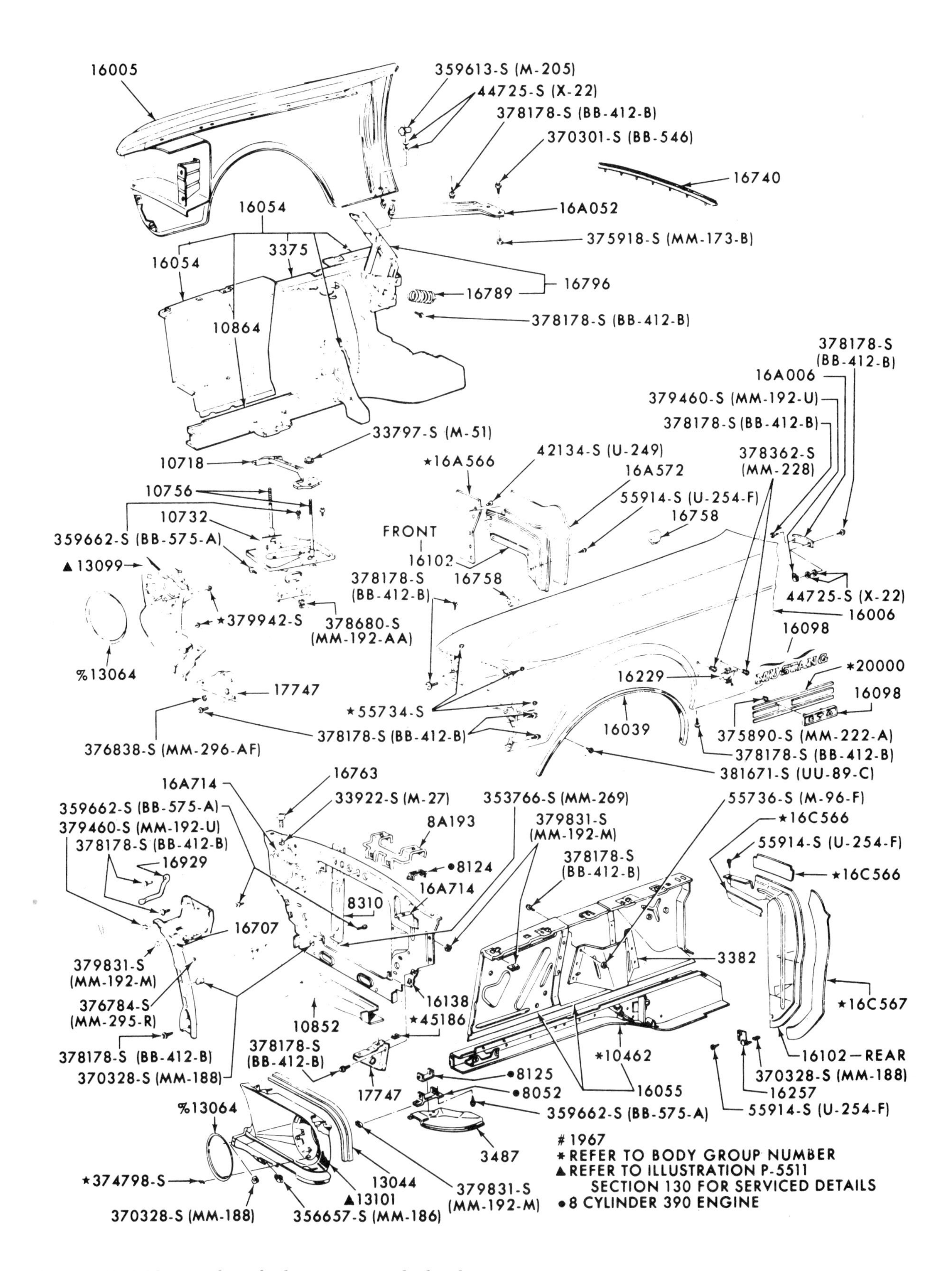

The 1967-1968 Mustang front fenders, aprons, and related parts.

All 1965-1968 Mustangs (except 1968 Shelby Mustangs) equipped with a radio were outfitted with a standard round-base telescoping antenna mounted on the right front fender. The 1968 Shelby Mustang antennas were mounted on the right rear fender.

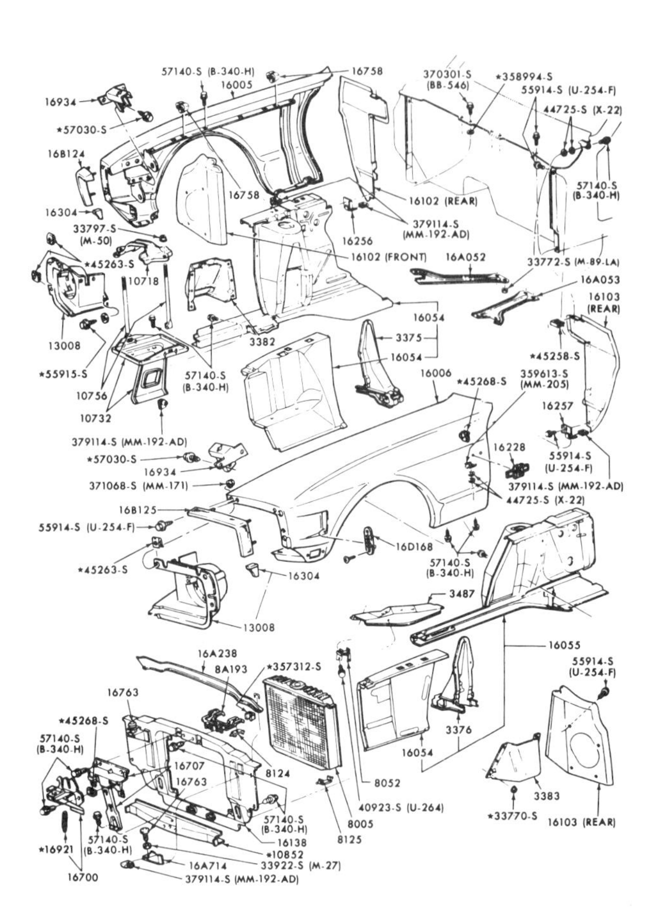

The 1969 Shelby Mustang front fenders, aprons, core support, and related parts.

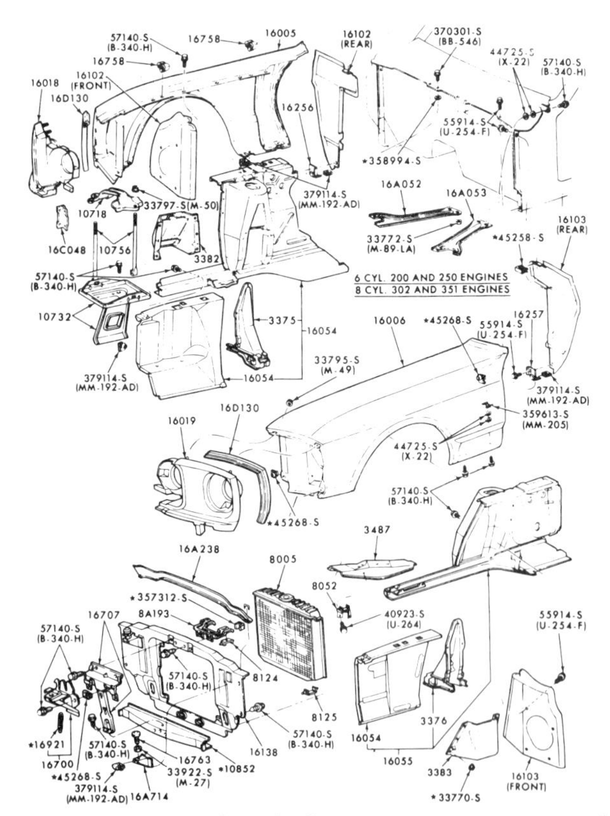

The 1969 Mustang front fenders, aprons, core support, and related parts.

A rectangular base telescoping antenna was installed on all 1969-1970 Mustangs equipped with a radio. Deleting the standard radio was an option for all years and models. The lower shaft of OEM rectangular-base antennas is teardrop shaped (viewed from above) to create an aerodynamic image. Replacement units have a round mast.

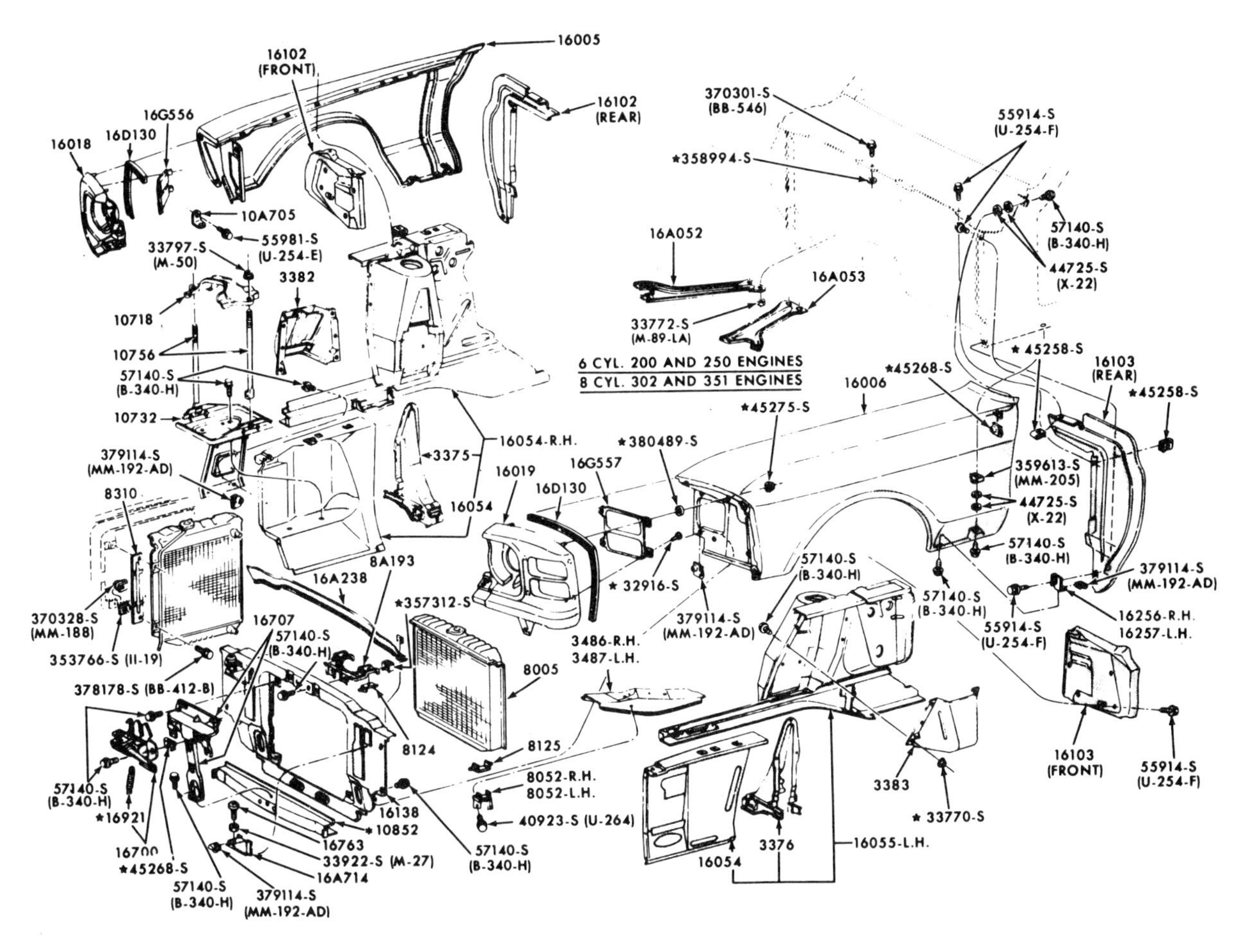

The 1970 Mustang front fenders, aprons, core support, and related parts.

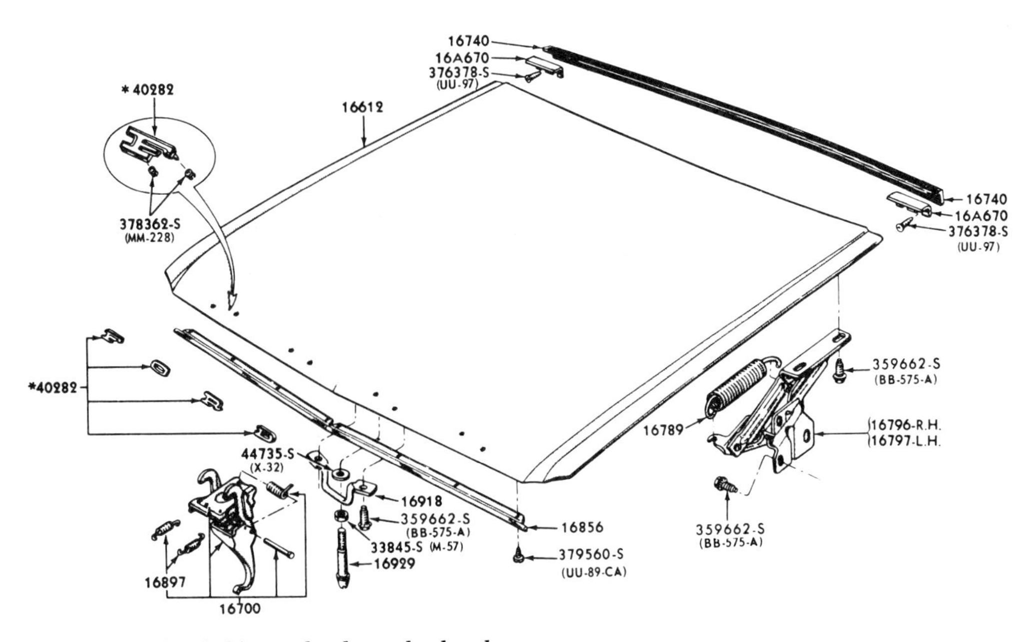

The 1965-1966 Mustang hood, hinges, latch, and related parts.

Hood louvers with integrated turn signal indicators were a popular 1967-1968 Mustang option.

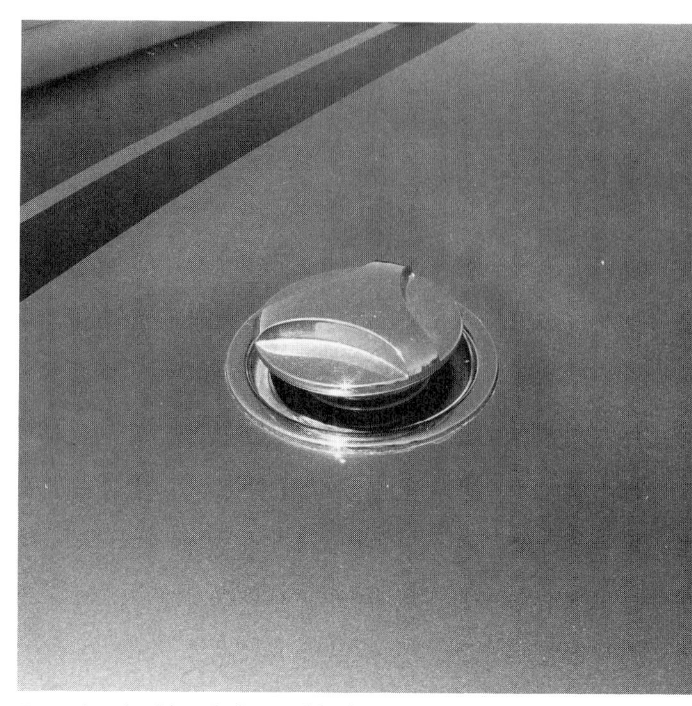

Race-inspired hood pins and locks were popular in the late 1960s. Chrome hood posts with linchpin clips were standard equipment on Mach 1s for 1969, and flush-mounted twist-locks (shown) took their place in 1970.

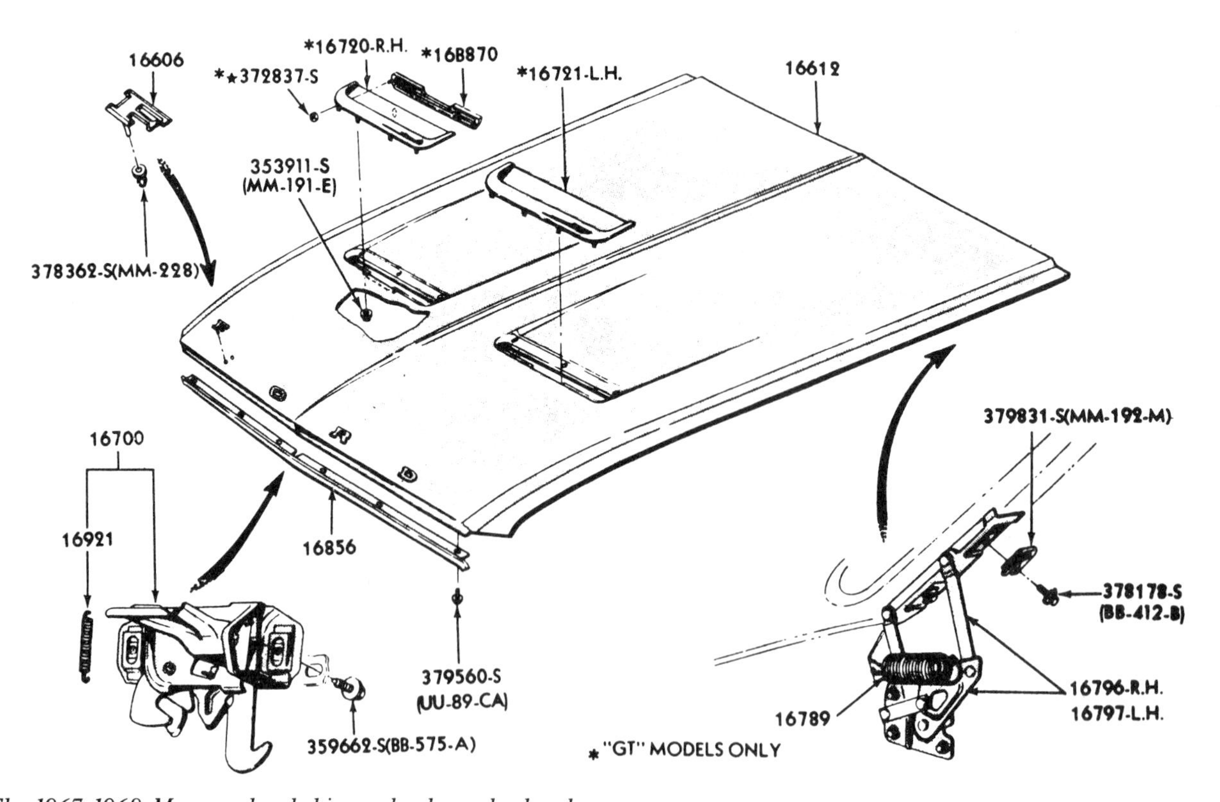

The 1967-1968 Mustang hood, hinges, latch, and related parts.

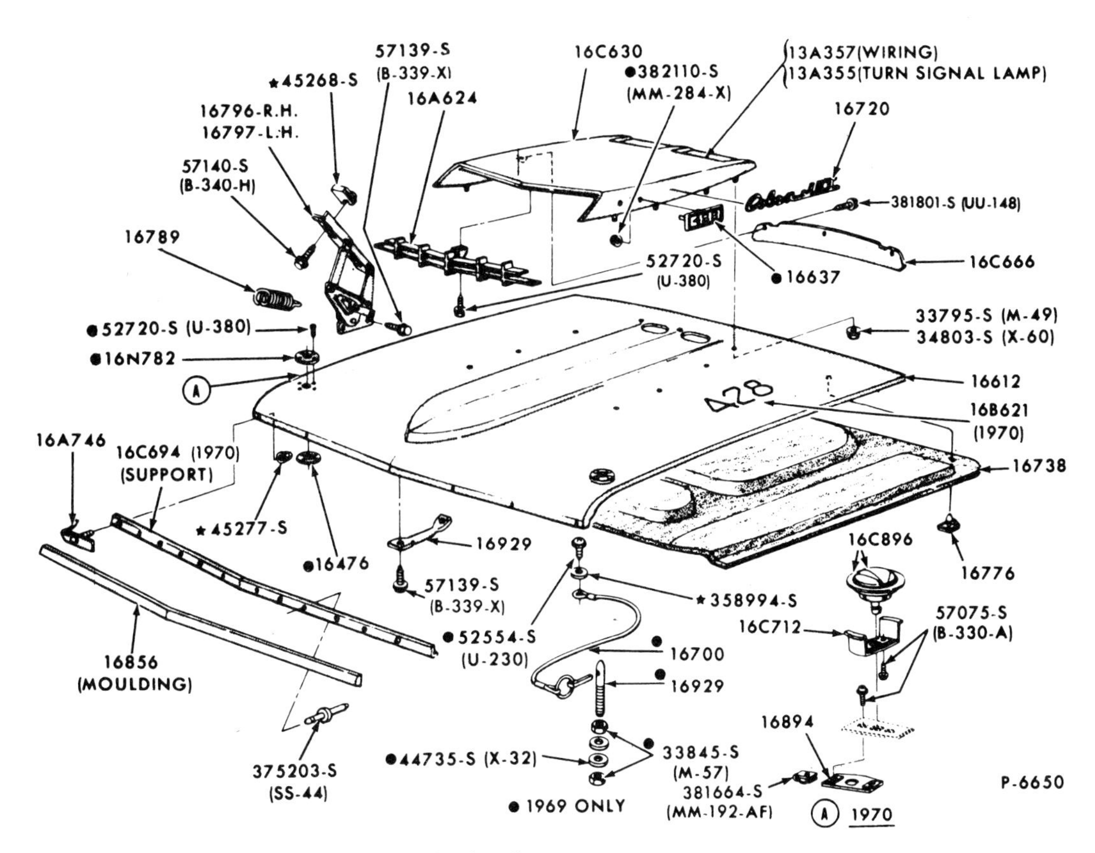

The 1969-1970 Mustang hood, hinges, scoop, and related parts.

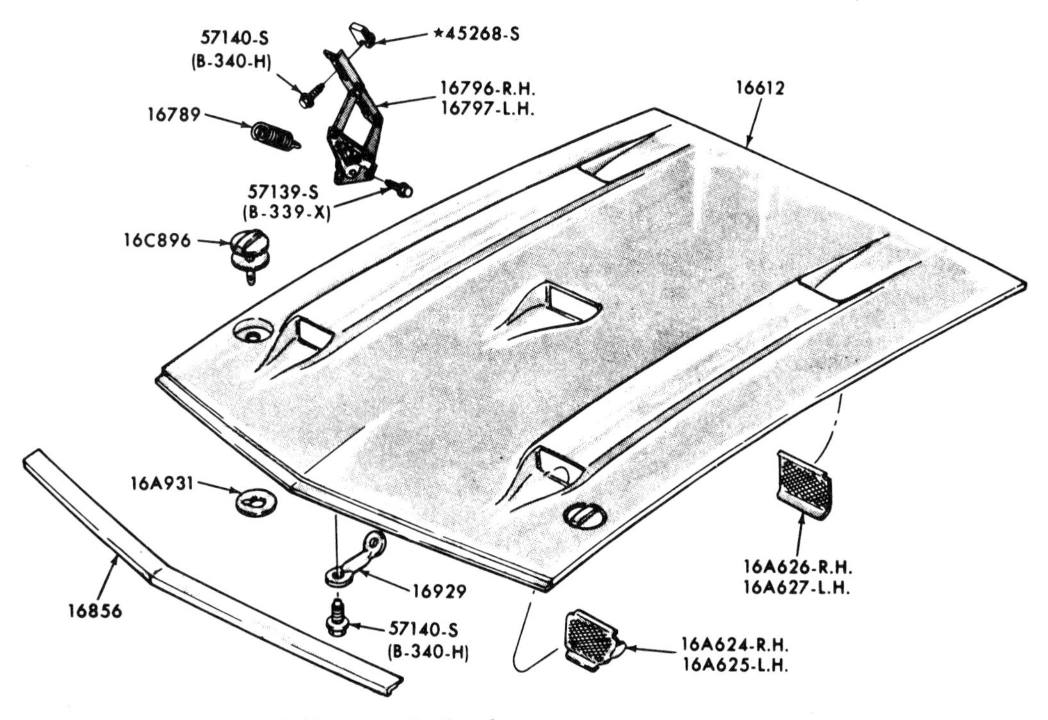

The 1969-1970 Shelby Mustang hood, hinges, and related parts.

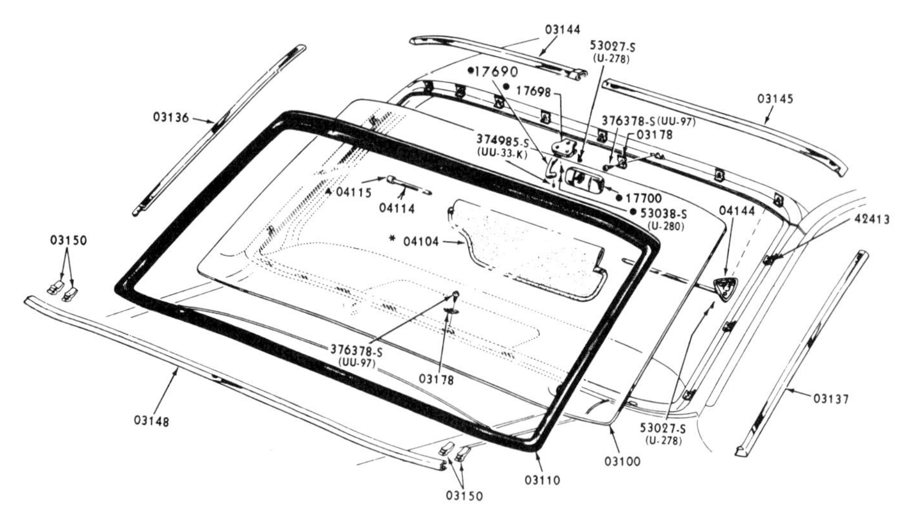

The 1965-1968 Mustang hardtop and fastback windshield, gasket, trim, and related parts.

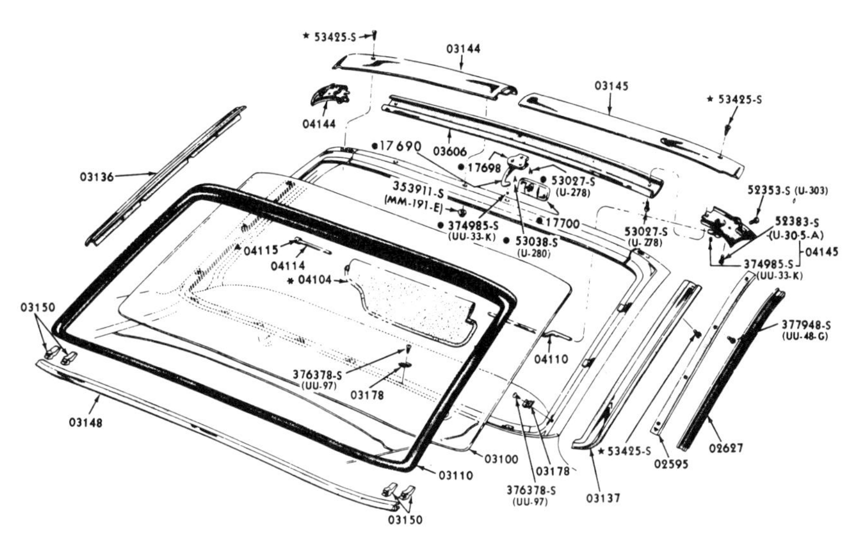

The 1965-1968 Mustang convertible windshield, gasket, trim, and related parts.

MUSTANG INSIDE REAR VIEW MIRROR APPLICATION CHART

YEAR	PART NUMBER	DESCRIPTION
65/66	C7OZ-17700-C	Day/night - 2.38" x 10" - 1/4" flat across
		shank for set screw mounted - vinyl covered
67	C7AZ-17700-F1A	Day/night - 2.42" x 10" - 1/4" flat arm for
		bolt mounted - black vinyl covered
67	C7AZ-17700-F1B	Less mounting arm - breakaway type - blue
		Day/night - 2.45" x 10" - diam. 1/2" - ball
		mounted - black vinyl covered less mounting arm
68	C7AZ-17700-E1K	Less mounting arm - cement-on type
70		Day/night - 2.45" x 10" - undetachable mounting
		arm - black vinyl covered

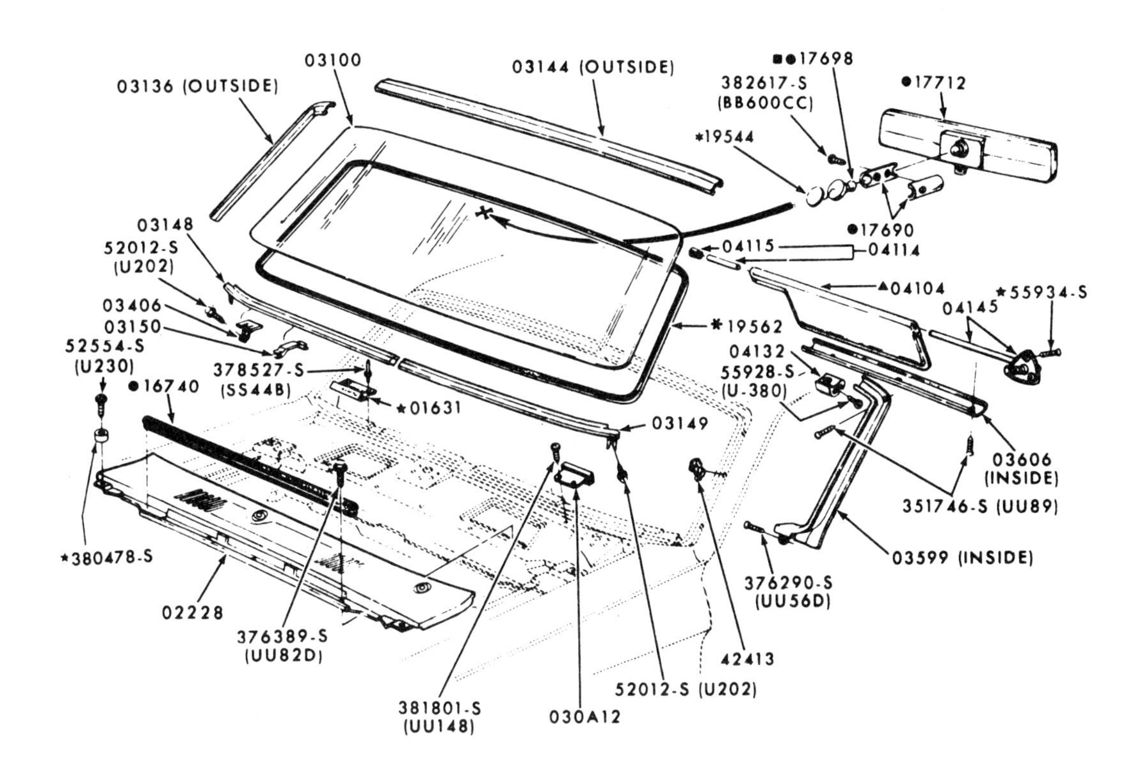

The 1969-1970 Mustang hardtop and SportsRoof windshield, gasket, trim, and related parts.

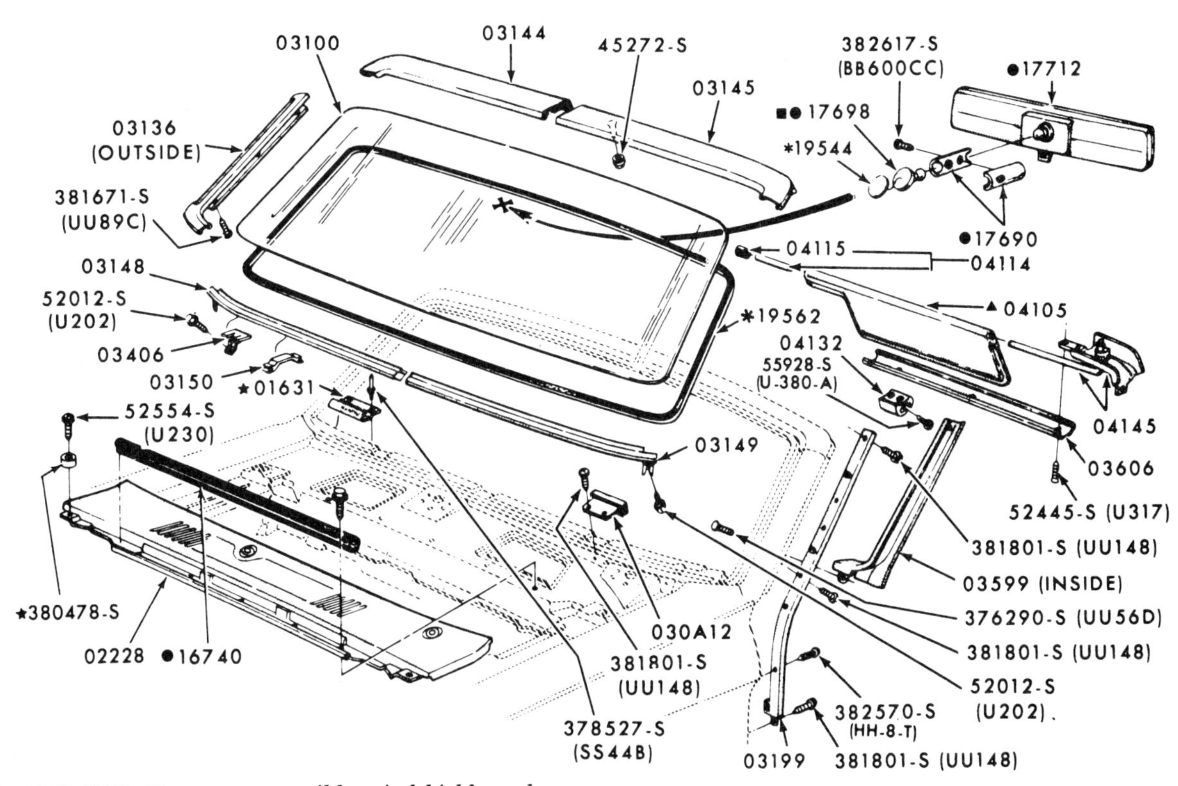

The 1969-1970 Mustang convertible windshield, gasket, trim, and related parts.

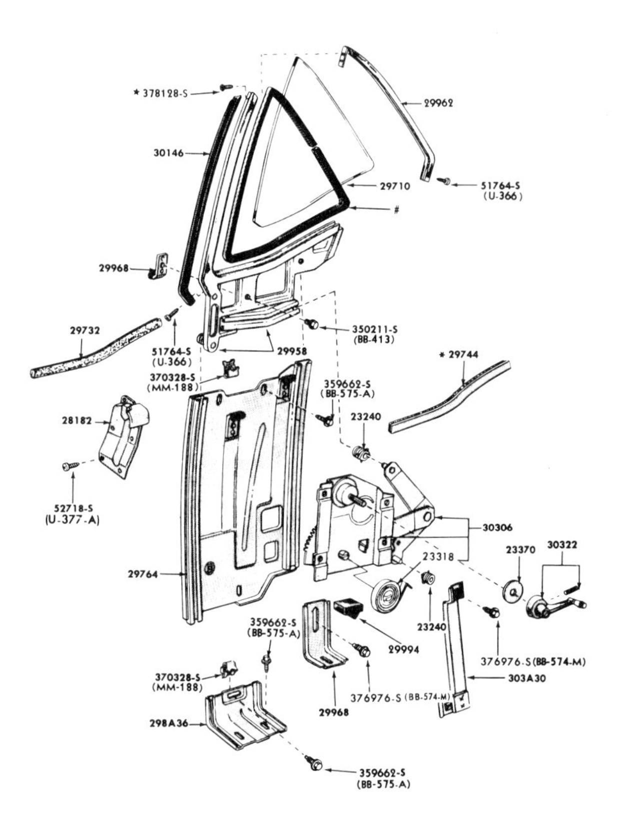

The 1965-1966 Mustang hardtop and convertible quarter window, regulator, and related parts.

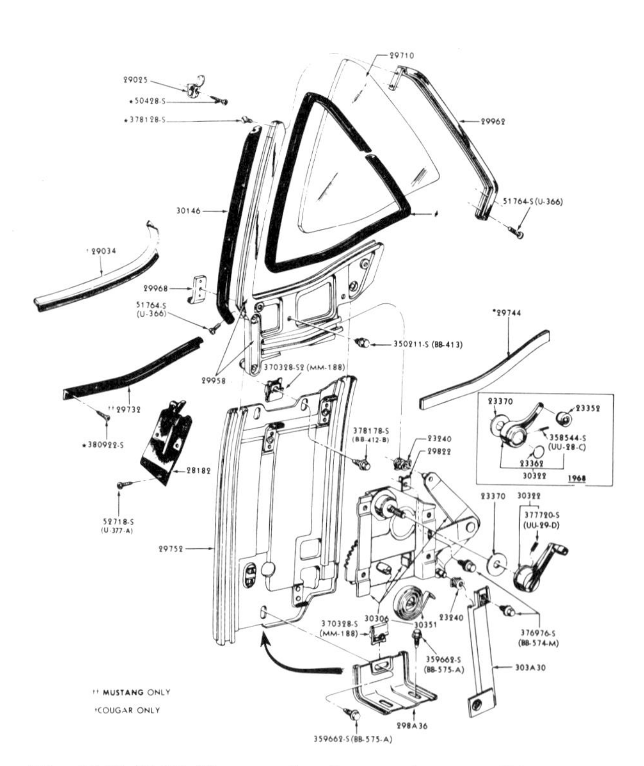

The 1967-1968 Mustang hardtop and convertible quarter window, regulator, and related parts.

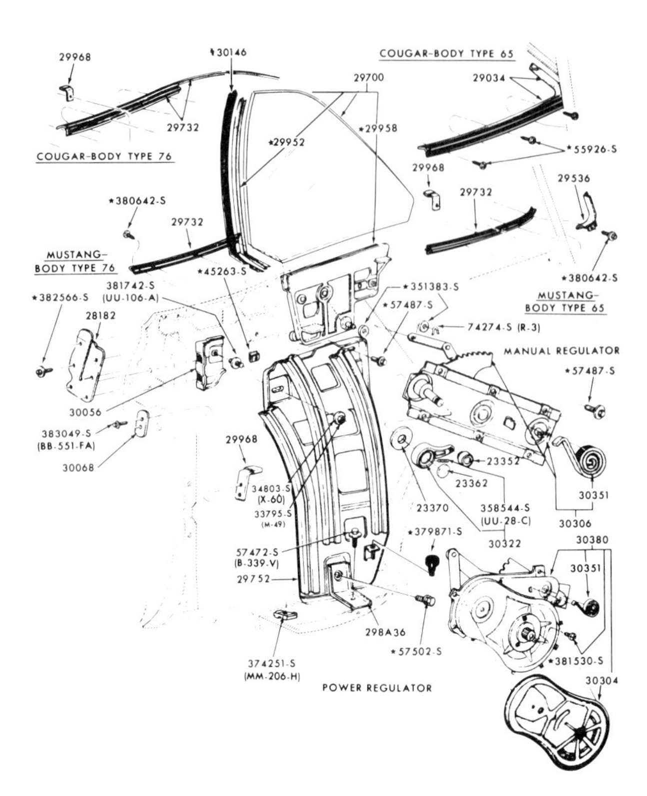

The 1969-1970 Mustang hardtop and convertible quarter window, regulator, and related parts.

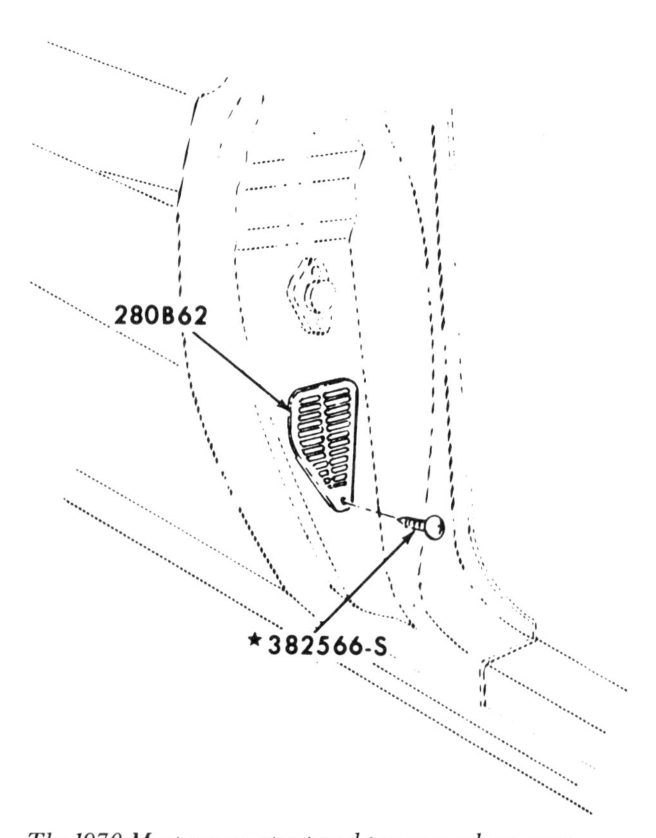

The 1970 Mustang quarter-panel pressure-release vent.

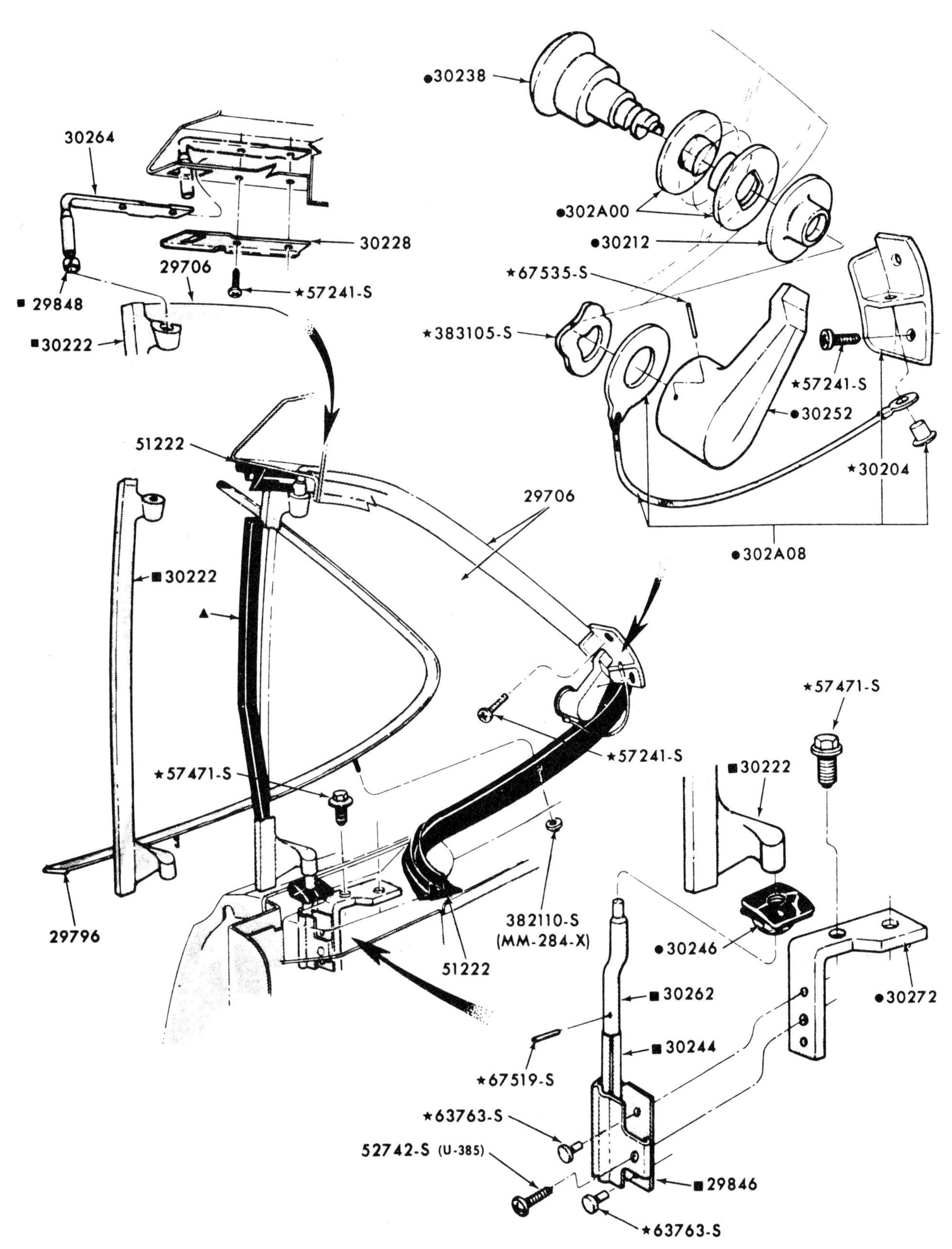

The 1969-1970 Mustang SportsRoof quarter window, latch, weatherstrip, and related parts.

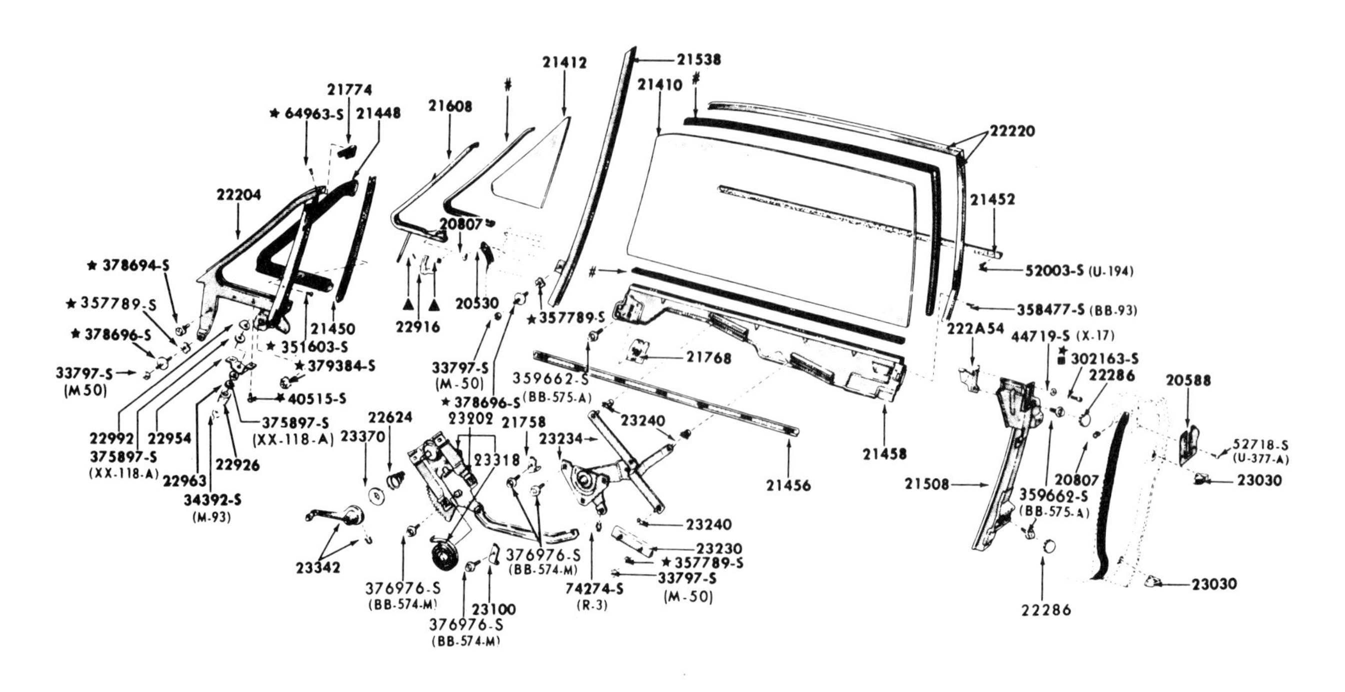

The 1965-1966 Mustang door window, window regulator, vent wing, and related hardware.

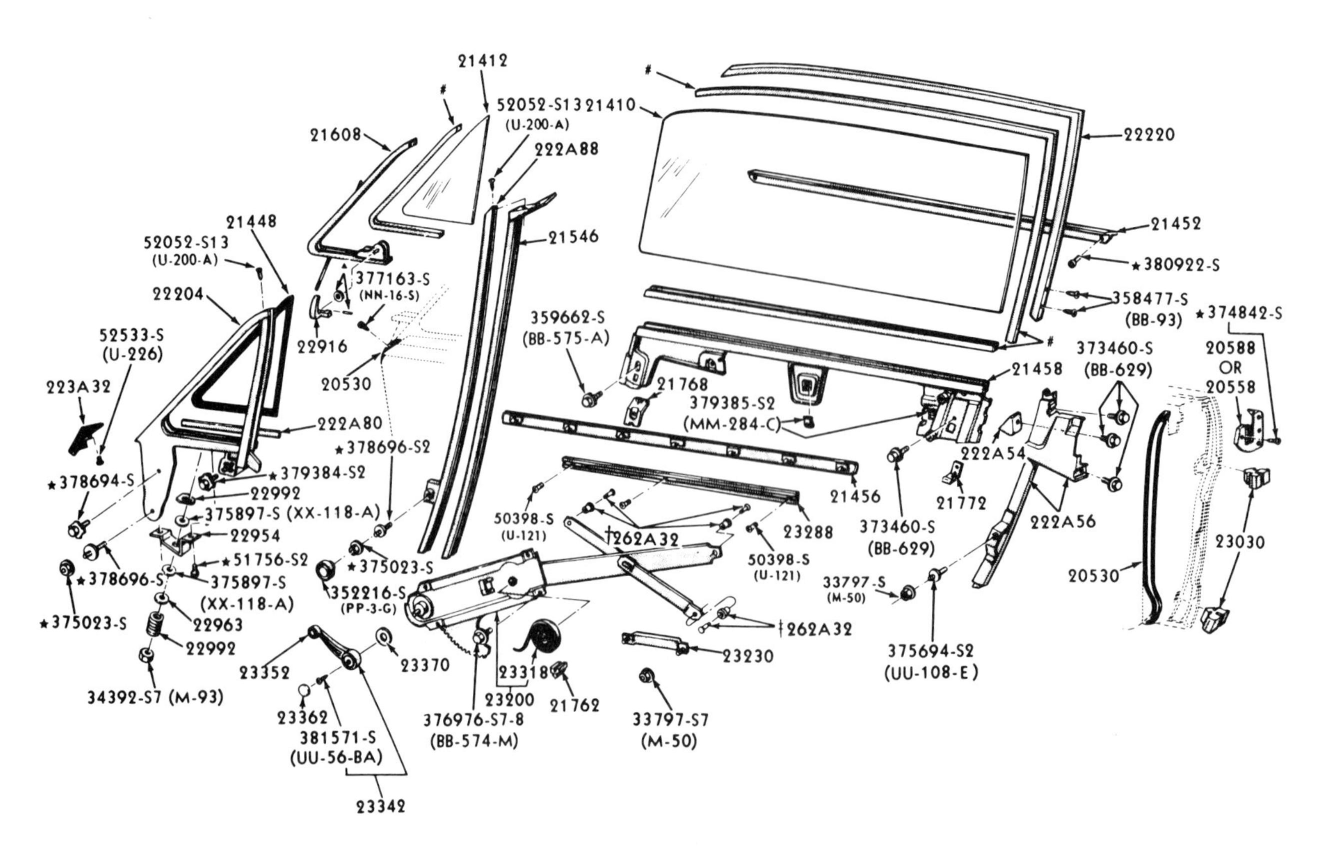

The 1967-1968 Mustang door window, window regulator, vent wing, and related hardware.

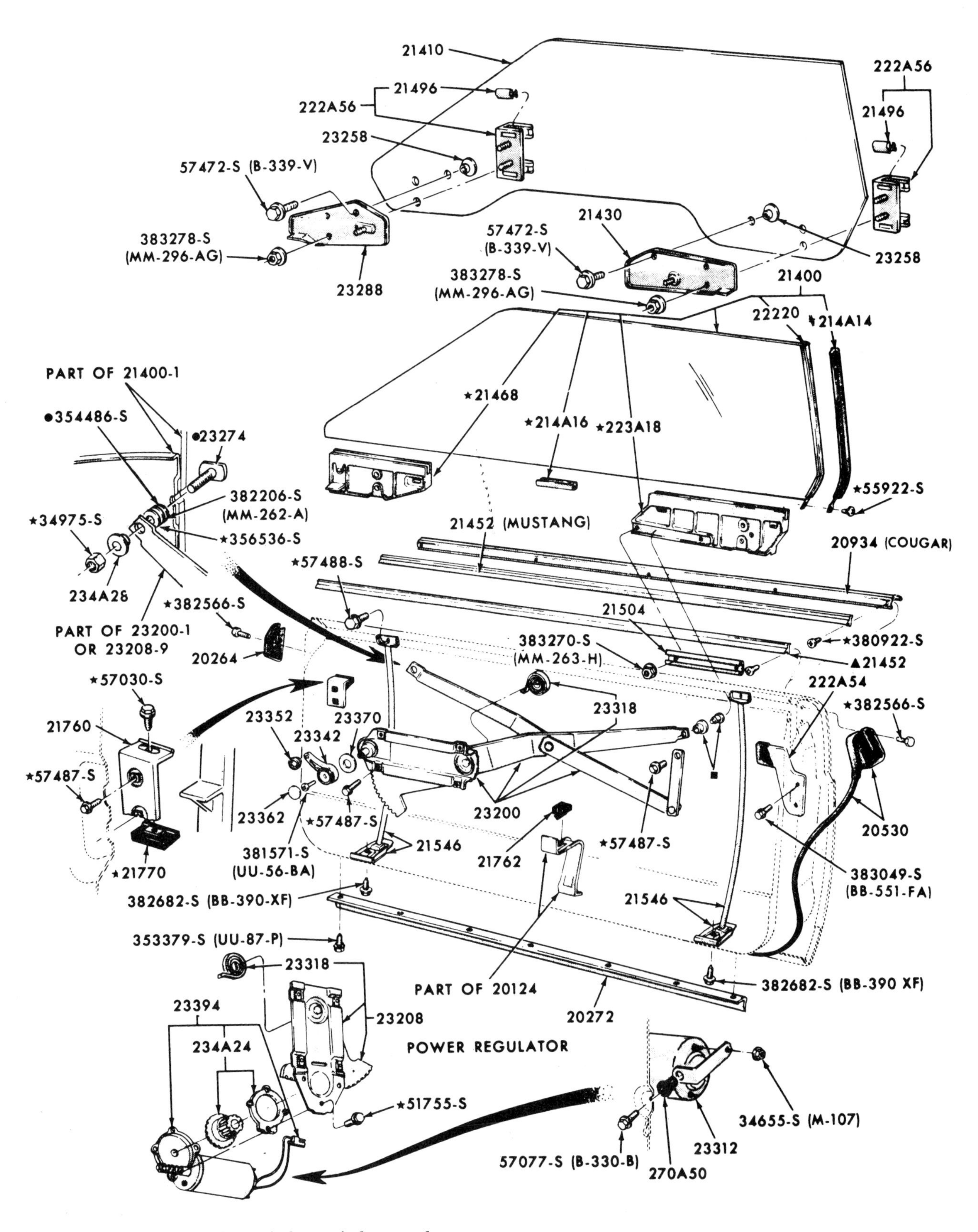

The 1969-1970 Mustang door window, window regulator, and related hardware.

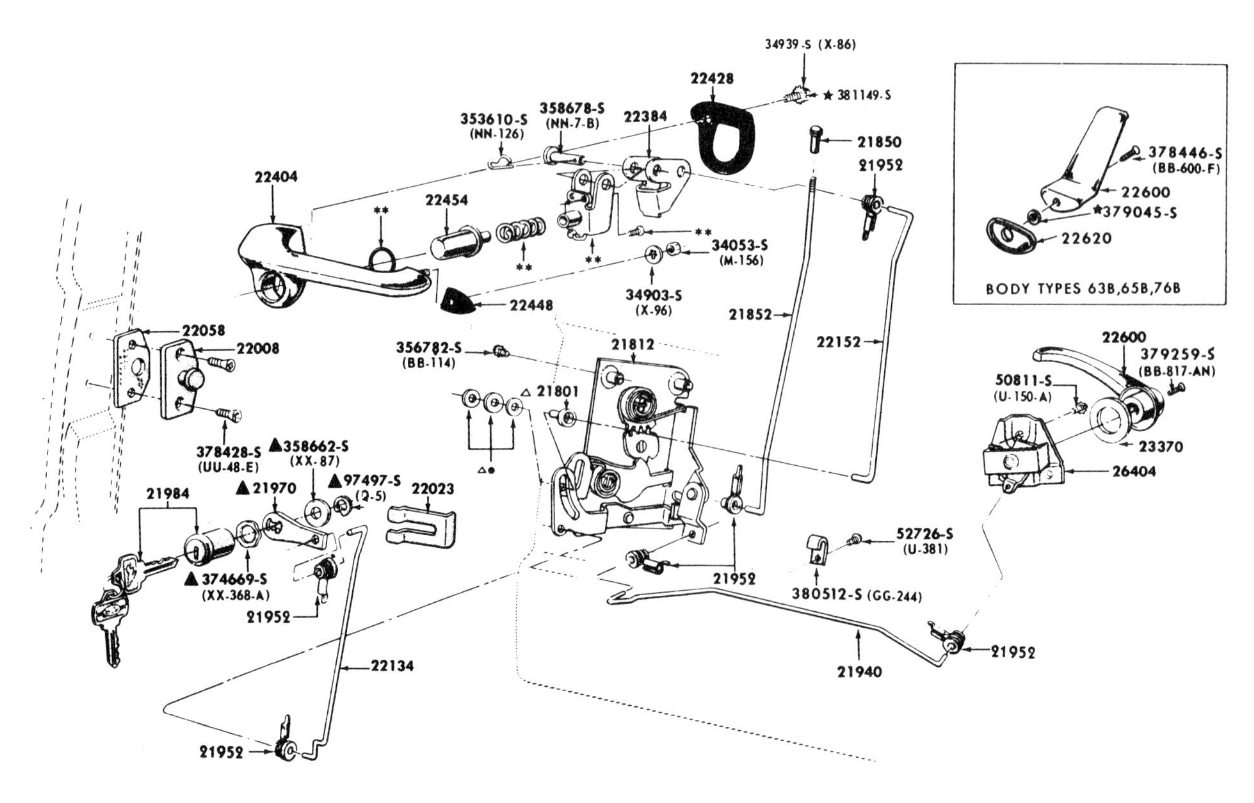

The 1965-1966 Mustang door handles, latch, and related hardware.

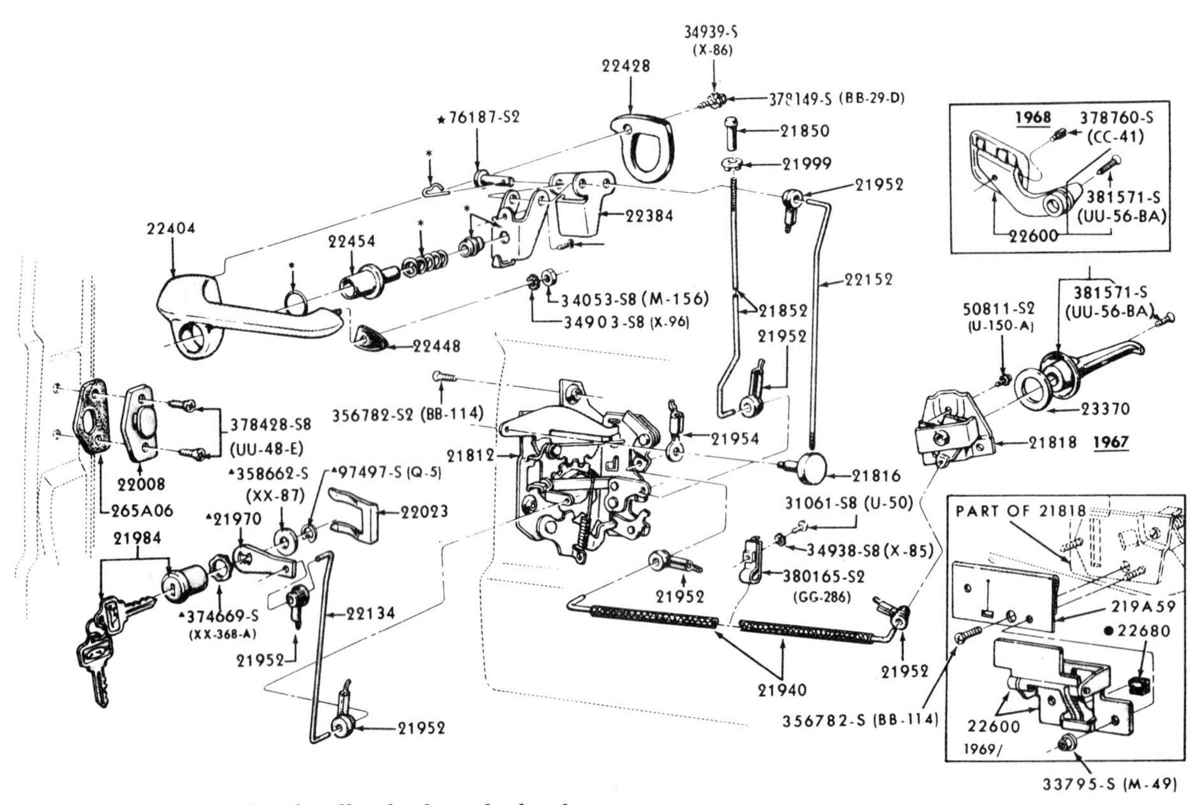

The 1967-1970 Mustang door handles, latch, and related hardware.

Functional rear-facing fresh air vents adorn the sides of all 1965-1968 Mustang fastbacks. The 1965-1966 louvers (shown) have five large openings.

The 1967-1968 fastback louvers are more numerous than the 1965-1966 models, and less pronounced.

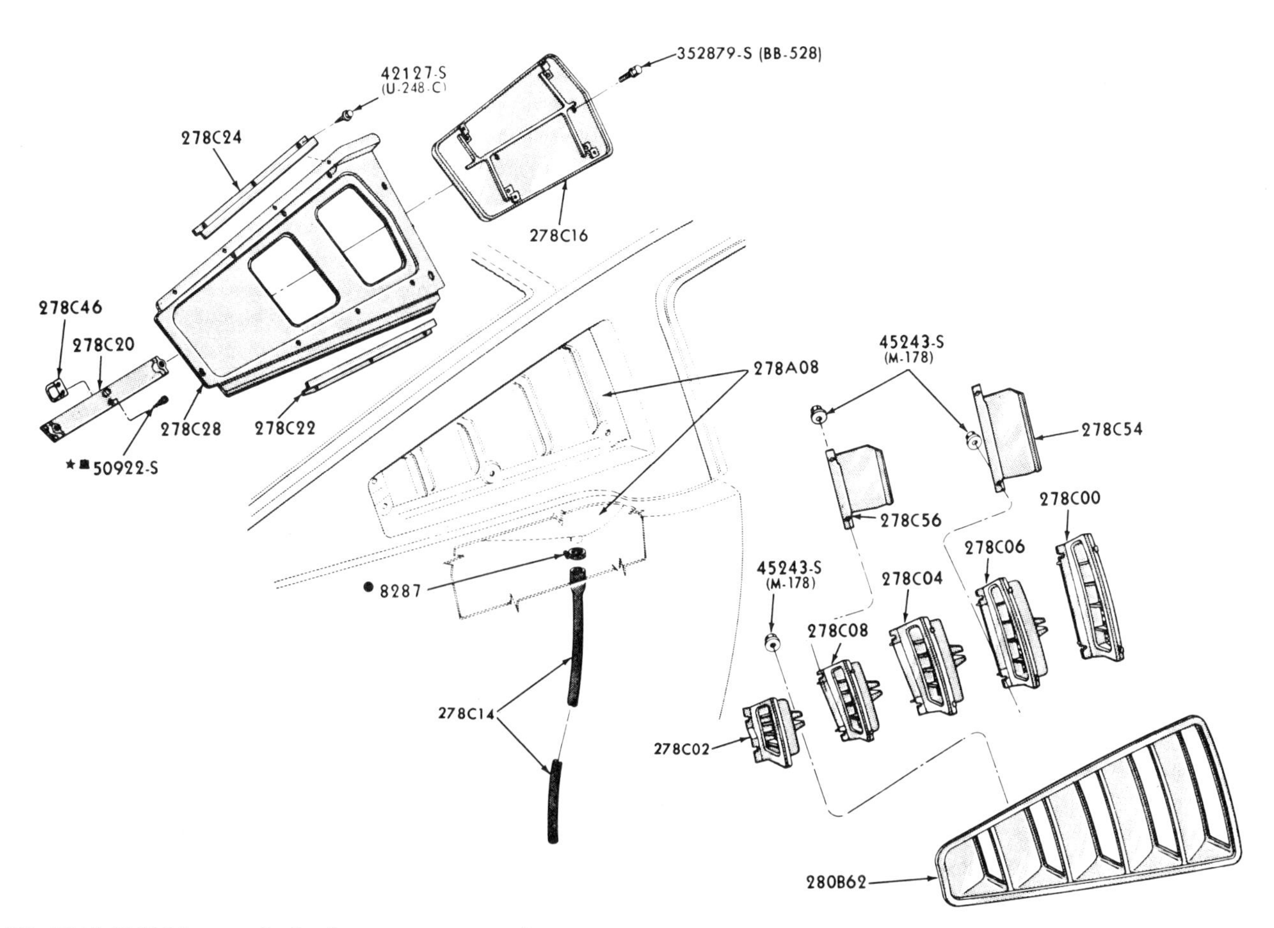

The 1965-1966 Mustang fastback rear-quarter vent louvers, hardware, and trim.

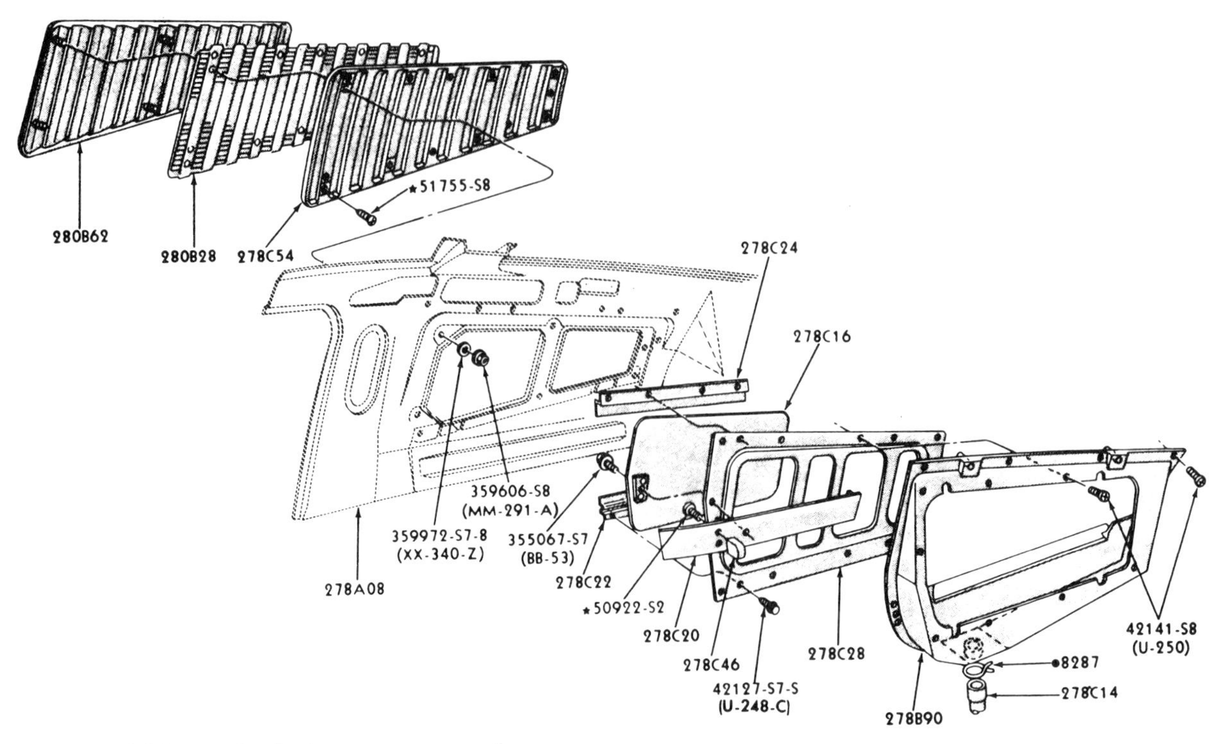

The 1967-1968 Mustang fastback rear-quarter vent louvers, hardware, and trim.

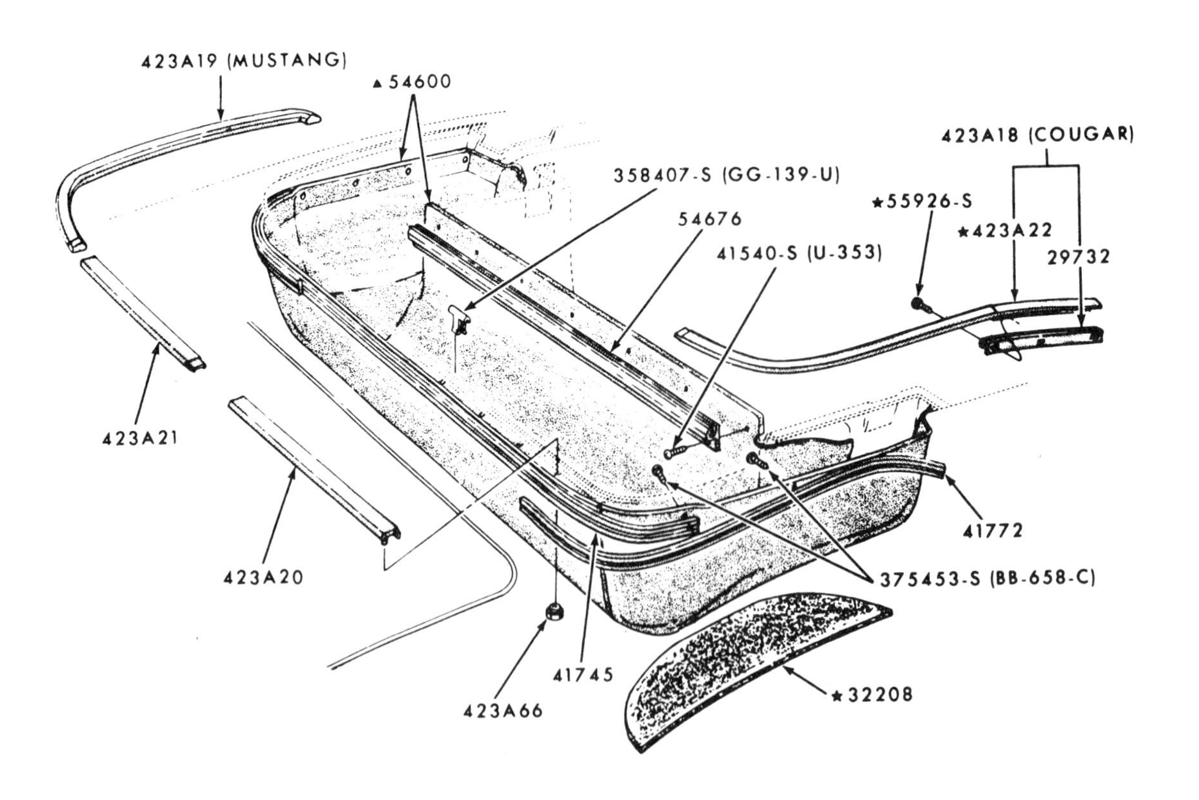

The 1969-1970 Mustang fastback rear window, gasket, and window molding.

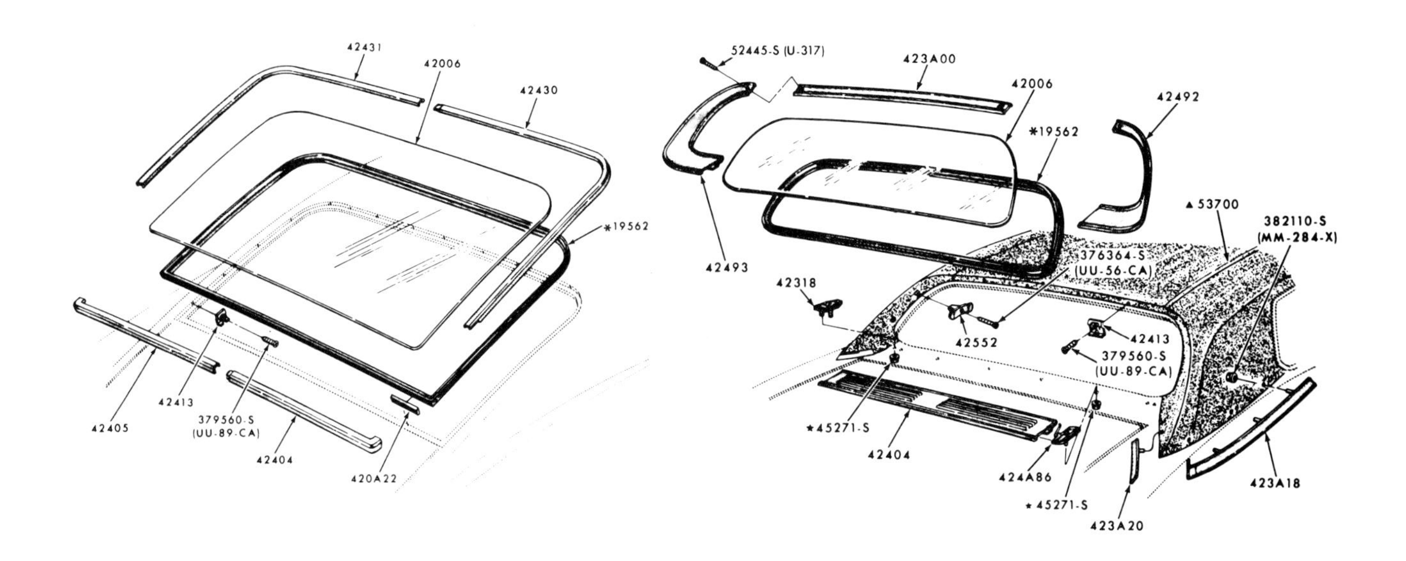

The 1969-1970 Mustang convertible-top well liner, rear belt moldings, and related parts.

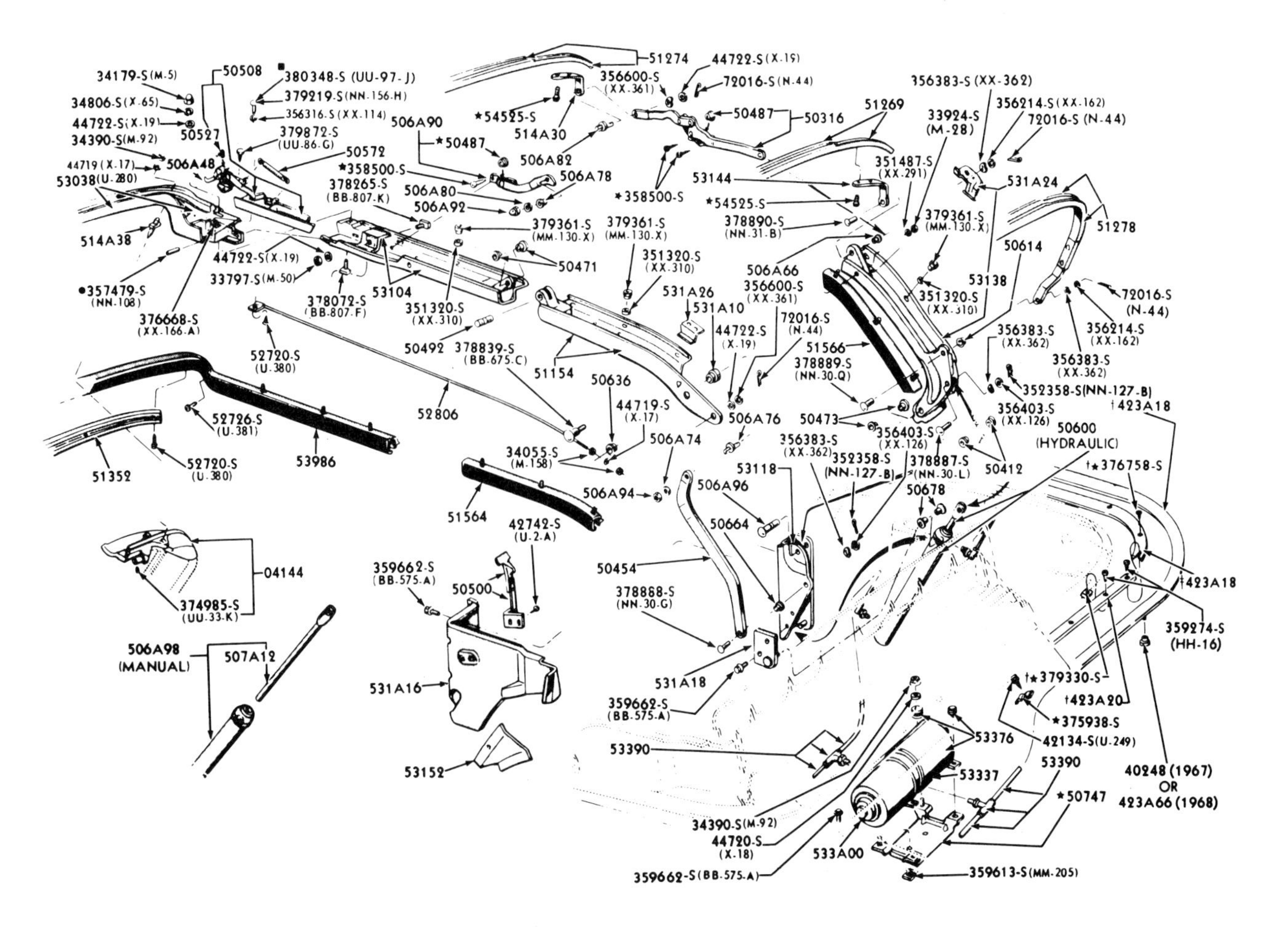

The 1965-1968 Mustang manual and power folding-top components, brackets, and hardware.

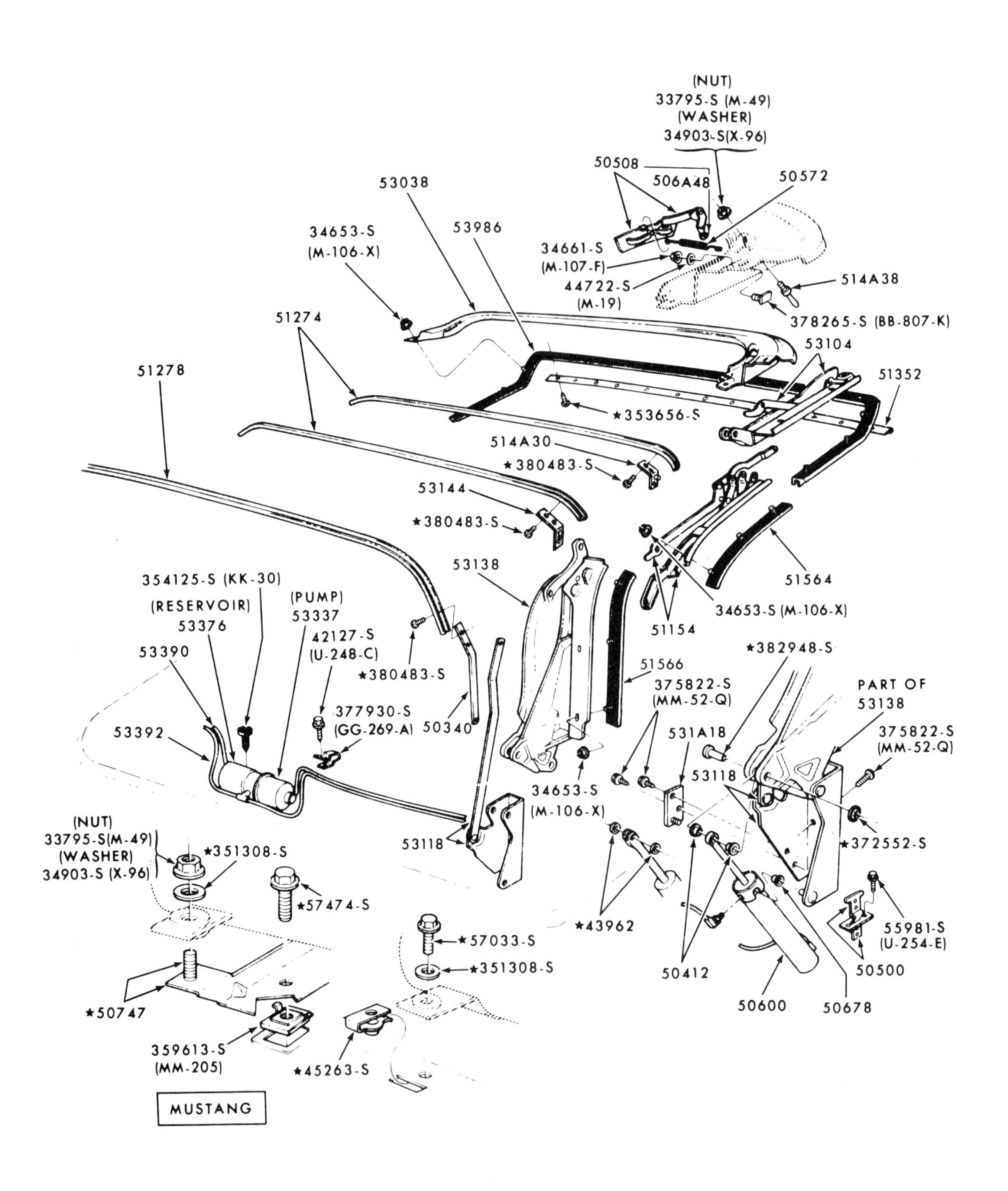

The 1969-1970 Mustang power folding-top components, brackets, and hardware.

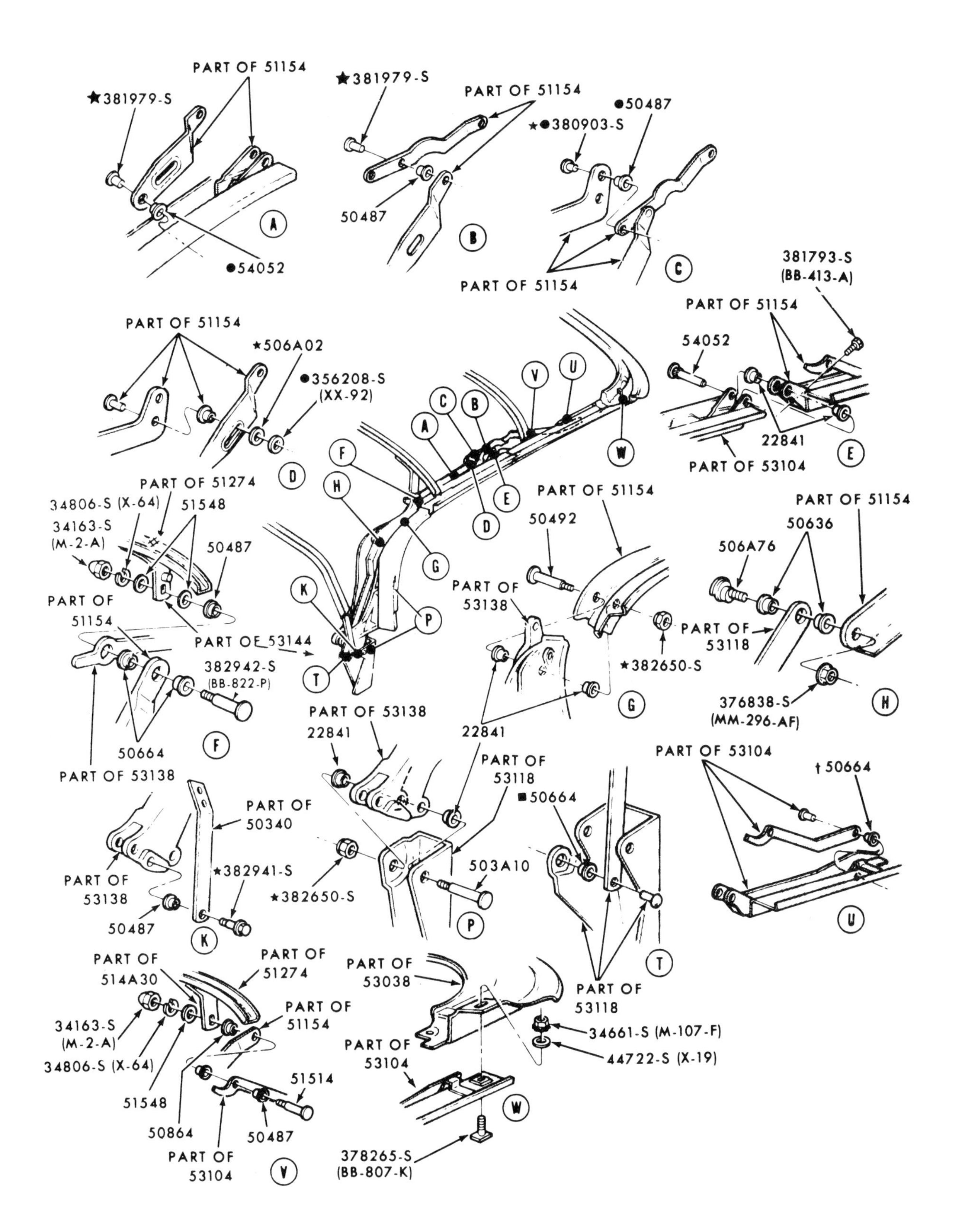

The 1969-1970 Mustang power folding-top components, brackets, and hardware (continued).

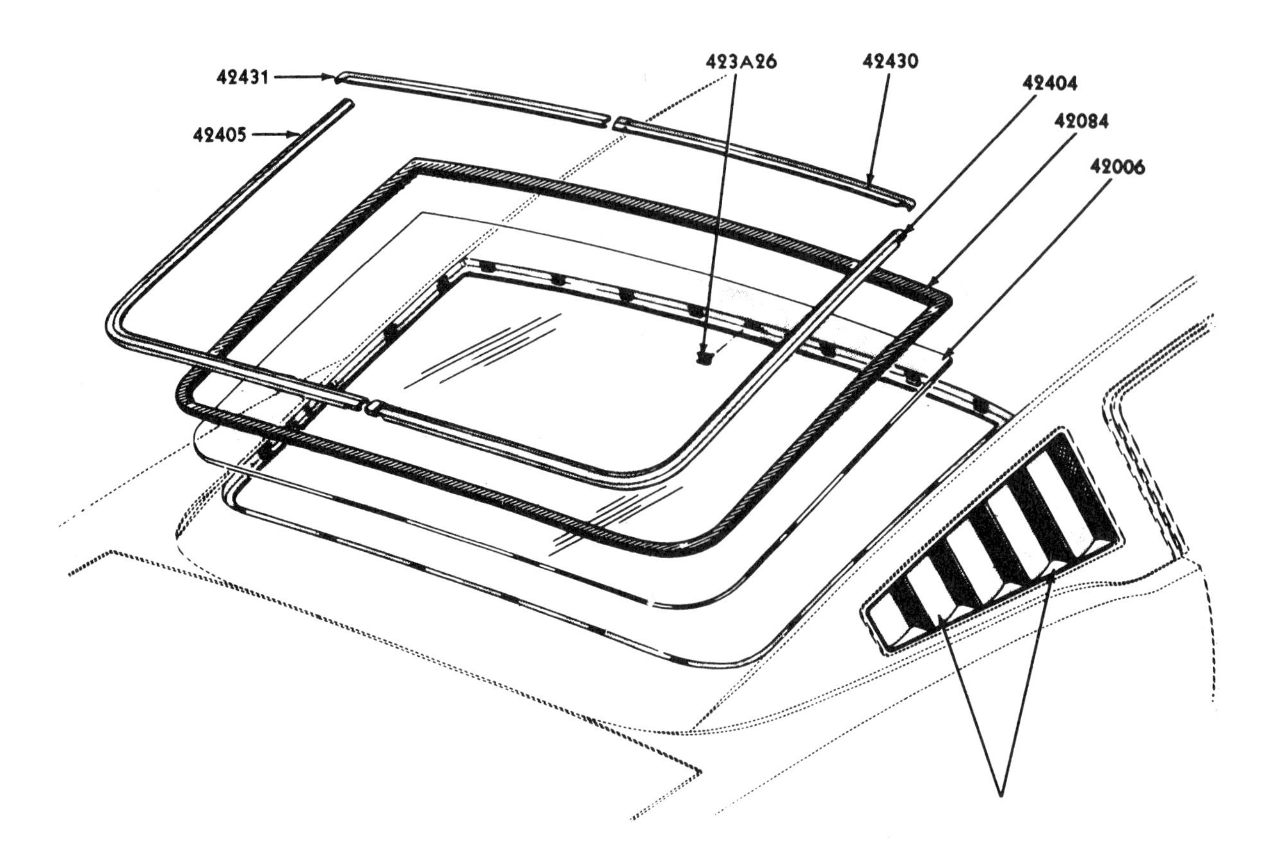

The 1965-1966 Mustang SportsRoof rear window, gasket, and window molding.

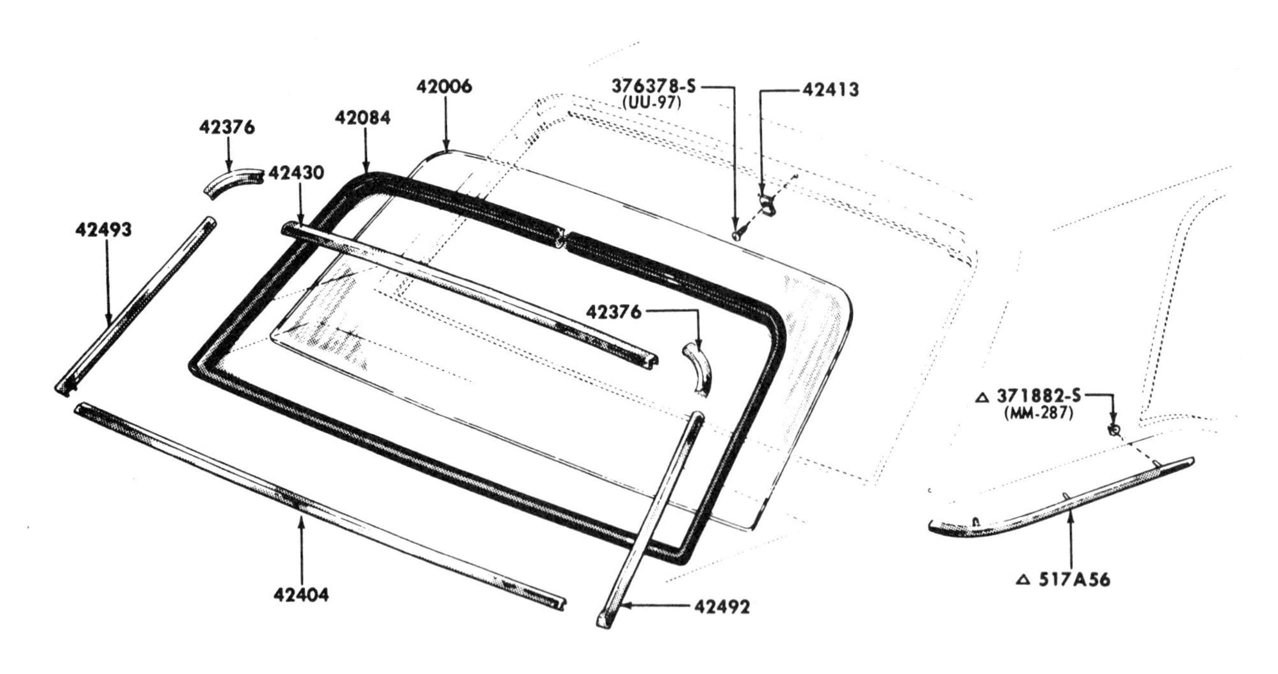

The 1965-1966 Mustang hardtop rear window, gasket, and window molding.

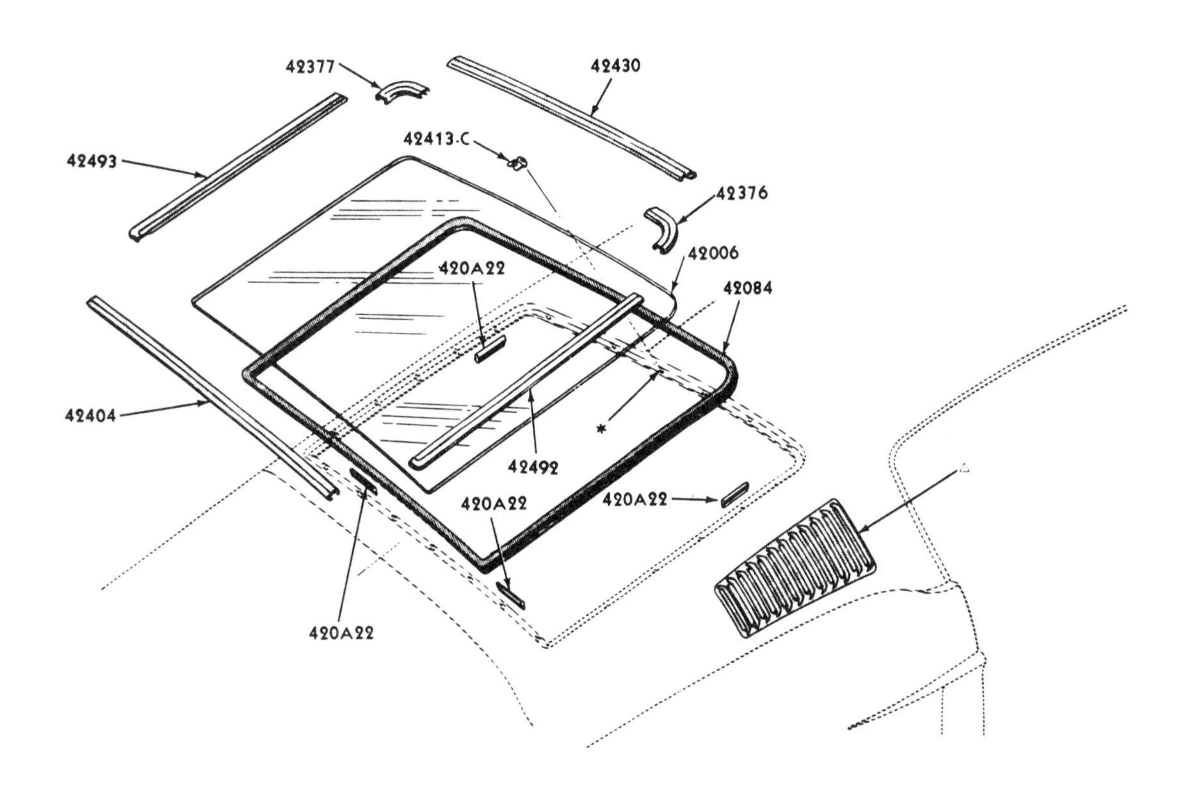

The 1967-1968 Mustang fastback rear window, gasket, and trim.

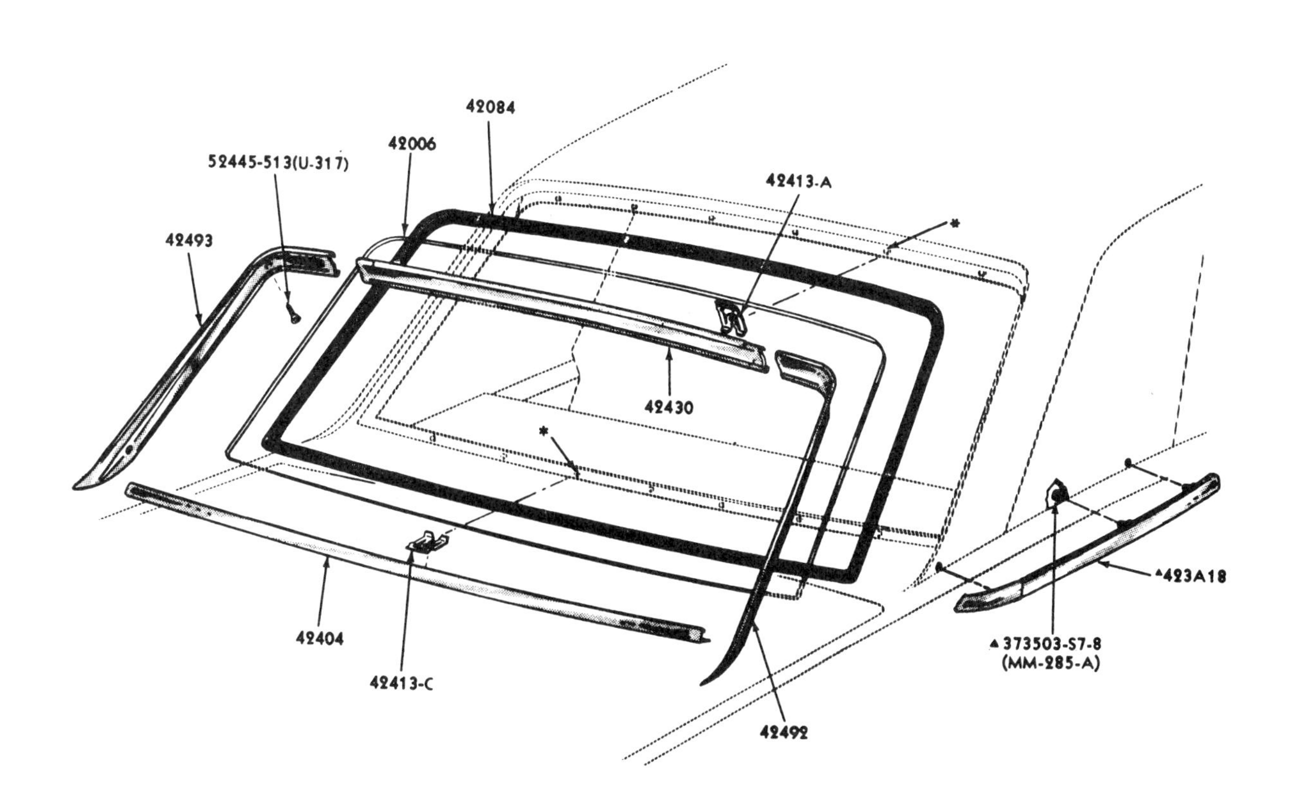

The 1967-1968 Mustang hardtop rear window, gasket, and trim.

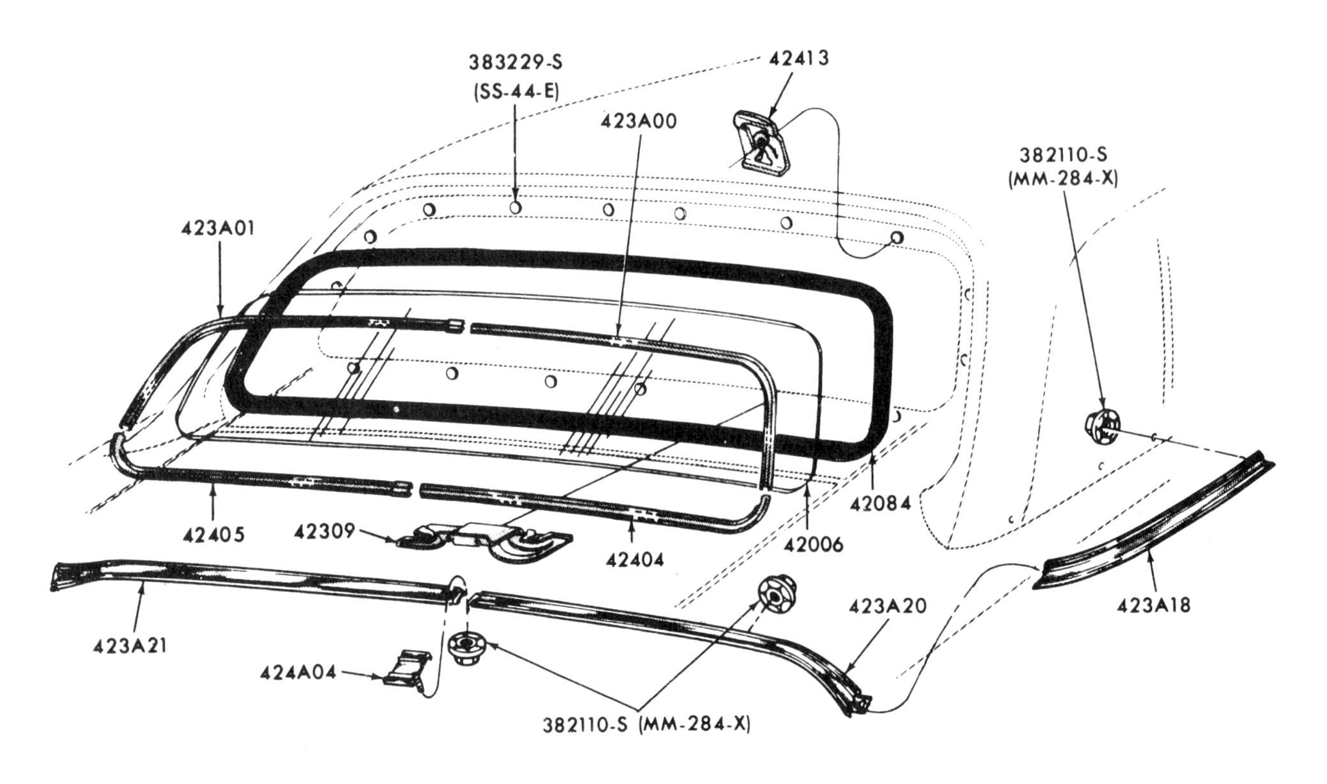

The 1970 Mustang hardtop rear window, gasket, and trim.

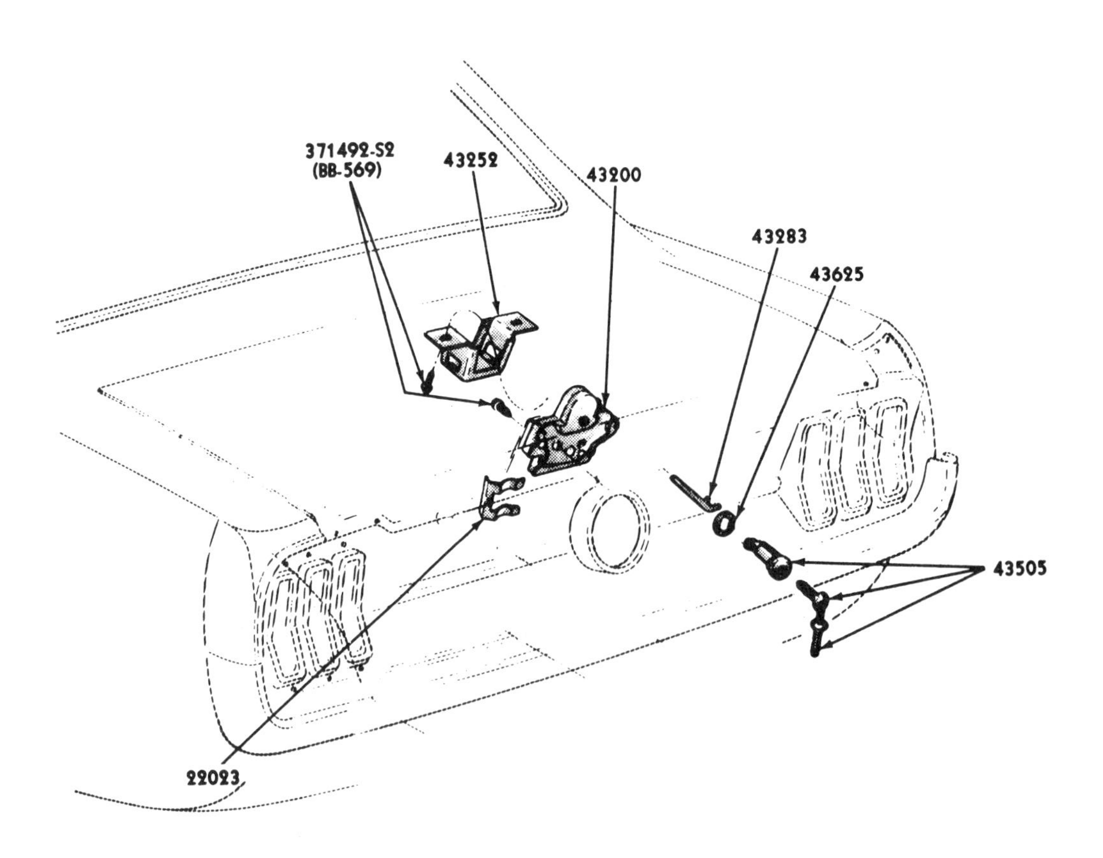

The 1967-1970 Mustang luggage compartment door latch and related parts.

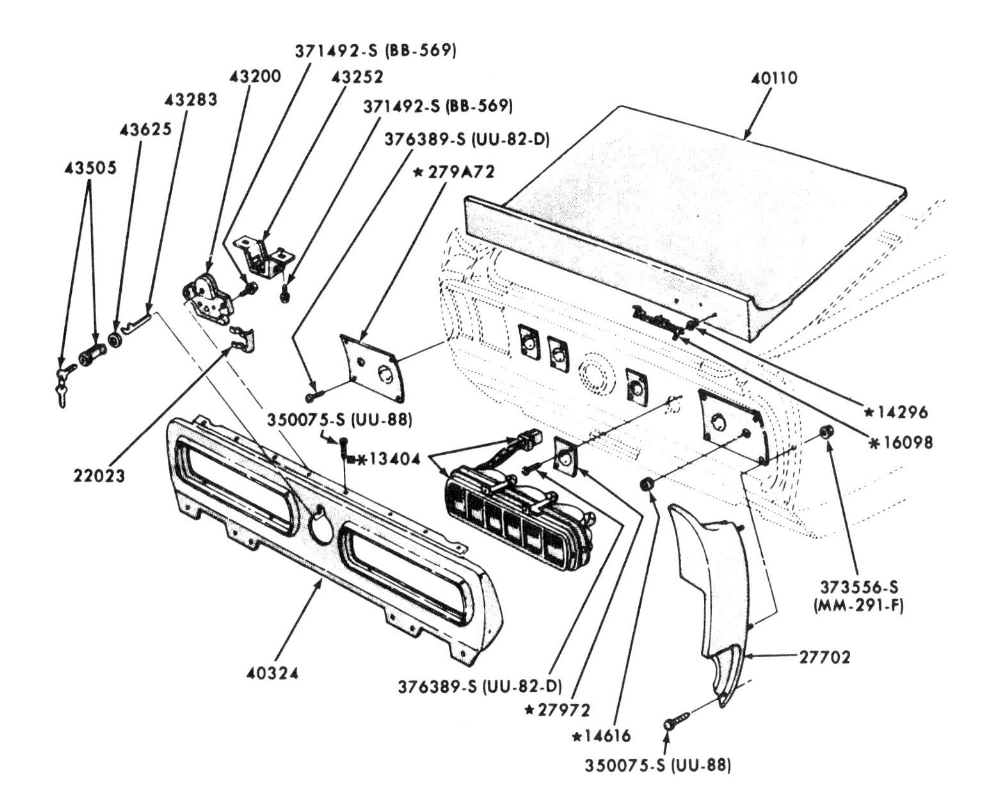

The 1968 California Special GT Mustang luggage compartment lid, fender extension, taillight, rear tail panel, luggage compartment latch, and related hardware.

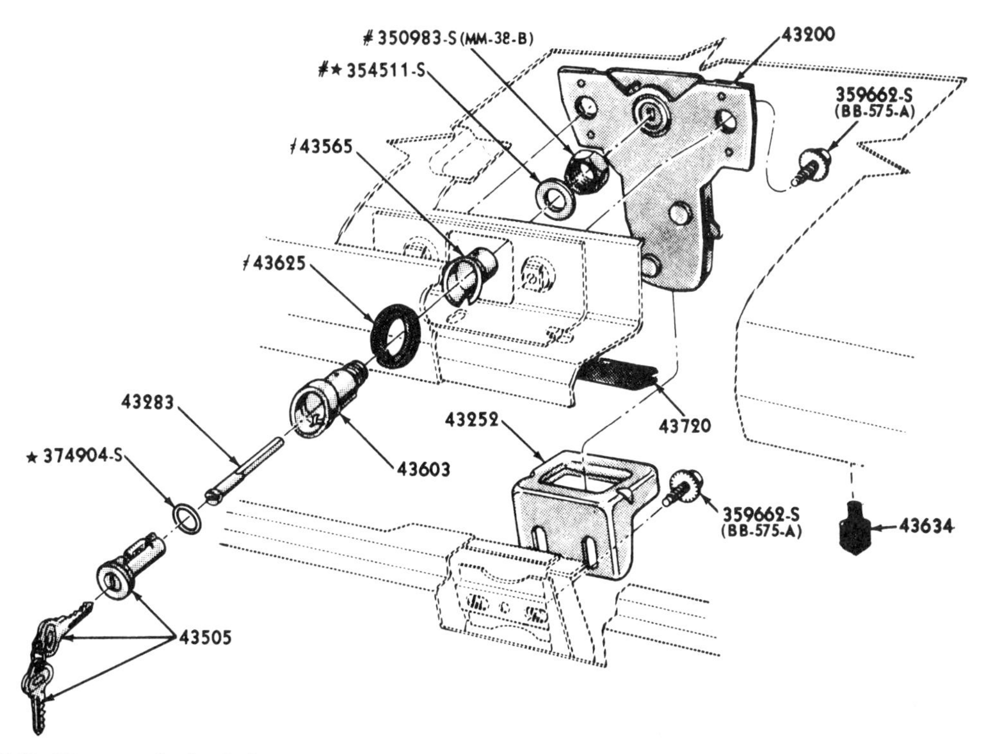

The 1965-1966 Mustang fastback luggage compartment door latch and related parts.

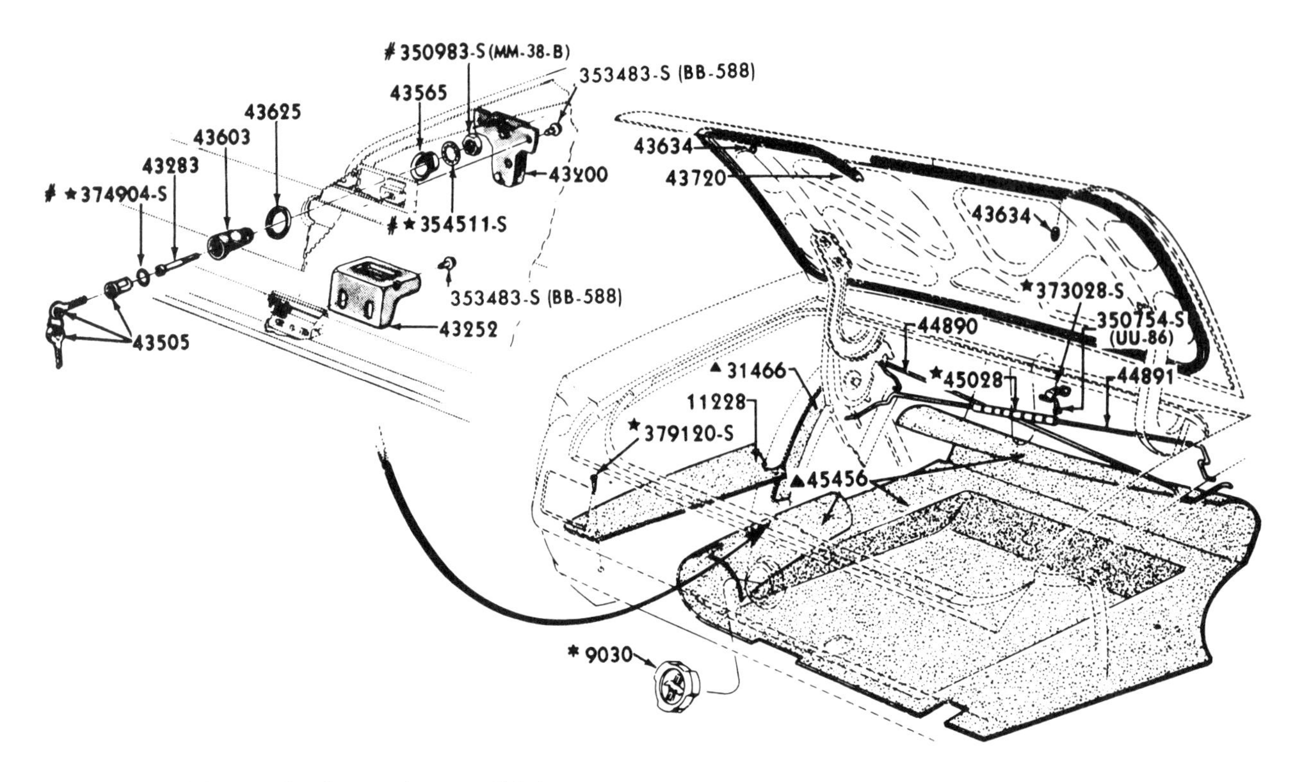

The 1965-1966 Mustang hardtop and convertible luggage compartment door latch, trunk floor covering, and related parts.

All 1966 Mustangs equipped with the GT package received a special gas cap (the 1965 GT received a standard Mustang gas cap). On this 1966 version the recessed letters in the Ford name were painted black while the word Mustang was in red. The raised GT letters are centered in a spun-aluminum disc and the remainder of the cap is chrome.

The three-fingered standard twist-off gas caps installed on all 1967 non-GT equipped Mustangs resembled a wheel-cover spinner. The intricate plastic center retains the FORD MUSTANG inscription encircling the traditional running horse logo.

The 1967 was the first year for the optional pop-open gas cap. On GT Mustangs, the embossed running horse logo is replaced by the letters GT. The 1967 GT pop-open cap also was used on 1968 Mustang GTs.

A non-GT pop-open cap was available in 1968. The unique embossed logo is centered on a brushed-aluminum disc.

An optional ribbed rear grille panel was available on 1967 Mustangs equipped with the Exterior Decor Group.

The same standard twist-off gas cap is installed on all 1969 and 1970 Mustang body styles except Mach 1, GT, and Shelby.

The standard twist-off gas cap for 1968.

The 1969 GT Mustangs received their own unique pop-open gas cap with its gear-toothed center disc and traditional GT lettering.

Mach 1 pop-open gas caps were outfitted with the running horse logo set atop a tri-colored bar.

The 1970 Mach 1 pop-open cap is similar to the model used in 1969. A flat tail panel (between the taillamps) was brought back for 1970. It was the first year since 1966 that the tail panel was not concave. Mach 1s and cars optioned with the Exterior Decor Group received a honeycomb rear panel.

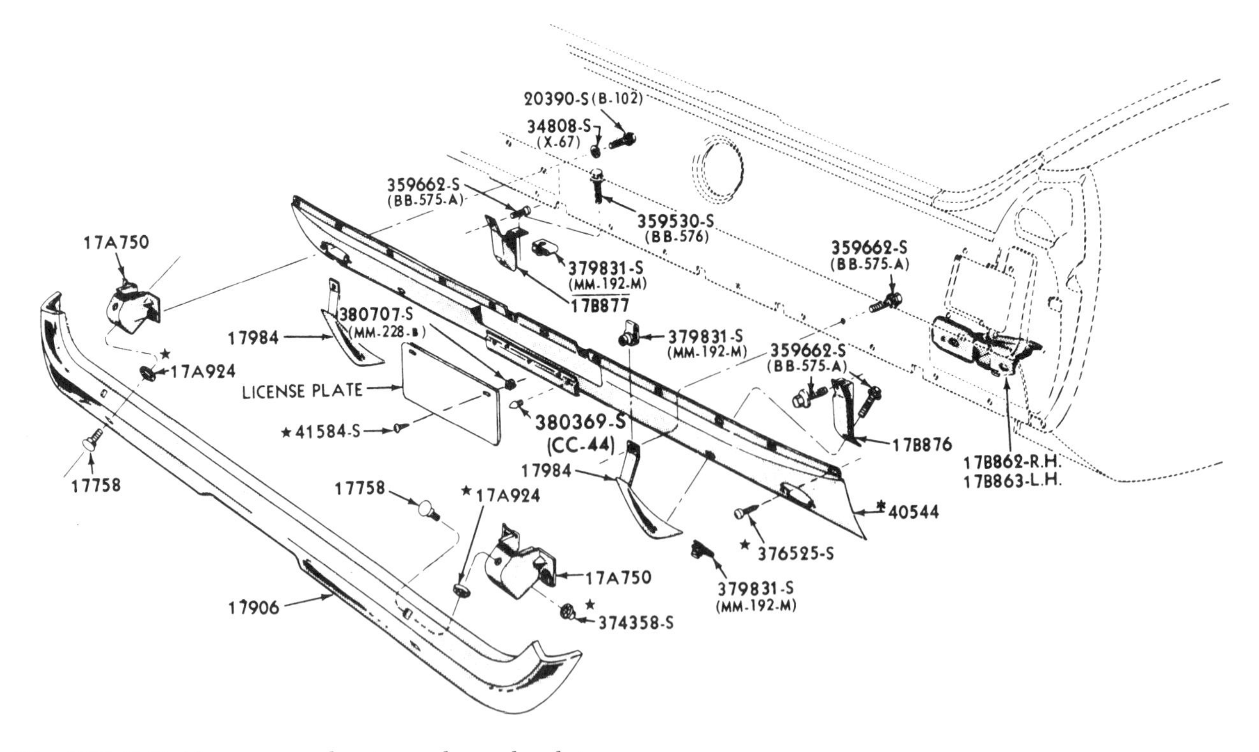

The 1965-1966 Mustang rear bumper, valance, brackets, and related mounting hardware.

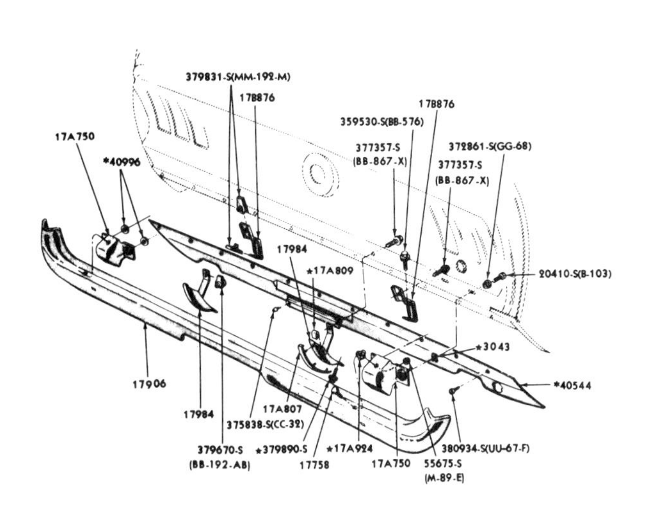

The 1967-1968 Mustang rear bumper, valance, brackets, and related mounting hardware.

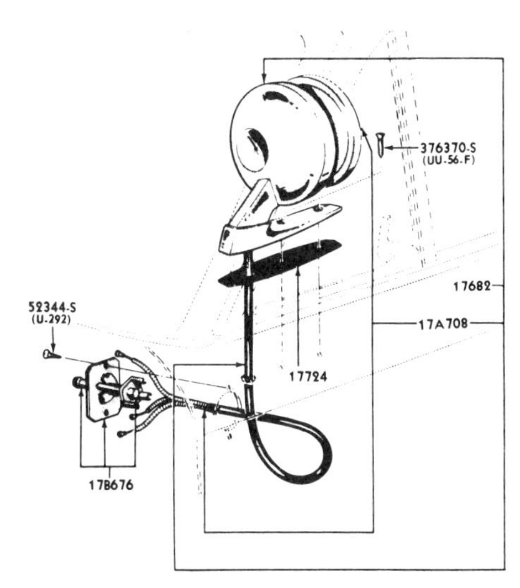

The 1965-1966 Mustang driver's side remote-control rear-view mirror.

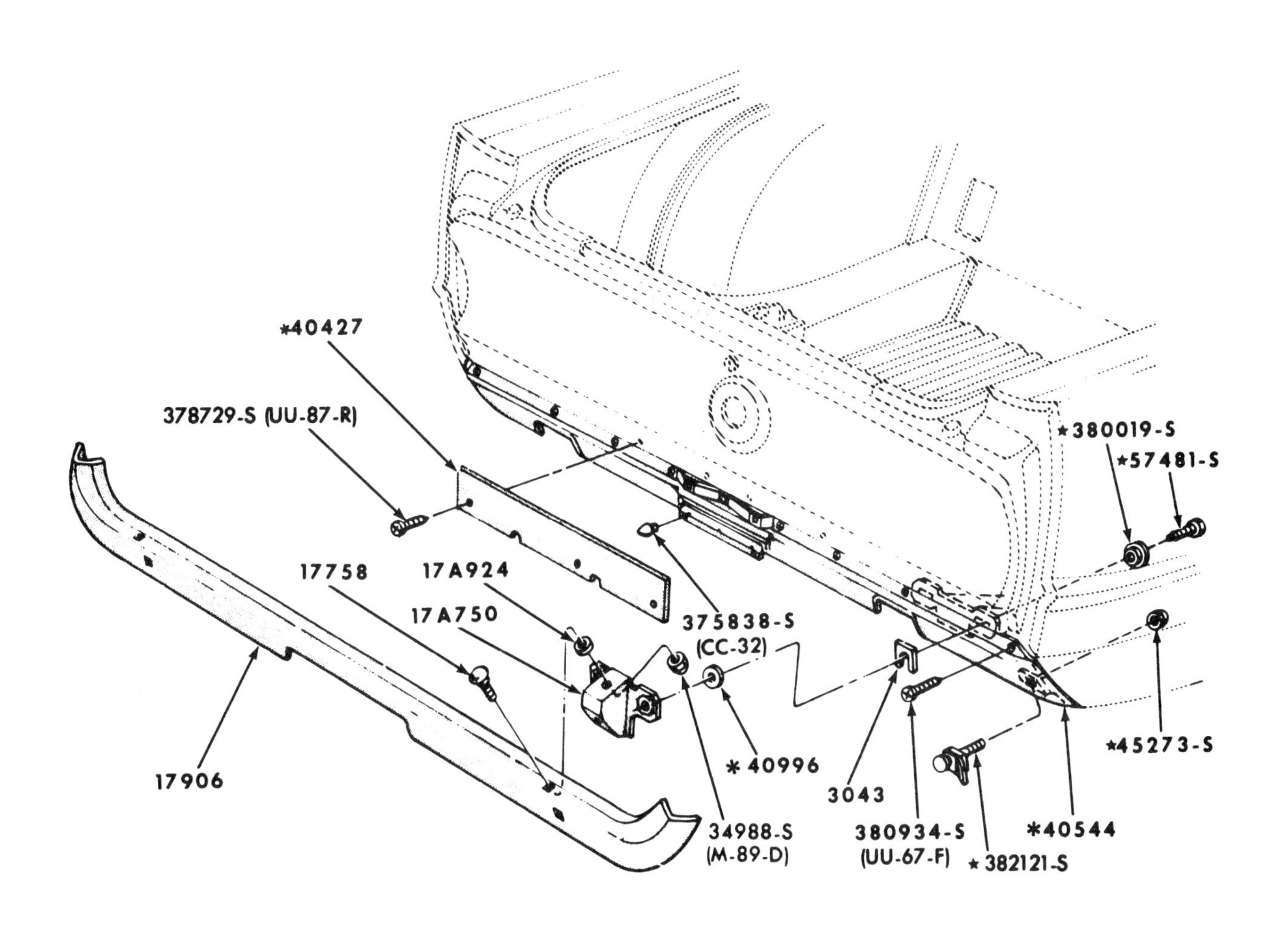

The 1969-1970 Mustang rear bumper, valance, brackets, and related mounting hardware.

MUSTANG OUTSIDE REAR VIEW MIRROR ASSEMBLY APPLICATION CHART

Year	Type	Part Number	Description	I.D. Number
65	1	B5AZ-17696-A	Conical shaped, not replaced	C3RA-17682-D
65/66	2	C1SZ-17696-B	Universal type, round head	C5DB-17682-C
65/66	5	C5AZ-17696-F	Rectangular head	C5AB-17682-C,F C5OB-17682-C
65/66	6	C3RZ-17696-A	L.H., universal type, round head. Kit.	
67	12	C7AZ-17696-D	Trailer towing mirror, R.H. or L.H. fender mounted. Kit.	
65	12	D3AZ-17696-B	Trailer towing mirror, R.H. or L.H. fender mounted. Kit.	
68	9	D4AZ-17696-A	Kit. R.H. or L.H., manual, rectangular head.	
67/68	9	D4AZ-17696-B	L.H., manual, rectangular head.	C7AB-17683-A C7TB-17683-U C7TB-17683-AB C8AB-17683-B
67	5	C7AZ-17696-E	L.H. - remote control - door mounted	C7AB-17683-B,D C7AB-17683-G,H C7AB-17683-J C7OB-17683-B
67/68	5	C8AZ-17696-A	R.H., rectangular head, companion to L.H. remote control mirror.	C7AB-17682-D,E C8AB-17682-C C8DB-17682-A C8GB-17682-A,B C8GB-17682-C,D
68	5	C8AZ-17682-D	L.H., remote control, rectangular head.	C8AB-17683-D
69	5	C9ZZ-17682-B	L.H., remote control, rectangular head for non-Mach 1 SportsRoof & Conv.	C9ZB-17683-D C9ZB-17683-F
69	9	C9ZZ-17682-D	L.H., remote control, rectangular head except Mach 1 & Grandé	C9ZB-17683-A
69	13	C9ZZ-17682-C	L.H., remote control, racing type head. Before 2/24/69	C9ZB-17683-CW C9ZB-17683-EW
69	13	C9ZZ-17682-E	L.H., remote control, racing type head. from 2/24/69	C9ZB-17683-GW
69/70	13	C9ZZ-17682-A	R.H., manual, racing type head.	C9ZB-17682-AW C9ZB-17682-BW
70	5	DOZZ-17696-A Consists of: DOZZ-17682-A C9ZZ-17724-B C8SZ-17B733-B C7SZ-17B732-B (2) 382879-S45 382031-S100 57259-S2 382930-S	L.H., remote control, rectangular head. Mirror assembly Gasket Plate Nut No. 10 screws (24 x 1") Clip assembly No. 8 screw/washer (32 x .75") Clip-wiring	
70	13	DOZZ-17682-B	L.H., remote control, racing type head.	DOZB-17683-BW
70	5	DOZZ-17682-A	L.H., remote control, rectangular head.	DOZB-17683-C

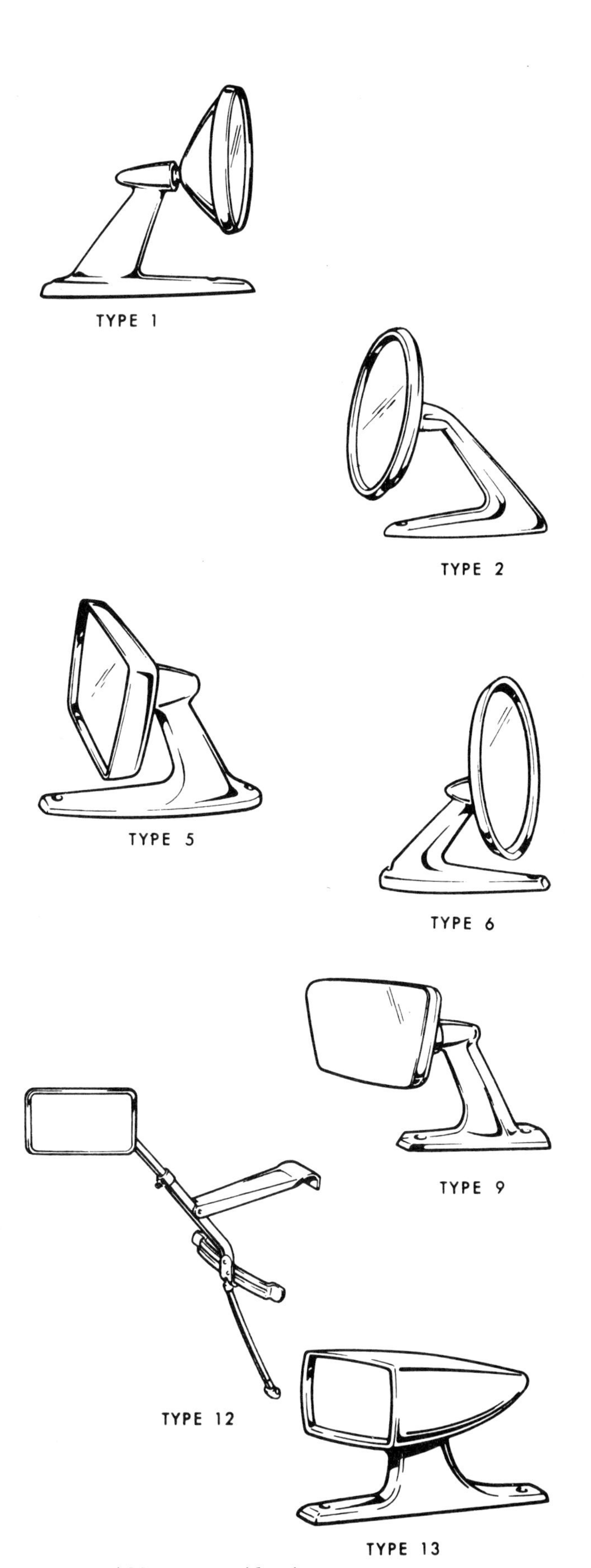

The 1965 Mustang outside mirrors.

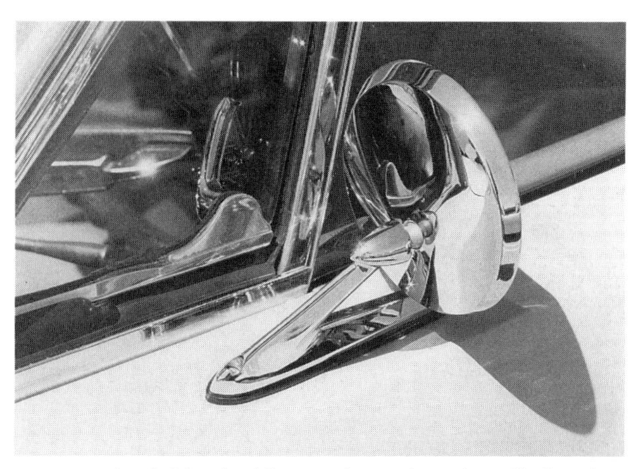

The standard driver's side rearview mirror installed on late 1965 and all 1966 Mustangs was an adaptation of a common mirror used on many other Ford cars. A conical-shaped mirror was installed on early 1965 Mustangs.

A passenger's side non-remote mirror was an inexpensive option on 1965 and 1966 Mustangs. It is actually a driver's side mirror mounted on the passenger's side door.

The base of the optional remote driver's side mirror mounts more toward the rear of the door compared to the standard non-remote mirror. The vent wing should just clear the mirror's head when opened.

In keeping with the times, standard mirrors went rectangular for 1967. This is the optional passenger's side non-remote mirror available for all 1967-1968 Mustangs. As with earlier standard mirrors, this was simply a driver's side mirror mounted on the passenger's side.

The optional Deluxe passenger's side mirror for 1967-1968 Mustangs resembled the driver's side remote mirror but was adjusted manually.

Standard outside mirrors for 1969-1970 Mustangs were similar in design to the 1967-1968 rectangular part. The base and stem were contoured to accommodate the change in body style. Unlike earlier standard mirrors, this design will not mount on the passenger's side door. A driver's side remote-control mirror with a rectangular head was also an option for 1969-1970 Mustangs.

Bullet-shaped racing mirrors color-keyed to the exterior body color were a popular Mustang option in 1969-1970. The package included both driver's side and matching passenger's side mirrors, and was available on all body styles. The driver's side mirror is remotely controlled.

The unique passenger's side racing mirror is a visual reverse image of the driver's side mirror, but is adjusted manually.

The familiar running horse over the red, white, and blue striped bar, and block letters spelling MUSTANG made up the standard lower front fender logo for all late 1965 and 1966 Mustangs. Early 1965 lettering measured $4^{3}/_{8}$in long, and the later emblem was lengthened to 5in.

A GT badge and large separated letters spelling MUSTANG bordered by a racing stripe replaced the traditional lower fender logo on models optioned with the Sports Handling Package.

The 1965 and 1966 non-GT fastback Mustangs were identified as 2 + 2 models. A special 2 + 2 fender emblem accompanied the Mustang fender lettering.

The 1965 and 1966 Mustangs equipped with V-8 engines boasted their cubic inch displacements with special upper front fender badges. The displacement, 260 or 289, is positioned above a wide chrome V, Ford's longtime symbol for a V-8 engine.

A black and chrome checked placard with the words HIGH PERFORMANCE engraved along its upper edge indicated a 271hp, 289ci engine was nestled under the hood.

Combining the 289 HiPo engine with the popular GT Sports Handling Package option makes for a desirable and collectible package. Interest goes up if the car is a convertible or fastback.

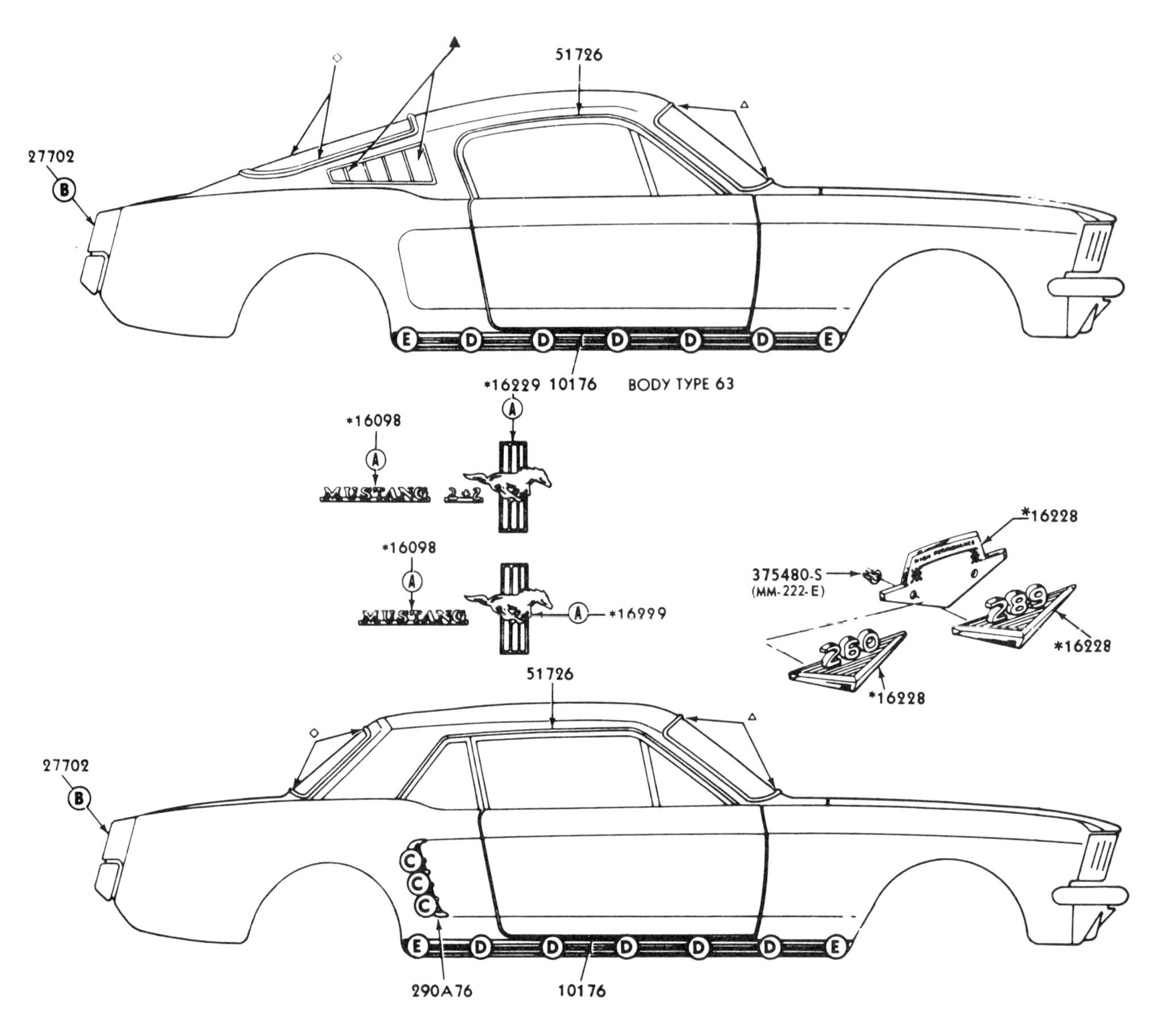

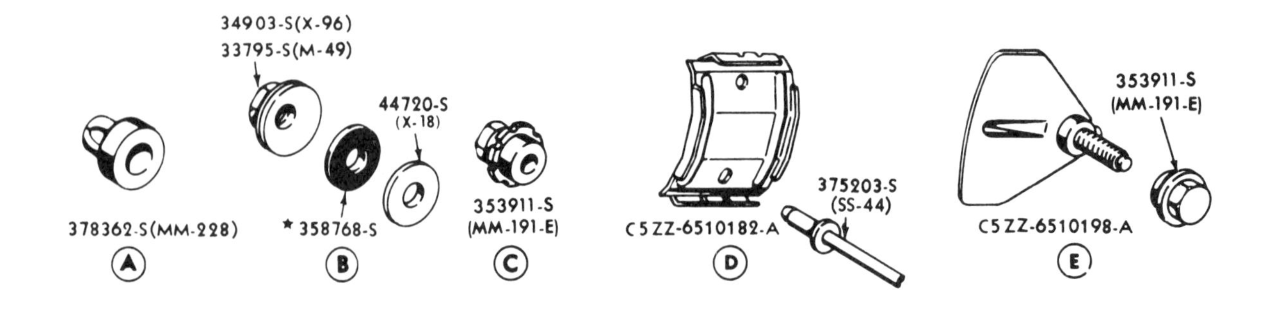

The 1965-1966 Mustang fastback and hardtop body side molding, trim, and attaching hardware.

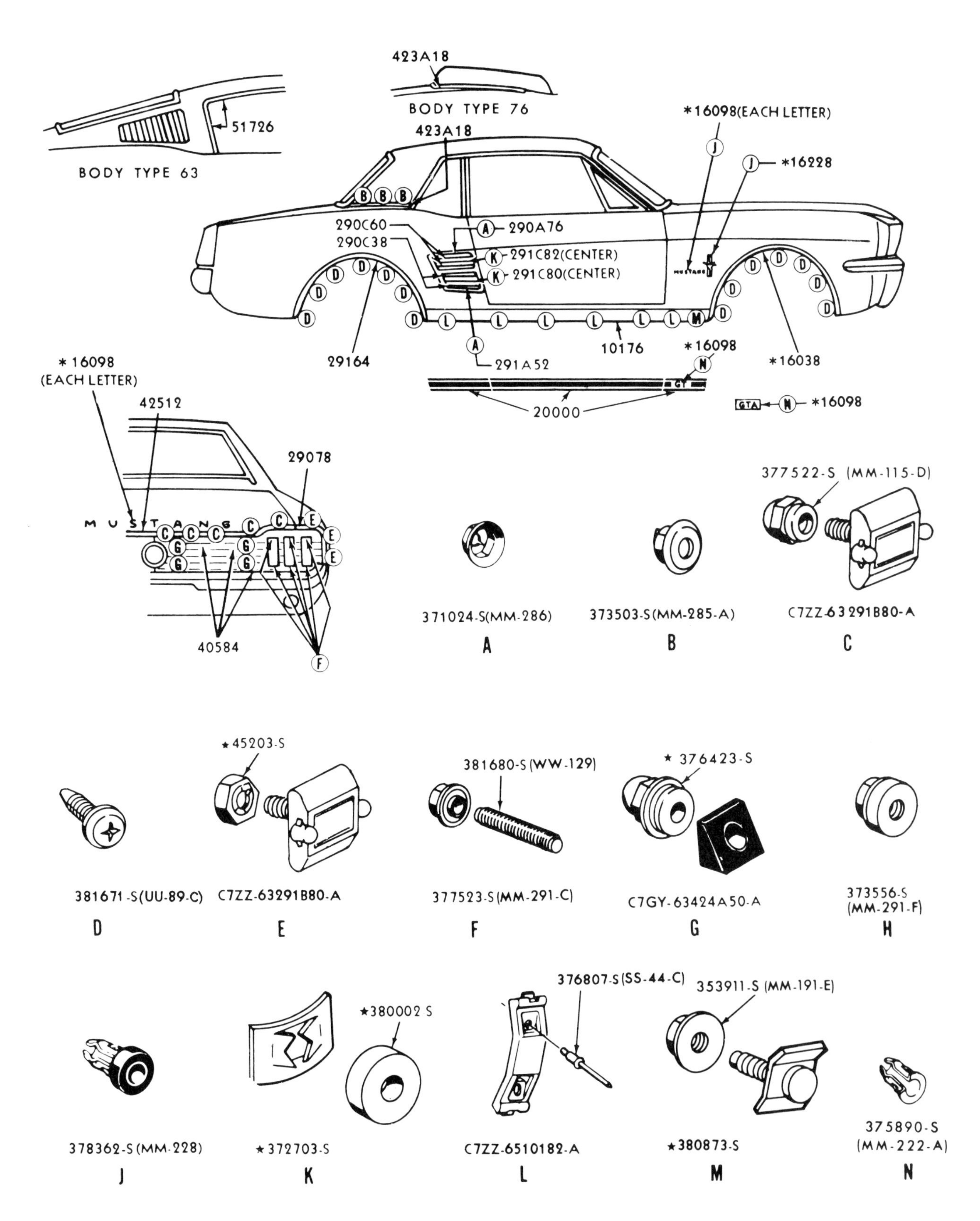

The 1967 Mustang body side molding, trim, and attaching hardware.

The block letters changed again in 1967, but the running horse logo remained the same as the standard lower fender treatment for Mustangs equipped with six-cylinder engines.

The cubic inch displacement was displayed above the tri-colored bar on 1967 and 1968 Mustangs equipped with V-8 engines.

The 1967 Mustangs equipped with the GT option received a special lower fender badge bordered by a racing stripe. In 1967 cars designated simply GT had manual transmissions. In that year only, the designation GTA was given a GT with an automatic transmission.

The straight GT stripe went away in 1968, as did the GTA designation. The Sports Handling Package option was indicated by a simple rectangular GT badge on each front fender.

The 1968 was the first year the Mustang fender emblem was written in script. The cubic inch displacement of V-8 powered Mustangs was still perched above the tri-colored bar.

Ford engineers added 0.130in to the 289 engine's 2.87in stroke and created a 302ci engine. Then, in the middle of the production year, it was installed in 1968 Mustangs. The 302 had completely replaced the 289 by the beginning of the 1969 model year.

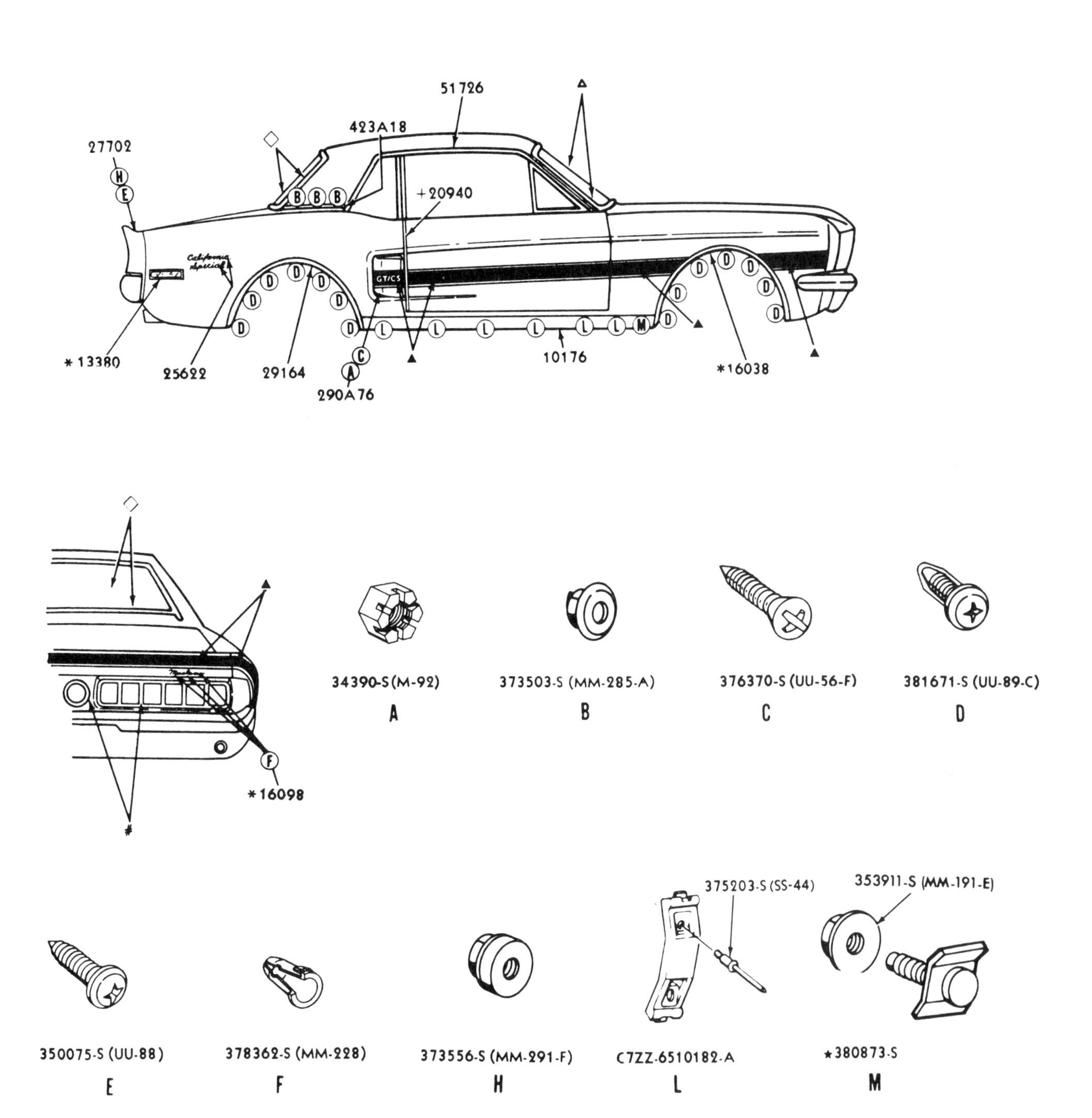

The 1968 California Special GT Mustang body side and rear molding, trim, strips, and attaching hardware.

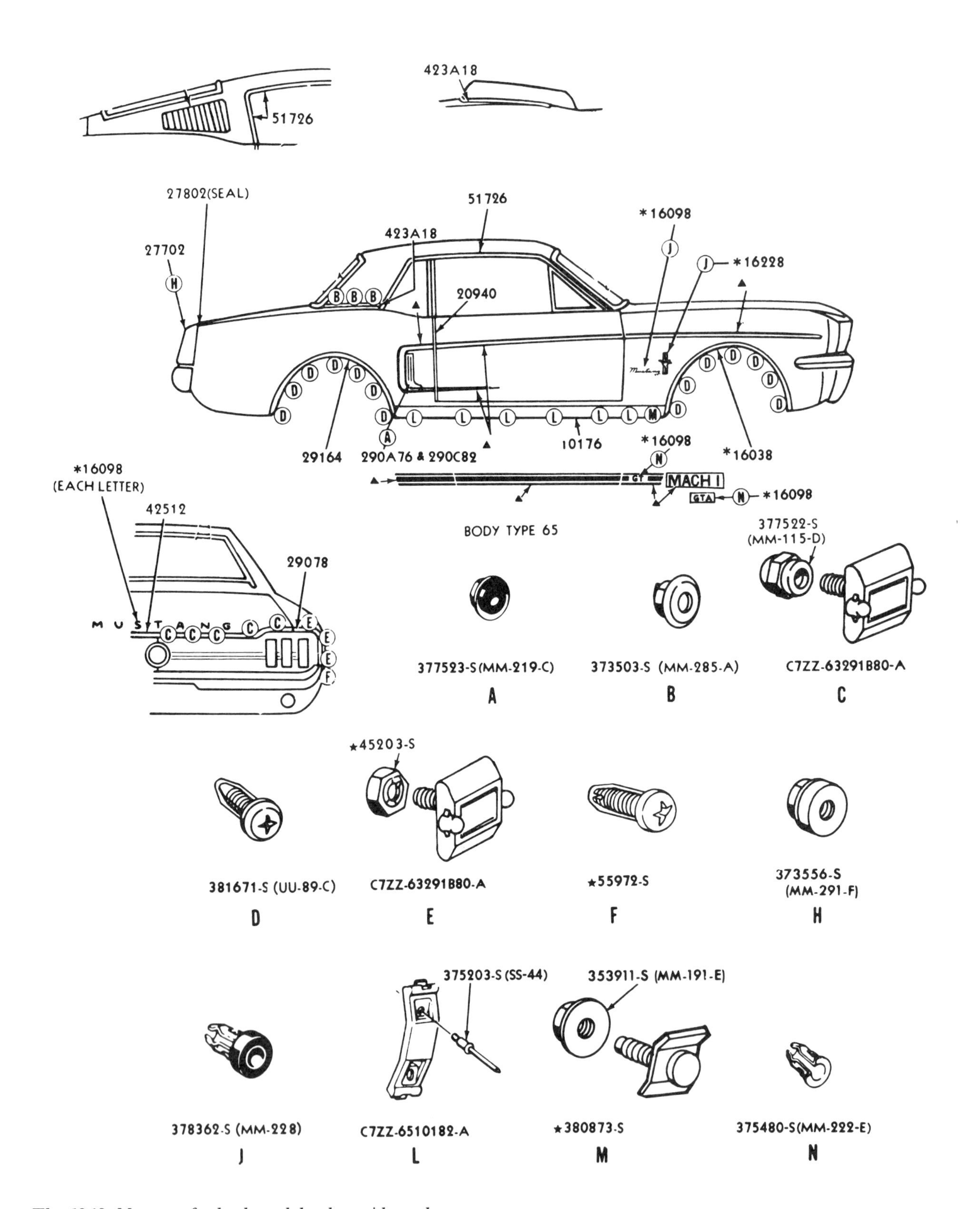

The 1968 Mustang fastback and hardtop side and rear molding, trim, strips, and attaching hardware. Note: Mach 1 was proposed in 1968 but never built.

The Mustang script was hollowed out and painted for 1969. The 351 Windsor engine also made its debut in 1969. The 351 and the 390 engines were designated with a numeric placard located just below the front fender script on standard models.

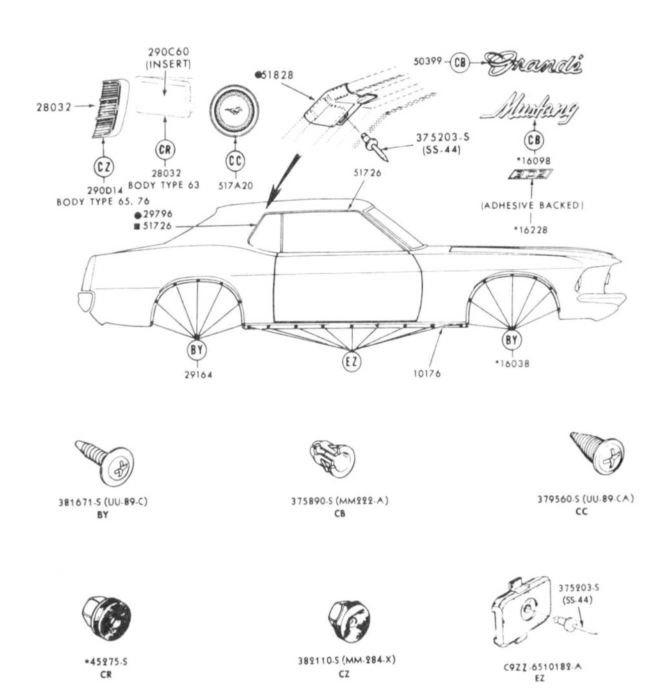

The 1969 Mustang body side molding, trim, and attaching hardware.

The 1969 Mustangs equipped with the Sport Handling Package did not have a traditional GT badge on the lower front fenders, but the straight lower body stripes reappeared.

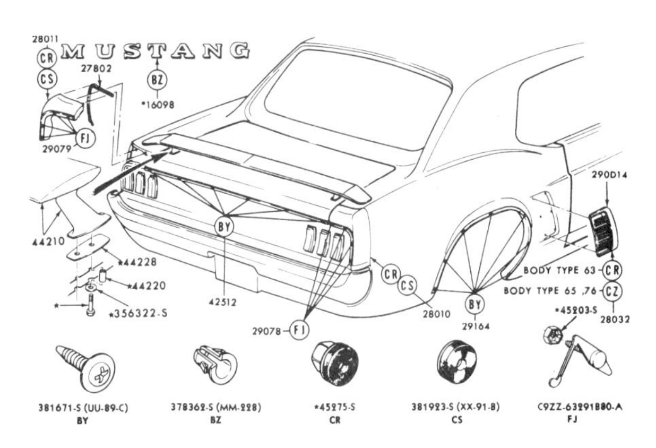

The 1969 Mustang rear spoiler, trim, and related hardware.

Ford copied the center of a 1969 twist-off gas cap to make this SportsRoof quarter panel medallion. Pieces are unique to each side of the car in order to keep the running horse always facing toward the front.

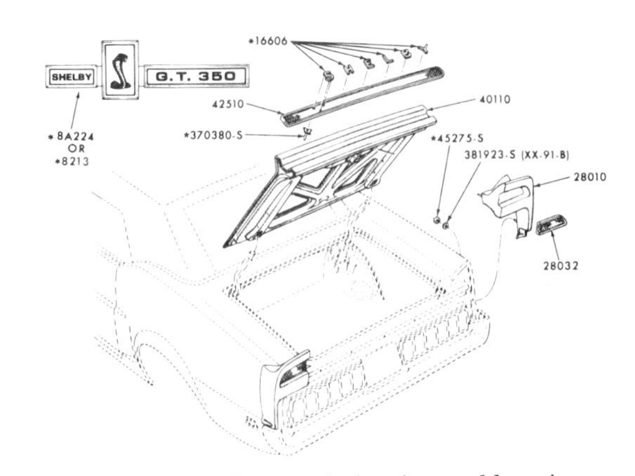

The 1969-1970 Shelby rear deck, trim, and lettering.

The Grande option converted a 1969-1970 Mustang hardtop into a luxury car.

Boss 302 Mustangs were labeled with taped letters in conjunction with tape stripes.

In 1970, a Landau vinyl roof was standard equipment on the Grande. A full vinyl roof was optional.

In 1969, "Boss 302" appeared in a C-stripe located on the car's sides. In 1970 (shown), "BOSS 302" was included in the Z-stripe that started at the leading edge of the hood.

In 1970, the Mustang script appeared just above the wide louvered rocker panel that was standard on the Mach 1.

The mighty Boss 429 had the most subdued insignia when compared to other Mustangs. An outlined "Boss 429" tape decal located on the front fenders was the only warning.

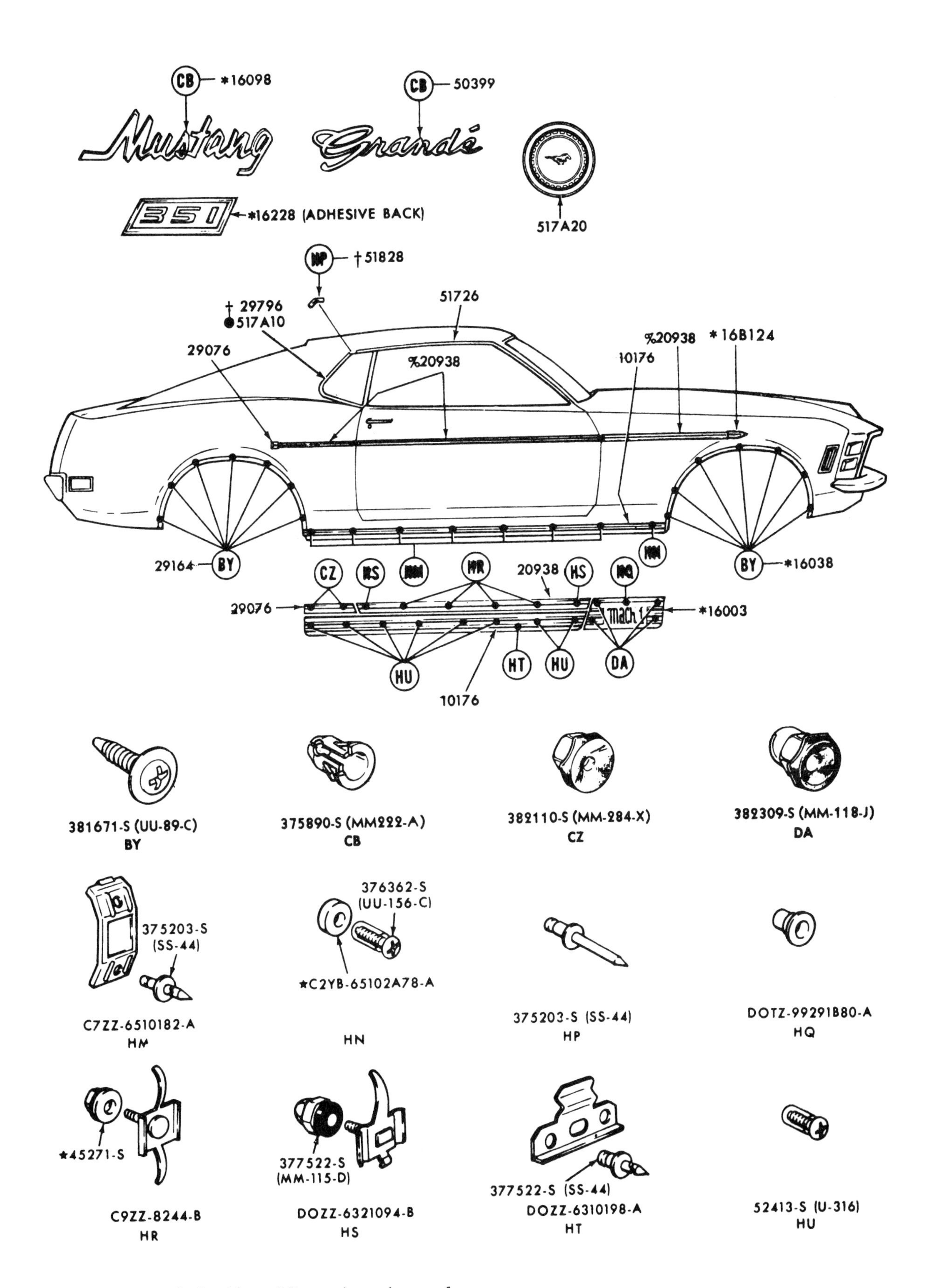

The 1970 Mustang body side molding, trim strips, and attaching hardware.

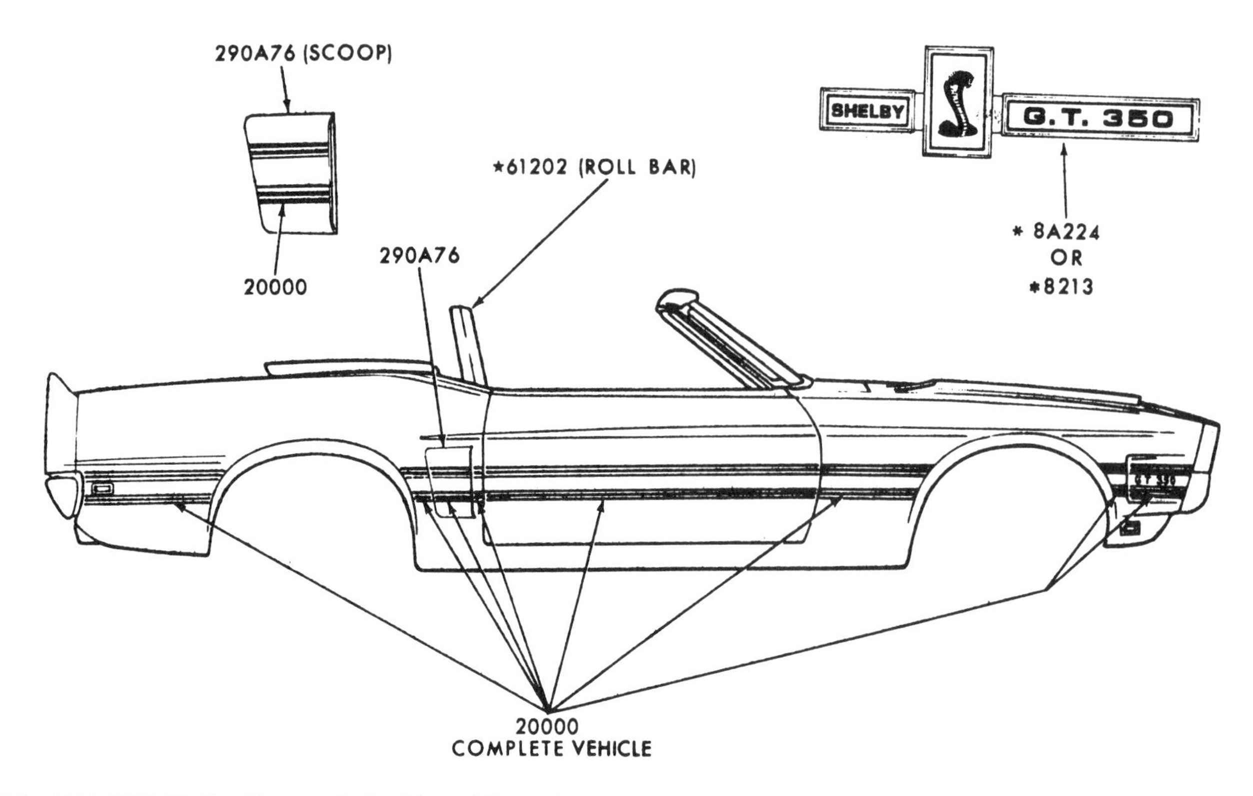

The 1969-1970 Shelby Mustang body side molding, trim, stripes, and lettering.

Rear quarter or body side trim was changed year by year, and therefore provided a good way to determine the model year of the car. The 1965 Mustangs sported a scalloped chrome strip; 1966 Mustangs (shown) had a ribbed bar with three fingers extending toward the front of the vehicle. Chrome rear quarter ornaments were deleted from fastbacks and all GT models in 1965 and 1966.

In 1967 the chrome rear quarter trim gave way to painted louvers that resembled scoops. These louvers were not deleted on GT equipped cars.

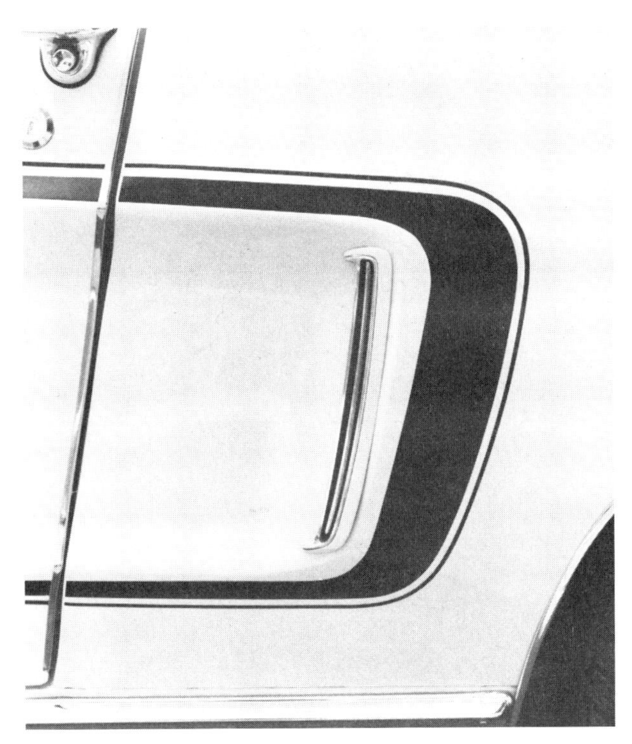

A thin vertical bar set in a body-colored bezel decorated Mustang rear quarters in 1968. These ornaments were also included on GT cars.

Ford dealers in Colorado and its adjacent states offered their own version of the popular California Special in 1968. The High Country Special carried a 51 DSO code and was available only as a hardtop. The High Country Special, primarily hardtops, also had been available through DSO 51 in 1966 and 1967, with special factory colors and ornamentation.

The GT C-stripe was a popular option in 1968. Bright rocker panel molding was standard on all 1968 Mustang models.

One of the Mustang trademarks, the sculptured body side recess, disappeared in 1969. A chrome simulated rear-facing scoop set in a body-colored bezel took its place.

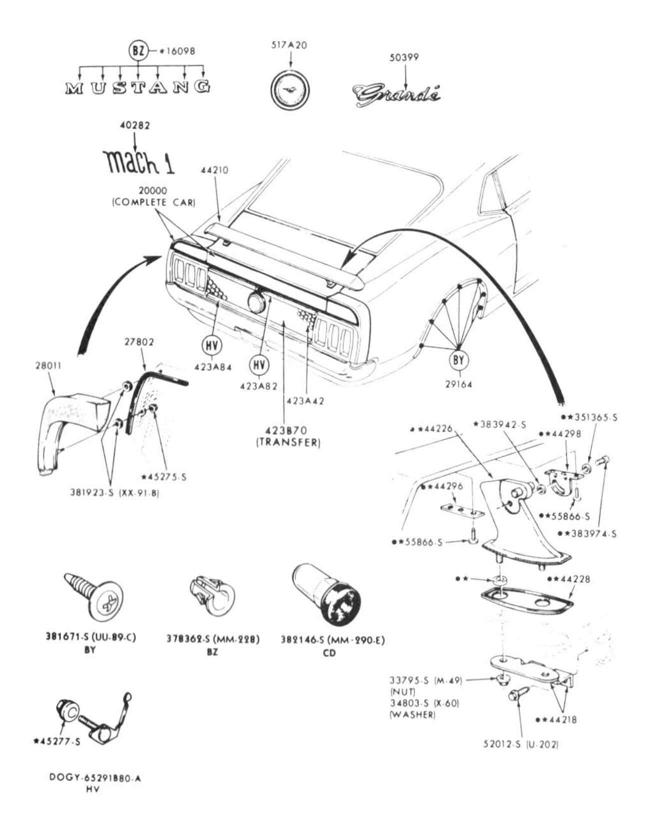

The 1970 Mustang rear spoiler, trim, moldings, and lettering.

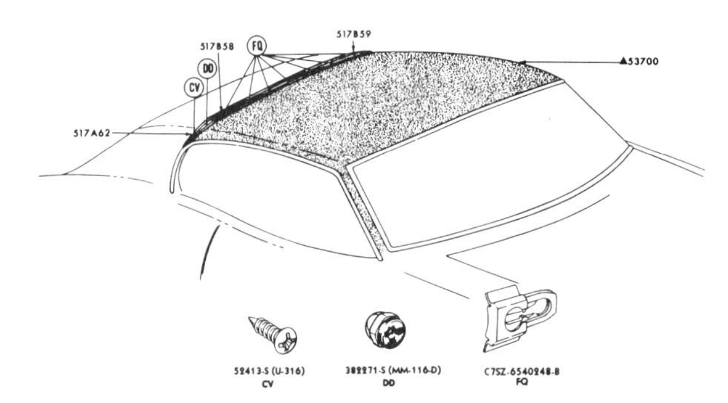

The 1970 Mustang SportsRoof vinyl roof molding and attaching hardware.

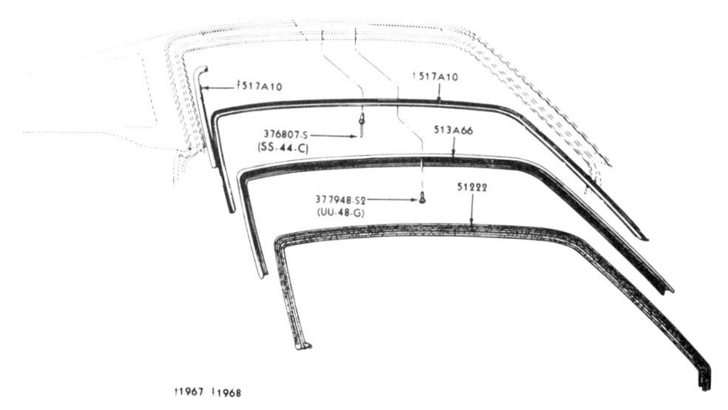

The 1965-1968 Mustang fastback roof side moldings and weatherstrips.

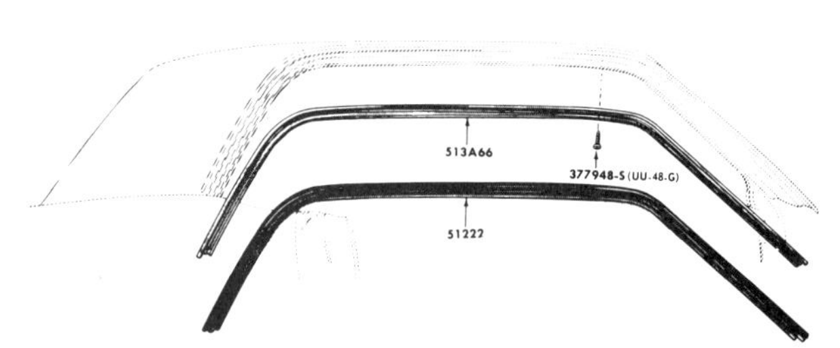

The 1965-1968 Mustang hardtop roof side moldings and weatherstrips.

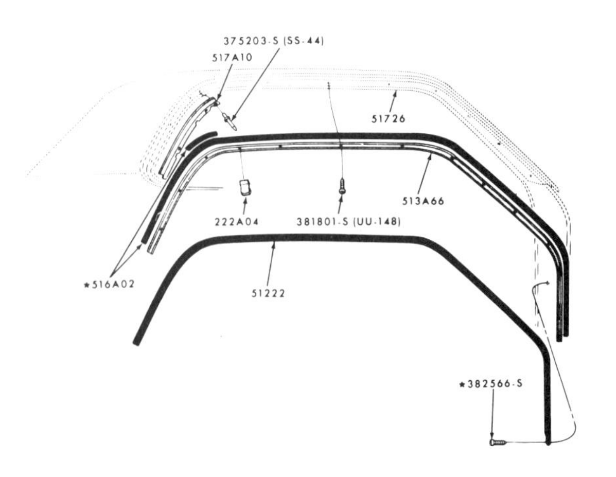

The 1969-1970 Mustang hardtop roof side moldings and weatherstrips.

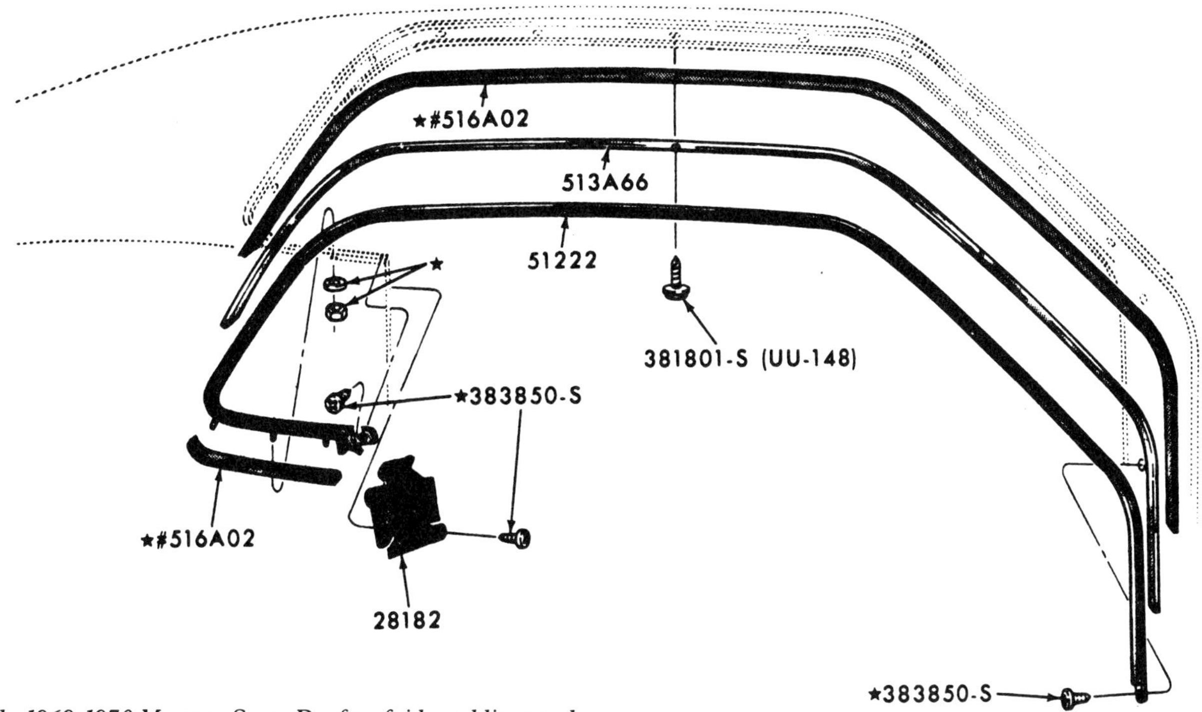

The 1969-1970 Mustang SportsRoof roof side moldings and weatherstrips.

Chapter 3

Interior

Few have openly recognized that a major appeal of the Mustang to its owners over the years has been its interior design. Basically grouped like Mustang exteriors in their 1965-66, 1967-68, and 1969-1970 similarities, the interiors always have held a personality and uniqueness all their own. From a plain bench seat decor with standard upholstery and an AM radio in 1965 to a full luxury pleated and woven bucket seat Mach 1 interior with woodgrain trim and piping on the floor mats, Mustang interiors have made it possible for owners to feel truly at home in their own cars.

The reader should note that this chapter will not include shifters (to be found in Chapter 5 on Drivetrain), steering wheels (to be found in Chapter 6 on Suspension), or instrument clusters (to be found in Chapter 7 on Wiring).

A huge percentage of the money spent in Mustang restoration is devoted to rejuvenating interiors. The restorer is always wise to check with concours experts and reputable parts vendors before spending money for interior components and upholstery. Small details such as the texture and grain of vinyl upholstery, the composition of the carpet material, the accuracy of seat belt retrofitting, and proper colors of interior paint are best researched before a mistake is made. It is suggested that anyone restoring a Mustang for the first time obtain the California Mustang (800-854-1737) publication, *How To Restore Your Mustang*, available from Classic Motorbooks (800-826-6600) in order to view color chips for carpet and upholstery.

Use this chapter for the correct placement of hardware, the proper assembly sequence, model-specific components, authentication of swap meet and used parts, restoration continuity, and selection of optional upgrades.

MUSTANG INSTRUMENT CLUSTER HOUSING APPLICATION CHART

Year	Model	Description	Part Number
65		Only w/ clusters w/ charge indicator light and oil pressure warning light.	C5ZZ-10838-A
65		Black camera case finish; only w/ clusters w/ ammeter and oil pressure gauge.	C5ZZ-10838-C
65,66		Wood grain finish; only w/ clusters w/ ammeter and oil pressure gauge.	C5ZZ-10838-B
66		Black camera case finish; only w/ clusters w/ ammeter and oil pressure gauge.	C6ZZ-10838-A
67	Std. interiors	Black camera case finish; includes lenses.	C7ZZ-10838-C
67	Deluxe interiors	Brushed aluminum; includes lenses.	C7ZZ-10838-B
68		Camera case.	C8ZZ-10838-A
68		Wood grain.	C8ZZ-10838-B
69	Except Mach 1 & Grandé.	Black leathergrain	C9ZZ-10838-A
69	Mach 1 & Grandé.	Light teakwood, not replaced.	C9ZZ-10838-B
69	Mach 1 & Grandé.	Dark teakwood.	C9ZZ-10838-C
70	Except Mach 1 & Grandé.		D0ZZ-10838-A
70	Mach 1 & Grandé.	Dark teakwood.	D0ZZ-10838-B

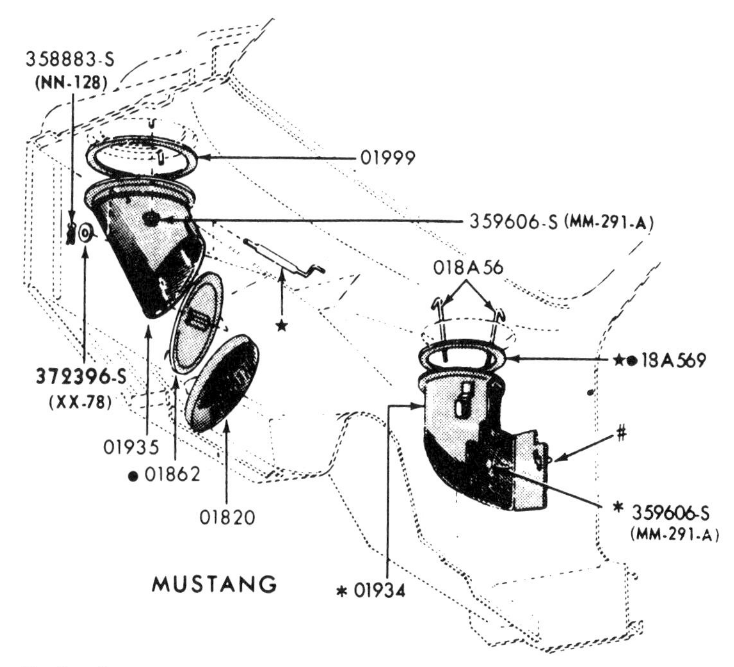

The 1965-1968 Mustang ventilating ducts.

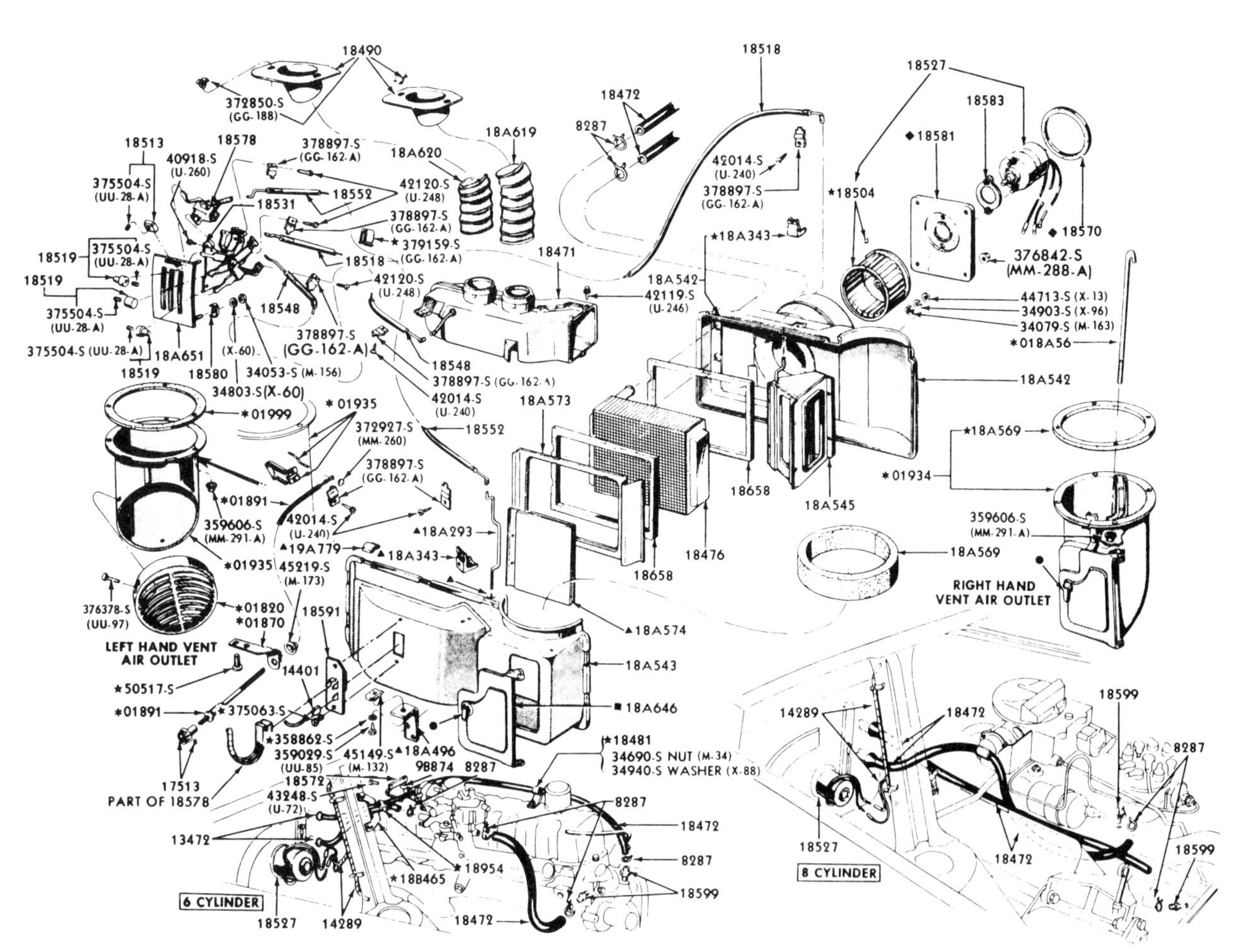

The 1965-1966 Mustang heater, fan, controls, fresh-air vents, and related hardware.

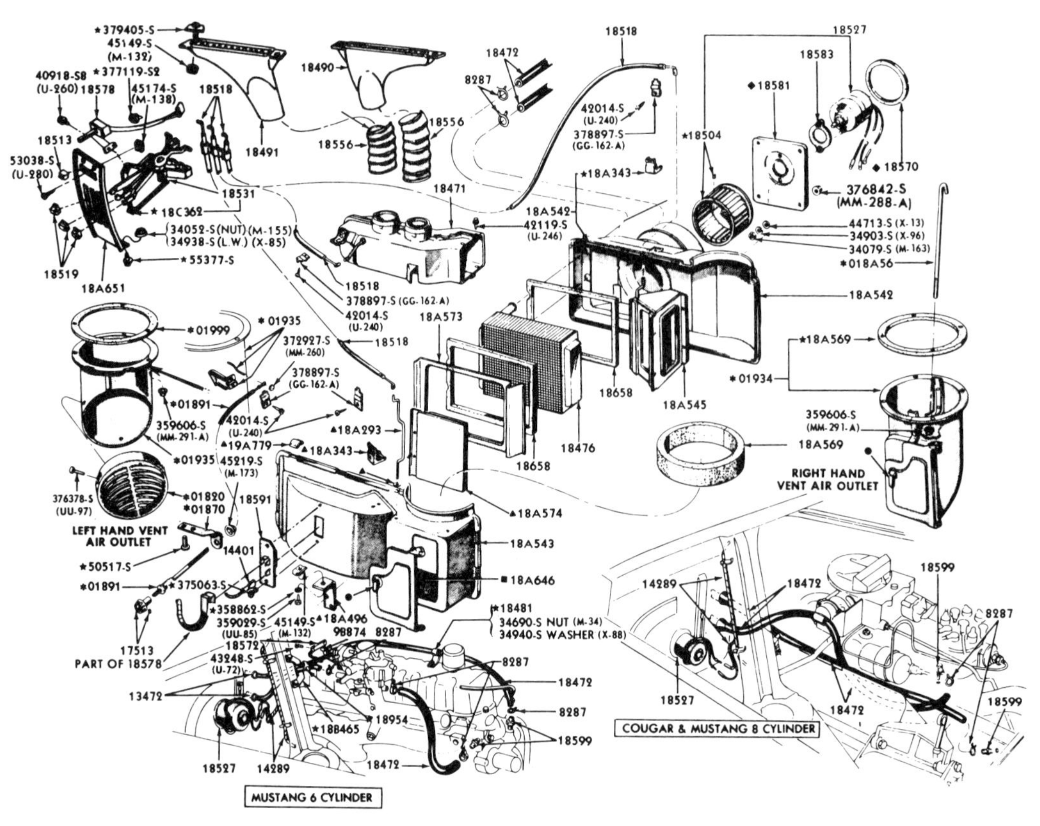

The 1967-1968 Mustang heater, fan, controls, fresh-air vents, and related hardware.

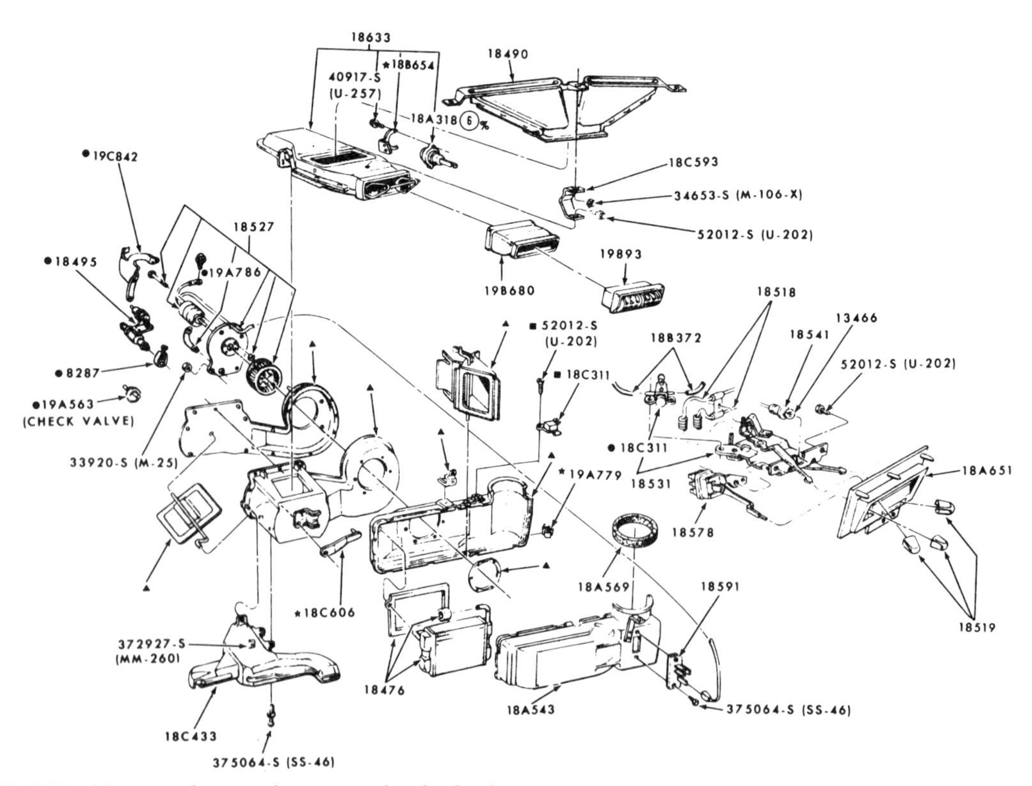

The 1969-1970 Mustang heater, fan, controls, fresh-air vents, and related hardware.

MUSTANG INSTRUMENT PANEL GLOVE COMPARTMENT APPLICATION CHART

YEAR	NOTES	PART NUMBER	DESCRIPTION
65/66		C5ZZ-6506010-A	
67/68		C7ZZ-6506010-A	Without hang-on air conditioning
69/70		C9ZZ-6506010-A	Without air conditioning
65/66		C5ZZ-6506024-B	With air conditioning
69			Includes retainer, spacer, socket and rivet
		C9ZZ-6506010-A2K	Rear - Aqua
		C9ZZ-6506010-A2B	- Blue
		C9ZZ-6506010-A1G	- Ivy gold
		C9ZZ-6506010-A1Y	- Nugget gold
		C9ZZ-6506010-A1D	- Red
70	1	DOZZ-6306010-D	Rear - Black
		DOZZ-6306010-B	- Blue
		DOZZ-6306010-F	- Ginger
		DOZZ-6306010-A	- Green
		DOZZ-6306010-E	- Red
		DOZZ-6306010-C	- Tobacco
70	2	DOOZ-6306010-K	Rear - Black
	3	DOOZ-6306010-H	- Blue
		DOOZ-6306010-M	- Ginger
		DOOZ-6306010-G	- Green
		DOOZ-6306010-L	- Red
		DOOZ-6306010-J	- Tobacco
70	4	DOOZ-6306010-R	Rear - Black
		DOOZ-6306010-P	- Blue
		DOOZ-6306010-T	- Ginger
		DOOZ-6306010-N	- Green
		DOOZ-6306010-S	- Red
		DOOZ-6306010-Q	- Tobacco

Notes:

1. Before 9/2/69
2. From 9/2/69
3. Before 12/1/69
4. From 12/1/69

The AM radio was the most popular option in 1965 and 1966. An AM/FM radio and AM with Stereophonic 8-track tape player were also available.

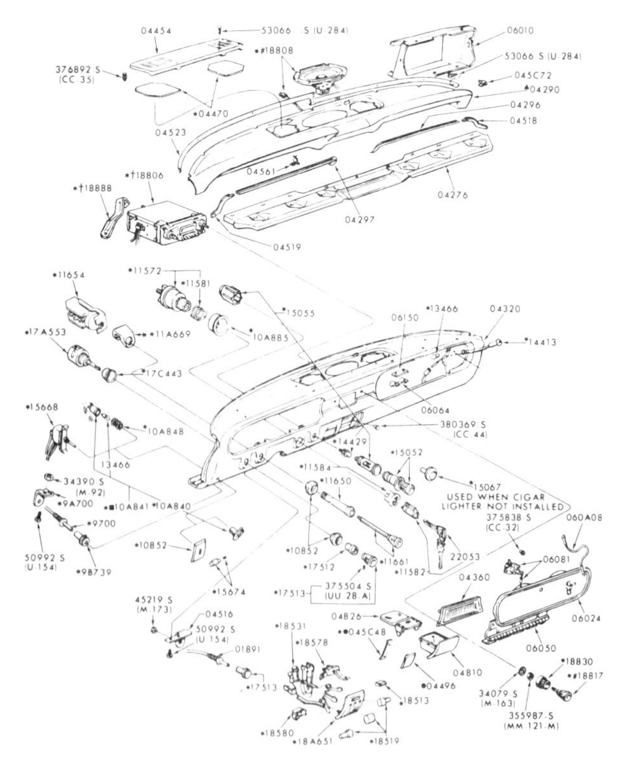

The 1965 Mustang instrument panel and related parts with indicator lights.

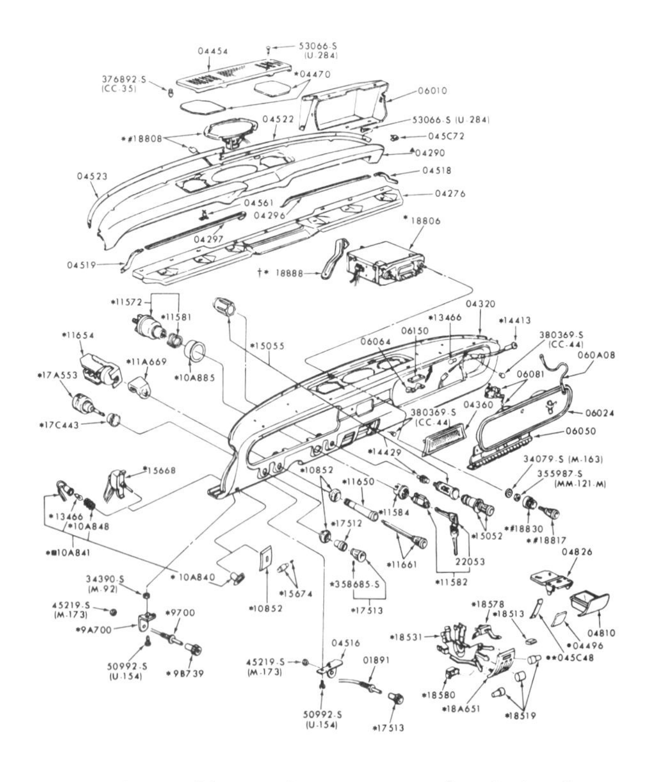

The 1965-1966 Mustang instrument panel and related parts with gauges.

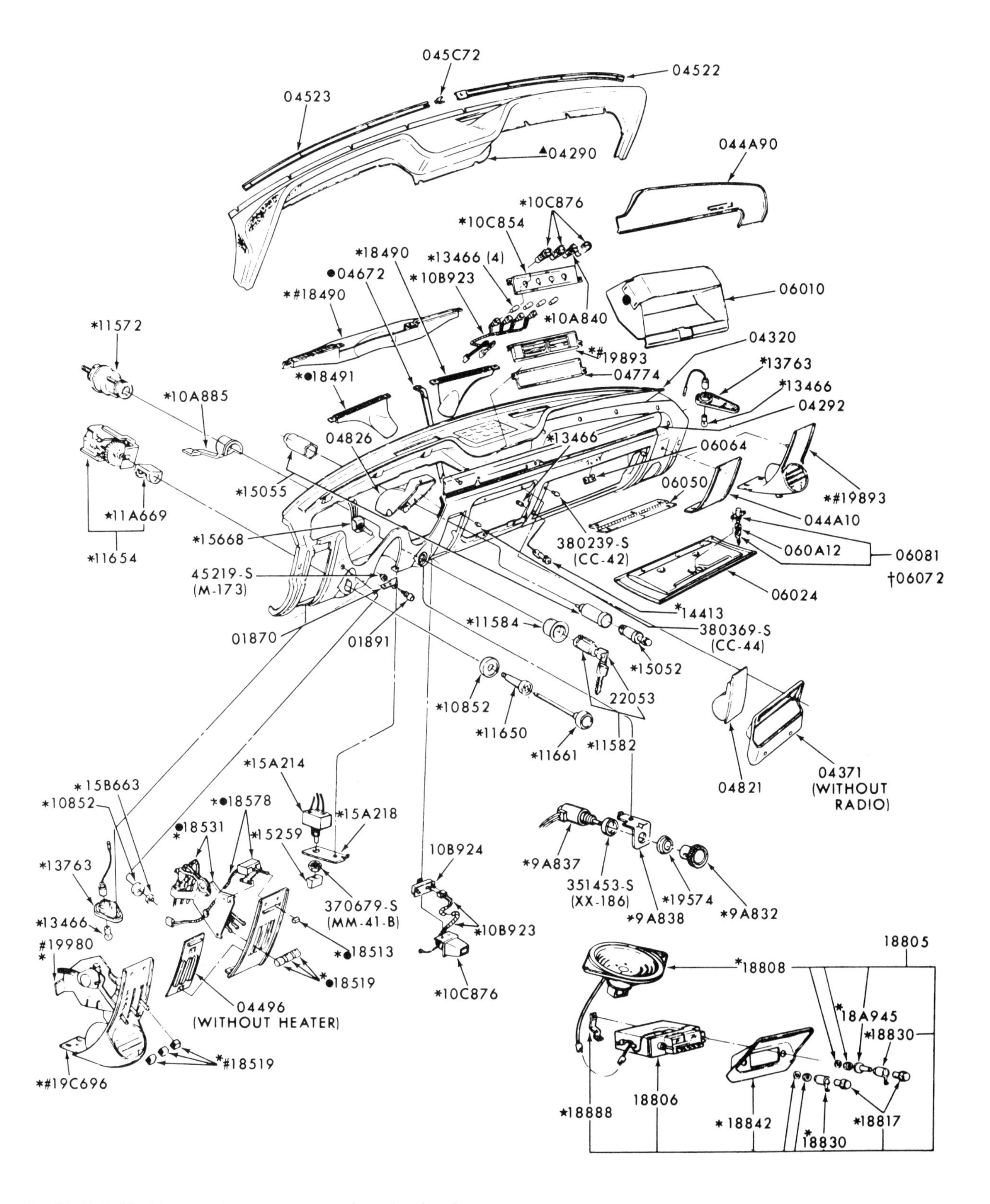

The 1967-1968 Mustang instrument panel and related parts.

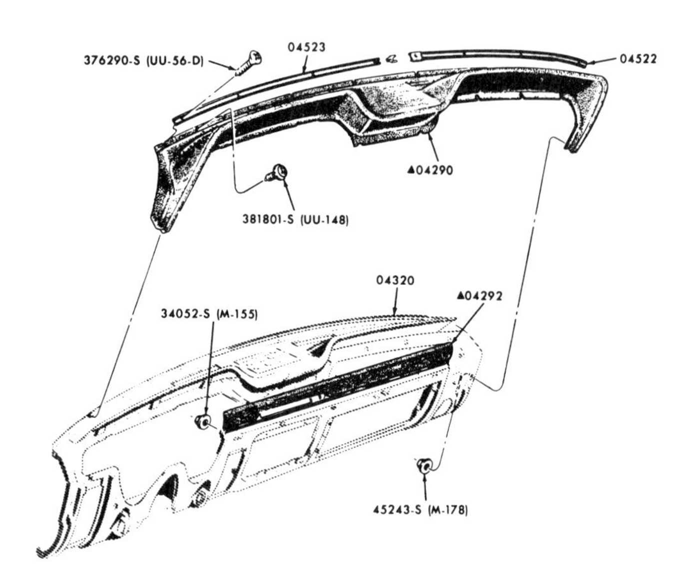

The 1967-1968 Mustang dash pad and attaching hardware.

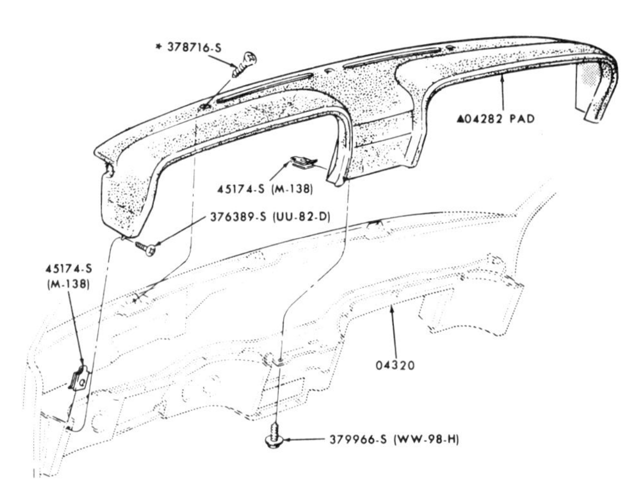

The 1969-1970 Mustang dash pad and attaching hardware.

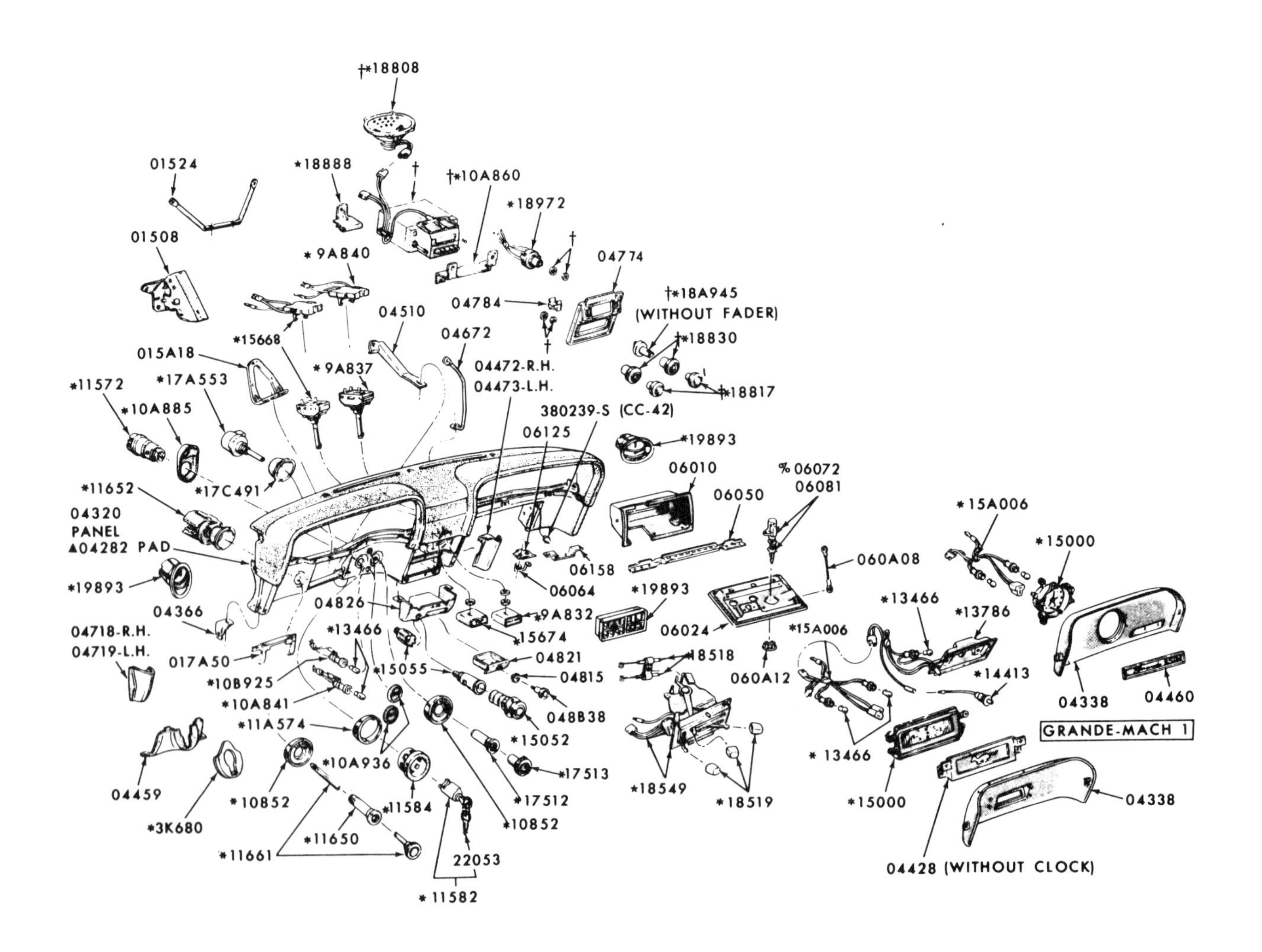

The 1969-1970 Mustang instrument panel and related parts.

Sculptured door panels and Thunderbird door handles were part of the Interior Decor option.

Early Shelby Mustang interiors, like this 1966 model, are simple and business-like. Except for a woodgrain wheel and dash-mounted tachometer, 1965-1966 Shelbys retained the standard black Mustang interior.

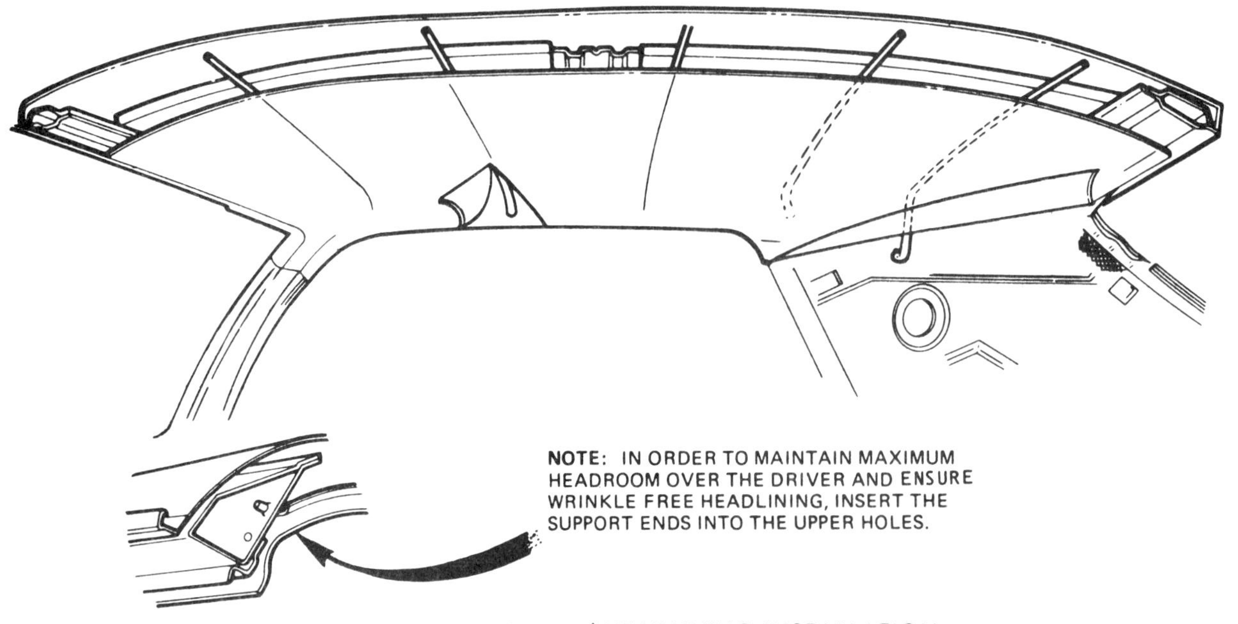

A typical Mustang hardtop roof bow and headlining installation.

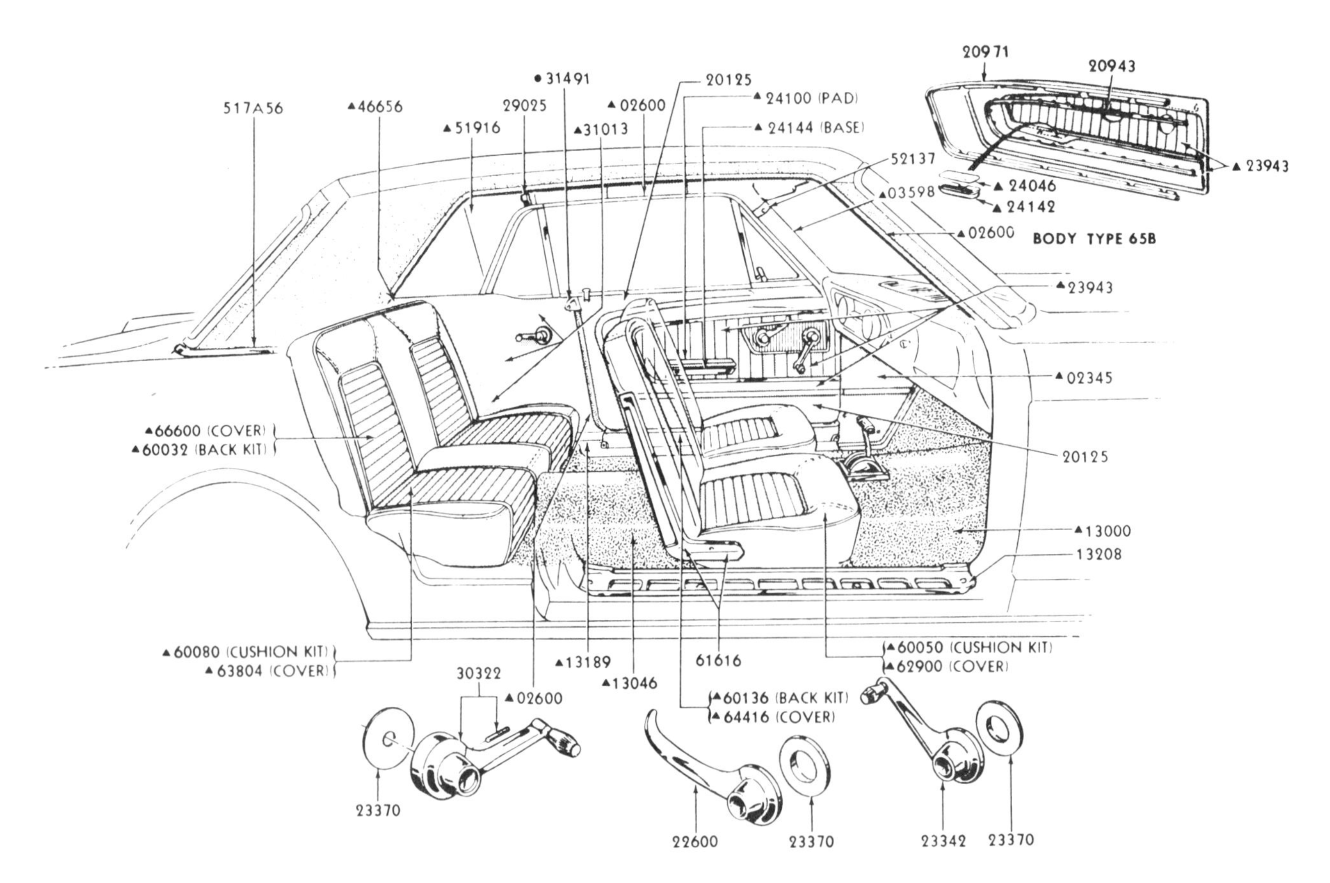

Typical 1965-1966 Mustang hardtop interior trim and related parts.

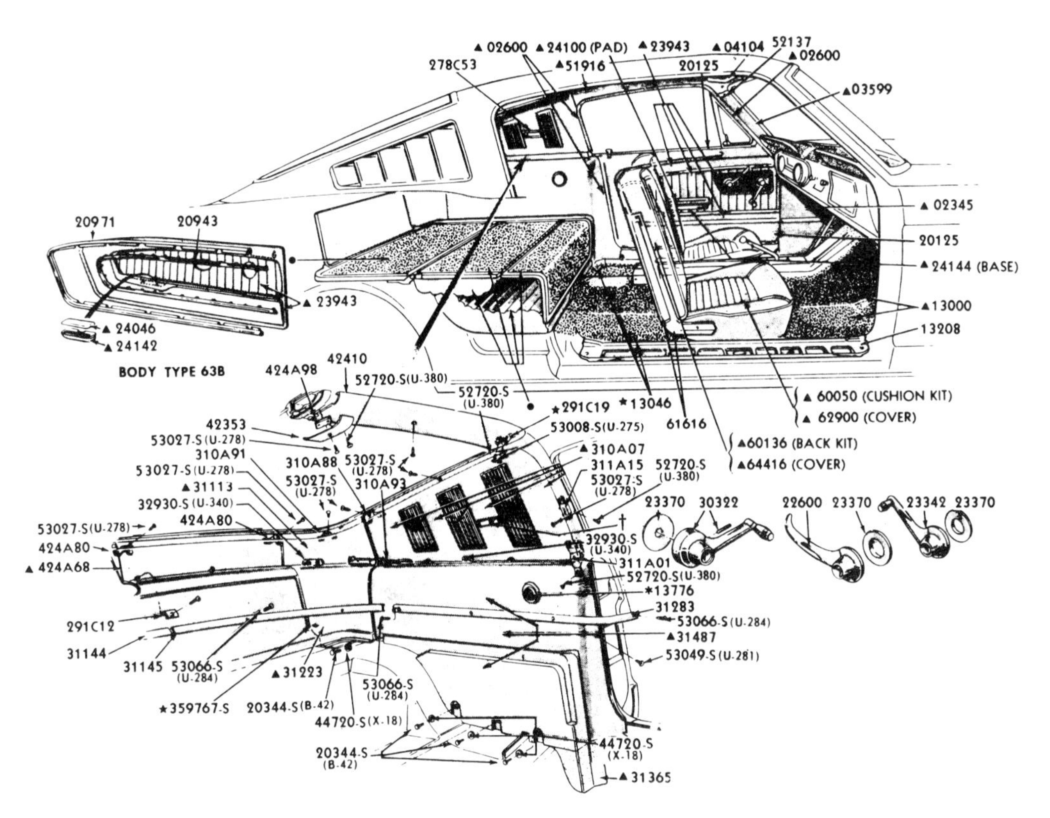

Typical 1965-1966 Mustang fastback interior trim and related parts.

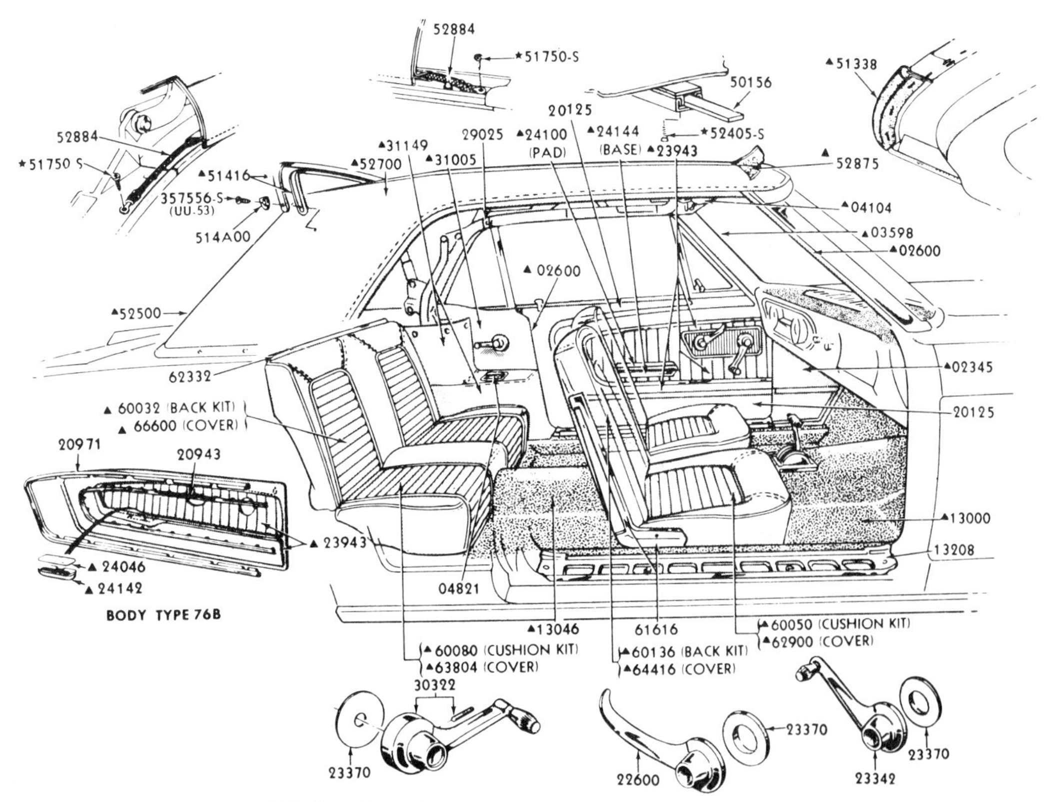

Typical 1965-1966 Mustang convertible interior trim and related parts.

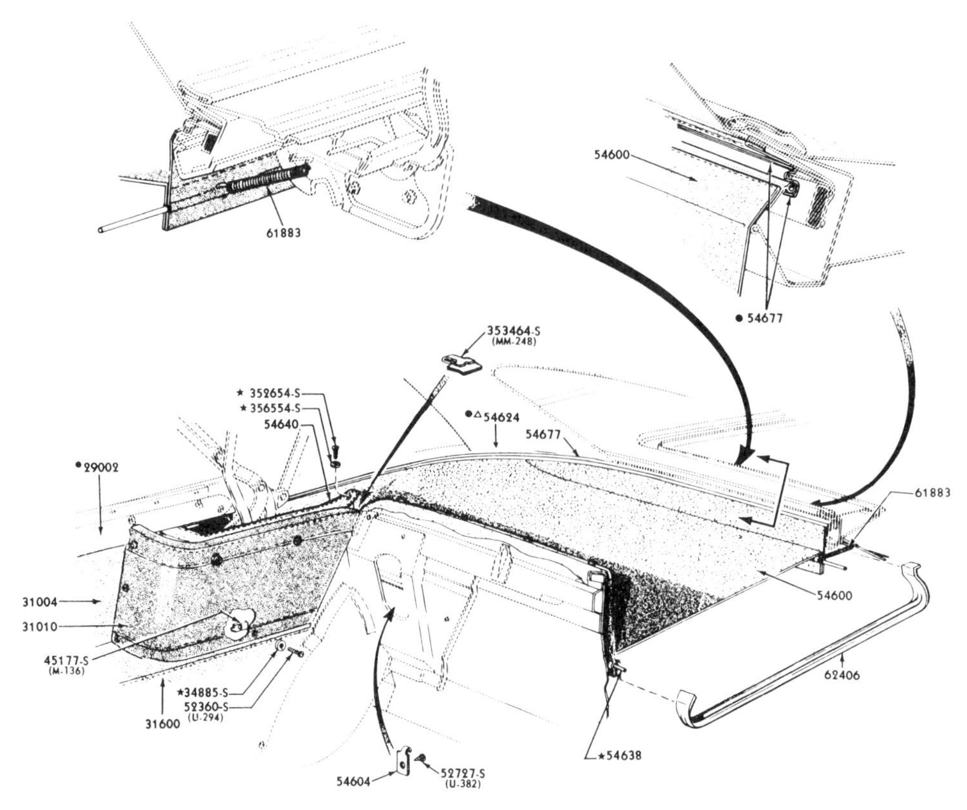

The 1965-1966 Mustang convertible top compartment trim and related parts.

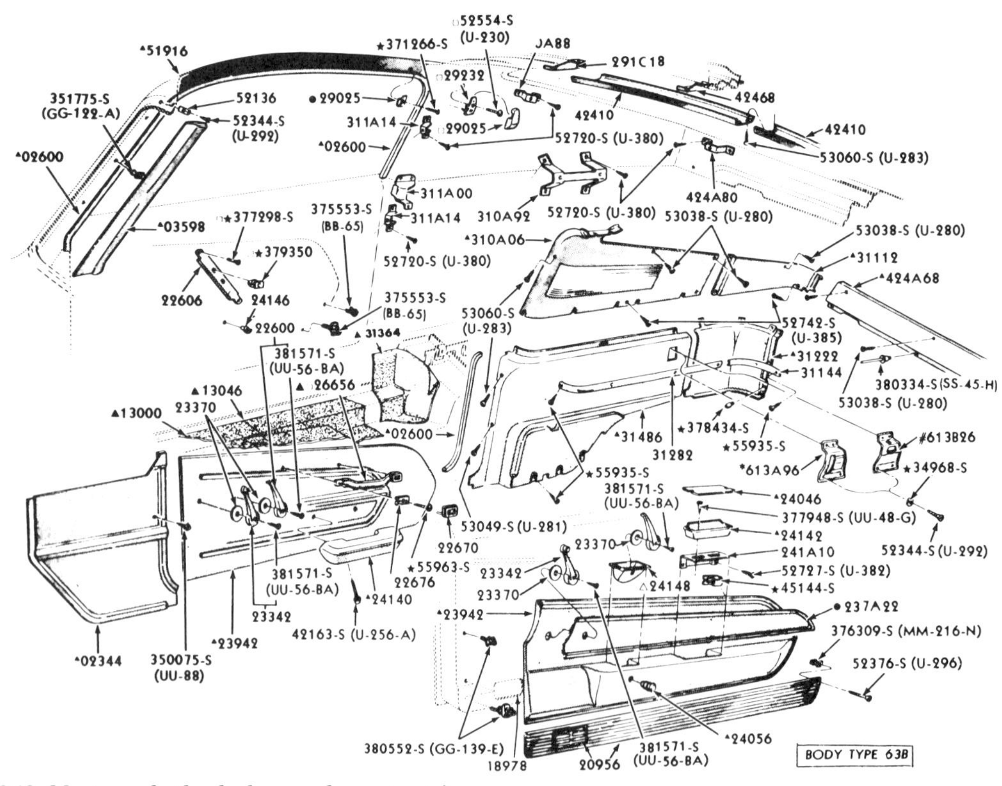

The 1967-1968 Mustang fastback door and quarter trim panels and related parts.

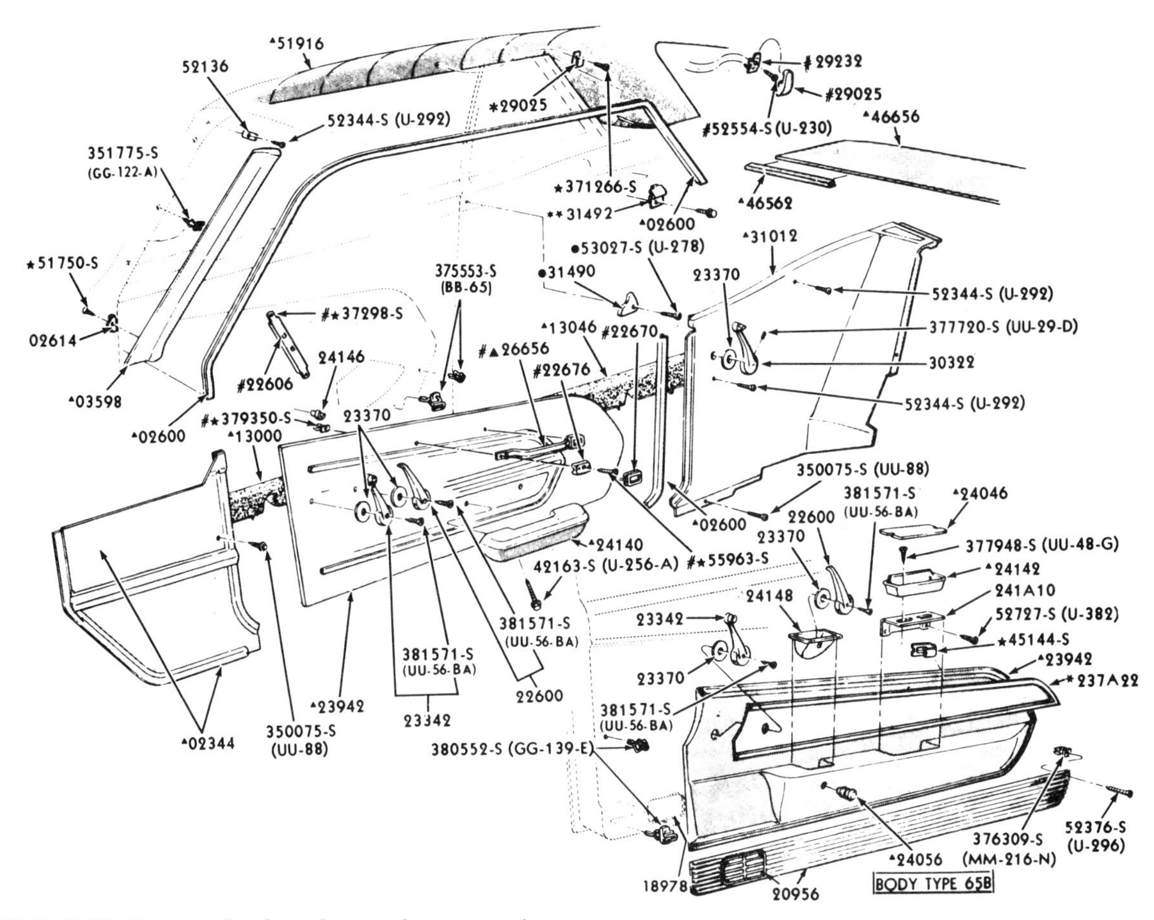

The 1967-1968 Mustang hardtop door and quarter trim panels and related parts.

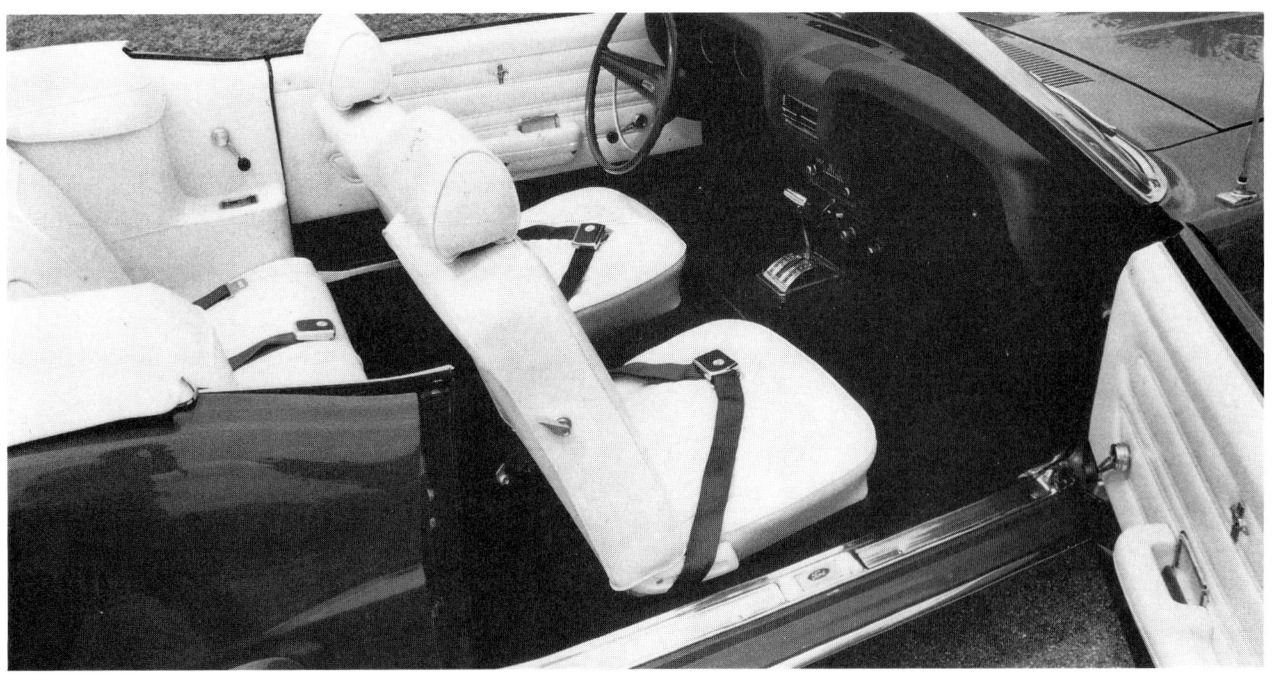

A white standard interior provides an attractive contrast in this beautifully restored 1969 Mustang convertible.

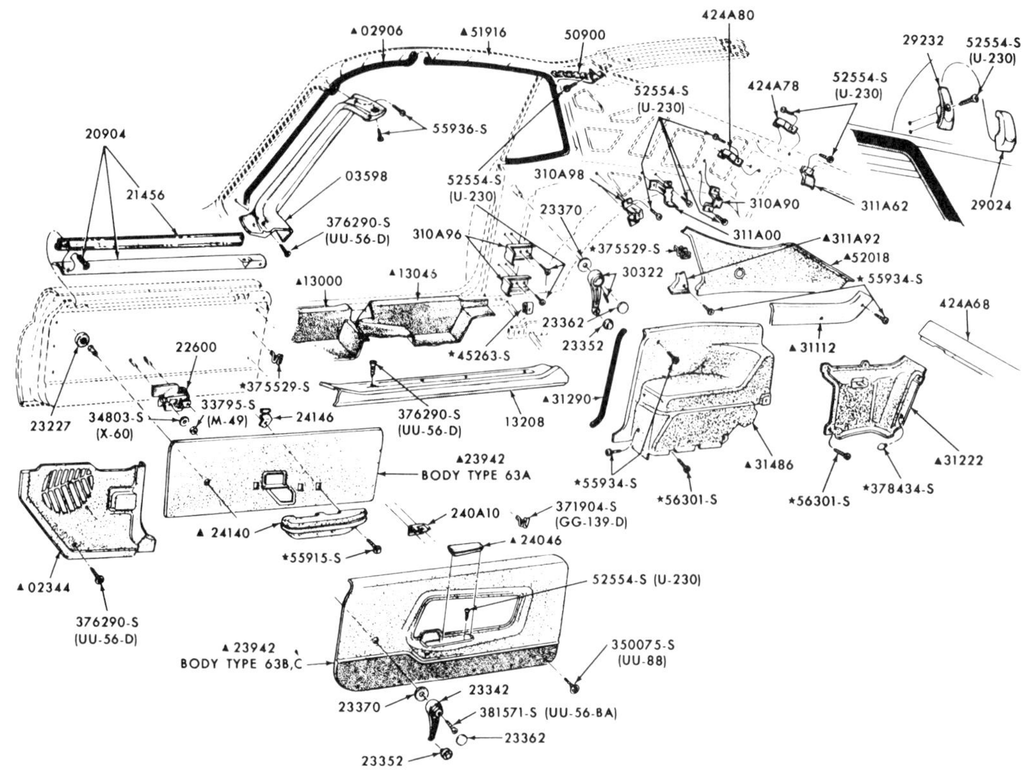

The 1969-1970 Mustang SportsRoof door and quarter trim panels and related parts (with fold-down rear seat).

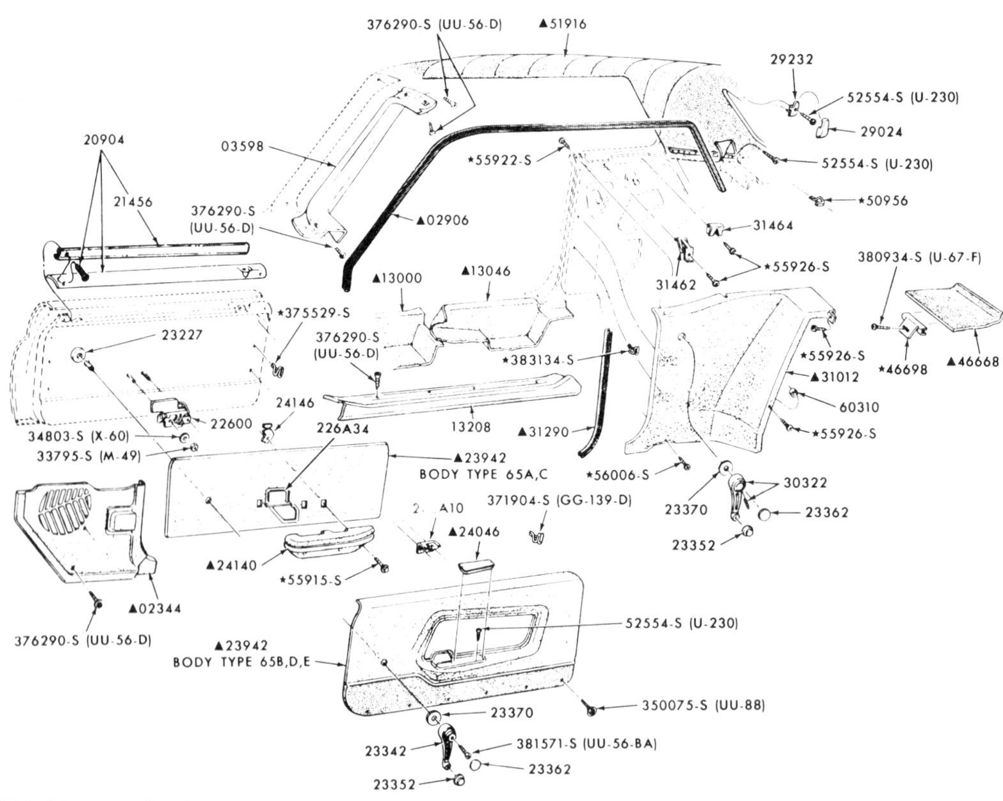

The 1969-1970 Mustang hardtop door and quarter trim panels and related parts.

The Deluxe interior was standard in 1969-1970 Shelby Mustangs. This 1969 model is also equipped with a tilt wheel and air conditioning.

A Deluxe interior, which included molded door panels with woodgrain inserts and a carpeted kick panel, was standard equipment on all 1969-1970 Mustang Mach 1s.

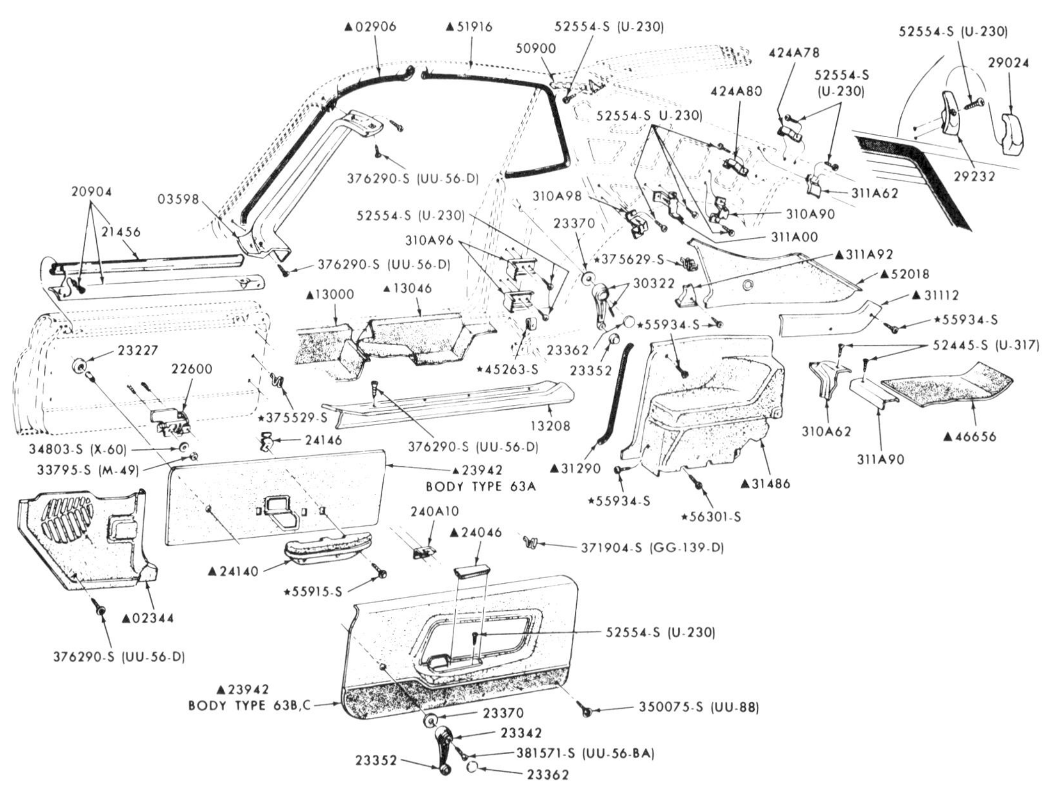

The 1969-1970 Mustang SportsRoof interior trim and related parts (with fixed rear seat).

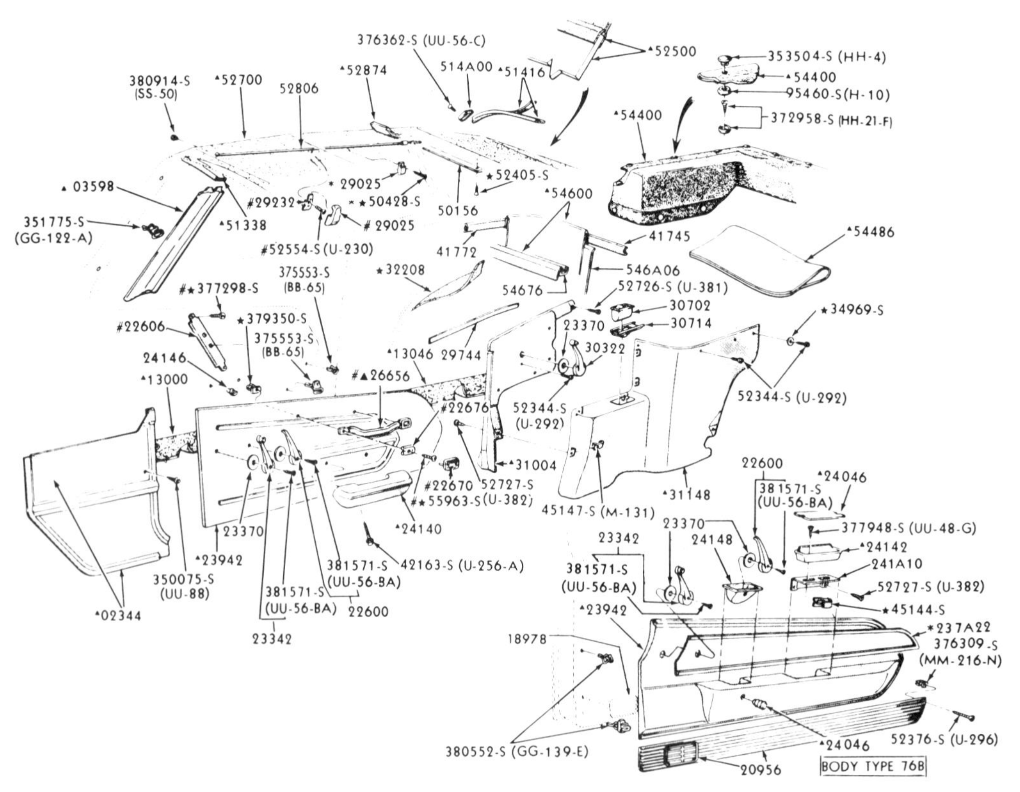

The 1967-1968 Mustang convertible door and quarter trim panels and related parts.

Mustang Interior Trim Scheme Codes, 1965

Code	Body Type	Trim Scheme
22	63A,65A	Medium blue crinkle all vinyl
22	76A	Medium blue crinkle all vinyl
25	63A,65A	Red crinkle all vinyl
25	76A	Red crinkle all vinyl
26	63A,65A	Black crinkle all vinyl
26	76A	Black crinkle all vinyl
28	63A,65A	Light ivy gold crinkle all vinyl
28	76A	Light ivy gold crinkle all vinyl
29	63A,65A	Medium palomino crinkle all vinyl
29	76A	Medium palomino crinkle all vinyl
32	65C	Medium blue crinkle vinyl all vinyl
32	76C	Medium blue crinkle vinyl all vinyl
35	65C	Red crinkle all vinyl
35	76C	Red crinkle all vinyl
36	65C	Black crinkle all vinyl
36	76C	Black crinkle all vinyl
39	65C	Medium palomino crinkle all vinyl
39	76C	Medium palomino crinkle all vinyl
42	63A,65A	White crinkle all vinyl with blue
42	76A	White crinkle all vinyl with blue
45	63A,65A	White crinkle all vinyl with red
45	76A	White crinkle all vinyl with red
46	63A,65A	White crinkle all vinyl with black
46	76A	White crinkle all vinyl with blue
D9	63A,65A	White crinkle all vinyl with palomino
D9	76A	White crinkle all vinyl with palomino
F2	63B,65B	White crinkle all vinyl with blue
F2	76B	White crinkle all vinyl with blue
F5	63B,65B	White crinkle all vinyl with red
F5	76B	White crinkle all vinyl with red
F6	63B,65B	White crinkle all vinyl with black
F6	76B	White crinkle all vinyl with black
F7	63B,65B	White crinkle all vinyl with turqouise
F7	76B	White crinkle all vinyl with turqouise
F8	63B,65B	White crinkle all vinyl with ivy gold
F8	76B	White crinkle all vinyl with ivy gold
F9	63B,65B	White crinkle all vinyl with palomino
F9	76B	White crinkle all vinyl with palomino

Body Type Codes
63A Fastback, Standard interior
63B Fastback, Deluxe interior
65A Hardtop, Standard interior
65B Hardtop, Deluxe interior
65C Hardtop, Bench seat option
76A Convertible, Standard interior
76B Convertible, Deluxe interior
76C Convertible, Bench seat option

1966 Mustang Interior Trim Scheme Codes

Code	Body Type	Trim Scheme
22	63A,65A,76A	Light blue crinkle vinyl and medium blue rosette pattern vinyl
22A	65A,76A	Light blue crinkle vinyl and medium blue crinkle vinyl
25	63A,65A,76A	Red crinkle vinyl and red rosette pattern vinyl
25A	65A,76A	Red crinkle all vinyl
26	63A,65A,76A	Black crinkle vinyl and black rosette pattern vinyl
26A	65A,76A	Black crinkle all vinyl
27	63A,65A,76A	Light aqua crinkle vinyl and medium aqua rosette pattern vinyl
27A	65A,76A	Light aqua crinkle vinyl and medium aqua crinkle vinyl
2DA	65A,76A	Parchment crinkle all vinyl
32	65C,76C	Light blue crinkle vinyl and medium blue rosette pattern vinyl
35	65C,76C	Red crinkle vinyl and red rosette pattern vinyl
36	65C,76C	Black crinkle vinyl and black rosette pattern vinyl
62	63B, 65B	White crinkle vinyl and light blue crinkle vinyl
62A	65B	White crinkle vinyl and light blue crinkle vinyl
64	63B,65B,76B	Parchment crinkle vinyl and medium emberglo crinkle vinyl
64A	65B	Parchment crinkle vinyl and medium emberglo crinkle vinyl
65	63B,65B,76B	Red crinkle all vinyl
65A	65B	Red crinkle all vinyl
66	63B,65B,76B	Black crinkle all vinyl
66A	65B	Black crinkle all vinyl

Code	Body Type	Trim Scheme
67	63B,65B,76B	White crinkle vinyl and light aqua crinkle vinyl
67A	65B	White crinkle vinyl and light aqua crinkle vinyl
68	63B,65B,76B	White crinkle vinyl and light ivy gold crinkle vinyl
68A	65B	White crinkle vinyl and light ivy gold crinkle vinyl
C2	65C,76C	Parchment crinkle vinyl and parchment rosette pattern vinyl
C3	65C,76C	Parchment crinkle vinyl & parchment rosette patt. vinyl w/ burgundy
C4	65C,76C	Parchment crinkle vinyl & parchment rosette patt. vinyl w/ emberglo
C6	65C,76C	Parchment crinkle vinyl and parchment rosette patt. vinyl with black
C7	65C,76C	Parchment crinkle vinyl and parchment rosette pattern vinyl with aqua
C8	65C,76C	Parchment crinkle vinyl & parchment rosette patt. vinyl w/ ivy gold
C9	65C,76C	Parchment crinkle vinyl & parchment rosette patt. vinyl & parchment
D2	63A,65A,76A	Parchment crinkle vinyl & parchment rosette patt. vinyl with blue
D2A	65A,76A	Parchment crinkle all vinyl with blue
D3	63A,65A,76A	Parchment crinkle vinyl & parchment rosette patt. vinyl w/ burgundy
D3A	65A,76A	Parchment crinkle all vinyl with burgundy
D4	63A,65A,76A	Parchment crinkle vinyl & parchment rosette patt. w/ emberglo
D4A	65A,76A	Parchment crinkle all vinyl with emberglo
D6	63A,65A,76A	Parchment crinkle vinyl and parchment rosette patt. vinyl w/ black
D6A	65A,76A	Parchment crinkle all vinyl with black
D7	63A,65A,76A	Parchment crinkle vinyl & parchment rosette patt. vinyl with aqua
D7A	65A,76A	Parchment crinkle vinyl with aqua
D8	63A,65A,76A	Parchment crinkle vinyl & parchment rosette patt. vinyl w/ ivy gold
D8A	65A,76A	Parchment crinkle vinyl with ivy gold
D9	63A,65A,76A	Parchment crinkle vinyl & parchment rosette patt. vinyl w/ palomino
D9A	65A,76A	Parchment crinkle vinyl with palomino
F2	63B,65B,76B	Parchment crinkle all vinyl with blue
F2A	65B	Parchment crinkle all vinyl with blue
F3	63B,65B,76B	Parchment crinkle all vinyl with burgundy
3A	65B	Parchment crinkle all vinyl with burgundy
F4	63B,65B,76B	Parchment crinkle all vinyl with emberglo
F4A	65B	Parchment crinkle all vinyl with emberglo
F6	63B,65B,76B	Parchment crinkle all vinyl with black
F6A	65B	Parchment crinkle all vinyl with black
F7	63B,65B,76B	Parchment crinkle all vinyl with aqua
F7A	65B	Parchment crinkle all vinyl with aqua
F8	63B,65B,76B	Parchment crinkle all vinyl with ivy gold
F8A	65B	Parchment crinkle all vinyl with ivy gold
F9	63B,65B,76B	Parchment crinkle all vinyl with palomino
F9A	65B	Parchment crinkle all vinyl with palomino

Body Type Codes

63A Fastback, Standard interior
63B Fastback, Deluxe interior
65A Hardtop, Standard interior
65B Hardtop, Deluxe interior
65C Hardtop, Bench seat option
76A Convertible, Standard interior
76B Convertible, Deluxe interior
76C Convertible, Bench seat option

1967 Mustang Interior Trim Scheme Codes

Code	Body Type	Trim Scheme
2A	63A,65A,76A	Charcoal black crinkle all vinyl
2B	63A,65A,76A	Light blue crinkle vinyl and dark blue crinkle vinyl
2D	63A,65A,76A	Red crinkle all vinyl
2D1	65A	Red crinkle all vinyl
2F	63A,65A,76A	Medium saddle haircell all vinyl
2F1	65A	Medium saddle haircell all vinyl
2G	63A,65A,76A	Light ivy gold crinkle vinyl & medium ivy gold crinkle vinyl

Code	Body Type	Trim Scheme
2G1	65A	Light ivy gold crinkle vinyl & medium ivy gold crinkle vinyl
2K	63A,65A,76A	Light aqua crinkle vinyl and dark aqua crinkle vinyl
2K1	65A	Light aqua crinkle vinyl and dark aqua crinkle vinyl
2U	63A,65A,76A	Paster parchment crinkle all vinyl
2U1	65A	Paster parchment crinkle all vinyl
4A	65C,76C	Charcoal black crinkle all vinyl
4B	65C	Blue all vinyl
4D	65C	Blue all vinyl
4U	65C,76C	Pastel parchment crinkle all vinyl
4U1	65C	Pastel parchment crinkle all vinyl
5A	63B,65B	Charcoal black crinkle vinyl & charcoal black knitted vinyl
5U	63B,65B	Pastel parchment crinkle vinyl & pastel parchment knitted vinyl
6A	63B,65B,76B	Charcoal black crinkle all vinyl
6B	63B,65B,76B	Light blue crinkle vinyl and dark blue crinkle vinyl
6D	63B,65B,76B	Red crinkle all vinyl
6F	63B,65B,76B	Medium saddle haircell all vinyl
6G	63B,65B,76B	Light ivy gold crinkle vinyl & medium ivy gold crinkle vinyl
6K	63B,65B,76B	Light aqua crinkle vinyl and dark aqua crinkle vinyl
6U	63B,65B,76B	Pastel parchment crinkle all vinyl
7A	63A,65A	Charcoal black crinkle vinyl & charcoal black knitted vinyl
7U	63A,65A	Pastel parchment crinkle vinyl & pastel parchment knitted vinyl

Body Type Codes
63A Fastback, Standard interior
63B Fastback, Deluxe interior
65A Hardtop, Standard interior
65B Hardtop, Deluxe interior
65C Hardtop, Bench seat option
76A Convertible, Standard interior
76B Convertible, Deluxe interior
76C Convertible, Bench seat option

1968 Mustang Interior Trim Scheme Codes

Code	Body Type	Trim Scheme
2A	63A,65A,76A	Charcoal black crinkle vinyl and charcoal black kiwi pattern vinyl
2AA	63A,65A,76A	Charcoal black crinkle vinyl and charcoal black kiwi pattern vinyl
2B	63A,65A,76A	Light blue crinkle vinyl and dark blue kiwi pattern vinyl
2BA	63A,65A,76A	Light blue crinkle vinyl and dark blue kiwi pattern vinyl
2D	63A,65A,76A	Dark red crinkle vinyl and dark red kiwi pattern vinyl
2DA	63A,65A,76A	Dark red crinkle vinyl and dark red kiwi pattern vinyl
2F	63A,65A,76A	Medium saddle crinkle vinyl and medium saddle kiwi pattern vinyl
2FA	63A,65A,76A	Medium saddle crinkle vinyl and medium saddle kiwi pattern vinyl
2G	63A,65A,76A	Light ivy gold crinkle vinyl and medium ivy gold kiwi pattern vinyl
2GA	63A,65A,76A	Light ivy gold crinkle vinyl and medium ivy gold kiwi pattern vinyl
2K	63A,65A,76A	Light aqua crinkle vinyl and dark aqua kiwi pattern vinyl
2KA	63A,65A,76A	Light aqua crinkle vinyl and dark aqua kiwi pattern vinyl
2U	63A,65A,76A	Pastel parchment crinkle vinyl & pastel parch. kiwi patt. vinyl & black
2UA	63A,65A,76A	Pastel parchment crinkle vinyl & pastel parch. kiwi patt. vinyl & black
2Y	63A,65A,76A	Light nugget gold crinkle vinyl & light nugget gold kiwi pattern vinyl
2YA	63A,65A,76A	Light nugget gold crinkle vinyl & light nugget gold kiwi pattern vinyl
5A	63B,65B	Charcoal black crinkle vinyl and charcoal black clarion knitted vinyl
5AA	63B,65B	Charcoal black crinkle vinyl and charcoal black clarion knitted vinyl
5B	63B,65B	Light blue crinkle vinyl and dark blue clarion knitted vinyl
5BA	63B,65B	Light blue crinkle vinyl and dark blue clarion knitted vinyl
5D	63B,65B	Dark red crinkle vinyl and dark red clarion knitted vinyl
5DA	63B,65B	Dark red crinkle vinyl and dark red clarion knitted vinyl
5U	63B,65B	Pastel parch. crinkle vinyl & pastel parch. clarion knit. vinyl w/ black
5UA	63B,65B	Pastel parch. crinkle vinyl & pastel parch. clarion knit. vinyl w/ black
6A	63B,65B,76B	Charcoal black crinkle vinyl and charcoal black kiwi pattern vinyl

Code	Body Type	Trim Scheme
6AA	63B,65B,76B	Charcoal black crinkle vinyl and charcoal black kiwi pattern vinyl
6B	63B,65B,76B	Light blue crinkle vinyl and dark blue kiwi pattern vinyl
6BA	63B,65B,76B	Light blue crinkle vinyl and dark blue kiwi pattern vinyl
6D	63B,65B,76B	Dark red crinkle vinyl and dark red kiwi pattern vinyl
6DA	63B,65B,76B	Dark red crinkle vinyl and dark red kiwi pattern vinyl
6F	63B,65B,76B	Medium saddle crinkle vinyl and medium saddle kiwi pattern vinyl
6FA	63B,65B,76B	Medium saddle crinkle vinyl and medium saddle kiwi pattern vinyl
6G	63B,65B,76B	Light ivy gold crinkle vinyl and medium ivy gold kiwi pattern vinyl
6GA	63B,65B,76B	Light ivy gold crinkle vinyl and medium ivy gold kiwi pattern vinyl
6K	63B,65B,76B	Light aqua crinkle vinyl and dark aqua kiwi pattern vinyl
6KA	63B,65B,76B	Light aqua crinkle vinyl and dark aqua kiwi pattern vinyl
6U	63B,65B,76B	Pastel parchment crinkle vinyl & pastel parch. kiwi patt. vinyl & black
6UA	63B,65B,76B	Pastel parchment crinkle vinyl & pastel parch. kiwi patt. vinyl & black
6Y	63B,65B,76B	Light nugget gold crinkle vinyl & light nugget gold kiwi pattern vinyl
6YA	63B,65B,76B	Light nugget gold crinkle vinyl & light nugget gold kiwi pattern vinyl
7A	63B,65B	Charcoal black crinkle vinyl and charcoal black clarion knitted vinyl
7AA	63B,65B	Charcoal black crinkle vinyl and charcoal black clarion knitted vinyl
7B	63A,65A	Light blue crinkle vinyl and dark blue clarion knitted vinyl
7BA	63A,65A	Light blue crinkle vinyl and dark blue clarion knitted vinyl
7D	63A,65A	Dark red crinkle vinyl and dark red clarion knitted vinyl
7DA	63A,65A	Dark red crinkle vinyl and dark red clarion knitted vinyl
7U	63A,65A	Pastel parch. crinkle vinyl & pastel parch. clarion knit. vinyl w/ black
7UA	63A,65A	Pastel parch. crinkle vinyl & pastel parch. clarion knit. vinyl w/ black
8A	63C,65C	Charcoal black crinkle vinyl and charcoal black clarion knitted vinyl
8AA	63C,65C	Charcoal black crinkle vinyl and charcoal black clarion knitted vinyl
8B	63C,65C	Light blue crinkle vinyl and dark blue clarion knitted vinyl
8BA	63C,65C	Light blue crinkle vinyl and dark blue clarion knitted vinyl
8D	63C,65C	Dark red crinkle vinyl and dark red clarion knitted vinyl
8DA	63C,65C	Dark red crinkle vinyl and dark red clarion knitted vinyl
8U	63C,65C	Pastel parch. crinkle vinyl & pastel parch. clarion knit. vinyl w/ black
8UA	63C,65C	Pastel parch. crinkle vinyl & pastel parch. clarion knit. vinyl w/ black
9A	63D,65D	Charcoal black crinkle vinyl and charcoal black clarion knitted vinyl
9AA	63D,65D	Charcoal black crinkle vinyl and charcoal black clarion knitted vinyl
9B	63D,65D	Light blue crinkle vinyl and dark blue clarion knitted vinyl
9BA	63D,65D	Light blue crinkle vinyl and dark blue clarion knitted vinyl
9D	63D,65D	Dark red crinkle vinyl and dark red clarion knitted vinyl
9DA	63D,65D	Dark red crinkle vinyl and dark red clarion knitted vinyl
9U	63D,65D	Pastel parch. crinkle vinyl & pastel parch. clarion knit. vinyl w/ black
9UA	63D,65D	Pastel parch. crinkle vinyl & pastel parch. clarion knit. vinyl w/ black

Body Type Codes

63A Fastback, Standard interior
63B Fastback, Deluxe interior
63C Fastback, Bench seat option
63D Fastback, Bench seat luxury option
65A Hardtop, Standard interior
65B Hardtop, Deluxe interior
65C Hardtop, Bench seat option
65D Hardtop, Bench seat luxury option
76A Convertible, Standard interior
76B Convertible, Deluxe interior

1969 Mustang Interior Trim Scheme Codes

Code	Body Type	Trim Scheme
1A	65E	Charcoal black corinthian vinyl & charcoal black hopsack pattern cloth
1AA	65E	Charcoal black corinthian vinyl & charcoal black hopsack pattern cloth
1B	65E	Light blue corinthian vinyl and light blue hopsack pattern cloth
1BA	65E	Light blue corinthian vinyl and light blue hopsack pattern cloth

Code	Body Type	Trim Scheme
1G	65E	Dark ivy gold corinthian vinyl and dark ivy gold hopsack pattern cloth
1GA	65E	Dark ivy gold corinthian vinyl and dark ivy gold hopsack pattern cloth
1Y	65E	Light nugget gold corinthian vinyl & light nug. gold hopsack patt. cloth
1YA	65E	Light nugget gold corinthian vinyl & light nug. gold hopsack patt. cloth
2A	63A,65A,76A	Charcoal black corinthian vinyl and charcoal black kiwi vinyl
2B	63A,65A,76A	Light blue corinthian vinyl and light blue kiwi vinyl
2D	63A,65A,76A	Dark red corinthian vinyl and dark red kiwi vinyl
2G	63A,65A,76A	Dark ivy gold corinthian vinyl and dark ivy gold kiwi vinyl
2Y	63A,65A,76A	Light nugget gold corinthian vinyl and light nugget gold kiwi vinyl
3C	63C	Charcoal black corinthian vinyl and charcoal black clarion kitted vinyl
3AA	63C	Charcoal black corinthian vinyl and charcoal black clarion kitted vinyl
3D	63C	Dark red corinthian vinyl and dark red clarion knitted vinyl
3DA	63C	Dark red corinthian vinyl and dark red clarion knitted vinyl
3D1	63C	Dark red corinthian vinyl and dark red clarion knitted vinyl
3W	63C	White corinthian vinyl and white clarion knitted vinyl
3WA	63C	White corinthian vinyl and white clarion knitted vinyl
3W1	63C	White corinthian vinyl and white clarion knitted vinyl
4A	63A,65A	Charcoal black corinthian vinyl and charcoal black clarion knitted vinyl
4D	63A,65A	Dark red corinthian vinyl and dark red clarion knitted vinyl
4D1	63A	Dark red corinthian vinyl and dark red clarion knitted vinyl
5A	63B,65B	Charcoal black corinthian w/ charcoal black ruffino & clarion knit vinyl
5AA	63B,65B	Charcoal black corinthian w/ charcoal black ruffino & clarion knit vinyl
5B	63B,65B	Light blue corinthian vinyl w/ light blue ruffino & clarion knit vinyl
5BA	63B,65B	Light blue corinthian vinyl w/ light blue ruffino & clarion knit vinyl
5D	63B,65B	Dark red corinthian vinyl and dark red ruffino and clarion knitted vinyl
5DA	63B,65B	Dark red corinthian vinyl and dark red ruffino and clarion knitted vinyl
5G	63B,65B	Dark ivy gold corinthian w/ dark ivy gold ruffino & clarion knit vinyl
5GA	63B,65B	Dark ivy gold corinthian w/ dark ivy gold ruffino & clarion knit vinyl
5W	63B,65B	White corinthian w/ white ruffino & clarion knitted vinyl w/ black
5WA	63B,65B	White corinthian w/ white ruffino & clarion knitted vinyl w/ black
5Y	63B,65B	Lt. nugget gold corinthian w/ lt. nug. gold ruffino & clarion knit vinyl
5YA	63B,65B	Lt. nugget gold corinthian w/ lt. nug. gold ruffino & clarion knit vinyl
6A	76A	Charcoal black corinthian vinyl and charcoal black clarion knitted vinyl
6D	76A	Dark red corinthian vinyl and dark red clarion knitted vinyl
7A	76B	Charcoal black corinthian w/ charcoal black ruffino and corinth. vinyl
7AA	76B	Charcoal black corinthian w/ charcoal black ruffino and corinth. vinyl
7B	76B	Light blue corinthian vinyl with light blue ruffino and corinthian vinyl
7BA	76B	Light blue corinthian vinyl with light blue ruffino and corinthian vinyl
7D	76B	Dark red corinthian vinyl with dark red ruffino and corinthian vinyl
7DA	76B	Dark red corinthian vinyl with dark red ruffino and corinthian vinyl
7G	76B	Dk. ivy gold corinthian vinyl w/ dk. ivy gold ruffino & corinthian vinyl
7GA	76B	Dk. ivy gold corinthian vinyl w/ dk. ivy gold ruffino & corinthian vinyl
7W	76B	White corinthian vinyl with white ruffino and corinthian vinyl w/black
7WA	76B	White corinthian vinyl with white ruffino and corinthian vinyl w/black
7Y	76B	Lt. nugget gold corinthian w/ lt. nugget gold ruffino & corinth. vinyl
7YA	76B	Lt. nugget gold corinthian w/ lt. nugget gold ruffino & corinth. vinyl
8A	65C	Charcoal black corinthian w/ charcoal black ruffino & clarion knit vinyl
8AA	65C	Charcoal black corinthian w/ charcoal black ruffino & clarion knit vinyl
8B	65C	Light blue corinthian vinyl w/ light blue ruffino & clarion knit vinyl
8BA	65C	Light blue corinthian vinyl w/ light blue ruffino & clarion knit vinyl
8D	65C	Dark red corinthian vinyl and dark red ruffino and clarion knitted vinyl
8DA	65C	Dark red corinthian vinyl and dark red ruffino and clarion knitted vinyl
8Y	65C	Lt. nugget gold corinthian w/ lt. nug. gold ruffino & clarion knit vinyl
8YA	65C	Lt. nugget gold corinthian w/ lt. nug. gold ruffino & clarion knit vinyl
9A	65D	Charcoal black corinthian w/ charcoal black ruffino & clarion knit vinyl
9AA	65D	Charcoal black corinthian w/ charcoal black ruffino & clarion knit vinyl
9B	65D	Light blue corinthian vinyl w/ light blue ruffino & clarion knit vinyl
9BA	65D	Light blue corinthian vinyl w/ light blue ruffino & clarion knit vinyl

Code	Body Type	Trim Scheme
9D	65D	Dark red corinthian vinyl and dark red ruffino and clarion knitted vinyl
9DA	65D	Dark red corinthian vinyl and dark red ruffino and clarion knitted vinyl
9Y	65D	Lt. nugget gold corinthian w/ lt. nug. gold ruffino & clarion knit vinyl
9YA	65D	Lt. nugget gold corinthian w/ lt. nug. gold ruffino & clarion knit vinyl
DA	63B,65B	Charcoal black corinthian vinyl and charcoal black clarion knitted vinyl
DAA	63B,65B	Charcoal black corinthian vinyl and charcoal black clarion knitted vinyl
DD	63B,65B	Dark red corinthian vinyl and dark red clarion knitted vinyl
DDA	63B,65B	Dark red corinthian vinyl and dark red clarion knitted vinyl
DD1	63B	Dark red corinthian vinyl and dark red clarion knitted vinyl
DW	63B,65B	White corinthian vinyl and white clarion knitted vinyl with black
DWA	63B,65B	White corinthian vinyl and white clarion knitted vinyl with black
DW1	63B	White corinthian vinyl and white clarion knitted vinyl with black

Body Type Codes

63A SportsRoof, Standard interior
63B SportsRoof, Deluxe interior
63C SportsRoof, Mach 1 option
65A Hardtop, Standard interior
65B Hardtop, Deluxe interior
65C Hardtop, Bench seat option
65D Hardtop, Bench seat luxury option
65E Hardtop, Grandé option
76A Convertible, Standard interior
76B Convertible, Deluxe interior

1970 Mustang Interior Trim Scheme Codes

Code	Body Type	Trim Scheme
3A	63C	Charc. black corinth. vinyl w/ charc black clarion knit & corinth. vinyl
3B	63C	Medium blue corinthian w/ medium black clarion knit & corinth. vinyl
3E	63C	Vermillion corinthian w/ vermillion clarion knit & corinth. vinyl
3F	63C	Medium ginger corinthian w/ med. ginger clarion knit & corinth. vinyl
3G	63C	Med. ivy green corinth. w/ med. ivy green clarion knit & corinth vinyl
3W	63C	White corinthian w/ white clarion knit & corinth. vinyl w/ black
AA	65E	Charcoal black corinthian vinyl & black houndstooth pattern cloth
AA1	65E	Charcoal black corinthian vinyl & black houndstooth pattern cloth
AB	65E	Medium blue corinthian vinyl & medium blue houndstooth pattern cloth
AB1	65E	Medium blue corinthian vinyl & medium blue houndstooth pattern cloth
AE	65E	Vermillion corinthian vinyl and vermillion houndstooth pattern cloth
AE1	65E	Vermillion corinthian vinyl and vermillion houndstooth pattern cloth
AF	65E	Medium ginger corinthian vinyl & med. ginger houndstooth pattern cloth
AF1	65E	Medium ginger corinthian vinyl & med. ginger houndstooth pattern cloth
AG	65E	Med. ivy green corinthian & med. ivy green houndstooth patt. cloth
AG1	65E	Med. ivy green corinthian & med. ivy green houndstooth patt. cloth
BA	63A,65A,76A	Charcoal black corinthian vinyl and charcoal black ruffino vinyl
BAA	63A,65A,76A	Charcoal black corinthian vinyl and charcoal black ruffino vinyl
BA1	63A	Charcoal black corinthian vinyl and charcoal black ruffino vinyl
BB	63A,65A,76A	Medium blue corinthian vinyl and medium blue ruffino vinyl
BBA	63A,65A,76A	Medium blue corinthian vinyl and medium blue ruffino vinyl
BE	63A,65A,76A	Vermillion corinthian vinyl and vermillion ruffino vinyl
BEA	63A,65A,76A	Vermillion corinthian vinyl and vermillion ruffino vinyl
BF	63A,65A,76A	Medium ginger corinthian vinyl and medium ginger ruffino vinyl
BFA	63A,65A,76A	Medium ginger corinthian vinyl and medium ginger ruffino vinyl
BG	63A,65A,76A	Medium ivy green corinthian vinyl & medium ivy green ruffino vinyl
BGA	63A,65A,76A	Medium ivy green corinthian vinyl & medium ivy green ruffino vinyl
BW	63A,65A,76A	White corinthian vinyl and white ruffino vinyl with black
BWA	63A,65A,76A	White corinthian vinyl and white ruffino vinyl with black
BW1	63A	White corinthian vinyl and white ruffino vinyl with black
CE	63B,76B	Vermillion corinthian vinyl and vermillion spectrum stripe vinyl
CE1	63B,76B	Vermillion corinthian vinyl and vermillion spectrum stripe vinyl

Code	Body Type	Trim Scheme
CF	63B,76B	Medium ginger corinthian vinyl & medium ginger spectrum stripe vinyl
CF1	63B,76B	Medium ginger corinthian vinyl & medium ginger spectrum stripe vinyl
EA	63B,76B	Charcoal black corinthian w/ charcoal black clarion knit & ruffino vinyl
EA1	63B,76B	Charcoal black corinthian w/ charcoal black clarion knit & ruffino vinyl
EB	63B,76B	Medium blue corinthian vinyl w/ med. blue clarion knit & ruffino vinyl
EB1	63B,76B	Medium blue corinthian vinyl w/ med. blue clarion knit & ruffino vinyl
EG	63B,76B	Med. ivy green corinth. w/ med. ivy green clarion knit & corinth. vinyl
EG1	63B,76B	Med. ivy green corinth. w/ med. ivy green clarion knit & corinth. vinyl
EW	63B,76B	White corinthian vinyl w/ white clarion knit & corinth. vinyl w/ black
EW1	63B,76B	White corinthian vinyl w/ white clarion knit & corinth. vinyl w/ black
TA	63B,65B	Charcoal black corinthian w/ charc. black clarion knit & corinth. vinyl
TAA	63B,65B	Charcoal black corinthian w/ charc. black clarion knit & corinth. vinyl
TA1	63B,65B	Charcoal black corinthian w/ charc. black clarion knit & corinth. vinyl
TB	63B,65B	Medium blue corinthian vinyl w/ med. blue clarion knit & corinth. vinyl
TBA	63B,65B	Medium blue corinthian vinyl w/ med. blue clarion knit & corinth. vinyl
TB1	63B,65B	Medium blue corinthian vinyl w/ med. blue clarion knit & corinth. vinyl
TGA	63B,65B	Med. ivy green corinth. w/ med. ivy green clarion knit & corinth. vinyl
TG1	63B,65B	Med. ivy green corinth. w/ med. ivy green clarion knit & corinth. vinyl
TW	63B,65B	White corinthian vinyl w/ white clarion knit & corinth. vinyl w/ black
TWA	63B,65B	White corinthian vinyl w/ white clarion knit & corinth. vinyl w/ black
TW1	63B,65B	White corinthian vinyl w/ white clarion knit & corinth. vinyl w/ black
UE	63B,65B	Vermillion corinthian vinyl and vermillion spectrum stripe vinyl
UEA	63B,65B	Vermillion corinthian vinyl and vermillion spectrum stripe vinyl
UE1	63B,65B	Vermillion corinthian vinyl and vermillion spectrum stripe vinyl
UF	63B,65B	Med. ginger corinthian & med. ginger spectrum stripe & corinth. vinyl
UFA	63B,65B	Med. ginger corinthian & med. ginger spectrum stripe & corinth. vinyl
UF1	63B,65B	Med. ginger corinthian & med. ginger spectrum stripe & corinth. vinyl

Body Type Codes
63A SportsRoof, Standard interior
63B SportsRoof, Deluxe interior
63C SportsRoof, Mach 1 option
65A Hardtop, Standard interior
65B Hardtop, Deluxe interior
65E Hardtop, Grandé option
76A Convertible, Standard interior
76B Convertible, Deluxe interior

The 1965 standard Mustang interior features unique molded door panels and all-vinyl seat upholstery. Front seat belts were optional in 1965.

Standard 1966 interiors, like this example in Parchment, were simple yet comfortable. Seat belts were standard equipment in 1966.

A folddown rear seat was included as standard equipment on all fastback models through 1968. This 1968 Mustang fastback has a black standard interior.

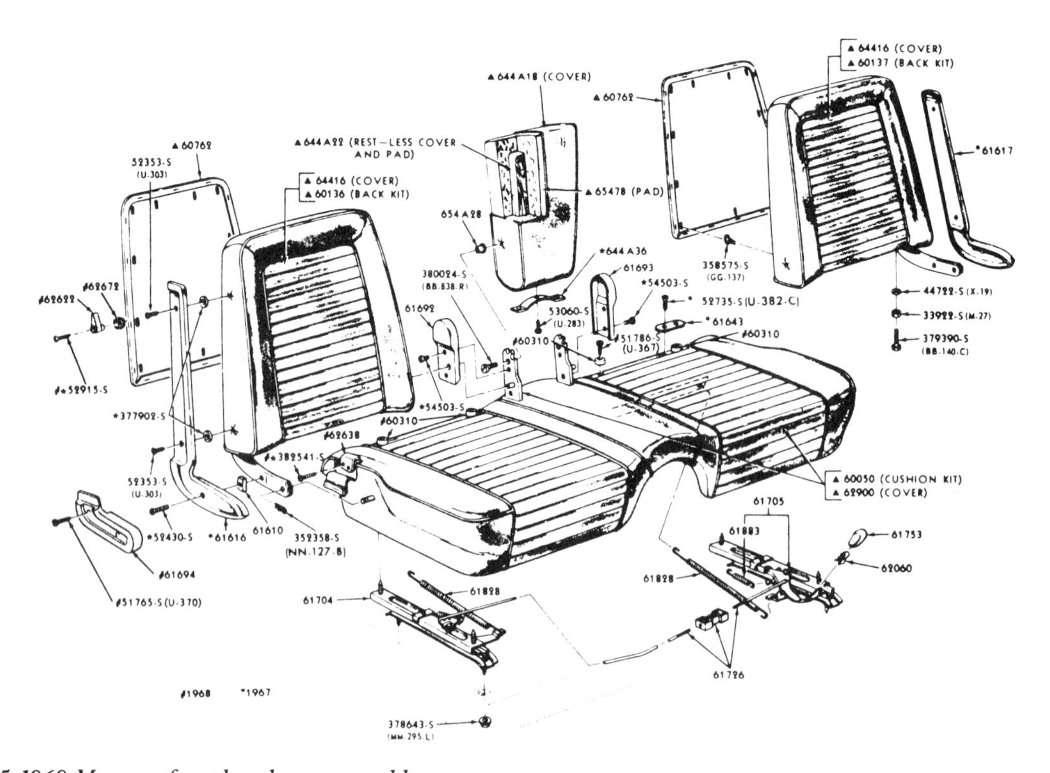

The 1965-1969 Mustang front bench seat assembly.

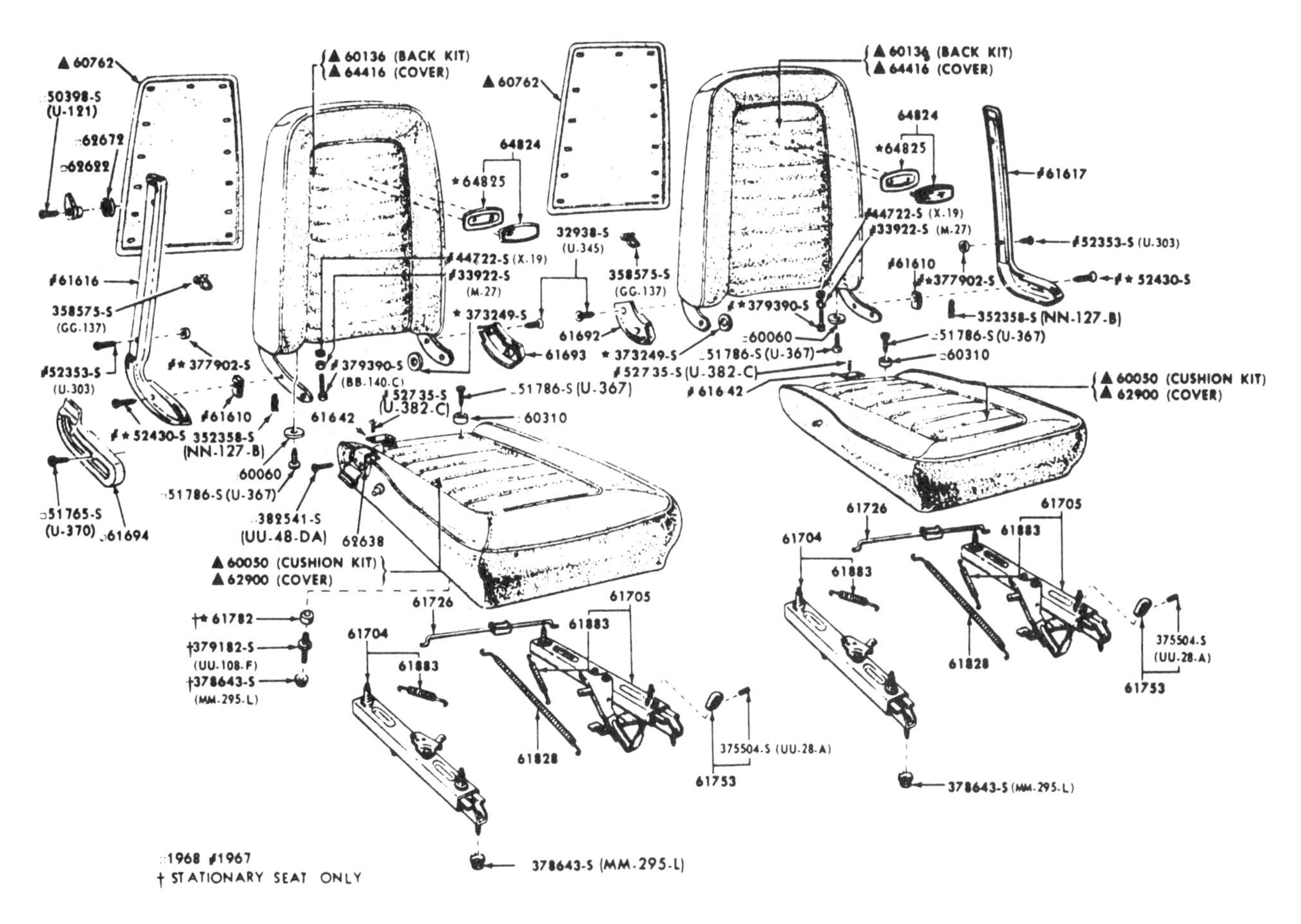

The 1965-1966 Mustang, 1967 Mustang (standard interior), and 1968-1969 Mustang front bucket seat assembly.

The 1968 standard seats have more padding and a seatback lock—yet another safety feature.

This 1966 Mustang convertible is equipped with the Interior Decor Group and air conditioning.

The optional Interior Decor Group included these unique seatbacks embossed with running horses.

This 1965 hardtop has a white standard interior, optional seat belts with retractors, and a Rally-Pac.

Standard interior for 1965 placed a pleated seat cushion inside smooth vinyl bolsters. Rear seat belts were a dealer-installed option in 1965.

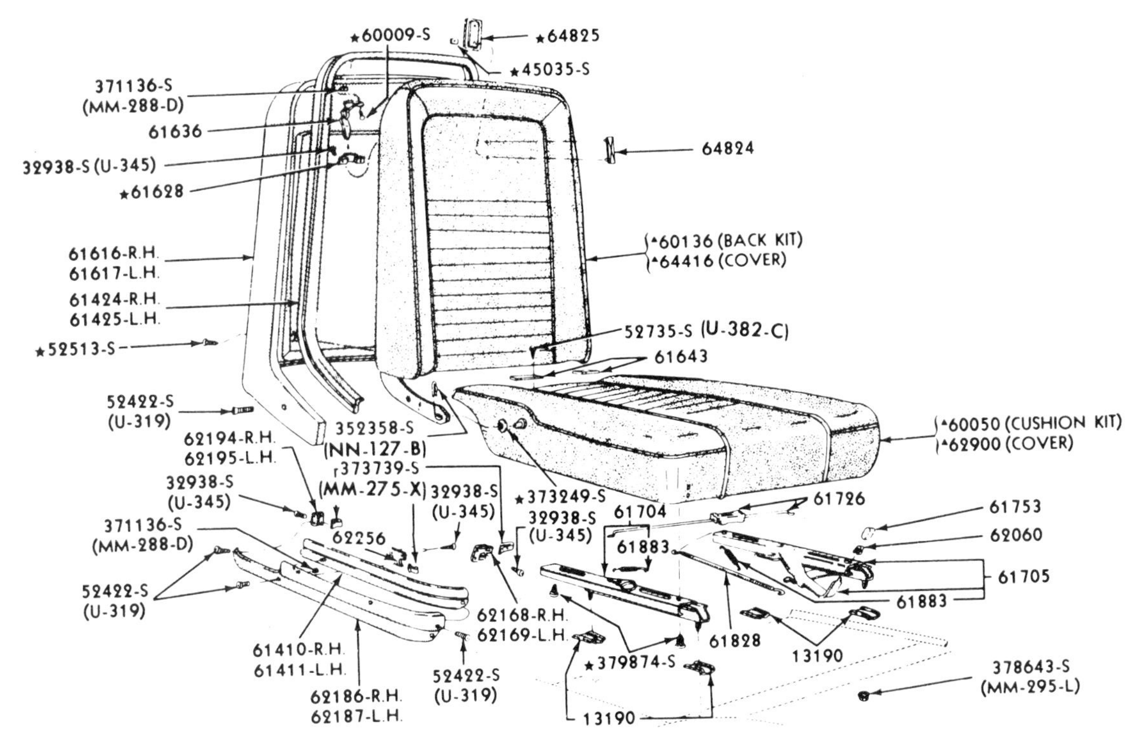

The 1967 Mustang front bucket seat assembly.

A Parchment front bench seat in a 1966 Mustang.

A full-width bench seat with center armrest was a Mustang option through 1969. This is a front bench seat in a 1967 convertible. Until 1968, a bench seat was available only with standard interior. In 1968 and 1969 a Deluxe bench seat option was offered.

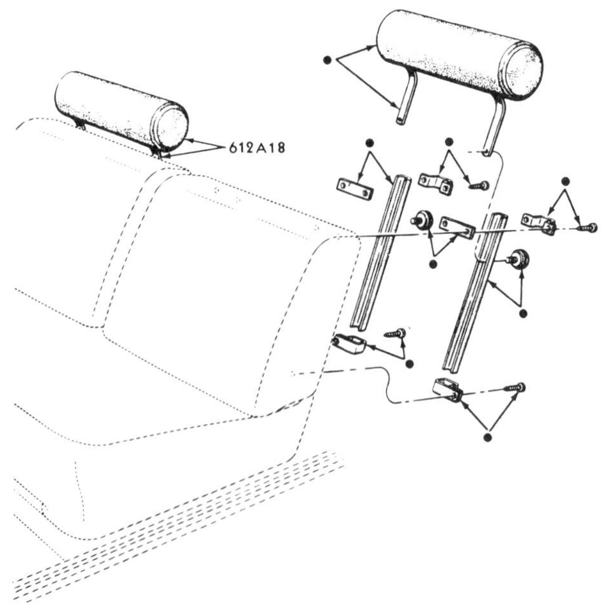

Optional 1966-1967 Mustang front seat headrests and related parts.

This two-tone Deluxe interior is in a 1967 convertible.

The full-length console for 1967 is attractive and functional. Its front compartment has a roll-top door and inside courtesy lamp.

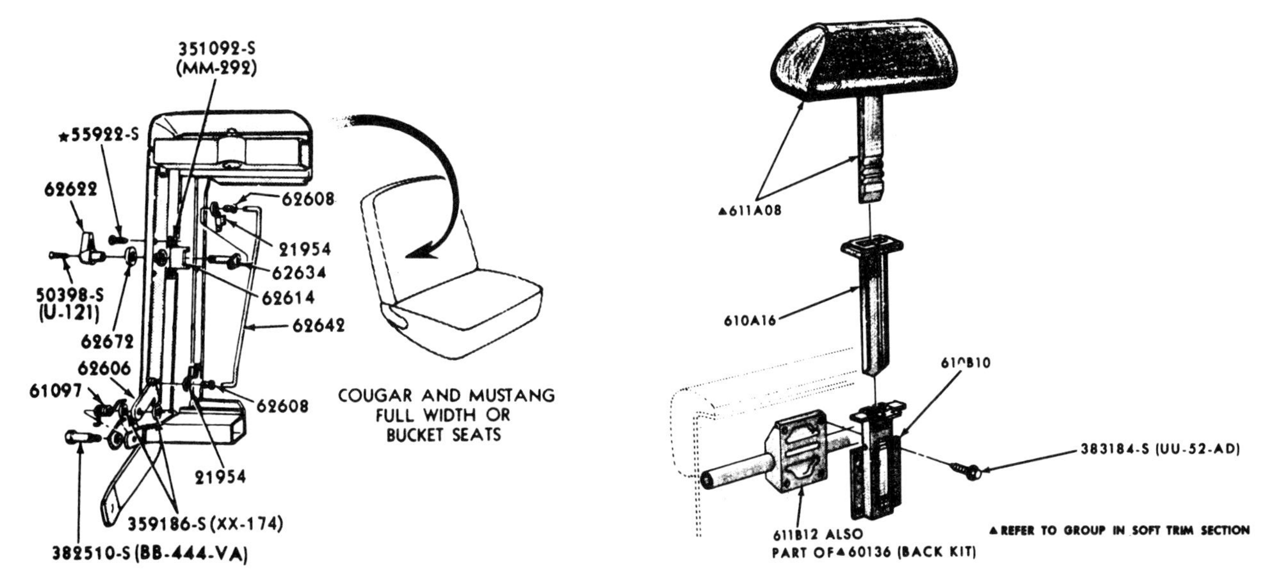

The 1968 Mustang front seatback latch assembly.

The 1969-1970 Mustang front seat headrest assembly.

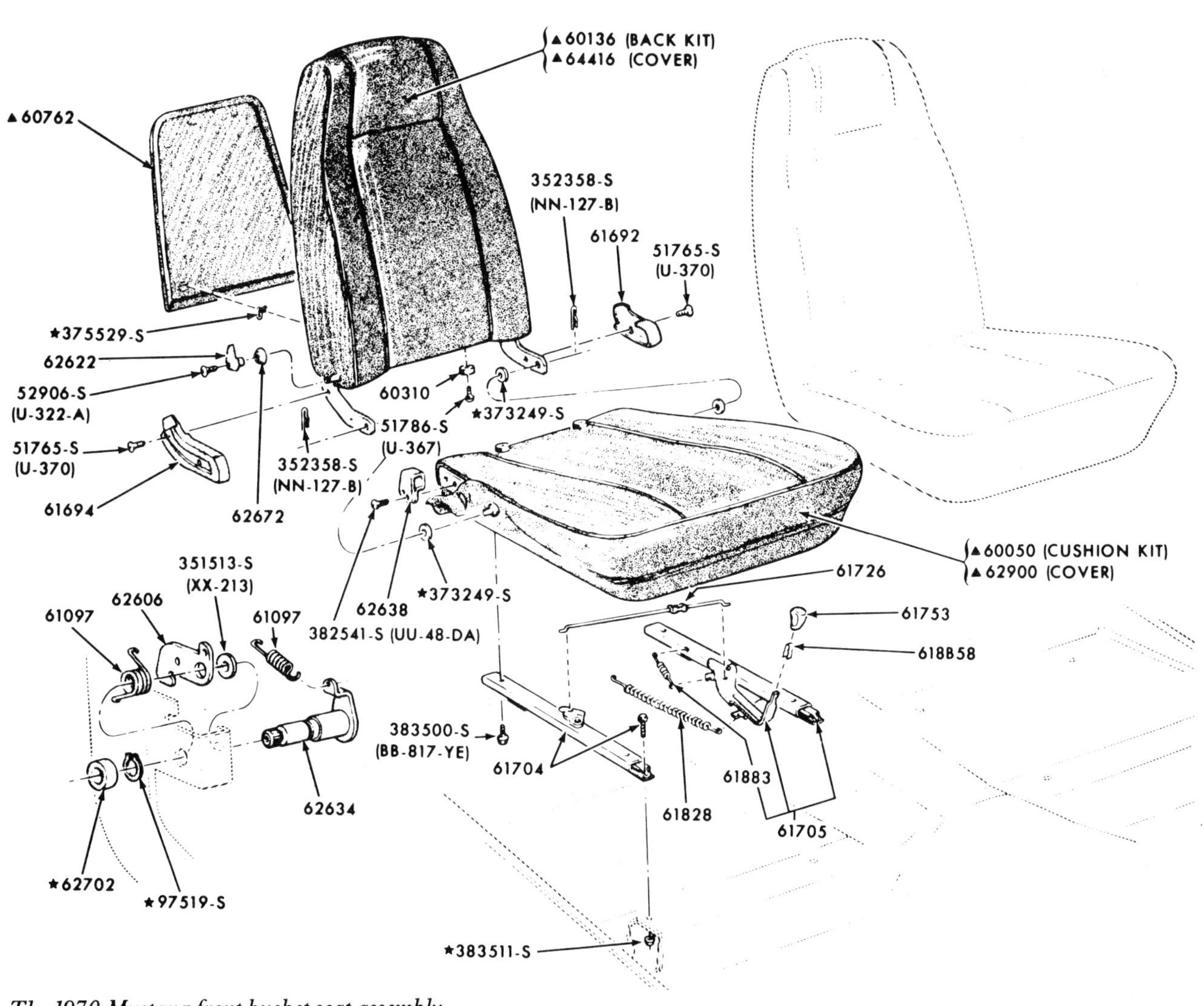

The 1970 Mustang front bucket seat assembly.

MUSTANG FRONT SEAT BELT ASSEMBLY APPLICATION CHART

BODY TYPE	PART NUMBER (less suffix)	R/N	COLORS and APPPLICABLE PART NUMBER SUFFIXES							
1966 *			AQUA	BLACK	BLUE	BUR-GUNDY	EMBER GLO	IVY GOLD	PARCH-MENT	RED
63,65	^C6OZ-62611A72	N	◊	GAB	◊	◊	◊	◊	◊	CAC
	^C7OZ-62611A72	R		B1A	B1B				◊	
Right	#C7OZ-76611A72	R		A1A	◊					
Left	#C7OZ-76611A73	R	DAF	A1A	◊			◊		
Before 3/28/66	#C6OZ-62611A72	R	DAF	DAB	GAD		GAG		◊	GAC
3/28 - 5/16/66	#C6OZ-62611A72	R	◊	GAB					◊	GAC
From 5/16/66	#C6ZZ-65611A72	R	DAF	DAB	◊		DAG	DAE	DAJ	DAC
76	^ C6ZZ-76611A72	N	◊	DAB	◊		◊	◊	◊	◊
Before 5/16/66	#C6ZZ-76611A72	N	◊	EAB	◊		◊	◊	◊	◊
From 5/16/66	#C6ZZ-76611A72	N	◊	EAB	◊		◊	◊	Grandé	◊

1967 *			AQUA	BLACK	BLUE	SADDLE		IVY GOLD	PARCH-MENT	RED
63, 65	#C7ZZ-65611A72	R	◊	D1A	◊	◊		◊	◊	◊
76	#C7ZZ-76611A72	N	◊	A1A	◊	◊		◊	◊	◊
	^ C7ZZ-76611A72	N		B1A	◊				◊	
63,65 exc. Shelby GT350/500	^ C7ZZ-65611A72	R		B1A	B1B				◊	
63 Shelby GT350/500	S7MS-65611A72-A		Includes emblem							

1968			AQUA	BLACK	BLUE		NUGGET GOLD	IVY GOLD	PARCH-MENT	RED
63,65 exc. Shelby GT350/500	+C8WZ-65611A72		B1K	B1A	B1B			B1G	B1U	B1D
76 exc. Shelby GT350/500	+C8ZZ-76611A72	N	◊	E1A	◊		◊	◊	◊	◊
	@D1TZ-10611B60			E						

1969			AQUA	BLACK	BLUE	SADDLE	NUGGET GOLD	IVY GOLD		RED
63,65	+C9ZZ-65611A72	R	F2K	F2A	F2B	F2F	F2Y	F2G		F2D
76	+C9ZZ-76611A72	N	◊	B1A	◊	◊	◊	◊		◊
Ctr. & rear seat	+C9AZ-62611A72	N	K1K	K1A	K1B		K1Y	K1G		K1D

1970				BLACK	BLUE	GINGER	GREEN		PARCH-MENT	
63,65	DOZZ-65611A72			J	G	H	K		F	
	«DOZZ-65611A72	R		N						
76	DOZZ-76611A72	N		B	A	E	C			

R/N: Retractable or Nonretractable harness

* Includes attaching parts

^ With fingertip release

\# With push button release

◊ Use black

\+ Chrome and plastic buckle

@ Plastic buckle

« Standard buckle - plastic covered

Body Type Codes:

63 Fastback/SportsRoof

65 Hardtop (incl. Grandé)

76 Convertible

MUSTANG REAR SEAT BELT ASSEMBLY APPLICATION CHART

BODY TYPE:	PART NUMBER LESS SUFFIX	COLORS and PART NUMBER SUFFIXES:							
1966			EMBER-GLO	BLUE	IVY GOLD	PARCH-MENT	RED		TOB-ACCO
Before 1/3/66 *	^C6AZ-62613B84				DAE		DAC		DAK
From 1/3/66 to 5/16/66 *	#C6AZ-63613B84		@	EAD		EAJ			EAK
From 5/16/66 *	#C6ZZ-65613B84		BAG	BAD		BAJ			
1967		AQUA	BLACK	BLUE	IVY GOLD	PARCH-MENT	RED	SADDLE	TOB-ACCO
Exc. Shelby	#C7AZ-62613B84	A1K	A1A	A1B	A1G	A1U	&	A1F	&
GT 350/500	+C7AZ-62613B84		B1A	B1B		&			
1968		AQUA	BLACK	BLUE	IVY GOLD	PARCH-MENT	RED	SADDLE	NUGGET GOLD
Exc. Shelby	=C8AZ-62613B84		C1A	C1B		&			
GT 350/500	~C8WZ-65613B84	A1K	A1A	A1B	A1G	A1U	A1D	A1F	A1Y
1969									
Rear & aux. seat	Refer to group 611A72-3								
1970		GINGER	BLACK	BLUE	GREEN	PARCH-MENT			
	DOMZ-76613B84	E	A	B	H	J			
	>DOAZ-62613B84		A						

* Includes attaching parts

@ Use C7AZ-62613B84-A1D

^ With fingertip release

\# With push button release

= Plastic buckle

& Use black

~ Chrome and plastic buckle

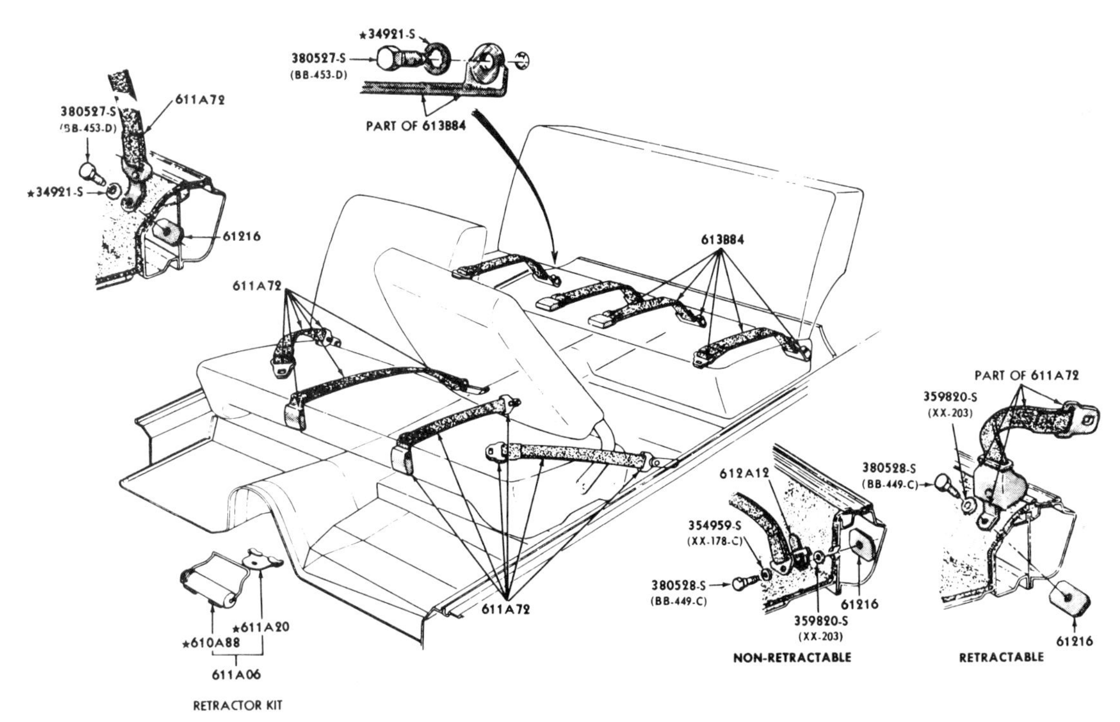

Typical 1966-1969 Mustang front and rear seat belt installation.

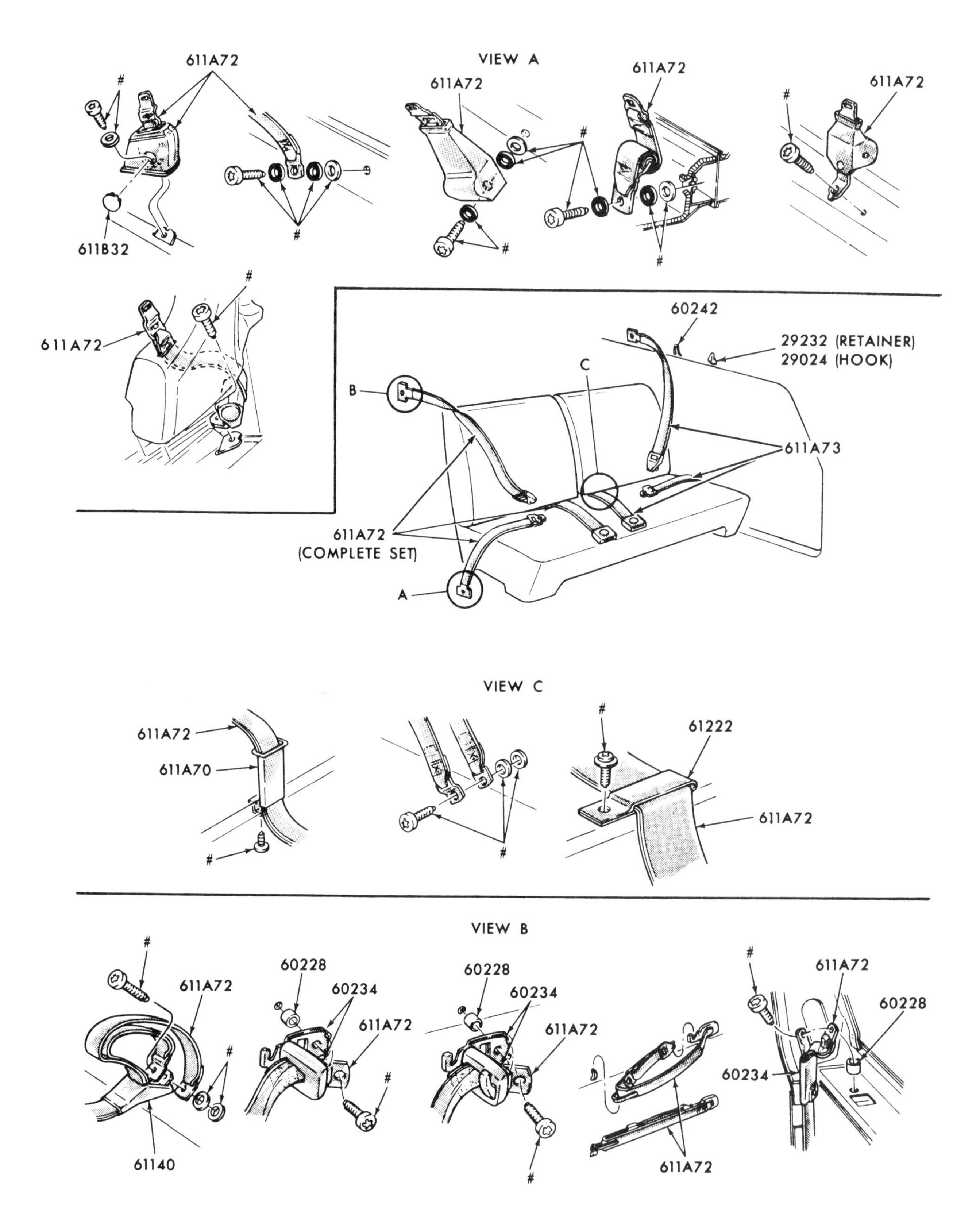

Typical 1970 Mustang seat belt and shoulder harness installation.

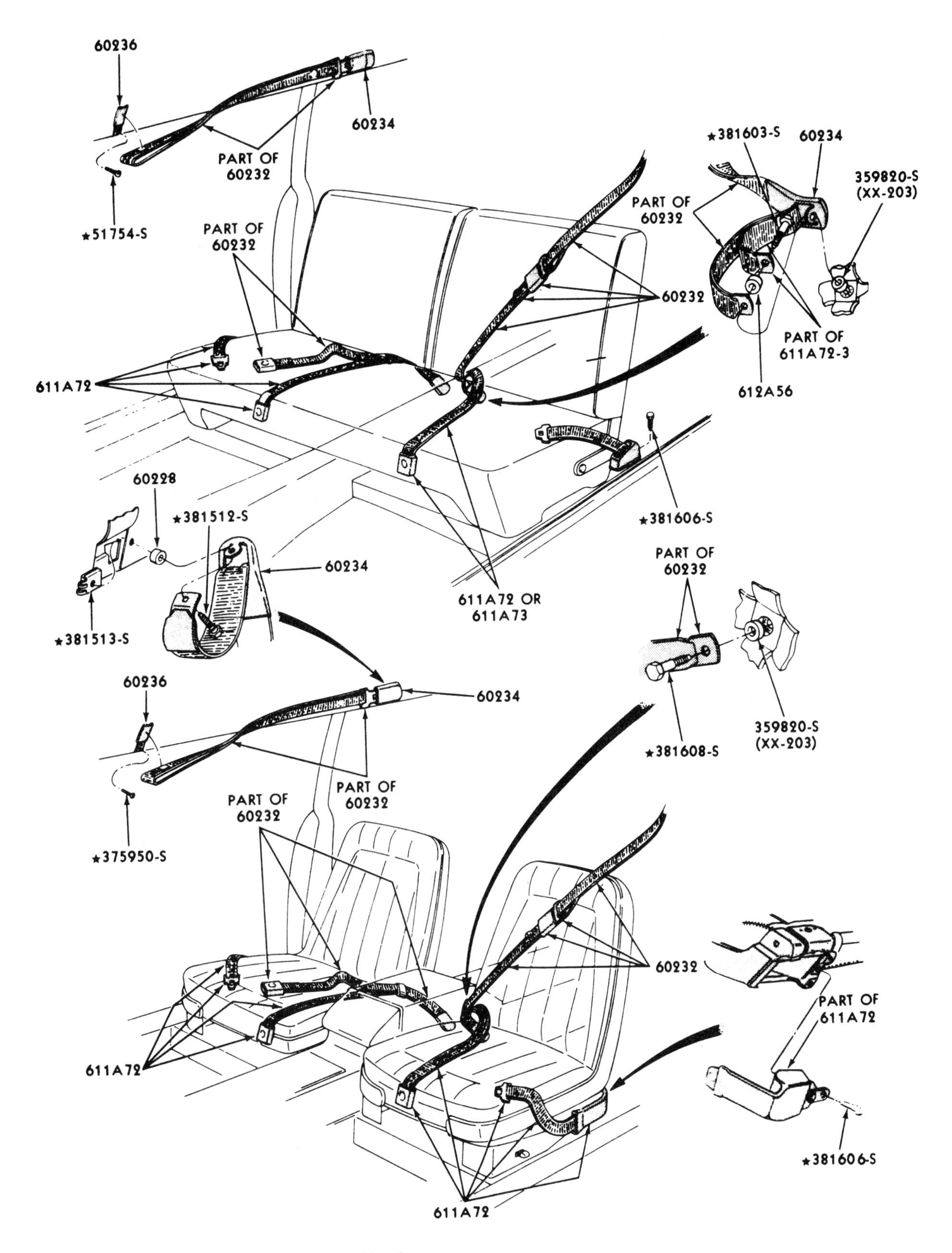

Typical 1967-1969 Mustang front seat shoulder harness installation.

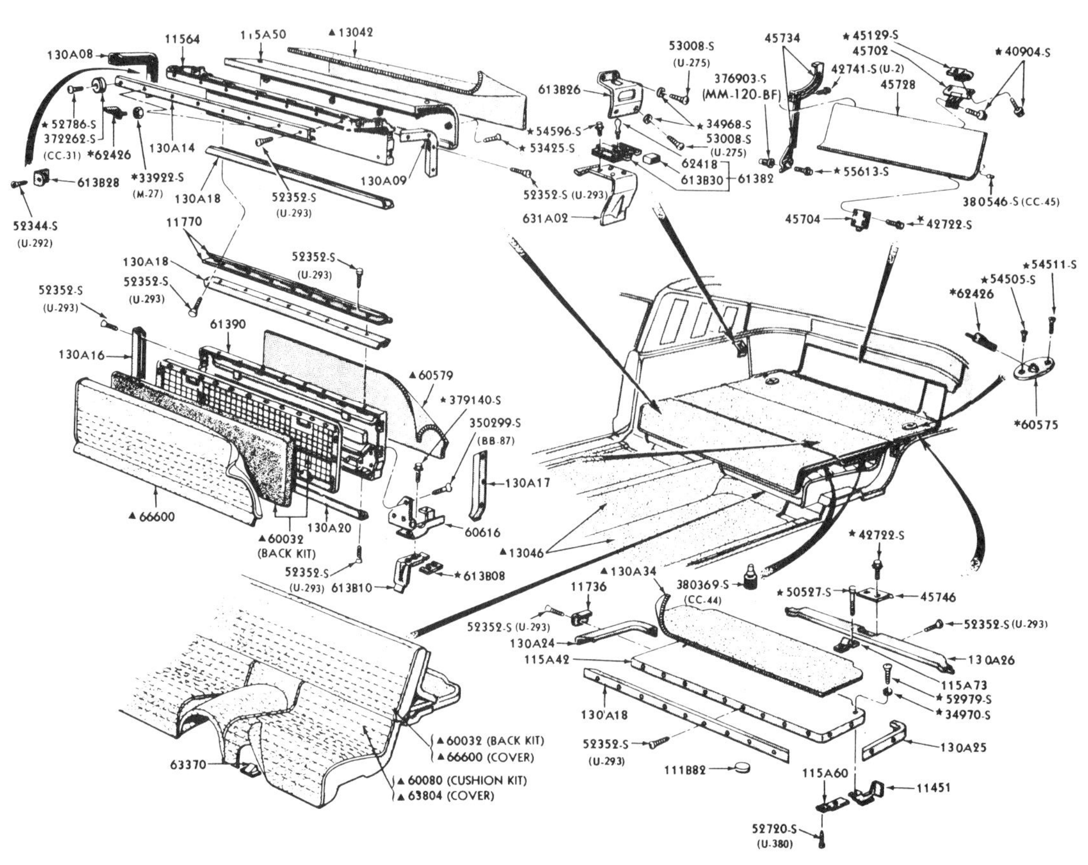

The 1965-1966 Mustang fastback fold-down rear seat assembly.

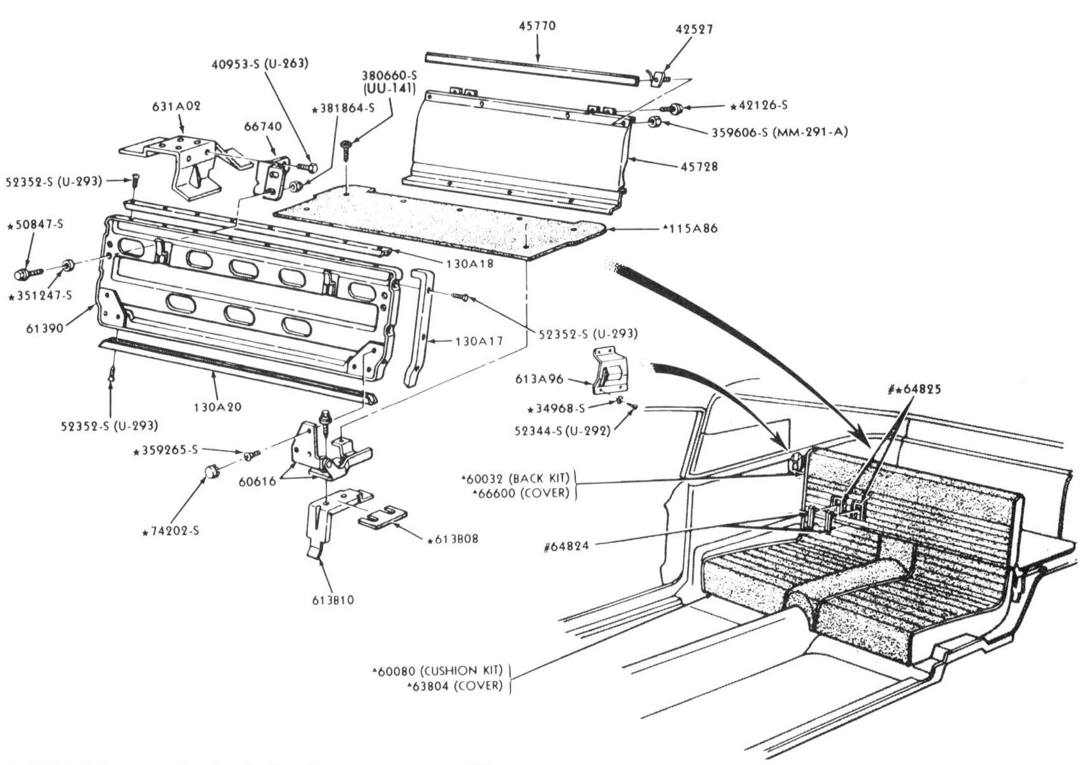

The 1967-1970 Mustang fastback fixed rear seat assembly.

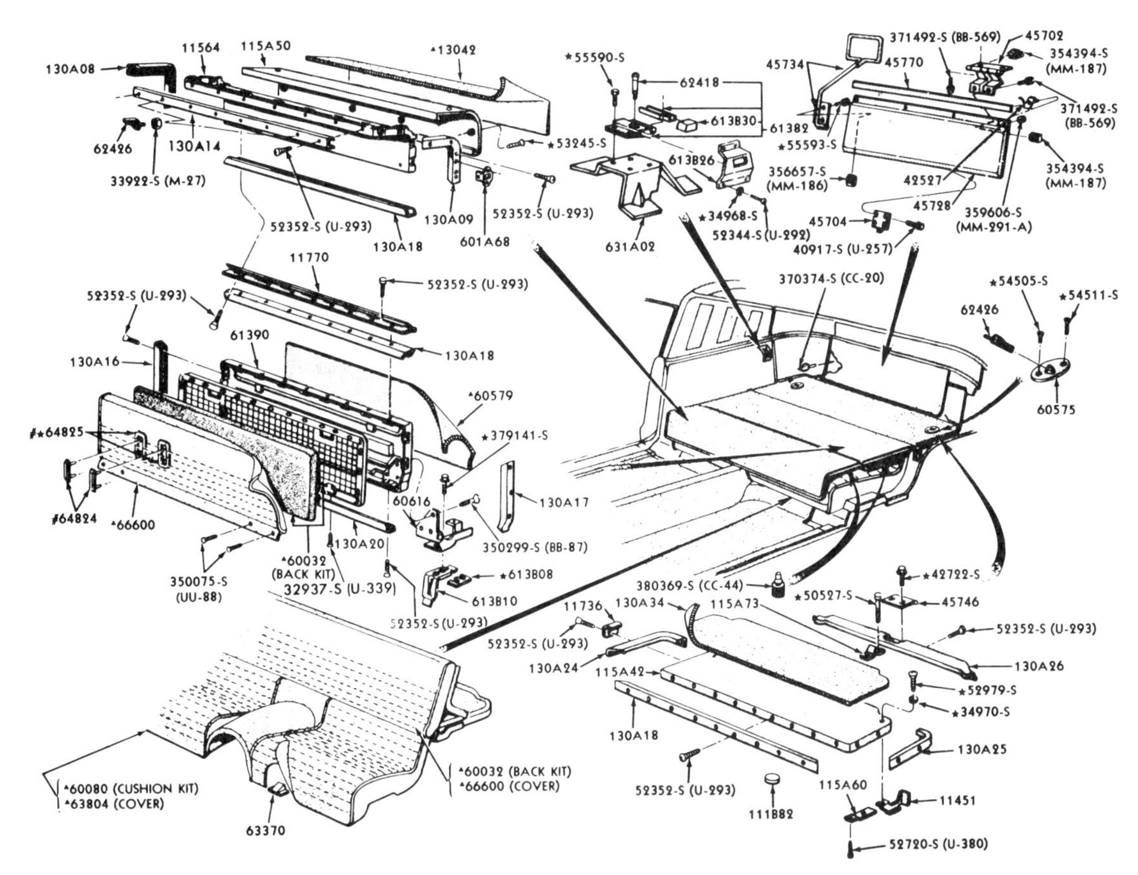

The 1967-1970 Mustang SportsRoof fold-down rear seat assembly.

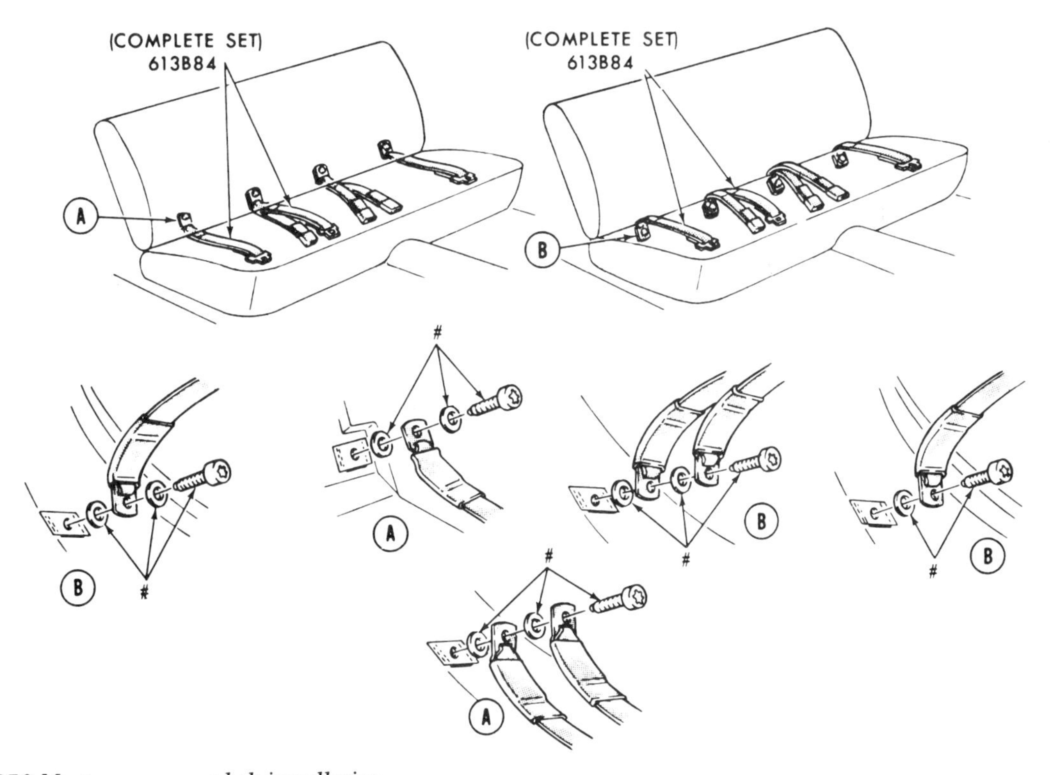

Typical 1970 Mustang rear seat belt installation.

Mustang Console Panel Application Chart

Colors & Applicable Part Number Suffixes

Body Type	Part Number Less Suffix	White	Black	Blue	Aqua	Ivy Gold	Nugget Gold	Red	Ginger
1965									
65,76 w/o A/C Before 12/1/64	C5OZ-65045A36	CAF	—	—	—	—	—	—	—
65,76 w/A/C Before 8/20/64	C5ZZ-65045A36	CAF	—	—	—	—	—	—	—
63,65,76 w/A/C From 12/1/64	C5ZZ-65045A36	CAF	CAB	—	—	—	—	—	—
1966									
	C5ZZ-65045A36	—	CAB	—	—	—	—	—	—
1967									
	C7ZZ-65045A36	—	A1A	A1B	—	—	—	—	—
1968									
	C8ZZ-65045A36	—	A1A	A1B	—	—	—	—	—
1969									
	C9ZZ-65045A36	—	C1A	—	C1K	C1G	C1Y	C1D	—
1970									
	DOZZ-65045A36	—	G	—	—	—	—	—	H

Body Type Codes
63: Fastback
65: Hardtop
76: Convertible

MUSTANG CENTER INSTRUMENT PANEL APPLICATION CHART

YEAR	BODY TYPE	DESCRIPTION	PART NUMBER
67/68	All, Std. Interior	Camera case finish	C7ZZ-6504774-B
68	All	Wood grain finish	C8ZZ-6504774-A
69	Mach 1 & Grandé	Light teakwood finish, w/o radio, Incl. attaching parts	C9ZZ-6504774-C
	Before 1/6/69.	Light teakwood finish, with radio, Incl. attaching parts	C9ZZ-6504774-D
69/70	Mach 1 & Grandé	Dark teakwood finish, w/o radio, Incl. attaching parts	C9ZZ-6504774-E
	From 1/6/69	Dark teakwood finish, with radio, Incl. attaching parts	C9ZZ-6504774-F
69/70	Exc Mach 1 & Grandé	Black cameracase finish, w/o radio, Incl. attach. parts	C9ZZ-6504774-A
69/70	Exc Mach 1 & Grandé	Black Corinthian grain finish - with radio	C9ZZ-6504774-B

MUSTANG INSTRUMENT PANEL GLOVE COMPARTMENT DOOR APPLICATIONS

YEAR	PART NUMBER	DESCRIPTION
65/66	C5ZZ-6506024-D	All Deluxe interiors; Wood grain finish
67/68	C7ZZ-6506024-A	
69/70	C9ZZ-6506024-A	

MUSTANG CONSOLE PANEL GLOVE COMPARTMENT DOOR APPLICATIONS

YEAR	PART NUMBER	DESCRIPTION
65/66	C5ZZ-6506024-B	With Camera Case finish
65/66	C5ZZ-6506024-C	With wood grain finish, From 3/8/65
67/68	C8ZZ-6506024-A	
	379162-S(CC34A)	Bumper
69	C9ZZ-6506024-B1Y	Nugget gold - less hinge - no catch - plain
	C9ZZ-6506024-B1D	Red
69/70	(4) 52720-S8(U380)	No. 8 pan head screws - 18 x 3/8"
69/70	C9ZZ-6506024-B1A	Black, less hinge, no catch, plain, Before 9/2/69
	C9ZZ-6506024-B2B	Blue
70	DOZZ-6506024-C	Red, Before 9/2/69
70	DOZZ-6306024-K	Black, less hinge, magnetic catch, plain, From 9/2/69
	DOZZ-6306024-B	Blue, to 12/1/69
	DOZZ-6306024-F	Ginger
	DOZZ-6306024-A	Green
	DOZZ-6306024-D	Red
	DOZZ-6306024-C	Tobacco
70	DOZZ-6306024-K	Black, less hinge, positive catch, plain, From 12/1/69
	DOZZ-6306024-H	Blue
	DOZZ-6306024-M	Ginger
	DOZZ-6306024-G	Green
	DOZZ-6306024-L	Red

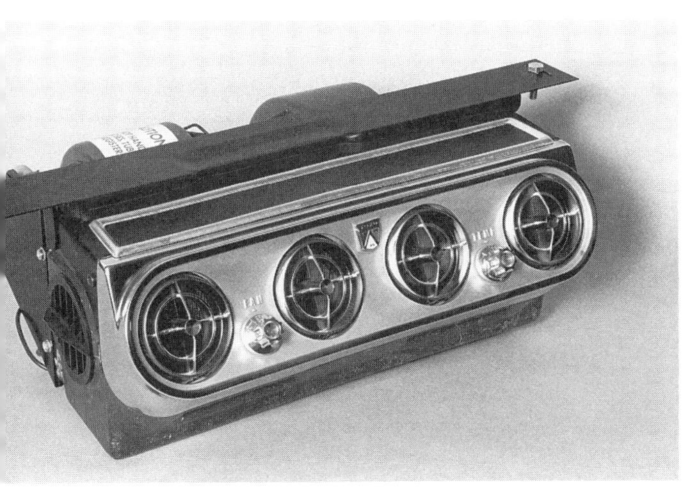

Air conditioning in a Mustang was a bolt-on option (even if it was factory installed) until 1967 when it became an integrated part of the instrument panel. This compact 1965 model air handler contains the evaporator core, a fan, a thermostatic control valve, and the two control switches. Hang-on air conditioning can only recycle air in the passenger compartment because there are no provisions for a fresh-air duct.

The 1966 air conditioning unit has a black camera-case finish and smaller, less intricate vents, and is slightly smaller in size.

A full-length console with front compartment and rear ashtray was and remains a popular Mustang option.

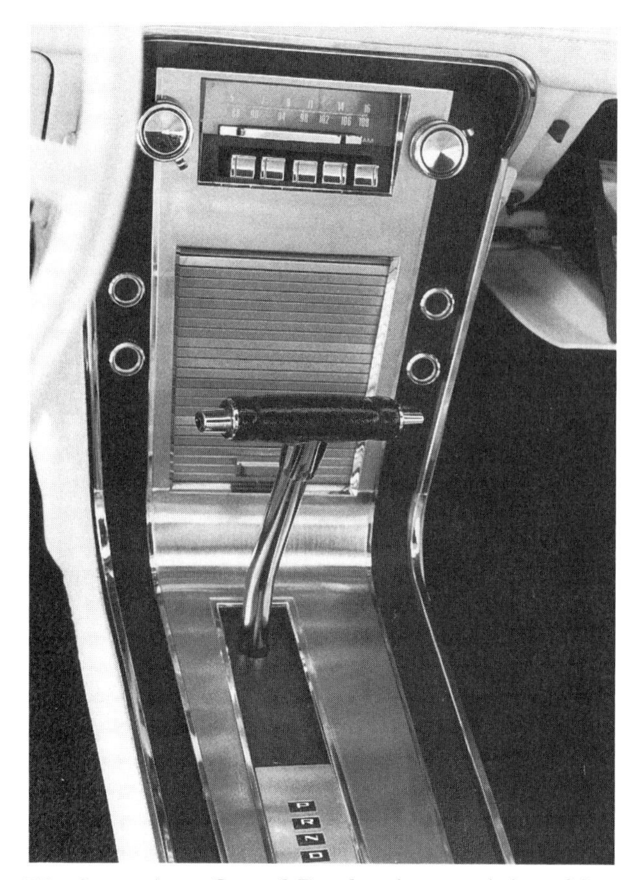

The Convenience Control Panel option, consisting of four warning lights, is located in the console when air conditioning is installed. Otherwise, the lights are positioned in the center of the dash panel just above the radio.

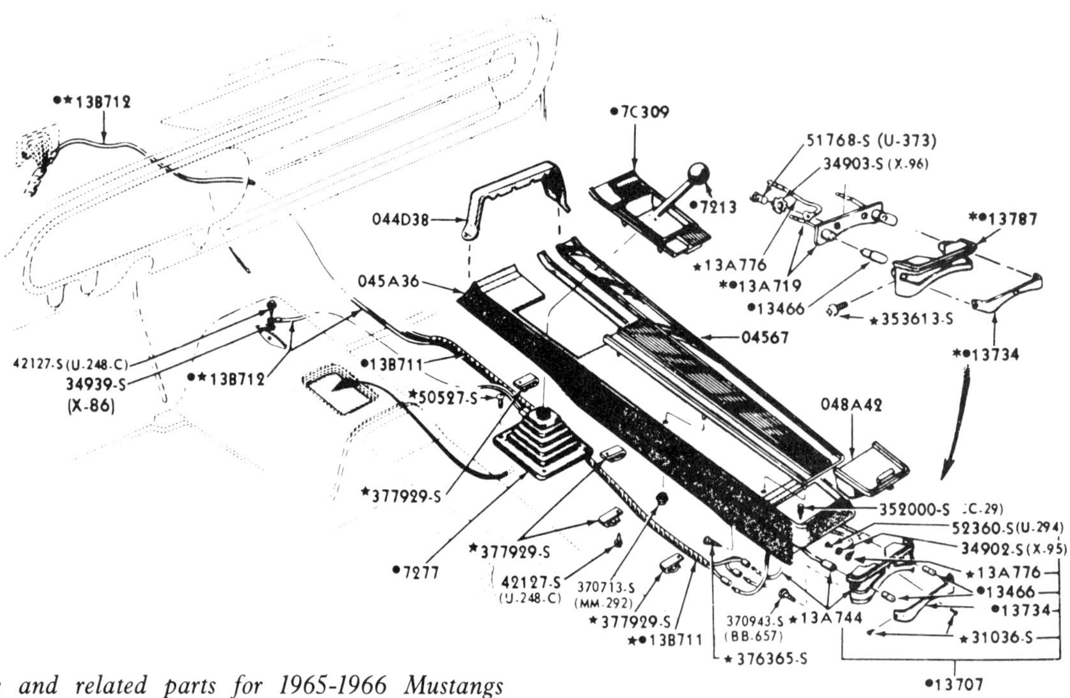

Console and related parts for 1965-1966 Mustangs equipped with air conditioning.

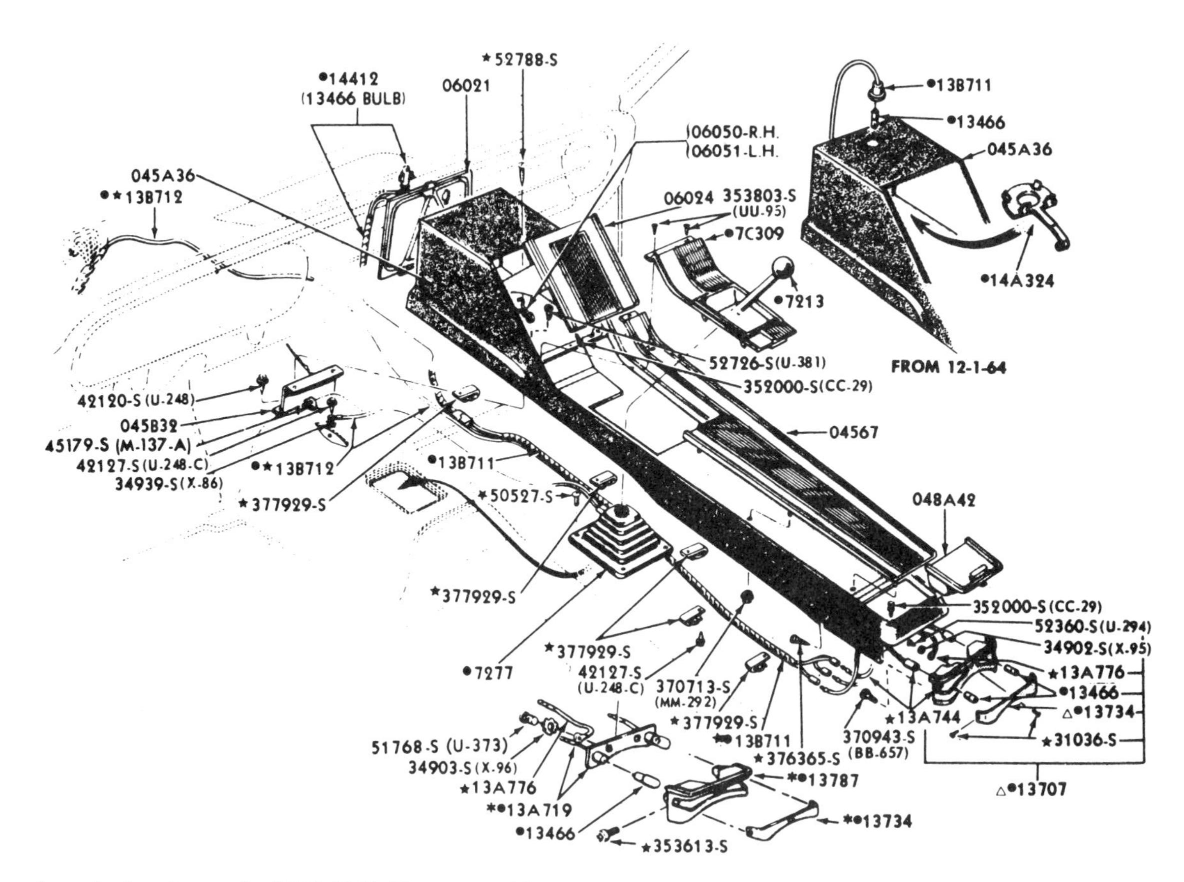

Console and related parts for 1965-1966 Mustangs without air conditioning.

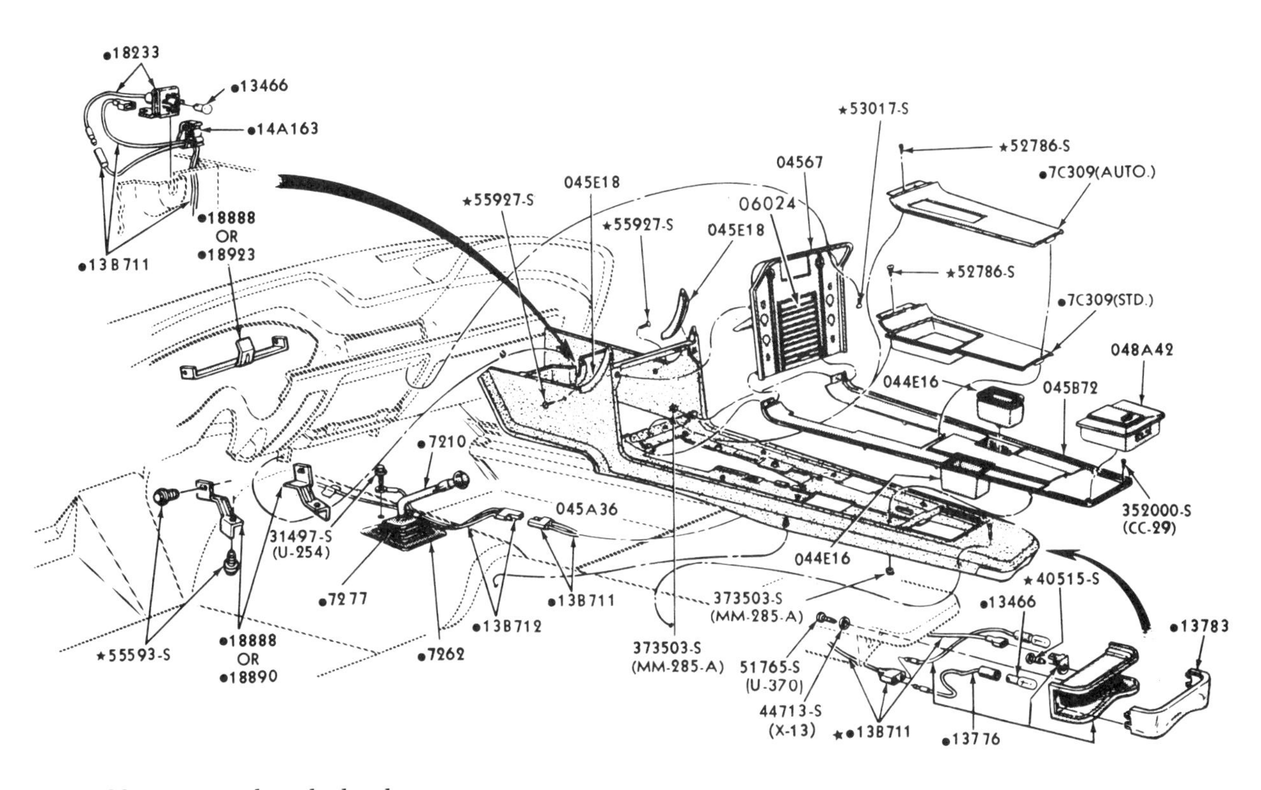

The 1967 Mustang console and related parts.

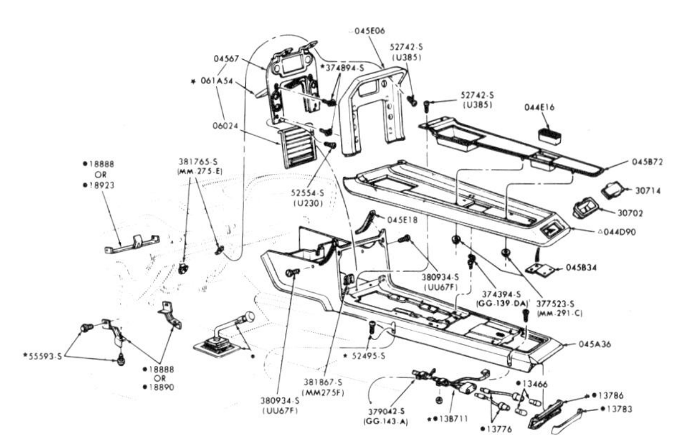

The 1968 Mustang console and related parts.

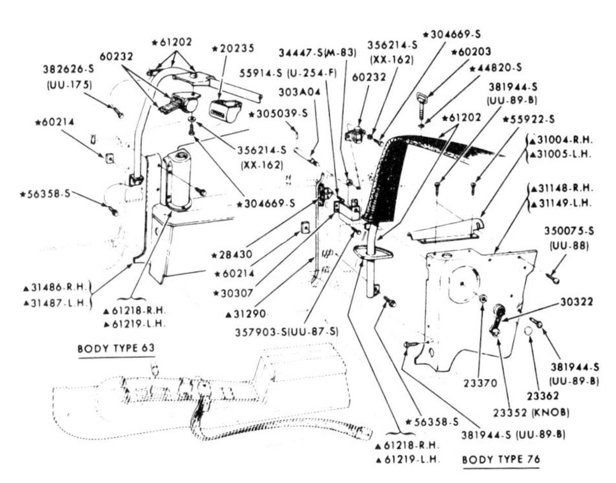

The 1969-1970 Shelby Mustang rollbar and related parts.

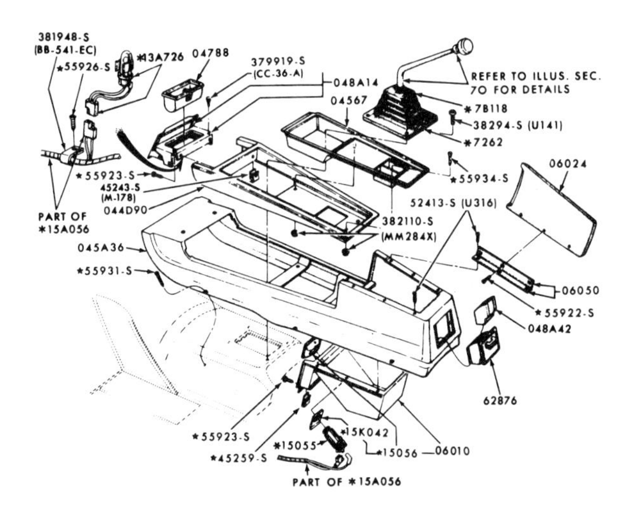

The 1969-1970 Mustang console and related parts.

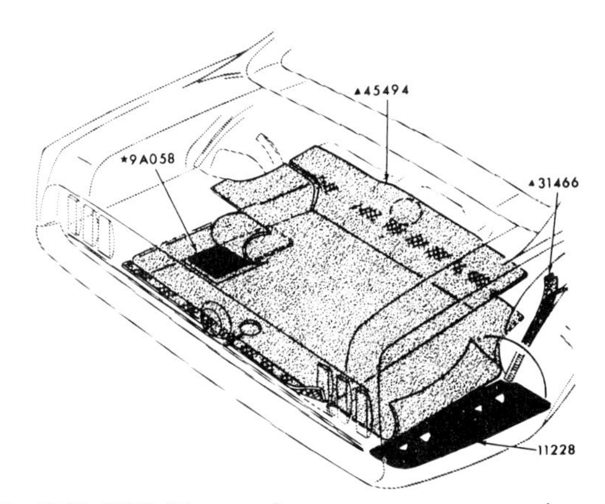

The 1969-1970 Mustang luggage compartment trim.

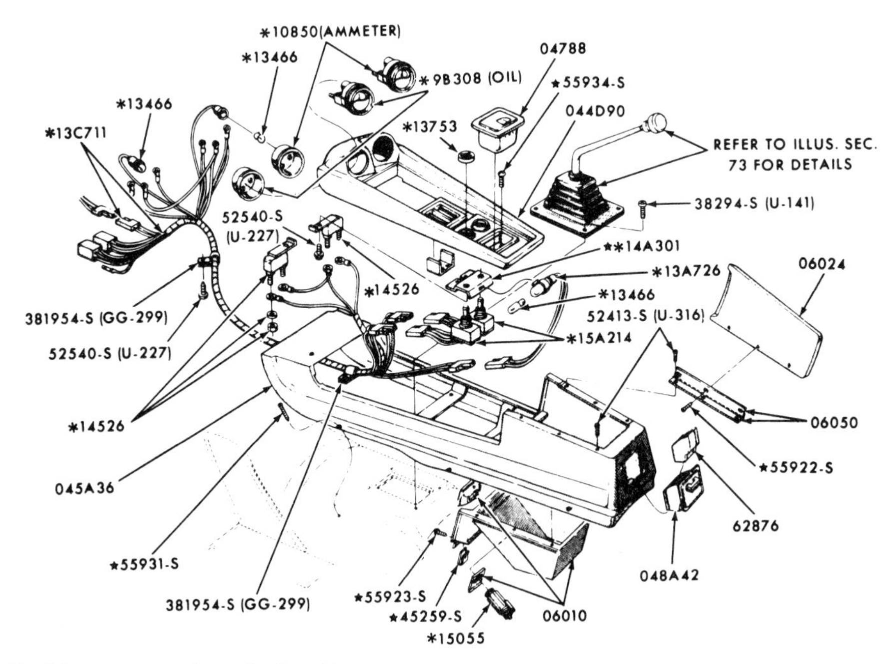

The 1969-1970 Shelby Mustang console and related parts.

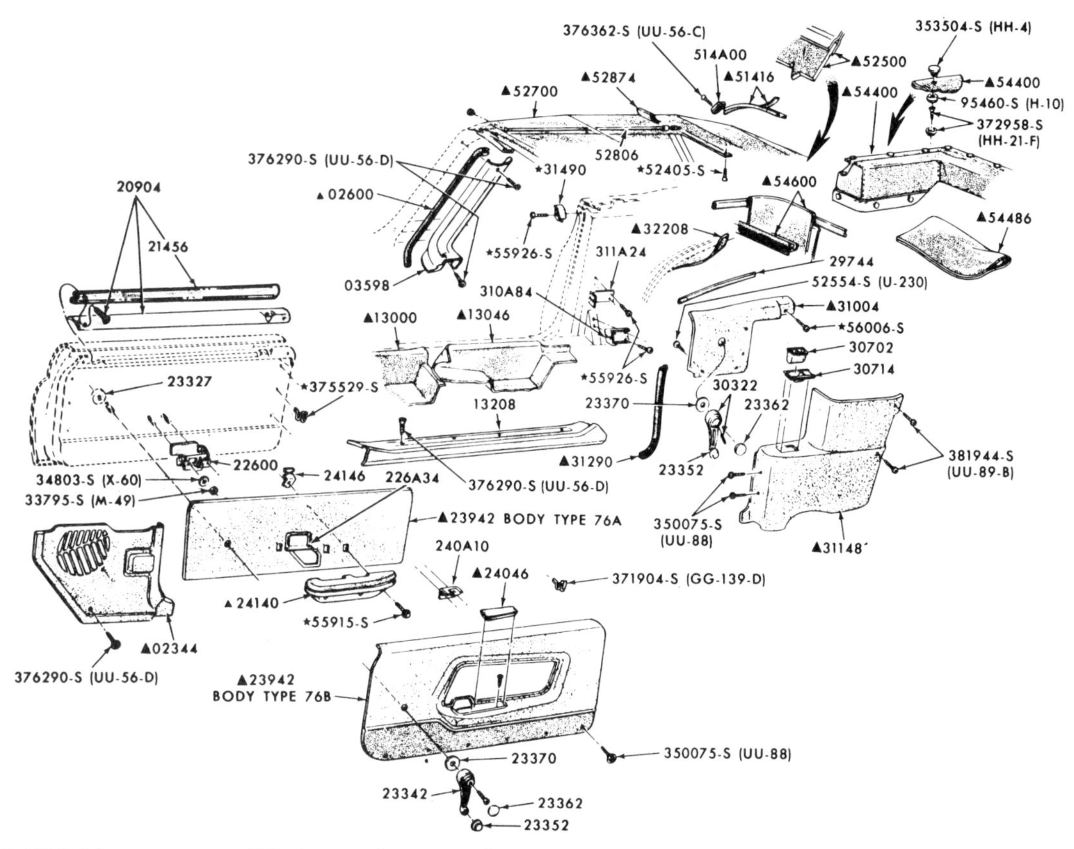

The 1969-1970 Mustang convertible door and quarter trim panels and related parts.

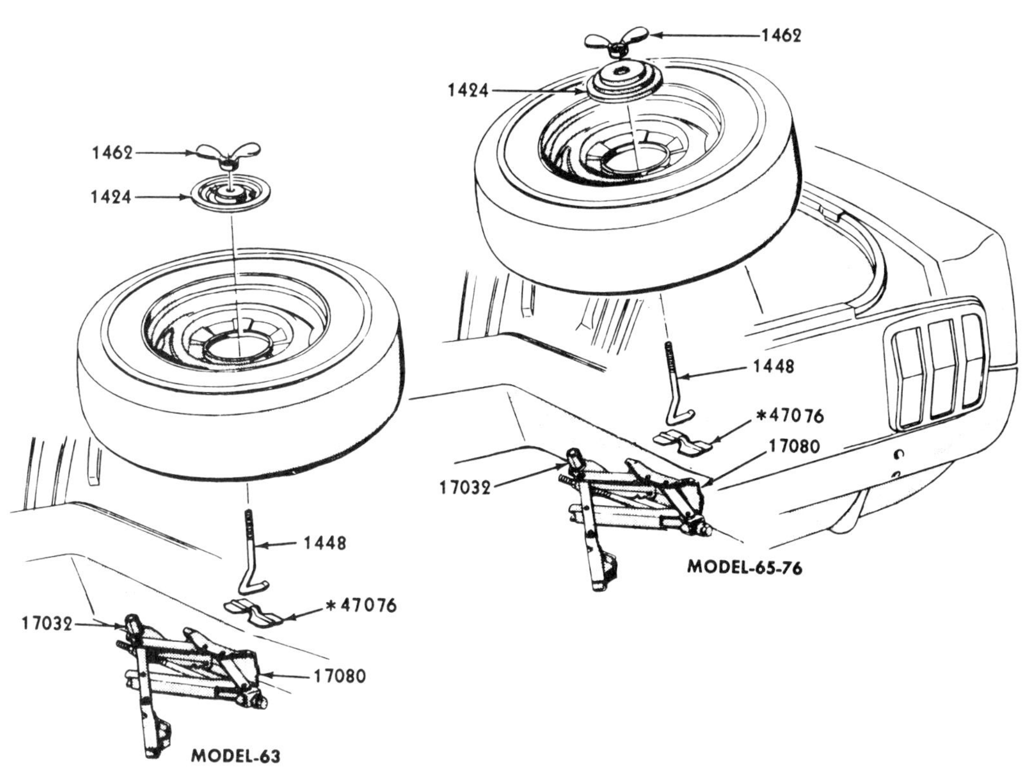

The 1965-1966 Mustang spare wheel, jack, and related parts.

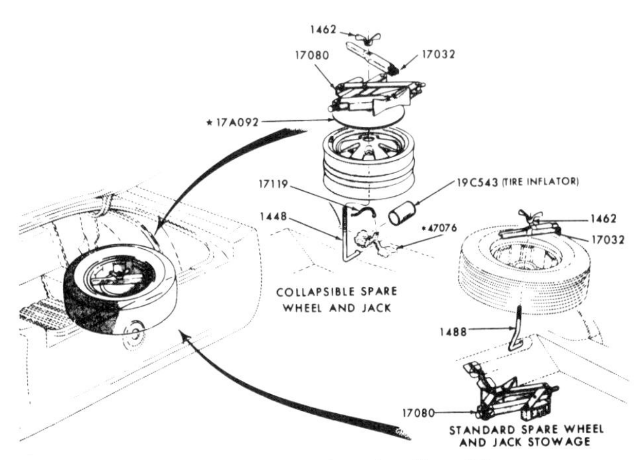

Both a standard and an optional collapsible spare were offered in 1969 and 1970 Mustangs.

A full-sized spare tire in a 1968 Mustang hardtop model.

Correct placement of a Space Saver spare tire in a 1969 Mustang SportsRoof.

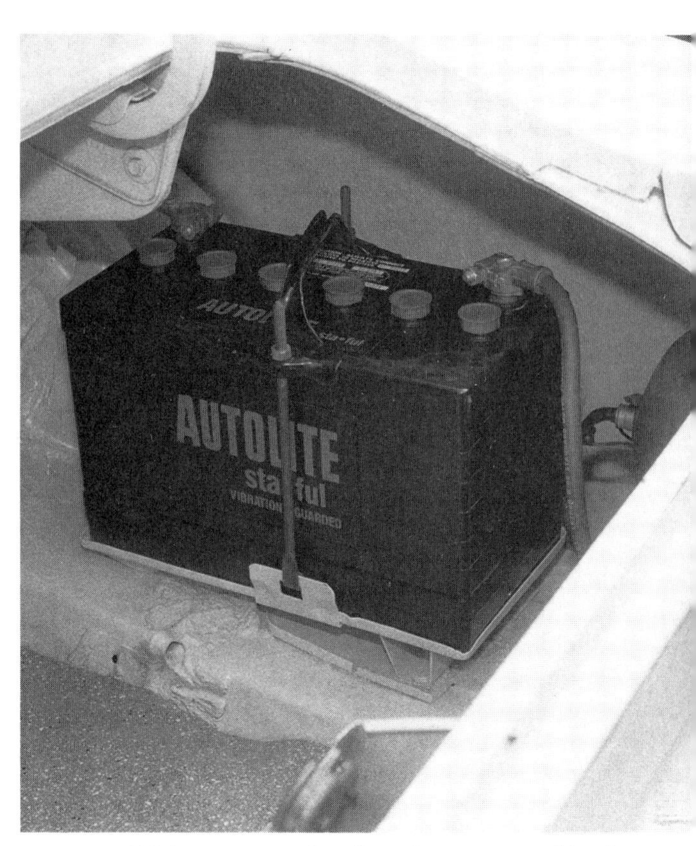

The Boss 429 battery was placed on the passenger side of the trunk floor for better distribution of weight. The spare tire was moved to the driver's side of the car.

The correct storage for a full-size spare tire in a 1970 Mustang SportsRoof.

Chapter 4

Engine

From 1964 through 1970 surprisingly few engines powered the Ford Mustangs. Several shared exterior componentry, but applications of accessories varied according to the weight of the car, the options included (especially air conditioning), and performance packages. Included in this chapter will be not only the engines and their internal components, but assemblies and subassemblies either located in the engine compartment or in the category of engine-driven accessories, most notably air conditioning.

From the base 170ci six-cylinder engine to the 429ci NASCAR engine in the special edition 1969 and 1970 Boss 429s, the sources of power evolved along with the Mustang itself. Due in part to the popularity of the Mustang and the nature of auto sales competition, especially with GM, in the 1960s, Ford involved itself in remarkable levels of research and development of performance engines. This pleased not only the potential Mustang buyer, but the performance enthusiast in general; the appreciation of the more powerful Ford V-8 engines of that era continues into the 1990s. It is remarkable that today's 5.0 liter engines are based on the 302ci introduced in 1968—which, in turn, derived from the 260ci V-8 and 221ci V-8 of the early 1960s.

Today, enthusiasts are deeply involved in a quest to authenticate many engines, accessories, and subassemblies, but especially the more uncommon "muscle" parts of the past. Many original components, especially equipment related to emissions controls, were discarded or swapped with no regard for their future value, rarity, or difficulty in identification and replacement. In this chapter we will attempt to provide guidelines for identification based upon production dates, original dimensions, tie-ins to related componentry, and cross-references to various applications and options.

Much of this material is drawn from vintage Ford references, though the reader should note that Ford always has made running changes in its supply system and there is no guarantee that all part numbers will relate directly to the original factory-installed equipment. The reason for this is that Ford, as time passed, tended to consolidate applications of many parts in order to reduce inventories and costs. A replacement part may be less specialized than the original piece, but will work satisfactorily, according to Ford. That doesn't necessarily help the concours restorer in search of assembly-line perfection.

But the part numbers themselves are not the goal here; the value of this document lies in the description, dimension, and application of Ford replacement componentry. Those facts can help anyone to a better understanding of what was original equipment or what was accepted in the past as original equipment.

The 289 HiPo Engine

A K-Code High Performance 289, nicknamed HiPo, is an A-Code 289 4-V that has been modified and strengthened to handle high-rpm operation. The engine block is basically unchanged. It is necessary to shift the torque and horsepower curve to a higher-rpm level when, as in this case, cubic inches are in limited supply. As stresses on internal components increase dramatically with higher rpm, it is necessary to beef up pieces and areas that are known to fail. An engine will not make more high-rpm power just by making some internal components stronger. Other modifications, such as free-flowing induction and exhaust systems and a radical camshaft profile are needed to successfully shift the powerband upward.

K-Code 289 High Performance Nomenclature

To better contain the forces generated by the HiPo crankshaft, special heavy-duty main bearing caps were installed. All 289 HiPo engines used a two-bolt block, meaning the main bearing caps are held in place with just two bolts. The oil passages

continued on next page

supplying the lifter, cam, and main bores are sealed with screw-in pipe plugs instead of pressed-in caps for increased durability. The 4-V flat-tappet hydraulic camshaft is replaced with a solid-lifter cam, which promotes durability and enables the engine to breathe at higher rpm. The cylinder heads were exclusive HiPo pieces.

Specifications

Gross brake horsepower	271 @ 6000rpm
Gross torque	312 @ 3400rpm
Bore	4.00in
Stroke	2.87in
Compression ratio	10.5:1 (1964-1968)
Combustion chamber volume (cc)	47.7-50.7cc
Bore spacing	4.38in
Crankshaft material	Nodular iron
Main journal diameter	2.2486in
Rod journal diameter	2.1232in
Block deck height	8.206in
Deck height clearance	0.016in
Compression height	1.60in
Connecting rod length, center to center	5.155in
Intake valve head diameter	1.665in (1964) 1.778in (1965-1968)
Exhaust valve head diameter	1.450in
Valve spring pressure (closed)	83.5-92.5lb @ 1.77in
Valve spring pressure (open)	234.5-259.5lb @ 1.32in
Valve lash (mechanical)	0.020in (hot)
Rocker arm ratio	1.60:1
Firing order	15426378

280 HiPo Part Numbers

Part number	Description
C4OZ-6010-C	Engine Block (5-bolt bellhousing) 1964-1965
C5OZ-6010-C	Engine Block (6-bolt bellhousing) 1966-1968
C5OZ-12127-E	Distributor (dual point, mechanical advance)
C3OZ-6250-C	Camshaft (solid lifter) Marked #VE
C9ZZ-6200-C	Connecting Rod (5.155in long)
C3OZ-6500-A	Lifter (solid, 0.8742in O.D. 2in long)
C3OZ-6A527-A	Rocker Arm Stud ($2^1/_2$in long, $^3/_8$x24 top, $^7/_{16}$x14 bottom)
C4OZ-6A574-A	Valve Cover Kit (COBRA, aluminum)
C3OZ-6211-M	Rod Bearings (std.)
C3AZ-6333-A	Upper Main Bearings (std. red 0.0956in) 1963-1965
C3AZ-6333-B	Upper Main Bearings (std. blue 0.0960in) 1963-1965
C3AZ-6333-P	Upper Main Bearings (std. 0.0959in) 1966
C3AZ-6333-H	Lower Main Bearings (std. red 0.0956in) 1963-1965
C3AZ-6333-J	Lower Main Bearings (std. blue 0.0960in) 1963-1965
C3AZ-6333-AA	Lower Main Bearings (std. 0.0959in) 1966
C3AZ-6337-A	Upper Main Thrust Bearing (std. 0.0956in) 1963-1965
C3AZ-6337-B	Upper Main Thrust Bearing (std. 0.0960in) 1963-1965
C3AZ-6337-P	Upper Main Thrust Bearing (std. 0.0959in) 1966
C3AZ-6337-H	Lower Main Thrust Bearing (std. 0.0956in) 1963-1965
C3AZ-6337-J	Lower Main Thrust Bearing (std. 0.0960in) 1963-1965
C3AZ-6337-AA	Lower Main Thrust Bearing (std. 0.0959in) 1966
C3OZ-6303-B	Crankshaft (nodular iron)
C5OZ-6049-A	Cylinder Head
C3AZ-6051-C	Cylinder Head Gasket
C3OZ-6505-A	Valve (exhaust)
C4OZ-6507-A	Valve (intake)
C3OZ-6A511-A	Spring (valve)
C3OZ-6514-A	Valve Spring Retainer
7HA-6518-A	Valve Stem Lock
C3OZ-6571-B	Valve Stem Seals
C2OZ-6065-A	Head Bolts (short)
C2OZ-6065-B	Head Bolts (long)
C3OZ-6108-L	Piston (includes pin)
C2OZ-6261-A	Bearing Camshaft (front)
C2OZ-6267-A	Bearing Camshaft (#2 journal)
C2OZ-6262-A	Bearing Camshaft (#3 journal)
C2OZ-6270-A	Bearing Camshaft (#4 journal)
C2OZ-6263-A	Bearing Camshaft (#5 journal)
C3OZ-6265-A	Spacer (camshaft sprocket)
C2OZ-6266-A	Plug (camshaft rear bearing)
C3OZ-6268-A	Timing Chain (58 links)
C5OZ-6269-A	Plate (camshaft thrust)
C3OZ-6306-A	Camshaft Sprocket (21 teeth)
C2OZ-6600-A	Oil Pump
C5AZ-6312-B	Crankshaft Pulley (marked #C5AE-A single sheave $6^{23}/_{64}$in diameter $^3/_8$in belt)
C5OZ-6316-A	Crankshaft Damper
C3OZ-6375-C	Flywheel (3- or 4-speed, marked #C3OE-6380-B, used before 7/1/66)
C3OZ-6375-D	Flywheel (3- or 4-speed, marked #A, 160 tooth ring gear, used after 7/1/66)
C4OZ-6375-C	Flywheel (C-4 automatic trans)
C4OZ-6980-A	Engine Dress Up Kit (chrome)
Consists Of:	C2OZ-2162-A Master Cylinder Cap C4OZ-6A547-A Valve Covers C4GE-6750-A Dip Stick C3DE-6766-C Oil Filler Cap C2SZ-8100-A Radiator Cap C4DZ-9600-C Air Cleaner

The 428ci Engines

The first 428ci engine was engineered for and installed in 1966 Thunderbirds. It is the last and youngest member of Ford's FE engine family. Thunderbird 428s were comparatively docile engines incorporating a mild hydraulic cam profile that generated more than enough low-end torque to get the heavy car away from a stoplight.

The 428ci engine became part of Mustang's option list on April 1, 1968. Capitalizing on years of high-performance development and experience, Ford engineers made numerous modifications that transformed the slow-pulling Thunderbird engine into a tire-smoking brute. The Medium Riser 427/390 Police Interceptor hybrid was thereafter called Cobra Jet, or CJ for short.

Later improvements to the engine's bottom end identified the big block as a Super Cobra Jet, abbreviated as Super CJ or simply SCJ.

Drag Pack was a race option available with most big-block high-performance engine packages.

Cobra Jet

The 428ci is the only FE engine identified as a Cobra Jet or Super Cobra Jet. As compared to other FEs, it closely resembles a bored and stroked 390 GT or PI (Police Interceptor) engine.

The internal features that distinguish a Cobra Jet from a Thunderbird 428ci engine include a high-strength, nodular-cast-iron crankshaft, forged-steel Police Interceptor-type connecting rods with larger 13/32in nut and bolt, and 427-type low-riser cylinder heads. A Thunderbird 428-4V combustion chamber measures from 68.1cc to 71.1cc as compared to the larger-chambered CJ head which measures from 72.8cc to 75.8cc.

Stock intake runners were 1.34in wide and 1.93in tall. CJ runners measured 1.34in by 2.34in. CJ 2.085in intake valves were 0.060in larger than stock while the 1.658in exhaust valves were 0.100in larger.

The hydraulic CJ cam profile is hotter than the stock 428 version and the 735cfm carburetor is 135cfm larger.

The 428 is the only FE engine that is externally balanced. All other FEs are internally balanced.

Super Cobra Jet

A Super Cobra Jet is a Cobra Jet with some stronger reciprocating pieces added. CJ connecting rods are replaced with forged 427 rods, which use stronger cap screws instead of bolts and nuts to secure the bearing caps.

SCJ engines built before December 26, 1968, were equipped with pistons weighing 692 grams. SCJs built after that date were outfitted with heavier 712 gram pistons.

Each assembly had its own unique crankshaft, flywheel, harmonic balancer, and spacer counterweight balanced to suit the individual application. A heavy-duty oil cooler positioned directly behind the

continued on next page

This special oil filter adapter bolted to the SCJ block, and directed oil to the oil cooler. A system relief valve and pressure gauge port were part of the factory engineered adapter.

The 428ci Cobra Jet engine could be ordered with a little known option called Drag Pack. Among the numerous internal and external drivetrain components included with the Drag Pack option was this heavy-duty engine oil cooler. A 428ci engine equipped with Drag Pack had a stronger lower end and was known as a Super Cobra Jet (SCJ).

grille is also part of the Super CJ equipment package. The air-to-oil heat exchanger is attached to the radiator-core support with factory brackets and plumbed with high-pressure hose from a special oil filter adapter.

Drag Pack

Drag Pack adds an optional V-Code 3.91:1 or W-Code 4.30:1 rear-axle ratio to a 428 SCJ performance package. Low ratio gear sets are available only with Traction-Loc, Ford's own positive-locking differential. In some years, an aftermarket Detroit Locker differential was an available option.

The 351 Engines

A few facts about 351 engines installed in 1969 and 1970 Mustangs:

- Cleveland engines are part of the 335 Series of Ford engines which include the 351C, the 351M (Modified), and the 400ci. The 351M and 400 were never installed in a Mustang. These engines were manufactured at the Cleveland, Ohio, plant, hence the C designation.
- Windsor engines, manufactured at the Windsor plant in Ontario, Canada, include the 221, 260, 289, 302, and 351W versions. Windsor engines are not identified by a commonly known series number.
- In 1969, all 351ci engines with two- or four-barrel induction were Windsor engines.
- In 1970, both Cleveland and Windsor engines were installed in Mustangs.
- Only the Cleveland was available with both two- and four-barrel induction in 1970.
- The 351 Windsor engine was not available with a four-barrel carburetor in 1970.
- The engine code for the two-barrel 351 was H and the code for the four-barrel was M.
- The codes were the same for both engines in 1969 and 1970 regardless of engine type.
- Both engines utilized a 4.00in bore and 3.50in stroke.
- Both engines used ten headbolts to secure each cylinder head.
- Cylinder heads were unique to each engine and are not interchangeable although the patterns for the headbolts were the same.
- Windsor and Cleveland intake manifolds, camshafts, and crankshafts were not interchangeable.
- Stock Windsor connecting rods are longer than Cleveland rods.
- Coolant passes through the intake manifold of a Windsor engine, but not in a Cleveland.
- The thermostat housing attaches to the engine block on a Cleveland with two vertical bolts, and to the intake manifold on a Windsor with two bolts positioned horizontally.
- The Cleveland has wide cylinder heads incorporating canted valves and open-type combustion chambers. The Windsor uses narrower cylinder heads with in-line valves and wedge-shaped combustion chambers.
- The two bolts holding the fuel pump on a Cleveland engine are vertical in relationship to each other; on a Windsor, they are horizontal.
- Rocker arms are unique to each engine.
- The Boss 302 is a Windsor engine that uses Cleveland-type cylinder heads.
- Cleveland and Boss 302 engines use eight bolts to secure each valve cover. All other Windsor engines, including the standard 302, use six bolts per valve cover.
- Cleveland four-barrel heads differ from the two-barrel heads. Four-barrel heads have large intake and exhaust runners and quench-type combustion chambers. Two-barrel heads have small intake and exhaust runners and open combustion chambers.
- Two-barrel and four-barrel 351 Windsor heads are basically the same.
- Cleveland engines have round intake and exhaust ports; Windsor engines have rectangular-shaped ports.
- The intake manifold bolts on a Windsor engine are all vertical. Only four intake bolts, two on each side of the carburetor, on a Cleveland are vertical. The rest are angled toward each cylinder head.
- The Boss 351, which is a Cleveland engine, did not appear until 1971.

Mustang Data Plate Engine Codes, 1964-1970

Code	Years	Engine	Remarks
A	'65-'67	289-4V	
C	'65-'68	289-2V	
D	'64½	289-4V	
F	'64½	260-2V	
	'68-'73	302-2V	
G	'69-'70	Boss 302 (4V)	
H	'69-'70	351-2V	
J	'68	302-4V	The '71 Mustang had a 429-4V Ram Air engine.
K	'65-'67	289-4V (HP)	The solid lifter High Performance, or Hi-Po engine.
L	'69-'73	250-1V	
M	'69-'71	351-4V	A Windsor in '69 & '70; a Cleveland in '70 & '71.
Q	'69-'70	428-4V	The '71-'73 Mustangs had a Q-code 351-4V engine.

Code	Years	Engine	Remarks
R	'68-'70	428-4V	Ram Air Cobra Jet
S	'67-'69	390-4V	
T	'65-'69	200-1V	
U	'64½	170-1V	
W	'68	427-4V	Yes, one is known to exist. A former drag car.
X	'68	302-2V	Uncommon.
Z	'69-'70	Boss 429 (4V)	

Years of Production, Mustang Engines

6-Cylinder Engines

CID	1965	1966	1967	1968	1969	1970
170	x					
200	x	x	x	x	x	x
250					x	x

8-Cylinder Engines

CID	1965	1966	1967	1968	1969	1970
260	x					
289	x	x	x	x		
302				x	x	x
351					x	x
390			x	x	x	
427				x		
428				x	x	x
429					x	x

Note: The 170 cid and 260 cid engines were primarily 1964-1/2 production.

The Falcon's largest contribution to the Mustang was its drivetrain. Base engine for the 1964½ Mustang was the 170ci inline six-cylinder, which was designed for and made its debut in the first production Falcon. Generating 101hp, this economical engine offered reasonably good performance while delivering over 22mpg of regular grade gasoline. Transmission and rear-end components were also Falcon carryovers on the first Mustang. Baseline 170ci engines and three-speed manual transmissions had their drawbacks which led Ford to upgrade the pair. Early 170ci engines incorporated a solid-lifter camshaft and constant-mesh, helical-cut cam gears, which made the little engine noisy. Three-speed transmissions were considered inconvenient because first gear was not synchronized. The car had to be brought to a complete stop before first gear could be engaged without grinding. The code for the 1964½ vintage 170ci engine is U.

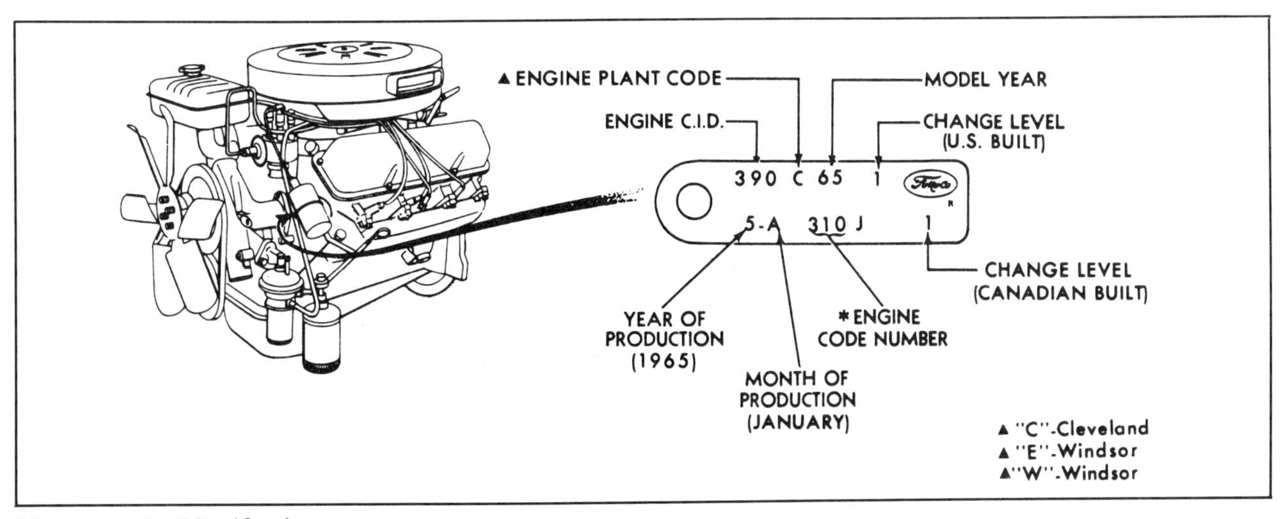

Mustang engine identification tags.

Mustang Engine I.D. Tag Code Identifier Chart, Part One.

Tag Code	Cyl.	CID	Carb.	Data Code
5	6	170	F-1	U
6	6	170	F-1	U
7	6	170	F-1	U
8	6	170	F-1	U
11	6	170	F-1	U
12	6	170	F-1	U
16	6	170	F-1	U
17	6	170	F-1	U
21	6	200	F-1	T
23	6	200	F-1	T
24	6	200	F-1	T
25	6	200	F-1	T
26	6	170	F-1	U
26	6	200	F-1	T
27	6	200	F-1	T
28	6	170	F-1	U
29	6	170	F-1,S/E	U
32	6	200	F-1	T
33	6	200	F-1	T
34	6	200	F-1	T
35	6	200	F-1	T
36	6	200	F-1	T
40	6	200	F-1	T
41	6	200	F-1	T
42	6	200	F-1	T
45	6	250	F-1	L
46	6	250	F-1	L
47	6	250	F-1	L
48	6	250	F-1	L
50	6	250	F-1,S/E	L
200	8	351W	F-2	H
201	8	351W	F-2	H
202	8	351W	F-2	H
203	8	351W	F-2	H
205	8	351W	F-2	H
206	8	351W	F-2	H
207	8	351W	F-2	H
208	8	351W	F-4	M
209	8	351W	F-4	M

Tag Code	Cyl.	CID	Carb.	Data Code
210	8	351W	F-4	M
211	8	351W	F-4	M
212	8	351W	F-4	M
213	8	351W	F-4	M
231	8	289	F-2	C
232	8	289	F-2	C
233	8	289	F-2	C
236	8	289	F-2	C
237	8	289	F-2	C
239	8	289	F-2	C
240	8	289	F-2	C
241	8	289	F-2	C
245	8	289	F-4	K
246	8	289	F-4	K
250	8	289	F-4,P/F	A
252	8	289	F-2	C
253	8	289	F-2	C
257	8	289	F-2	C
271	8	302	F-2	F,G
272	8	302	F-2	F,G
273	8	302	F-2	F,G
274	8	302	F-2	F,G
275	8	302	F-2	F
276	8	302	F-2	F
277	8	302	F-2	F
279	8	302	F-2	F,G
280	8	302	F-2	F,G
281	8	302	F-2	F
282	8	302	F-2	F
283	8	302	F-4	J
284	8	302	F-4	J
285	8	302	F-2	F
287	8	302	F-2	F
288	8	302	F-2	F
296	8	302	F-2	G
299	8	302HP	H-4	G
300	8	302HP	H-4	G
318	8	390GT	F-4	S

Carburetor Identification:

F-1	Ford 1V
F-2	Ford 2V
F-2,P/F	Ford 2V, Premium Fuel
F-4	Ford 4V
F-4,P/F	Ford 4V, Premium Fuel
H-4	Holley 4V

Mustang Engine I.D. Tag Code Identifier Chart, Part Two.

Tag Code	Cyl.	CID	Carb.	Data Code
319	8	390GT	F-4	S
321	8	390GT	F-4	S
322	8	390GT	F-4	S
324	8	390GT	F-4	S
341	8	390	F-4	Z
343	8	390	F-4	Z
350	8	427	H-4	W
353	8	427	H-8	R
357	8	390	F-4	Z
359	8	427	F-4	W
360	8	427	H-4	W
361	8	427	H-8	W
364	8	427	H-4	W
382	8	390GT	F-4	S
383	8	390GT	F-4	S
384	8	390GT	F-4	S
385	8	390GT	F-4	S
386	8	390GT	F-4	S
400	8	428	F-4	Q
401	8	428	F-4	Q
407	8	428CJ	H-4	R
408	8	428CJ	H-4	R
410	8	428	F-4	Q
418	8	428CJ	H-4	R
419	8	428CJ	H-4	R
420	8	428CJ	H-4	R
421	8	428CJ	H-4	R
422	8	428SCJ	H-4	R
423	8	428SCJ	H-4	R
424	8	428SCJ	H-4	R
425	8	428SCJ	H-4	R
426	8	428SCJ	H-4	
491	8	260	F-2	F
492	8	260	F-2	F
500	8	260	F-2	F
502	8	260	F-2	F
504	8	260	F-2	F
506	8	260	F-2	F

Tag Code	Cyl.	CID	Carb.	Data Code
534	8	260	F-2	F
536	8	260	F-2	F
538	8	260	F-2	F
540	8	260	F-2	F
548	8	289	F-2	C
549	8	289	F-2	C
550B	8	289	F-4	D
550J	8	289	F-2	C
551B	8	289	F-4	D
551J	8	289	F-2	C
552	8	289	F-4	A
554	8	289	F-4	K
557	8	289	F-2	C
558	8	289	F-2	C
561	8	289	F-2	C
562	8	289	F-2	C
563	8	289	F-4	K
564	8	289	F-4	K
566	8	289	F-4,P/F	A
567	8	289	F-4,P/F	A
600	8	351C	F-2	H
601	8	351C	F-2	H
602	8	351C	F-2	H
604	8	351C	F-2	H
606	8	351C	F-2	H
608	8	351C	F-4	M
609	8	351C	F-4	M
610	8	351C	F-2	H
611	8	351C	F-2	H
612	8	351C	F-4	M
613	8	351C	F-4	M
614	8	351C	F-2	H
615	8	351C	F-2	H
616	8	351C	F-4	M
617	8	351C	F-4	M
630	8	351C	F-4	M
632	8	351C	F-2	H

Carburetor Identification:

F-1	Ford 1V
F-2	Ford 2V
F-2,P/F	Ford 2V, Premium Fuel
F-4	Ford 4V
F-4,P/F	Ford 4V, Premium Fuel
H-4	Holley 4V
H-8	Holley 2-4V

The 170ci six-cylinder was replaced with a 200ci version in the fall of 1964. The 30ci and 19hp gains presented no obvious visual changes; it was the internal modifications that made the 200ci more durable. The 170 had four main bearings; the 200 had seven main bearings, which complemented the new improved crankshaft design. All six-cylinder Mustangs produced through the end of the 1970 model year were equipped with four-lug wheels even though drivetrain and suspension components were upgraded over the six-year period. Early six-cylinder engines were painted black except for the long, scalloped valve cover and snorkeled air cleaner lid which were bright red. T is the engine code for the 200.

The earliest first-generation Mustangs could have been equipped with an optional 260ci V-8 engine. These two-barrel low-compression Windsor engines were borrowed from the Fairlane, Ford's only mid-size car for 1964. Sporting 164hp, 260ci base V-8s offered a real boost to the Mustang's performance image. The engine generates lots of low-end torque, which translates into an impressive seat-of-the-pants performance feel. Early V-8 Mustang engines also used a Fairlane air cleaner, which can be identified by its small-diameter lid that fits inside an opening in the top of the housing. Later V-8 air cleaner lids were the same diameter as the lower housing. The earliest V-8s had cast-iron water-pump pulleys and dual generator drivebelts on cars not equipped with power steering. A three-speed Cruise-O-Matic automatic was the only transmission used with the 260ci engine. The engine block and heads were painted black while the valve covers and air cleaner were gold. Color pictures of the very first Mustang (00001) show its 260 engine painted light blue. The total number of 260 engines that were painted light blue is unknown. Production code was F designating a 260ci two-barrel engine.

Mustang Engine Power Comparison Chart

Year	Engine	Data Code	Comment	Brake Horsepower	Torque
'65-67	289-4V	A		225 @ 4800 in 1965-6-7	305 @ 3200
'64-1/2	289-2V	C		195 @ 4400 in 1964	282 @ 2400 in 1964
'65-67	289-2V	C		200 @ 4400 in 1965-6-7	282 @ 2400
'68	289-2V	C		195 @ 4400 in 1968	288 @ 2400 in 1968
'64-1/2	289-4V	D		210 @ 4400	300 @ 2400
'64-1/2	260-2V	F		164 @ 4400	285 @ 2200
'68-69	302-2V	F		210 @ 4400	300 @ 2600
'69-70	Boss 302	G		290 @ 5800	290 @ 4300
'69-73	351-2V	H		250 @ 4600	355 @ 2600
'68	302-4V	J		230 @ 4800	310 @ 2800
'65-68	289-4V	K	HiPo	271 @ 6000	312 @ 3400
'69-73	250-1V	L		155 @ 4400	240 @1600
'69	351-4V	M		290 @ 4800 Windsor, 1969	385 @ 3200
'70	351-4V	M		300 @ 5400 Cleveland, 1970	380 @ 3400
'69-70	428CJ	Q		335 @ 5200	440 @ 3400
'68-70	428CJ	R	Ram Air	335 @ 5200	440 @ 3400
'67-69	390-4V	S		320 @ 4800	427 @ 3200
'65-70	200-1V	T		120 @ 4400	190 @ 2400
'64-1/2	170	U		101 @ 4400	158 @ 2400
'69-70	Boss 429	Z		375 @ 5600	450 @ 3400

At first introduction, the 289ci two-barrel engine shared the optional spotlight with the 260 but offered an additional 46hp and 29ci. This over-bored version of the 260 generated 210 (gross) horsepower. Early production V-8 engines were equipped with generators instead of alternators and had an oil filler tube protruding from the timing cover. Both features were changed by the end of the calendar year though not at the same time. This engine has a generator but the oil filler opening has been relocated to the driver's side valve cover, a change that became synonymous with all small-block Ford engines. Early V-8 water pumps were made out of aluminum as was the timing cover. The long block was painted black, the valve covers gold. The air cleaner housing and lid were gold but the round snorkel extending toward the right shock tower was gloss black. This engine is in a 1964½ Mustang equipped with optional power steering. Early power steering pumps, like the example shown, were cast-iron units manufactured by Eaton Corporation and are identified by the small cast-iron pulley and the large oil reservoir perched on top. Production code for the 1964½ 289ci two barrel engine is C.

The D-Code 289ci four-barrel engine could easily be the rarest engine option since it was available only during April through September of 1964. Essentially, it was a four-barrel version of the C-code 289 two-barrel. The only obvious identifying mark was the black, white, and red "289 Cubic Inch 4-V Premium Fuel" decal on the air cleaner lid. Early D-code engines should be equipped with generators and may or may not have the timing cover mounted oil filler tube.

The K-code 289ci High Performance engine, or HiPo for short, became available in June 1964 and remained an option until the end of the 1967 production year. It was a popular lightweight performance engine that produced 271 (gross) horsepower with help from a solid-lifter flat-tappet camshaft, 480cfm carburetor, and special free-flowing header-type cast-iron exhaust manifolds. The optional Dress-Up Kit, which includes chromed valve covers and an open-element air cleaner, was standard on HiPo engines. The vacuum-advance diaphragm is deleted from the fixed breaker plate, dual points distributor, and the alternator sports a large-diameter pulley that allows it to survive at high engine rpms.

By the beginning of the 1965 model year, the 170ci six-cylinder and the 260ci V-8 had been dropped from the optional engine list. The D-code 289 had evolved into an A-Code 225hp 289ci four-barrel engine. Generators were replaced with alternators when the traditional five-dial instrument cluster took the place of the horizontal speedometer bezel with dual warning lights that was a carryover from the Falcon. Ford installed its own compact power steering pumps in 1965. The small cast-iron pump was surrounded by a tin cover that acted as the reservoir. Over the next few years, the 4in spout mounted on top of the reservoir would be relocated many times to accommodate numerous other engine-driven accessories. The A-Code 289ci engine became the standard engine for the popular GT Equipment Group offered in April of 1965. All V-8 engines produced in 1965 were black with gold valve covers and air cleaners.

The C-code 289 two-barrel remained an option until January 1968 when all 289s were replaced with 302s. Early production C-code 289s were rated 210hp but by the 1965 model year, it was detuned to deliver 200hp. There was no significant difference between the two engines. Changing the compression ratio enabled the factory to vary horsepower figures. The hexagonal snout protruding from the vacuum-advance canister houses the diaphragm return spring. The metal vacuum tube is connected at each end by a threaded fitting. Because the method for connecting the tube changed at some point early in 1966, this threaded tube was used on all 1964½ and 1965 engines. There is no mechanical reason for the vertical loop in the tube that is traditionally formed just in front of the air cleaner.

The Sprint option was a cosmetic package used to enhance the marketability of base-level hardtops. Among the visual effects was an engine dress-up kit consisting of a chrome air cleaner, radiator, oil and master cylinder caps. A "Sports Sprint" or "Sport Sprint 200" garnished the air cleaner top. The T-Code 200ci engine was rated at 120hp in 1966. The Sports Sprint option added nothing to enhance performance.

All engines starting with the 1966 model year were painted a medium shade of blue. A similar color, later referred to as Corporate Blue, became part of Ford's trademark when it was used as the background color for the famous Ford oval. Almost every part sold over dealership counters contained the Ford script printed in Corporate Blue. Engine options for the 1966 model year were simple and easy to remember. There was the base level T-Code 200ci six-cylinder and three optional 289ci V-8s. Pictured is a 1966 Sprint 200 equipped with air conditioning and power steering.

Aside from the blue paint, the base 200ci six-cylinder engine hadn't changed much since 1965. This 1967 model shows the fuel filter canister has disappeared from the top of the fuel pump and the snorkle is missing from the air cleaner lid. The small-diameter metal tube following the route taken by the fuel line is the vacuum advance tube. The fuel filter, which is really just a nylon screen housed in a cylinder, can be seen protruding from under the front edge of the air cleaner.

This C-code 289ci two-barrel could be considered the base-level V-8. Its mild hydraulic-lifter camshaft and dual-plane intake manifold delivered smooth, even power and good gas mileage. The rounded valve covers are another trademark of pre-1967 V-8 engines. The additional tubular firewall-to-fender apron braces were used to strengthen the unibody construction of the convertible. This low-optioned 1966 convertible is common to the era. It has manual brakes, steering, and transmission. The C-code engine was an inexpensive option.

The T-code 1965 200ci six-cylinder was outfitted with the latest accessories. This is a late $1964^{1}/_{2}$ because it has an oil-pressure switch threaded into the left rear corner of the block indicating the instrument panel is the Falcon-type with low oil-pressure warning light. The later 1965 bezel utilized an oil-pressure gauge, which required a large-diameter gold-anodized sender. It is also equipped with a Ford power steering pump and an alternator. Note the location of the filler tube for the power steering reservoir. The fuel filter is located on top of the inverted fuel pump and the vacuum advance tube is clipped to the fuel line.

Not a HiPo, this is a 1966 A-code 289 four-barrel engine equipped with an optional Cobra Dress Up Kit. Cobra kits were commonly sold over-the-counter by Ford dealers. Bare aluminum valve covers cast with raised ribs and "Cobra Powered By Ford" on the top surface were part of the kit, which also included a chrome, open-element HiPo air cleaner. This particular engine is in a low-optioned GT as indicated by the large master cylinder with snap-on lid and adjustable proportioning valve used with front disc brakes.

An unusual option package by today's standards but common for the no-frills 1960s. A K-code HiPo engine in a stripped down non-GT. It's a late 1964 model because it has an oil-pressure switch needed for the Falcon-styled instrument cluster but instead of a generator, it is equipped with an alternator. It has a heater but no windshield washer.

C-code 289s could have been ordered in any body style or with any other option through 1967 except for the GT Equipment Group, which required an A-code four-barrel. Internally, all 289 A- and C-code engines were the same, except each had its own camshaft profile. The engine's compression ratio was dictated by the size of the combustion chamber in the cylinder heads. Windsor engines built before the middle of 1967 used slotted pushrod holes in the cylinder heads to maintain rocker-to-valve stem alignment. The exact date of the change is unknown. These engines were equipped with conventional, rounded, non-lettered valve covers. The air cleaner snorkel slips over a hot air duct and contains a temperature-controlled trap door. The bottom of the hot air duct wraps around the right side exhaust manifold. To enhance cold weather driveability, the trap door would be held open for the first few minutes after startup. The door would block the open end of the snorkel forcing the engine to draw air, warmed by the exhaust manifold, through the hot air tube and into the carburetor. After the engine achieved normal operating temperature, the door would close allowing the engine to breathe fresh, relatively cool air from the engine compartment.

By the end of the 1966 production year Chevrolet was a distant third in the pony car race. The Camaro was to debut in 1967 with their performance engine, the 295hp 350ci, designed to blow away a 289 HiPo and impress Ford-bound new car buyers. The Camaro performance car would be called—what else?—a Super Sport. But Henry's boys dealt the bow-tie brigade a tough hand in 1967 by introducing the first Mustang powered by a big-block engine. The S-code 390ci is a successor to the first FE Series engine introduced in 1958. This version produced 320hp. Chevrolet was caught off guard and in a last minute effort to meet Ford head on, stuffed a 396ci engine in the Camaro and shoveled it out the gate to do battle with the Mustang. A 1967 390 GT was impressive indeed, but the true potential of a big-block Mustang wouldn't be realized for another year. A 390-powered Mustang was hard to work on and even harder to keep cool but there was no stopping 427lb-ft of torque once the tachometer needle cleared 2800rpm. This heavily optioned GT is equipped with power brakes, power steering, and air conditioning.

In 1967 and 1969, the S-code GT engine was rated 320hp. In 1968 it drew a rating of 325hp. If the owner did not opt for Ram Air, the engine received a chrome air cleaner lid complete with High Performance decal. The LeMans finned aluminum valve covers shown here are not factory installed but were available through Ford Parts Departments.

Ford took a giant leap ahead of the competition when, in the middle of the 1968 production year, it built the first Mustangs equipped with a Cobra Jet 428ci engine. Cobra Jet and Super Cobra Jets were nicknames that stayed with the 428 until its demise at the end of the 1970 production year. This big-block beast, like the 390, is a member of the FE engine family. Early R-code Mustangs, referred to as 68½ Cobra Jets because they were not introduced until April 1, were conservatively rated at 335hp. Pictured is a 1969 Q-code 428 CJ. The Q signifies non-Ram Air.

Because big-block engines dominated the performance market, small-blocks were considered by many to be economy or passenger-car units, a stigma that prevailed even though in 1967 the popular GT package included any engine as long as it had a four-barrel carburetor. In addition to being offered one year only, those conditions make the 230hp 1968 J-code 302 four-barrel engine somewhat rare. Economy-minded buyers could choose between the T-code 200ci six-cylinder, the C-code 289ci two-barrel V-8, or the F-code 302ci two-barrel V-8, each offering good performance and efficiency. Serious performance buyers, obviously giving no thought to fuel economy, chose either the S-code 390ci or the R-code 428ci V-8 engine. Both were only available with a four-barrel carburetor. The J-code 302 replaced the K-code 289ci HiPo, though that is where the similarity ends. The J-code was a four-barrel version of the F-code 302. It didn't have a special block, head castings, or flat-tappet camshaft. The ignition system was also common to other Windsor engines. Its hydraulic cam profile, 10.0:1 compression ratio, and four-barrel induction system gave it 35 more horsepower than the two-barrel version.

The Mustang was a longer, heavier car in 1969. For that reason, an optional L-code 250ci six-cylinder engine was offered as a step up from the base level T-code 200ci six-cylinder. The 250ci was highly refined compared to all other sixes and offered smooth economical performance. Generating 155hp at 4000rpm and 240lb-ft of torque at 1600rpm, the big six-cylinder could perform almost as well as the entry level F-code 302ci V-8. With two-barrel induction, the 302 engine made 220hp at 4600rpm and 300lb-ft of torque at 2600rpm. For $39 a customer could upgrade from a 200ci to a 250ci engine.

A G-code Boss 302 is similar in many respects to the 289 HiPo in that it has special canted valve heads with large intake and exhaust runners, a special block with four-bolt main bearing caps and screw-in core plugs, and a solid-lifter camshaft. Horsepower output was conservatively rated at 290. Boss 302 engines were engineered to compete in Trans-Am racing during the 1969-1970 season. Few Boss 302 components were designed for or will work with standard production 302 Windsor engines. The free-flowing heads and induction system provided extra-quick throttle response and the strong bottom end kept the engine together during sustained high-rpm operation. Boss 302s are still considered one of the most desirable performance engines of the decade.

There were nine different engine options available for Mustang in 1969, more than in any other year. Cubic inch options were: 200, 250, 302, 351, 390, 428, and 429. The 200ci six was standard for a non-performance car. There were two individual 302s, and two 428ci engines that were each given a separate code. The entry level V-8 was this F-code 302ci two-barrel engine.

Except for some necessary external changes that mostly involved the accessory drive assemblies, the 200ci six-cylinder remained mostly unchanged throughout its six-year tenure. It was replaced by the L-code 250ci six-cylinder as the base engine beginning with 1971 production. This basic 200 with optional power steering is in a 1967 convertible.

Even though the 428 Cobra Jet and High Performance 390 are both members of the FE engine family, the 428 shares a closer relationship to the race-bred 427. Cobra Jet engines received harder nodular-iron crankshafts and Police Interceptor rods. Both pieces are stronger and more durable than the stock Thunderbird 428 that was never installed in a Mustang by the factory. All CJs were equipped with dished, cast-aluminum pistons with double valve reliefs. CJ cylinder heads closely resemble 406ci and 427ci four-barrel low-riser pieces, and feature larger-than-stock valve sizes. A 735cfm carburetor replaced the stock 600cfm units used on non-CJ engines. It's unusual to see a 428 equipped with air conditioning, power steering, and power brakes. The R engine code for 1968, 1969, and 1970 indicates the 428ci engine is equipped with a Ram Air hood scoop. In 1968 the fixed scoop was permanently attached to the hood. Ram Air scoops in 1969 and 1970, like the one shown, were bolted to the engine.

Cleveland 351ci engines appeared in 1970. If a two-barrel 351ci engine was ordered in a new 1970 Mustang, there was no way to tell whether it would be delivered with a Cleveland or a Windsor block. Only the Cleveland was offered with a four-barrel induction system in 1970. M-code four-barrel Cleveland engines developed 300hp and 380lb-ft of torque. With an 11.0:1 compression ratio, premium fuel was required and indicated on the air cleaner decal. Ram Air was an option if a 351-4V was installed in a Mach 1 but, because the basic engine specifications did not change, it was not given its own code (as were 428ci Ram Air engines).

The Z-code Boss 429 was the biggest and baddest big-block of them all. It was only installed in 1969 and 1970 Mustangs. Ford had to legitimize the engine so NASCAR would allow it in Grand National racing. To meet the requirement, at least 500 of the Hemi-styled engines had to be installed in passenger cars and sold to the public. Virtually every part of the Boss 429 is special. The spark plug holes in the center of the massive heads required a one-of-a-kind valve cover. The canted valves were activated by two different-length rocker arms. Even the intake and exhaust manifolds were particular to the application. The Z-code 429 incorporates a crescent-shaped combustion chamber, which, according to the Ford engineers, "generates low-end torque commonly associated with wedge engines while combining the high rpm breathing capability of a hemi."

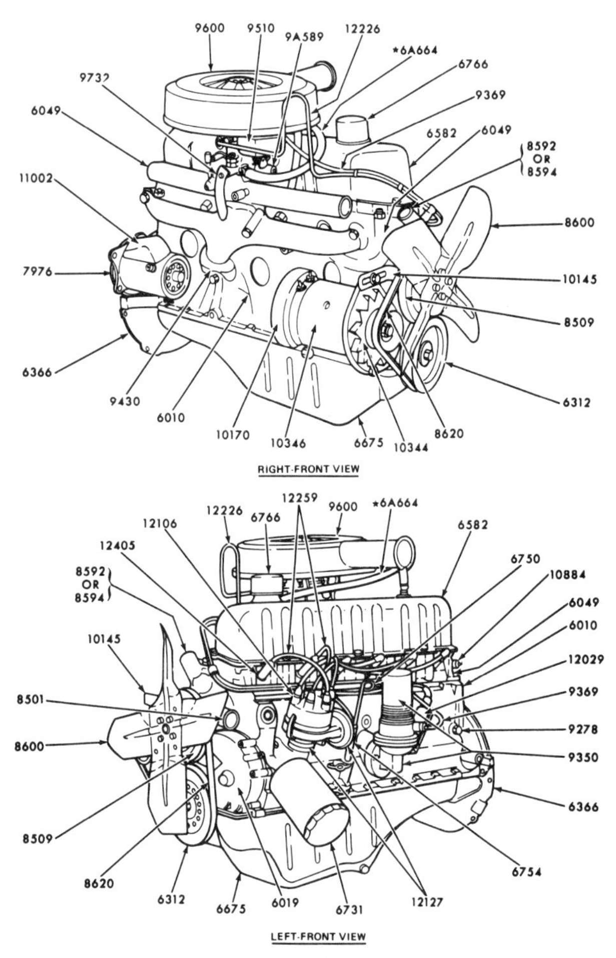

Right and left front views of a typical 170, 200, or 250 engine.

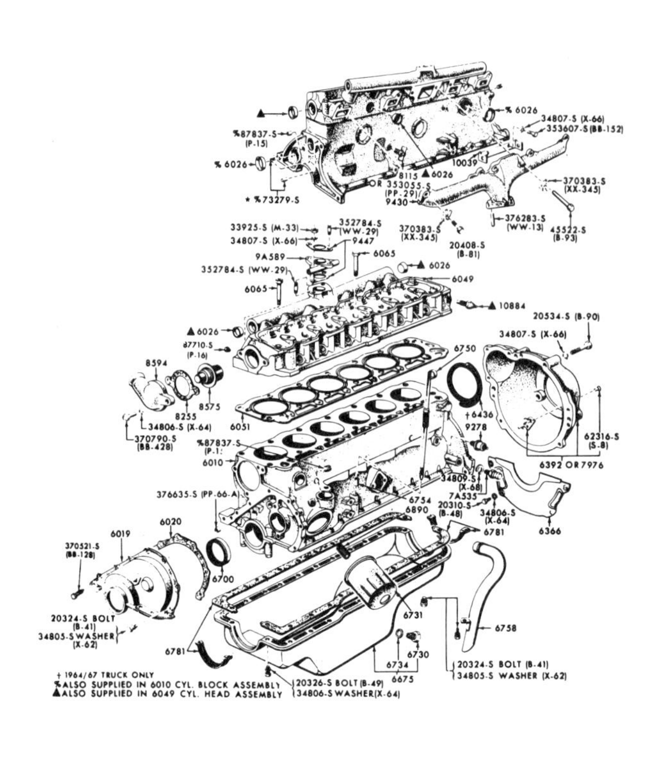

Typical 170, 200, and 250ci engine block and related parts.

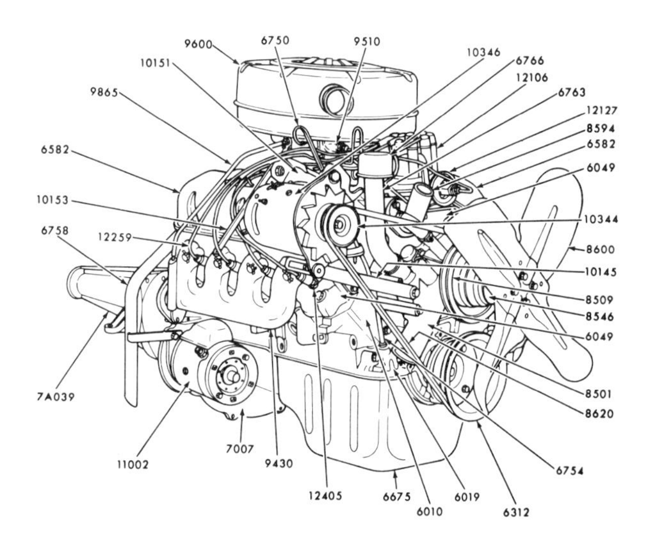

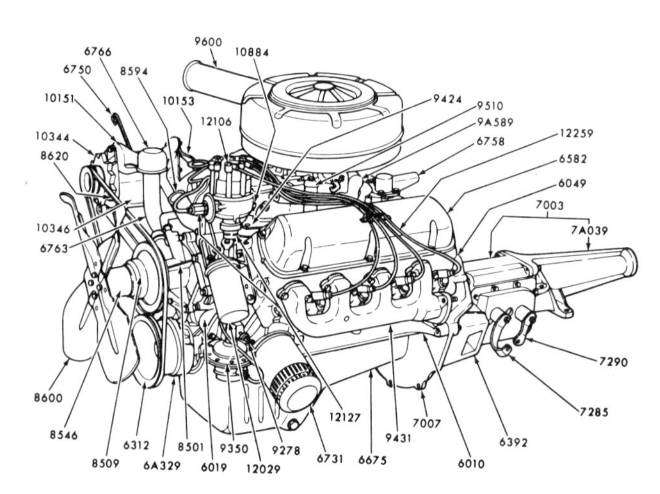

Right and left front views of a typical 260ci engine and transmission assembly.

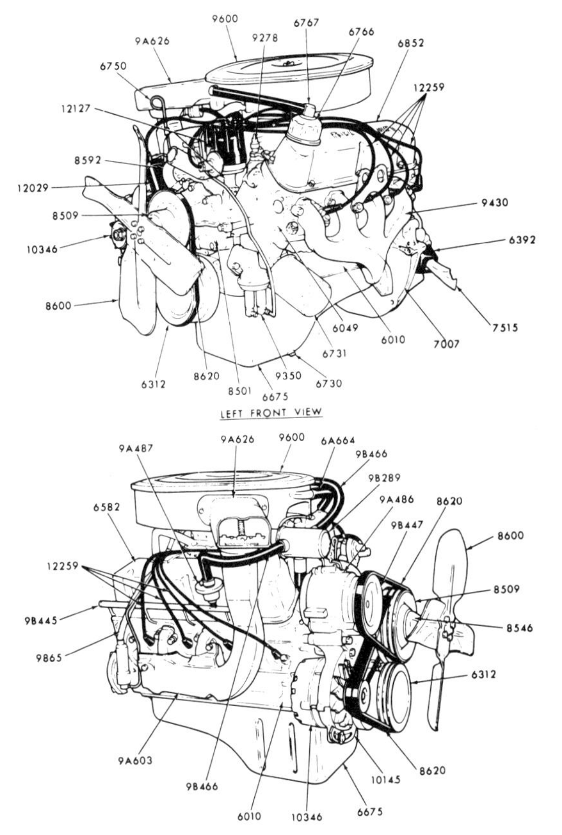

Right front and left front views of 302 and 351 Windsor engines.

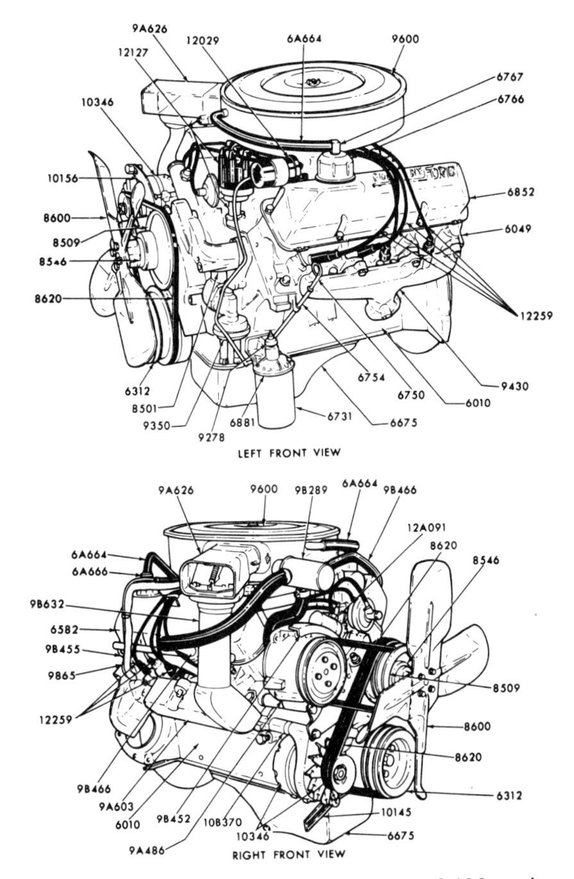

Right front and left front views of 390 and 428 engines.

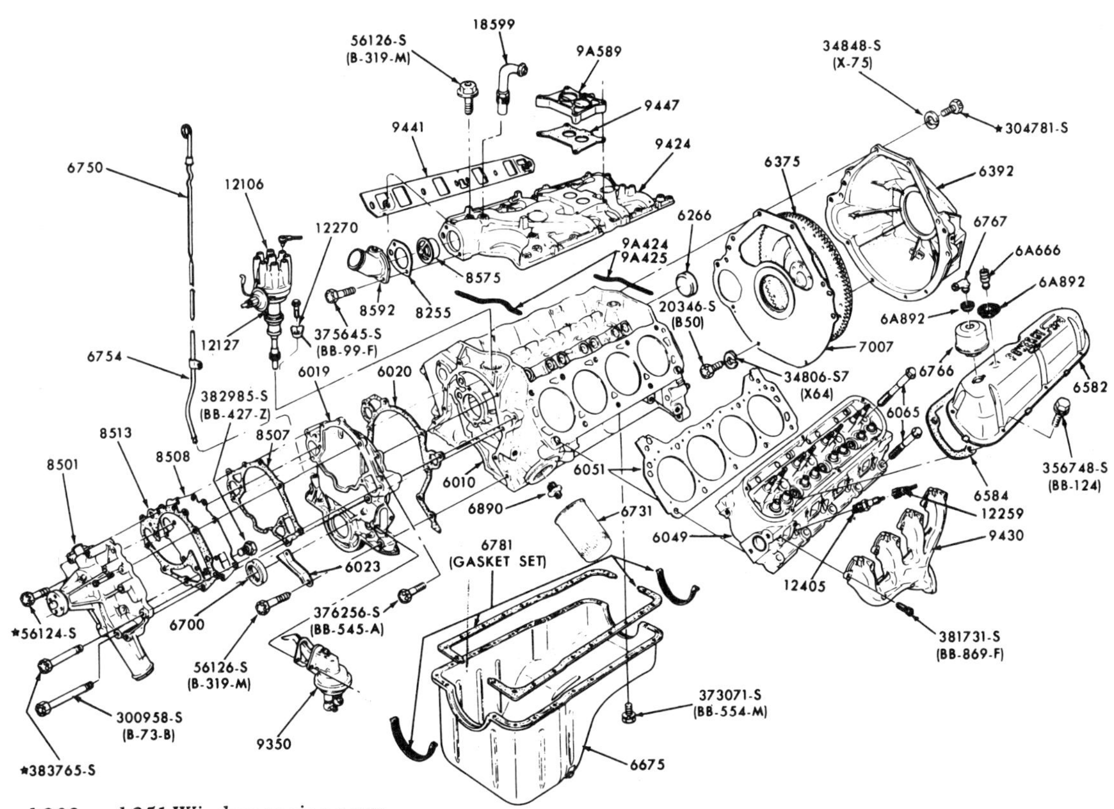

External 302 and 351 Windsor engine parts.

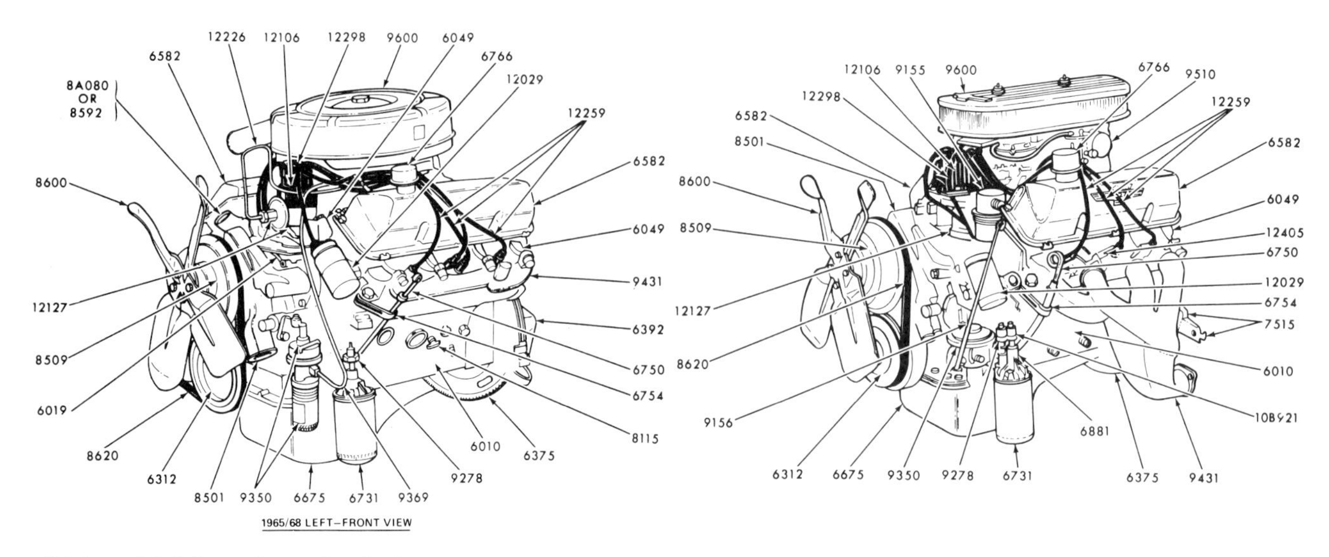

Right and left front views of typical 390 and 428ci engines.

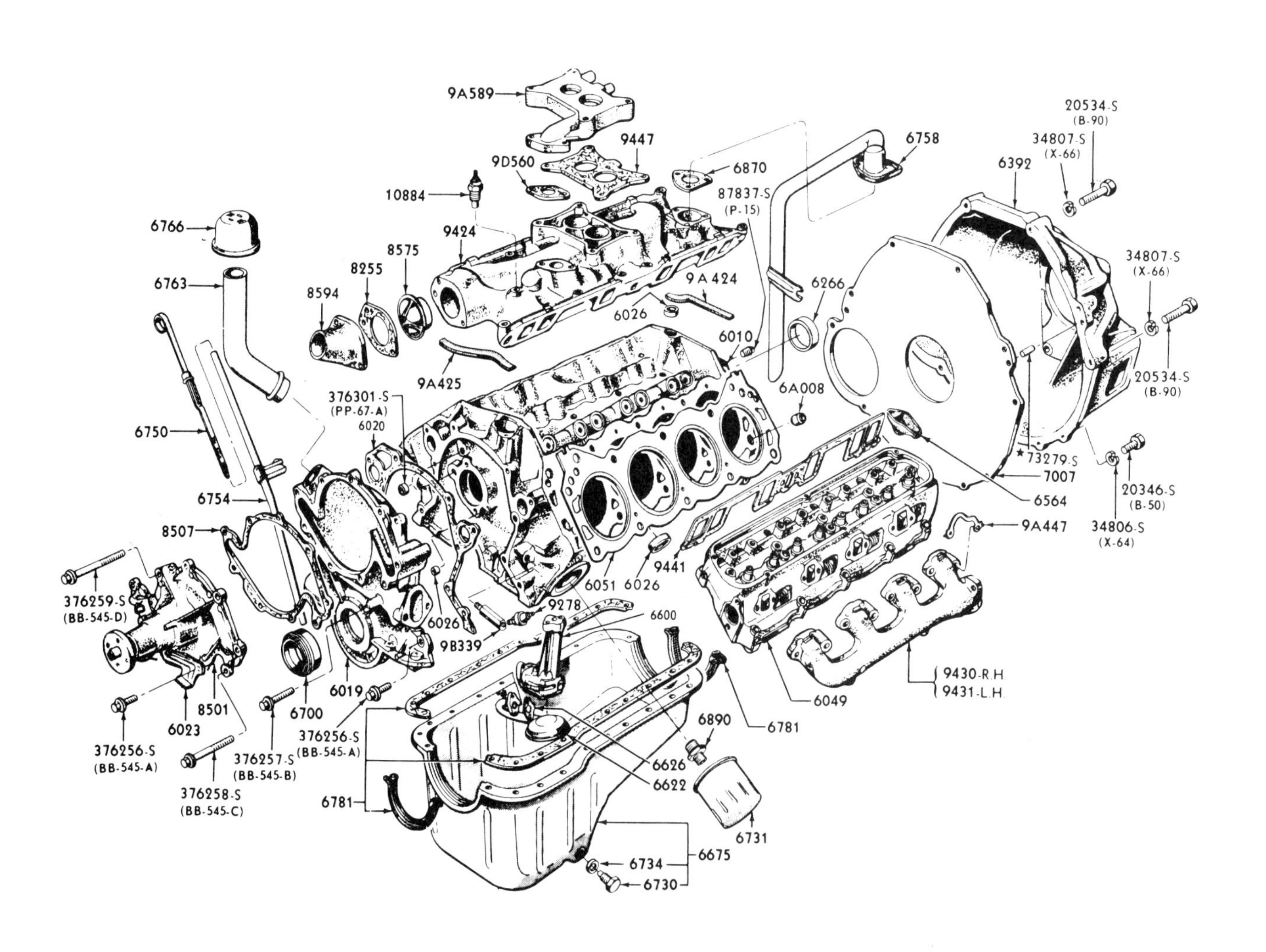

Typical 260 and 289ci engine block and related parts.

Mustang Spark Plug Application Chart, 14MM - 18MM, 8-Cylinder Engines

Year	Model/Description	C.I.D.	LSA	Part Number	NSA	Part Number	HDSA	Part Number
65		260	BF42	B8A-12405-A	BF42	B8A-12405-A	BF42	B8A-12405-A
65/69	2/B,4/B R.F OR P.F	289	BF42	B8A-12405-A	BF42	B8A-12405-A	BTF42	C6TZ-12405-A
65/69	Special (4/B)	289	B432	C0AZ-12405-A	BF32	C0AZ-12405-A	BF32	C0AZ-12405-A
65/68	2/B, 4/B	352,390	BF42	B8A-12405-A	BF-42	B8A-12405-A	BF-42	B8A-12405-A
68/69	4/B	427,428	BF42	B8A-12405-A	BF-42	B8A-12405-A	BF-42	B8A-12405-A
		429						
68/70		302	BF42	B8A-12405-A	BF42	B8A-12405-A	BF42	B8A-12405-A
68/69		427	BF32	C0AZ-12405-A	BF32	C0AZ-12405-A	BF32	C0AZ-12405-A
69/70	Boss	302,429	AF32	C9ZZ-12405-A	AF32	C9ZZ-12405-A	AF32	C9ZZ-12405-A
69/70	Boss	302,429	ARF32	C9PZ-12405-A	ARF32	C9PZ-12405-A	ARF32	C9PZ-12405-A
69/70	Boss	351-2V	BF42	B8A-12405-A	BF42	B8A-12405-A	BF42	B8A-12405-A
69/70		351-4V	BF32	C0AZ-12405-A	BF32	C0AZ-12405-A	BF32	C0AZ-12405-A
		390, 428						
69/70		390	BF42	B8A-12405-A	BF42	B8A-12405-A	BF42	B8A-12405-A
69/70		351	AF42	C9ZZ-12405-B	AF42	C9ZZ-12405-B	ARF42	C9PZ-12405-B
70	Before 11/5/69	351	AF32	C9ZZ-12405-A	AF32	C9ZZ-12405-A	ARF32	C9PZ-12405-A
70	From 11/15/69	351	AF42	C9ZZ-12405-B	AF42	C9ZZ-12405-B	ARF42	C9PZ-12405-A
70	4/B	351	AF42	C9ZZ-12405-B	AF42	C9ZZ-12405-B	AF42	C9ZZ-12405-B

LSA: Light Service Autolite NSA: Normal Service Autolite HDSA: Heavy Duty Service Autolite

MUSTANG DISTRIBUTOR APPLICATION AND IDENTIFICATION CHART, Part One.

				CALIBRATION CODE					
			SERV. REPL.	DPHRGM.	CAM	Springs (12191-2)			STOP
YEAR	MODEL	I.D.	DISTRIBUTOR	(12370)	(12210)	PRIM.	SEC.	VAC.	(12202)
6-cylinder 170 CID									
65	C4	C4DF-12127-B C8UF-12127-J	C4DZ-12127-E	302		231	211		
66	S/T	C4ZF-12127-A	C4DZ-12127-D	302		226	331		
66	3/S/T with T/E	C6DF-12127-A C7TF-12127-J,L	C6DZ-12127-A	304 304	110 110	240 237	229 225	2 14	
6-cylinder 200 CID									
65/66	C4 except T/E	C8DF-12127-H	C40Z-12127-F	302		226	213		
66	C4 with T/E	C6DF-12127-E C6DF-12127-K	C6DZ-12127-K	304 304	110 110	234 234	240 240	218 217	7 3
66	3/4/S/T with T/E	C6DF-12127-C	C6DZ-12127-C	304	110	234	240	217	3
65/67	S/T except T/E	C5DF-12127-E C8DF-12127-G	C4DZ-12127-D	302 302		226 226	215 212		
65/67	C4 with T/E	C5DF-12127-F C5DF-12127-K C8DF-12127-H	C4OZ-12127-F	302 302 302		212 226 226	212 213 213		
67	C4 with T/E	C7DF-12127-D	C7DZ-12127-F	304	111	234	233	217	3
67	3/S/T with T/E	C7DF-12127-C	C7DZ-12127-F	304	111	238	244	217	3
67	C4 with Imco	C6DF-12127-D	C6DZ-12127-D	304	110	234	240	229	2
67	C4 with Imco	C7DF-12127-F	C6DZ-12127-F	304	111	236	242	216	14
68/69	3/S/T with Imco	C8DF-12127-C	C8DZ-12127-C	374	111	238	244	245	16
68/69	C4 with Imco	C8DF-12127-D	C8DZ-12127-D	374	111	237	246	245	11
70	S/T	DODF-12127-J	DODZ-12127-C	304	110	236	235	217	14
70	A/T	DODF-12127-C	DODZ-12127-C	360	110	236	235	241	3

MUSTANG DISTRIBUTOR APPLICATION AND IDENTIFICATION CHART, Part Two.

				CALIBRATION CODE					
YEAR	MODEL	I.D.	SERV. REPL. DISTRIBUTOR	DPHRGM. (12370)	CAM (12210)	Springs (12191-2) PRIM.	SEC.	VAC.	STOP (12202)
6-cylinder 250 CID									
69	S/T	C9OF-12127-R	C9OZ-12127-A	304	111	234	244	217	14
69	A/T	C9OF-12127-U C9OF-12127-V	C9OZ-12127-V	304 304	110 110	236 243	244 240	248 225	1 3
70	A/T,S/T,Imco	D0OF-12127-A	D2OZ-12127-P	360	110	243	241	218	13
8-cylinder 260 CID									
65	S/T,C4	C4OF-12127-B C4ZF-12127-B,E C5ZF-12127-J C5JF-12127-B	C4OZ-12127-B	 301 303 303	103 103 108 108	223 222 236 237	206 230 240 240	218 211 218 218	1 3 6 1
8-cylinder 289 CID									
65/67	4/B carb. w/ T/E 4/B S/T exc. T/E	C50F-12127-E	C50Z-12127-E		107	236	235		
65	Special 4/B carb.	C3OF-12127-D,F	C50Z-12127-E		104	222	232		
65	4/B S/T exc. Hi-Po	C4ZF-12127-C Z5AF-12127-B	C7ZZ-12127-A	301	103	222	204	216	10
65	Special	C4ZF-12127-D	C5OZ-12127-E						
65	C4 exc. Special	C4GF-12127-B C5GF-12127-C	C7ZZ-12127-A	301 303	103 107	228 241 210	201 233	216 216	10 10
65/66	2/B S/T exc. T/E	C5AF-12127-M	C7ZZ-12127-A	303	107	241	240	216	1
65/66	2/B,C4,C/M exc. T/E	C5AF-12127-N	C7ZZ-12127-A	303	107	234	233	216	11
65/66	4/B exc. T/E	C5GF-12127-A	C7ZZ-12127-A	303	108	234	235	218	1
66	2/B,3/4/S/T,T/E	C6AF-12127-J	C7OZ-12127-D	303	108	243	233	218	11
66	2/B,C/M,C4,T/E	C6AF-12127-AK	C6AZ-12127-S	303	108	236	219	216	11
66	4/B, 3/4/S/T & T/E	C6ZF-12127-A	C7ZZ-12127-C	303	108	243	235	216	6
66	4/B, C4 & T/E	C7ZF-12127-B	C7OZ-12127-E	303					
67	2/B & C4 exc. T/E	C7AF-12127-AE	C7OZ-12127-B	303	107	243	235	216	2
67	2/B & C4 exc. T/E	C7OF-12127-B	C7ZZ-12127-A	303	107	234	244	216	11
67	2/B, C4 & T/E	C7AF-12127-AH	C7AZ-12127-A	303	107	236	244	216	10
67	2/B, C4 & T/E	C7OF-12127-J	C7AZ-12127-A	303	107	236	244	217	9
67	2/B,3/4/S/T exc. T/E	C7OF-12127-A	C7OZ-12127-A	303	107	243	233	216	4
67	2/B, C4 & T/E	C7OF-12127-E	C7OZ-12127-E	303	107	236	244	217	9
67	4/B, C4 exc. T/E	C7ZF-12127-B	C7ZZ-12127-A	303					
67	4/B, 3/4/S/T, T/E	C7ZF-12127-C,E	C7ZZ-12127-C	303	107	243	235	216	11, 2
67	4/B,C4,T/E	C7ZF-12127-D	C7ZZ-12127-C	303	107	234	235	218	2
67	4/B,C4,T/E	C7ZF-12127-G	C7ZZ-12127-H	303	107	241	244	225	2
67	4/B,C4,T/E	C7ZF-12127-H	C7OZ-12127-E	303	107	234	235	216	2
67	Special 4/B,T/E	C7ZF-12127-J	C7ZZ-12127-K		107	236	235		
67 68	2/B,3/4/S/T,T/E 2/B,3/S/T,T/E	C7OF-12127-D	C7AZ-12127-U	303	108	237	240	218	2
67/68	Special 4/B,T/E	C7OF-12127-K	C7ZZ-12127-K		107	236	235		
67/68	4/B, 3/4/S/T exc. T/E	C7ZF-12127-A C7ZF-12127-F	C7ZZ-12127-A	303 303	108 107	234 243	235 234	218 218	1 2
68	2/B,A/T,Imco	C8OF-12127-C	C8OZ-12127-D	35	107	236	233	229	2
8-cylinder 302 CID									
68	2/B,A/T,Imco	C8OF-12127-C	C8OZ-12127-C	35	107	236	233	229	2
68	4/B,A/T,Imco	C8ZF-12127-B	C8ZZ-12127-A	307					
68	4/B,A/T,Imco	C8ZF-12127-D	C8ZZ-12127-D	303	108	238	235	217	11
68	4/B,3/S/T,T/E	C8ZF-12127-A	C8ZZ-12127-A	307	108	236	235	248	2
69	S/T, Imco	C8AF-12127-E	C8AZ-12127-E	356	107	236	233	217	11

MUSTANG DISTRIBUTOR APPLICATION AND IDENTIFICATION CHART, Part Three.

YEAR	MODEL	I.D.	SERV. REPL. DISTRIBUTOR	CALIBRATION CODE DPHRGM. (12370)	CAM (12210)	Springs (12191-2) PRIM.	SEC.	VAC.	STOP (12202)
8-cylinder 289 CID continued									
69	2/B,A/T w/A/C, Imco w/o A/C	C9AF-12127-N	C9AZ-12127-M	303	107	234	235	229	2
69	A/T with A/C	C9AF-12127-R	C9AZ-12127-R	356	107	234	235	245	9
69	Boss 4/B,S/T w/ T/E	C9ZF-12127-B	C9ZZ-12127-E	311	112		233	218	16
69/70	Boss 4/B,S/T w/ T/E	C9ZF-12127-E	D1ZZ-12127-A	311	112	238	235	225	15
70	2/B,A/T,Imco	DOAF-12127-T	DOOZ-12127-AL	310	113	236	252	217	3
70	2/B,S/T,Imco	DOAF-12127-Y	DOAZ-12127-Y	356	113	236	235	217	1
70	2/B,A/T,Imco	DOOF-12127-AC	D2OZ-12127-H	375	113	236	252	245	7
8-cylinder 351 CID									
69	2/B,A/T,S/T,Imco	C9OF-12127-M	DOOZ-12127-M	303	107	241	233	218	13
69	2/B,A/T,S/T,Imco	DOOF-12127-M	DOOZ-12127-M	303	107	241	233	248	5
69	A/T,4/B	C9OF-12127-T	C9OZ-12127-Z	303	107	236	235	218	6
69	A/T,4/B	C90F-12127-Z	DOOZ-12127-R	303	107	241	244	245	11
69	S/T,4/B,Imco	C90F-12127-N	DOOZ-12127-N	303	107	237	235	220	6
69	S/T,4/B,Imco	DOOF-12127-N	DOOZ-12127-N	303	107	237	235	229	6
69	4/B,A/T,Imco	DOOF-12127-R	DOOZ-12127-R	303	107	241	244	245	11
70	2/B,S/T,Imco	DOAF-12127-H	DOAZ-12127-H	307	107	241	233	248	#7
70	Shelby GT350 w/ S/T								
70	2/B,A/T,Imco	DOAF-12127-V	DOAZ-12127-V	354	107	241	233	248	#7
		DOAF-12127-AC		354	113	236	235	248	#16
70	Shelby GT350 w/ A/T	DOAF-12127-V	DOAZ-12127-V	354	107	241	233	248	#7
		DOAF-12127-AC		354	113	236	235	248	#16
70	4/B,S/T,A/T,Imco	DOOF-12127-Z	DOOZ-12127-G	357	107	241	235	217	2
70	2/B,S/T,Imco	DOOF-12127-E	D2ZZ-12127-A	361	119	236	233	250	
		DOOF-12127-T		355	107	251	240	245	9
70	2/B,A/T,Imco	DOOF-12127-U	DOOZ-12127-U	303	107	241	233	217	1
70	S/T,4/B,Imco	DOOF-12127-V	DOOZ-12127-V	356	107	238	233	217	5
8-cylinder 390 CID									
67	4/B,3/4/S/T exc. T/E	C7OF-12127-H	C7AZ-12127-U	303	107	243	244	216	11
67	4/B,C6,3/4/S/T,T/E	C7OF-12127-F,G	C7OZ-12127-F	303	108	243	233	216	12
68/69	A/T with T/E								
68	4/B,T/E	C8OF-12127-D	C8OZ-12127-D	357	108	243	233	216	11
68	2/B,C6 with Imco	C8WF-12127-B	C8WY-12127-B	356	107	236	233	216	10
68	2/B,S/T,T/E	C8AF-12127-M	C8AZ-12127-M	355	107	236	233	245	9
68/69	2/B,A/T,Imco	C8AF-12127-N	C7AZ-12127-D	356	107	241	233	216	8
8-cylinder 428 CID									
67/69	CJ A/T, Imco	C7OZ-12127-F	C7OZ-12127-F						
68/69	C6,S/T,T/E	C8OF-12127-D	C8OZ-12127-D	357	108	243	233	216	11
69	A/T,T/E	C8OF-12127-J	C8OZ-12127-J	303	107	243	244	216	12
69	CJ S/T with T/E	C8OF-12127-H	D8OZ-12127-H	357	107	243	244	216	11
70	CJ 4/B,S/T,T/E	DOZF-12127-C	DOZZ-12127-C	322	118	234	244	218	17
70	CJ 4/B,A/T,T/E GT350	DOZF-12127-G	DOZZ-12127-D	303	107	236	233	217	11
8-cylinder 429 CID									
69	Boss 4/B Spec. Before 2/28/69	C9AF-12127-U	C9ZZ-12127-D	311	112	237	244	229	11
69	Boss 4/B Spec. From 2/28/69	C9ZF-12127-D	C9ZZ-12127-D	307	112	238	233	225	10

* For Shelby GT350 with A/T, change stop #7 or #16 to #14.

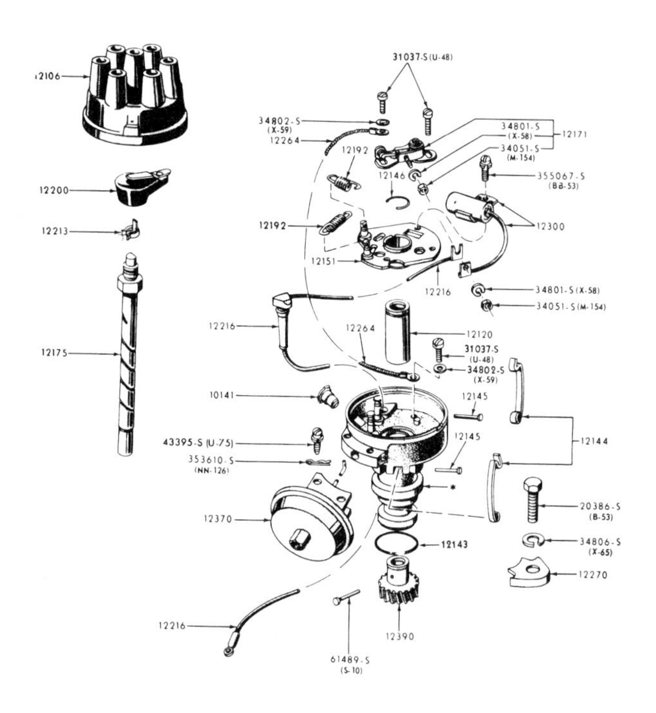

Exploded view of a six-cylinder distributor.

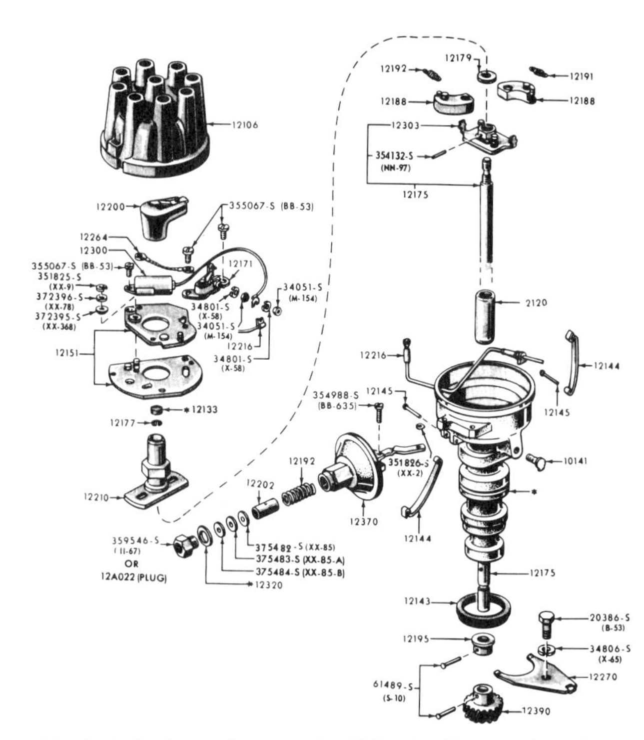

Exploded view of an early V-8 distributor with forged centrifugal weights.

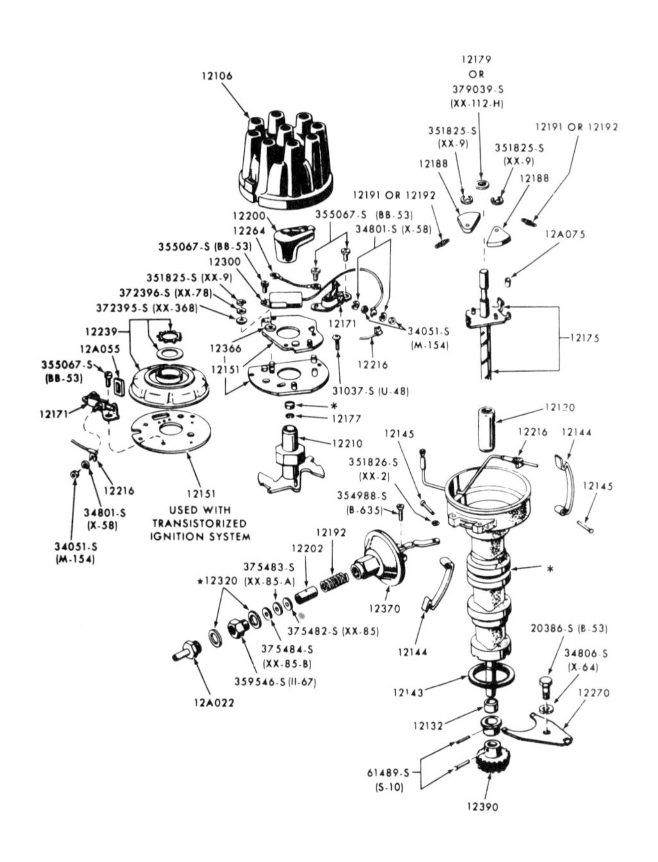

Exploded view of a V-8 distributor with stamped centrifugal weights.

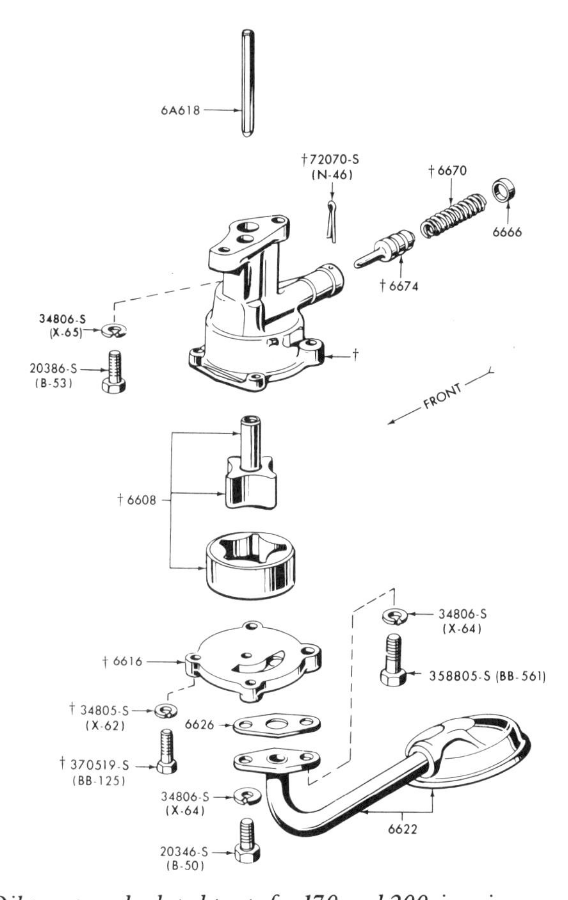

Oil pump and related parts for 170 and 200ci engines.

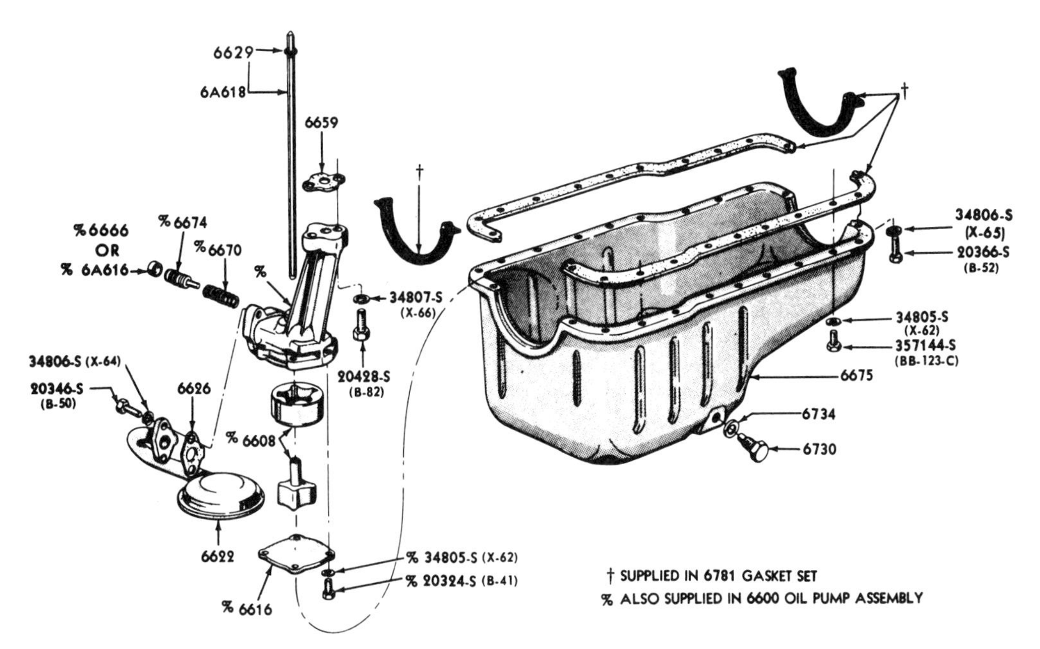

Typical oil pump, pan and related parts for 260, 289, 302, 351, and 429ci engines.

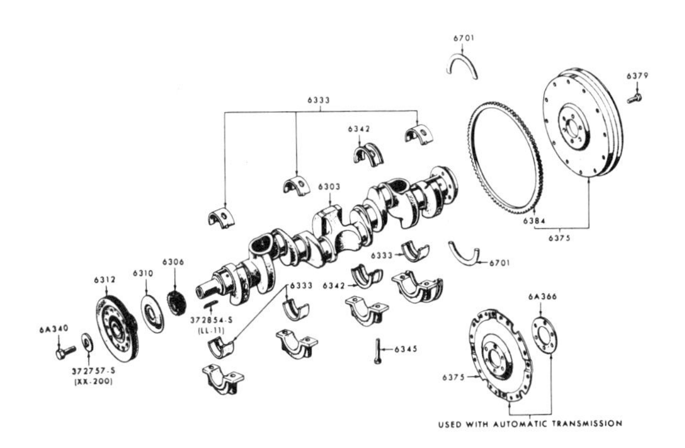

Typical crankshaft, flywheel, main bearings, and related parts for early 1965 model 170ci with four main bearings.

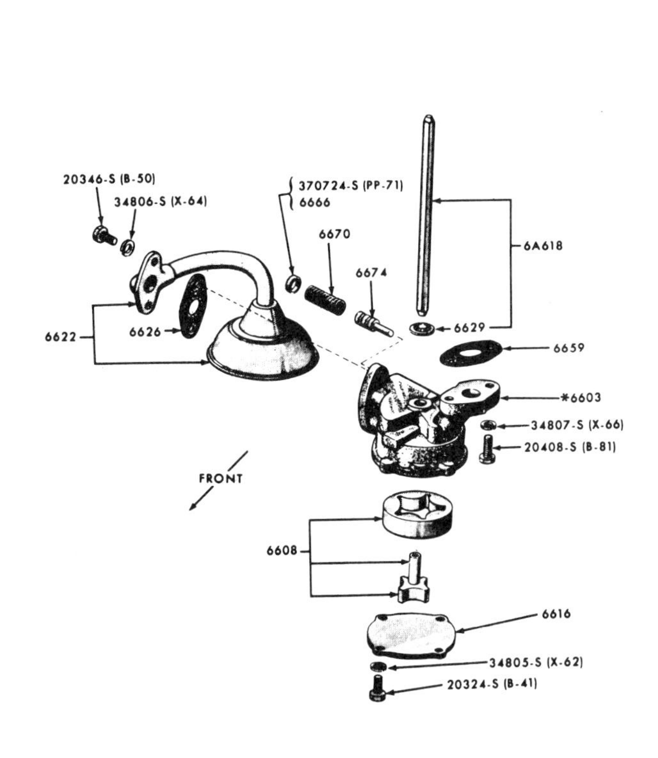

Oil pump and related parts for 390 and 428ci engines.

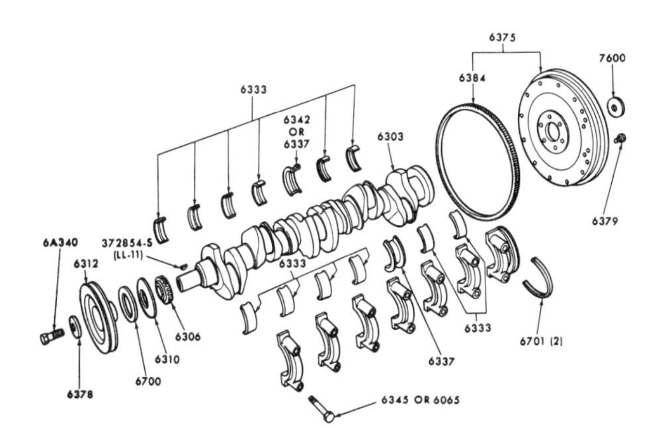

Typical crankshaft, flywheel, main bearings, and related parts for 200 and 250ci engines with seven main bearings.

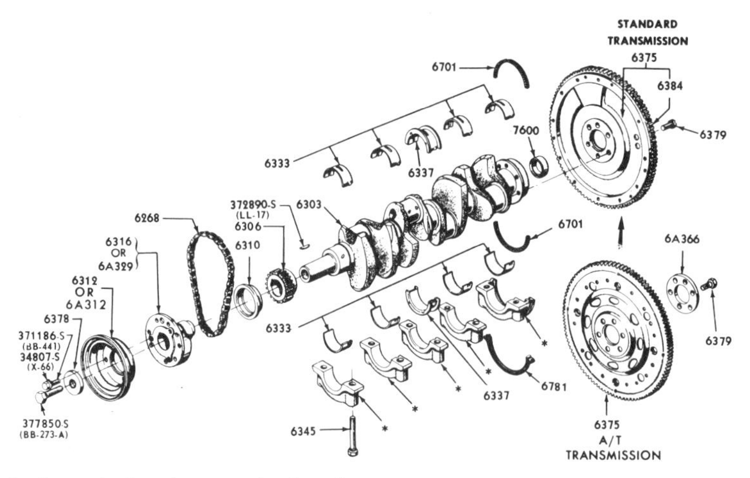

Typical crankshaft, flywheel, main bearings, and related parts for 260, 289, 302, 351, and 429ci engines.

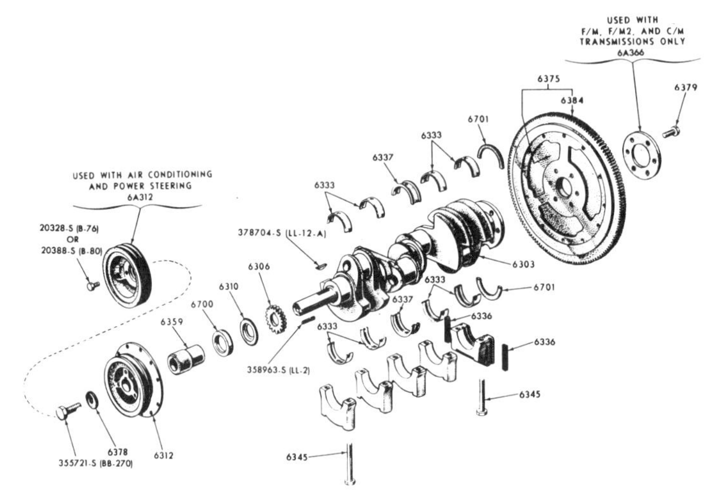

Crankshaft, flywheel, main bearings, and related parts for 390 and 428ci engines.

HiPo engines utilize thicker main bearing caps as compared to a standard 289. The block, however, is virtually unchanged.

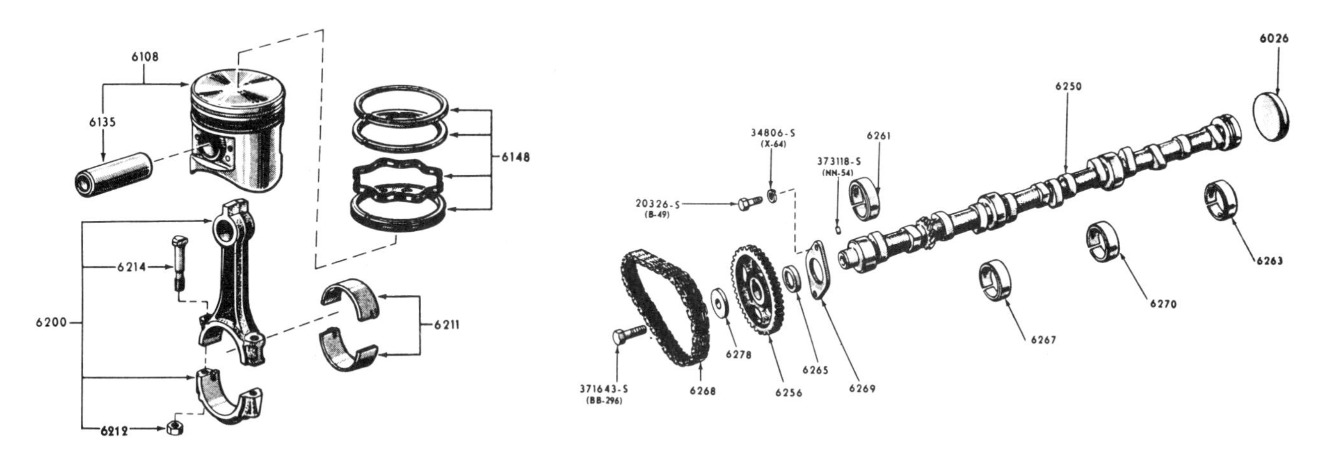

Typical piston, connecting rod, rings, engine camshaft, gear, bearings, and related parts for 170, 200, and 250ci engines.

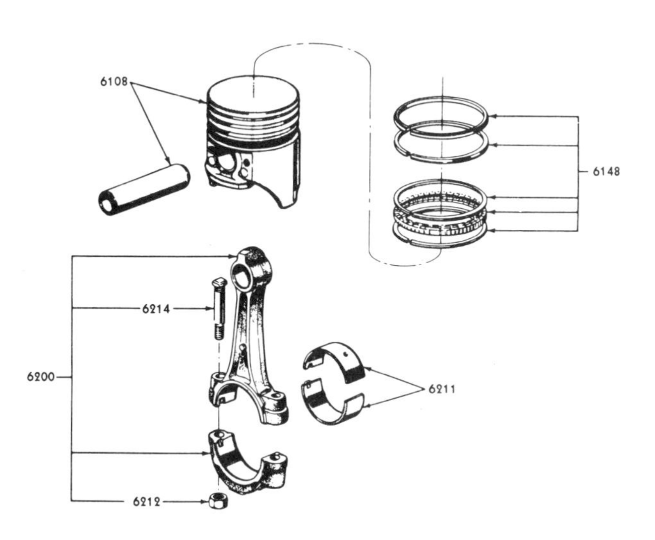

Typical piston, connecting rod, rings, and related parts for 260, 289, 302, 351, 400, and 429ci engines.

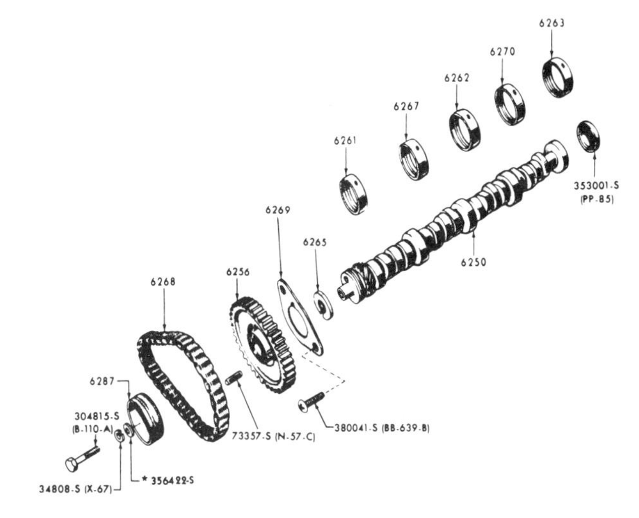

Typical camshaft, gear, bearings, and related parts for 390 and 428ci engines.

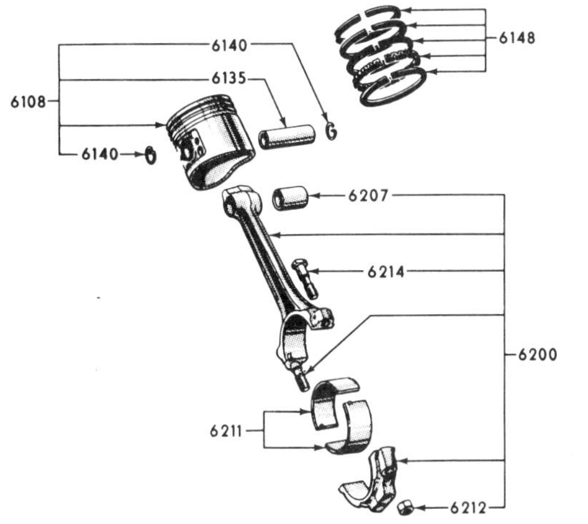

Typical piston, connecting rod, and related parts for 390 and 428ci engines.

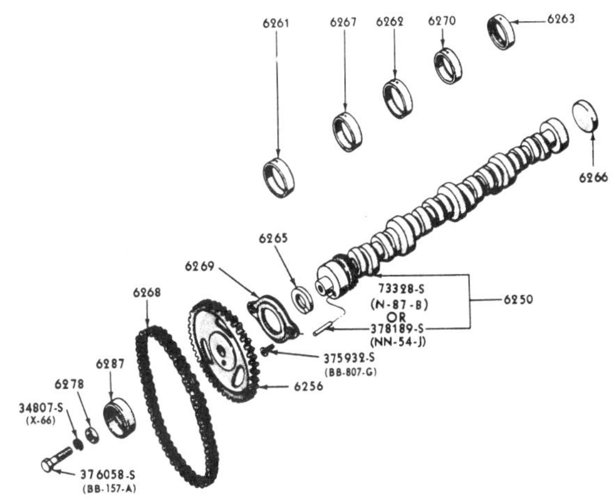

Typical camshaft, gear, bearings, and related parts for 289, 351, and 429ci engines.

Mustang Exhaust Manifold Application Chart, 1965-1970

For six-cylinder applications:

Years	CID	Part Number
65/67	200	C3OZ-9430-A
65/67	170	C3OZ-9430-A
67	200	D3BZ-9430-A
68	200	D3BZ-9430-A
69/70	200	D3BZ-9430-A
69	250	D3BZ-9430-A

For eight-cylinder applications:

Years	CID	Right Hand	Left Hand	Notes
65/67	289 K	C5ZZ-9430-B	C3OZ-9431-B	
65/68	260,289	D4DZ-9430-E	D0OZ-9431-D	
67	390	C6OZ-9430-A	C7OZ-9431-A	
68	390	C6OZ-9430-A	C7OZ-9431-A	Before 2/15/68
68	390	C8OZ-9430-B	C7OZ-9431-A	From 2/15/68
68	427	C6OZ-9430-A	C7OZ-9431-A	
68	428 CJ	C8OZ-9430-C	C8OZ-9431-B	
68/70	302	DOAZ-9430-C	D0OZ-9431-D	
		r/b D4DZ-9430-E &		
		D5DZ-9A603-C(2/75)		
69	351	C9AZ-9430-A	C9OZ-9431-A	
69	390	C9LZ-9430-B	C7OZ-9431-A	
69	428 CJ	C8OZ-9430-C	C8OZ-9431-B	
69/71	Boss 302	C9ZZ-9430-B	C9ZZ-9431-A	
69/70	Boss 429	C9AZ-9430-B	C9AZ-9431-A	
70	351W-2V	C9AZ-9430-A	C9OZ-9431-A	
70	351C-2V	D2OZ-9430-B	D0AZ-9431-A	
70	351	D0AZ-9430-B	D0AZ-9431-B	
70	428 CJ	C9OZ-9430-C	C8OZ-9431-B	
70	428 SCJ	C9OZ-9430-C	C8OZ-9431-B	

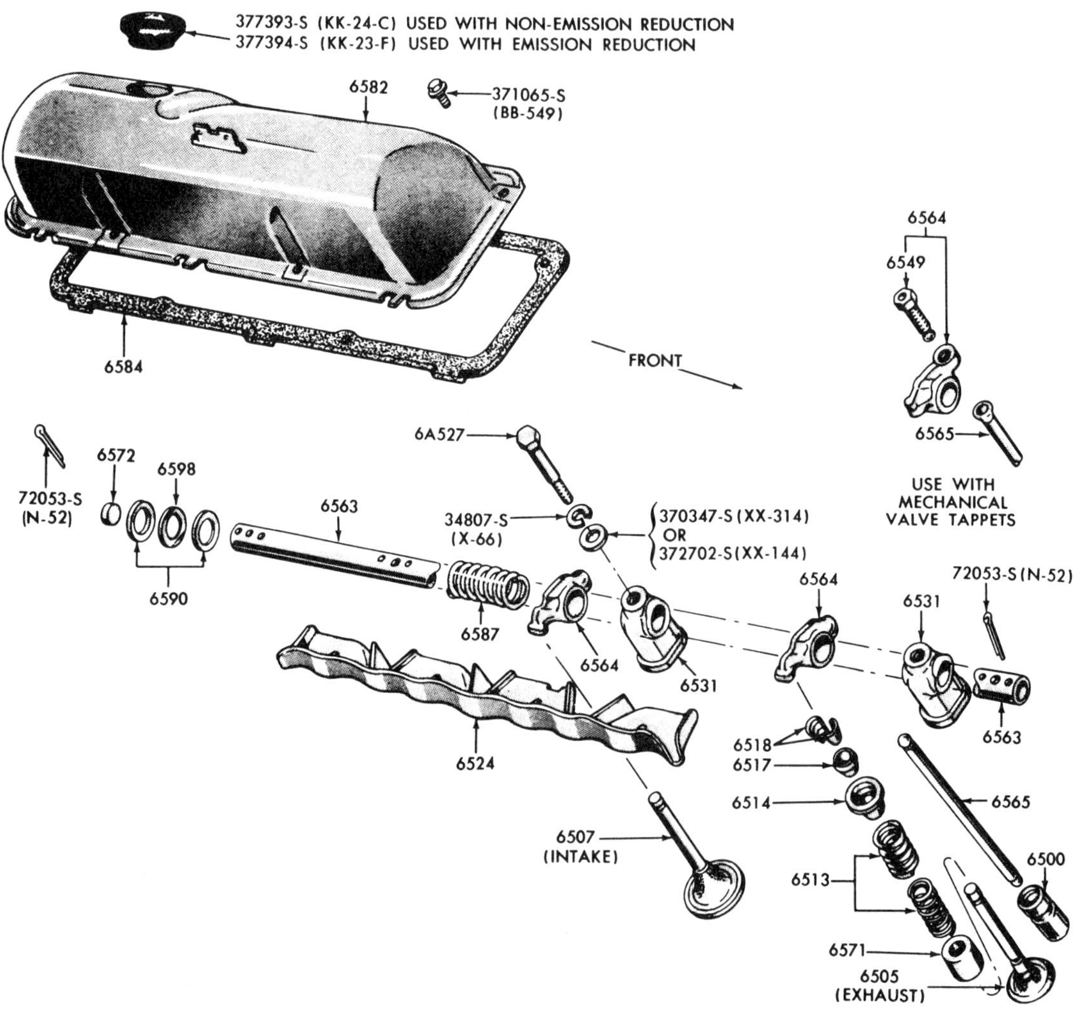

Valves, springs, rocker arms, cover, and related parts for 390 and 428ci engines.

Boss 302 engines were dressed out and finely detailed. A dress-up kit which included either chrome valve covers for 1969 or polished finned aluminum valve covers for 1970 were standard equipment. Here you can see the mangled mess of wires above the coil that incorporates the rev limiter and the 1970 vintage decal proclaiming the engine a Boss 302 built by Ford.

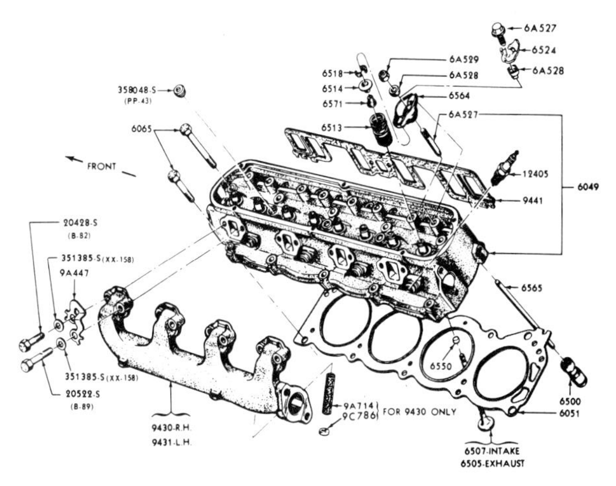

Typical cylinder head, exhaust manifold, gaskets, and related parts for all V-8 engines.

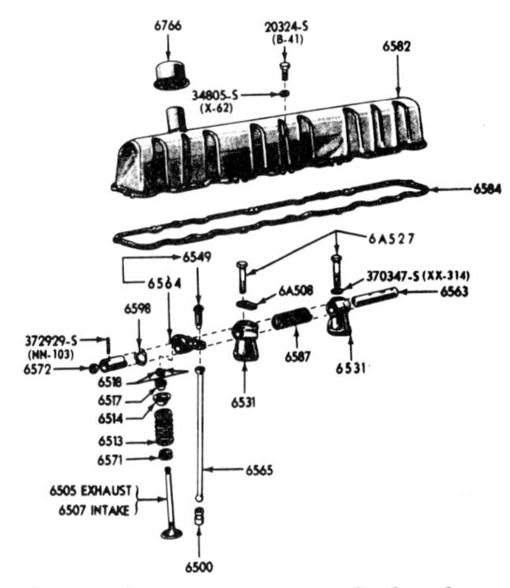

Valves, springs, rocker arms, cover and related parts for 170, 200, and 250ci engines.

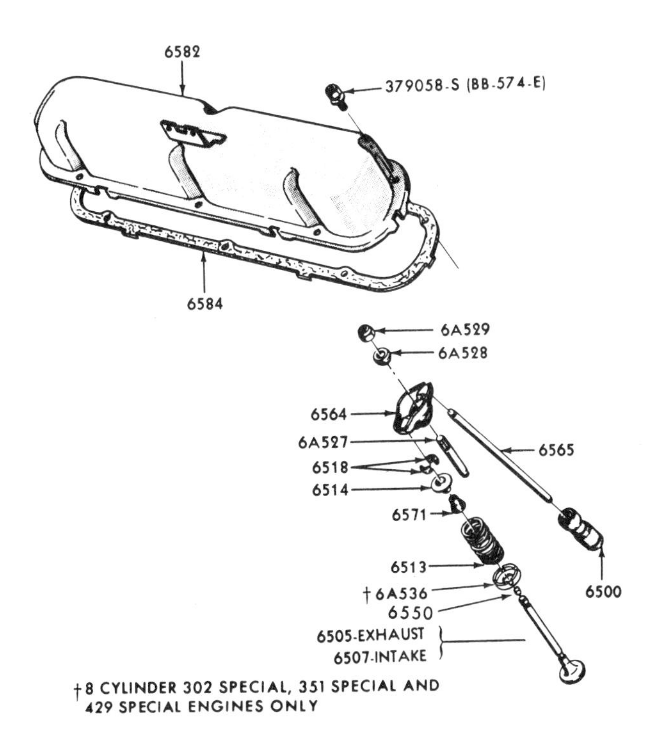

Typical valves, springs, rocker arms, cover, and related parts for all V-8 engines.

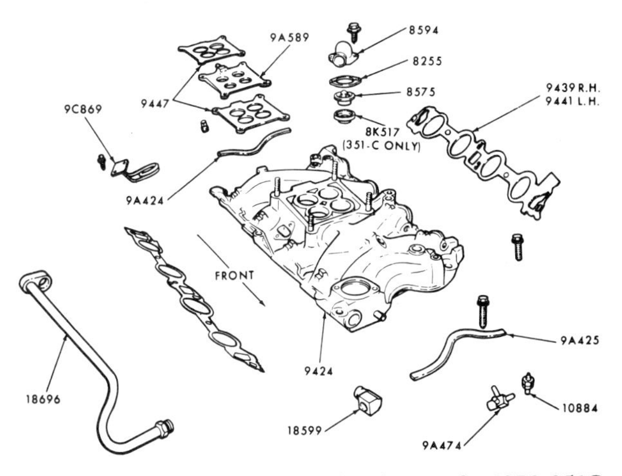

Typical intake manifold and related parts for 1970 351C and 1969-1970 429ci engines. Note: the thermostat and housing are not attached to the intake manifold on 351C engines.

MUSTANG INTAKE MANIFOLD APPLICATIONS CHART

Year	Cyl/CID	Part Number	Description
65	8/260 (Emission Reduction)	C4OZ-9424-E	Pkg. Includes one
			C5AZ-8592-C Connection Assy.
65	8/260 (Exc. w/ Emission Reduction)	C3AZ-9424-G	Pkg. Includes one
			C3AZ-10911-A Adapter
65	8/289 (4/B Carb. & Exc. Emission	C3OZ-9424-C	Pkg. Includes one
	Reduction		C3AZ-10911-A Adapter
			Before 8/20/65
65/67	8/289 (2B Carb)	C4AZ-9424-H	Pkg. Includes one
			C6AZ-8592-C Connection Assy.
66/68	8/289,302 (4B Carb.)	C4OZ-9424-H	
65	8/289 (4/B Carb)	C4OZ-9424-H	
65	8/289 (4/B Carb. & Emis. Reduction)	C4OZ-9424-H	Before 8/20/65
65/69	8/289,302 High-Perf. (4/B Carb)	C9OZ-9424-D	
65/68	8/289,302 (4/B Carb.)	C9OZ-9424-D	
68	8/427 (K, C6, T/E & 4/B Carb.)	C6ZA-9424-H	
66	8/390 High-Perf. (4/B Carb.)	C6AZ-9424-H	
66/67	8/390 (4/B Carb.)	C6AZ-9424-N	
68/70	8/427CJ (4/B Carb.)	C8OZ-9424-B	Cast Iron 1.60" Primary Openings,
			1.70" Secondary
66/70	8/390 High-Perf., 4/B Carb.	C8OZ-9424-B	Cast Iron 1.60" Primary Openings,
			1.70" Secondary
68	8/390 (2/B Carb.)	C9AZ-9424-E	
68	8/390 (4/B Carb.)	C8AZ-9424-C	Pkg Includes one 8A7617-B Plug
68	8/289 (2/B Carb.)	C4AZ-9424-J	
68/70	8/302 (2/B Carb.)	C4AZ-9424-J	
69	8/351 (4/B Carb.)	C9OZ-9424-B	
69	8/390 (4/B Carb.)	C9ZZ-9424-A	
69/70	8/351W High-Perf. (4/B Carb.)	C9OZ-9424-E	Aluminum High-Riser
69/70	8/429 Boss (4/B Carb.)	C9AZ-9424-D	Aluminum
69/70	8/302 Boss (4/B Carb.)	C9ZZ-9424-C	Aluminum
69	8/251 (2/B Carb.)	C9OZ-9424-A	
70	8/351W (2/B Carb.)	C9OZ-9424-A	
69/70	8/302 Boss	D0ZC-9425-A	Aluminum "Cross-Boss" Base Only uses
	Off-Highway Units		D0ZX-9C484-A Cover Gasket
			Inline 4V Carb.
			Use D0ZX-9C483-A Cover
70	8/351C (4/B Carb.)	D0AZ-9424-C	
70	8/351C 2/B Carb.)	D1AZ-9424-D	

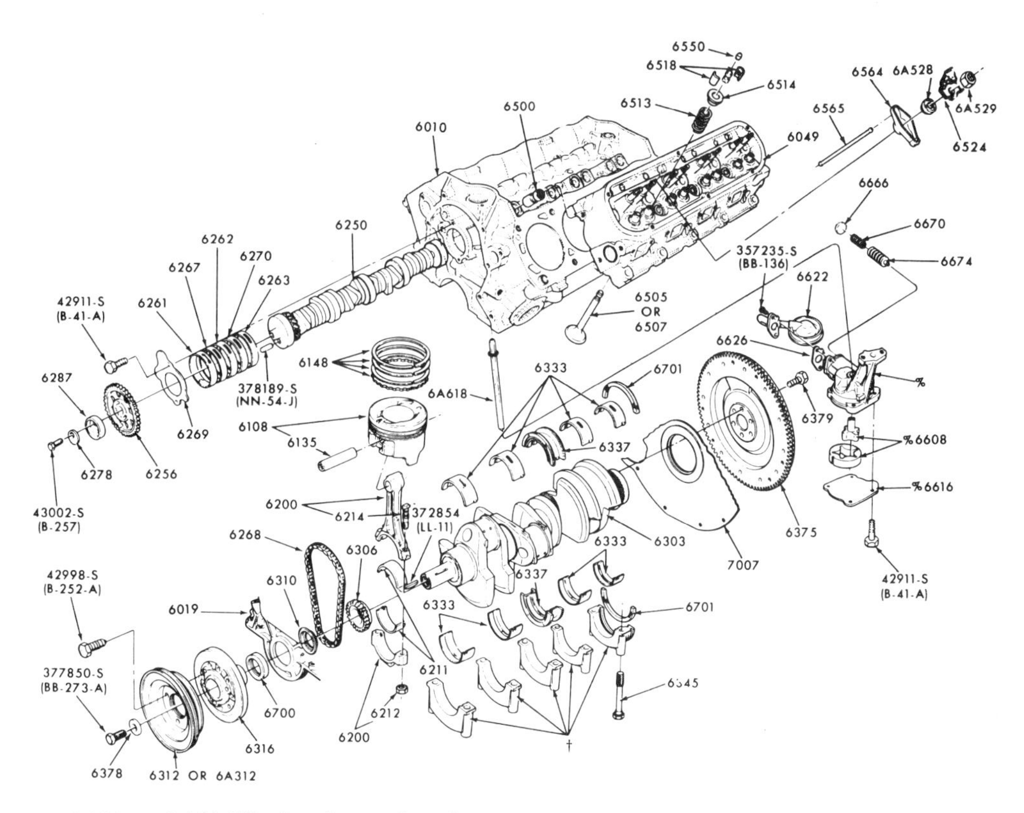

Exploded view of 302 and 351 Windsor internal engine parts.

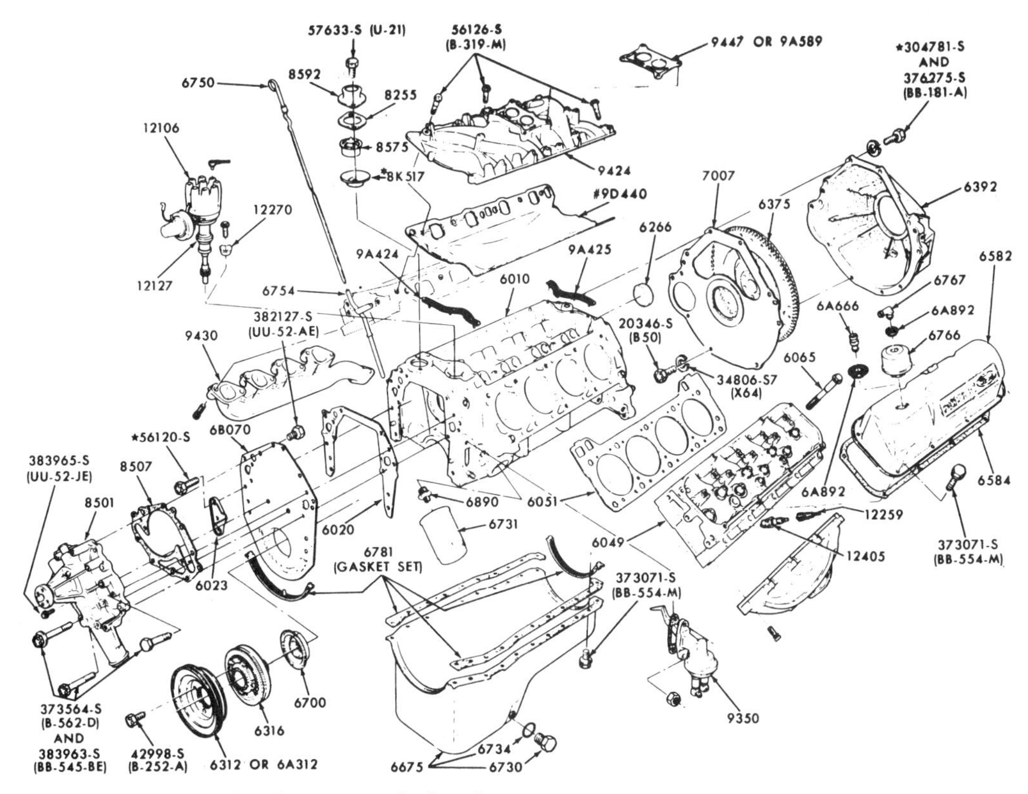

The 351 Cleveland engine external engine parts and related hardware.

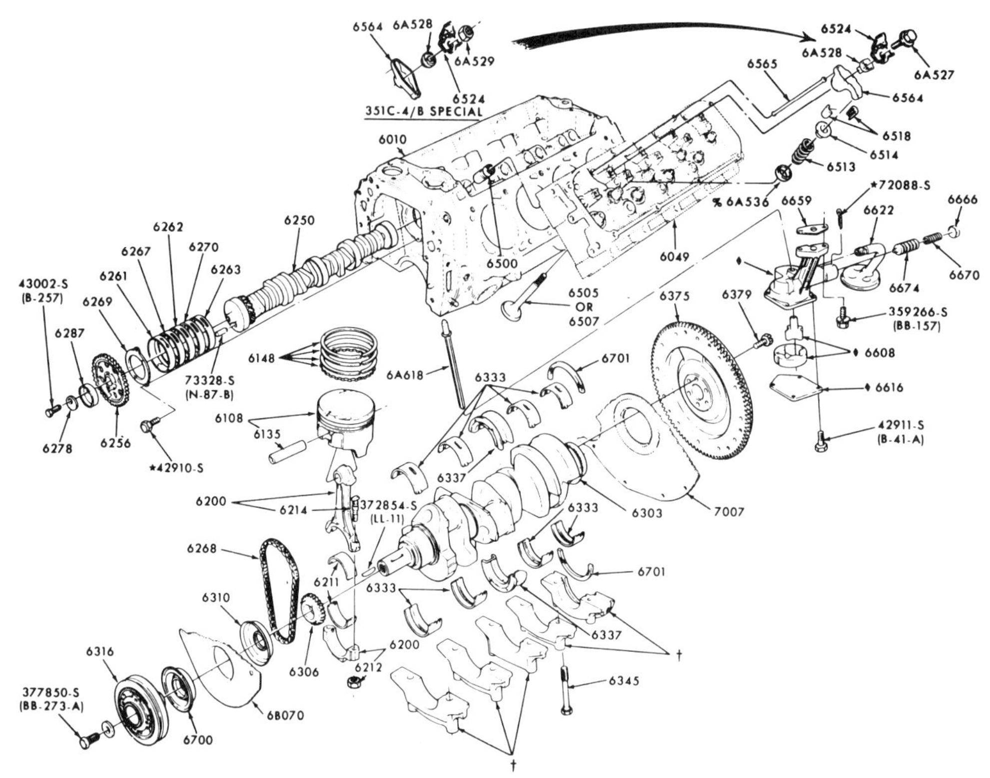

An exploded view of a 351 Cleveland long-block assembly.

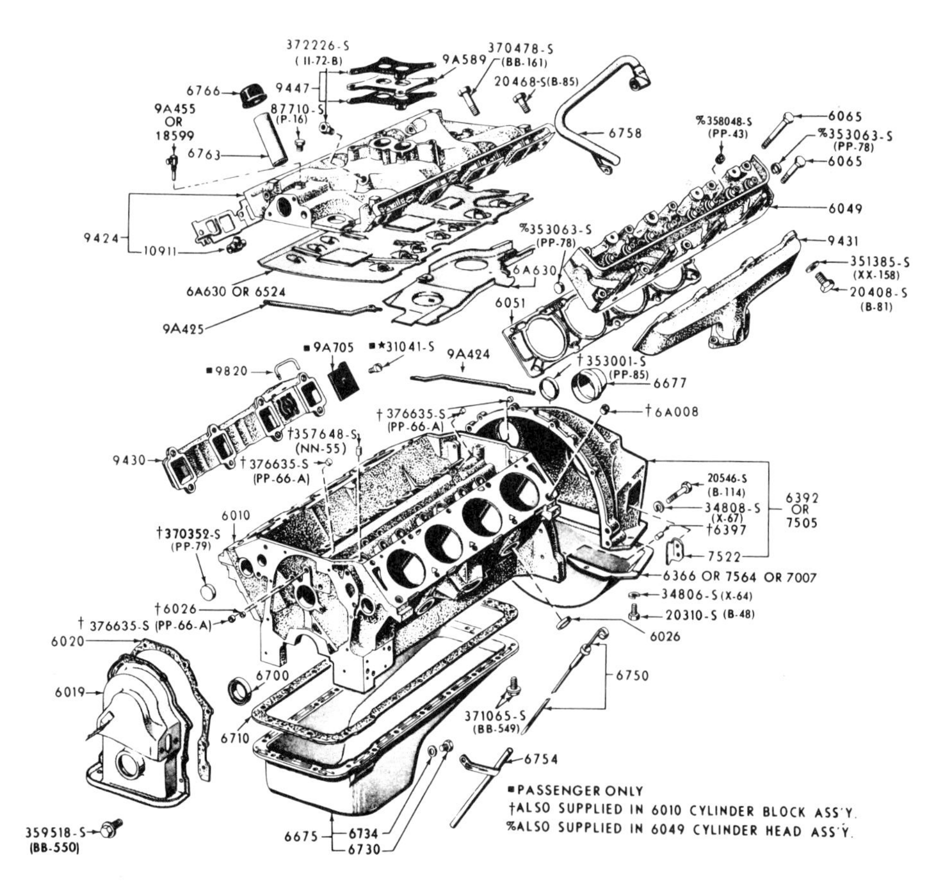

The 390 and 428 engine block and related parts.

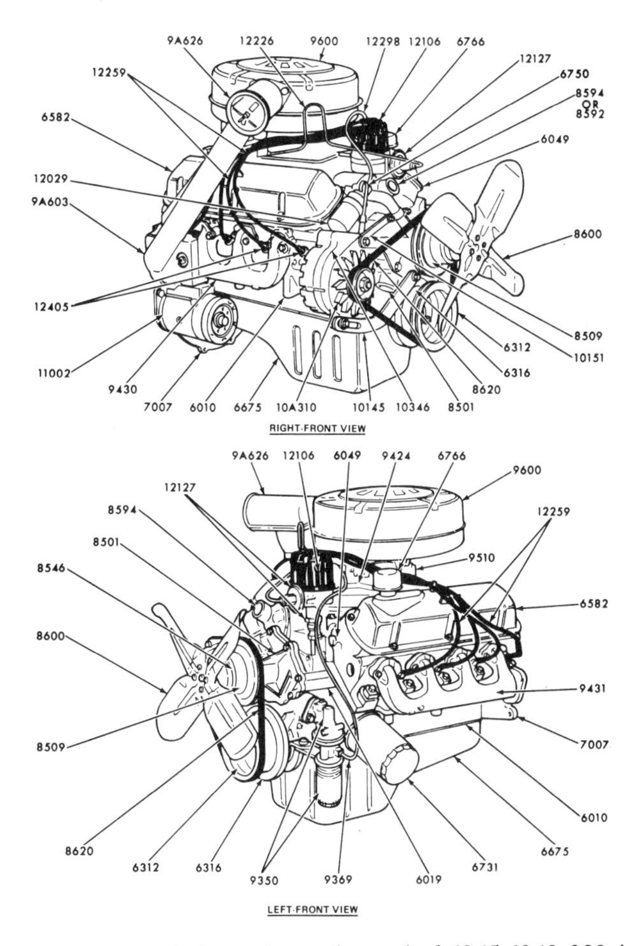

Right and left front views of a typical 1965-1968 289ci engine.

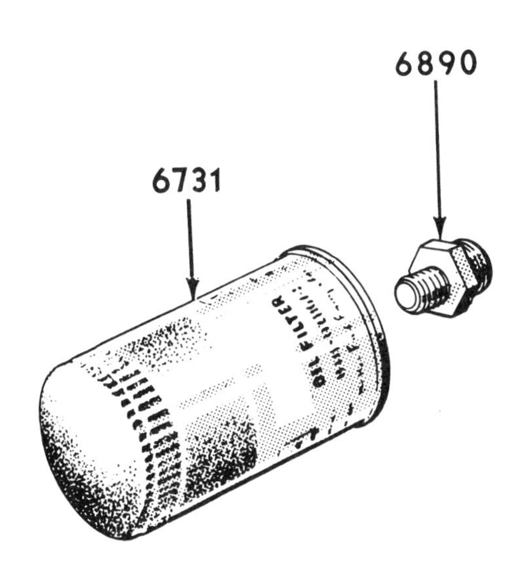

Engine oil filter and adapter fitting for all engines except 390 and 428.

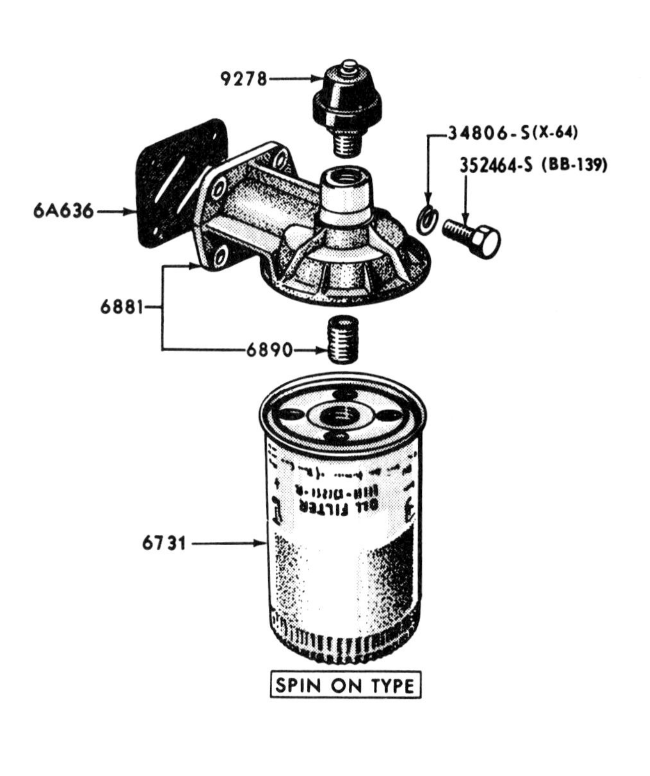

Engine oil filter, adapter, and related parts for 1968-1970 model 390 and 428ci engines.

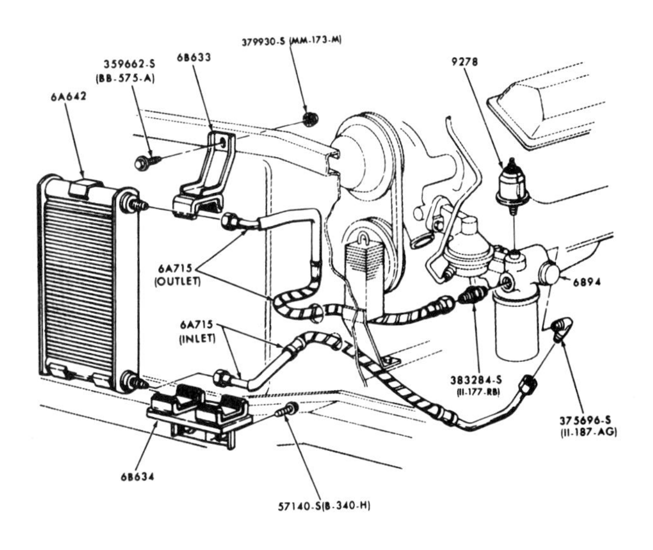

Engine oil cooler assembly and attaching hardware for 1969-1970 Boss 302, Boss 429, and 428 Super Cobra Jet engines.

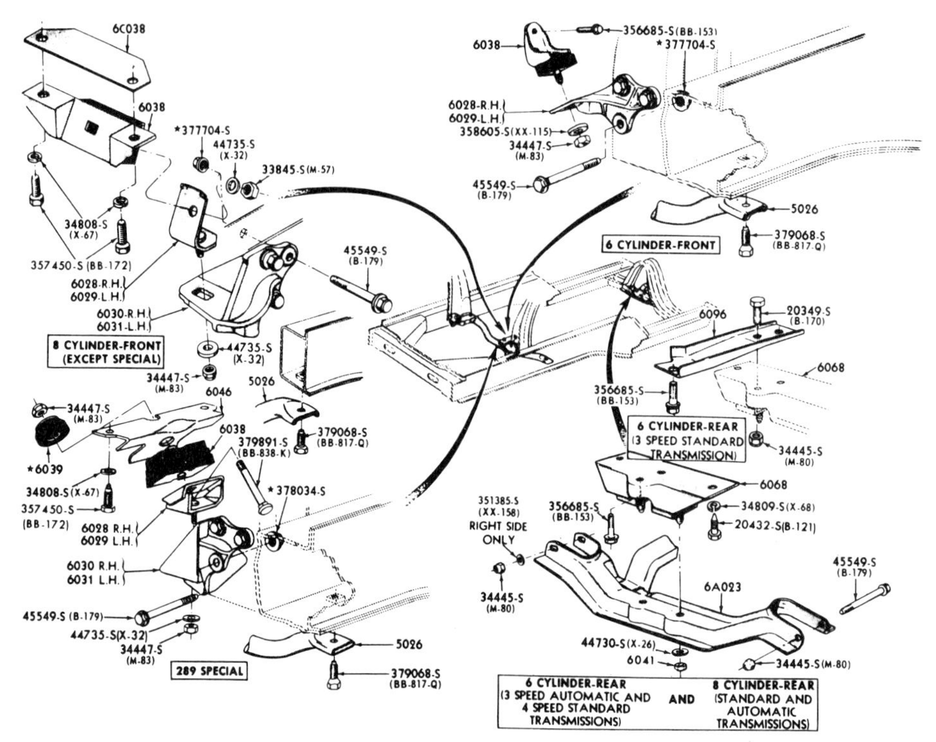

Engine and transmission mounts, accessories, and hardware for 1965-1966 Mustang (all engines and all transmissions).

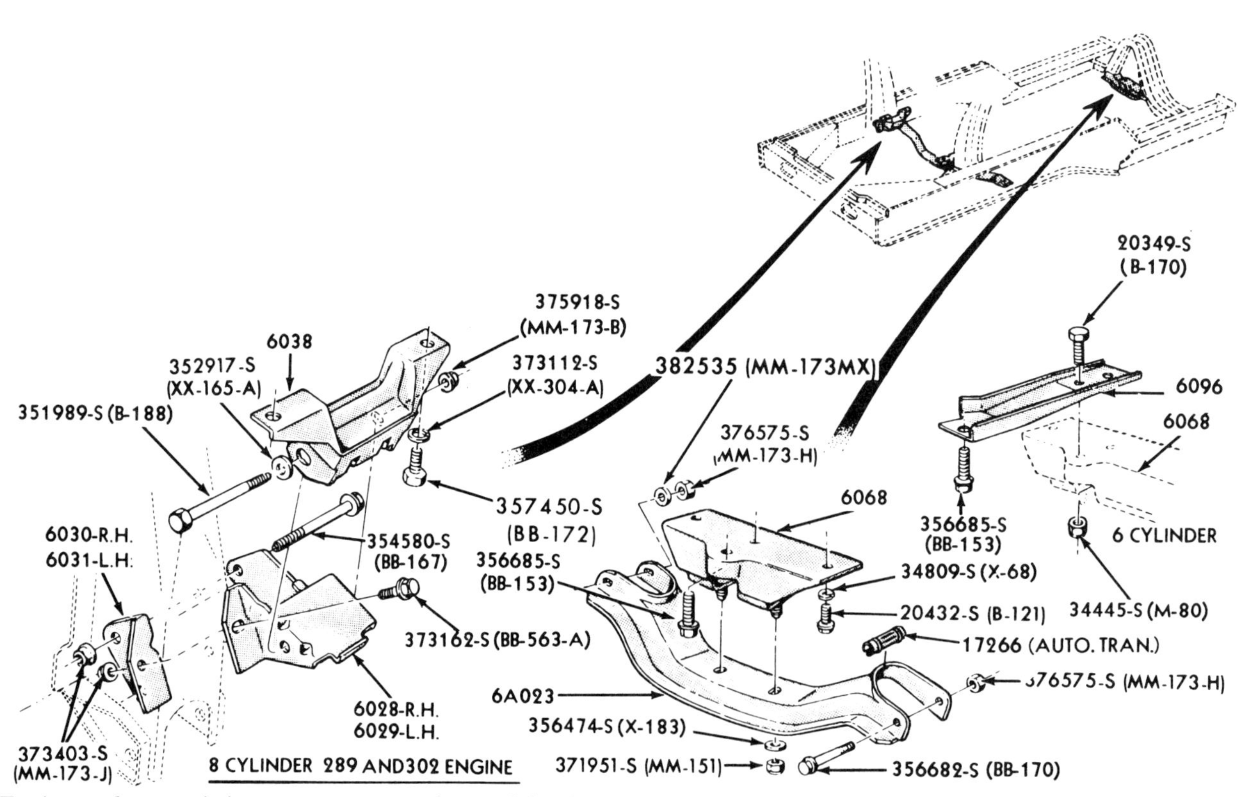

Engine and transmission mounts, accessories, and hardware for 1967-1970 Mustang equipped with six-cylinder or small-block V-8 engines.

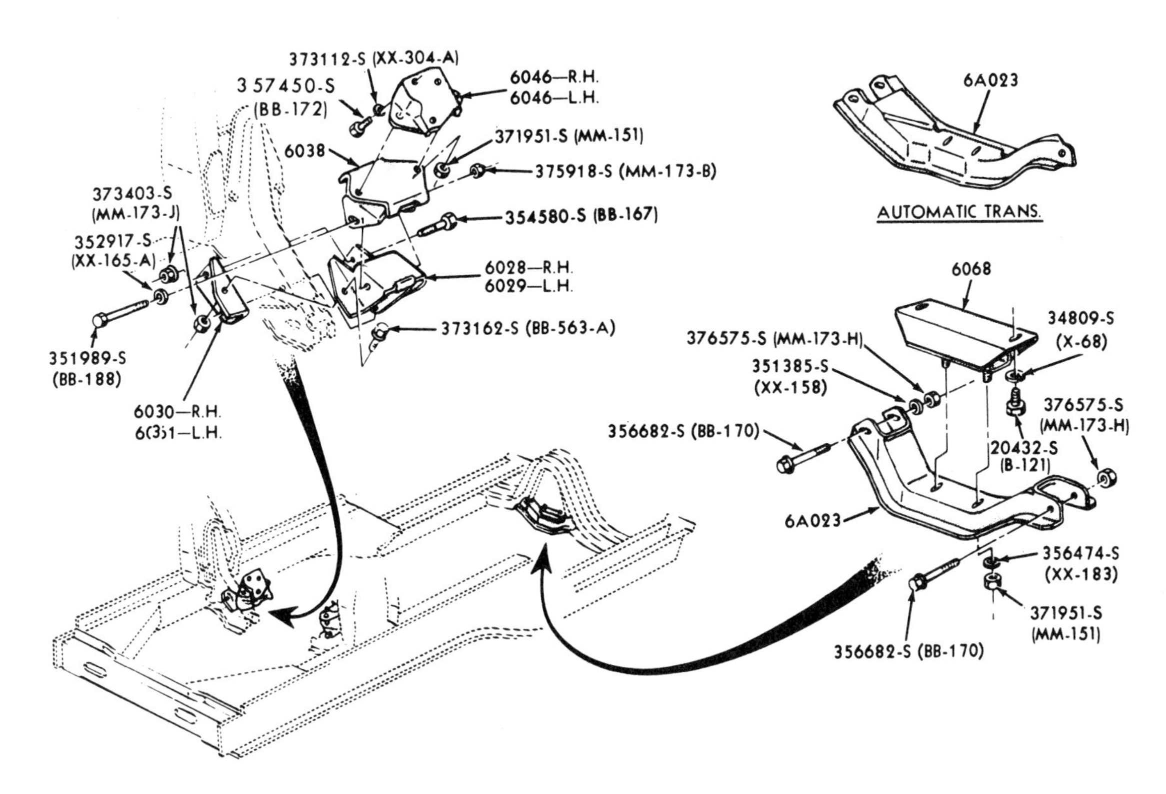

Engine and transmission mounts, accessories, and hardware for 1967-1970 Mustang equipped with 390 or 428ci engines.

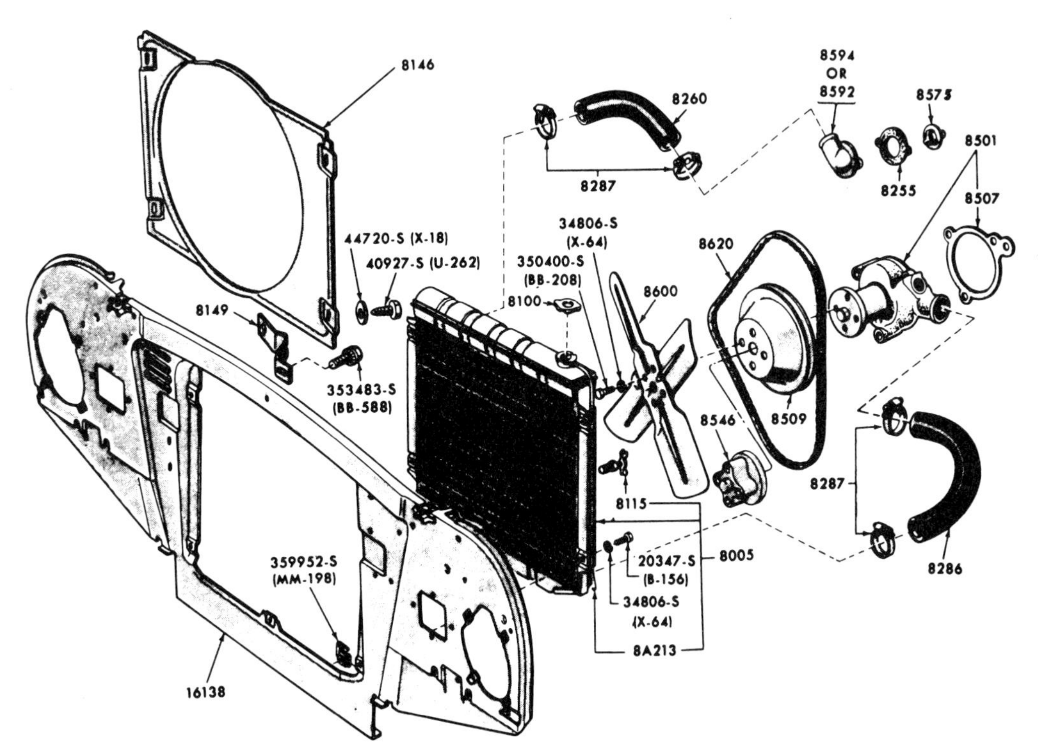

Typical cooling system and related parts for 1965-1970 Mustang equipped with all six-cylinder and V-8 engines except 390 or 428.

MUSTANG WATER PUMP ASSEMBLY APPLICATION CHART

Year	Notes	Engine	Description	Part Number
65		170,200		D4DZ-8501-A
65	2	260,289	Aluminum; uses C4OZ-8512-B &	C5AZ-8501-K
			C4OZ-8530-C	
65	2,5	289		D3UZ-8501-A
65	1	289		C5OZ-8501-A
65	1,4	289		D3UZ-8501-A
66		200		D4DZ-8501-A
66	7	289		D3UZ-8501-A
66	7	289	Aluminum; uses C4OZ-8530-C Bearing	C5AZ-8501-K
			with C4AZ-6019-B	
66	6	289		D3UZ-8501-A
66	1	289		C5OZ-8501-A
67		200		D4DZ-8501-A
67	2	289		D3UZ-8501-A
67	1	289		C5OZ-8501-A
67		390		D0AZ-8501-D
68		200		D4DZ-8501-A
68	1	289		C5OZ-8501-A
68	2,8	289,302		D3UZ-8501-A
68		390		D0AZ-8501-D
69		200		D4DZ-8501-A
69		250		D0DZ-8501-A
69	8	302-2V,351	Except Boss 302	D3UZ-8501-A
69		Boss 302		C9ZZ-8501-A
69		390,428CJ		D0AZ-8501-D
69		Boss 429	Pump incl. C8VZ-8507-A & C8SZ-8513-A	C9VZ-8501-A
70		200		D4DZ-8501-A
70		250		D0DZ-8501-A
70		Boss 302		D0ZZ-8501-B
70		351C	Incl. D0AZ-8513-A gasket; r/b D2SZ-8501-A	D0AZ-8501-E
70		428CJ		D0AZ-8501-D
70		302,351W	Except Boss 302	D0OZ-8501-C

Notes:

1 Used with High Performance 289
2 Except with High Performance 289
3 With generator or alternator
4 With alternator
5 With alternator, with A/C
6 Except with High Performance 289, with Thermactor
7 Except with High Performance 289, without Thermactor
8 Change water bypass tube for 289, 302 when necessary.

MUSTANG RADIATOR SHROUD APPLICATION CHART

Year	Engine	Notes	Description	Part Number
65	200	1	One-piece type	C3DZ-8146-C
65	260,289	1	One-piece type	C3DZ-8146-D
65	289	2	One-piece type	C3DZ-8146-D
66	200	1	One-piece type	C3DZ-8146-C
66	289	1	One-piece type; use w/ C3DZ-8005-K core	C6ZZ-8146-A
66	289	1	One-piece type; use w/ C3DZ-8005-K core	C3DZ-8146-D
66	289	2	One-piece type	C3DZ-8146-D
67	200	1	One-piece type-plastic	C7ZZ-8146-C
67	289	1	One-piece type-plastic, marked C7ZE-B	C9ZZ-8146-B
67	289	2	One-piece type-plastic, marked C7ZE-B	C9ZZ-8146-B
67	390	1,3,5,11	One-piece type-plastic, marked C7ZE-A	C7ZZ-8146-A
68	200	1,3		C7ZZ-8146-C
68	289	1,12	Plastic, marked C9OE-F	C9OZ-8146-A
68	289	4,11	Plastic, marked C9OE-F	C9OZ-8146-A
68	302	1	Plastic, marked C9OE-F	C9OZ-8146-A
68/70	390	1,3,5		C8ZZ-8146-A
68/70	428 CJ			C8ZZ-8146-A
68/70	302 (GT350)			C9OZ-8146-A
69	200	1 or 3		C7ZZ-8146-C
69	250	1 or 3,6,9	Plastic, use on 23-1/4 inch wide radiator	C9ZZ-8146-A
69	250	1 or 3,10	Plastic, use on 20-1/4 inch wide radiator	C9ZZ-8146-C
69	302,351	11	Plastic, marked C9ZE-C, w/ 20-inch radiator	C9ZZ-8146-B
69	302,351	3,6	Plastic, marked C9ZE-C, w/ 20-inch radiator	C9ZZ-8146-B
69	302,351	1,3	Plastic, marked C9OE-F, w/ 24-inch radiator	C9OZ-8146-A
69	302,351	3,6	Plastic, marked C9OE-F, w/ 24-inch radiator	C9OZ-8146-A
69	Boss 302		Plastic, marked C9OE-F, w/ 24-inch radiator	C9OZ-8146-A
70	200			C7ZZ-8146-C
70	250		Plastic	C9ZZ-8146-C
70	302		Plastic, marked D0ZE-B	D0ZZ-8146-B
70	351-2V		Plastic, marked D0ZE-B	D0ZZ-8146-B
70	351-4V		Plastic, marked D0ZE-B	D0ZZ-8146-B
70	302,351		Plastic, marked D0ZE-A or E	D0ZZ-8146-A
70	Boss 302		Plastic, marked D0ZE-A or E	D0ZZ-8146-A
70	351-4V		Plastic, marked D0ZE-A or E	D0ZZ-8146-A
70	428 CJ			C8ZZ-8146-A

Notes:

1 with Air Conditioning
2 High Performance
3 with Extra Cooling
4 with IMCO (Improved Combustion
5 with Thermactor
6 with Hang-on A/C
7 with 3.50:1 rear axle
8 without 3.50:1 rear axle
9 Before 12/16/68
10 From 12/16/68
11 with Standard transmission
12 except w/ High Performance

MUSTANG RADIATOR APPLICATION CHART

Year	Engine	Notes	Core Width	Core Height	Core Thickness	Part Number
65/66	170,200	1,2,4	17-1/4	16-1/2	1-1/4	C5ZZ-8005-C
65/66	260,289	1,2,3,4	17-1/4	17-1/2	1-1/4	C3DZ-8005-K
67	200	1,2,4	20-1/4	16-7/16	1-1/4	C9ZZ-8005-A
67	289	1,2,4	20-1/4	16-7/16	1-1/4	C9ZZ-8005-B
67	390	1,2,4	23-1/4	16	2	C7ZZ-8005-C
68	200	1,2,4	20-1/4	16-7/16	1-1/4	C9ZZ-8005-A
68	289-2V	1	20-1/4	16-7/16	1-1/4	C9ZZ-8005-B
68	289-2V	2,4	24-3/16	16	1-1/2	C9ZZ-8005-C
68	289 HiPo	1,2,4	24-3/16	16	1-1/2	C9ZZ-8005-C
68	302	1	20-1/4	16-7/16	1-1/4	C9ZZ-8005-B
68	302	2,4	24-3/16	16	1-1/2	C9ZZ-8005-C
68	390	1,2,4	24-3/16	16	2-1/4	C8ZZ-8005-C
68	428	1,2,4	24-3/16	16	2-1/4	C8ZZ-8005-C
69	200	1,2,4	20-1/4	16-7/16	1-1/4	C9ZZ-8005-A
69	250	1,2,4,5	23-1/4	15-7/16	1-1/4	C9ZZ-8005-D
69	250	1,2,4,6	20-1/4	16-7/16	1-1/4	C9ZZ-8005-E
69	302-2V,351	1	20-1/4	16-7/16	1-1/4	C9ZZ-8005-B
69	351	2,4	24-3/16	16	1-1/2	C9ZZ-8005-C
69	Boss 302	1	24-3/16	16	1-1/2	C9ZZ-8005-C
69	Boss 302	2,4	24-3/16	16	1-1/2	C9ZZ-8005-C
69	390 GT	1,2,4	24-3/16	16	2-1/4	C8ZZ-8005-C
69	428 CJ	1,2,4	24-3/16	16	2-1/4	C8ZZ-8005-C
69	Boss 429	1,2,4	24-3/16	16	2-1/4	C8ZZ-8005-C
70	200	1,4	20-1/4	16-7/16	1-1/4	C9ZZ-8005-A
70	200	2	20-1/4	16-7/16	1-1/4	C9ZZ-8005-E
70	250	1,2,4	20-1/4	16-7/16	1-1/4	C9ZZ-8005-E
70	302	1	20-1/4	16-7/16	1-1/4	C9ZZ-8005-E
70	351-2V	1	20-1/4	16-7/16	1-1/4	C9ZZ-8005-E
70	351-4V	2,7	20-1/4	16-7/16	1-1/4	C9ZZ-8005-E
70	Boss 302	1,2	24-3/16	16	2-1/4	D0ZZ-8005-C
70	428 CJ	1,2,4	24-3/16	16	2-1/4	C8ZZ-8005-C
70	Boss 429	1,2,4	24-3/16	16	2-1/4	C8ZZ-8005-C
70	302	2,4	24-3/16	16	2-1/4	D0ZZ-8005-D
70	351-2V & 4V	2,4	24-3/16	16	2-1/4	D0ZZ-8005-D
70	351-4V	1,2,8	24-3/16	16	2-1/4	D0ZZ-8005-D

Notes:

1. Standard Cooling
2. Extra Cooling
3. Super Extra Cooling
4. With Air Conditioning
5. Before 12/16/68
6. From 12/16/68
7. Without 3.50:1 rear axle
8. With 3.50:1 rear axle

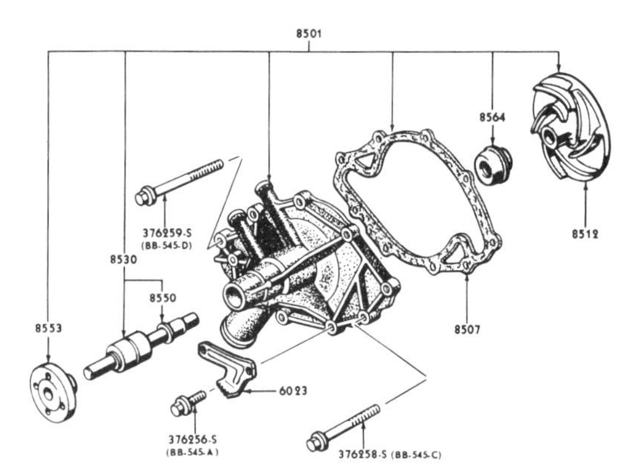

Aluminum water pump (without cover) assembly and attaching hardware for 1965 260 and 289ci engines.

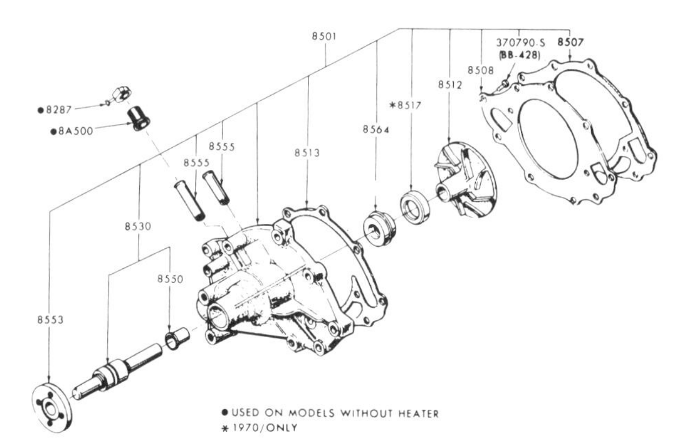

Iron water pump (with cover) for 1965-1970 289, 302, and 351W engines.

An information tag explaining antifreeze warranty and service specifications was attached to the radiator filler neck or overflow hose by the dealer during the standard predelivery check.

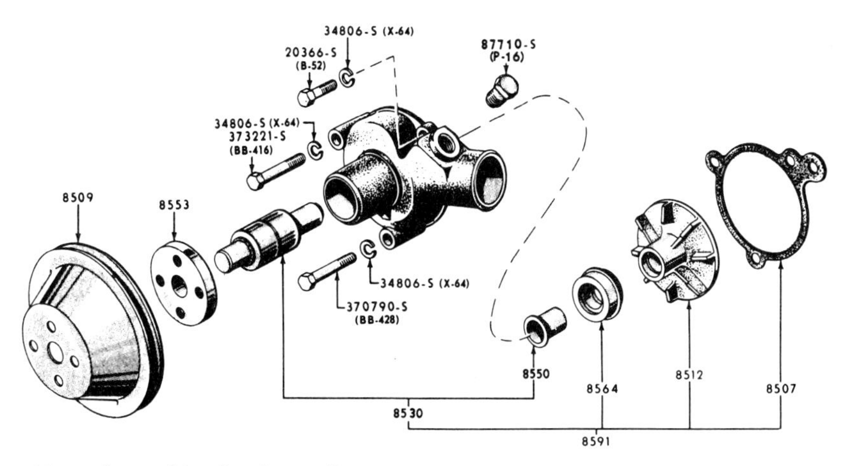

Engine water pump assembly and attaching hardware for 170, 200, and 250ci engines.

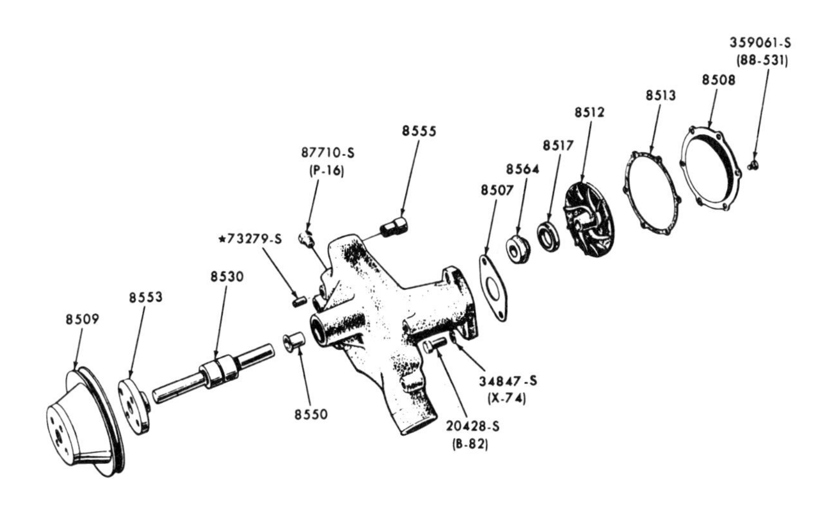

Water pump assembly and attaching hardware for 390 and 428ci engines.

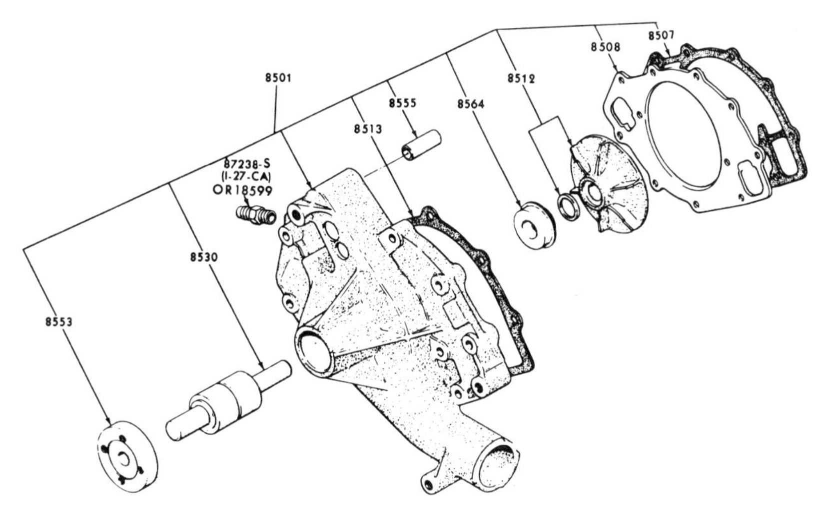

Water pump assembly and attaching hardware for 351 Cleveland and 429ci engines.

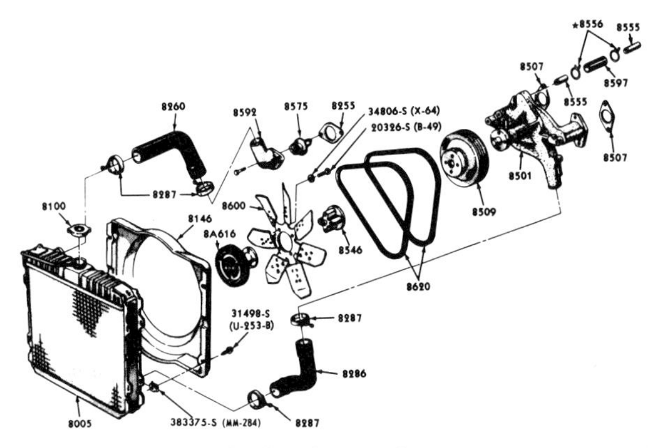

Cooling system and related parts for 1967-1970 Mustang equipped with 390 or 428 engines.

MUSTANG FAN CLUTCH ASSEMBLY APPLICATION CHART

Year	Engine	Notes	Description	Part Number
65	260,289	1		C6OZ-8A616-D
66	289	2,3,5	3-3/16" depth, marked C6ZE-A	C6ZZ-8A616-A
66	289	4,6		C6OZ-8A616-D
66	289	7		C6OZ-8A616-D
67	289,390	8	2.91" depth, marked C7ZE-B	C7OZ-8A616-C
67	289	9	3-3/16" depth, marked C6ZE-A	C6ZZ-8A616-A
67	428		for GT-500, not replaced	
68	390 GT	10	marked C8OE-A	C8OZ-8A616-A
68	200,289 302,428CJ	11	marked C8OE-B or C	C8OZ-8A616-B
68	428		for GT-500	C8ZX-8A616-A
69	250		marked C9DE-A or C9ZE-C	C9DZ-8A616-A
69	390GT	1	marked C8OE-A	C8OZ-8A616-A
69	390GT	12	marked C8OE-B or C	C8OZ-8A616-B
69	428CJ, or SCJ		marked C9ZE-B	C9ZZ-8A616-A
70	250	1	marked C9DE-A or C9ZE-A	C9DZ-8A616-A
70	428CJ	1,13	marked C9ZE-B	C9ZZ-8A616-A
70	428SCJ	13	not used from 9/1/69	C9ZZ-8A616-A
70	428CJ	1,14		D0ZZ-8A616-A

Notes:
1 with factory installed A/C
2 with hang-on A/C, w/ or w/o Thermactor
3 Before 1/5/66
4 From 1/5/66
5 Use with C6ZZ-8805-A, 2-1/4" core.
6 Use with C3DZ-8805-K, 1-1/4" core.
7 Integral A/C, w/ or w/o Thermactor
8 except hang-on A/C
9 with hang-on A/C
10 with Thermactor
11 without A/C
12 Standard cooling package
13 Before 9/1/69
14 From 9/1/69

MUSTANG FAN BLADE ASSEMBLY APPLICATION CHART, PART ONE, 1965-1968

Year	Engine	Notes	Description	Blades	Diam.	Hole	Part Number
65	170,200	1		6	15-1/16	5/8	C9DZ-8600-A
65	170,200	3	Replaces C2DA-A,C2OA-A	4	15-1/2	5/8	C2DZ-8600-D
65	260,289	4,5,6		4	17	5/8	C2OA-8600-B
65	260,289	2,7		5	17-1/4	2-3/8	C3OZ-8600-D
65	260,289	1,11	Aluminum, replaces 6-blade fan	5	17	5/8	C5AZ-8600-A
65	289K	10		4	17-1/8	5/8	C6OZ-8600-D
65	289	2,8		7	17-1/2	2-3/8	C4OZ-8600-D
66	200	3	Replaces C2DA-A,C2OA-A	4	15-1/2	5/8	C2DZ-8600-D
66	200	12		6	15-1/16	5/8	C9DZ-8600-A
66	289K	10	marked C3OA-C	4	17-1/8	5/8	C6OZ-8600-D
66	289	4,9,12	marked C6OE-B,C6AE-E,C1VV-D	4	17-1/2	5/8	C6OZ-8600-A
66	289	5,6,9,13	marked C5AE-B	5	17	5/8	C5AZ-8600-A
66	289	1	marked C6OE-G	7	17-1/2	2-3/8	C4OZ-8600-D
67	200	3,5,14	Replaces C2DA-A,C2OA-A	4	15-1/2	5/8	C2DZ-8600-D
67	200	12		6	15-1/16	5/8	C9DZ-8600-A
67	289K	10	marked C6OE-D	4	17-1/8	5/8	C6OZ-8600-D
67	289	3,9,15,17	marked C6AE-E or C1VV-D	4	17-1/2	5/8	C6OZ-8600-A
67	289	4,9,12	marked C6AE-E or C1VV-D	4	17-1/2	5/8	C6OZ-8600-A
67	289	5,6,9,13,16	marked C5AE-B	5	17	5/8	C5AZ-8600-A
67	289	9,12,18	marked C5AE-B	5	17	5/8	C5AZ-8600-A
67	289	12	marked C6DA-A	5	17-1/2	5/8	C6DZ-8600-A
67	289	12	marked C6OE-G	7	17-1/2	2-3/8	C4OZ-8600-D
67	390	5	marked C6OE-F	7	18-1/4	2-3/8	C6OZ-8600-F
68	200	3,5 or 6 or 12		6	15-1/16	5/8	C9DZ-8600-A
68	289	3,5		4	17-1/8	5/8	C6OZ-8600-D
68	289,302	12		5	17-1/2	5/8	C8SZ-8600-B
68	302	3,5		4	17-1/2	5/8	C6OZ-8600-A
68	351	5,6,17		4	17-1/2	5/8	C6OZ-8600-A
68	390	5,6,12	marked C8OE-B	7	18-1/4	2-3/8	C8OZ-8600-A
68	GT350		marked C6OE-G	7	17-1/2	2-3/8	C4OZ-8600-D

Notes:
1 with Hang-on A/C
2 with factory-installed Hang-on A/C
3 Except with A/C
4 Except Hi-Po
5 Standard Cooling
6 Extra Cooling
7 with factory generator
8 with factory alternator
9 with Thermactor
10 with Hi-Po engine
11 with generator or alternator
12 with A/C
13 with Power Steering
14 includes IMCO
15 with Auto. Transmission
16 before 11/14/66
17 from 11/14/66
18 with Manual Transmission

MUSTANG FAN BLADE ASSEMBLY APPLICATION CHART, PART TWO, 1969-1970

Year	Engine	Notes	Description	Blades	Diam.	Hole	Part Number
69	200	2	marked C9ZE-D	4	16	5/8	C9ZZ-8600-A
69	200	3		6	15-1/16	5/8	C9DZ-8600-A
69	200,250	1,3	marked C5AE-B	5	17	5/8	C5AZ-8600-A
69	250	4		4	17-1/8	5/8	C9OZ-8600-A
69	250	4 or 9	marked C9ZE-C	6	17	2-5/8	C9ZZ-8600-B
69	302-2V,351	4		5	17-1/2	5/8	C8SZ-8600-B
69	302-2V	2,3		4	17-1/2	5/8	C6OZ-8600-A
69	351	2,3	marked C8SE-A	4	17-9/16	5/8	C9OZ-8600-B
69	302-2V,351	1,3		5	18	5/8	C5AZ-8600-K
69	Boss 302		marked C9WE-A	5	17-1/2	2-3/8	C9WZ-8600-A
69	Boss 302			4	17-1/8	5/8	C6OZ-8600-D
69	Boss 302			5	17-1/2	5/8	C8SZ-8600-B
69	390 GT		marked C8OE-B	7	18-1/4	2-3/8	C8OZ-8600-A
69	390	1,3	marked C8AE-D	5	19	5/8	C8AZ-8600-D
69	428CJ,SCJ	2,4	marked C9ZE-E	7	18-1/4	2-5/8	C9ZZ-8600-C
69	428 SCJ	2	marked C9OE-H	6	18	5/8	C9OZ-8600-D
69	Boss 429		marked C9ZE-F	7	18	3/4	C9ZZ-8600-D
70	200	2,10	marked C9ZE-D	4	16	5/8	C9ZZ-8600-A
70	200	3,10		6	15-1/16	5/8	C9DZ-8600-A
70	200	2,3,11	marked C8YE-8600-A	5	16-1/2	5/8	C8YZ-8600-A
70	200,250	1,3,14	marked C5AE-B	5	17	5/8	C5AZ-8600-A
70	250	2,3		4	17-1/8	5/8	C9OZ-8600-A
70	250	4	marked C9ZE-C	6	17	2-5/8	C9ZZ-8600-B
70	302	4		5	17-1/2	5/8	C8SZ-8600-B
70	351	2 or 3,4		5	17-1/2	5/8	C8SZ-8600-B
70	Boss 302			5	17-1/2	5/8	C8SZ-8600-B
70	351	2 or 3	marked C8SE-A	4	17-9/16	5/8	C9OZ-8600-B
70	302-2V	2 or 3		4	17-9/16	5/8	C9OZ-8600-B
70	428CJ	2,12	marked C9ZE-E	7	18-1/4	2-5/8	C9ZZ-8600-C
70	428CJ,SCJ	4		7	18-1/4	2-5/8	C9ZZ-8600-C
70	428 SCJ	2,13		7	18-1/4	2-5/8	C9ZZ-8600-C
70	428 SCJ	12		6	18	5/8	C9OZ-8600-D
70	428 CJ	13		7	18	5/8	D0TZ-8600-B
70	Boss 429		marked D0ZE-A	5	18	1	D0ZZ-8600-A

Notes:
1 with Hang-on A/C
2 Standard Cooling
3 Extra Cooling
4 with A/C
5 with Power Steering
6 includes IMCO
7 with Auto. Transmission
8 with Manual Transmission
9 Economy 1V model
10 Before 8/27/69
11 From 8/27/69
12 Before 9/1/69
13 From 9/1/69
14 Slot existing holes away from center hole (4 places).

MUSTANG RADIATOR HOSE APPLICATION CHART, PART ONE, 1965-1967

Year	Engine	Notes	Type	Interior Diameter	Length	Part Number
65	170,200	M	Upper	1-1/4		C5TZ-8260-A
		M	Lower			D2PZ-8286-H
		F	Upper			C9PZ-8260-F
65	260,289	M,1	Upper			C3DZ-8260-J
		M,1	Lower	1-3/4	14-7/8	C3DZ-8286-F
		F,1	Upper			C9PZ-8260-K
65	260,289	M,2	Upper			C5TZ-8260-A
		M,2	Lower	1-3/4	14-7/8	C3DZ-8286-F
65	289	M,3	Upper			D2PZ-8260-H
		M,3	Lower	1-3/4	14-7/8	C3DZ-8286-F
		F,3	Upper			C9PZ-8260-K
65	289	M,4	Upper			C5TZ-8260-A
		M,4	Lower	1-3/4	14-7/8	C3DZ-8286-F
66	200	M	Upper			C2OZ-8260-H
		M	Lower			D2PZ-8286-H
		F	Upper			C9PZ-8260-F
66	289	M,5	Upper			D2PZ-8260-H
		M,5	Lower	1-3/4	14-7/8	C3DZ-8286-F
		F,5	Upper			C9PZ-8260-K
66	289	M,6	Upper			C5TZ-8260-A
		M,6	Lower	1-3/4	14-7/8	C3DZ-8286-F
66	289	M,8	Upper			D2PZ-8260-H
		M,8	Lower	1-3/4	14-7/8	C3DZ-8286-F
		F,8	Upper			C9PZ-8260-F
		F,8	Lower			C9PZ-8286-R
67	200	M	Upper			C2OZ-8260-H
		M	Lower			D2PZ-8286-H
		F	Upper			C9PZ-8260-F
67	289	M,7	Upper			D2PZ-8260-H
		M,7	Lower	1-3/4	14-7/8	C3DZ-8286-F
		F,7	Upper			C9PZ-8260-K
67	289	M,6	Upper			C5TZ-8260-A
		F,6	Lower	1-3/4	14-7/8	C3DZ-8286-F
67	390	M	Upper	1-3/4	14-1/2	C5AZ-8260-G
		M	Lower			D2PZ-8286-J
		F	Upper			C9PZ-8260-R

Notes:
- M Moulded Hose
- F Flexible Hose; to be used when moulded hose not available.
- 1 with generators, except with 4 (2V) Carb. Kit.
- 2 with generators & 4 (2V) Carb. Kit.
- 3 with alternators, except with 4 (2V) Carb. Kit.
- 4 with alternators & 4 (2V) Carb. Kit.
- 5 without A/C, except with 4 (2V) Carb. Kit.
- 6 with 4 (2V) Carb. Kit.
- 7 except with 4 (2V) Carb. Kit.
- 8 with A/C

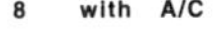

The Tecumseh cast-iron air conditioning compressor (left) was used primarily from 1965 through 1967, but not always. The York aluminum case compressor (right) was used intermittently through the years 1965 through 1970.

MUSTANG RADIATOR HOSE APPLICATION CHART, PART TWO, 1968-1969

Year	Engine	Notes	Type	Interior Diameter	Length	Part Number
68	200	M	Upper			C2OZ-8260-H
		M	Lower			D2PZ-8286-H
		F	Upper			C9PZ-8260-F
68	289-4V	M,9	Upper	1-1/2		C5AZ-8260-F
		M,9	Lower	1-3/4		C8ZZ-8286-A
		F,9	Upper			C9PZ-8260-L
68	289-2V,302	M,9	Upper	1-1/2	13	C8ZZ-8260-C
		M,9	Lower	1-3/4	14-7/8	C3DZ-8286-F
		F,9	Upper			C9PZ-8286-T
68	289-2V,302	M,10	Upper	1-1/2		C5AZ-8260-F
		M,10	Lower	1-3/4		C8ZZ-8286-A
		F,10	Upper			C9PZ-8260-L
68	390-4V,428CJ	M	Upper	1-3/4	14-1/2	C5AZ-8260-G
		M	Lower			D2PZ-8286-J
		F	Upper			C9PZ-8260-R
69	200	M	Upper			C2OZ-8260-H
		M	Lower			D2PZ-8286-H
		F	Upper			C9PZ-8260-F
69	250	M,11	Upper			C9ZZ-8260-D
		M,11	Lower	1-7/8	18	C9ZZ-8286-A
69	250	M,12	Upper	1-1/2	11	C9ZZ-8260-A
		M,12	Lower	1-7/8	18	C9ZZ-8286-A
		F,12	Upper			C9PZ-8260-N
69	302,351	M,9	Upper			D2PZ-8260-H
		M,9	Lower	1-3/4	14-7/8	C3DZ-8286-F
		F,9	Upper			C9PZ-8260-K
69	351,Boss 302	M,10	Upper			D2PZ-8260-H
		M,10	Lower	1-3/4	14-7/8	C3DZ-8286-F
		F,10	Upper			C9PZ-8260-K
69	390,428CJ	M	Upper	1-3/4	14-1/2	C5AZ-8260-G
		M	Lower			D2PZ-8286-J
		F	Upper			C9PZ-8260-R
69	Boss 429	M	Upper	1-1/2	11-1/4	B9A-8286-D
		M	Lower			C9AZ-8286-D

Notes:
- M Moulded Hose
- F Flexible Hose; to be used when moulded hose not available.
- 9 Standard cooling
- 10 Extra Cooling or A/C
- 11 Before 12/16/68
- 12 From 12/16/68

MUSTANG RADIATOR HOSE APPLICATION CHART, PART THREE, 1970

Year	Engine	Notes	Type	Interior Diameter	Length	Part Number
70	200	M	Upper			C2OZ-8260-H
		M	Lower			D2PZ-8286-H
		F	Upper			C9PZ-8260-F
70	250	M	Upper	1-1/2	11	C9ZZ-8260-A
		M	Lower	1-7/8	18	C9ZZ-8286-A
		F	Upper			C9PZ-8260-N
70	351C	M,9,10	Upper			D2PZ-8260-M
		M,9,10	Lower	1-3/4	14-7/8	D0ZZ-8286-A
		F,9,10	Upper			C9PZ-8260-L
		F,9,10	Lower	1-3/4	16-1/2	C9PZ-8286-R
70	302,351W	M,9,10	Upper			D2PZ-8260-H
		M,9,10	Lower	1-3/4	14-7/8	D0ZZ-8286-A
		F,9,10	Upper			C9PZ-8260-T
		F,9,10	Lower	1-3/4	16-1/2	C9PZ-8286-R
70	Boss 302	M	Upper			D2PZ-8260-H
		M	Lower	1-3/4	14-7/8	D0ZZ-8286-A
		F	Upper			C9PZ-8260-T
		F	Lower	1-3/4	16-1/2	C9PZ-8286-R
70	428CJ	M	Upper	1-3/4	14-1/2	C5AZ-8260-G
		M	Lower			D2PZ-8286-J
		F	Upper			C9PZ-8260-R
70	Boss 429	M	Upper	1-1/2	11-1/4	B9A-8286-D
		M	Lower			C9AZ-8286-D

Notes:
- M Moulded Hose
- F Flexible Hose; to be used when moulded hose not available.
- 8 with A/C
- 9 Standard cooling
- 10 Extra Cooling or A/C

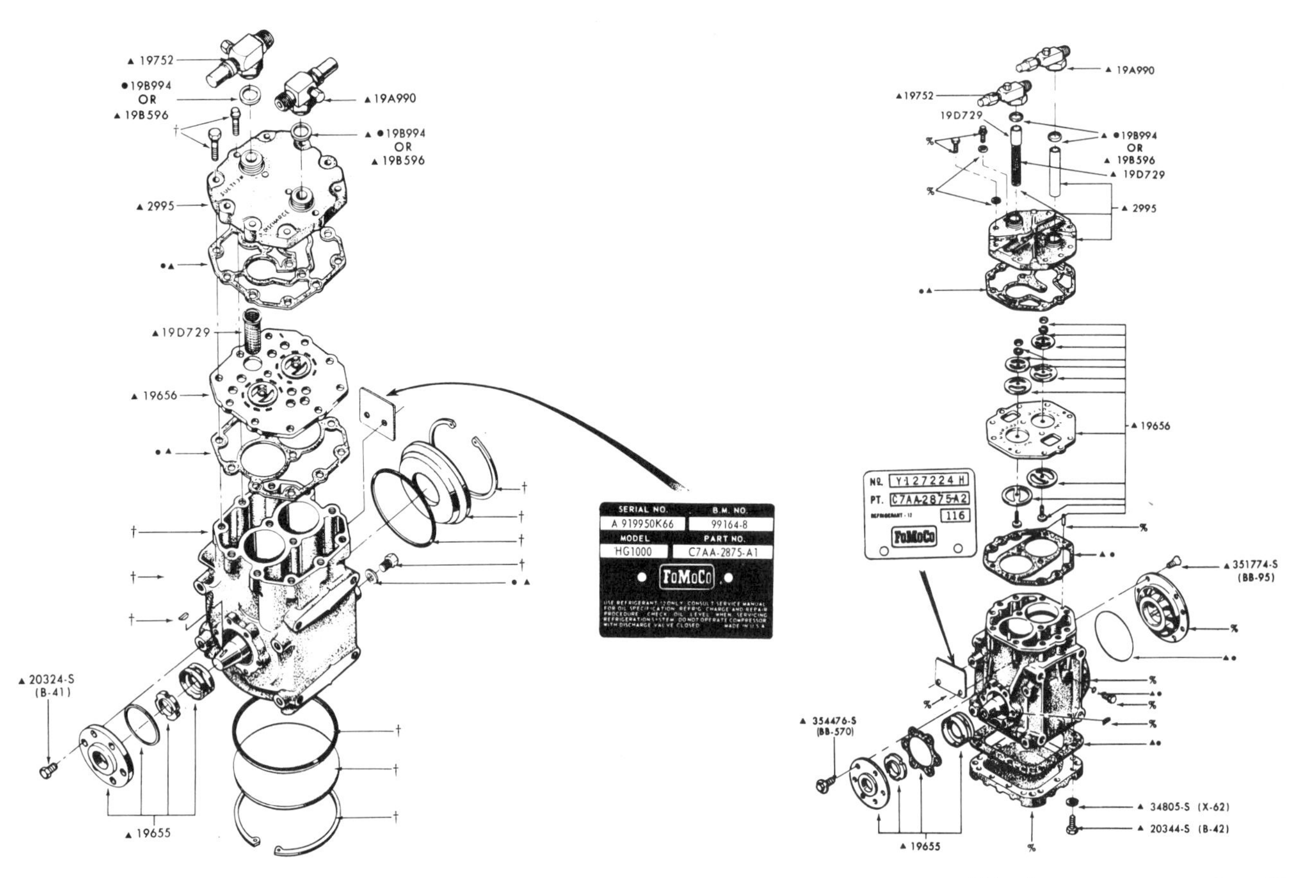

The Tecumseh cast-iron air conditioning compressor used on all 1965-1970 Mustangs.

The York aluminum air conditioning compressor used on all 1965-1970 Mustangs.

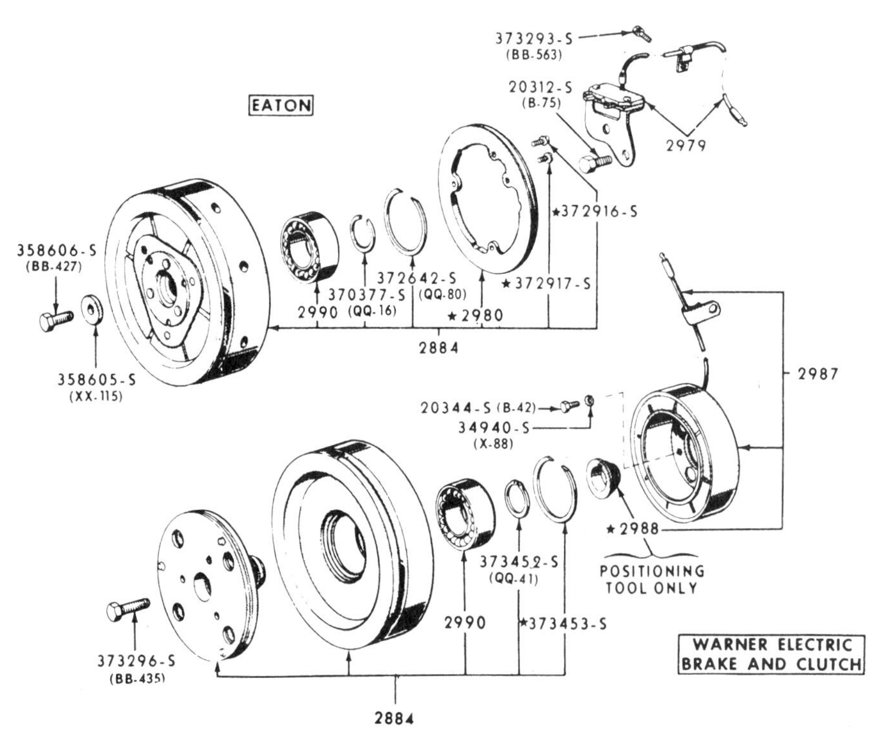

Eaton and Warner air conditioner compressor clutches used on all 1965-1970 Mustangs.

MUSTANG AIR CONDITIONER COMPRESSOR CLUTCH AND PULLEY APPLICATION CHART

YEAR	MODEL	TYPE	IDENTIFICATION	PART NUMBER
65/66	6 cylinder	2C	C4DA-2981-B, B1	C8OZ-2884-G
65/66	6 cylinder	6C	C4DA-2981-B2,	
65/66	6 cylinder	4C	C4DA-2981-C	C5DZ-2884-A
65/66	8 cylinder 260, 289	2B	C3AA-2981-A, A1	C3AZ-2884-A
65/66	8 cylinder 260, 289	6B	C3AA-2981-A2	
65	8 cylinder 260,289	5A	C4OA-2981-A	C3AZ-2884-C
65	8 cylinder 260,289	4A	C3AA-2981-B	C3AZ-2884-C
65/66	8 cylinder 260,289	4A	C2AA-2981-C	C8OZ-2884-G
65/66	8 cylinder 289	2B	C2AA-2981-A, A1	C8OZ-2884-B
65/66	8 cylinder 289	6B	C2AA-2981-A2	
66	8 cylinder 390	5B	C4AA-2981-A	C6AZ-2884-A
67	6 cylinder	2C	C7AA-2981-B	C7AZ-2884-B
67	6 cylinder	6C	C7OA-2981-B	C7OZ-2884-B
67	6 cylinder	2C	C4DA-2981-B, B1	C8OZ-2884-G
67	6 cylinder	6C	C4DA-2981-B2	
67	6 cylinder	4C	C7ZA-2981-C	C7ZZ-2884-B
67	8 cyl. 289 (exc. GT-350)	2C	C7AA-2981-B	C7AZ-2884-B
67	8 cyl. 289 (exc. GT-350)	3B	C7AA-2981-F	C9AZ-2884-A
67	8 cyl. 289 (exc. GT-350)	2B	C2AA-2981-A, A1	C8OZ-2884-B
67	8 cyl. 289 (exc. GT-350)	6B	C2AA-2981-A2	C8OZ-2884-B
67	8 cyl. 289 GT-350	5B	C7SA-2981-B	C7AZ-2884-D
67	8 cylinder 390	6B	C7AA-2981-H,	C9AZ-2884-B
67	8 cylinder 390	6B	C7SA-2981-A,	C9AZ-2884-B
67	8 cylinder 390	6B	C9AA-2981-G	C9AZ-2884-B
67	8 cylinder 390	2B	C2AA-2981-A, A1	C8OZ-2884-B
67	8 cylinder 390	6B	C2AA-2981-A2	C8OZ-2884-B
67	8 cylinder 390	2B	C7AA-2981-C	C7AZ-2884-C
68	6 cylinder	5C	C8OA-2981-A	C8OZ-2884-A
68	8 cylinder 289, 302	3B	C7AA-2981-F	C9AZ-2884-A
68	8 cylinder 289, 302	3B	C9AA-2981-F	C9AZ-2884-A
68	8 cylinder 289, 302	5B	C7AA-2981-G	C9AZ-2884-H
68	8 cylinder 390	6B	C7AA-2981-H,	C9AZ-2884-B
68	8 cylinder 390	6B	C7SA-2981-A,	C9AZ-2884-B
68	8 cylinder 390	6B	C9AA-2981-G	C9AZ-2884-B
68	8 cylinder 390, 428	5B	C7AA-2981-J	C8AZ-2884-F
69	6 cylinder 250	5C	C9OA-2981-E	C9OZ-2884-B
69	8 cylinder 302, 351	3B	C7AA-2981-F,	C9AZ-2884-A
69	8 cylinder 302, 351	3B	C9AA-2981-F	C9AZ-2884-A
69	8 cylinder 302, 351	5B	C9AA-2981-D, D1, D2	C9AZ-2884-D
69		6B	C7AA-2981-H,	C9AZ-2884-B
69	8 cylinder 390, 428	6B	C7SA-2981-A,	C9AZ-2884-B
69	8 cylinder 390, 428	6B	C9AA-2981-G	C9AZ-2884-B
69	8 cylinder 390, 428	5B	C7AA-2981-J	C8AZ-2884-F
70	6 cylinder	5C	C9OA-2981-E	C9OZ-2884-B
70	8 cyl. 302-2B, 351, 428	5B	C9AA-2981-D, D1, D2	C9AZ-2884-D
70	8 cylinder 351	3B	C7AA-2981-F	C9AZ-2884-A
70	8 cylinder 351	3B	C9AA-2981-F	C9AZ-2884-A

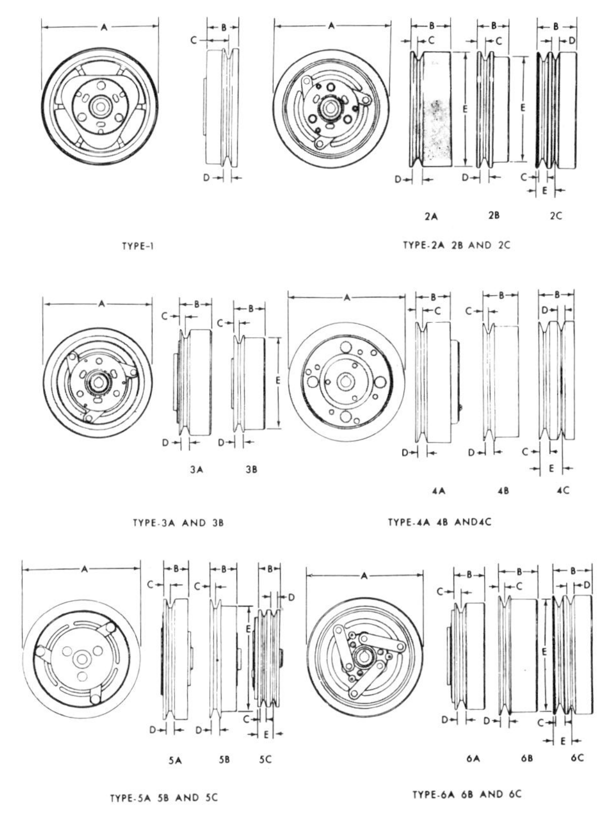

Mustang air-conditioner compressor clutch and pulley reference.

MUSTANG AIR CONDITIONER COMPRESSOR CLUTCH AND PULLEY I.D. CHART

PART NUMBER	TYPE	A	B	C	D	E	*	VENDOR	STAMPED NO.	REMARKS
C3AZ-2884-A	2B	6-3/4	1-7/8	7/16	1/2	6-1/16	1	Eaton	C3AA-2981-A,A1	Has integral field-use w/C8SZ-2979-A.
C3AZ-2884-A	6B	6-3/4	1-7/8	7/16	1/2	6-1/16	1	Eaton	C3AA-2981-A2	Has integral field-use w/C8SZ-2979-A.
C3AZ-2884-C	4A	6-7/8	1 7/16	7/16	1/2		1	Warner	C3AA-2091-B	Less field-use w/C3SZ-2987-A.
C3AZ-2884-C	5A	6-3/4	1-5/8	7/16	1/2		1	Pitts	C4OA-2981-A	Less field-use w/C3SZ-2987-A.
C5DZ-2884-A	4C	6-7/8	2	1/2	3/8	1	2	Warner	C4DA-2981-C	Less field-use w/C3SZ-2987-A.
C6AZ-2884-A	5B	6-11/16	2-1/8	2/8	1/2	6-3/8	1	Pitts	C4AA-2981-A	Less field-use w/C3SZ-2987-A.
C7AZ-2884-B	2C	6-1/4	1-1/2	1/2	1/2	1-3/16	2	Eaton	C7AA-2981-B	Has integral field-use w/C8SZ-2979-A.
C7AZ-2884-C	2B	6-3/4	1 3/4	3/8	1/2	5-5/16	1	Eaton	C7AA-2981-C	Has integral field-use w/C8SZ-2979-A.
C7AZ-2884-D	5B	6-11/16	1-5/8	3/8	1/2	6	1	Pitts	C7SA-2981-B	Less field-use w/C3SZ-2987-A.
C7OZ-2884-B	6C	7	1-1/2	1/2	3/8	1-1/16	2	Eaton	C7OA-2981-B	Has integral field-use w/C8SZ-2979-A.
C7ZZ-2884-B	4C	6-7/8	2-1/16	1/2	3/8	1-1/16	2	Warner	C7ZA-2981-C	Less field-use w/C7AZ-2987-A.
C8AZ-2884-F	5B	6-11/16	1-5/8	3/8	1/2	6	1	Pitts	C7AA-2981-J	Less field-use w/C8AZ-2987-C.
C8OZ-2884-A	5C	6-3/4	1-7/16	3/8	3/8	1-1/8	2	Pitts	C8OA-2981-A	Less field-use w/C8AZ-2987-C.
C8OZ-2884-B	2B	6-3/4	2-1/4	3/8	1/2	6-3/8	1	Eaton	C2AA-2981-A,A1	Used on 1967-has integral field-use with C8SZ-2979-A.
	6B	6-3/4	1-7/8	7/16	1/2	6-1/16	1	Eaton	C3AA-2981-A2	Has integral field-use w/C8SZ-2979-A.
	6B	6-3/4	2-1/4	3/8	1/2	6-3/8	1	Eaton	C7OA-2981-F	Has integral field-use w/C8SZ-2979-A.
C8OZ-2884-G	2C	6-3/4	2-5/16	1/2	3/8	1-1/16	2	Eaton	C4DA-2981-B,B1	Has integral field-use w/C8SZ-2979-A.
	4A	6-7/8	2-1/8	3/8	1/2		1	Warner	C2AA-2981-C	Has wire and field assembly.
	6C	6-3/4	2-5/16	1/2	3/8	1-1/16	2	Eaton	C4DA-2981-B2	Has integral field-use w/C8SZ-2979-A.
C9AZ-2884-A	3B	6-5/16	2-1/16	3/8	1/2	6	1	Eaton	C7AA-2981-F	Has integral field-use w/C8SZ-2979-A.
									C9AA-2981-F	
C9AZ-2884-B	6B	6-3/4	2	5/8	1/2	6	1	Eaton	C7AA-2981-H	Has integral field-use w/C8SZ-2979-A.
	6B	6-3/4	2-1/16	3/8	1/2	6	1	Eaton	C7SA-2981-A	Has integral field-use w/C8SZ-2979-A.
	6B	6-3/4	2-1/16	5/8	1/2	6	1	Eaton	C9AA-2981-F	Has integral field-use w/C8SZ-2979-A.
C9AZ-2884-D	5B	6-3/8	1-11/16	3/8	1/2	5-3/16	1	Pitts	C9AA-2981-D	Less field-use w/C8AZ-2987-D.
	5B	6-5/16	1-1/2	3/8	1/2	5-3/16	1	GPD	C9AA-2981-D1	
	5B	6-5/16	1-1/2	3/8	1/2	5-3/16	1	Pitts	C9AA-2981-D2	
C9AZ-2884-H	5B	6-1/4	5/8	5/16	1/2	6	1	Pitts	D2AA-2981-G	Less field-use w/C8AZ-2987-D.
C9OZ-2884-B	5C	6-1/4	1-7/16	3/8	3/8	1	2	Pitts	C9OA-2981-E	Less field-use w/C9AZ-2987-A.

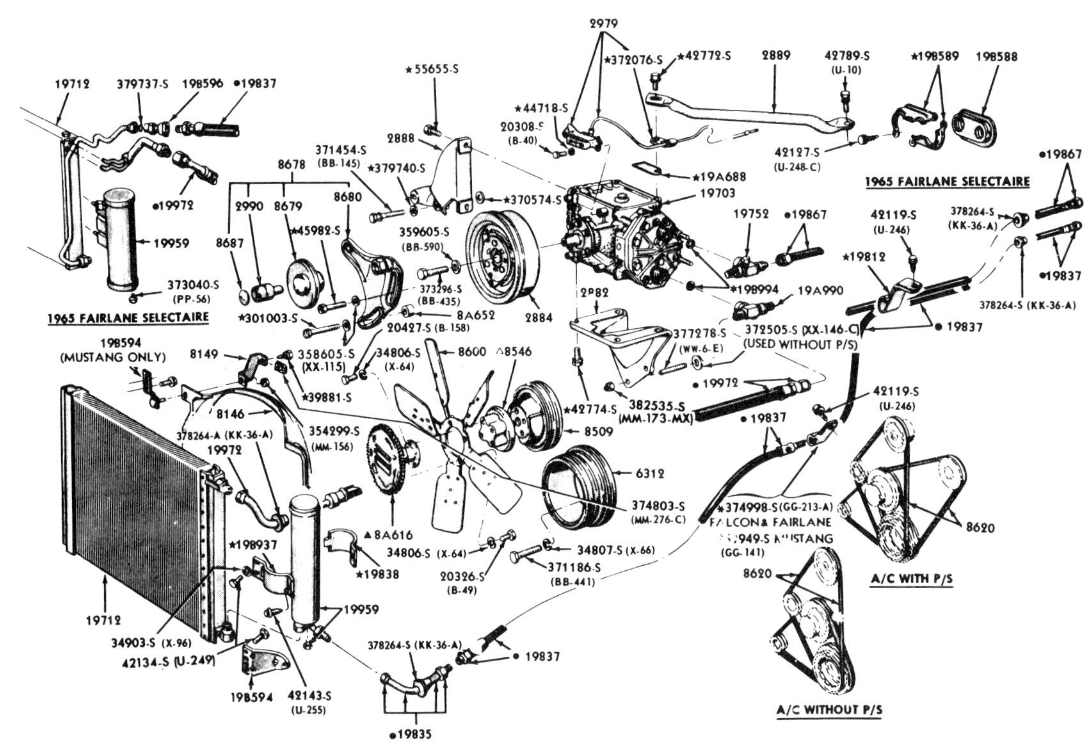

Air-conditioning compressor, condenser, brackets, hoses, and related hardware for 1965-1966 Mustangs equipped with eight-cylinder engines (with alternator).

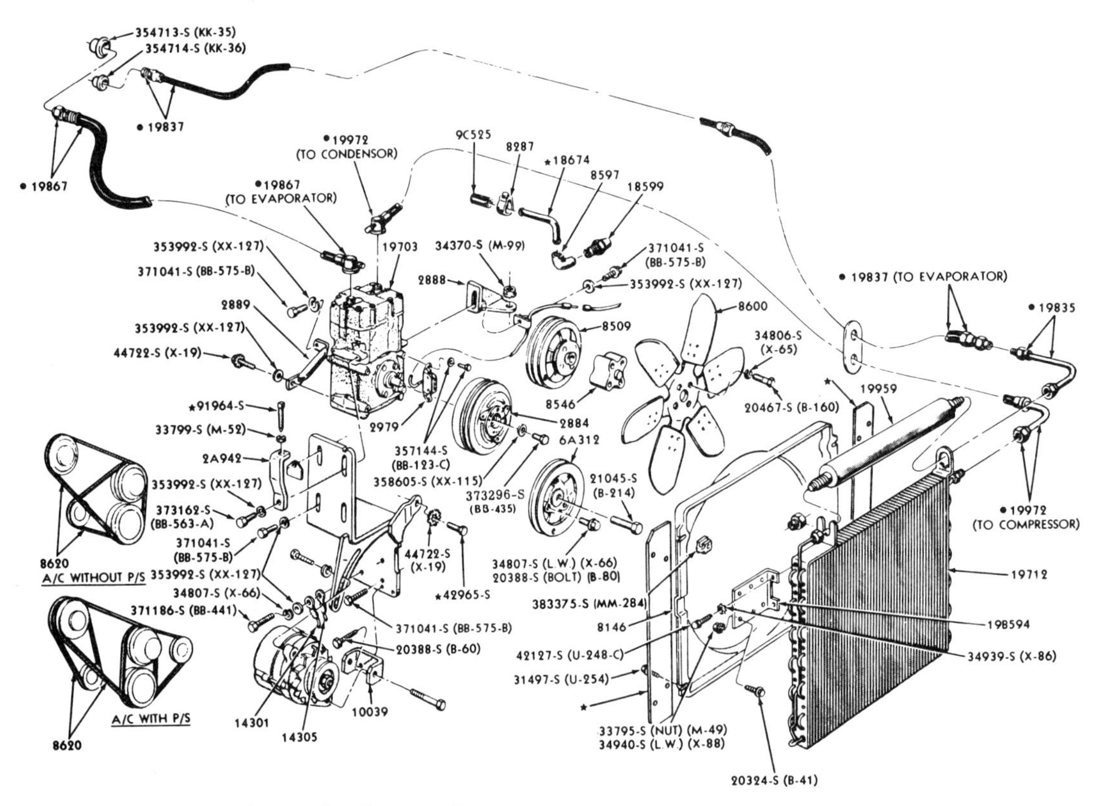

Air-conditioner compressor, condenser, brackets, and related hardware for hang-on system installed on 1967 Mustangs equipped with six-cylinder engines.

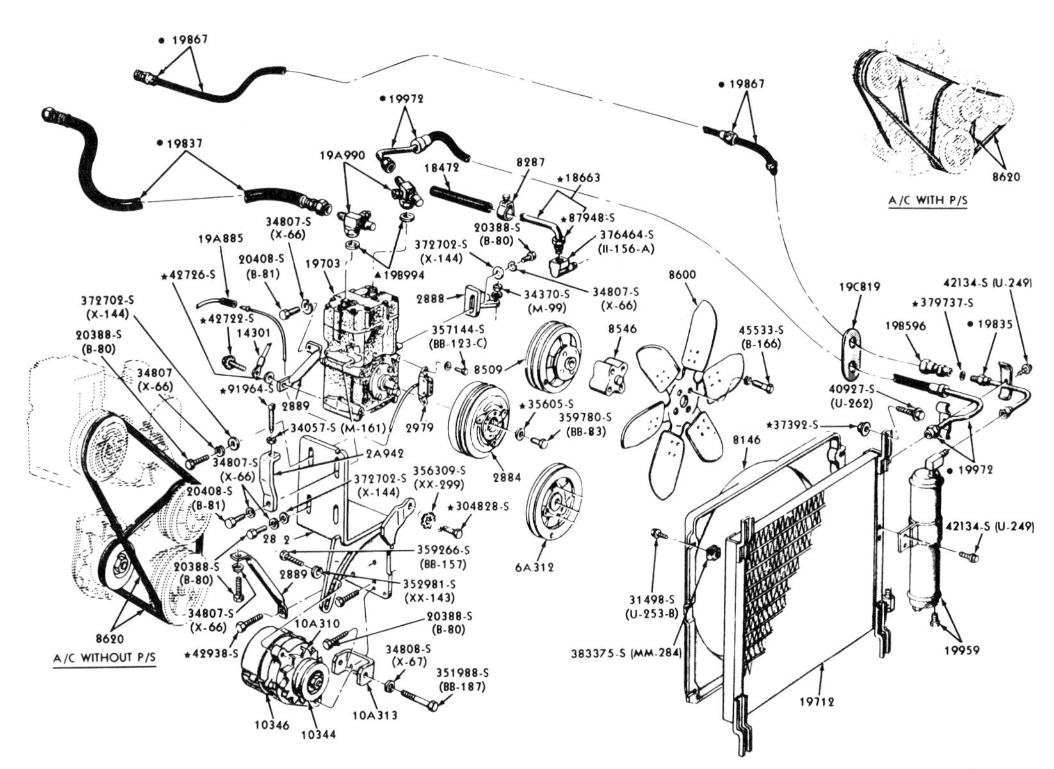

Air-conditioning compressor, condenser, brackets, hoses, and related hardware for 1967 Mustangs equipped with six-cylinder engines and factory-installed integral air conditioning.

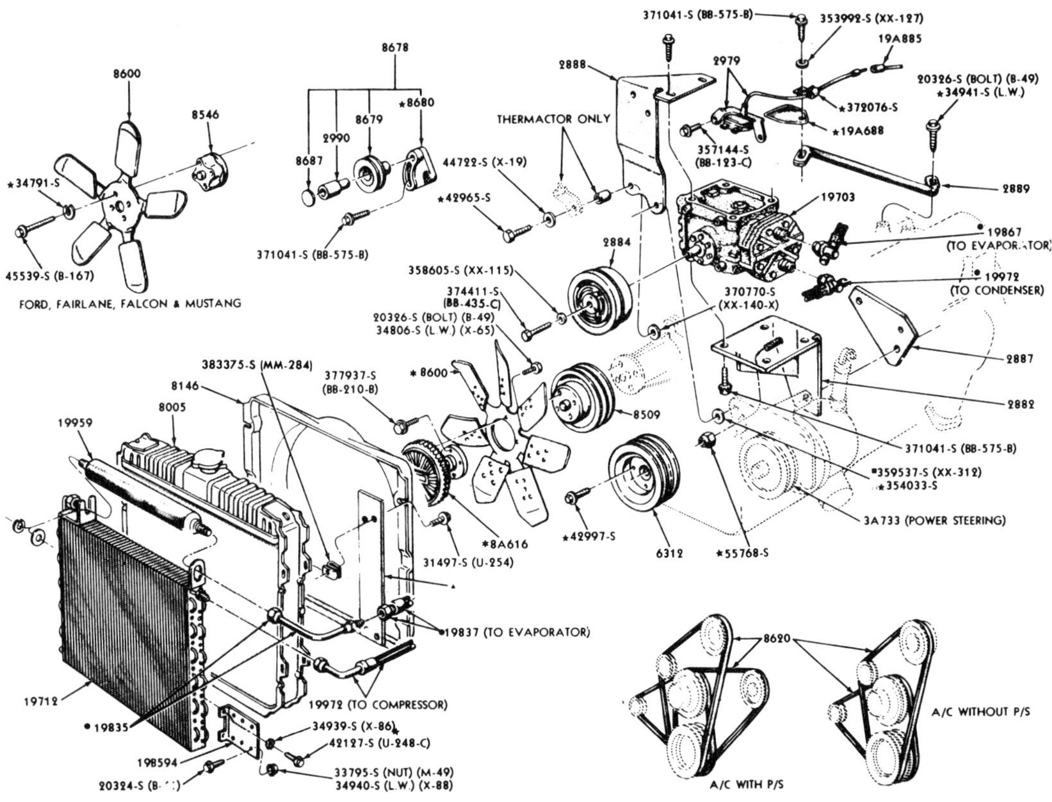

Air-conditioning compressor, condenser, brackets, hoses, and related hardware for 1967 Mustangs equipped with a 298ci engine and hang-on or dealer-installed air conditioning.

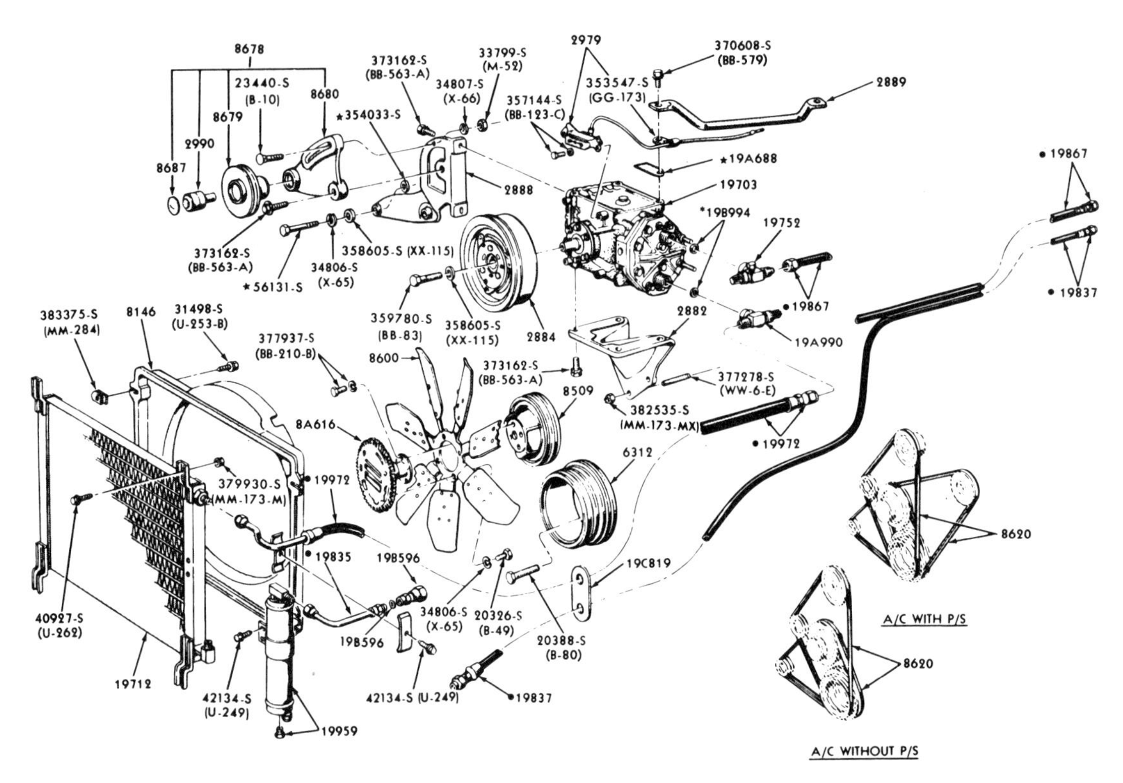

Air-conditioning compressor, condenser, brackets, hoses, and related hardware for 1967 Mustangs equipped with 289ci engines and factory-installed integral air conditioning.

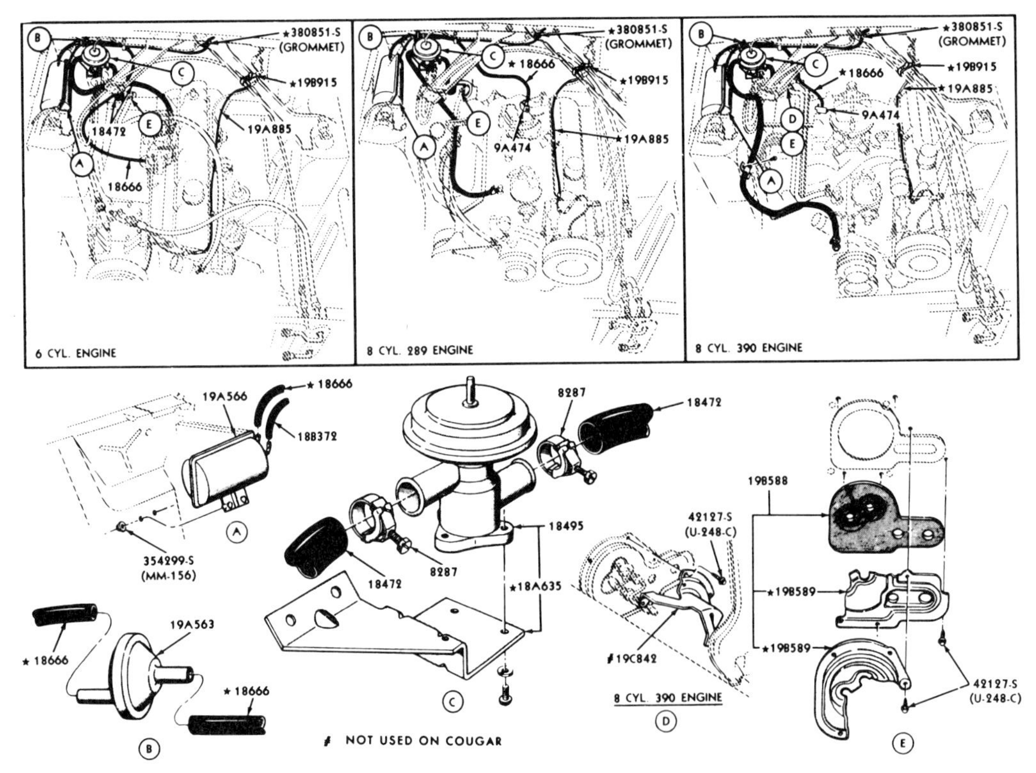

Vacuum reservoir, hose connections, vacuum-operated heater control valve, and heater hoses for 1967 Mustangs equipped with factory-installed integral air conditioning.

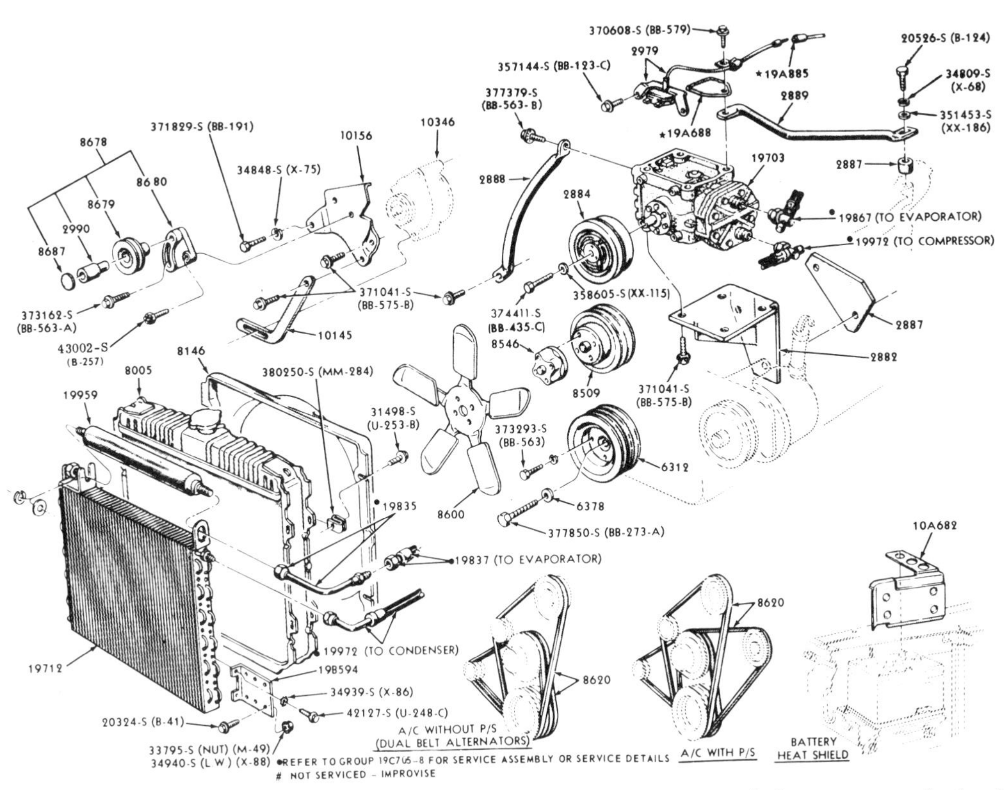

Air-conditioning compressor, condenser, brackets, hoses, and related hardware for 1967 Mustangs equipped with a 390ci engine and hang-on or dealer-installed air conditioning.

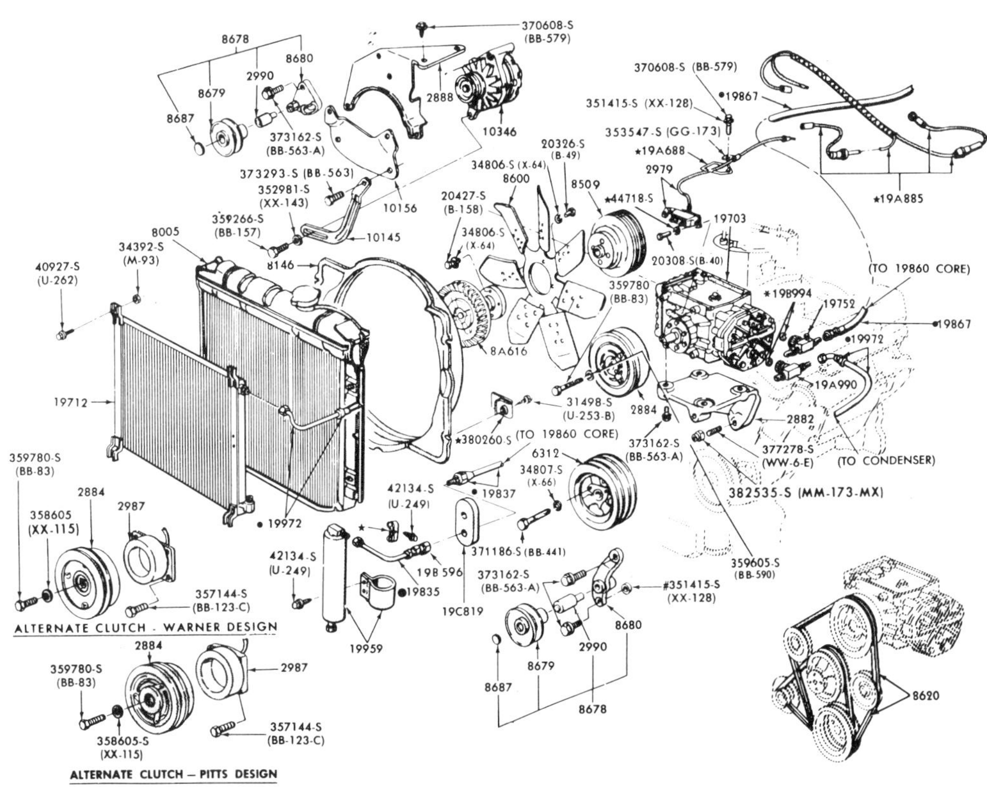

Air-conditioning compressor, condenser, brackets, hoses, and related hardware for 1967-1970 Mustangs equipped with a 390 or a 428ci engine and factory-installed integral air conditioning.

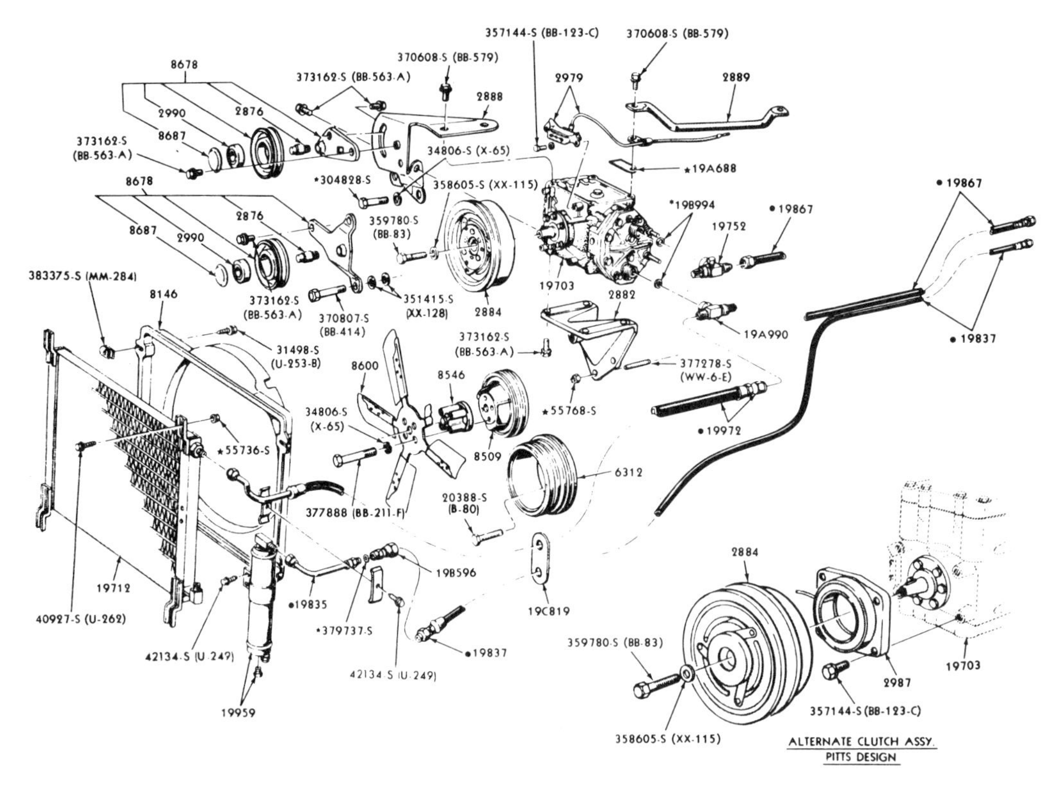

Air-conditioning compressor, condenser, brackets, hoses, and related hardware for 1967-1970 Mustangs equipped with a 302 or 351ci engine and factory-installed integral air conditioning.

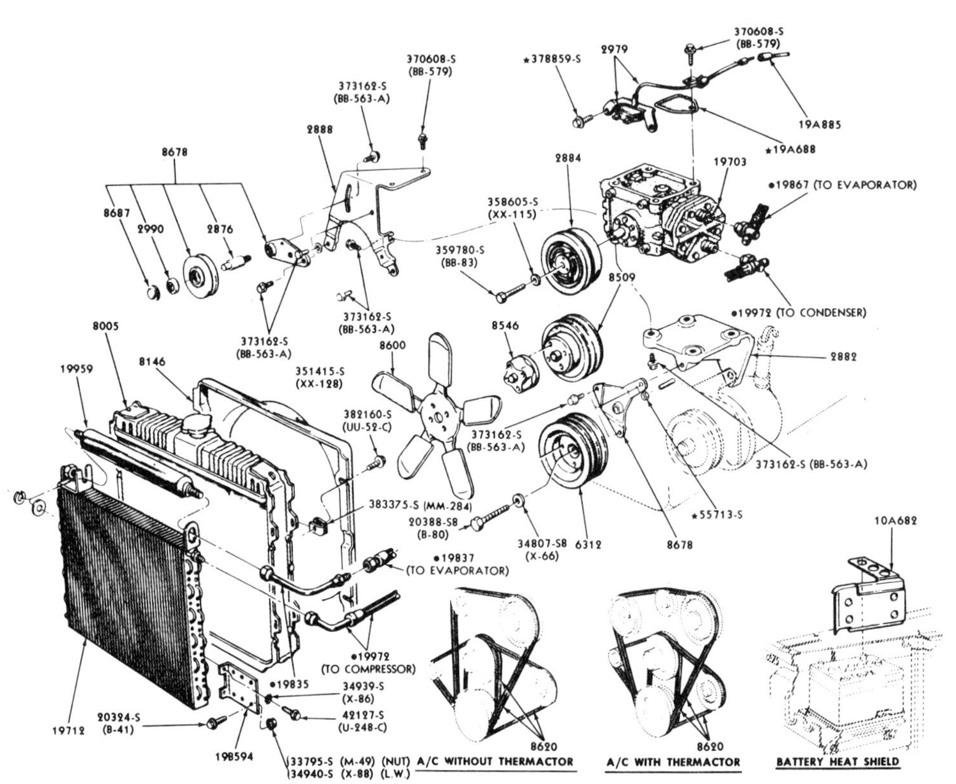

Air-conditioning compressor, condenser, brackets, hoses, and related hardware for 1968 Mustangs equipped with a 390ci engine and hang-on or dealer-installed air conditioning.

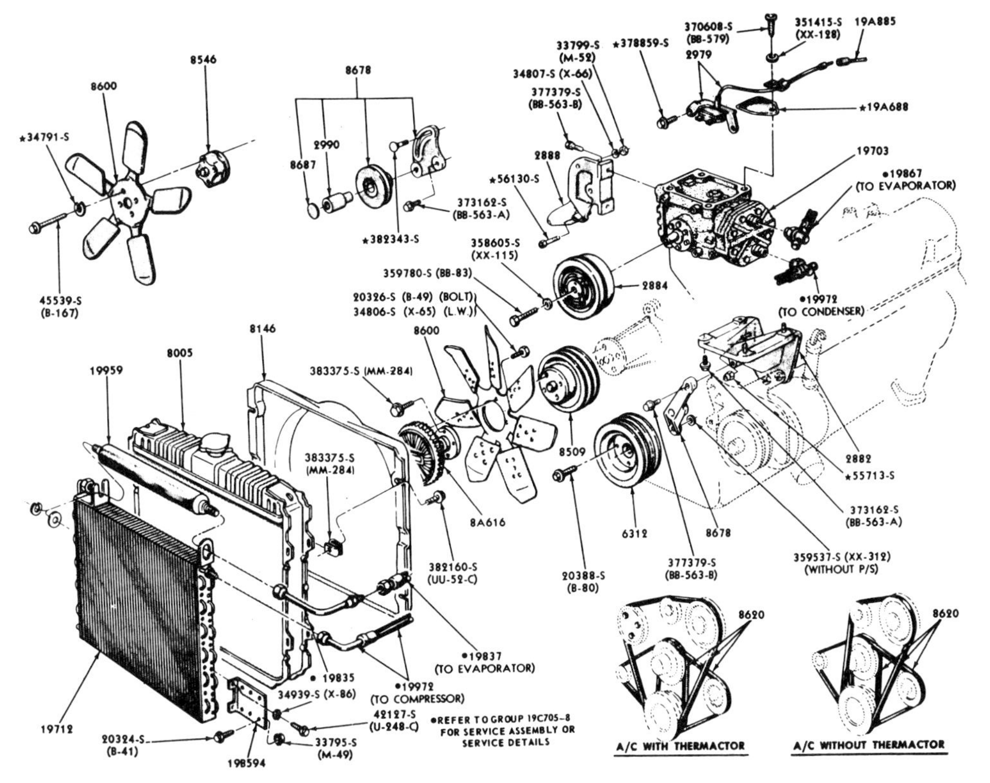

Air-conditioning compressor, condenser, brackets, hoses, and related hardware for 1968 Mustangs equipped with a 289 or a 302ci engine and hang-on or dealer-installed air conditioning.

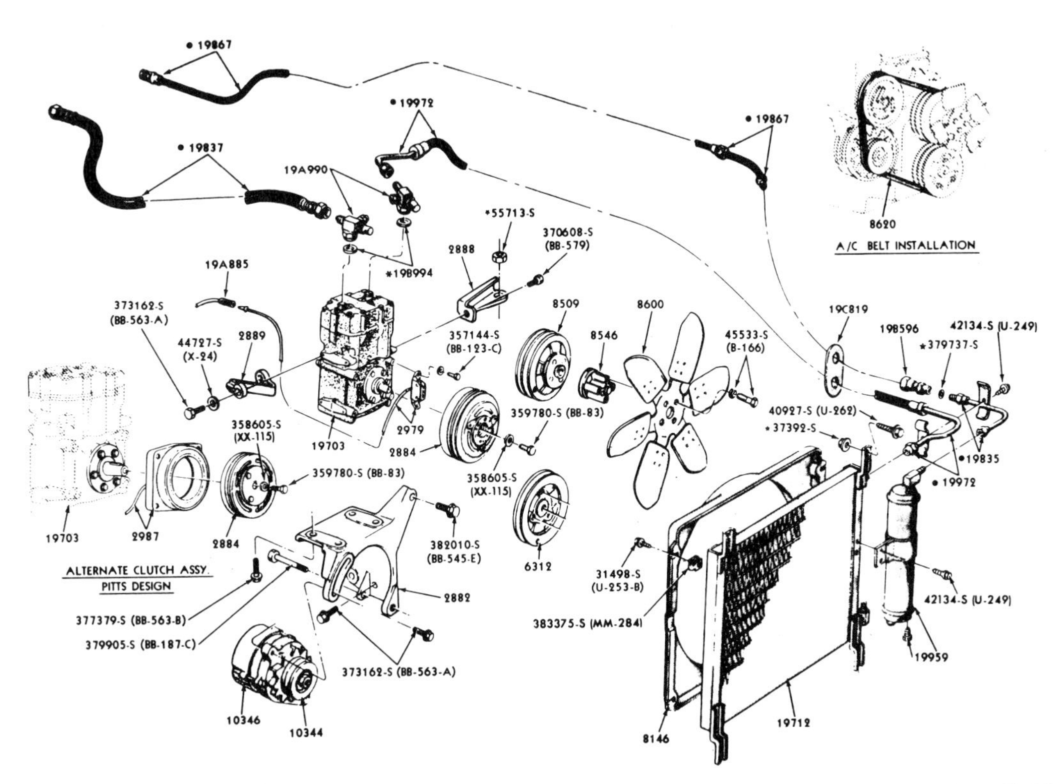

Air-conditioning compressor, condenser, brackets, hoses, and related hardware for 1968-1970 Mustangs equipped with six-cylinder engines and factory-installed integral air conditioning.

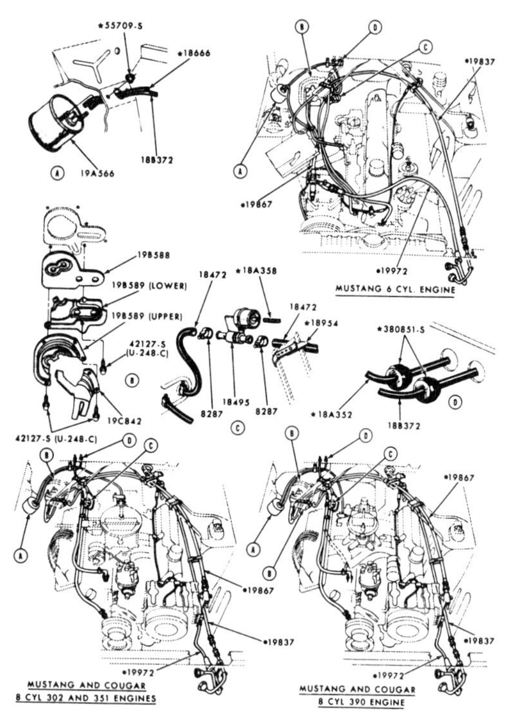

Vacuum reservoir, hose connections, vacuum-operated heater control valve, and heater hoses for 1969-1970 Mustangs equipped with factory-installed integral air conditioning.

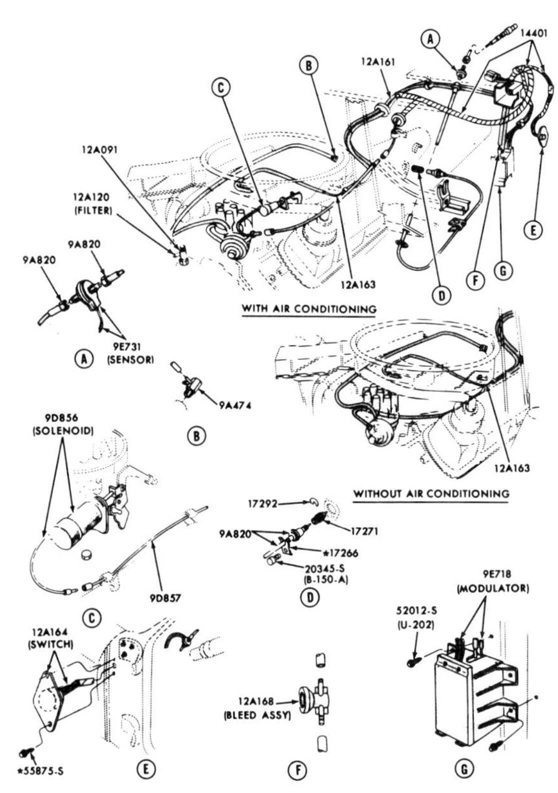

Distributor modulator for 1970 Mustang equipped with a 351ci engine and Ram Air.

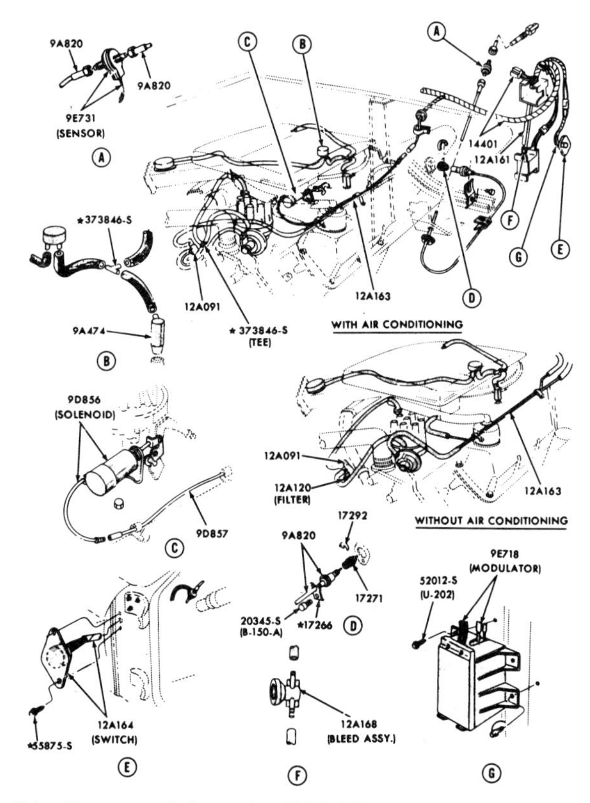

Distributor modulator for 1970 Mustang equipped with a 302ci engine.

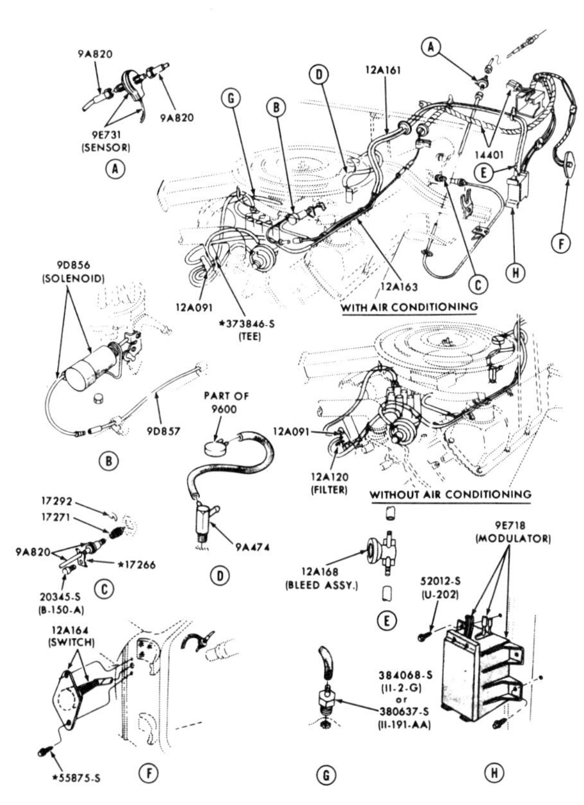

Distributor modulator for 1970 Mustang equipped with a 351ci engine without Ram Air option.

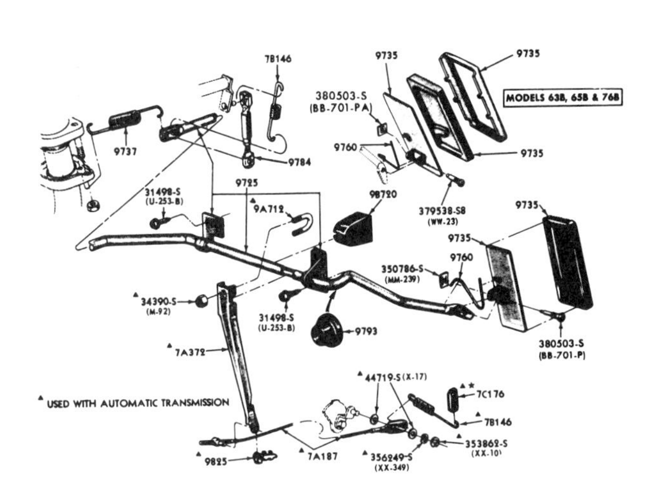

Accelerator linkage for 1965-1968 Mustangs equipped with six-cylinder engines.

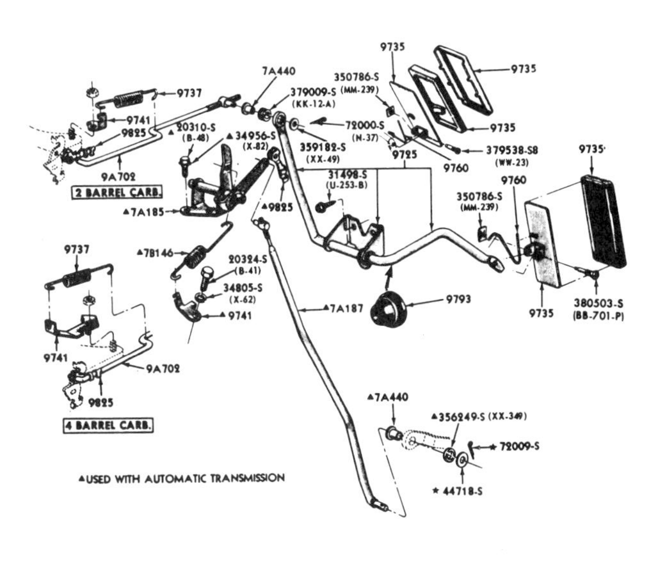

Accelerator linkage for 1965 Mustangs equipped with V-8 engines.

Few sales promotions included or bothered with a particular engine or drivetrain combination. The Sports Sprint package introduced in the spring of 1966 was a cosmetic option designed around the base Mustang hardtop. Besides numerous exterior appearance upgrades, the base 200ci six-cylinder also received a few dress-up items. A chrome air cleaner lid replaced the stock Corporate Blue version. A special decal proclaimed the cubic inch displacement and identified the package as a Sprint. The oil filler cap, radiator cap, and master cylinder cap were also chrome. Internally the engine and drivetrain were identical to the non-Sprint regular production models, though to this day rumors of special cam grinds, unique cylinder heads, and higher output persist.

MUSTANG ACCELERATOR PEDAL to CARB THROTTLE CABLE ASSEMBLY CHART

1969 and 1970

Year	Engine	Notes	Description	Part Number
69	All 8s		22.71" long	C9ZZ-9A758-A
69	All 6s		30.50" long	C9ZZ-9A758-C
70	All 8s	Except Boss 429	22.83" long	C9ZZ-9A758-E
70	All 6s		30.00" long	C9ZZ-9A758-D

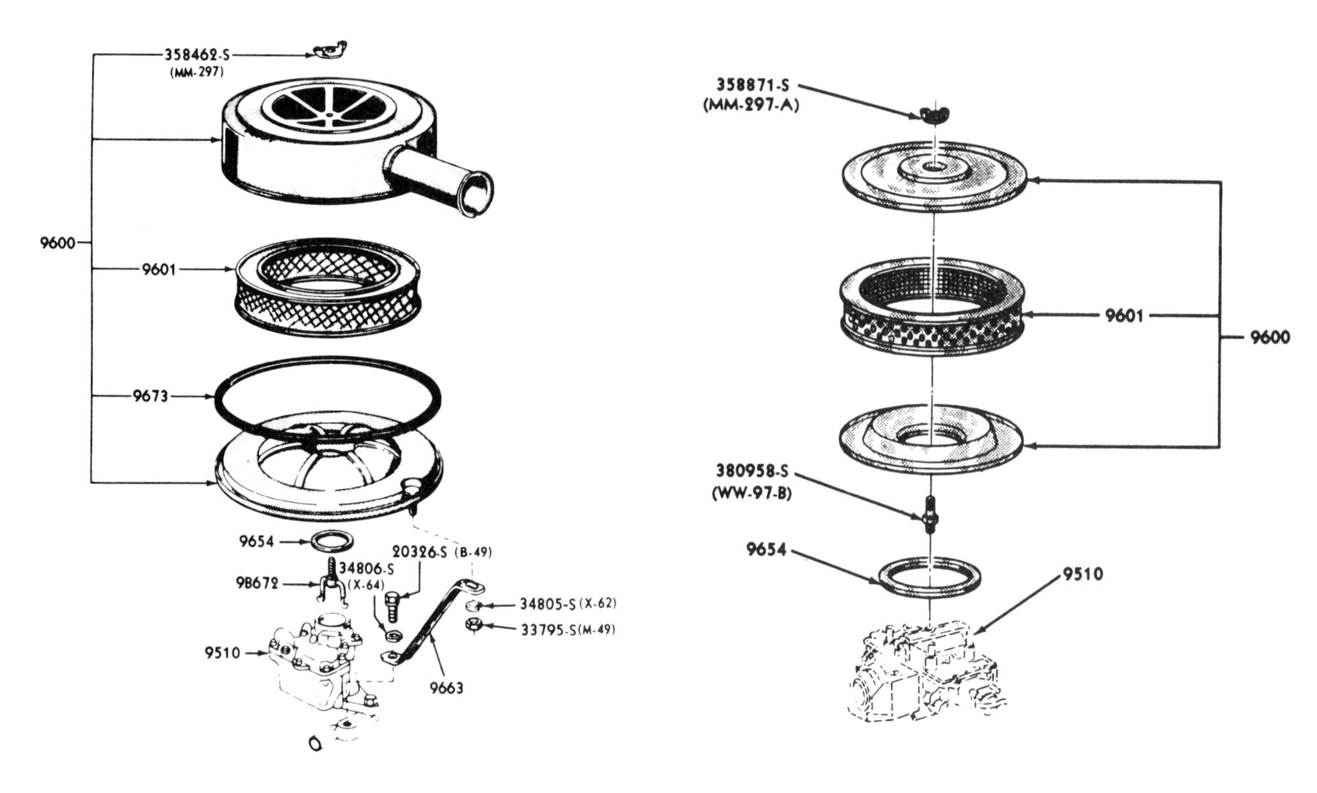

Air cleaner assembly for 1965-1967 Mustangs equipped with a six-cylinder engine.

Air cleaner for 1965-1967 Mustangs equipped with 289 High Performance (HiPo) engine.

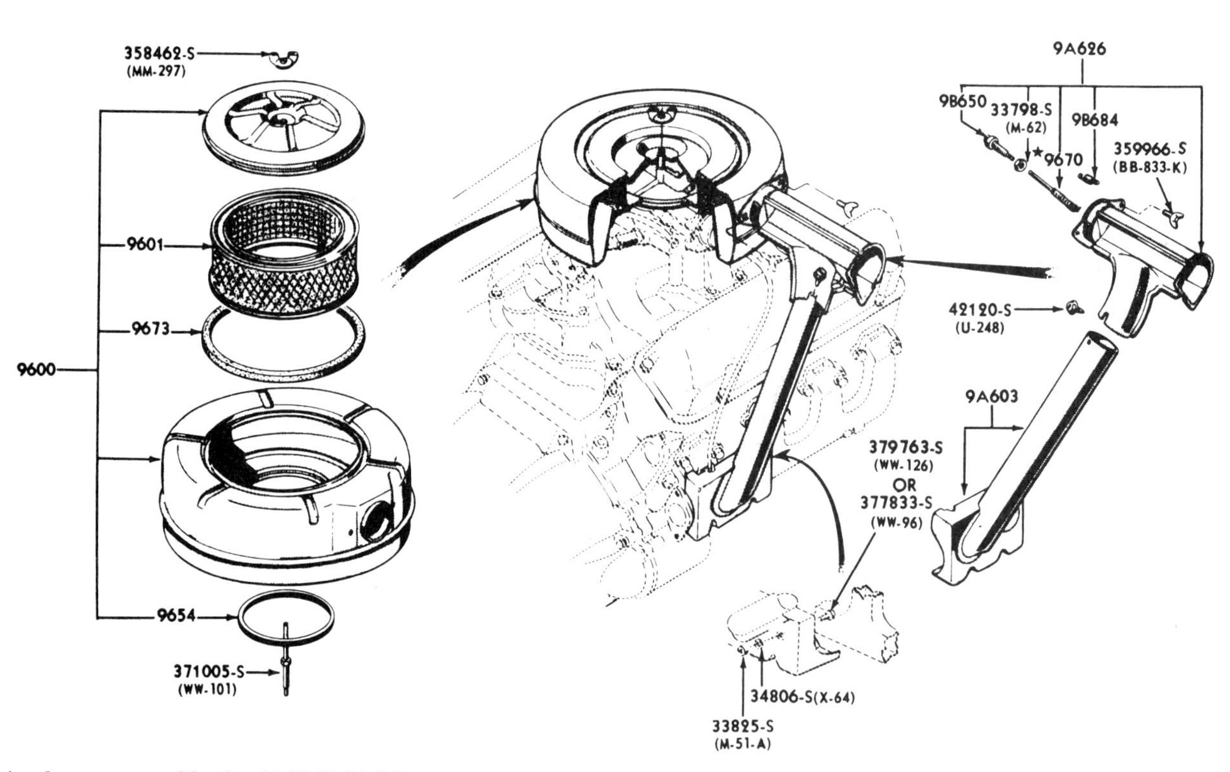

Air cleaner assembly for 1965-1967 Mustangs equipped with a V-8 engine (except 289 HiPo).

MUSTANG CARBURETOR AIR CLEANER ELEMENT APPLICATION CHART

Year	Engine	Model	Description	Part Number
68	289	Except Hi-Po	11.32" I.D.; 13.30" O.D., 2.80" high. Used with Air Cleaner Assemblies C8AF-D, C8AZ-A,D;C8OZ-A,B;C8SZ-A,B;D0OZ-A;	C8AZ-9601-A
68/70	302			
68	390	w/ 2V carb.		
69,70	351	w/o Ram Air		
65/67	260,289	Except Hi-Po	7.5" I.D.; 10.04" O.D., 2.38" high. Used with Air Cleaner Assembly C5ZZ-C,D	C5ZZ-9601-B
68	289	2V, open emission		
68	302	4V, open emission		
67	390		11.32" I.D.; 13.30" O.D., 2.22" high. Used with Air Cleaner Assemblies C7SZ-A, C7ZZ-F;C5ZZ-W;C8OZ-C;C8ZZ-B;C9OZ-E,H; C9ZZ-B,F,H,J;D0ZZ-C,F,G,N;	B7SZ-9601-A
68	390	4V & Thermactor		
69	390	4V, Exc. GT & Ram Air		
68,69	428CJ	w/ 4V carb.		
69	390	GT w/ 4V carb.		
69,70	302	Boss 302 w/ 4V		
69	351,390	w/ Ram Air		
69,70	429	Boss 429 w/ 4V		
65/67	289	High Performance		
68	289	Canada only.		
70	351	w/ 2V, Ram Air		
70	428CJ			
70	302	Boss 302 w/ Ram Air		
70	351	4V, Ram Air	11.32" I.D.; 13.30" O.D., 2.50" high. Used with Air Cleaner Assembly D0OZ-C	D0GY-9601-A
67	428	GT500	w/ 4V or 8V Carb.	EDJ-9601-A
68	302,428	GT350/500 Exc. KR		
65 65/67	170 170,200		7.96" I.D.; 9.90" O.D., 2.26" high. Used with Air Cleaner Assemblies C3DZ-C, C6ZZ-M;C7ZZ-A	C1KE-9601-A
69,70	250		9.86" I.D.; 11.75" O.D., 2.00" high. Used with Air Cleaner Assemblies C9ZZ-A; D0DZ-A	C9ZZ-9601-A
70	200		8.94" I.D.; 10.90" O.D., 2.12" high. Used with Air Cleaner Assembly D0DZ-C	C8TZ-9601-A
67/69	200	w/ Imco & C4	8.35" I.D.; 10.28" O.D., 2.25" high. Used with Air Cleaner Assemblies C7DZ-A; C8ZZ-A;C7ZZ-B	C8ZZ-9601-A
67,68	289,302	GT350 w/ P/S	w/ Supercharger, 9" cylindrical	S2MS-9601-A
67,68	289,302	GT350 w/o P/S	w/ Supercharger, 9" cylindrical	S7MS-9601-H

MUSTANG CARBURETOR AIR CLEANER ASSEMBLY APPLICATION CHART, Part One.

Year	Engine	Model	Description	Part Number
65/67	170,200	Except Imco with C4.	Color "Blue."	C3DZ-9600-C
66	200		Chrome.	C6ZZ-9600-M
67	200	With Imco.	12.25" dia., 2.52" high, color "Blue" tray, and Chrome cover; use w/ Carter carb.	C7ZZ-9600-B
67	200	With Imco & C4.	12.25" dia., 2.52" high, color "Blue."	C7DZ-9600-A
67	200	With Thermactor	12.25" dia., 2.52" high, color "Blue" tray, and Chrome cover; use w/ Ford carb.	C7ZZ-9600-A
68	200	w/ Ford carb; US only	Dry type, 12.4" dia., 3.48" high	C8ZZ-9600-A
69	200	w/ Ford carburetor	Dry type, 12.4" dia., 3.48" high	C8ZZ-9600-A
68	200	Open emission	Canada only.	C6DF-9600-F
69	250		14.62" diameter; 3.25" high.	C9ZZ-9600-A
70	200		13.47" diameter; 3.63" high.	D0DZ-9600-C
70	250		14.62" diameter; 3.24" high.	D0DZ-9600-A
65/67	289	High Performance	14.05" diam., 2.61" high	C5ZZ-9600-W
68	289	w/ 4V, open emission.		
66/70	390	with 4V carburetor		
65/67	260,289		17.25 diam., 4.22" high, color "Blue" tray and Chrome cover.	C5ZZ-9600-D
68	289	w/ 2V, open emission.		
68	302	w/ 4V, open emission.		
67	289	Exc. High Peformance	17.25 diam., 4.22" high, color "Blue" tray and Chrome cover.	C5ZZ-9600-C
67	390	Except GT	16.79 diam., 3.66" high, color "Blue" tray and Chrome cover.	C7SZ-9600-A
67	390	GT	16.79 diam., 3.66" high.	C7ZZ-9600-F
68	390	GT, 4V, open emission		
67	428	w/ 8V carb.		S7MS-9600-A
68	302,428	w/ 8V carb.		S7MS-9600-A
67,68	289,302	w/ Supercharger	Cylindrical, 9" long.	S7MS-9600-H
68	390	GT w/ 4V	17" diam., 4.20" high	C8OZ-9600-B
68	390	Exc. GT w/ 4V	17" diam., 4.20" high	C8OZ-9600-A
68	302		17.84" diam., 4.20" high	C8AZ-9600-D
	351			
	289	Exc. High Peformance		
	390	w/ 2V carb.		
69,70	351,428	GT350/500	18" diam., 4.98" high	C9ZZ-9600-B
68	428CJ	w/ 4V & Ram Air		C8ZZ-9600-B
68,69	428CJ*	w/ 4V, w/o Ram Air	16.79" diam., 3.66" high	C8OZ-9600-C
69	390	GT w/ 4V		
69	302	Boss 302		
69	428CJ#	w/ 4V, w/o Ram Air	16.79" diam., 3.66" high, Chrome cover.	C9OZ-9600-H
69	351	w/ Ram Air	18.83 diam., 6.14" high, w/ door in cover	C9ZZ-9600-H
69	390	w/o Ram Air, Exc. GT	(Exc. GT) 16.79" diam., 3.66" high	C9OZ-9600-E
69	390	GT, 4V, Ram Air	18.83 diam., 5.82" high, w/ door in cover	C9ZZ-9600-J
69	428CJ*	w/ 4V & Ram Air	18.83 diam., 5.50" high	D0ZZ-9600-F
70	428CJ	w/ Ram Air		
69	428CJ#	w/ 4V & Ram Air	18.83 diam., 5.50" high, "Cobra Jet"	C9ZZ-9600-F
70	302	Boss 302 w/o Ram Air	16.79" diam., 3.66" high.	D0ZZ-9600-C
70	302	Boss 302 w/ Ram Air	18.83 diameter	D0ZZ-9600-N
70	351C	2V,4V,w/o Ram Air	17" diam., 4.20" high	D0ZZ-9600-M
70	302	2V, w/o Ram Air	17.84" diam., 4.20" high	D0ZZ-9600-B
70	351W			
70	351W	2V, w/ Ram Air	18.83 diam., 6.14" high	D0ZZ-9600-G
70	351C	2V,4V,w/ Ram Air	18.88 diam., 6.25" high	D0ZZ-9600-S
70	390	4V,w/o Ram Air	16.79" diam., 3.66" high.	D0ZZ-9600-D

Notes:

* Before 2/17/69

\# From 2/17/69

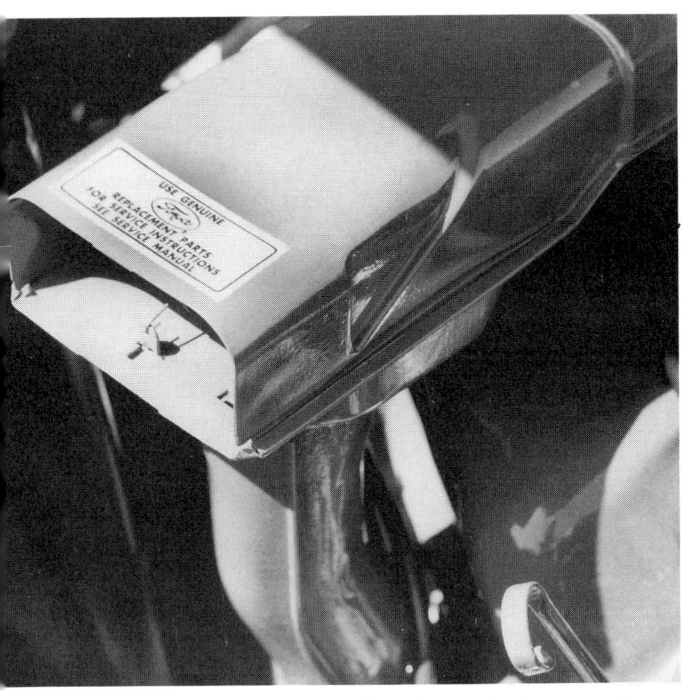

Because cars were being built as fast as the plants could turn them out, Ford became inconsistent with their assembly procedures. This 1967 302ci engine sports a decal promoting Genuine Ford replacement parts on its air cleaner snorkel, though this application was not universal with all cars and/or engines.

Specification and warning decals changed size, type, and location every year. Each engine option came with a different set of decals as well. This exhaust emissions decal, located on the air cleaner snorkel, lists tuning information for a 1970 Cobra Jet engine.

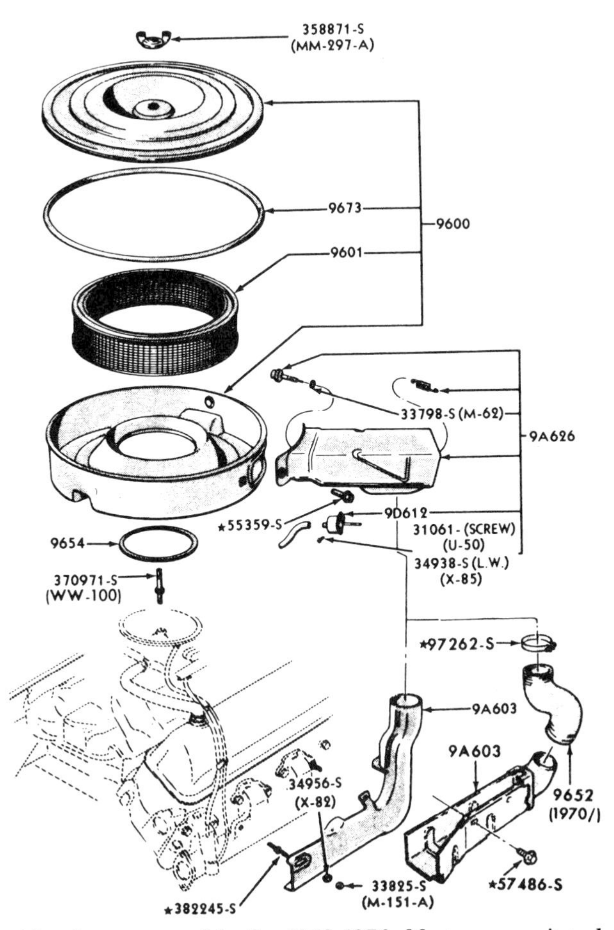

Air cleaner assembly for 1968-1970 Mustangs equipped with 289 or 302ci engines, and 1969 Mustangs equipped with 351ci engine.

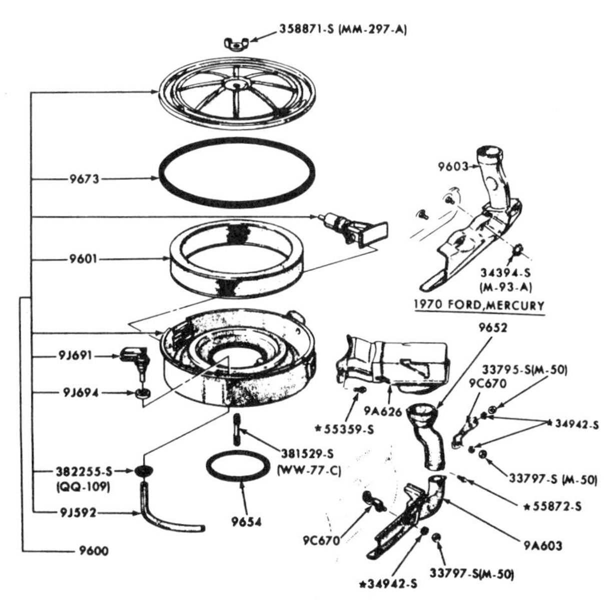

Air cleaner assembly (non-Ram Air) for 1969-1970 Mustangs equipped with 428ci engine.

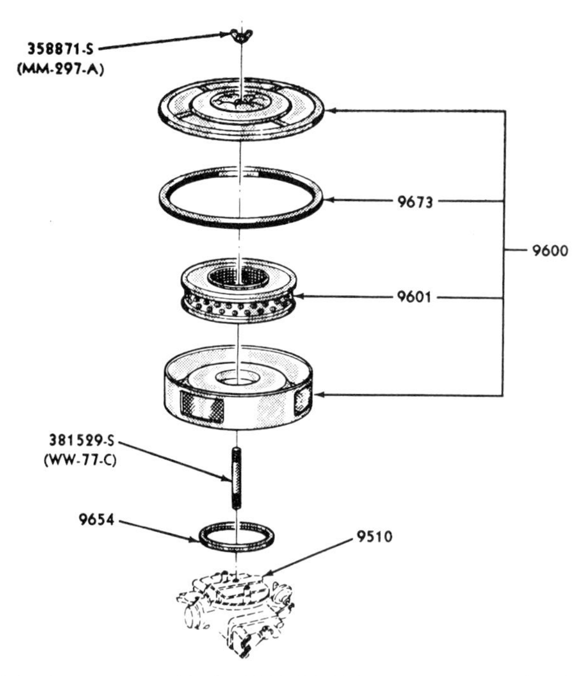

Air cleaner assembly for 1967 GT Mustangs equipped with 390ci engine.

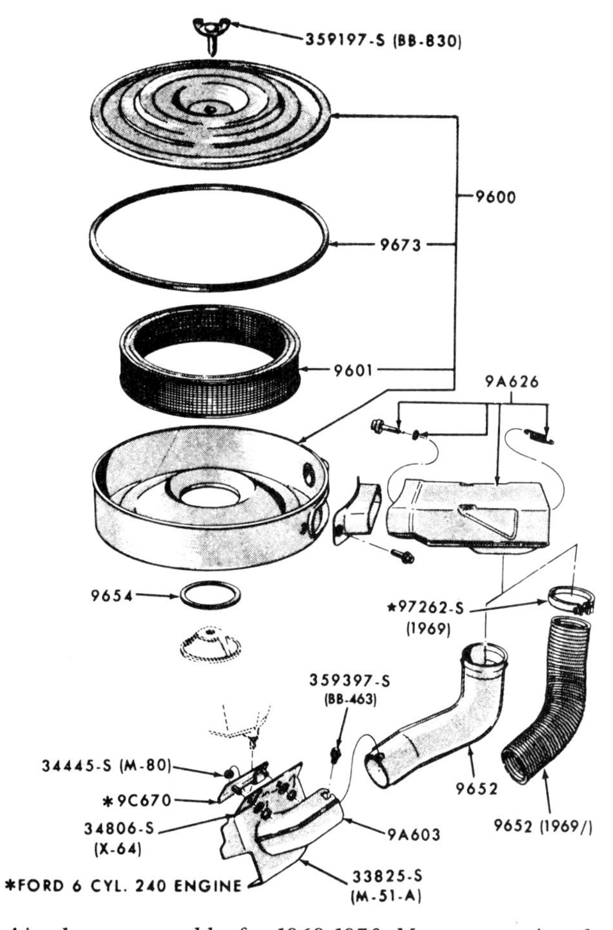

Air cleaner assembly for 1968-1970 Mustangs equipped with six-cylinder engines.

The Boss 429 air cleaner snorkel is unique because it required an extra-long extension to bridge the wide cylinder heads. Even the mighty Boss 429 had a hot air intake tube connecting the exhaust manifold stove to the air cleaner. The stove provided the carburetor a source of warm air to enhance cold-weather operation.

The Ram Air system in 1969 and 1970 consisted of a finned aluminum scoop integrated into the air cleaner lid. The functional scoop, which nearly doubled the height of the stock air cleaner, extended through a hole in the hood. Fresh air flowing over the hood was diverted to the air cleaner by the scoop. A vacuum-operated trap door located between the scoop and the air cleaner opened during wide-open-throttle operation and allowed fresh air to enter the engine. Otherwise the system was non-functional. It was commonly called a Shaker hood scoop because it vibrated or shook with the normal movement of the engine. Ram Air was available only with four-barrel engines.

Compare this snorkel attached to a 428ci air cleaner to the Boss 429 cleaner. Both are equipped with Ram Air and both were produced in 1970.

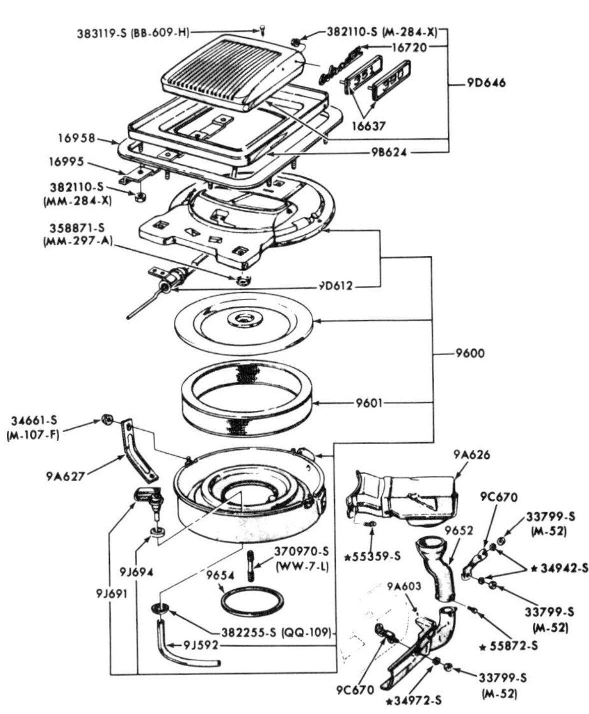

Ram Air (Shaker) scoop and air cleaner assembly for 1969 Mustangs equipped with 351, 390, and 428 engines, and 1970 Mustangs equipped with 428 engines.

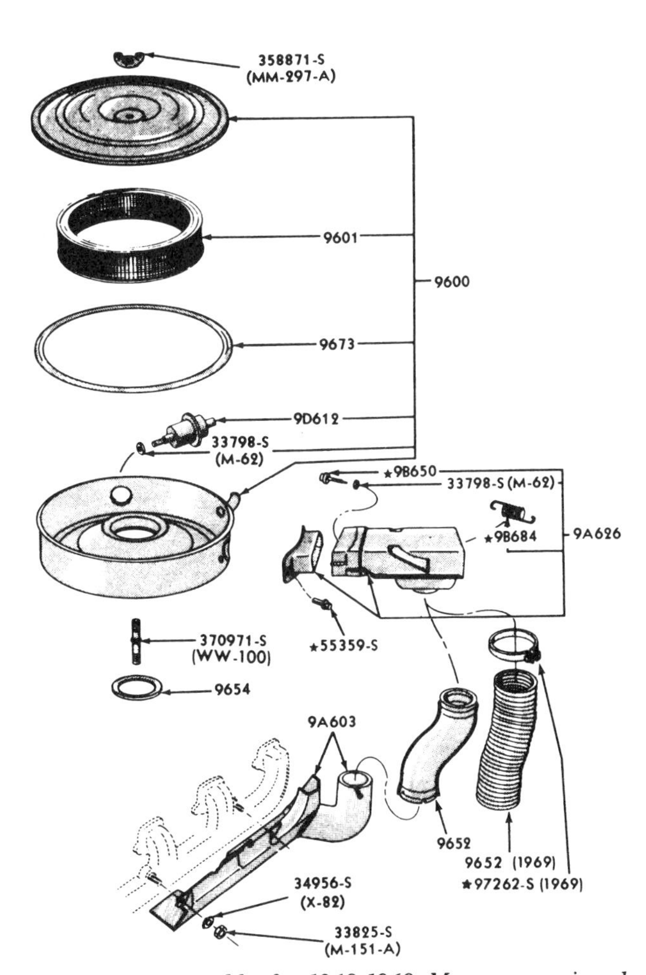

Air cleaner assembly for 1968-1969 Mustangs equipped with 390ci engines.

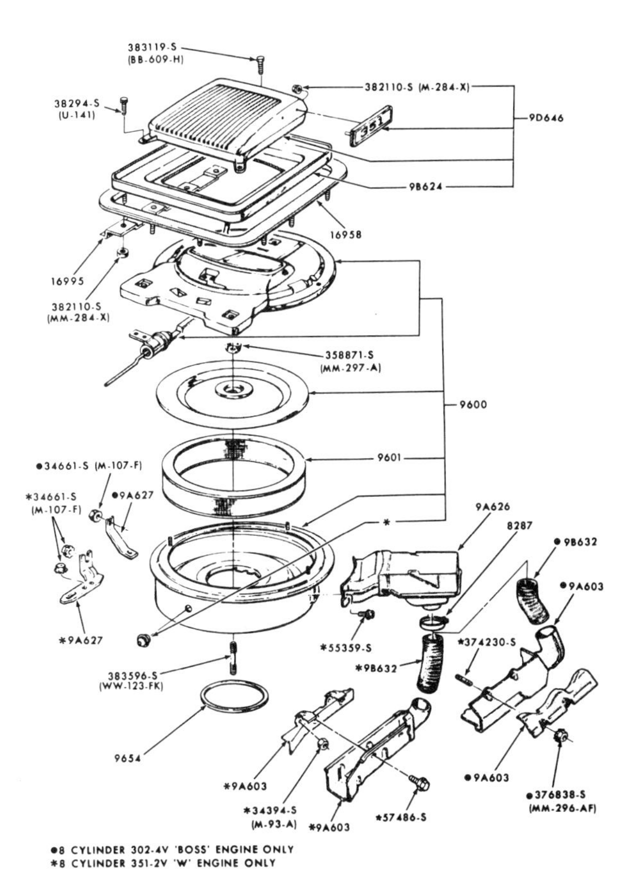

Ram Air (Shaker) scoop and air cleaner assembly for 1970 Mustangs equipped with 302-4V and 351 Windsor engines.

MUSTANG AIR CLEANER HOT AIR TUBE ASSEMBLY APPLICATION CHART

Year	Engine	Model	Description	Part Number
68	200		Steel, 2.51" diam. on (1) end & 2.74" dia.	C8ZZ-9652-A
69	200,250	before 10/15/68	Paper & Aluminum, 2.02" I.D., 18" long, cut.	C9UZ-9652-A
69	390	w/ 4V Carb.		
69,70	302	Boss		
70	429CJ			
70	200,250	All		
70	302,390	w/ 2V Carb.		
70	351W	w/ 2V Carb.		
70	302	Boss, w/ 4V		
70	351C			D0OZ-9652-A
68	390		Upper (steel) 2.03" dia. on (1) end &	C8AZ-9652-A
68,69	428CJ	w/ 4V Carb.	2.85" diameter on the other.	
69	390	w/ 4V & Ram Air	8.36" formed length.	
70	428CJ		Use w/o transistor ignition.	
70	351	W/ 2V & Ram Air		

MUSTANG CARBURETOR AIR CLEANER SCOOP & SEAL ASSEMBLY APPLICATION CHART

Year	Engine	Model	Description	Part Number
69/70	428CJ	w/ 4V & Ram Air	Marked "Cobra Jet"	C9ZZ-9D646-A
69	351,390	w/ Ram Air	Replace Cobra Jet name plates w/	C9ZZ-9D646-A
			(2) C9ZZ-16637-C for 351	
			(2) C9ZZ-16637-D for 390	
70	351	W/ 4V Carb.		C9ZZ-9D646-A
70	351	w/ 2V,4V & Ram Air	w/ White Hood Stripes	D0ZZ-9D646-A
70	302	Boss, w/ 4V	w/ White Hood Stripes	D0ZZ-9D646-A
70	351	w/ 2V,4V & Ram Air	w/ Black Hood Stripes	D0ZZ-9D646-B
70	302	Boss, w/ 4V	w/ Black Hood Stripes	D0ZZ-9D646-B

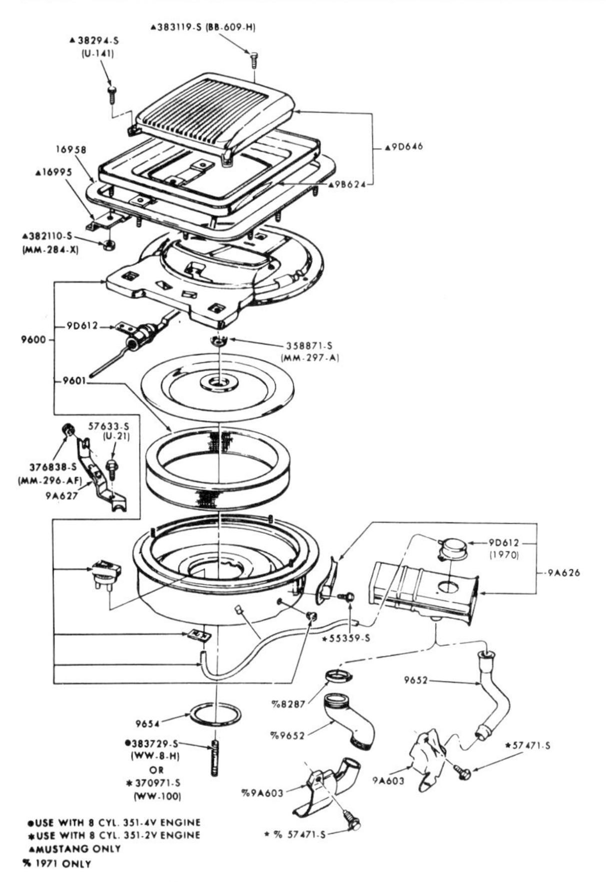

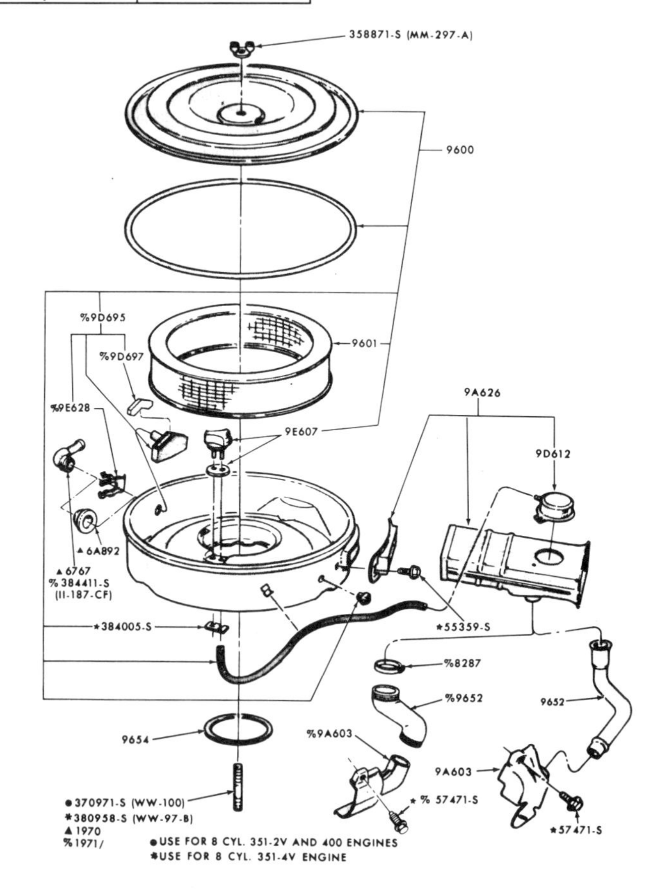

Ram Air (Shaker) scoop and air cleaner assembly for 1970 Mustangs equipped with 351 Cleveland engines.

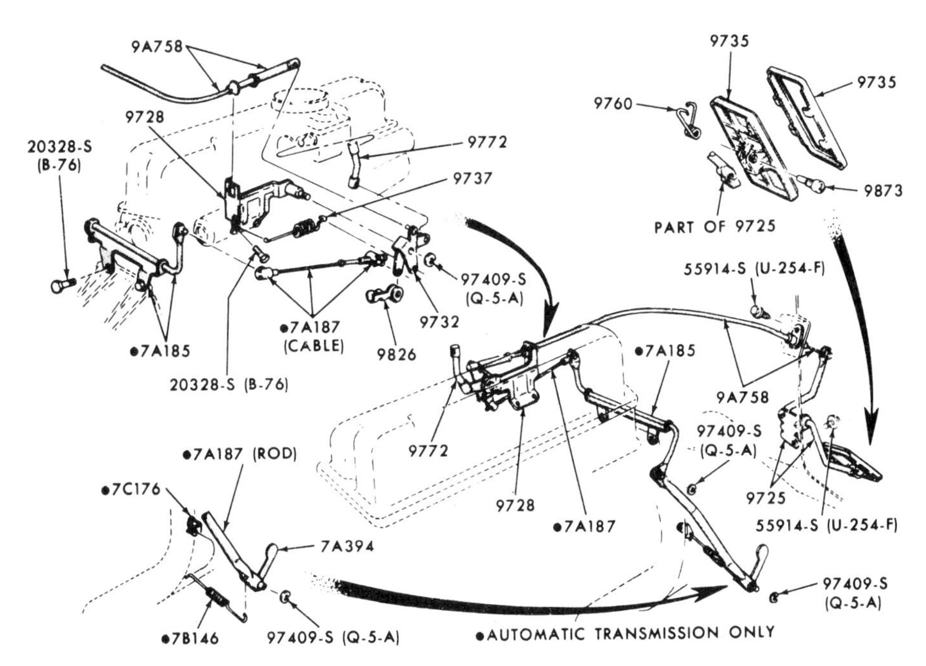

Accelerator linkage for 1969-1970 Mustangs equipped with six-cylinder engines.

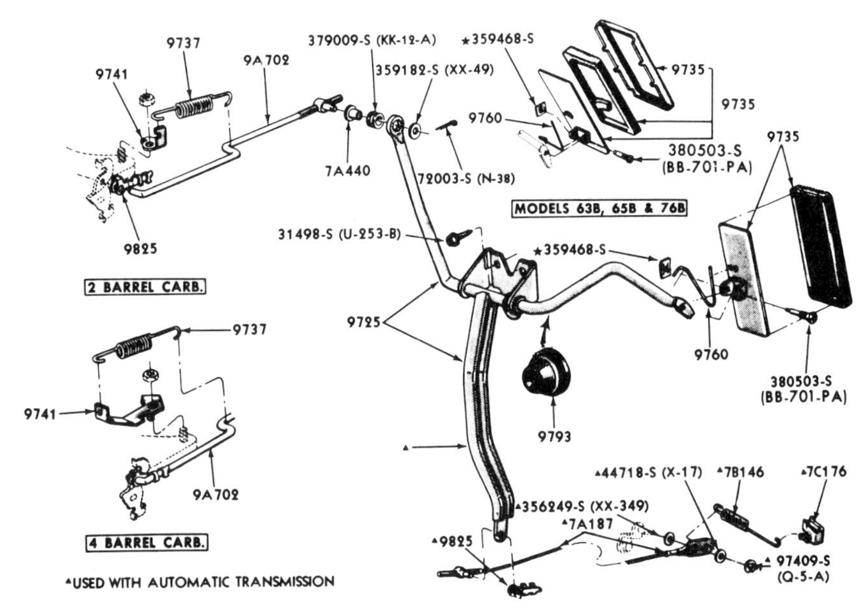

Accelerator linkage for 1967-1968 Mustangs equipped with V-8 engines.

MUSTANG ACCELERATOR PEDAL to CARB THROTTLE CABLE ASSEMBLY CHART

1969 and 1970

Year	Engine	Notes	Description	Part Number
69	All 8s		22.71" long	C9ZZ-9A758-A
69	All 6s		30.50" long	C9ZZ-9A758-C
70	All 8s	Except Boss 429	22.83" long	C9ZZ-9A758-E
70	All 6s		30.00" long	C9ZZ-9A758-D

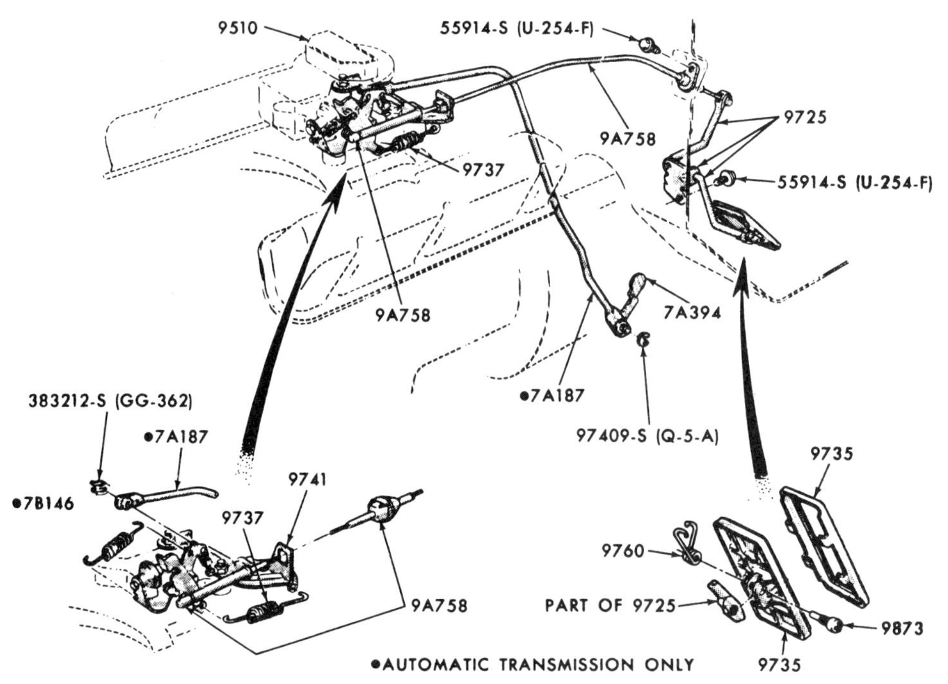

Accelerator linkage for 1969-1970 Mustangs equipped with V-8 engines.

MUSTANG CARBURETOR CHOKE CONTROL ASS'Y APPLICATION CHART, 1965.

Year	Engine	Description	Part Number
65/67	289 High Peformance	58.00" long, 7/16" -14 Thread	C5OZ-9700-A
69/70	Boss 302	57.62" long, 7/16" -14 Thread	C5ZZ-9700-C
70	Shelby GT350/500	44.00" long, 6 1/2" -20 Thread	C9ZZ-9700-C
70	Boss 429	41" long	C9ZZ-9700-B

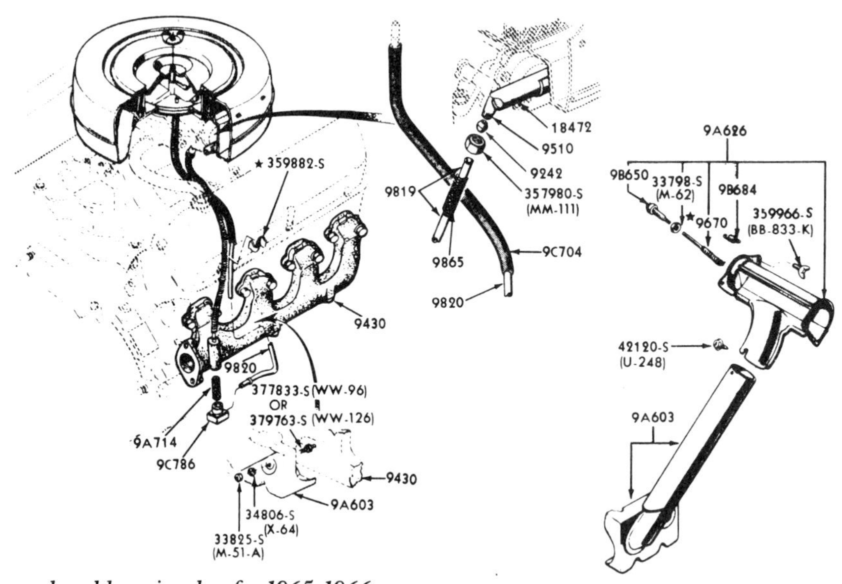

Thermostatic choke control and hot air tubes for 1965-1966 Mustang equipped with 260 or 289ci engines.

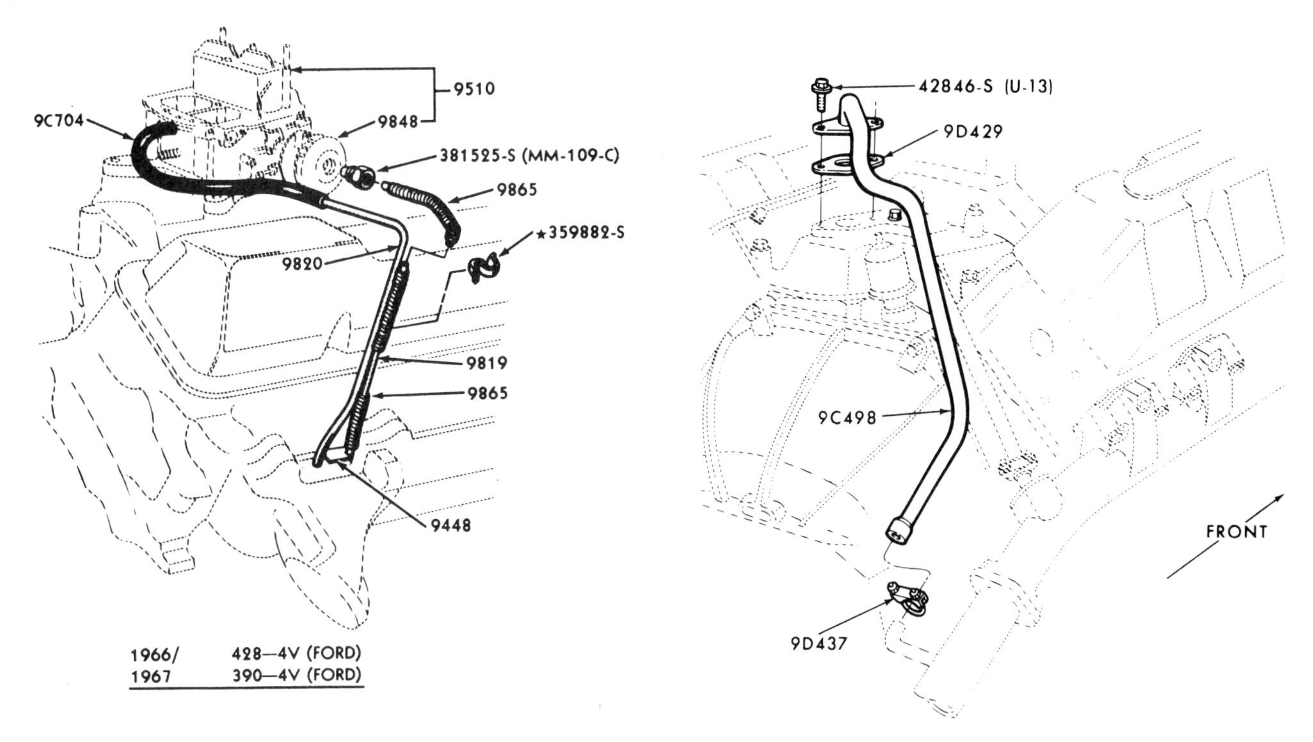

Thermostatic choke control tubes for 1967-1970 Mustang equipped with 390 or 428ci engines.

Boss 429 exhaust manifold heat tube assembly.

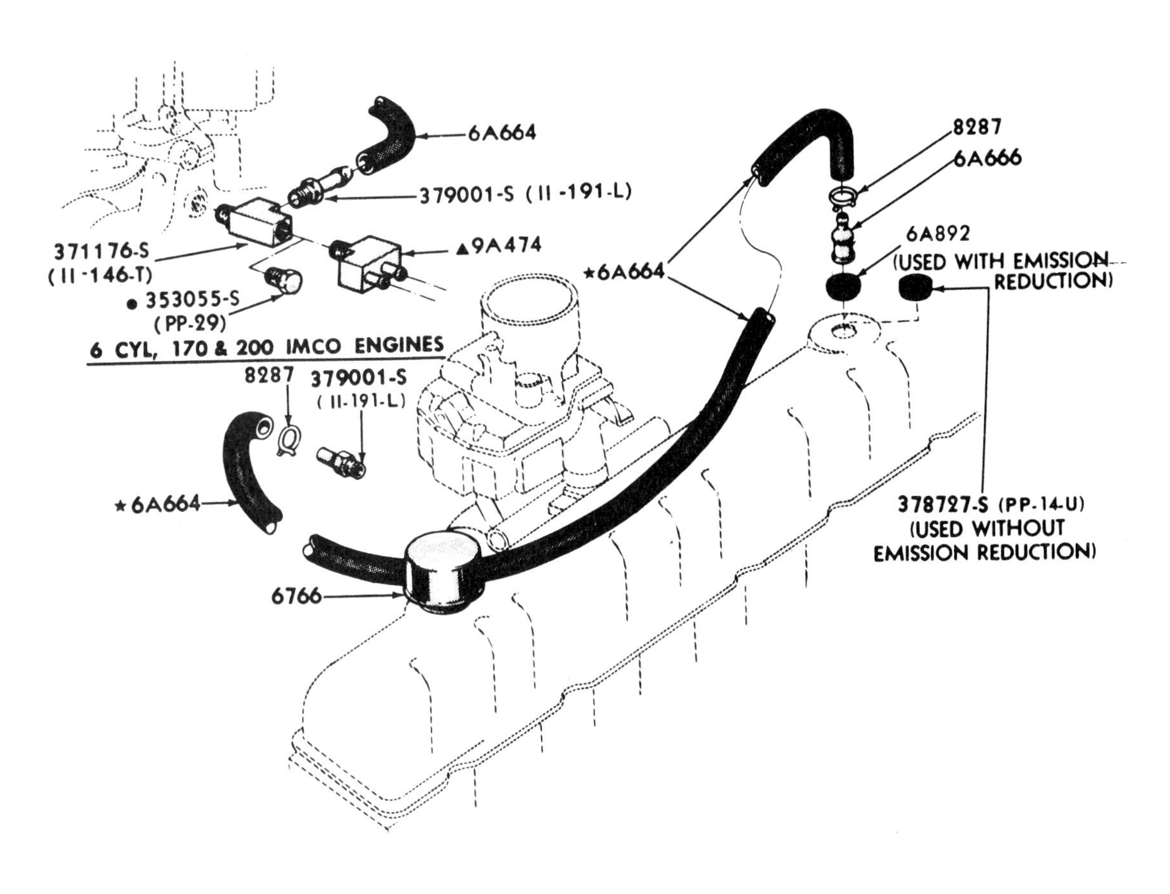

Crankcase emissions (open) evacuation system for 170 and 200ci engines.

Mustang Engine Emission Control Systems Application Chart

Year	Engine	Data Code	Emissions Equipment
'65-67	289-4V	A	Positive (Closed) Crancase Ventilation (PCV) - All 50 States Thermactor System in California
'65-68	289-2V	C	Positive (Closed) Crancase Ventilation (PCV) - All 50 States Thermactor System in California Improved Combustion Exhaust Emissions Control System (ICS) - With Automatic Transmission - Beginning in 1968
'64-1/2	289-4V	D	Positive (Closed) Crancase Ventilation (PCV) - All 50 States
'64-1/2	260-2V	F	Positive (Closed) Crancase Ventilation (PCV) - All 50 States
'68-73	302-2V	F	Positive (Closed) Crancase Ventilation (PCV) - All 50 States Thermactor System in California Improved Combustion Exhaust Emissions Control System (ICS) - with Automatic Transmission
'69-70	Boss 302	G	Positive (Closed) Crancase Ventilation (PCV) Thermactor System - All 50 States
'69-73	351-2V	H	Positive (Closed) Crancase Ventilation (PCV) - All 50 States Thermactor System in California Improved Combustion Exhaust Emissions Control System (ICS) - with Automatic Transmission
'68	302-4V	J	Positive (Closed) Crancase Ventilation (PCV) - All 50 States Thermactor System in California Improved Combustion Exhaust Emissions Control System (ICS) - with Automatic Transmission
'65-68	289-4V	K	Positive (Closed) Crancase Ventilation (PCV) - All 50 States Thermactor System in California Improved Combustion Exhaust Emissions Control System (ICS) - with Automatic Transmission Beginning In 1968
'69-73	250-1V	L	Positive (Closed) Crancase Ventilation (PCV) - All 50 States Thermactor System in California
'69-71	351-4V	M	Positive (Closed) Crancase Ventilation (PCV) Thermactor System in California Improved Combustion Exhaust Emissions Control System (ICS) -
'69-70	428CJ	Q	Positive (Closed) Crancase Ventilation (PCV) Thermactor System Improved Combustion Exhaust Emissions Control System (ICS) -
'68-70	428 CJ	R	Positive (Closed) Crancase Ventilation (PCV) Thermactor System Improved Combustion Exhaust Emissions Control System (ICS) -
'67-69	390-4V	S	Positive (Closed) Crancase Ventilation (PCV) - All 50 States Thermactor System In California
'65-70	200-1V	T	Positive (Closed) Crancase Ventilation (PCV) - All 50 States Thermactor System In California Improved Combustion Exhaust Emissions Control System (ICS) - with Automatic Transmission Beginning In 1968
'64-1/2	170	U	Positive (Closed) Crancase Ventilation (PCV)
'69-70	Boss 429	Z	Positive (Closed) Crancase Ventilation (PCV) Thermactor System

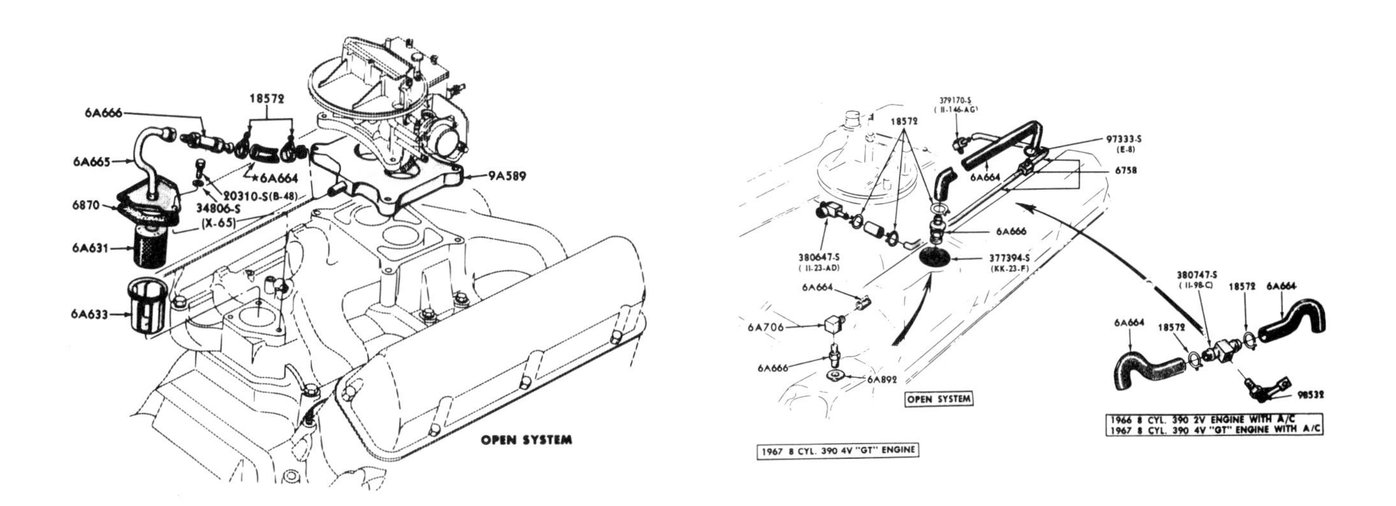

Intake manifold type crankcase evacuation (open) system for 1965 model 260 and 289 engines.

Open emissions reduction hot idle compensator and related parts for 1967 model 390ci engines.

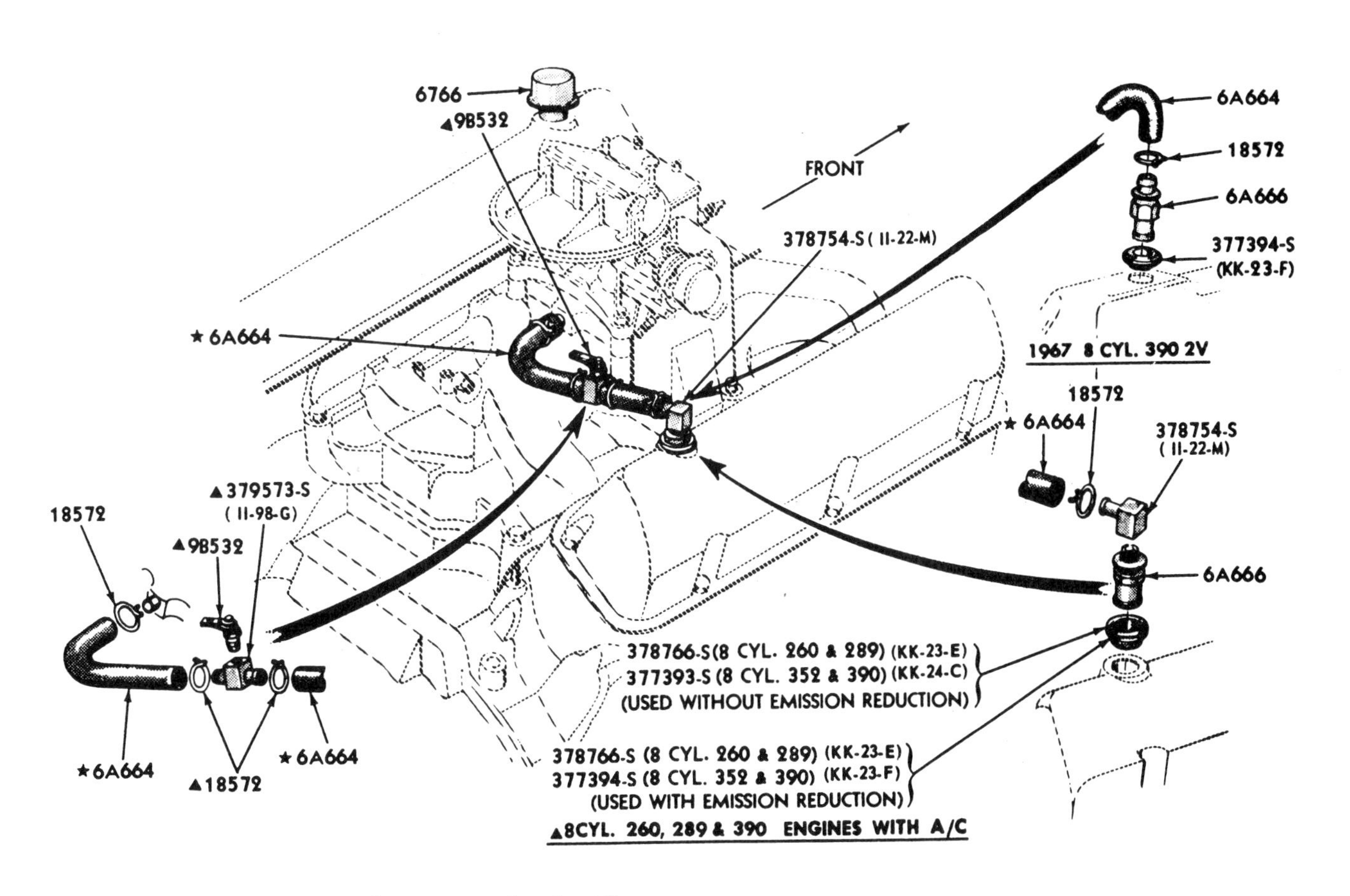

Engine emissions reduction, crankcase evacuation, hot idle compensator, and related parts for 1965-1970 260, 289, and 351W engines.

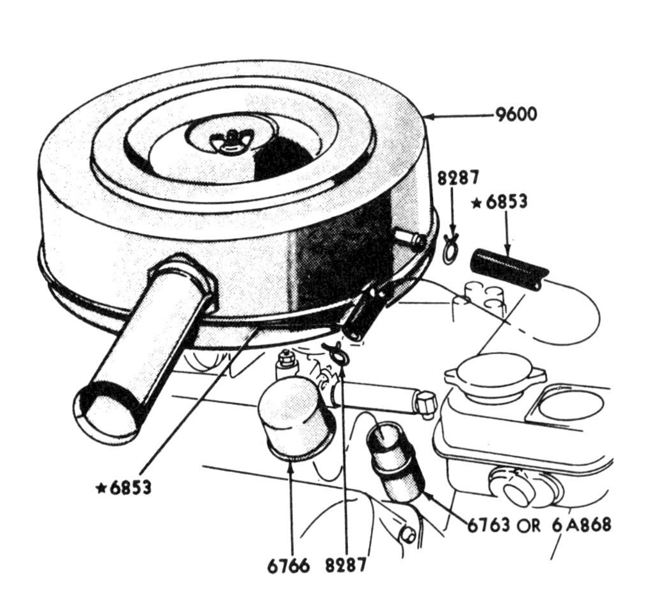

Closed emissions reduction air cleaner oil breather cap and related parts for all engines.

An engine equipped with a PCV (positive crankcase ventilation) system used a vented oil cap. Air entered the vented cap as the induction pulled crankcase fumes from the engine. Engines equipped with CCV (closed crankcase ventilation) systems had sealed oil caps. Venting the lower air cleaner housing was the factory's attempt to provide more air to a non-Ram Air-equipped big-block.

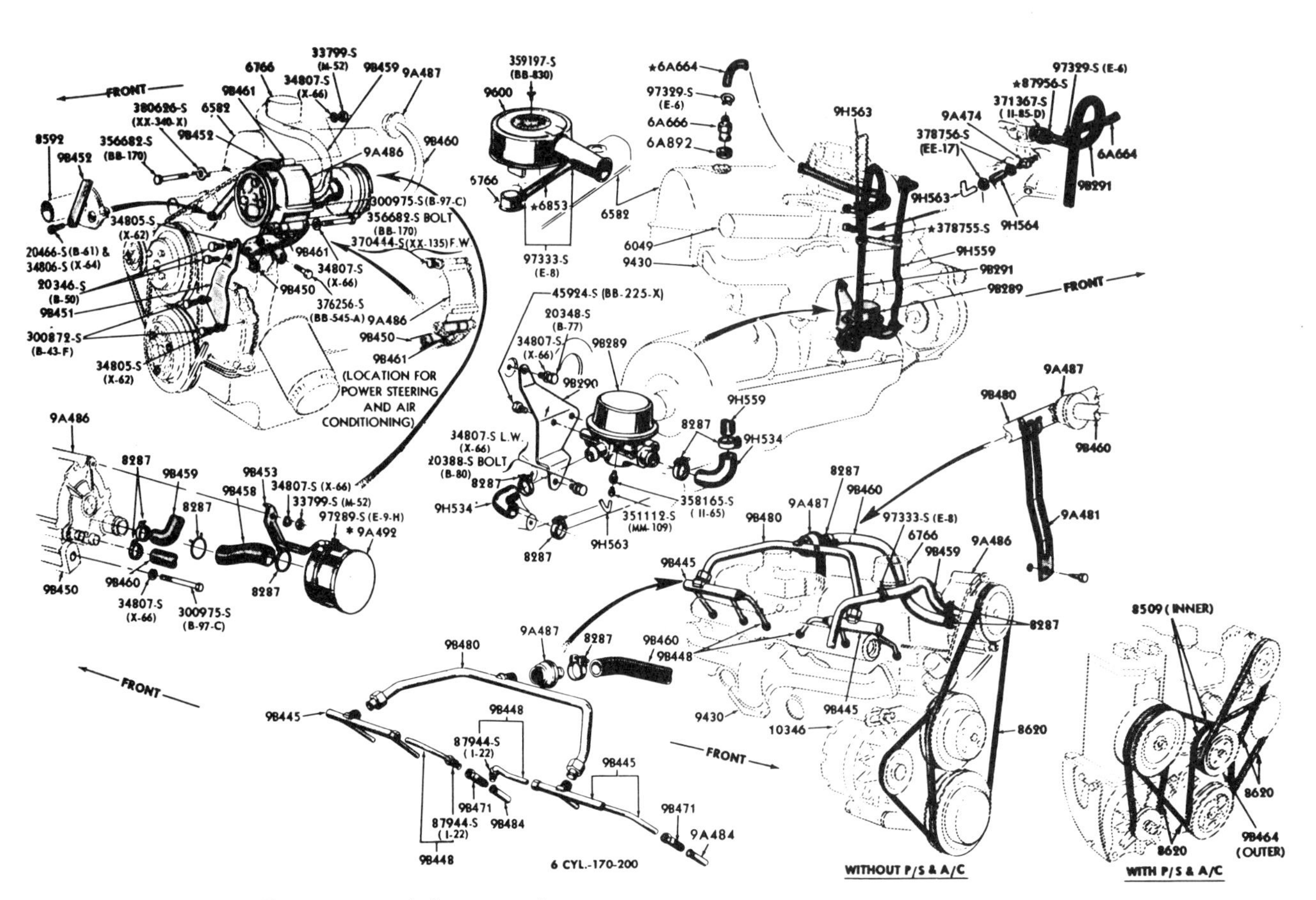

Intake manifold type Thermactor emissions control system for 1966 Mustang equipped with 200ci engine (California only).

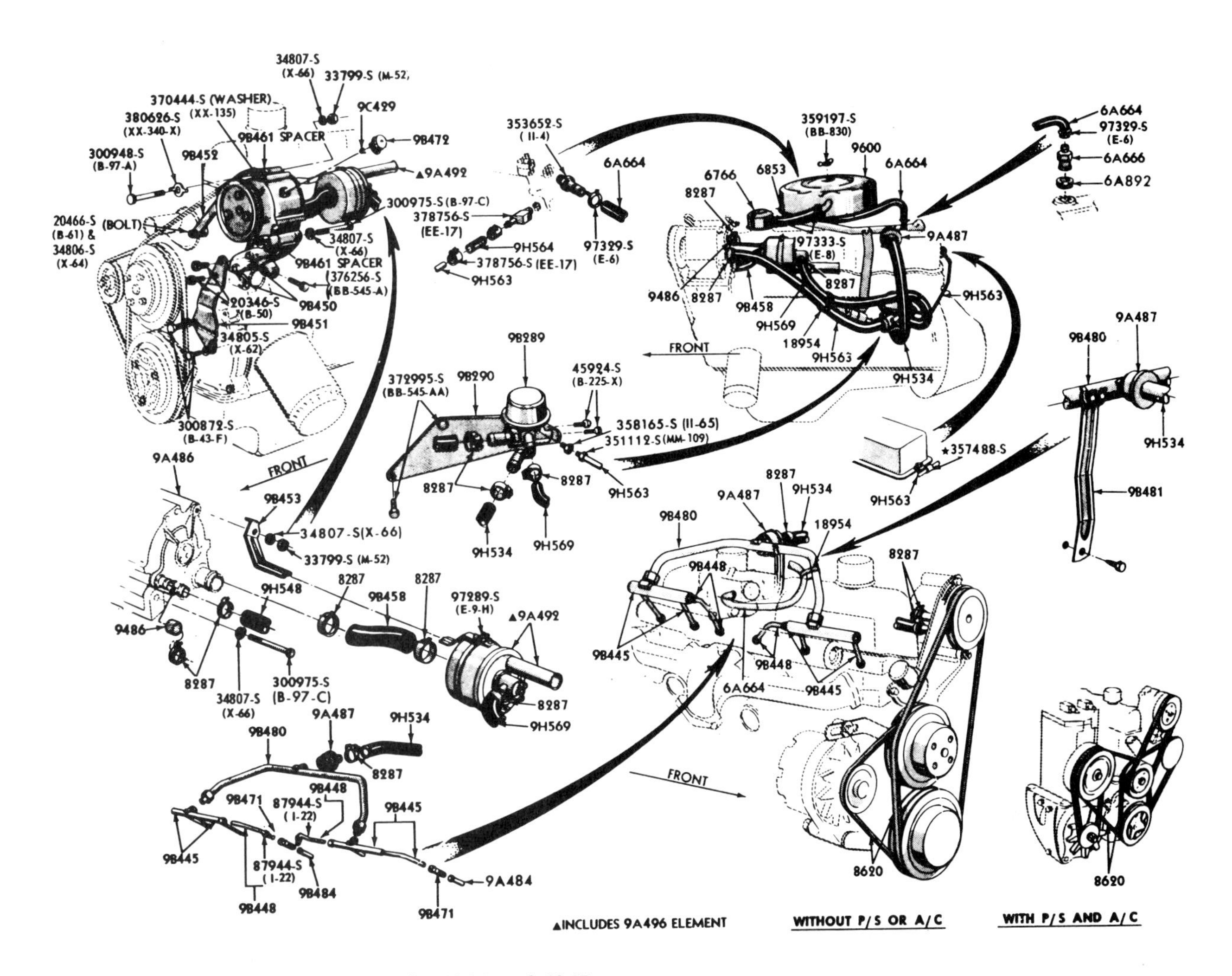

Thermactor emissions control system for 1966 and 1967 Mustang equipped with 200ci engine (California only).

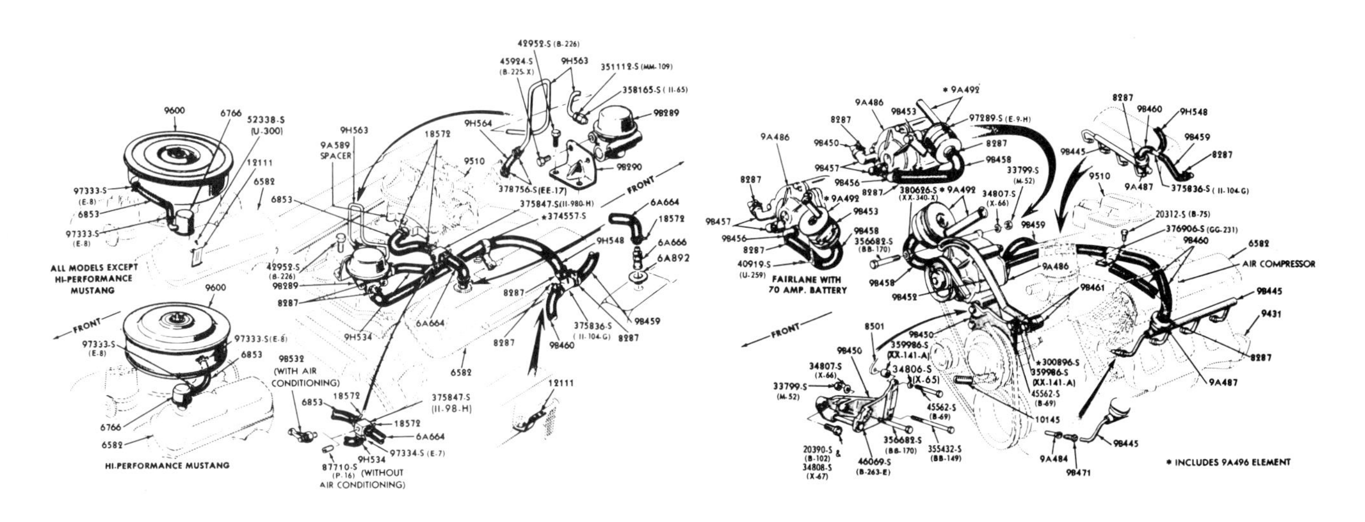

Intake manifold type Thermactor emissions control system for 1966 Mustang equipped with 289ci engine.

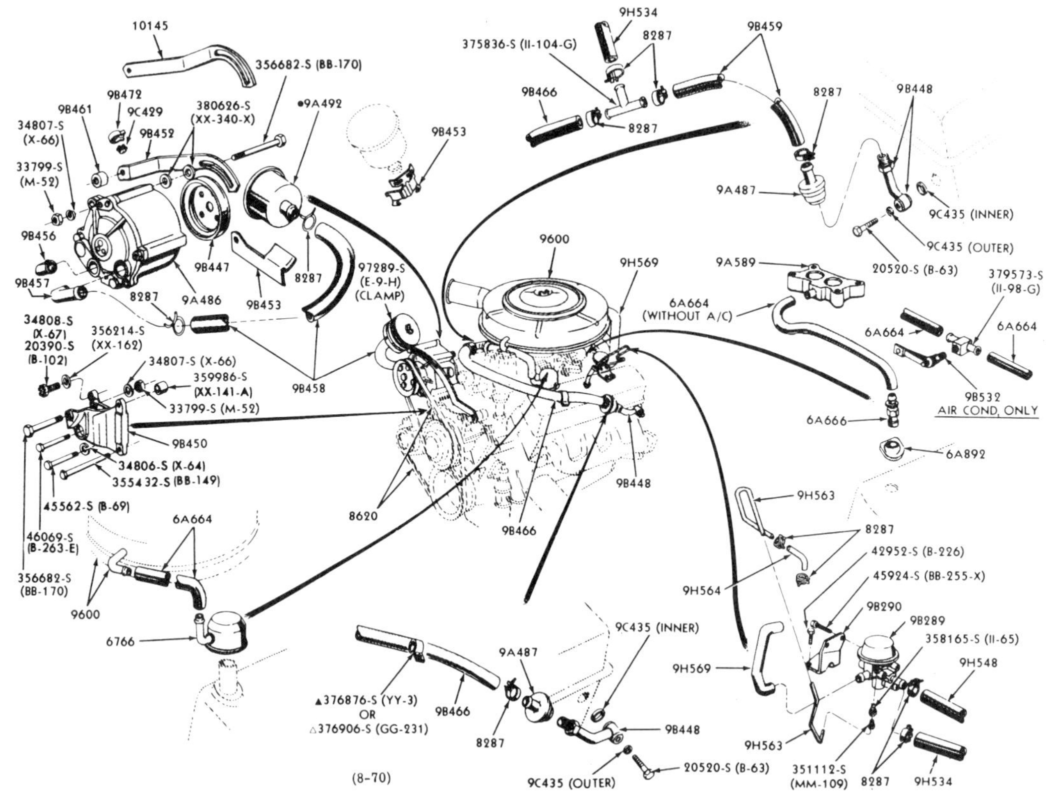

Thermactor emissions control system for 1967 Mustang equipped with a 289ci engine (except 289 HiPo).

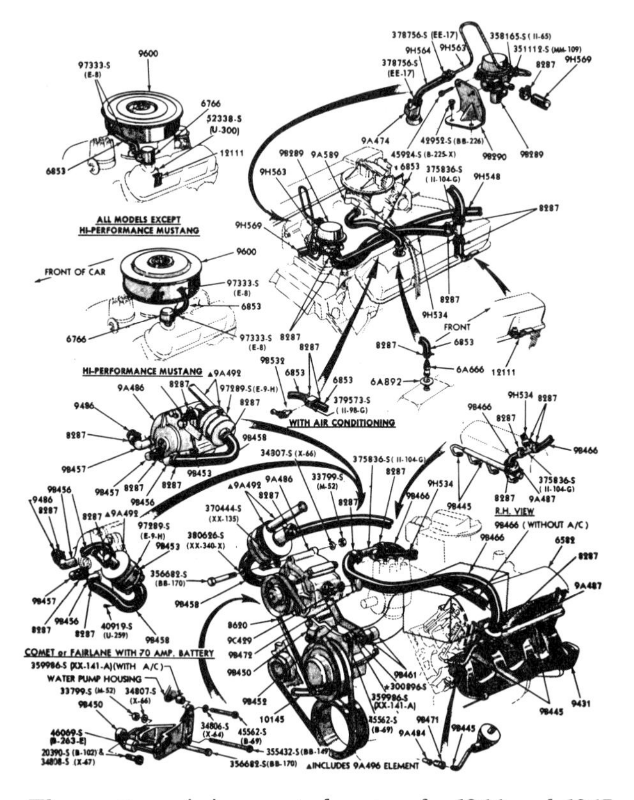

Thermactor emissions control system for 1966 and 1967 Mustang equipped with 289 HiPo engine.

By 1968 all Ford engines utilized some vacuum and/or electrical controls that greatly improved cold-weather startup and exhaust emission ratings. This was accomplished mostly by regulating the ignition timing based on the engine's operating temperature. The "vacuum tree," threaded into the thermostat housing, is really a temperature-activated vacuum switch that controls the dual-diaphragm vacuum advance canister attached to the distributor.

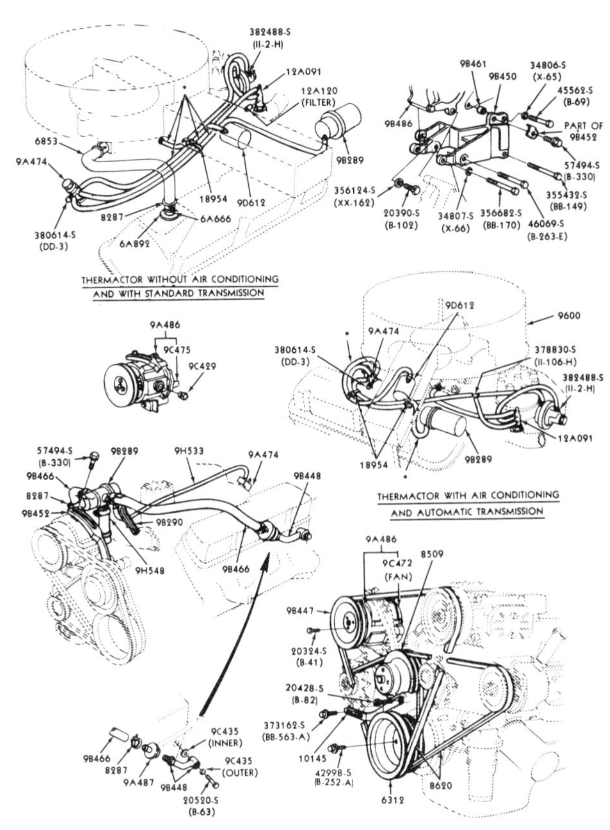

Thermactor emission control system used on 1968 Mustangs equipped with 289 engines.

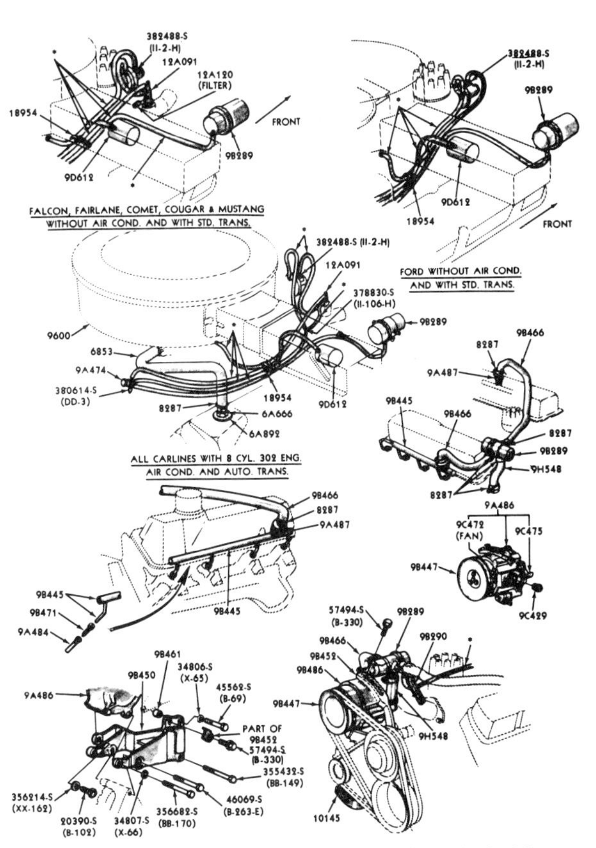

Thermactor emissions control system for 1968 Mustang equipped with 289 and 302ci engines.

In 1966, all Mustangs sold in California were equipped with Thermactor emissions reduction equipment designed to inject fresh air into the exhaust ports in an effort to promote more complete burning of the spent gasses.

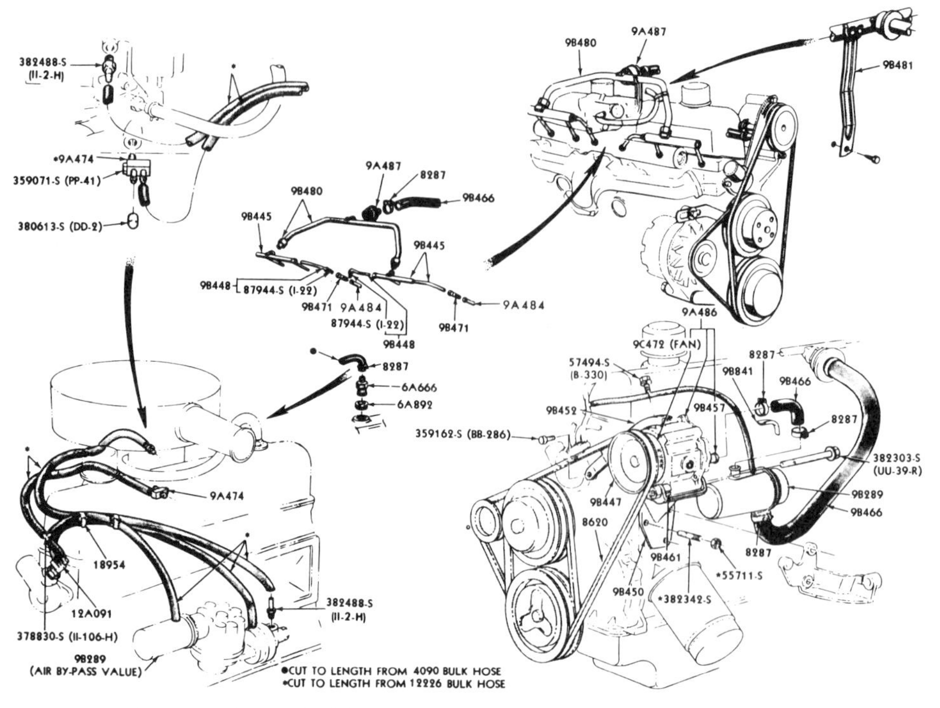

Thermactor emissions control system for 1968 Mustang equipped with 200ci engine.

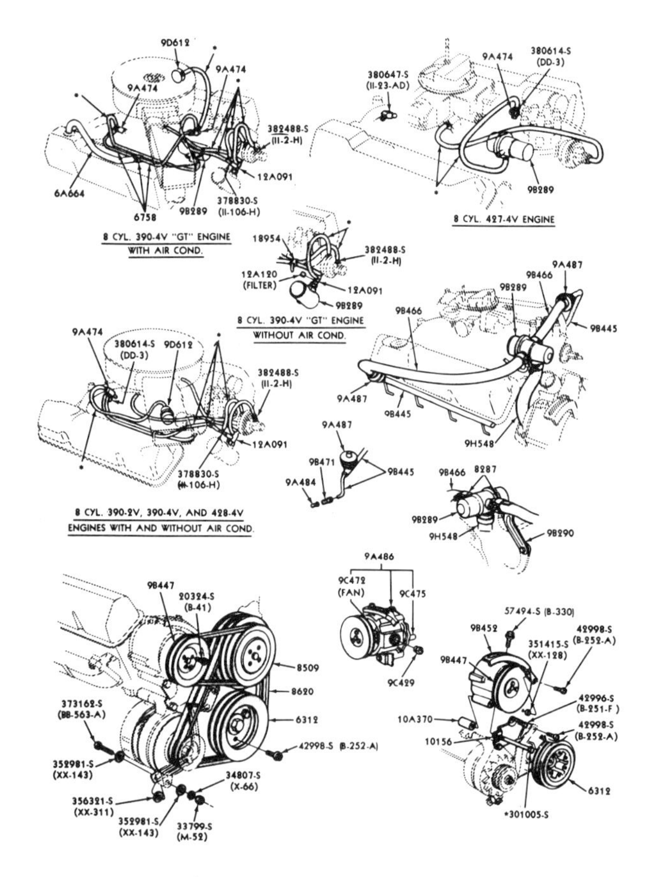

Thermactor emissions control system for 1968 through 1970 Mustang equipped with 390 and 428ci engines.

By 1968, most big-block Ford engines were equipped with some type of emissions device. The Thermactor system on this 428ci engine consists of a belt-driven air pump, several check valves, and a network of hoses needed to service the two air distribution manifolds connected to each set of exhaust ports. Simply put, the pump forces fresh air into the exhaust ports just above the valve. The presence of oxygen and moisture in the air combined with the hot exhaust temperatures promotes complete burning of any unused fuel. The air pump can be seen under the top radiator hose and the air bypass valve, identified with 424S. The two fresh air hoses leading to the manifolds can be seen above the thermostat housing.

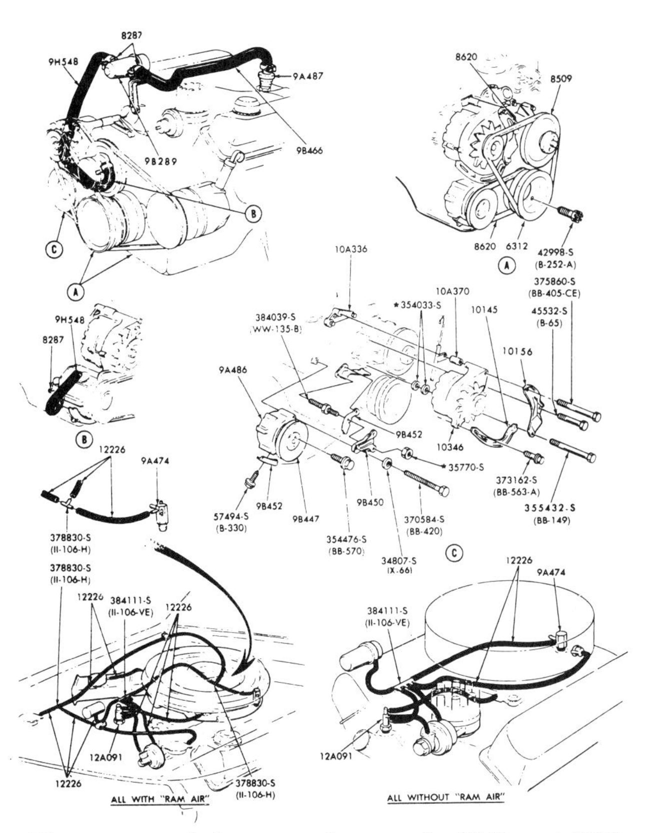

Thermactor emissions control system for 1969 and 1970 Boss 302 Mustangs.

Beginning in 1969, power steering filler necks were much smaller in diameter and the cap contained a little dipstick. This compact neck was located at the back of the pump and angled rearward. Other than rotating the canister which angled the neck from side to side, no other variation was necessary.

MUSTANG EXHAUST AIR SUPPLY MANIFOLD ASSEMBLY APPLICATION CHART

Years	Cyl/CID	Part Number	Description
66/68	6/200	C6DZ-9B445-A	FRONT
66/68	6/200	C6DZ-9B445-B	REAR
66/67	8/289	C6OZ-9B445-A	R.H.
67	8/289K	C6OZ-9B445-A	FROM 4/17/67
66/67	8/289	C6OZ-9B445-B	L.H.
67	8/289K (T/E)	C6OZ-9B445-B	L.H. FROM 4/17/67
68	8/289-2V/302	C8AZ-9B445-A	R.H. & L.H.
67	8/390 (T/E)	C6AZ-9B445-C	R.H. BEFORE 9/1/66
67/68	8/390 (T/E)	C9ZZ-9B445-A	R.H.
68	8/427 (T/E)	C9ZZ-9B445-A	R.H.
68/69	8/428CJ (T/E &4/B CARB)	C9ZZ-9B445-A	R.H.
70	8/428CJ (4/B CARB)	C9ZZ-9B445-A	
68/69	8/428CJ (4/BCARB) (T/E)	C9ZZ-9B445-B	L.H.
70	8/428CJ (4/B CARB)	C9ZZ-9B445-B	
67/68	8/390 (T/E)	C9ZZ-9B445-B	L.H.
68	8/427K (T/E)	C9ZZ-9B445-B	L.H.
69	8/429 BOSS (T/E)	C9AZ-9B445-B	R.H. & L.H. ID: C9AE-9B466-A
70	8/429 BOSS (T/E)	D0AZ-9B445-A	R.H. & L.H.

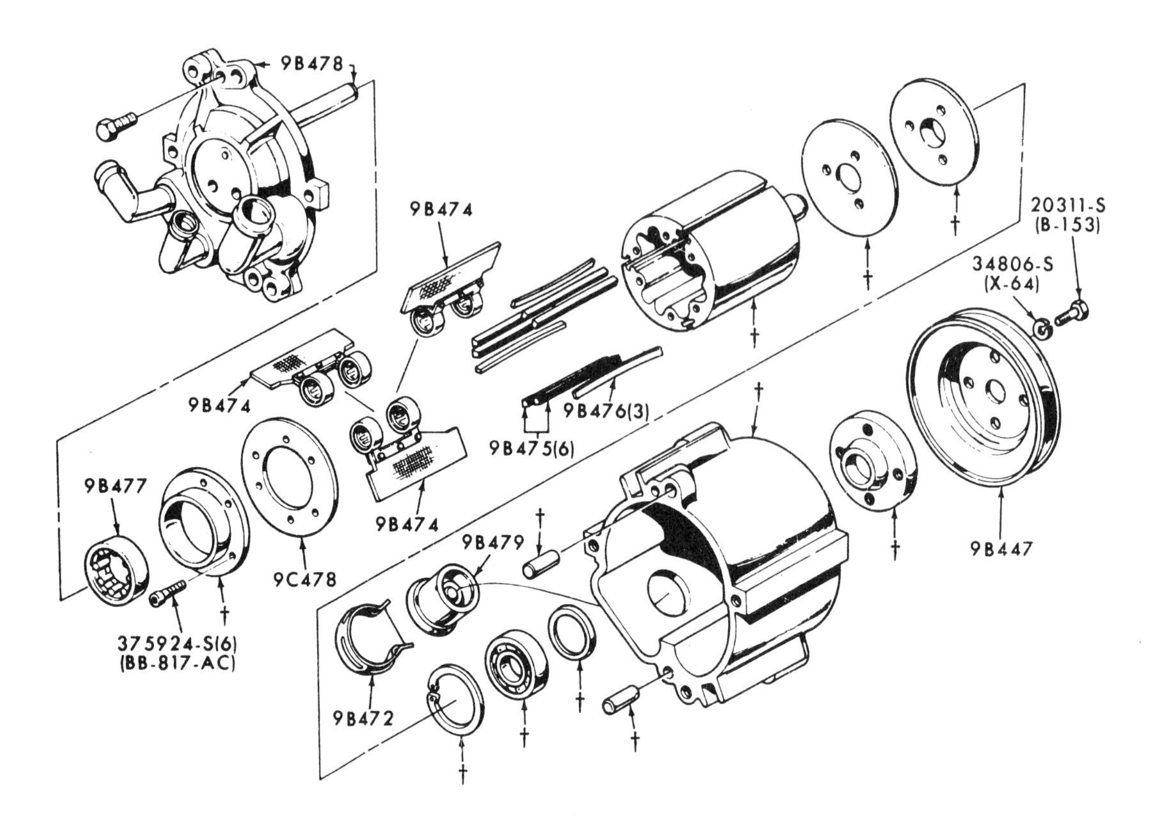

Thermactor exhaust air supply pump used on 1966 and 1967 Mustangs.

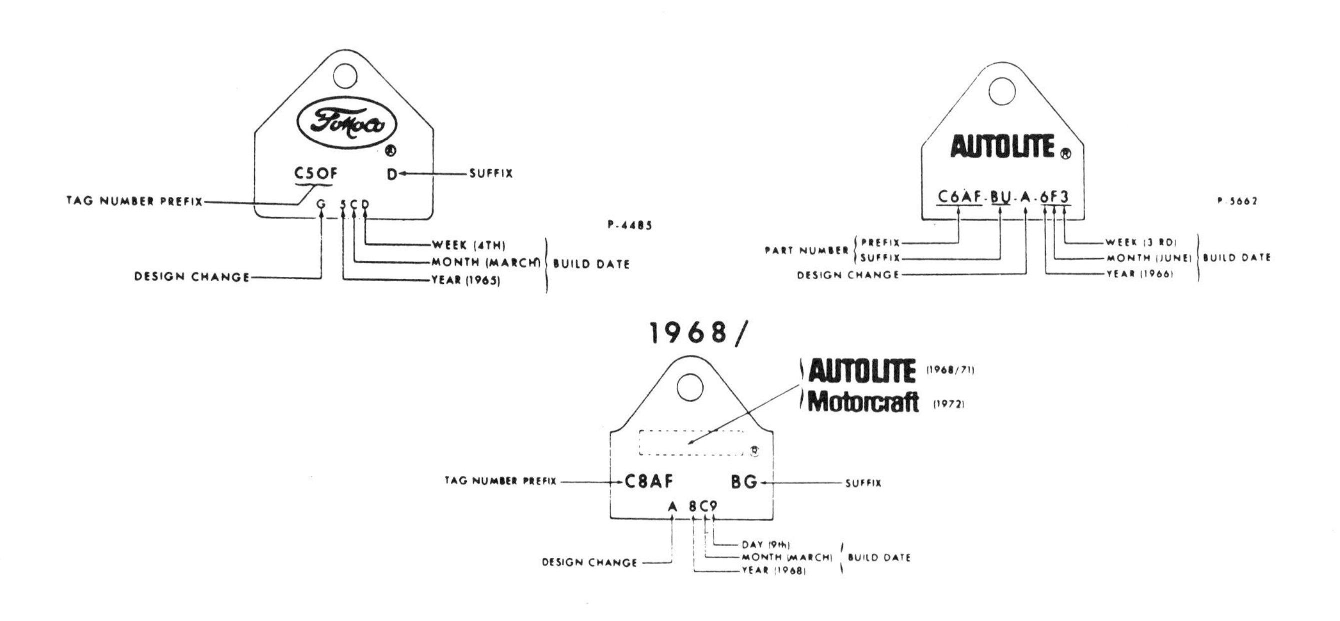

Mustang carburetor identification tags.

EXHAUST AIR SUPPLY PUMP PULLEY APPLICATION CHART

Years	Cyl/CID	Part Number	Notes	Description
66/67	6/170, 200	C6TZ-9B447-B	1	4.74" O.D.-ident. C6TE-9B447-B
66/67	6/170, 200	C6DZ-9B447-B	2	4.58" O.D.-ident. C6DE-9B447-B
67	8/390 (T/E)	C6AZ-9B447-E		5.95" O.D.-ident. C6AE-9B447-E
67	8/289 (w/ T/E & A/C)	C6AZ-9B447-E	3	5.95" O.D.-ident. C6AE-9B447-E
67	8/289	C6AZ-9B447-E	4	5.95" O.D.-ident. C6AE-9B447-E
67	8/289	C6AZ-9B447-D	5	5.65" O.D.-ident. C6AE-9B447-D
66/67	8/289 (w/ T/E, except A/C)	C6OZ-9B447-A	6	4.07" O.D.-ident. C6OE-9B447-B
67	8-289 (4/B carb. & T/E,	C6OZ-9B447-A	7	4.07" O.D.-ident. C6OE-9B447-B
	except A/C)			
67	8/289 (2/B carb., S/T,	C6OZ-9B447-A	7	4.07" O.D.-ident. C6OE-9B447-B
	& T/E, except A/C)			
66	8/289 (A/C) (T/E)	C6AZ-9B447-C		4.93" O.D.-ident. C6AE-9B447-C
66/67	6/170,200	C6TZ-9B447-A		5.33" O.D.-ident. C6TE-9B447-A
67	8/289 (T/E) (except A/C)	C7AZ-9B447-A	8	3.97" O.D.-ident. C7ZE-9B447-A
67	8/289 (S/T & T/E, exc. A/C)	C7AZ-9B447-A	9	3.97" O.D.-ident. C7ZE-9B447-A
67	8/289	C7AZ-9B447-E		4.75" Diam., ident. C7OE-9B447-B
67	8/289K, 4/B carb. & T/E	C7AZ-9B447-D	10	4.31" O.D. ident. C6AE-9B447-F
68	8/289,302 (AC &/OR P/S)	C8ZA-9B447-D		5.89" Diam., ident. C8AE-9C480-B
	(T/E)			
68	8/390 (4/B carb.) T/E A/C	C8OZ-9B447-B		4.981" Diam.-ident. C8OE-9C480-B
68	8/390 (P/S & 55 AMP.ALT.)	C8AZ-9B447-E		5.68" Diam.-ident. C8AE-9C480-D
	(T/E)			
69	8/302 BOSS (T/E)	C8AZ-9B447-E		5.68" Diam.-ident. C8AE-9C480-D
68	6 & 8/200, 289, 302, 390	C8AZ-9B447-B		4.75" Diam.-ident. C8AE-9C480-A
	(except A/C) (T/E)			
68	8/390 (GT 4/B carb..) T/E	C8AZ-9B447-B		4.75" Diam.-ident. C8AE-9C480-A
	except A/C			
68/69	8/428 (except A/C) (T/E)	C8AZ9B447-B		4.75" Diam.-ident. C8AE-9C480-A
70	8/428 CJ	C8AZ-9B447-B		4.75" Diam.-ident. C8AE-9C480-A
68	8/427 (4B & T/E)	C8OZ-9B447-A	11	5.49" Diam.-ident. C8OE-9C480-A
69	8/428 CJ (A/C) (T/E)	C9OZ-9B447-A		5.10" O.D.-ident. C9OE-9C480-A
70	8/428 CJ	C9OZ-9B447-A		5.10" O.D.-ident. C9OE-9C480-A
69/70	8/429 (T/E)	C9AZ-9B447-A		6.251" Dia.-ident. C9AE-9C480-A
70	8/302 BOSS (T/E)	D0ZZ-9B447-A		ident. D0ZE-9C480-A

Notes:

1 Except w/ P/S and A/C; except Imco w/ C4
2 With P/S, except A/C; with T/E
3 Before 8/29/66
4 With T/E and A/C,from 9/23/66
5 With T/E and A/C,from 8/29/66 to 9/23/66
6 Before 11/14/66
7 From 11/14/66
8 From 9/23/66 to 11/14/66
9 From 9/23/66
10 With Super Extra Cooling
11 With or Without Power Steering

Carburetor Application Chart
Mustang 170 & 200 6-Cylinder Engines, 1965-1970

Year	CID	Model	Type	Identification	Part Number	Replaced By:
65 (64)	170		F-1	C6PF-J; C8OF-F	C5OZ-9510-A	D0PZ-9510-A
65 (64)	170	1	F-1	C4ZF-J,L	C5ZZ-9510-E	D0PZ-9510-A
65 (64)	170	2	F-1	C4ZF-K,M	C5ZZ-9510-F	D0PZ-9510-A
65 (64)	170		F-1	C5DF-H,K,L,M	C5DZ-9510-C	D0PZ-9510-A
				C6DF-R,S		
65/67	200		F-1	C6PF-J; C8OF-F	C5OZ-9510-H	D0PZ-9510-A
65/67	200		H-1	D0PF-K	D0PZ-9510-A	
65	200		F-1	C5OF-E	C5OZ-9510-A	D0PZ-9510-A
65	200		F-1	C5OF-F	C5ZZ-9510-B	D0PZ-9510-A
65	200		F-1	C5OF-N	C5DZ-9510-E	D0PZ-9510-A
65/67	200		F-1	C5OF-R,Z,Y	C5DZ-9510-C	D0PZ-9510-A
				C6OF-AC,AD		
66	200		F-1	C6DF-C	C6DZ-9510-A	D0PZ-9510-G
66	200		F-1	C6OF-G	C6OZ-9510-D	D0PZ-9510-G
66/67	200		F-1	C7DF-J	C6DZ-9510-E	D0PZ-9510-G
66/67	200		F-1	C7DF-K	C6OZ-9510-M	D0PZ-9510-G
66/69	200		F-1	C8PF-L	C6DZ-9510-F	D0PZ-9510-G
67	200		F-1	C7OF-N	C7OZ-9510-J	D0PZ-9510-G
67	200		F-1	C7OF-R	C7OZ-9510-K	D0PZ-9510-G
67	200		C-1	C7DF-T	C7DZ-9510-T	D0PZ-9510-G
68	200		F-1	C8OF-A	C8OZ-9510-A	D0PZ-9510-G
68/69	200		F-1	C8OF-B	C8OZ-9510-B	D0PZ-9510-G
68	200		F-1	C6PF-J	C5OZ-9510-H	D0PZ-9510-A
				C8OF-E,F		
68	200		H-1	D0PF-K	D0PZ-9510-A	
69	200		F-1	C9DF-B	C9DZ-9510-B	D0PZ-9510-G
66/69	200		F-1	C9DF-B	C9DZ-9510-C	D0PZ-9510-G
66/69	200		H-1	D0PF-L	D0PZ-9510-G	
70	200		C-1	D0DF-C,M	D0DZ-9510-C	D2DZ-9510-G
70	200		C-1	D0DF-D,L	D0DZ-9510-D	D2DZ-9510-G
70	200		C-1	D0DF-G,T	D0DZ-9510-G	D2DZ-9510-G
70	200		C-1	D0DF-H,V	D0DZ-9510-H	D2DZ-9510-G
70	200		C-1	D0PF-AD	D0DZ-9510-J	D2DZ-9510-G

Model:

1 with Manual Transmission
2 with C4 Transmission
C-1 Carter 1V
F-1 Ford 1V
H-1 Holley 1V

Carburetor Application Chart
Mustang 250 cid 6-Cylinder Engines, 1969-1970

Year	CID	Model	Type	Identification	Part Number	Replaced By:
69	250	1	F-1	C9OF-A	C9OZ-9510-A	D0PZ-9510-T
69	250	2	F-1	C9OF-B	C9OZ-9510-B	
69	250	3	F-1	C9OF-J	C9OZ-9510-J	D0PZ-9510-T
69	250	4	F-1	C9OF-K	C9OZ-9510-K	D0PZ-9510-T
69	250	5	F-1	C9OF-M	C9OZ-9510-M	D0PZ-9510-T
69	250	6	F-1	C9OF-B	C9OZ-9510-L	D0PZ-9510-T
69	250		H-1	D0PF-D	D0PZ-9510-T	
70	250		C-1	D0PF-AB	D0ZZ-9510-AA	
70	250	7	C-1	D0ZF-C	D0ZZ-9510-C	D0ZZ-9510-AA
70	250	8	C-1	D0ZF-D	D0ZZ-9510-D	D0ZZ-9510-AA
70	250	9	C-1	D0ZF-E	D0ZZ-9510-E	D0ZZ-9510-C
70	250		C-1	D0ZF-F	D0ZZ-9510-F	D0ZZ-9510-AA

Model:

1 with C4 and IMCO, without A/C
2 with Manual Transmission and IMCO, without A/C
3 with Manual Transmission and IMCO, with A/C
4 with C4 and IMCO, with A/C
5 Economy model, with C4, M/T, and IMCO
6 with C4, M/T, and IMCO
7 with Manual Transmission and IMCO, cfm 215
8 with C4, IMCO, and A/C; cfm 215
9 with Manual Transmission and IMCO, without A/C, cfm 215

C-1 Carter 1V
F-1 Ford 1V
H-1 Holley 1V

Carburetor Application Chart
Mustang 260 cid 8-Cylinder Engines, 1965 (1964-1/2)

CID	Note:	Type	Identification	Part Number	Replaced By:
260	1	F-2	C4ZF-R,N;C4DF-AB	C5ZZ-9510-K	D2AZ-9510-D
			C4OF-BR,BS;C4DF-AC		
260	2	F-2	C9OF-B	C5ZZ-9510-L	D2AZ-9510-D
260	1	F-2	C9OF-J	C2AZ-9510-J	D2AZ-9510-D
260	1	F-2	C9OF-K	C5ZZ-9510-C	D2AZ-9510-D
260	1	F-2	C9OF-M	C2AZ-9510-K	D2AZ-9510-D
260	1	F-2	C9OF-B	C8ZZ-9510-H	D2AZ-9510-D
260		F-2	D0PF-D	D0AZ-9510-AC	D2AZ-9510-D

Notes:

1 Stamping identification located on edge of carburetor mounting flange.
2 Stamped identification on carburetor mounting flange is main body identification only. Carburetor identification appears on identification tag.

F-2 Ford 2V

Carburetor Application Chart, Part One
Mustang 289 cid 8-Cylinder Engines, 1965-1968

Year	Note:	Type	Identification	Part Number	Replaced By:
65/66	1	H-4	DOPF-U	C9AZ-9510-AE	
65/66	2	F-2	C6PF-F	C2AZ-9510-J	
65/67	2	F-2	C6PF-F	C2AZ-9510-K	D2AZ-9510-D
66	2,3	F-2	C6DF-E	C6DZ-9510-C	D2AZ-9510-D
66	2,4	F-2	C6DF-F	C6DZ-9510-D	D2AZ-9510-D
67	2,5	F-2	C7DF-E,S	C7DZ-9510-E	D2AZ-9510-D
67	2,3	F-2	C7DF-G	C7DZ-9510-G	D2AZ-9510-D
67	2,6	F-2	C7DF-F,R	C7DZ-9510-F	D2AZ-9510-D
67	2,4	F-2	C7DF-H,N,V	C7DZ-9510-H	D2AZ-9510-D
68	2,3	F-2	C8AF-AK	C8ZZ-9510-F	D2AZ-9510-D
68	2,7	F-2	C8ZF-G	C8ZZ-9510-G	D2AZ-9510-D
68	2,7	F-2	C8AF-L	C8AZ-9510-AR	D2AZ-9510-D
68	8	F-2	C8DF-E	C8DF-9510-E	D2AZ-9510-D
68	8	F-2	C8AF-AF;C8ZF-H	C8ZF-9510-H	D2AZ-9510-D
68	9	F-2	C8OF-S;C8ZF-E	C8OF-9510-S	D2AZ-9510-D
68	9	F-2	C8DF-F	C8DF-9510-F	D2AZ-9510-D
68	2,10	F-4	C6ZF-C,F	C4OZ-9510-F	C4OZ-9510-G
68	2,3	F-2	C8AF-BD;C8PF-T	C8ZZ-9510-H	D2AZ-9510-D
67	2,11	F-2	C8AF-BD;C8PF-T	C8ZZ-9510-H	D2AZ-9510-D
65	2,12	F-2	C4OF-AL,AT,BU,BT	C4OZ-9510-D	C4OZ-9510-G
65	2	F-2	C5ZF-A,B	C5ZZ-9510-H	D2AZ-9510-D
65	14	F-2	C5ZF-G,H;C6DF-A,B	C5ZZ-9510-L	D2AZ-9510-D
66	13,14	F-2	C5ZF-G,H;C6DF-A,B	C5ZZ-9510-L	D2AZ-9510-D
65	1,2	F-4	C5ZF-C,D,E,F	C5ZZ-9510-J	C9AZ-9510-AE
65	1,2	F-4	C5ZF-J,K,L,M	C5ZZ-9510-M	C9AZ-9510-AE
			C6ZF-A,B		
66	1,2	F-4	C5ZF-J,K,L,M	C5ZZ-9510-M	C9AZ-9510-AE
			C6ZF-A,B		
65	1	F-4	C6PF-H	C3AZ-9510-AU	C9AZ-9510-AE
66	1,13	F-4	C6PF-H	C3AZ-9510-AU	C9AZ-9510-AE

Notes:
1 Except High-Performance
2 Stamping identification located on edge of carburetor mounting flange.
3 With manual transmission and Thermactor.
4 With C4 transmission and Thermactor.
5 With manual transmission, except Thermactor.
6 With C4 transmission, except Thermactor.
7 With C4 transmission and Imco.
8 Canada only. Manual transmission; open emission.
9 Canada only. Automatic transmission, open emission.
10 Canada only. Open emission.
11 With Thermactor.
12 High Performance.
13 Except Thermactor.
14 Stamped identification on carb mounting flange is main body identifcation only. Carburetor identification appears on identification tag.

Carburetor Application Chart, Part Two
Mustang 289 cid 8-Cylinder Engines, 1965-1968

Year	Note:	Type	Identification	Part Number	Replaced By:
65/67	2,12	F-4	C5OF-L,M,T,U	C4OZ-9510-F	C4OZ-9510-G
65/67	12	F-4	DOPF-V	C4OZ-9510-G	
68	10	H-4	C6PF-F	C4OZ-9510-G	
65	1,2	F-4	C4GF-AE,AF,AZ,BA	C5ZZ-9510-D	C9AZ-9510-AE
			C4ZF-C		
65	1	H-4	DOPF-U	C9AZ-9510-AE	
65/68	8	H-4	DOPF-AN	DOPF-9510-U	not replaced
65/68	9	H-4	C9OF-R	C9OZ-9510-N	
66	1,2,3	F-4	C6ZF-D	C6ZZ-9510-A	C9AZ-9510-AE
66	1,2,4	F-4	C6ZF-E	C6ZZ-9510-B	C9AZ-9510-AE
67	1,2,5	F-4	C7DF-A	C7DZ-9510-A	D2AZ-9510-M
67	1,2,5	F-4	C7DF-L,AE	C7DZ-9510-L	D2AZ-9510-M
67	1,2,11,13	F-4	C8PF-V	C7AZ-9510-AT	D2AZ-9510-M
67	1,2,14	F-4	C7DF-C,AG	C7DZ-9510-C	D2AZ-9510-M
67	1,2,6	F-4	C7DF-B	C7DZ-9510-B	D2AZ-9510-M
67	1,2,6	F-4	C7DF-M,AF	C7DZ-9510-M	D2AZ-9510-M
67	1,2,4	F-4	C7DF-D,AH	C7DZ-9510-D	D2AZ-9510-M
67	2,3,12,15	F-4	C8ZF-K	C7DZ-9510-E	C7DZ-9510-C
67	1	F-4	D2PF-SA	D2AZ-9510-M	

Notes:
1 Except High-Performance
2 Stamping identification located on edge of carburetor mounting flange.
3 With manual transmission and Thermactor.
4 With C4 transmission and Thermactor.
5 With manual transmission, except Thermactor.
6 With C4 transmission, except Thermactor.
7 With C4 transmission and Imco.
8 600 cfm
9 600 cfm, center pivot.
10 Canada only. Open emission.
11 With or without Thermactor.
12 High Performance.
13 Manual or Automatic transmission.
14 With Thermactor.
15 4-barrel carburetor with manual choke.

F-2 Ford 2V
F-4 Ford 4V
H-4 Holley 4V

Carburetor Application Chart
Mustang 302 cid 8-Cylinder Engines, 1968-1970

Year	Note:	Type	Identification	Part Number	Replaced By:
68	1,2	F-2	C8AF-BD;C8PF-T	C8ZZ-9510-H	D2AZ-9510-D
69	1,2	F-2	C8AF-BD;C8PF-T	C2AZ-9510-J	D2AZ-9510-D
68	1,8	F-4	C8AF-AS;C8PF-V	C2AZ-9510-K	D2AZ-9510-M
69	1,14	F-2	C9AF-A,T	C9AZ-9510-T	D2AZ-9510-D
69	1,13	F-2	C9ZF-G	C9ZZ-9510-G	D2AZ-9510-D
68/69	10	F-4	C8ZF-M	C8ZF-9510-M	D2AZ-9510-M
68/71	11	F-4	C8ZF-L	C8ZF-9510-L	D2AZ-9510-M
68/71	12	F-4	C8ZF-A,B	C8ZF-9510-A	D2AZ-9510-M
68	1,3	F-4	C8ZF-D	C8ZZ-9510-D	D2AZ-9510-M
68	1,4	F-4	C8ZF-C	C8ZZ-9510-C	D2AZ-9510-M
68/71	6	H-4	DOPF-AN	DOPZ-9510-U	
68/71	7	H-4	C9OF-R	C9OZ-9510-N	
69	4,5,8	H-4	C9ZF-J	C9ZZ-9510-J	DOZZ-9510-Z
69/71	5,15	F-4		DOZX-9510-A	
69/71	5,16	F-4		DOZX-9510-B	
70	2	F-2	DOAF-C	DOAZ-9510-C	D2AZ-9510-D
68/70	9	F-2	DOAF-U	DOAZ-9510-U	D2AZ-9510-D
70	9	F-2	DOAF-D	DOAZ-9510-D	D2AZ-9510-D
69/70		F-2	DOAF-D	DOAZ-9510-AC	D2AZ-9510-D
68/71	12	F-4	D2PF-SA	D2AZ-9510-M	
70		H-4	DOZF-Z	DOZZ-9510-Z	

Notes:
1 Stamping identification located on edge of carburetor mounting flange.
2 With manual transmission and Imco.
3 With C4 transmission and Imco.
4 With manual transmission and Thermactor.
5 Boss.
6 600 cfm
7 600 cfm, center pivot.
8 4-barrel carburetor with manual choke.
9 Automatic transmission, IMCO, and A/C.
10 Canada only. Manual transmission; open emission.
11 Canada only. Automatic transmission, open emission.
12 Canada only. Open emission.
13 With C4 or FMX transmissions, and Imco.
14 With C6 or FMX transmissions, and Imco.
15 Off Highway vehicles, in-line carb, 1-11/16" dia. throat, 875 cfm.
16 Off Highway vehicles, in-line carb, 2-1/4" dia. throat, 1425 cfm.

F-2 Ford 2V
F-4 Ford 4V
H-4 Holley 4V

Carburetor Application Chart
Mustang 351 cid 8-Cylinder Engines, 1969-1970

Year	Note:	Type	Identification	Part Number	Replaced By:
69	1,2	F-2	C9ZF-A	C9ZZ-9510-A	D2AZ-9510-D
69	2	F-4	C9ZF-C	C9ZZ-9510-C	D2AZ-9510-M
69	1,3	F-2	C9ZF-B	C9ZZ-9510-B	D0AZ-9510-AB
69	3	F-4	C9ZF-D	C9ZZ-9510-D	D2AZ-9510-M
69	4	F-2	C9AF-J,Z; C9ZF-L	C9AZ-9510-AD	D2AZ-9510-D
			C9ZF-M		
69	1,3	F-4	C9ZF-D	C9AZ-9510-AC	D2AZ-9510-M
69	5	F-4	D0PF-AG; D1OF-AAA	D0OZ-9510-U	D2AZ-9510-M
70		F-4	D0PF-AG; D1OF-AAA	D0OZ-9510-U	D2AZ-9510-M
69/71	1,6	H-4	D0PF-AN (Windsor)	D0PZ-9510-U	
69/71	7	H-4	C9OF-R (Windsor)	C9OZ-9510-N	
70	8	F-4	D0OF-C,Y,AC	D0OZ-9510-C	D2AZ-9510-M
70	9	F-4	D0OF-D,Z,AB	D0OZ-9510-C	D2AZ-9510-M
70	10	F-4	D0OF-G	D0OZ-9510-G	D2AZ-9510-M
70	11	F-4	D0OF-H,AA,AD	D0OZ-9510-H	D2AZ-9510-M
70	12	F-2	D0AF-E	D0AZ-9510-E	D2AZ-9510-D
70	13	F-2	D0AF-F	D0AZ-9510-F	D2AZ-9510-D
70	14	F-2	D0AF-V	D0AZ-9510-V	D2AZ-9510-D
69/70	3	F-2	D0AF-J,AR,AS,AT,AU,AV	D0AZ-9510-AB	D2AZ-9510-D
70		F-2	D2AF-FB,FC,FD,GB,GC,GD	D2AZ-9510-D	D2AZ-9510-P
			D2PF-GA,GB; D4PE-FA		

Notes:
1 Stamping identification located on edge of carburetor mounting flange.
2 With manual transmission and Imco, w/ or w/o Ram Air.
3 With FMX transmission and Imco, w/ or w/o Ram Air.
4 With manual transmission or FMX, and Imco.
5 With Imco.
6 Manual, C4 or FMX transmission, 600 cfm.
7 Center pivot, 600 cfm.
8 With C4, FMX & Imco., w/o A/C, 605 cfm.
9 With manual transmission and Imco, w/o A/C, 605 cfm.
10 With manual transmission, Imco, and A/C, 605 cfm.
11 With C4, FMX, Imco, & A/C, 605 cfm.
12 With manual transmission and Imco, 351 cfm.
13 With FMX transmission and Imco, 351 cfm.
14 With FMX transmission, A/C, and Imco, 351 cfm.

F-2 Ford 2V
F-4 Ford 4V
H-4 Holley 4V

Carburetor Application Chart
Mustang 390 cid 8-Cylinder Engines, 1967-1969

Year	Model	Notes	Type	Identification	Part No.	Replaced By
67	w/ M/T exc. T/E	1,2	H-4	C7OF-A	C7OZ 9510-L	C9AZ 9510-U
67	w/ C/6 exc. T/E	1,2	H-4	C7OF-B	C7OZ 9510-M	C9AZ 9510-U
67	w/ M/T & T/E	1,2	H-4	C7OF-C	C7OZ 9510-N	C9AZ-9510-U
67	w/ C6 & T/E	1,2	H-4	C7OF-D	C7OZ 9510-P	C8OZ 9510-D
68	GT w/ M/T	1,2	H-4	C7OF-A	C7OZ 9510-L	
	open emission					
68	GT, w/ A/T	1,2	H-4	C7OF-B	C7OZ 9510-M	C9AZ 9510-U
	open emission					
68	w/ A/T	1,4	F-2	C8AF-AZ,BA,BB	C8AZ 9510-AP	D2AZ 9510-D
				C8OF-AE,AF		
				C8PF-U,C8WF-A		
68	w/ C6 &	1,2	F-2	C8OF-K	C8OZ 9510-K	D2AZ 9510-D
	Imco w/ P.F.					
67		2,3	H-4	C8OF C	C8OZ 9510-C	C9AZ 9510-U
68	GT w/ M/T & T/E	2,3	H-4	C8OF C	C8OZ 9510-C	
68	GT w/ C6 & M/T	2,3	H-4	C8OF-D	C8OZ 9510-D	
67	w/ C6 & T/E	2,3	H-4	C8OF-D	C8OZ 9510-D	
69	w/ M/T & Imco	2	F-4	C9ZF-E	C9ZZ 9510-E	D2AZ 9510-M
	w/ or w/o Ram Air					
69	GT W/ C6 & Imco	2	F-4	C9ZF-F	C9ZZ 9510-F	D2AZ 9510-M
	w/ or w/o Ram Air					
69	GT W/ C6 & Imco	5	F-4	D0PF-AG	C0OZ 9510-U	D2AZ 9510-M
	w/ or w/o Ram Air					
68		6	F-2	D0AF-J, AF, AS,	D0AZ 9510-AB	D2AZ 9510-D
				AT, AU, AV		

Notes:
1 Stamping indentification is located on edge of carburetor mounting flange
2 Service package includes mounting gasket
3 Stamping indentification is located on choke plate flange
4 Package includes gasket, limiter, cap gasket, adaptor, screw
5 Package includes C8SZ 9447-A, gasket, limiter, spacer, plug & screw
6 Package includes gasket, limiter, cap, gasket, hose, washer, adaptor, screw

F-2 Ford Two-barrel
F-4 Ford Four-barrel
H-4 Holley Four-barrel

Carburetor Application Chart
Mustang 428 cid 8-Cylinder Engines, 1968-1970

Year	Model	Notes:	Type	Identification	Part Number	Replaced By:
68/69		1,2	H-4	C8OF-AB	C8OZ-9510-AB	C9AZ-9510-U
68	CJ w/ C6 & T/E	1,2	H-4	C8OF-AB	C8OZ-9510-AB	C9AZ-9510-U
68	CJ w/ M/T & T/E	1,2	H-4	C9AF-U	C8OZ-9510-AA	C9AZ-9510-U
68/70	CJ w/ C6 & T/E	2	H-4		C9AZ-9510-U	
68	GT500 w/ T/E		H-4	C9AF-N,ED	C9ZX-9510-A	
70	GT500	2	H-4	D0ZF-G,H,U,T,	C9AZ-9510-N	C9AZ-9510-U
70	M/T & T/E	2	H-4	AA,AB,AC,AD	D0ZZ-9510-H	

Notes:
1 Stamping identification located on edge of carburetor mounting flange.
2 Service package includes mounting gasket
H-4 Holley 4V

Carburetor Application Chart
Mustang 429 cid 8-Cylinder Engines, 1969-1970

Year	Model	Notes	Type	Identification	Part No.	Replaced By
69	Boss w/ M/T	1,2,5	H-4	C9AF-S	C9AZ 9510-S	
	& Thermactor				(CA-708)	
70	CJ w/ M/T, C6	2,3	R-4	D0OF-B	D0OZ 9510-B	
	& IMCO				(CA-771)	
	w/ or w/o A/C					
70	Boss w/ M/T	1,2,4	H-4	D0OF-S	D0OZ 9510-S	D0ZZ 9510-H
	& Thermactor				(CA-776)	(CA-781)
70	Boss w/ M/T	2,4	H-4	D0ZF-G,H,U,T	D0ZZ 9510-H	
	& Thermactor			AA,AB,AC,AD	(CA-781)	

Notes:

1 w/ manual choke
2 Service package includes mounting gasket
3 715 cfm
4 735 cfm
5 750 cfm
H-4 Holley four-barrel
R-4 Rochester Quadra-jet

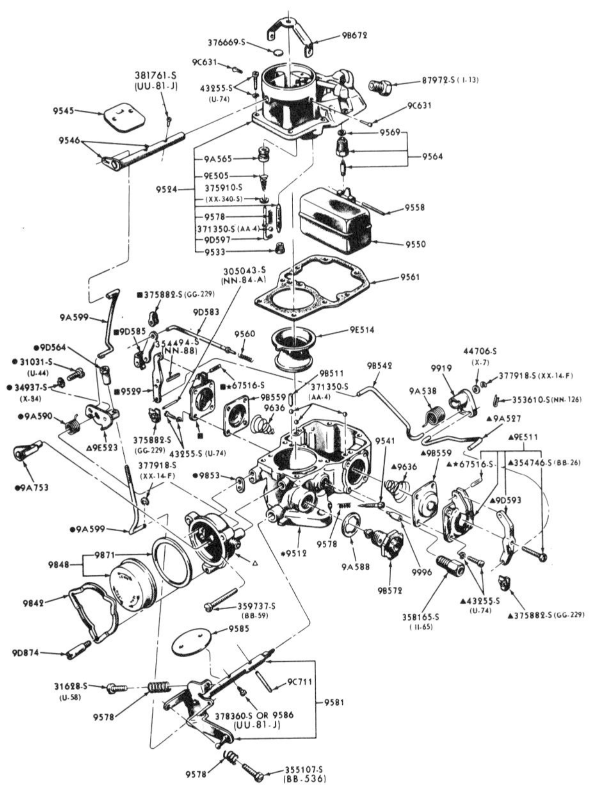

Exploded view of Ford IV carburetor for 1965 model 170 and 1966-1970 model 200ci engines. See chart for application.

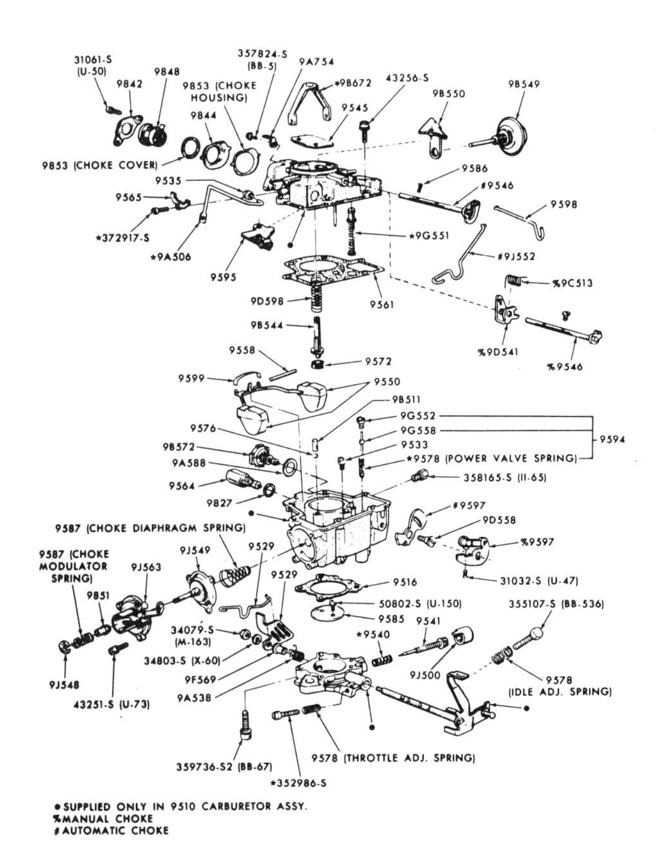

Exploded view of a Holley IV carburetor. See chart for application.

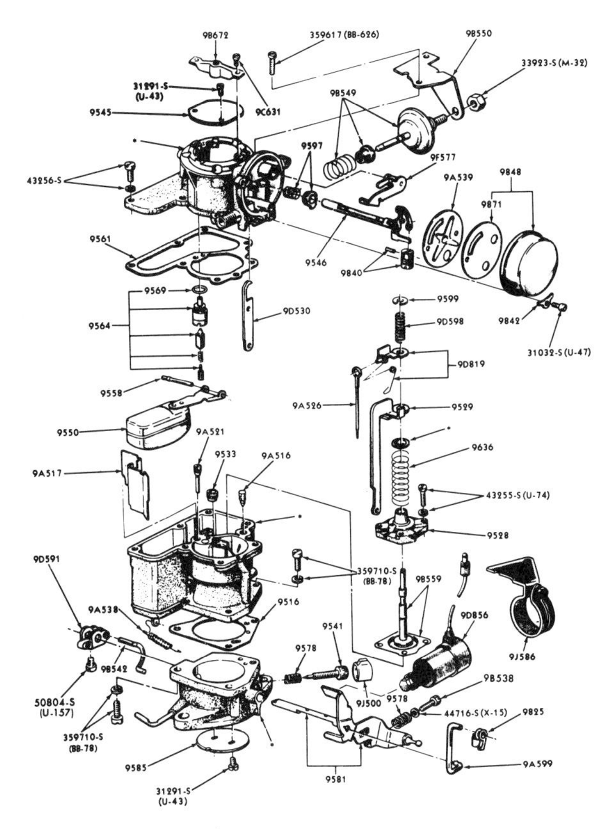

Exploded view of Ford 1V carburetor used on 1969-1970 250ci engines. See chart for application.

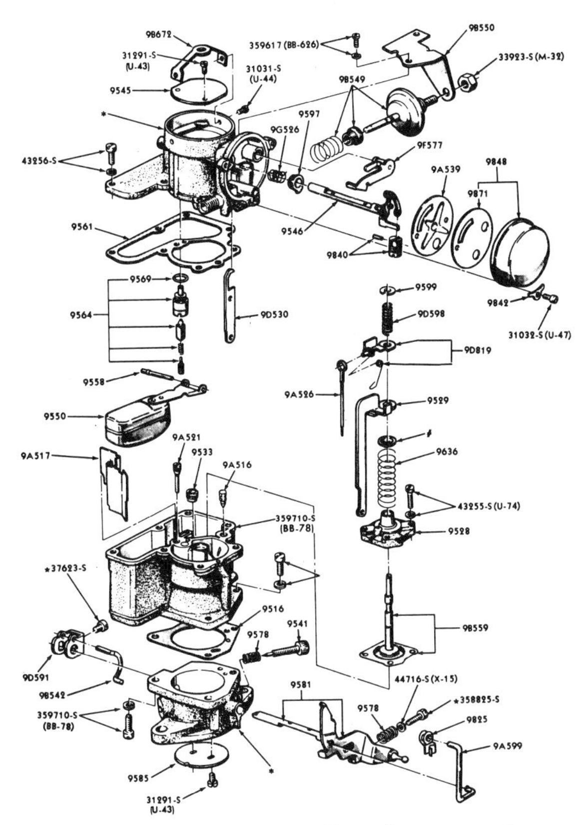

Exploded view of a Carter 1V carburetor used on 1967 200ci engines with IMCO. See chart for application.

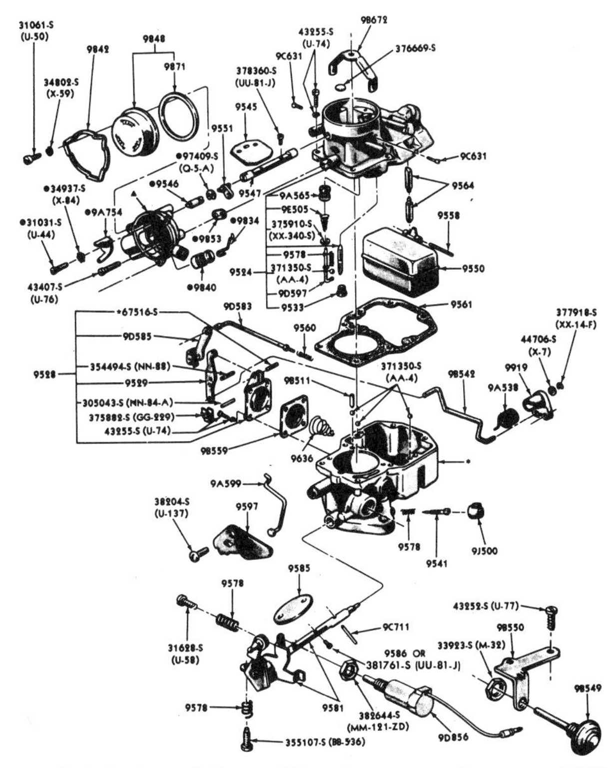

Exploded view of Carter 1V carburetor used on some 1970 model 200ci engines. See chart for application.

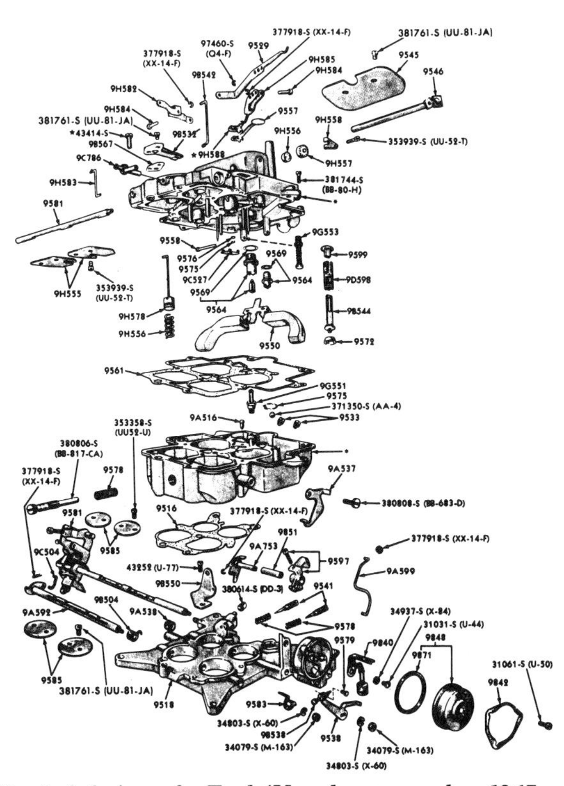

Exploded view of a Ford 4V carburetor used on 1967 and later engines. See chart for application.

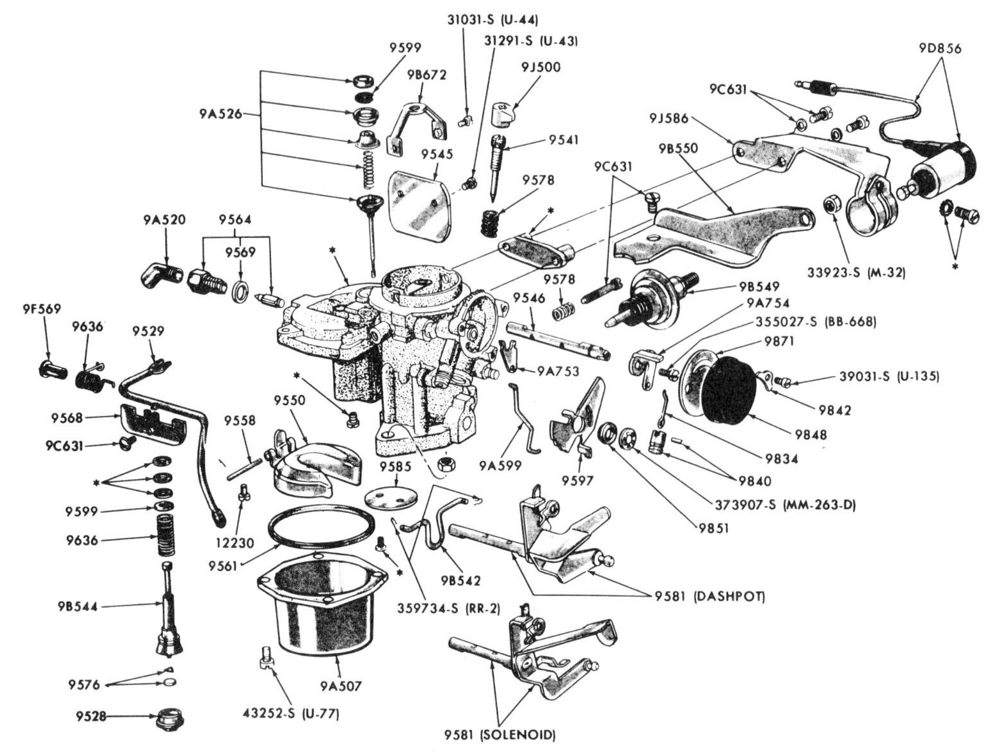

Exploded view of Carter IV carburetor used on 1970 250ci engines. See chart for application.

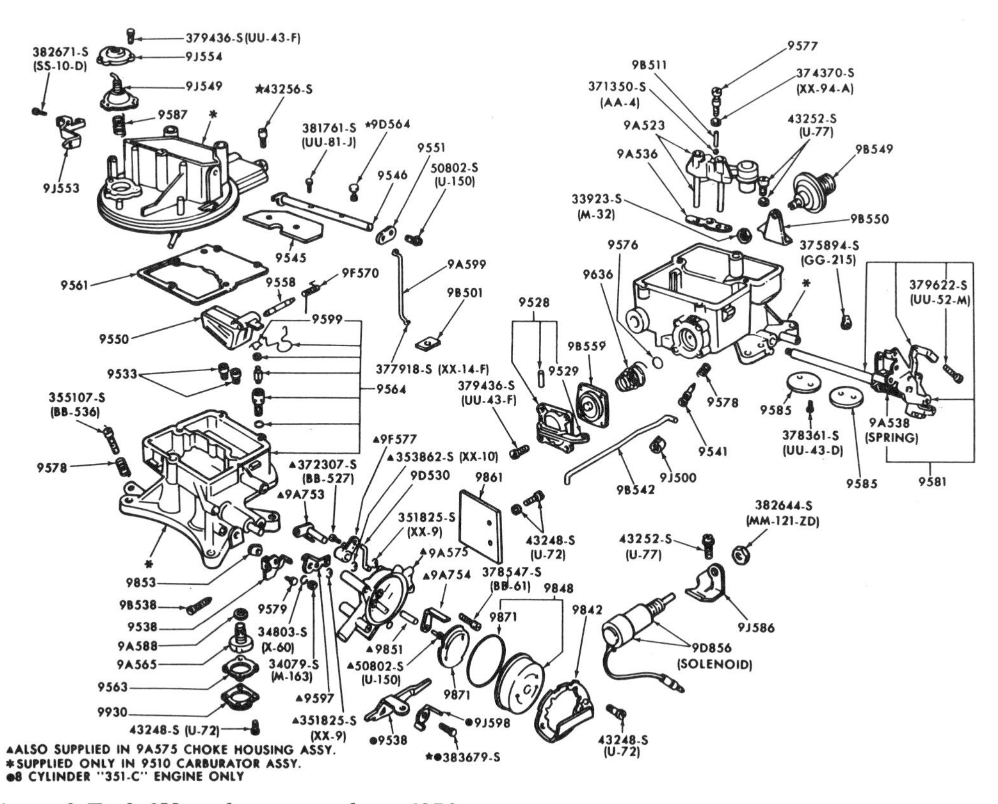

Exploded view of Ford 2V carburetor used on 1970 Mustangs equipped with V-8 engines. See chart for application.

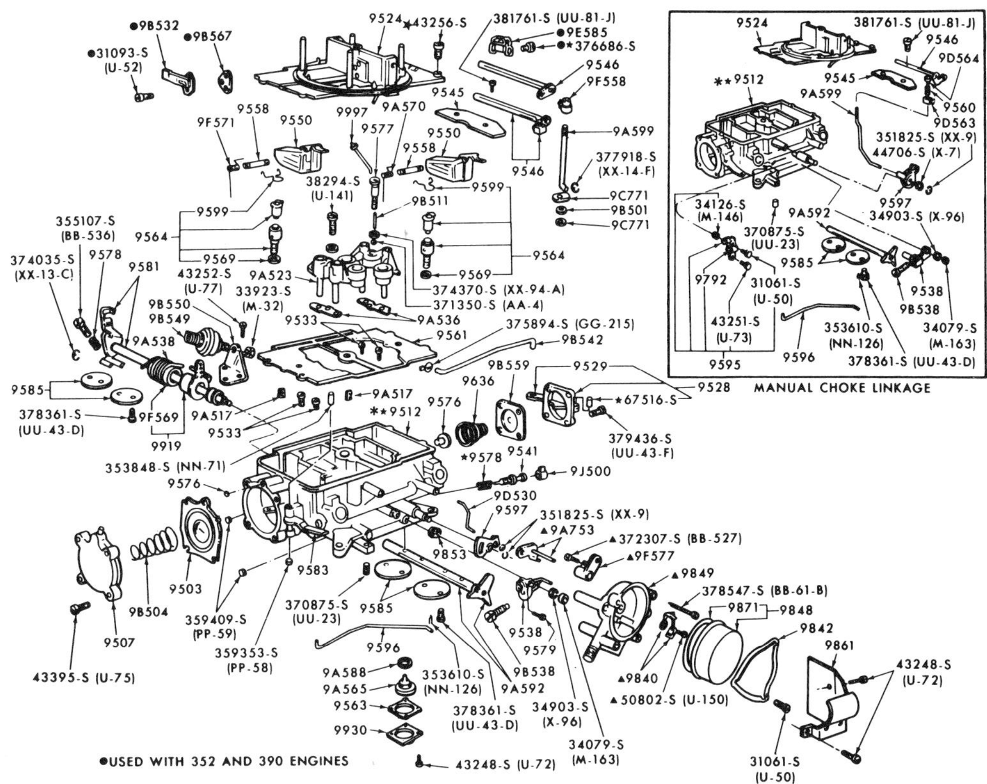

Exploded view of Ford 4V carburetor with piston-type choke and accelerator pump check valve. See chart for application.

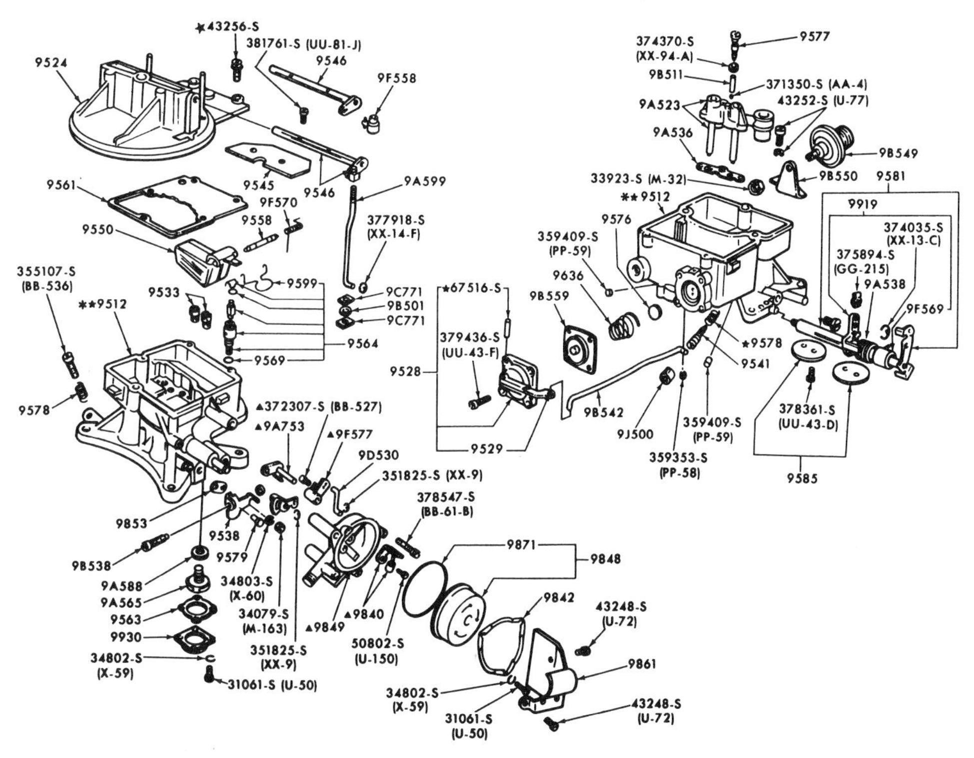

Exploded view of a Ford 2V carburetor with piston-type choke and accelerator pump check valve. See chart for application.

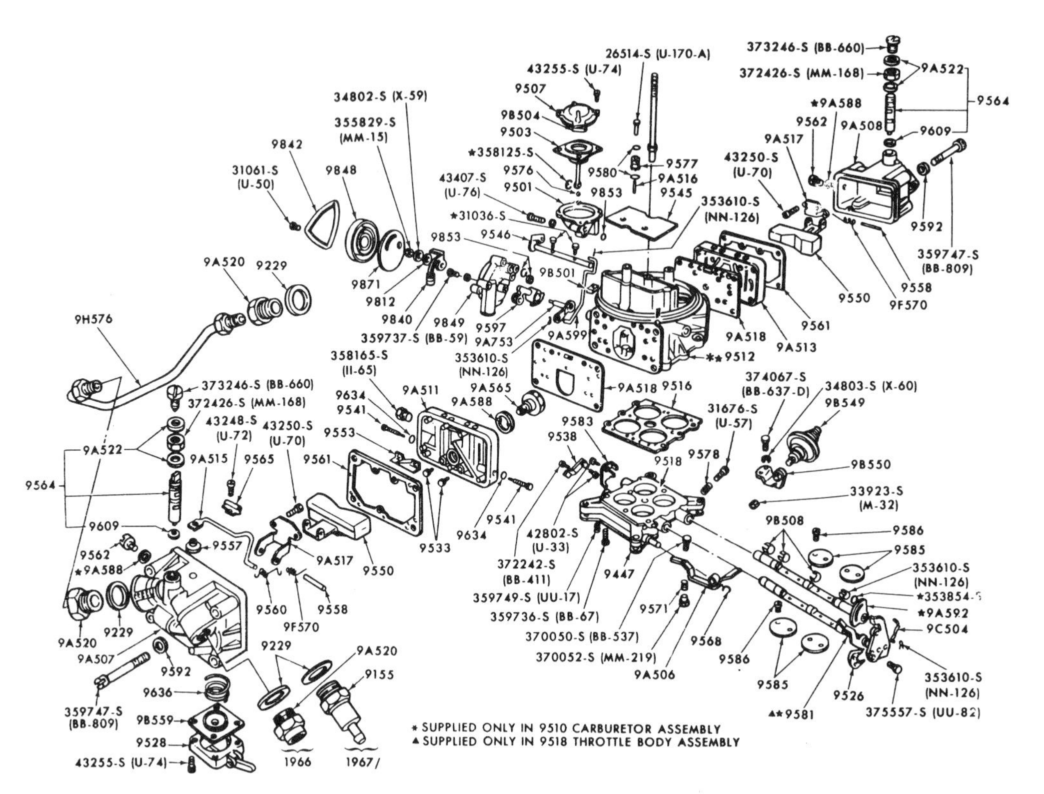

Exploded view of Holley 4V carburetor used on Boss 302, 429 CJ and SCJ, and Boss 429 engines. See chart for application.

FUEL PUMP ASSEMBLY CHART, for Six-Cylinder Mustangs

Year	Cyl/CID	Part Number	Description
65	6/170	C3DZ-9350-A	ident. by 6803 stamped
		r/b C3AZ-9350-Y	on pump mounting flange
65	6/170-200	C3AZ-9350-T	ident. by 0227 stamped
		r/b C3AZ-9350-Y	on fuel pump mounting flange
65	6/170-200	C3AZ-9350-V	ident. by 0485 stamped ON
		r/b C3AZ-9350-Y	on mounting flange
65	6/170, 200	C3AZ-9350-Y	CARTER - ident. by 3938-S
			stamped on mounting flange
65	6/170, 200	C5DE-99350-B	CARTER - ident. by 3913 stamped
		OR D2TE-9350-RA	on mounting flange
		r/b C3AZ-9350-Y	
66/67	6/200	C5UZ-9350-A	ident. by 0290 stamped on
		r/b C5UZ-9350-C	mounting flange
66/67	6/200	C5UZ-9350-C	CARTER - ident. by 4092 stamped
			on mounting flange
68/70	6/200	C8DZ-9350-A	CARTER - ident. by 4532-S or
		r/b D3TZ-9350-A	4531-S stamped on mounting flange
69	6/250	C8DZ-9350-A	CARTER - ident. by 4532-S or
		r/b D3TZ-9350-A	4531-S stamped on mounting flange
68/70	6/200	D3TZ-9350-A	identified D3TE-AA and also 6399-S
			on mounting flange - CARTER
69	6/250	D3TZ-9350-A	identified D3TE-AA and also 6399-S
			on mounting flange - CARTER

FUEL PUMP ASSEMBLY CHART, for Eight-Cylinder Mustangs

Year	Cyl./CID	Part Number	Description
65	8/260, 289	C3AZ-9350-M	CARTER-18 deg. inlet angle & 3/8"
			inlet tube, ident. by 3734-S
			Stamp on Mount. Flange
			used w/ oil pressure gauge
65	8/260, 289	C3AZ-9350-S	CARTER - ident. by 3911-S
		r/b C5OZ-9350-A	Stamp on Mount. Flange
65	8/289 (Exc. K)	C3AZ-9350-S	CARTER - ident. by 3911-S
		r/b C5OZ-9350-A	Stamp on Mount. Flange
65	8/289-4V K	C5OZ-9350-A	CARTER - ident. by 3939-S
			Stamp on Mount. Flange
67/68	8/390 (GT)	C6AZ-9350-B	CARTER - ident. by 4194-S
			Stamp on Mount. Flange
67/68	8/289, 302 (GT350)	S7MK-9350-B	Supercharger
	4/B CARB)		
66/67	8/289 (Exc. K)	C5AZ-9350-B	CARTER - ident. by 4193
			Stamp on Mount. Flange
66/67	8/289K (4/B Carb)	C6ZZ-9350-A	CARTER - ident. by 4201
			Stamp on Mount. Flange
68	8/289 (2/B Carb)	C8OZ-9350-A	CARTER - ident. by 4567
		r/b D0AZ-9350-B	Stamp on Mount. Flange
68/69	8/302	C8OZ-9350-A	CARTER - ident. by 4567
		r/b D0AZ-9350-B	Stamp on Mount. Flange
69	8/351	C8OZ-9350-A	CARTER - ident. by 4567
		r/b D0AZ-9350-B	Stamp on Mount. Flange
69/70	302-4V Boss	C9ZZ-9350-A	CARTER = ident. by 4910-S
			Stamp on Mount. Flange
68	8/427K	C7AZ-9350-A	CARTER = ident. by 4441-S
68/70	8/428CJ	C7AZ-9350-A	Stamp on Mount. Flange
68	8/390 (2/B Carb)	C7SZ-9350-A	CARTER - ident. by 4385
		r/b D4TZ-9350-A	Stamp on Mount. Flange
69	8/390	C7SZ-9350-A	CARTER - ident. by 4385
		r/b D4TZ-9350-A	Stamp on Mount. Flange
69/70	429 BOSS (4/B Carb)	C9AZ-9350-A	CARTER - ident. by 4842-S
			Stamp on Mount. Flange
70	8/351C	D0AZ-9350-A	CARTER - ident. by D0AE-E &
			also 4861-S
			Stamp on Mount. Flange
70	8/302 (2/B Carb)	D0AZ-9350-B	CARTER - ident. by 4896-S
		r/b D3OZ-9350-B	OR SA Stamp on Mount. Flange
70	8/302 (2/B Carb)	D3OZ-9350-B	CARTER - ident. by MS-6477-S
			Stamp on Mount. Flange
70	8/351W	D0AZ-9350-C	CARTER - ident. by 4888-S OR
			SA Stamp on Mount. Flange

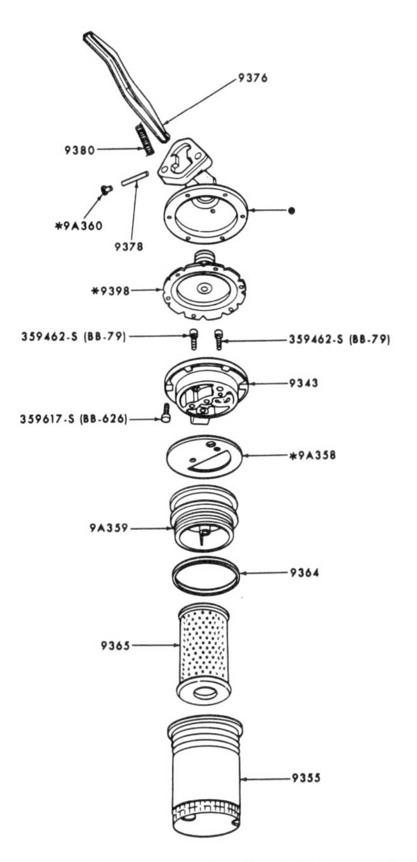

Fuel pump and filter assembly for 1965 model 260 and 289ci engines.

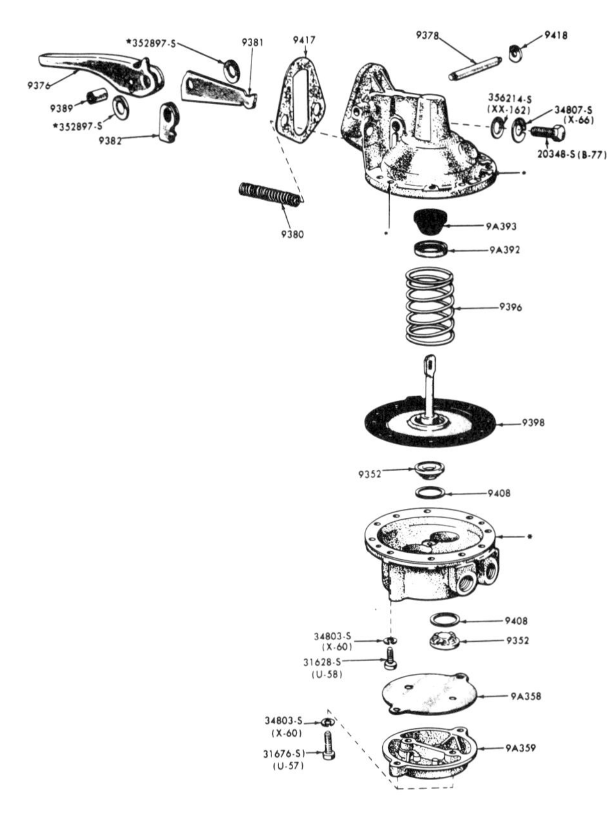

Fuel pump assembly for 1967-1970 Mustangs equipped with 390ci engine, and 1968-1970 Mustangs equipped with 428ci engine.

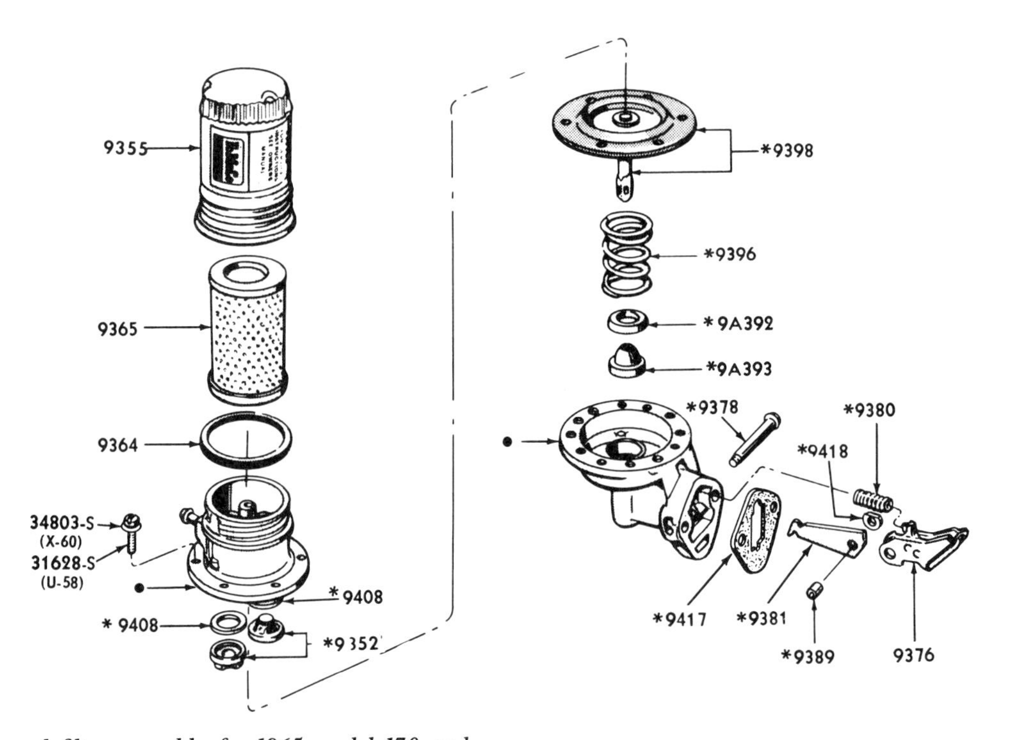

Fuel pump and filter assembly for 1965 model 170 and 200ci engines.

Mustang Fuel Tank Applications

Year	Model	Part Number	Description
65/68		C5ZZ-9002-D	16 GAL. CAPACITY
69		C9ZZ-9002-A	20 GAL. CAPACITY
70	WITHOUT EV/EM	D0ZZ-9002-A	22 GAL. CAPACITY
70	WITH EV/EM	D0ZZ-9002-B	20 GAL. CAPACITY

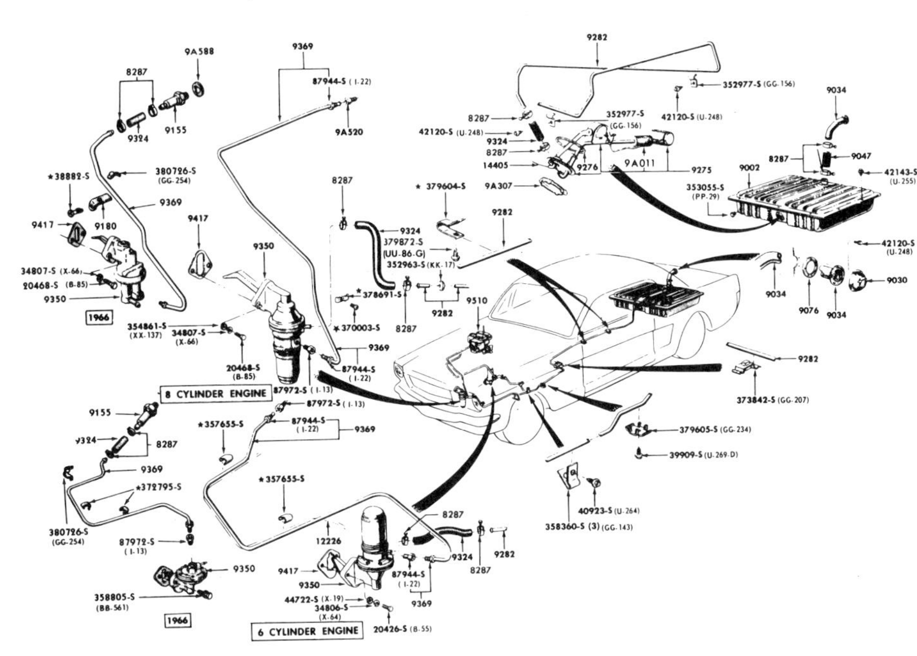

1965-1968 Mustang complete fuel delivery system and mounting hardware.

MUSTANG FUEL TANK FILLER CAP APPLICATION CHART

Year	Model	Part Number	Description
66/67	GT	C7ZZ-9030-C	Ident. by letters "GT"
68	GT/CS	C8ZZ-9030-C	Pop-Off type cap assy. -
			Ident. by Mustang Emblem
67	Exc. GT	C8ZZ-9030-C	Pop-Off type cap assy. -
			Ident. by Mustang Emblem
67	GT	C8ZZ-9030-D	Pop-Off type cap assy. -
			Ident. by letters "GT"
68	GT Exc. GT/CS	C8ZZ-9030-3	Pop-Off type cap assy. -
			Ident. by letters "GT"
69	GT	C9ZZ-9030-B	Pop-Off type cap assy. -
			Ident. by letters "GT"
69	Mach 1	C9ZZ-9030-C	Pop-Off type cap assy. -
			Ident. by Mustang Emblem
65		C9ZZ-9030-A	Ident. by Mustang Emblem
66/68		C9ZZ-9030-A	Ident. by Mustang Emblem
69	Exc. GT AND Mach 1	C9ZZ-9030-A	Ident. by Mustang Emblem
70	Exc. Mach 1	C9ZZ-9030-A	Ident. by Mustang Emblem
	w/o EV/EM		
70	Exc. Mach 1, w/ EV/EM	C0ZZ-9030-A	Ident. by Mustang Emblem
70	Mach 1 w/ EV/EM	D0ZZ-9030-D	Pop-Off type cap assy. -
			Ident. by Mustang Emblem
70	Mach 1 w/o EV/EM	D0ZZ-9030-E	Pop-Off type cap assy. -
			Ident. by Mustang Emblem

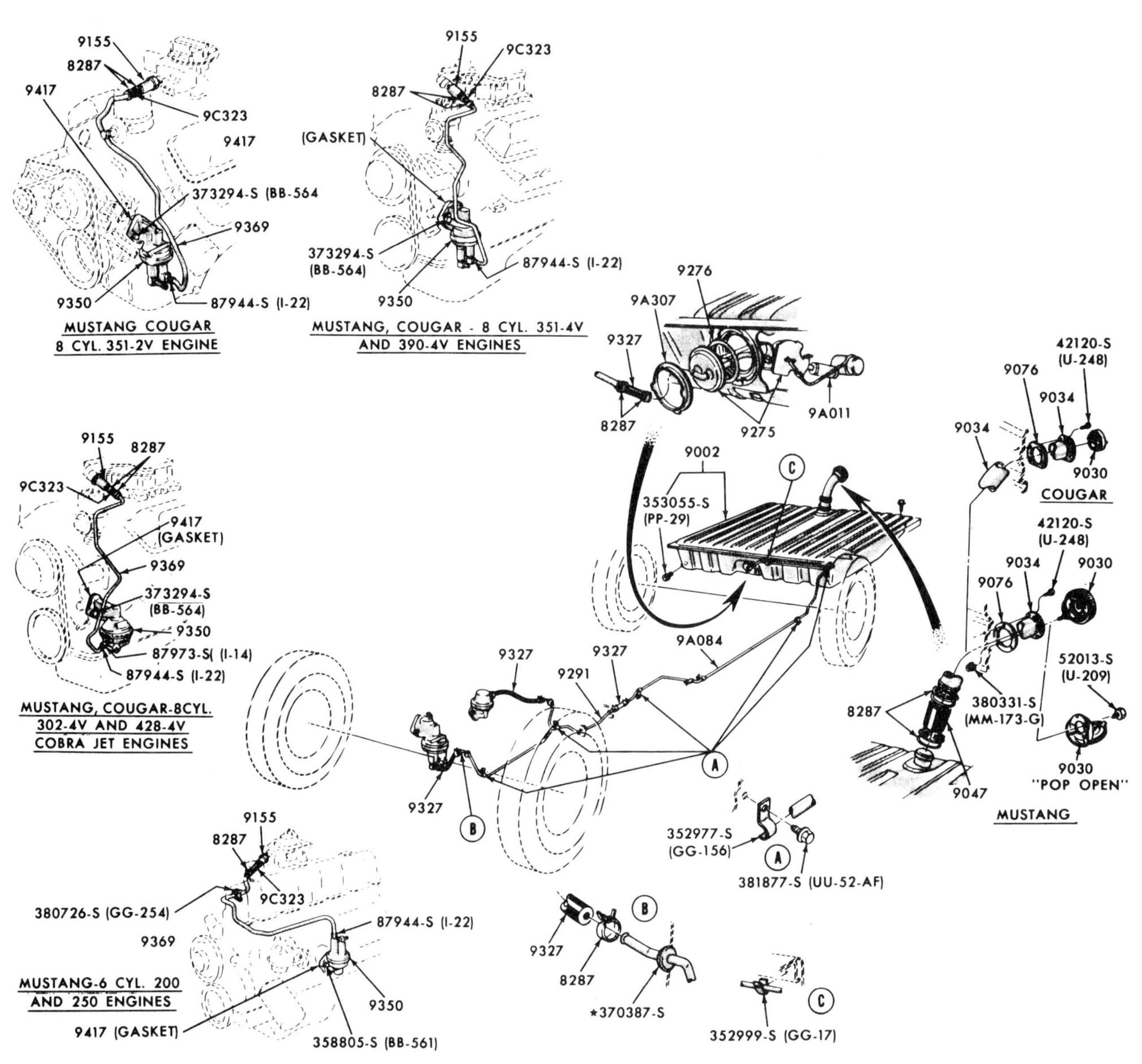

1969 Mustang fuel system and related parts. All models except Boss 429.

MUSTANG FUEL TANK UNIT GAUGE ASSEMBLY CHART

Year	Model	Part Number	Description
65/68	w/o low fuel warn system	C8ZZ-9275-C	includes C8ZZ-9275-B gauge assy.,
			C0AF-9276-A gasket, C8AF-9327-A
67/68	with low fuel warn system	C8ZZ-9275-D	includes C8ZZ-9275-B gauge assy.,
			C0AF-9276-A gasket, C8AF-9327-A
69	w/o low fuel warn system	C9WY-9275-A	gauge & outlet tube assy. - identified
			C9WF-9275-A 3/8" line-service pkg.
			includes C0AF-9276-A
70	with low fuel warn system	D0WY-9275-A	includes gauge assy. C0AF-9276-A
			gasket identified D0WF-9275-A

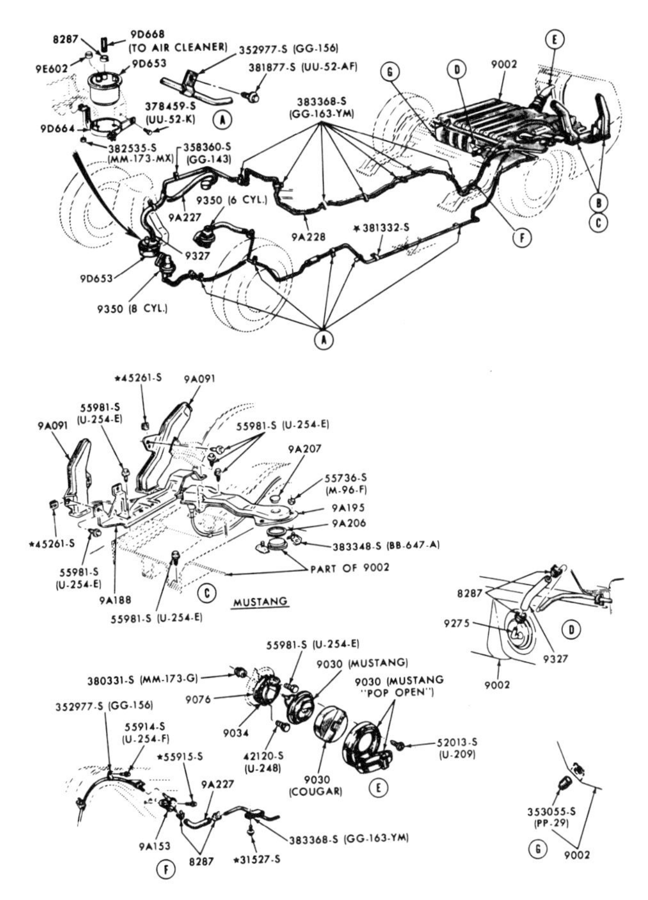

1970 Mustang fuel system without evaporative emission controls.

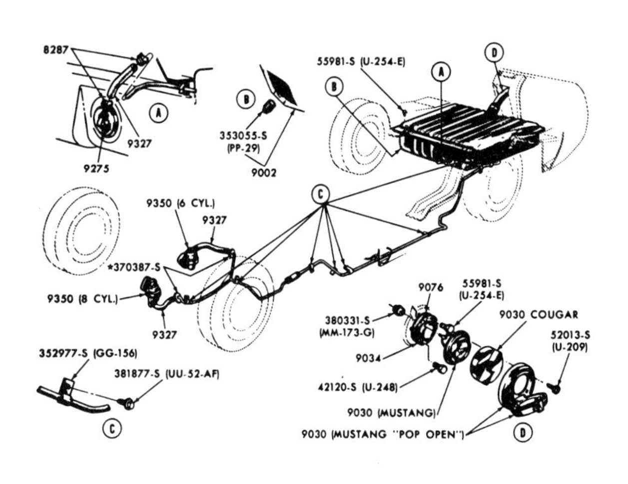

1970 Mustang fuel system with evaporative emission controls.

Chapter 5

Drivetrain

A Mustang is made up of hundreds of subassemblies. The steering wheel, the radio, and most of the various wiring harnesses and electrical pigtails are subassemblies that are composed of numerous smaller parts. Most minor subassemblies, such as the rearview mirrors, door panels, and window regulators, are manufactured and assembled by independent companies under contract to Ford Motor Company. Major subassemblies, such as the engine, transmission, and rear axle, are manufactured by Ford and built at different divisions within the company.

Each plant has its own internal systems and production procedures. To categorize components and to keep track of running changes that affect parts procurement, each subassembly is equipped with a small metal tag. Each tag is attached in the same place and in a similar manner. The coded information on the tag tells where the assembly was manufactured, what car line it was manufactured for (Ford, Torino, Mustang, and so on), the date it was assembled, the number of changes that have been implemented since its conception, and other miscellaneous facts that help describe the component in greater detail.

It may be that the exploded views of Mustang drivetrains in this chapter will save the restorer weeks of headaches and mysteries. Even the weekend mechanic will find that the disassemblies and reassemblies necessary for repairs and maintenance are made easier with a map. For that reason, suspension, brakes, steering, wheels, and exhaust are presented as a general running gear and underchassis chapter. Because most of the components in these subsystems are not in general day-to-day view of the average car owner, they tend to be more of a mystery to the novice than, say, interior parts or exterior trim.

Fortunately, the complications inherent in the mix-and-match nature of drivetrain components do not apply here. Through this chapter the reader will find that certain applications are quite narrow, and the possibilities for swaps are reduced by safety considerations or lack of interchangeability.

The real value of this chapter can be found in the exploded views, where the differences in model years and body styles become more evident. Someone rebuilding a steering system, for instance, can be more assured that the replacement components are compatible; that person also can take advantage of correlated part numbers when dealing with Ford or other vendors for such replacement parts.

Many Mustangers wish to upgrade their braking and steering systems to power assist, and the information herein will provide an exacting guide to necessary parts. But hobbyists are reminded that any work on the steering or braking systems requires a dedication to safety. Ford's original specifications for parts of those systems still apply as minimum requirements. If there are any questions as to brake parts or steering linkage compatibility, the advice of experts should be obtained.

Mustang Transmission Applications, 1964-1970

Three-Speed Manual Transmissions
Ford 2.77
Ford 3.03

Four-Speed Manual Transmissions
Dagenham
Ford Heavy Duty
Borg Warner T-10
Ford Close Ratio HD

Automatic Transmissions
C-4 Cruise-O-Matic
C-4 Select-Shift Cruise-O-Matic
C-6 Select-Shift Cruise-O-Matic
FMX Select-Shift

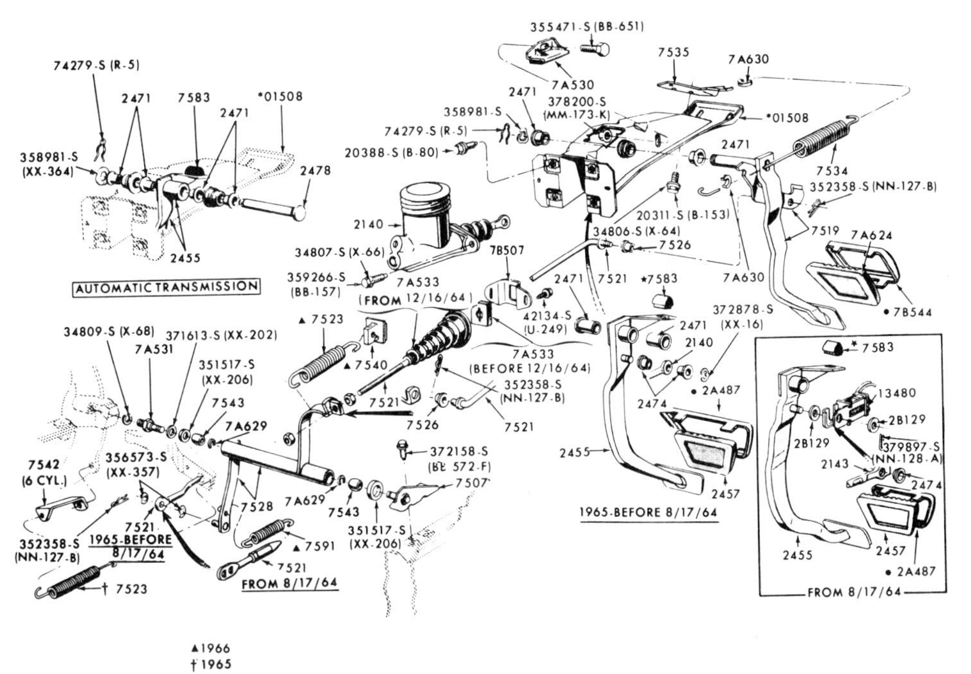

All first-generation Mustangs used a ridged mechanical linkage between the pedal and the clutch-release bearing. A rod (7521) connected to the pedal activates a lever (7528) called a bellcrank. The crank pushes an adjustable rod (also 7521) that pushes a clutch fork (7515) that is actually a lever. At the end of the fork is a sealed bearing (7580). The bearing presses the clutch fingers (part of pressure plate 7563) and disengages the clutch disc (7550). The size of the clutch disc and pressure plate and the configuration of the bellcrank differ between six- and eight-cylinder Mustangs. In later years, high-performance engines were outfitted with extra-heavy-duty clutches and bellcranks. Interestingly, the factory service manual provides instructions for lubricating the hood hinges but never addresses the clutch pedal bushings (2471), a high-wear item.

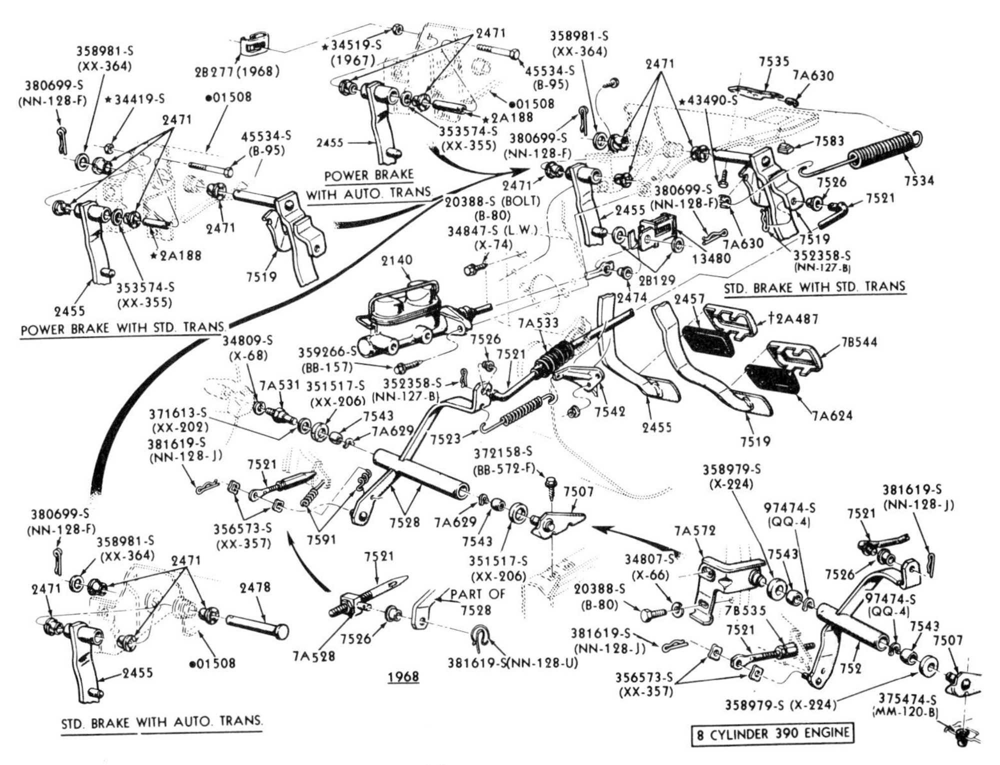

The 1967-1969 Mustang clutch and brake pedal assembly.

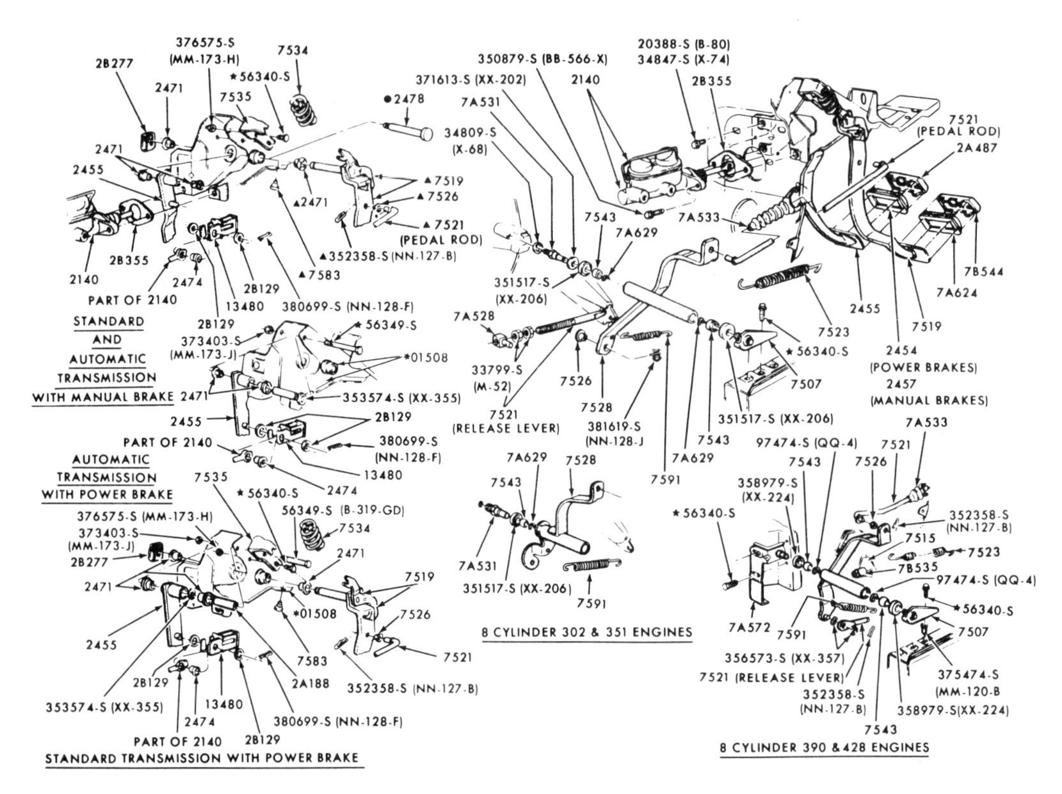

The 1969-1970 Mustang clutch and brake pedal assembly.

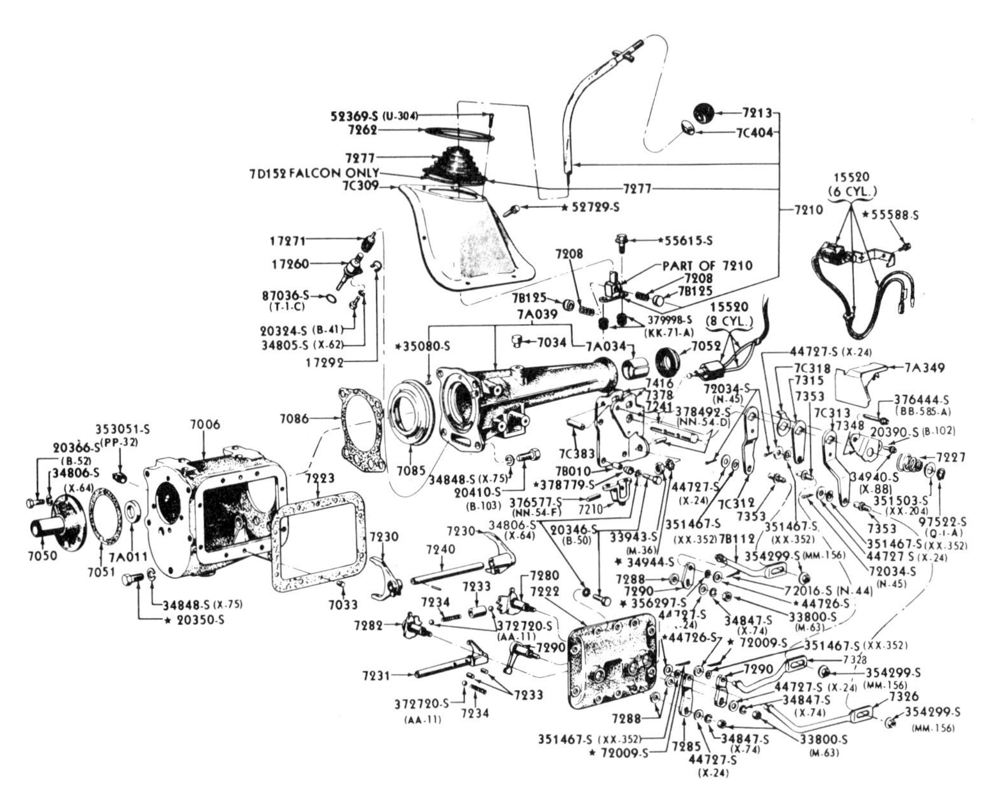

The Ford shifter for the Ford 4-speed transmission, 1965-1967. Factory shifters incorporate a long throw, in reference to the distance the lever must travel between gear selections. Long-throw shifters require less effort compared to short-throw shifters, and that accomplished one of Ford's primary goals: to make Mustangs equipped with manual transmissions fun and easy to drive.

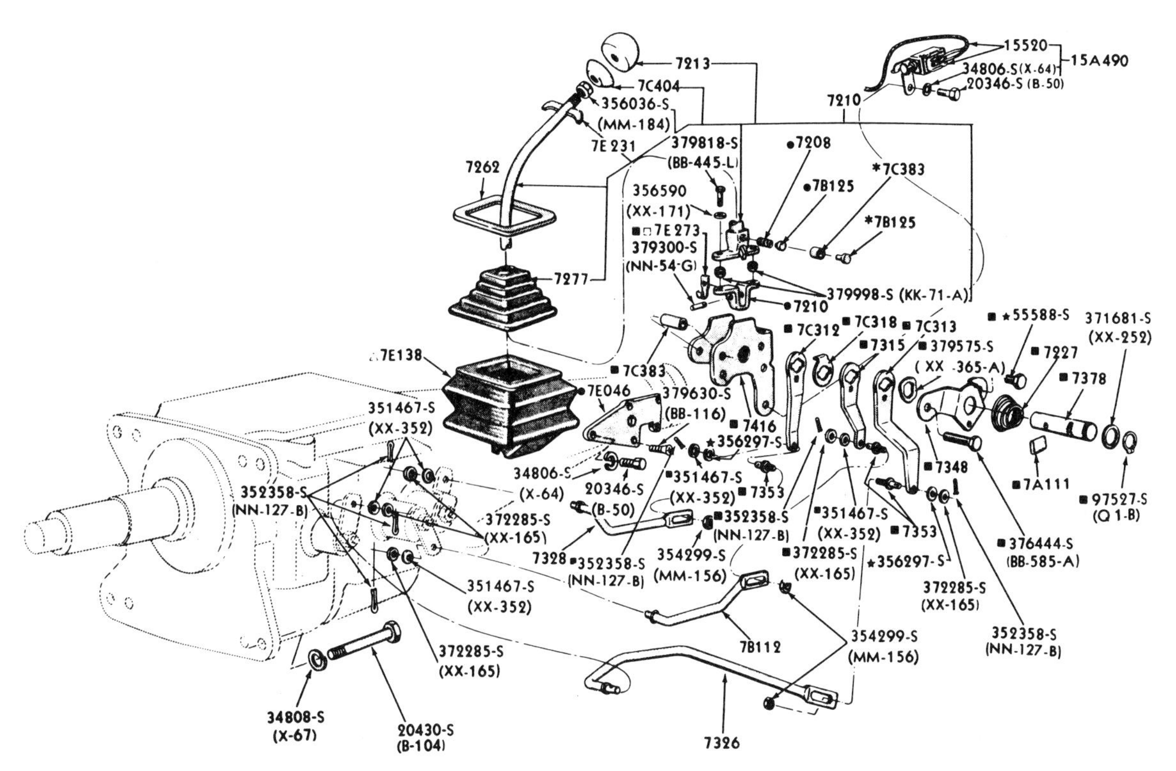

Four-speed shifter and linkage used from 1965 through 1968 on Ford-built Toploader transmissions installed in Mustangs.

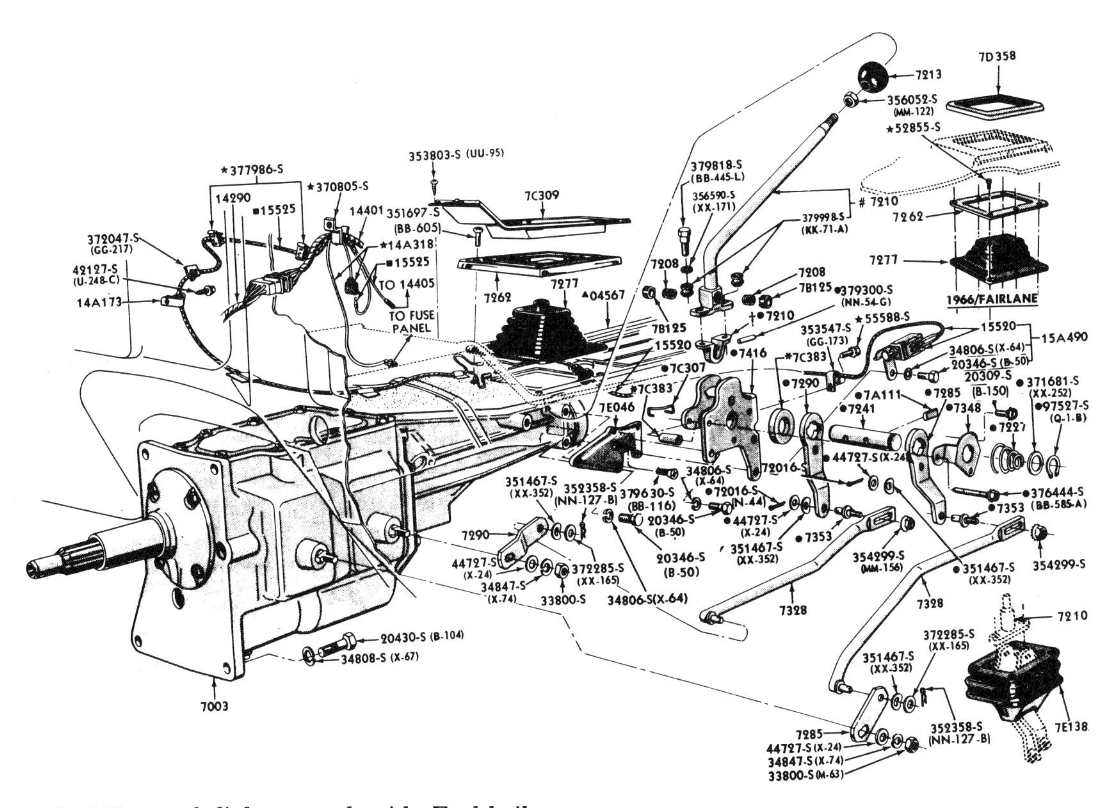

Three-speed shifter and linkage used with Ford-built transmissions between 1965 and 1968.

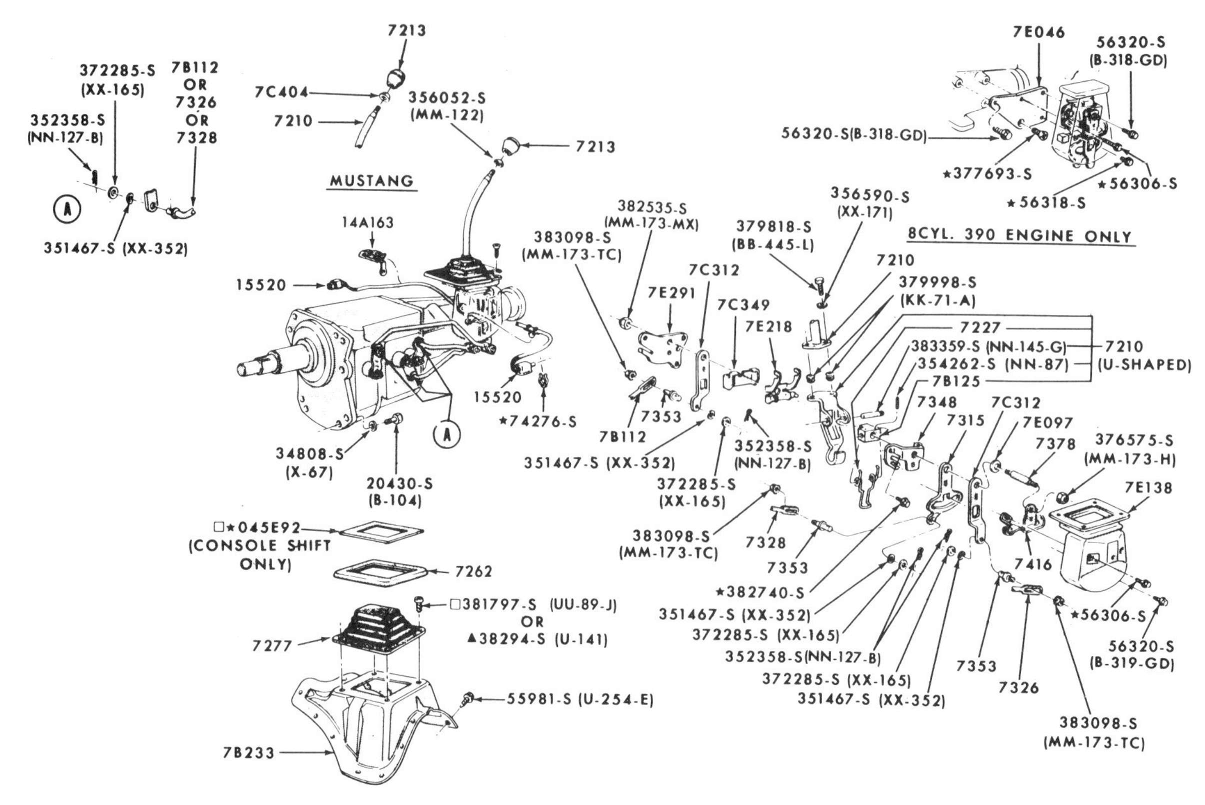

A typical view of a shifter and linkage installed in 1969 Mustangs equipped with Ford-built Toploader four-speed transmissions.

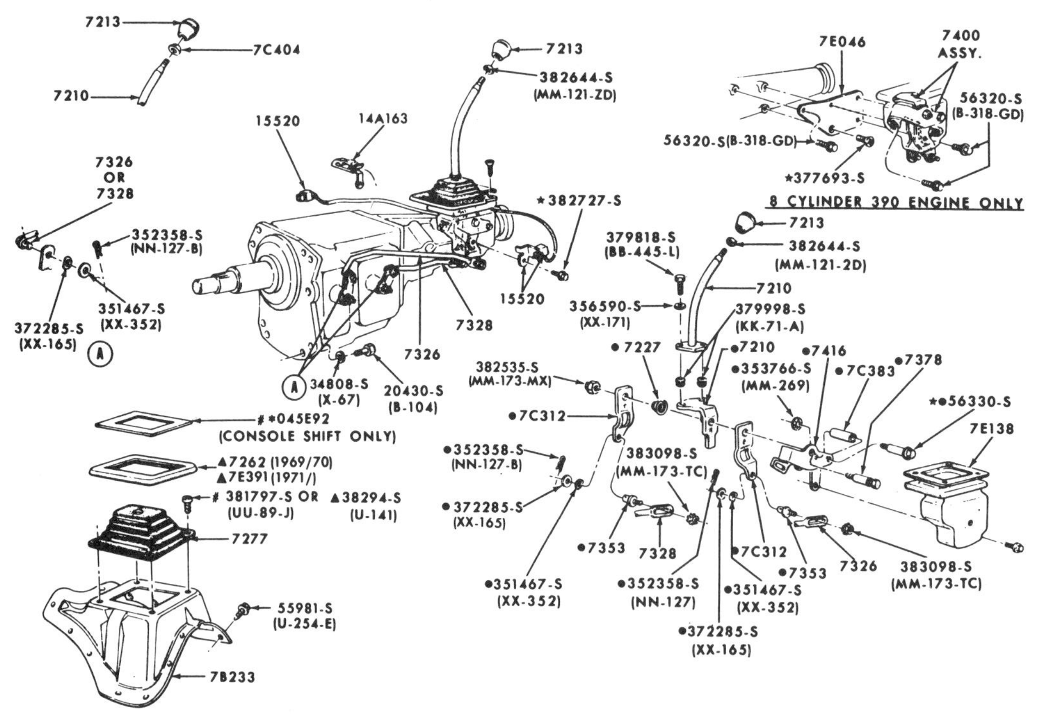

Mustang three-speed transmission shifter and linkage used with 1969-1970 Ford-built transmissions.

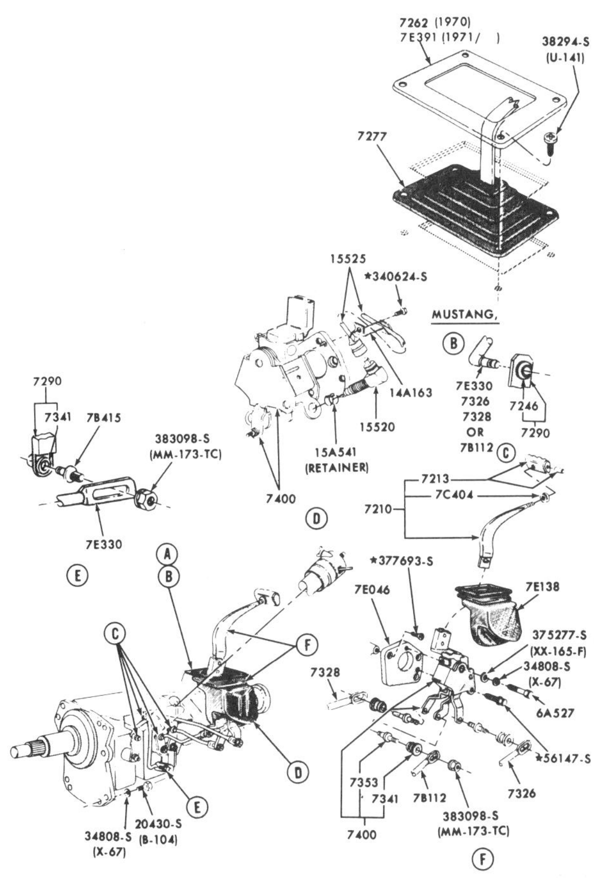

The four-speed Hurst shifter for a Ford transmission. In 1970, Ford began installing Hurst™ four-speed shift handles in Mustangs equipped with high-performance engines. The rectangular-shaped chrome handles are embossed with the letters H-U-R-S-T on each side. Using Hurst shift handles was a ploy by Ford's marketing experts to improve the company's image among high-performance enthusiasts. Aftermarket short-throw Hurst shifters are well known for their quick-shift engagement. But Ford installed only the Hurst shift handle and levers, not the entire assembly. The remainder of the linkage connecting the shifter to the transmission was Ford's own. The result is an improved Ford shifter that compromises Hurst's original performance standards.

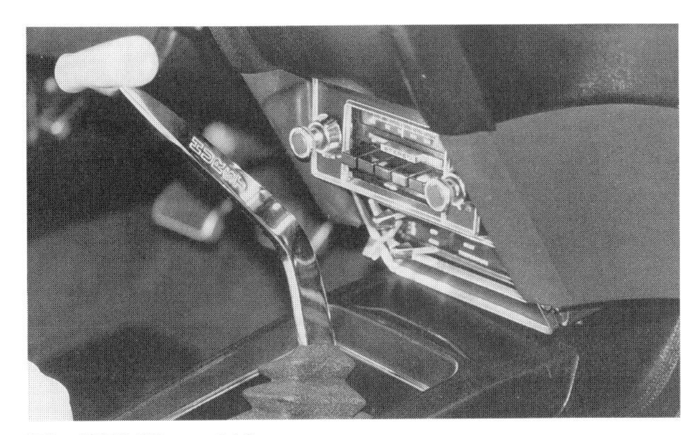

The 1970 Hurst shifter.

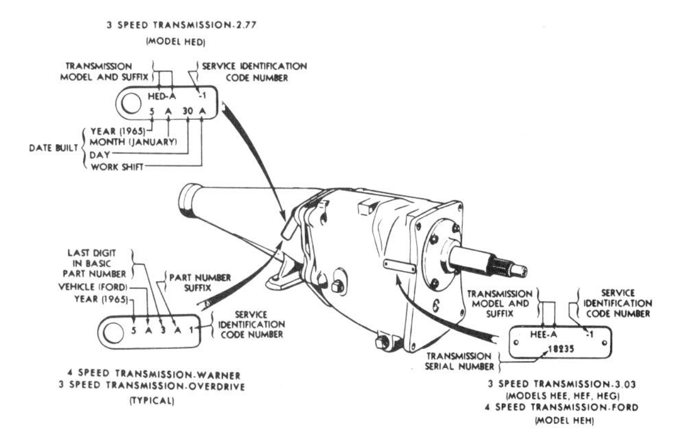

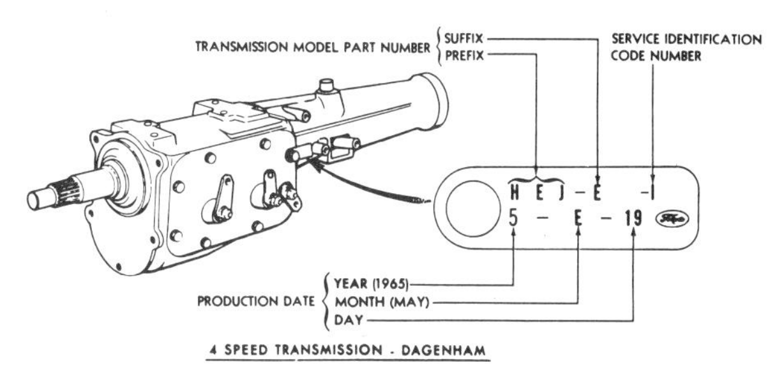

Manual transmission identification for 1965.

Mustangs were outfitted with one of five different manual transmissions: a 2.77 three-speed, a 3.03 three-speed, a Dagenham four-speed, a Ford four-speed, or a Warner built four-speed. The 2.77 and 3.03 are commonly thought to represent gear ratios but actually are the measurement (in inches) between the countershaft and the mainshaft. In other words, a transmission with a 3.03 designation indicates the center of the countershaft is 3.03in from the center of the mainshaft. A tag affixed to the gear housing or tailshaft extension contains coded information indicating the transmission's model number, service identification code number, and the date of manufacture.

Mustang Transmission Tag Codes By Year and Drivetrain, 1965 and 1966.

Year	Tag Code	Engine	Trans Type	Notes
1965	HED-AV	170	2.77	
	CHED-AV, C4Z,3,A	170	2.77	
	HEJ-B	170	D/4/S	
	HED-BD	200	2.77	
	CHED-BD, C5Z,3,A	200	2.77	
	HEJ-D	200	D/4/S	Before 7/1/65
	HEJ-E	200	D/4/S	After 7/1/65
	HEF-BB	260	3.03	
	HEF-BK	289	3.03	
	HEF-CS	289	3.03	
	HEF-CV	289	3.03	
	HEH-G	260, 289	F/4/S	Before 8/20/64
	HEH-P	260, 289	F/4/S	From 8/20/64 to 12/30/64
	HEH-BR	260, 289	F/4/S	From 12/30/64 to 2/1/65
	HEH-BT	289	F/4/S	From 2/1/65
	HEH-S	289 HiPo	F/4/S	Before 8/20/64
	HEH-T	289 HiPo	F/4/S	From 8/20/64 to 10/1/64
	HEH-BX	289 HiPo	F/4/S	After 10/1/64
	HEK-M	289	W/4/S	Before 11/23/64
	HEK-V	289	W/4/S	After 11/23/64
	HEK-R	289	W/4/S	
1966	HEJ-E	200	D/4/S	
	HED-BG	200	2.77	
	RED-B	200	2.77	
	HEF-CV	289	3.03	Before 11/1/65
	HEF-CW	289	3.03	After 11/1/65
	HEK-AD	289	W/4/S	
	HEH-BW	289	F/4/S	
	HEH-BX	289 HiPo	F/4/S	

KEY:
F - Ford-built transmission
W - Borg Warner-built transmission
D - Dagenham (English-built) transmission
3/S - Three forward gears (speeds)
4/S - Four forward gears (speeds)

Mustang Transmission Tag Codes By Year and Drivetrain, 1967 to 1969.

Year	Tag Code	Engine	Trans Type	Notes
1967	RAN-S	200	3.03	
	RAN-S1	200	3.03	
	RUG-E	289	F/4/S	
	RUG-E1	289	F/4/S	
	RUG-E2	289	F/4/S	
	RUG-N	289 HiPo	F/4/S	
	RUG-N1	289 HiPo	F/4/S	
	RUG-N2	289 HiPo	F/4/S	
	RAT-N	390	3.03	
	RAT-N1	390	3.03	
	RUG-M	390	F/4/S	
	RUG-M1	390	F/4/S	
	RUG-M2	390	F/4/S	
1968	RAN-S1	200	3.03	
	RAN-D1	289	3.03	
	RUG-E2	289-302	F/4/S	
	RAT-U	302	3.03	
	RAT-N1	390 GT	3.03	
	RUG-M2	390 GT	F/4/S	w/ 2.78 low gear
	RUG-AD	390 GT	F/4/S	w/ 2.32 low gear
	RUG-S	428	F/4/S	
	RUG-AE	428	F/4/S	
	RUG-AE1	428	F/4/S	
1969	RAN-AM	200-250	3.03	
	RAT-U1	302	3.03	
	RAT-AM	351	3.03	
	RUG-AG	302-351	F/4/S	w/ 2.32 low gear
	RUG-E3	302-351	F/4/S	w/ 2.78 low gear
	RUG-AD1	390	F/4/S	w/ 2.32 low gear
	RUG-M3	390	F/4/S	w/ 2.78 low gear
	RUG-AE2	428-429	F/4/S	

KEY:
F - Ford-built transmission
W - Borg Warner-built transmission
D - Dagenham (English-built) transmission
3/S - Three forward gears (speeds)
4/S - Four forward gears (speeds)

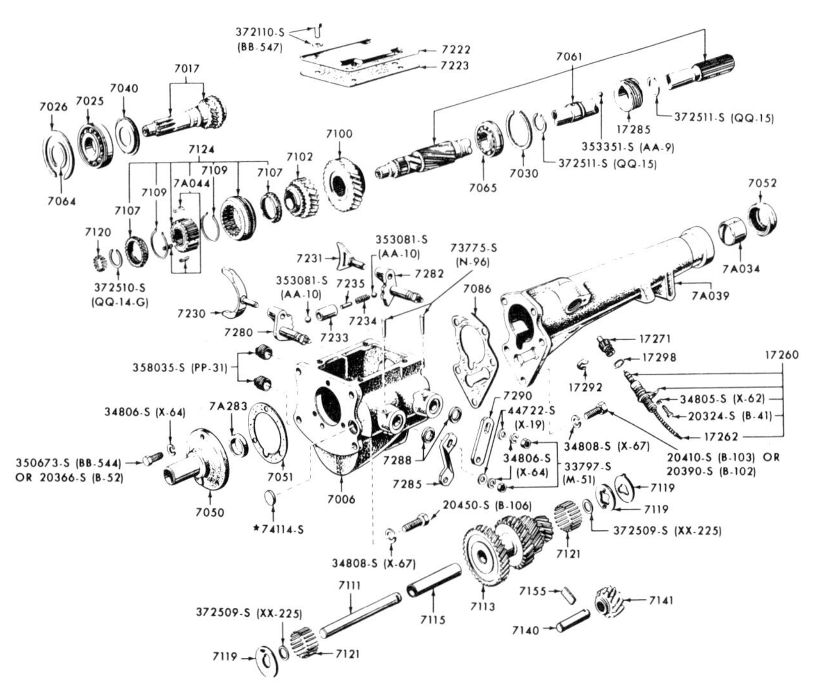

The shift levers extend through the side of the main case of a Ford transmission, and there are no shift rails. Once the mainshaft is assembled, it is lowered into the case through an opening in the top, hence the term Toploader. A stamped-tin cover is used to seal the opening. Ford-built 2.77 three-speed transmissions have a non-synchronized first gear, which means first gear cannot be engaged while the car is moving. These light-duty gearboxes were installed behind six-cylinder engines in 1965 and 1966 Mustangs. By 1967, the fully synchronized 3.03 three-speed transmission had replaced the 2.77.

Mustang Transmission Tag Codes By Year and Drivetrain, 1970.

Year	Tag Code	Engine	Trans Type	Notes
1970	RAN-AV	200-250	3.03	
	RAN-AV1	200-250	3.03	
	RAT-BA	302	3.03	
	RAT-BA1	302	3.03	
	RAT-BB	351	3.03	
	RAT-BB1	251	3.03	
	RAT-BB2	351	3.03	
	RUG-AV	302-351	F/4/S	w/ 2.78 low gear
	RUG-AV1	302-351	F/4/S	w/ 2.78 low gear
	RUG-AW	302-351	F/4/S	w/ 2.32 low gear
	RUG-AW1	302-351	F/4/S	w/ 2.32 low gear
	RUG-AZ	428-429	F/4/S	
	RUG-AZ1	428-429	F/4/S	

KEY:
F - Ford-built transmission
W - Borg Warner-built transmission
D - Dagenham (English-built) transmission
3/S - Three forward gears (speeds)
4/S - Four forward gears (speeds)

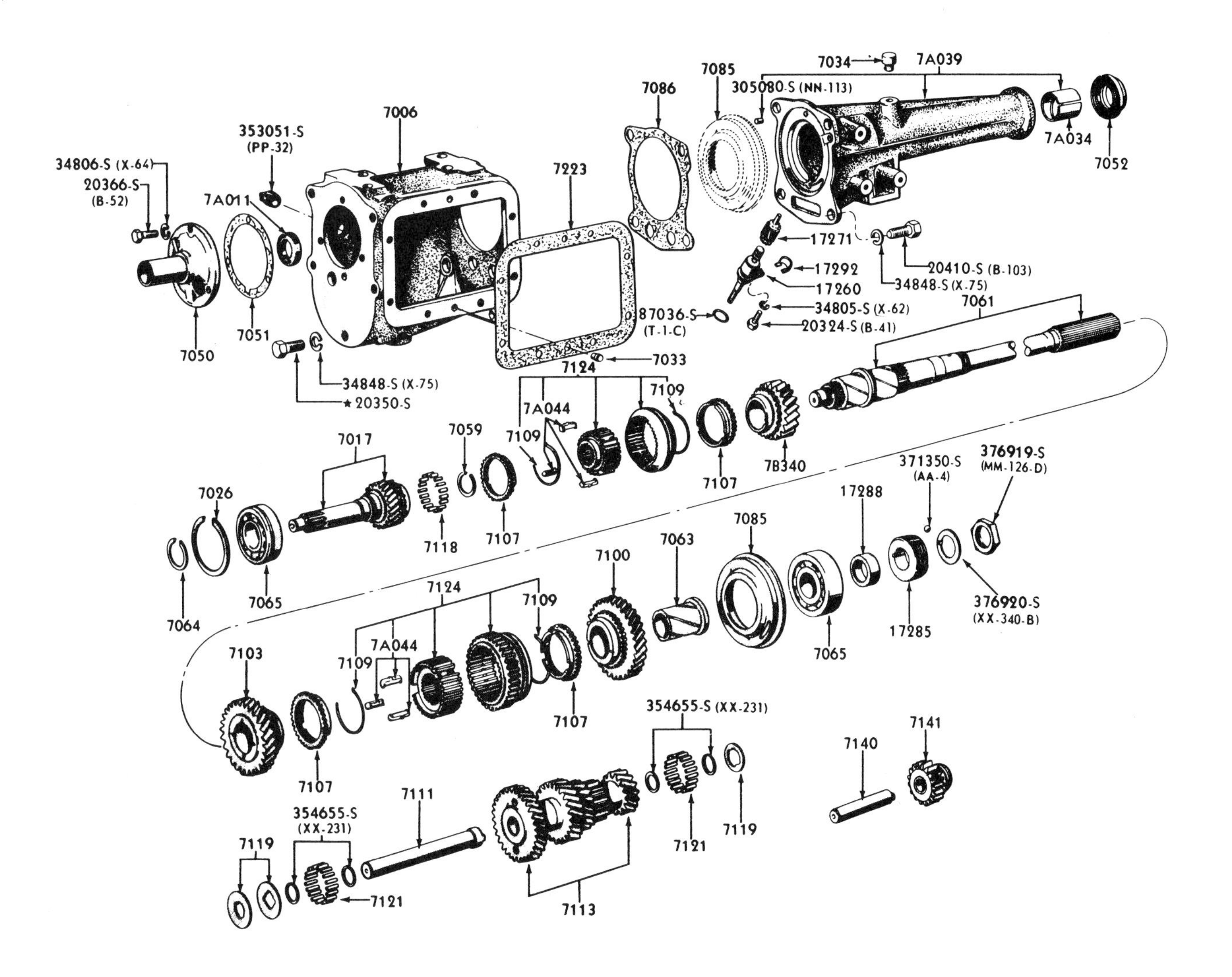

English-built Dagenham transmissions were lightweight units designed to offer the six-cylinder Mustang buyer an optional fourth gear. The transmission's durability was questionable at best, and they were quickly phased out early in 1966.

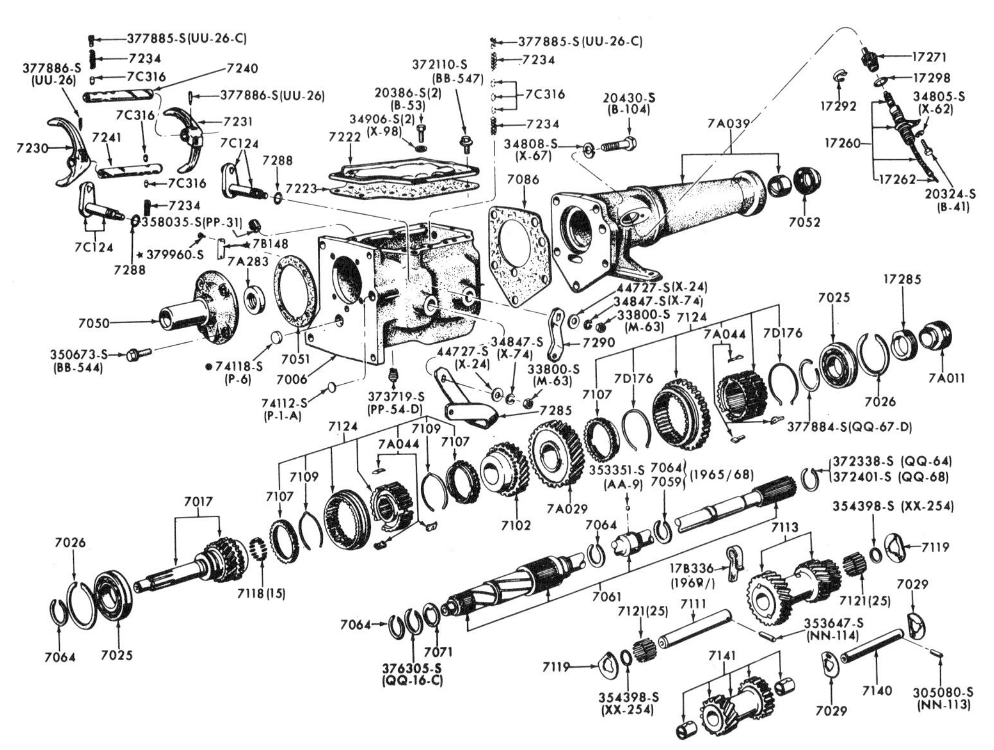

An exploded view of a fully synchronized Ford-built 3.03 transmission. The engineering term "3.03" is the distance in inches between the center of the countershaft and the center of the mainshaft.

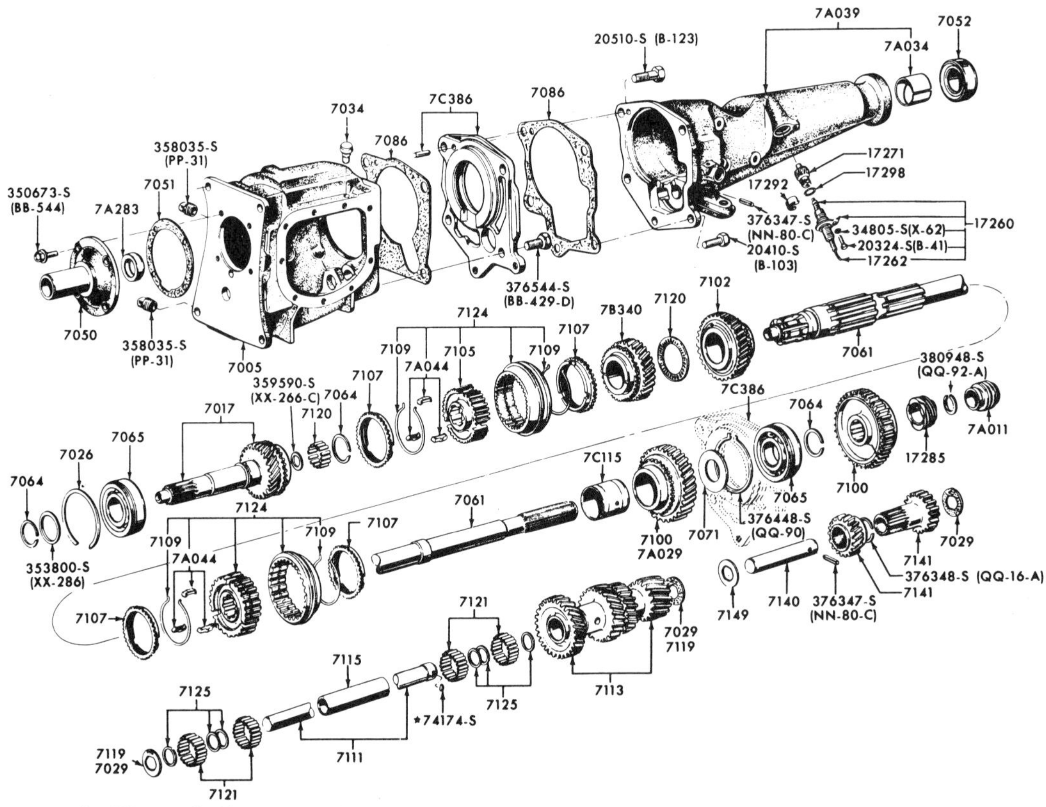

The side cover of a Warner-built transmission contains the shift rails, forks, and a shift rail interlock mechanism. A shift rail interlock will not allow two gears to be engaged at the same time. The side cover provides the only access to the main case once the transmission is assembled. Warner three-speed transmissions are fully synchronized, and were installed in some 1965 and 1966 V-8-powered Mustangs.

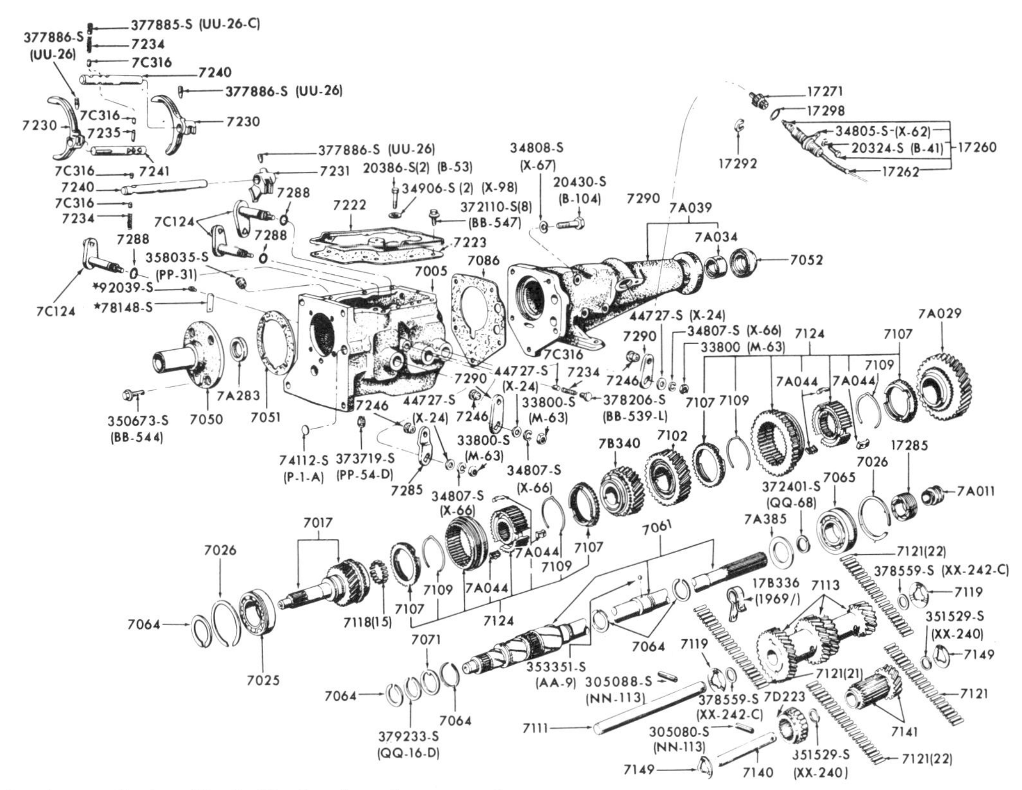

Exploded view of the Ford Toploader four-speed transmission.

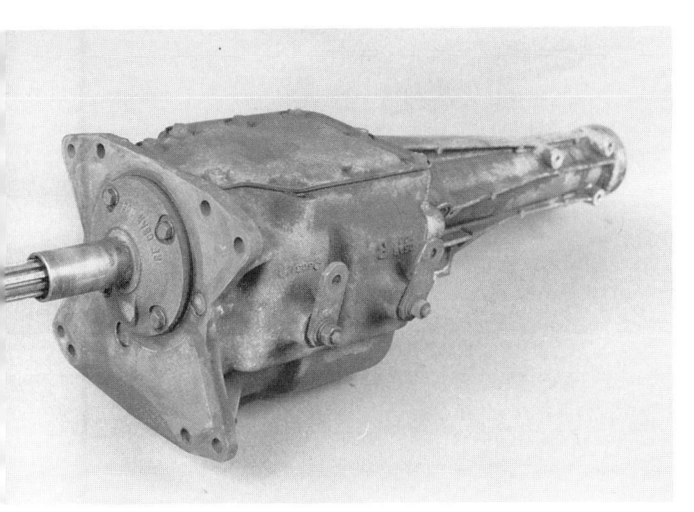

Ford's Toploader four-speed is still hailed as the strongest, most reliable manual transmission ever built. Numerous high-performance versions were built beginning in 1967 when Ford began installing big-block engines in Mustangs. Designed at a time when total vehicle weight and maximum fuel mileage were not major issues, more was better, or in this case, stronger. The Toploader's gear case and tailhousing were made of heavy cast iron, and each gear and shaft on average, 15 percent larger than necessary. Some rare race-bred Toploaders installed behind 428 Cobra Jet engines were outfitted with cast-aluminum tailhousings in an effort to save weight.

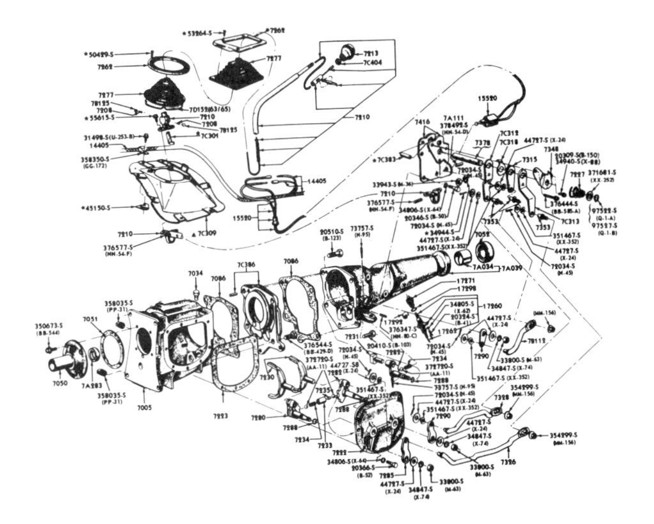

The four-speed Warner transmission shifter. Though they looked similar from inside the passenger compartment, Ford-built shifters were unique to a particular type and manufacturer of transmission. A four-speed lever designed to fit a Warner-built transmission would not fit a Toploader Ford unit, for example. All Ford-type four-speed shifters incorporate a manual release reverse lock-out lever or trigger located near the knob. Ford typically used slotted shift rods as a means for adjustment, a system designed for quick, efficient assembly at the factory. Aftermarket performance shifters normally use stronger, threaded rods.

MUSTANG CLUTCH DISC AND PRESSURE PLATE CHART

YEAR	MODEL AND DESCRIPTION	CID	TRANS.	*SPRINGS	DISC DIAM.	MATCHING DISC	MATCHING PLATE
65	6 Cyl.	170	All	6	8-1/2"	@C5DZ-7550-A	@C2OZ-7563-C
		200	All	6	8-1/2"	^@C5DZ-7550-A	@C5OZ-7563-B
	8 Cyl.	260	3.03	6L-Pink	10"	@C6OZ-7550-G	@C2OZ-7563-D
				6S-Orange			
	8 Cyl. exc. K	289	All	6L-4S	10-1/2"	@C5AZ-7550-F	@C2OZ-7563-D
	6 Cyl.	200	2.77	6	8-1/2"	~@C5DZ-7550-B	#@C5OZ-7563-B
66	6 Cyl.	200	D/4/S	6	8-1/2"		
65/66	8 Cyl., K-Code	289	All 4/S	8 Gray	10-1/2"	@C7ZZ-7550-B	@C7ZZ-7563-B
66	6 Cyl.	200	2.77	6 Alum.	9"	@C6OZ-7550-E	@C6DZ-7563-C
66/67	8 Cyl. exc. K	289	All	6L-6S	10"	@C6DZ-7550-C	@C2OZ-7563-D
	Before 5/1/67						
67	8 Cyl. exc. K	289	All	6L-6S	10"	@C6DZ-7550-C	@C2OZ-7563-F
	From 5/1/67						

* L=Large, S=Small

@ Supplied only as a remanufactured item

^ Before 5/8/65

~ From 6/8/65

\# From 2/4/65

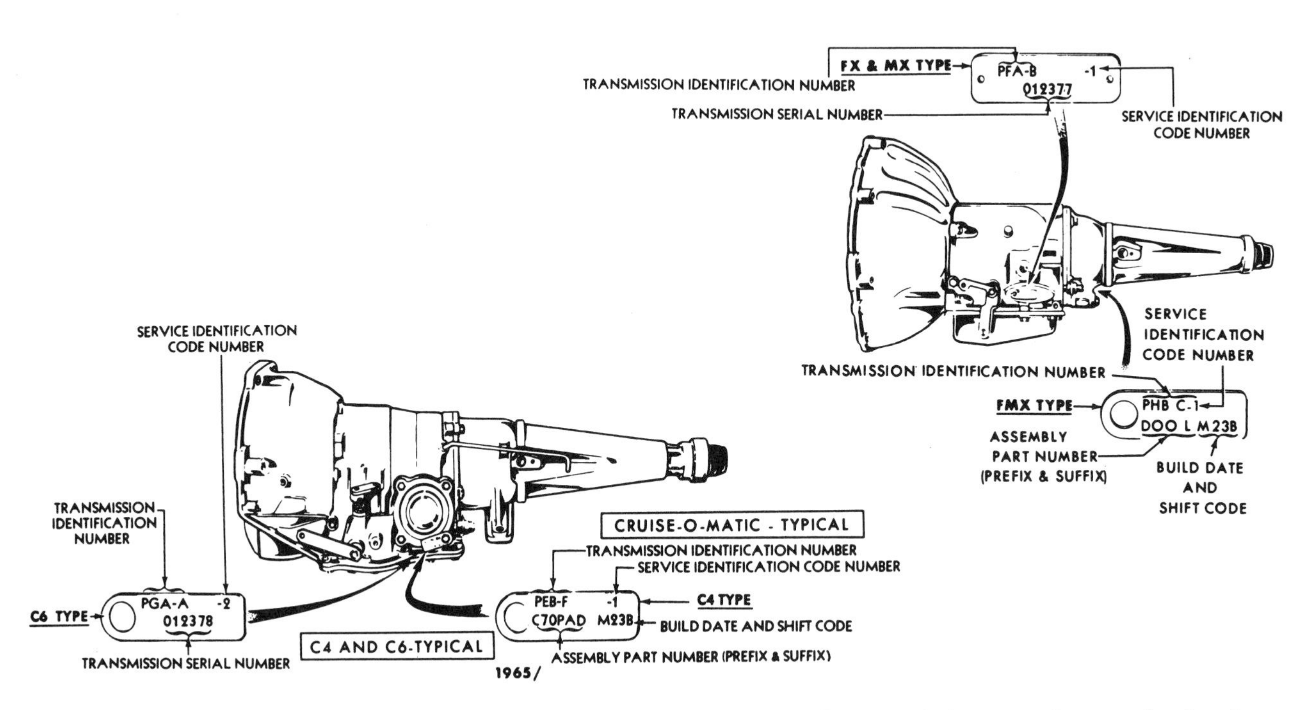

Three different automatic transmissions were installed in Mustangs from 1965 through 1970. All were designed, engineered, and built by Ford, and each has three forward speeds. All six-cylinder engines and 302ci and smaller V-8 engines received C4 transmissions. All Mustangs equipped with 351ci engines were outfitted with FMX transmissions, and all big-block engines were mated with a C6. The C4 transmissions are better known by their trade name, Cruise-O-Matic. A metal tag bolted to the driver's side servo cover on C4 and C6 transmissions or to the extension housing on FMX transmissions is used to identify each unit. The tag is stamped with the transmission's identification number as well as its service code number. Tags affixed to C6 transmissions list the unit's serial number. C4 and FMX tags show the prefix and suffix of the transmission's assembly part number in addition to the build date and shift code.

Mustang Automatic Transmissions by Year, Code, Engine Size and Type.

Production Year	Transmission Code	Engine CID	Transmission Type
1965	PCS-C	170	C4
	PCS-F	200	C4
	PCW-G	260	C4
	PCW-H	289	C4
	PCW-J	289	C4
1966	PCS-Y	200	C4
	PCW-AS	289	C4
	PCW-BA	289 HiPo	C4
1967	PEB-B	200	C4
	PEE-C	289	C4
	PEE-K	289 HiPo	C4
	PGA-P, P1	390GT	C6
1968	PEB-B1	200	C4
	PEE-C1	289	C4
	PEE-S	302	C4
	PGA-S	390-2V	C6
	PGA-P2	390-GT	C6
	PGB-AF	427/428CJ*	C6
1969	PEB-B1	200	C4
	PEA-AA	250 Economy	C4
	PEE-AD	250	C4
	PEE-AC	302	C4
	PHB-E	351-2V	FMX
	PHB-H	351-4V	FMX
	PGA-Y, AE	390	C6
	PGB-AF1	427/428CJ*	C6
1970	PEB-B3	200	C4
	PEE-AD1	250	C4
	PEE-AC1	302	C4
	PHB-E1	351-2V	FMX
	PHB-P	351-4V	FMX
	PGB-AF2	428CJ	C6

* A Regular Production Mustang equipped with a 427cid engine was proposed but never built.

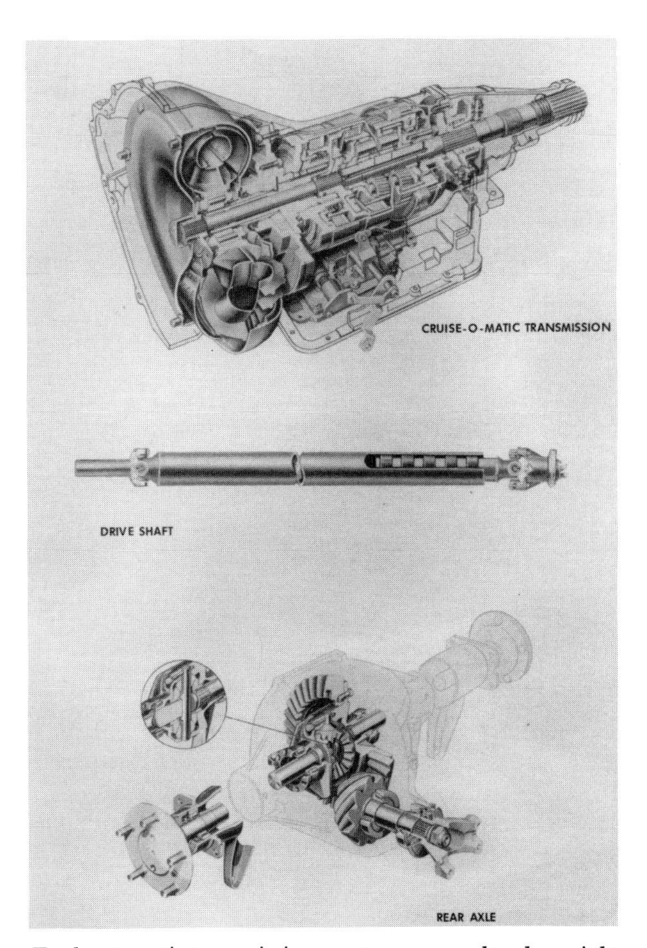

Ford automatic transmissions are programmed to the weight of the vehicle, torque output of the engine, and the rear-end gear ratio. Shift points are controlled by a combination of throttle position and load. Cruise-O-Matic C4 transmissions installed in 1965 Mustangs have a five-position shift selector. Placing the selector handle in the position marked with a red dot inside a black circle allows the transmission to upshift and downshift through all three forward gears automatically. Selecting the position marked with a black dot inside a black circle eliminates first gear from the automatic sequence. The factory believed starting out in second gear would limit tire spin during slippery driving conditions. The idea was abandoned in 1966.

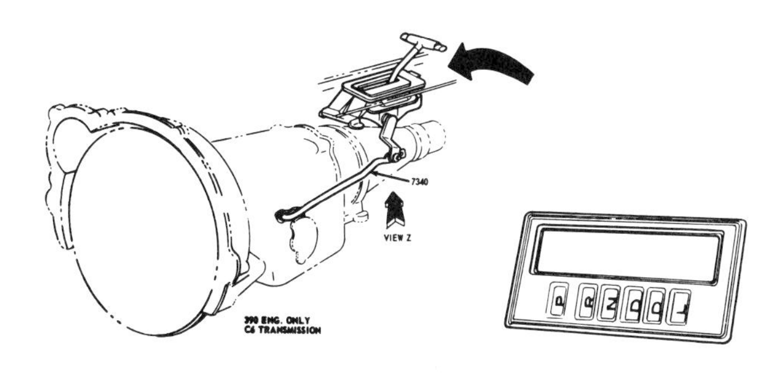

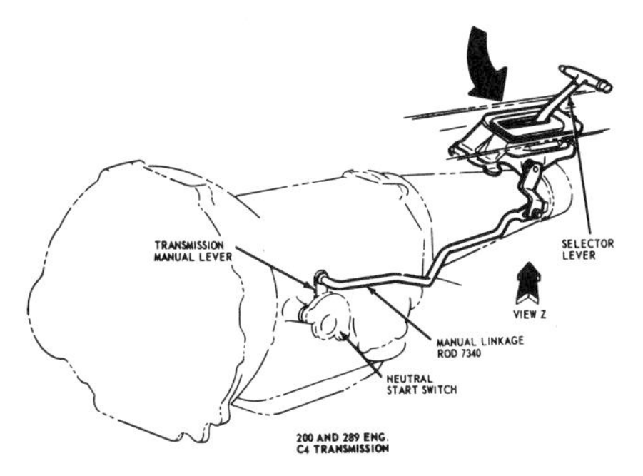

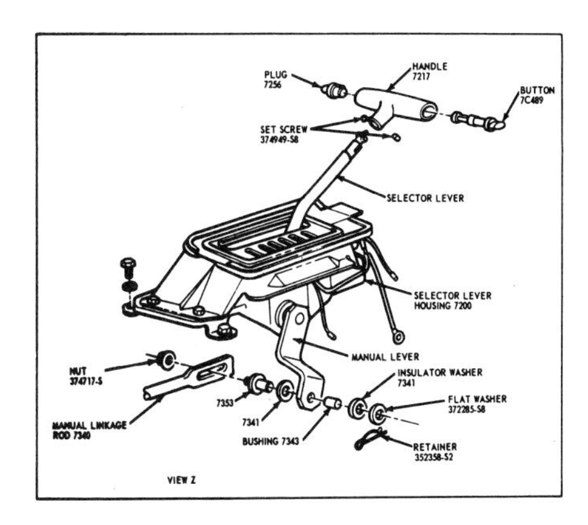

C4 and C6 transmission console linkages.

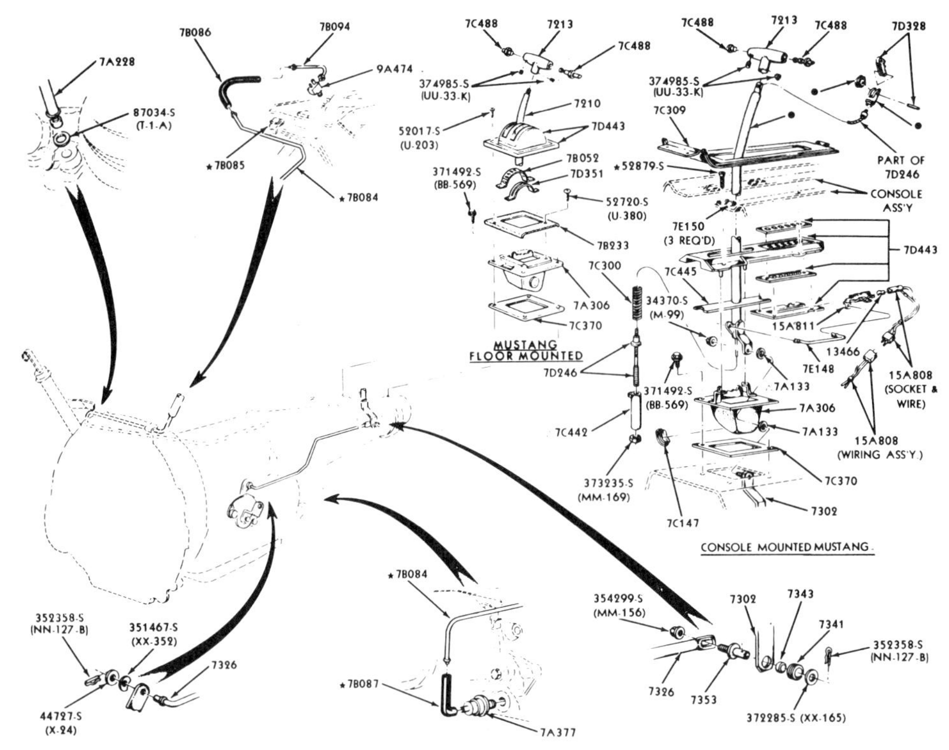

Automatic transmission floor shift control selector mechanisms for 1967-1968 Mustangs.

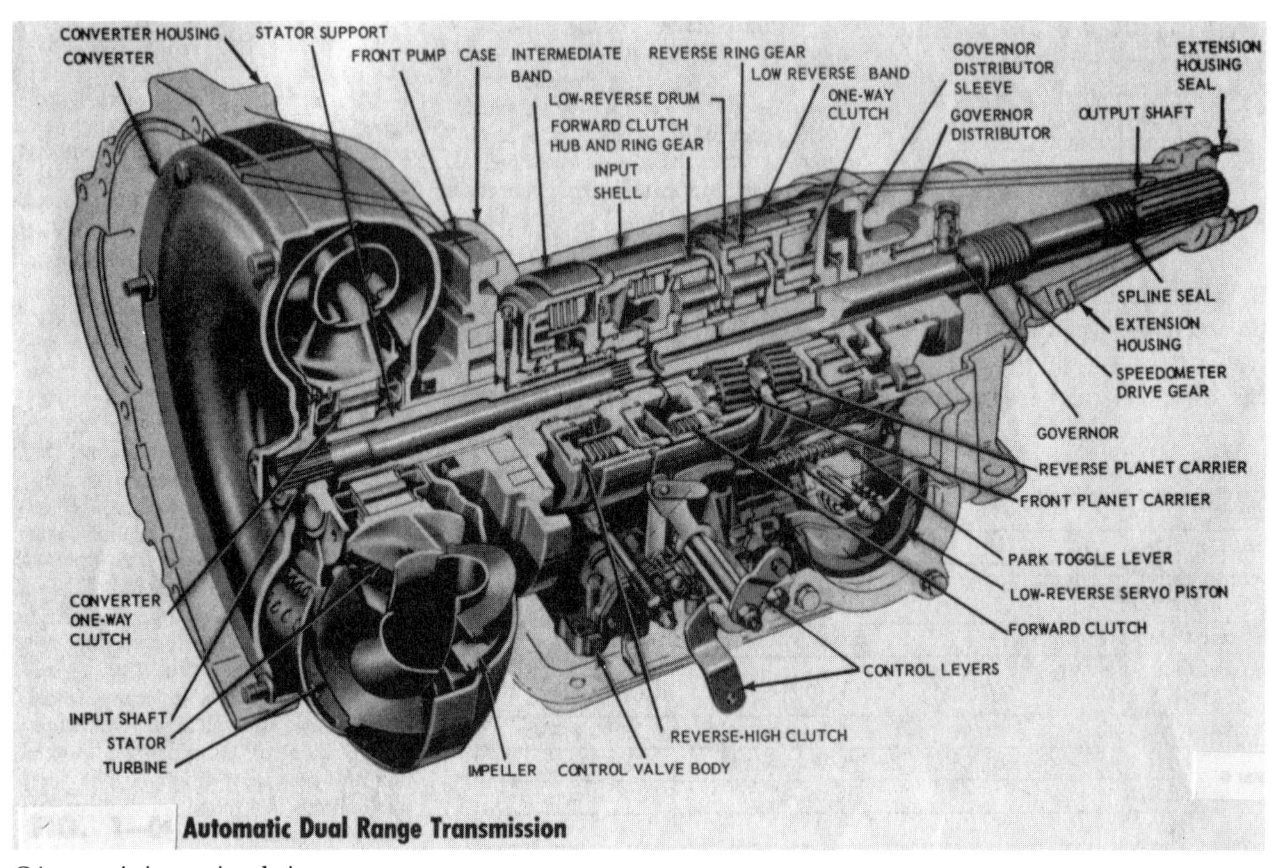

C4 transmission sectional view.

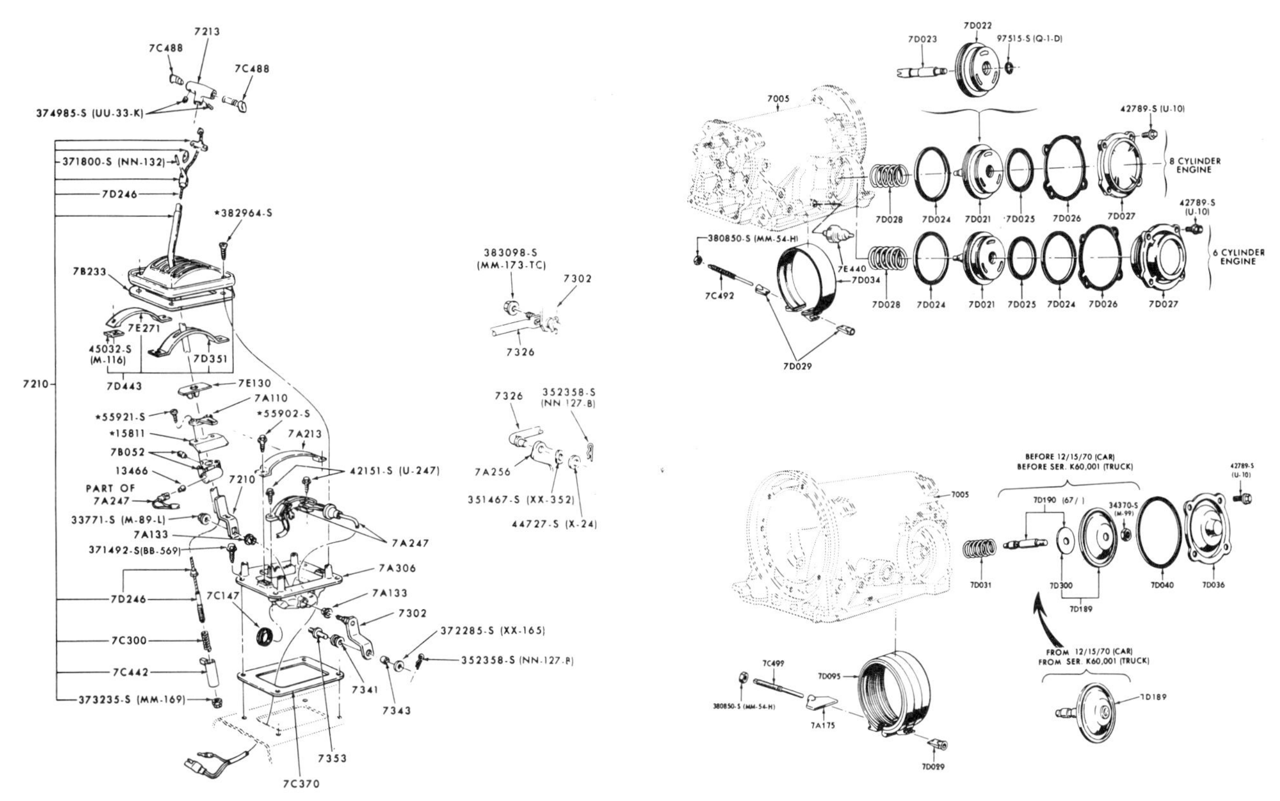

Automatic transmission floor shift control selector mechanisms for 1969-1970 Mustangs.

C4 automatic transmission front and rear servos and bands for 1965-1970 Mustangs.

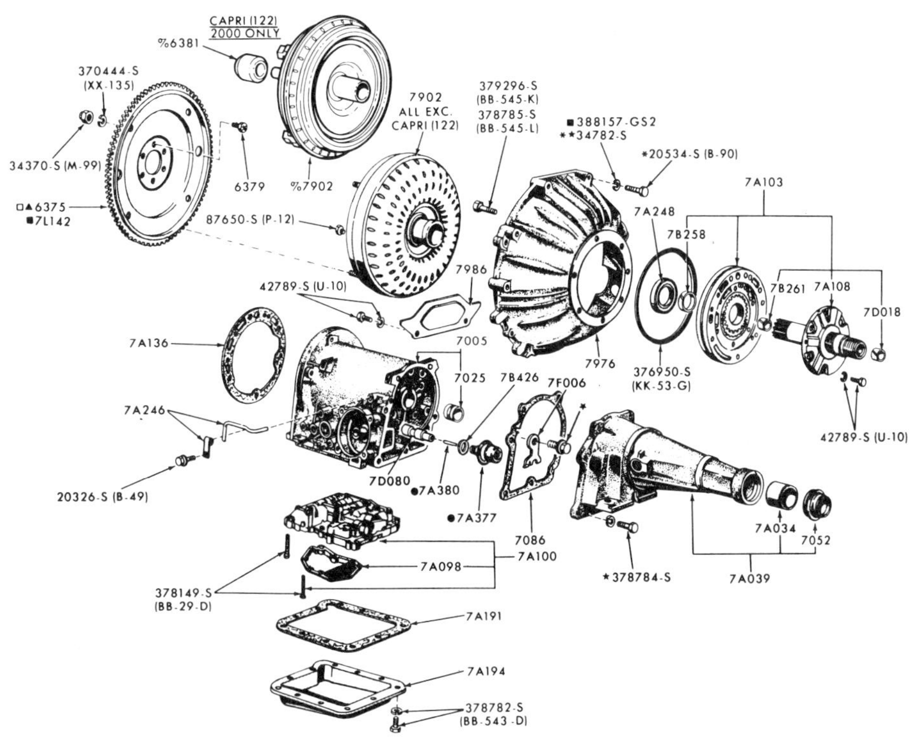

C4 automatic transmissions are lightweight and dependable. The bellhousing, main case, and extension housing are aluminum. All automatic-equipped Mustangs built in 1965 and 1966 were equipped with C4 transmissions. Different bellhousings (7976) were used to adapt the transmission to various six- and eight-cylinder engines. Different extension housings (7A039) were used to adapt the units to various body styles.

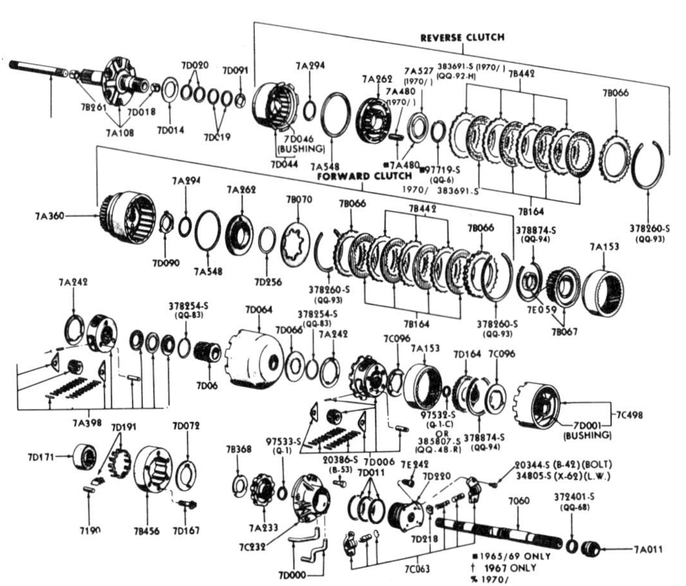

In an automatic transmission, different gear ratios are achieved by directing power through a series of compact planetary gear sets. The forward clutch is applied while the planet one-way clutch or low-reverse band prevents the low-reverse drum and reverse planet carrier from turning. The power flow is through the input shaft, the forward clutch, the ring gear, the planet gears, and the sun gears.

Counterclockwise rotation of the sun gear turns the reverse planet gear clockwise. With the reverse planet carrier held stationary, the clockwise rotation of the reverse planet gears rotates the reverse ring gear and hub clockwise. The hub of the reverse ring gear is splined to the output shaft which rotates clockwise at reduced speed, but with increased torque.

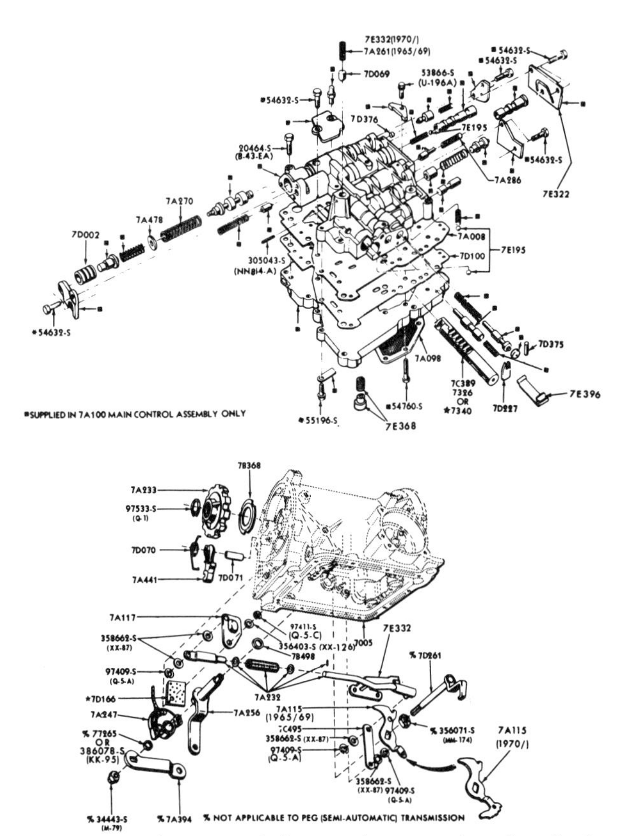

C4 automatic transmission main control valve body, parking pawl, throttle, and manual control linkage for 1965-1970 Mustangs.

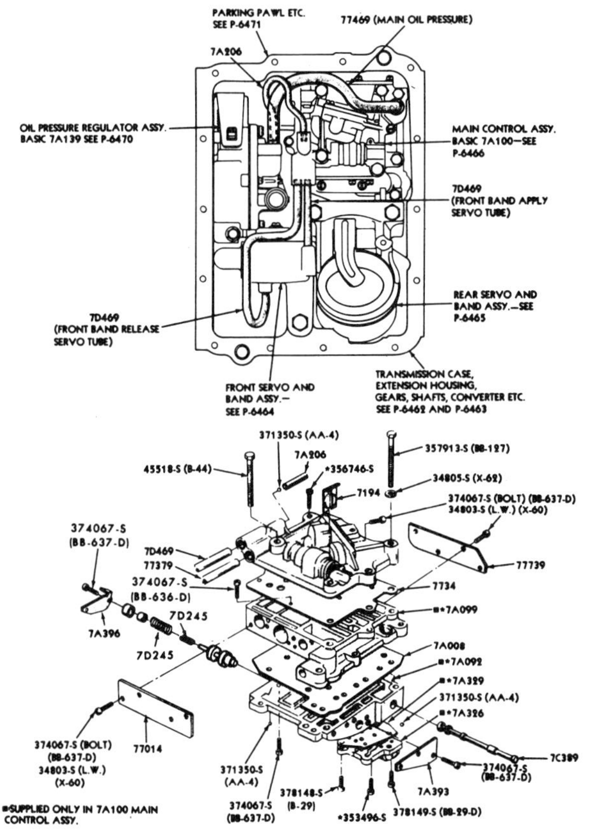

The FMX transmission valve body is completely different in operation compared to the C4 and C6 valve bodies.

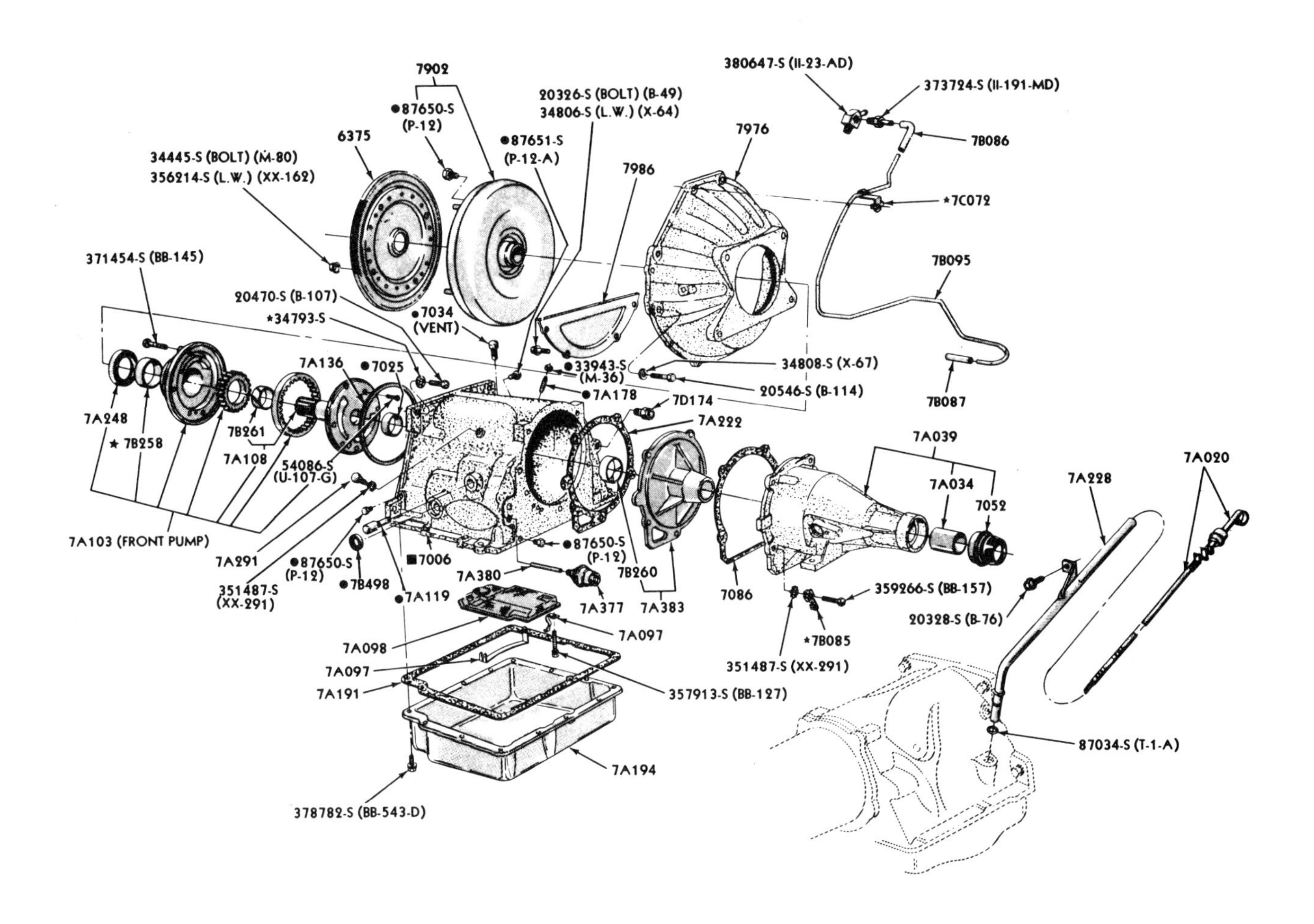

FMX transmissions were used behind mid-range torque engines such as the 351 Windsor in 1969-1970 and the 351 Cleveland in 1970; Cleveland engines were not installed in Mustangs in 1969. FMX transmissions have a cast-iron main case (7006), and an aluminum bellhousing (7976) and extension housing (7A039). The vacuum line (7B095) attaches to the intake manifold at fitting 380647-S and to the modulator valve, 7A337.

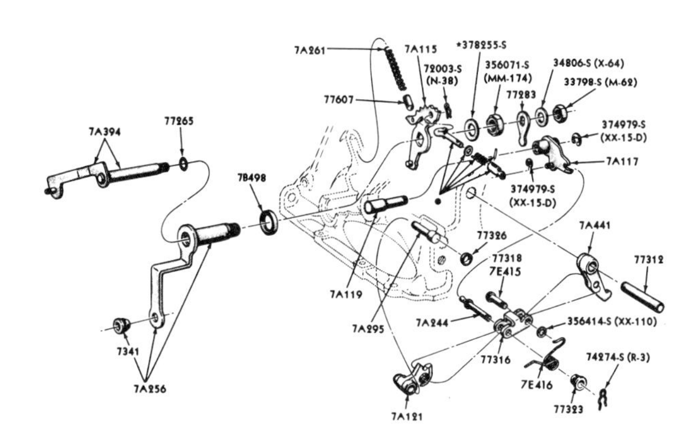

The FMX parking pawl, 1969-1970.

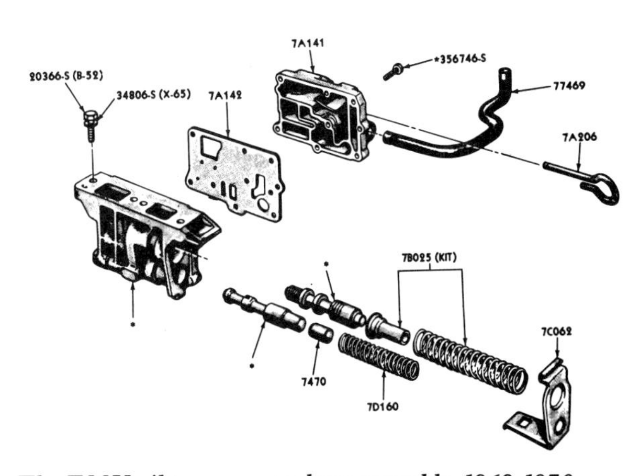

The FMX oil pressure regulator assembly, 1969-1970.

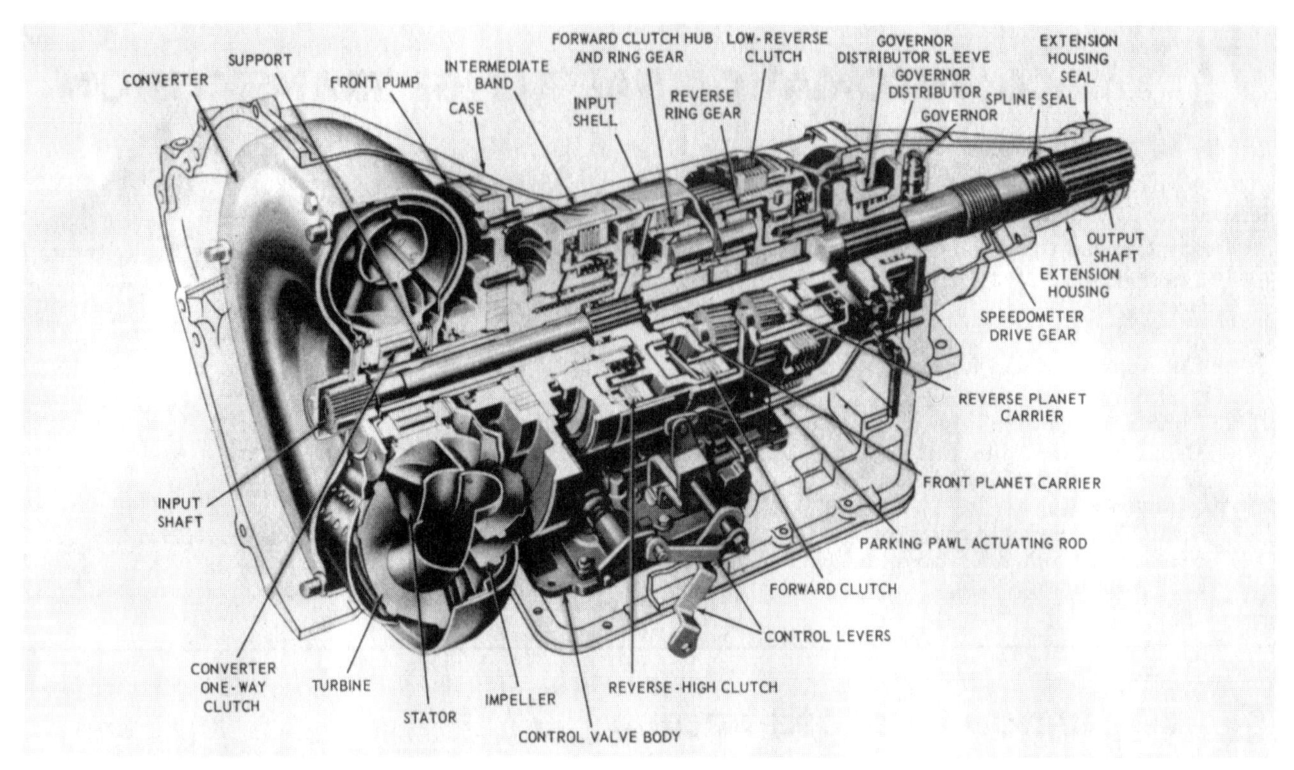

C6 transmission sectional view.

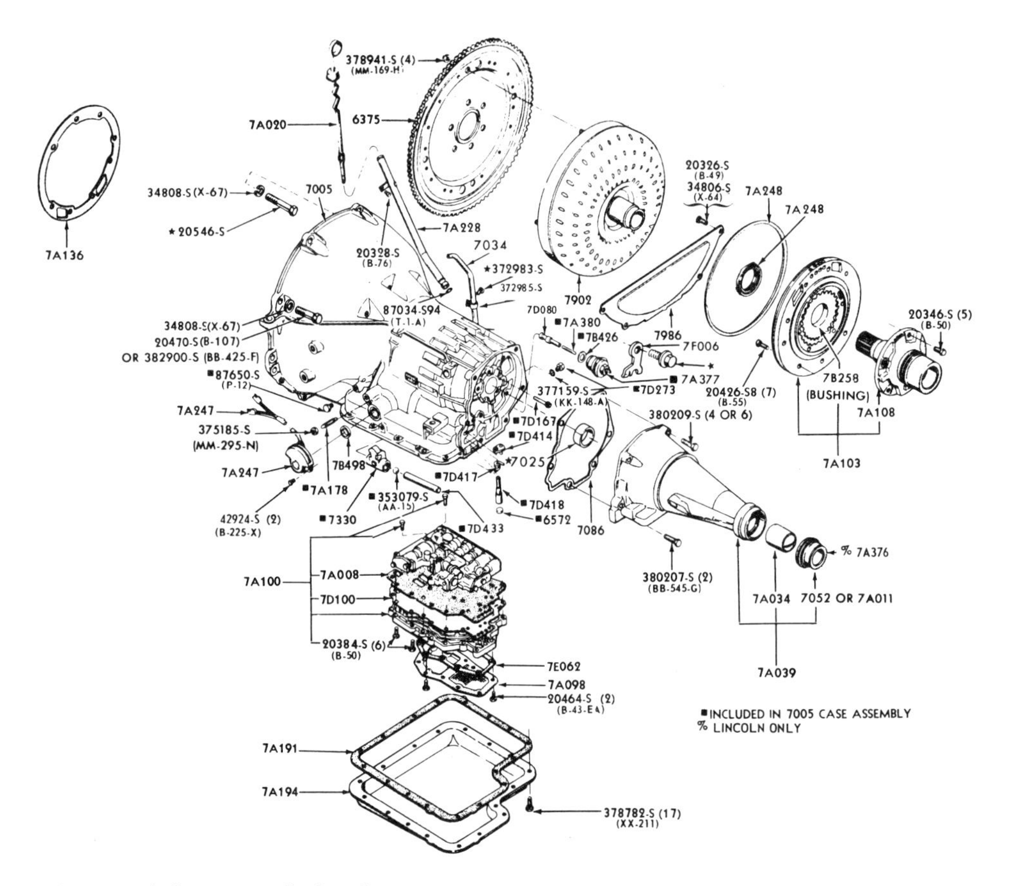

The C6 automatic transmission case and related parts.

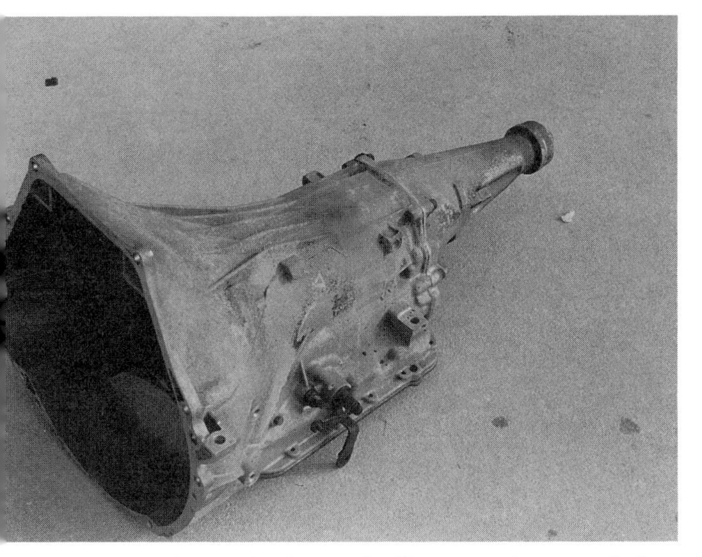

Ford's C6 is the brute of all automatic transmissions. Millions of C6 transmissions were installed in trucks and full-size Ford cars.

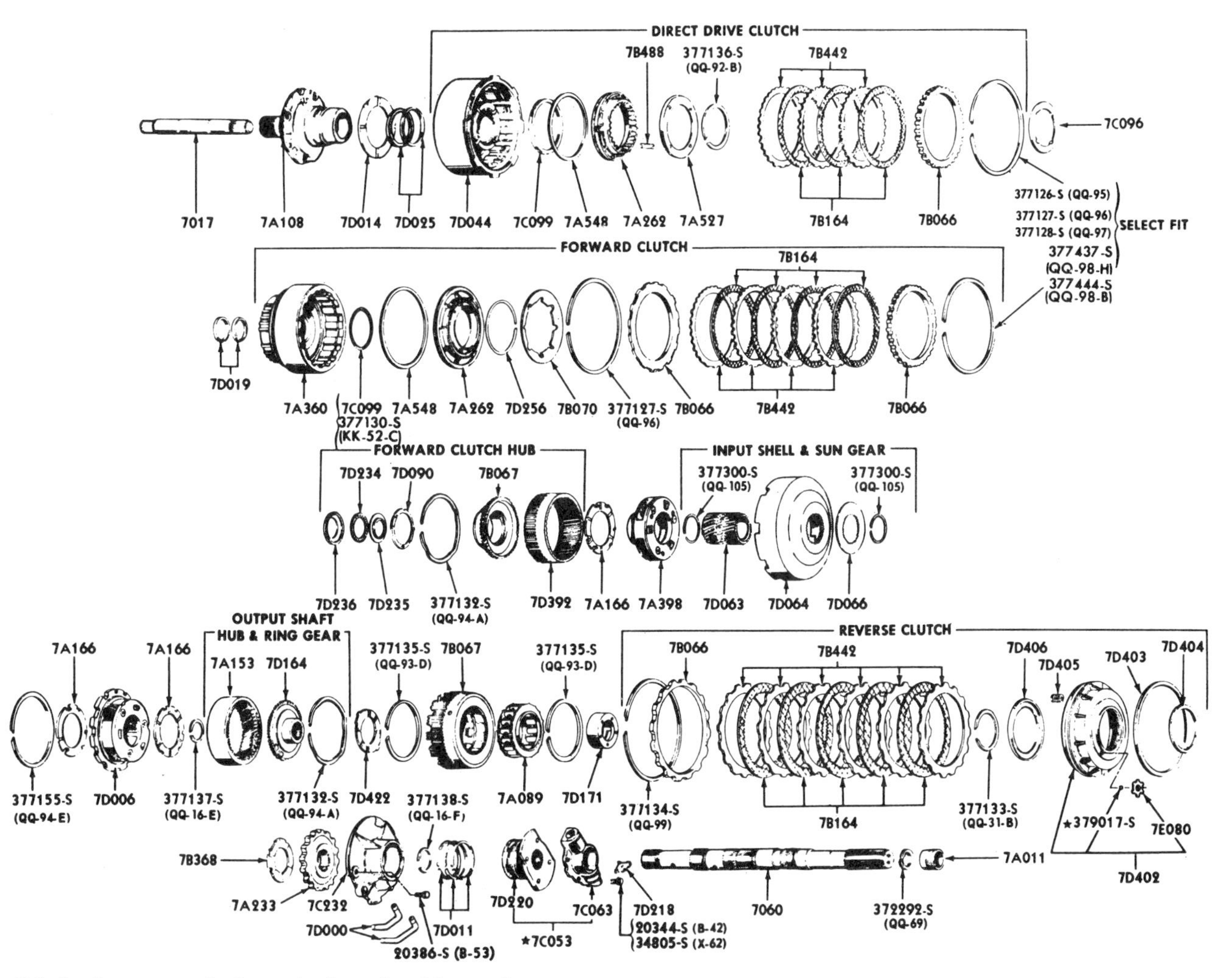

C6 clutches, gears, shafts, and other related internal parts.

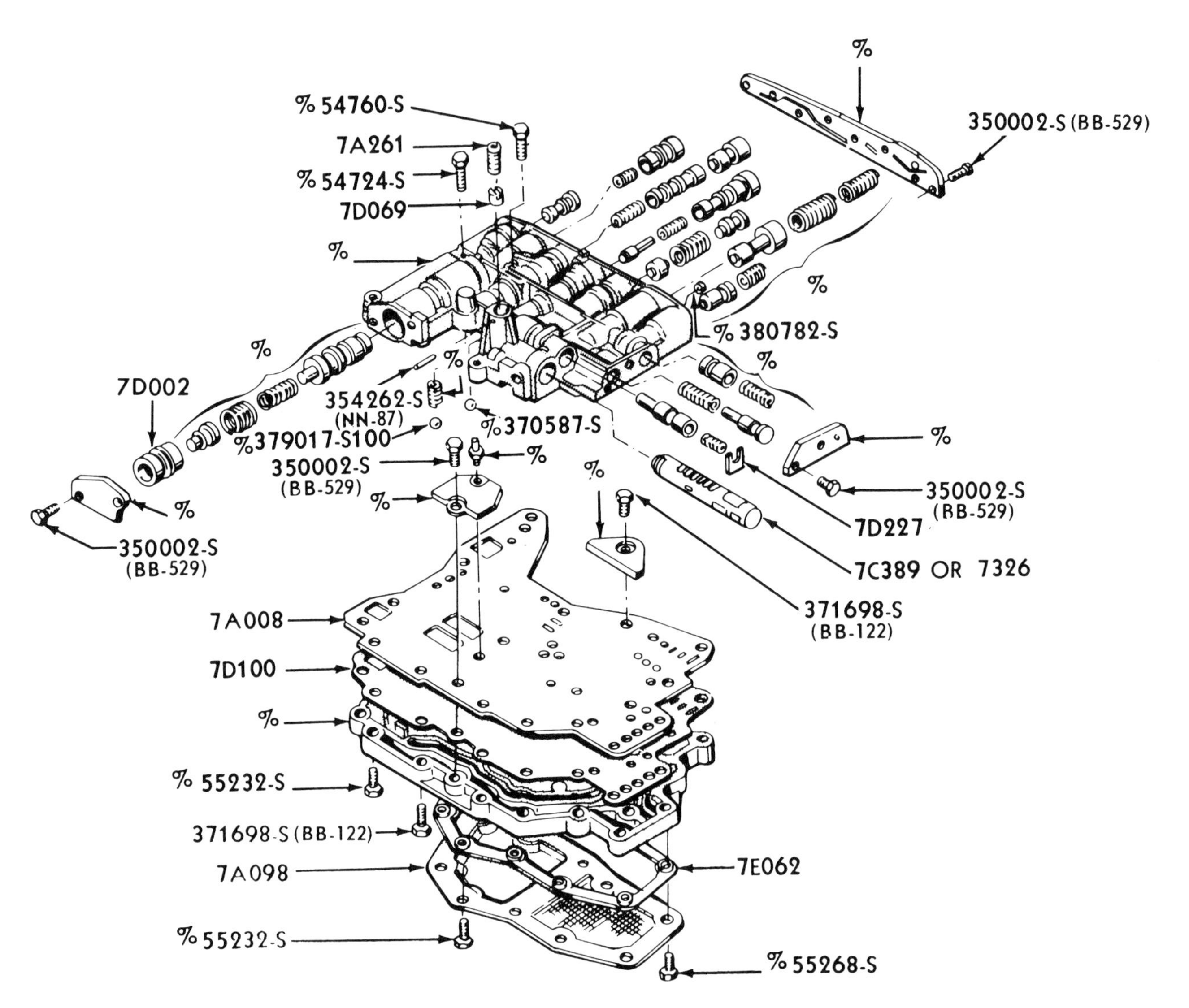

C6 automatic transmission main control valve body, parking pawl, throttle, and manual control linkage for 1967-1970 Mustangs.

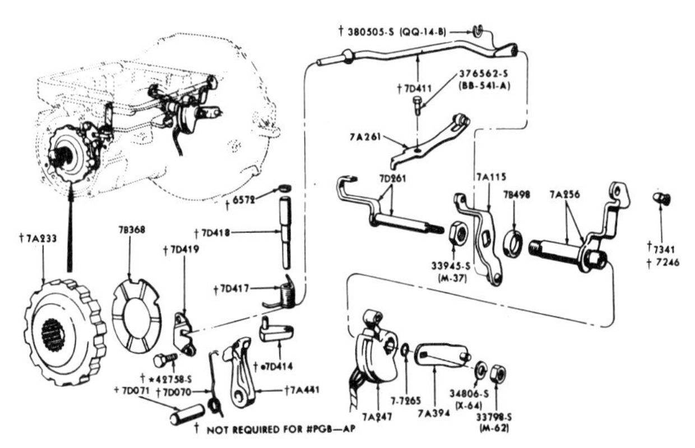

The C6 parking pawl, throttle, and manual control linkage, 1965-1970.

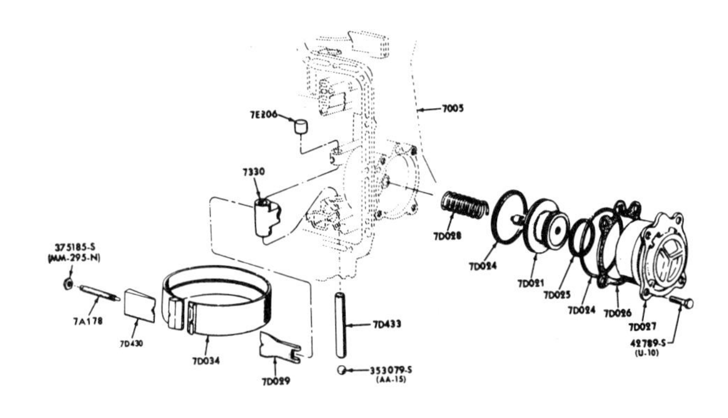

The C6 intermediate servo and band, 1965-1970.

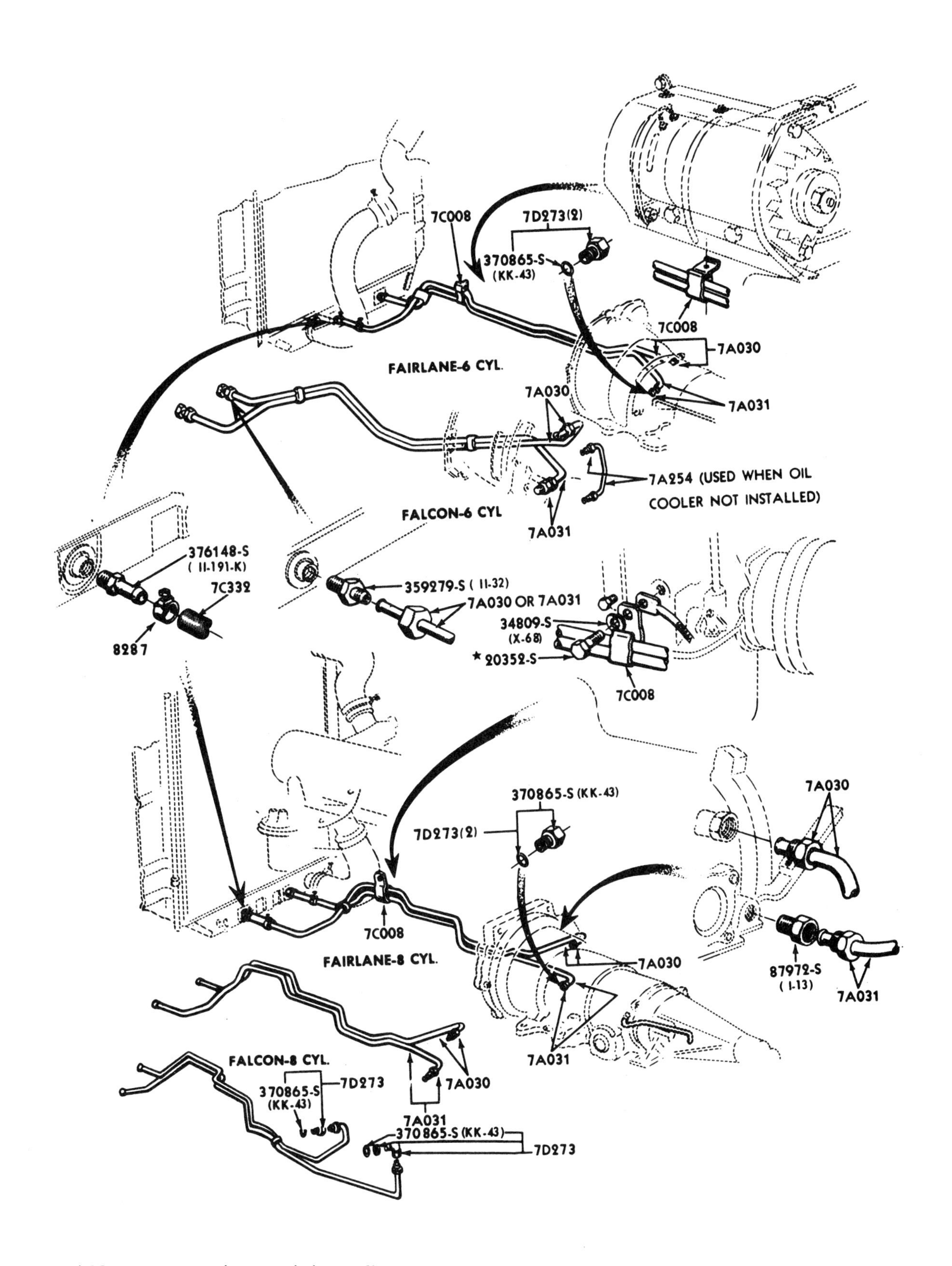

The 1965 Mustang automatic transmission cooling system.

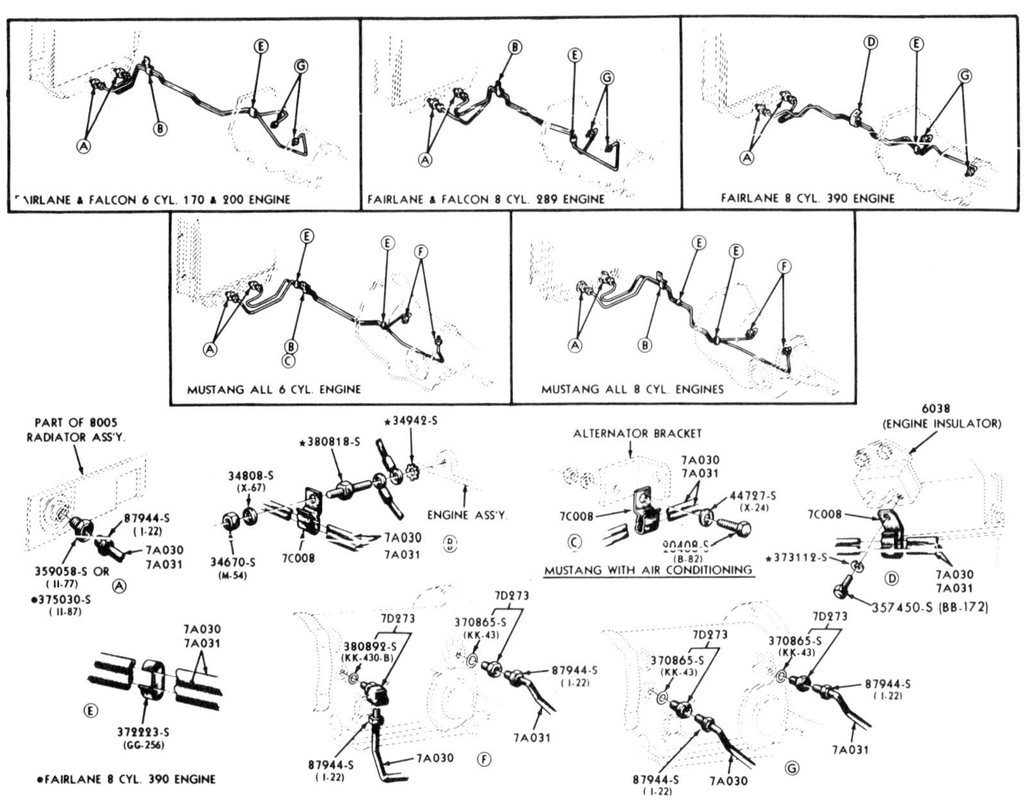
The 1966-1968 Mustang automatic transmission oil cooling system.

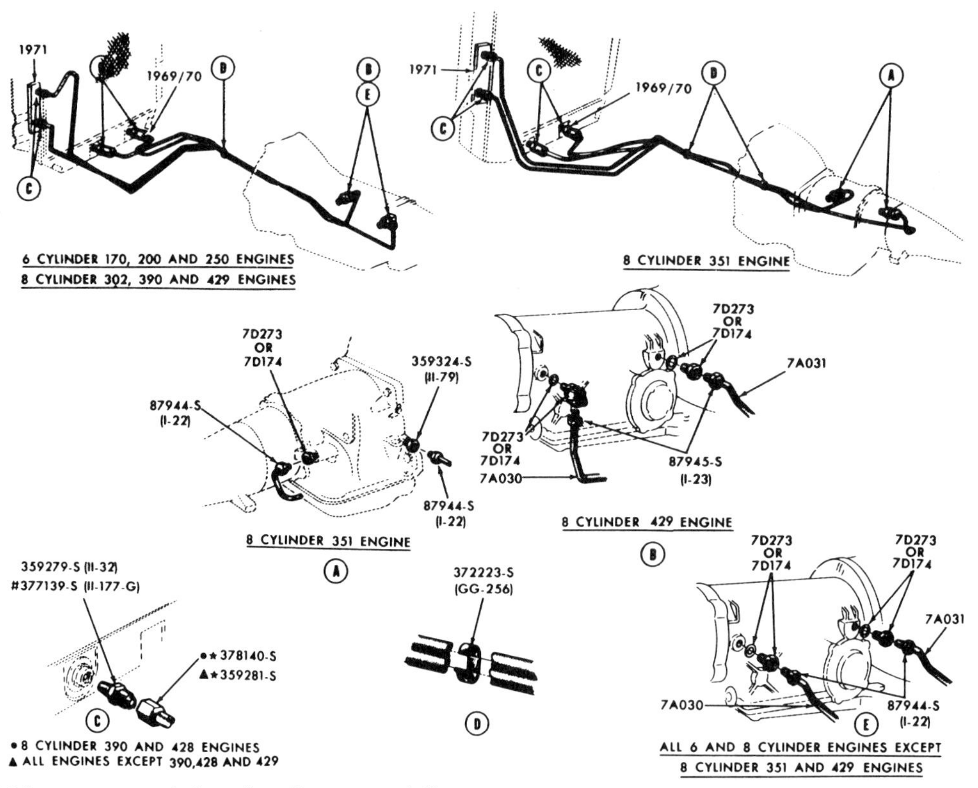
The 1969-1970 Mustang transmission oil cooling system (all engines).

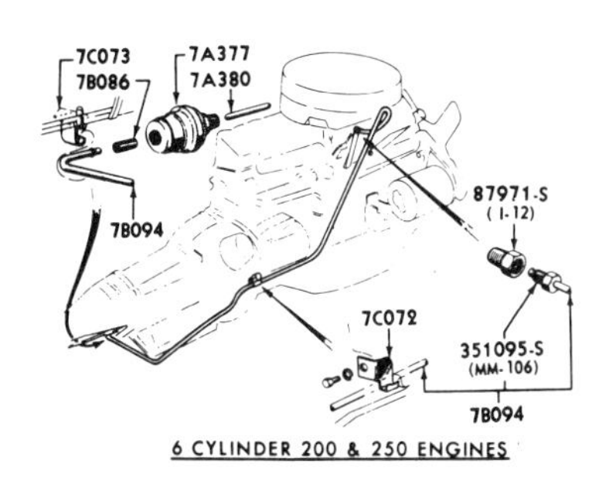

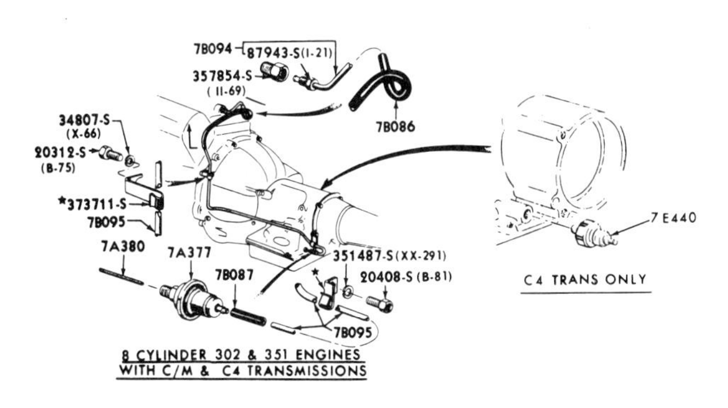

Automatic transmission vacuum throttle valve, lines, and fittings for 1965-1970 Mustangs.

MUSTANG Speedometer Cable Application Chart

Year	Model	Trans.	Speed Control Modulator or NOX Emission		Length	Remarks	Part Number
65/66		3/M/T	Yes	No	48-11/16		C5ZZ-17260-E
65/66	6 Cyl.	4/M/T			58-13/16		C5DZ-17260-A
65/66		C4			58-13/16		C5DZ-17260-A
65/66	8 Cyl.	4/M/T			64-15/16		C5OZ-17260-A
67		ALL	X		42-3/4	Pump to Speedo	C7ZZ-17260-H
67		ALL	X		36-7/16	Pump to Speedo	C7ZZ-17260-F
67		3/M/T, C4		X	63-1/16		C7ZZ-17260-D
67		C6		X	66-9/16		C7ZZ-17260-C
68		C4, C6		X	66-9/16	w/o 3.91 or 4.30 Axle	
67/68		4/M/T		X	77-1/16		C7ZZ-17260-J
68	390	3/M/T		X	77-1/16		C7ZZ-17260-J
68	Exc.390	3/M/T		X	58-13/16		C7DZ-17260-C
68		C6		X	62-3/4	w/ 3.91 or 4.30 Axle	C8ZZ-17260-F
						Adapter C8ZZ-17294-A	
						also required	
68		ALL	X		57-13/16	Pump to Speedo	C8ZZ-17260-E
69		ALL	X		56	Pump to Speedo	C8ZZ-17260-D
69		C6		X	63-1/2	w/o 3.91 or 4.30 Axle	C9ZZ-17260-B
69		Exc.4/M/T		X	56		C9ZZ-17260-C
69/70		4/M/T		X	66-1/2	w/o 3.91 or 4.30 Axle	C9ZZ-17260-A
69	428	4/M/T		X	66-1/2	w/ 3.91 or 4.30 Axle	C9ZZ-17260-A
69		4/M/T		X	70	w/ 3.91 or 4.30 Axle	C9ZZ-17260-F
						Adapter C9ZZ-17294-A	
						also Required	
						From 1/31/69	
69	Exc. 428	4/M/T/		X	70	w/ 3.91 or 4.30 Axle	C9ZZ-17260-F
						Adapter C9ZZ-17294-A	
						also Required	
						Before 1/31/69	
70		4/M/T/		X	70	w/ 3.91 or 4.30 Axle	C9ZZ-17260-F
						Adapter C9ZZ-17294-A	
						also Required	
						From 1/31/69	
69/70		C6		X	58	w/ 3.91 or 4.30 Axle	C9ZZ-17260-G
						Adapter C8ZZ-17294-1	
						also Required	
70		C4		X	56		D0OZ-17260-C
70		C/M		X	56		D0OZ-17260-C
70		C/M		X	61-13/16		D0OZ-17260-D
70		C/6		X	61-13/16	w/o 3.91 or 4.30 Axle	D0OZ-17260-D
70		3/M/T		X	53-13/16		D0OZ-17260-A

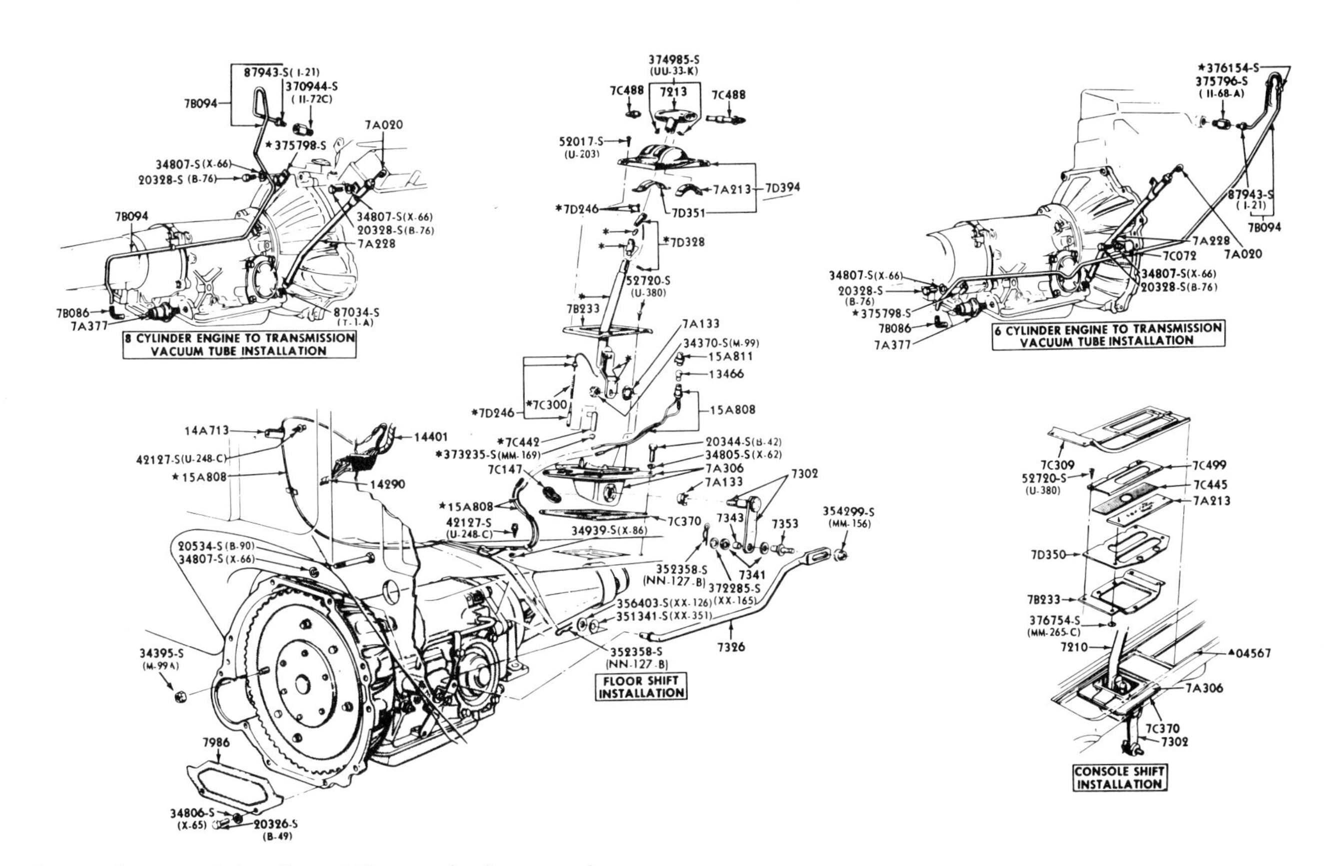

Automatic transmission floor shift control selector mechanisms for 1965-1966 Mustangs.

MUSTANG SPEEDOMETER DRIVEN GEAR IDENTIFICATION CHART
All years and models, 1965-1970.

DRIVEN GEAR	TYPE	MATERIAL	COLOR	TEETH
C0DD-17271-A	3a	Nylon	Wine	16
C0DD-17271-B	3a	Nylon	Gold or Yellow	18
C0DD-17271-C	3a	Nylon	Pink	19
C1DD-17271-A	3a	Nylon	Black	20
C2DZ-17271-F	3a	Nylon	Green or White	18
C2DZ-17271-G	3a	Nylon	Purple	17
C2DZ-17271-H	3a	Nylon	Blue	20
C2DZ-17271-J	3a	Nylon	Orange	16
C3DZ-17271-B	3a	Nylon	Green or White	17
C4DZ-17271-A	3a	Nylon	Pink-end only	19
C4OZ-17271-A	3a	Nylon	Red	21
C7SZ-17271-A	3	Nylon	Green	17
C7SZ-17271-B	3	Nylon	Gray	18
C7VY-17271-A	3	Nylon	Tan	19
C8SZ-17271-B	3	Nylon	Orange	20
DOOZ-17271-A	3	Nylon	Purple	21

MUSTANG SPEEDOMETER DRIVE GEAR IDENTIFICATION CHART
All years and models, 1965-1970.

				DIMENSIONS:			
TYPE	MATERIAL	COLOR	TEETH	I.D.	O.D.	Length	PART NUMBER
1	Steel	None	7	1-13/64"	2-3/16"	31/32"	C2DZ-17285-A
2	Nylon	None	6	1-5/32"	2-1/64"	11/16"	C0DR-17285-A
2a	Steel	None	6	1-3/8"	2"	11/16"	C2OZ-17285-C
2a	Nylon	Blue	7	1"	1-17/32"	1-1/8"	D0RY-17285-A
2a	Steel	None	7	1-1/32"	1-25/32"	7/16"	D1RY-17285-C
2a	Steel	None	8	1-1/8'	1-7/8"	7/16"	D1RY-17285-A
2b	Nylon	Yellow	7	1-3/8'	2"	11/16"	C3OZ-17285-C
3	Steel	None	6	1-3/32"	1-29/32"	35/64"	D1FZ-17285-A
3	Steel	None	6	1-19/64"	2"	39/64"	C3OZ-17285-A
3	Steel	None	7	1-3/8'	2"	11/16"	C8OZ-17285-B
3	Steel	None	8	1-19/64"	2"	39/64"	C5DZ-17285-C
3	Steel	None	8	1-19/64"	2-1/64"	39/64"	C5OZ-17285-B
4	Nylon	Green	6	1-3/8'	1-13/16"	11/16"	B7C-17285-A
4	Nylon	Black	7	1-1/4"	2"	11/16"	C6ZZ-17285-A
4	Nylon	Gray	8	1-3/8'	2-1/64"	11/16"	C5DZ-17285-B
4a	Nylon	Pink	6	1-3/8'	1-13/16"	11/16"	C5ZZ-17285-A
4a	Nylon	Black	7	1-3/8'	1-13/16"	11/16"	C4DZ-17285-A
4a	Nylon	Yellow	7	1-3/8'	2-1/64"	11/16"	C8AZ-17285-A
4a	Nylon	Brown	8	1-3/8'	2-1/64"	11/16"	C8OZ-17285-B
4b	Nylon	None	7	1-5/32"	2"	11/16"	C2OZ-17285-B
4b	Nylon	Pink or Yellow	7	1-3/8'	2"	11/16"	C3OZ-17285-C
4b	Steel	None	7	1-3/8'	2"	11/16"	C3OZ-17285-C

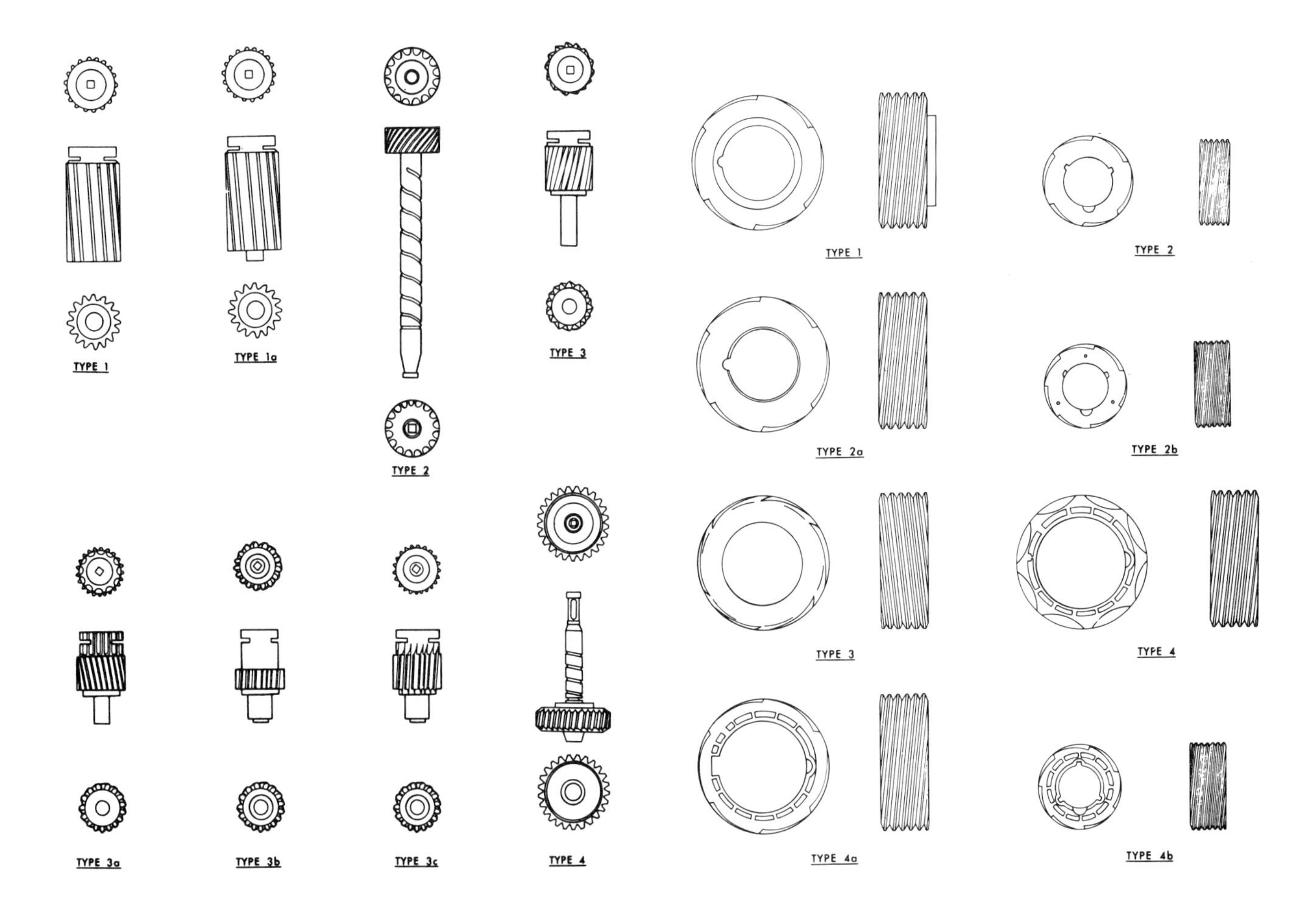

Speedometer drive gears, 1965.

MUSTANG SPEEDOMETER DRIVE GEAR APPLICATION CHART, 1965, Part One.

ENGINE CYL.	C.I.D.	TRANS.	TIRE SIZE	DRIVE GEAR	DRIVEN GEAR
			2.80 REAR AXLE RATIO		
6	200	C4	6.95x14	$	CODD-17271-A
8	289				
8	289	C4	6.50x13	$	CODD-17271-A
8	289	3/M/T	6.50x13		
	(Before 12/1/64)			C3OZ-17285-C	CODD-17271-A
	(From 12/1/64)			C8OZ-17285-B	CODD-17271-B
8	289	3/M/T	6.95x14		
	(Before 12/1/64)			C3OZ-17285-C	CODD-17271-A
	(From 12/1/64)			C8OZ-17285-B	CODD-17271-B
8	289	W/4/S	6.95x14		
	(HEK-L,R,V-Before 11/23/64)			*C5DZ-17285-C	C2DZ-17271-J
	(HEK-L,P,R,U,V-From 11/23/64)			C5DZ-17285-C	C2DZ-17271-F
8	289	W/4/S	6.50x13		
	(HEK-P,R,U,V)			C5DZ-17284-C	C4DZ-17271-A
8	289	W/4/S	6.95x14		
	(HEK-J Before 11/23/64)			*C5OZ-17285-B	CODD-17271-A
	(HEK-J,S From 11/23/64)			C5OZ-17285-B	CODD-17271-B
			2.83 REAR AXLE RATIO		
6	170,200	C4	6.50x13	$	CODD-17271-A
6	170,200	C4	6.95x14	$	CODD-17271-A
6	170	3/M/T	6.95x14	C2OZ-17285-B	CODD-17271-A
			3.00 REAR AXLE RATIO		
6	200	C4	6.95x14	$	C3DZ-17271-B
8	289				
8	289	C4	6.50x13	$	C3DZ-17271-B
8	289	3/M/T	6.50x13		
	(Before 12/1/64)			C3OZ-17285-C	C3DZ-17271-B
	(From 12/1/64)			C3OZ-17285-B	C1DD-17271-A
8	289	3/M/T	6.95x14		
	(Before 12/1/64)			C3OZ-17285-C	C3DZ-17271-B
8	289	W/4/S	6.95x14		
	(HEK-L,M Before 11/23/64)			*C5DZ-17285-C	C2DZ-17271-G
	(HEK-L,M,P,R,U,V From 11/23/64)			C5DZ-17285-C	C4DZ-17271-A
8	289	W/4/S	6.50x13		
	(HEK-M Before 11/23/64)			*C5DZ-17285-C	C2DZ-17271-J
	(HEK-M From 11/23/64)			C5DZ-17285-C	C4DZ-17271-A
			3.20 REAR AXLE RATIO		
8	289	W/4/S	6.95x14		
	HEK-J Before 11/23/64			*C5OZ-17285-B	C3DZ-17271-B
	HEK-J,S From 11/23/64			C5OZ-17285-B	CODD-17271-C

*If drive gear is replaced, driven gear must also be replaced
$ Part of transmission output shaft

MUSTANG SPEEDOMETER DRIVE GEAR APPLICATION CHART, 1965, Part Two.

ENGINE CYL.	C.I.D.	TRANS.	TIRE SIZE	DRIVE GEAR	DRIVEN GEAR
			3.20 REAR AXLE RATIO		
6	170,200	C4	6.50x13	$	CODD-17271-B
6	170,200	C4	6.95x14	$	CODD-17271-B
6	170,200	3/M/T	6.50x13	C2OZ-17285-B	CODD-17271-B
6	170,200	3/M/T	6.95x14	C2OZ-17285-B	CODD-17271-B
			3.25 REAR AXLE RATIO		
6	200	C4	6.95x14	$	CODD-17271-B
8	289	C/M	7.75x15	$	C3DZ-17271-B
6	200	3/M/T	6.95x14	C2OZ-17285-B	CODD-17271-B
8	289	3/M/T	6.95x14	C8OZ-17285-B	C4OZ-17271-A
8	289	W/4/S	6.95x14		
	Before 11/23/64			*C5DZ-17285-C	C2DZ-17271-F
	From 11/23/64			C5DZ-17285-C	C2DZ-17271-H
			3.50 REAR AXLE RATIO		
6	170	C4	6.50x13	$	C1DD-17271-A
8	289				
6	170,200	C4	6.95x14	$	C1DD-17271-A
8	289				
6	170,200	3/M/T	6.50x13	C2OZ-17285-B	C1DD-17271-A
6	170,200	3/M/T	6.95x14	C2OZ-17285-B	C1DD-17271-A
8	289	W/4/S	6.95x14		
	(HEK-L)			C5DZ-17285-C	C2DZ-17271-H
8	289	W/4/S	6.95x14		
	(HEK-J)			C5OZ-17285-B	C1DD-17271-A
8	289	W/4/S	6.95x14		
	(HEK-K)			C3OZ-17285-A	C3DZ-17271-B
			3.89 REAR AXLE RATIO		
8	289	C4	6.95x14	$	C4OZ-17271-A
8	289	C4	6.95x14	$	C4OZ-17271-A
8	289	W/4/S	6.95x14	C3OZ-17285-A	CODD-17271-A
8	289	F/4/S	6.95x14	C5ZZ-17285-A	C4DZ-17271-A
			4.11 REAR AXLE RATIO		
8	289	W/4/S	6.95x14	C3OZ-17285-A	C1DD-17271-A
8	289	F/4/S	6.95x14	C5ZZ-17285-A	C2DZ-17271-H

*If drive gear is replaced, driven gear must also be replaced
$ Part of transmission output shaft

MUSTANG SPEEDOMETER DRIVE GEAR APPLICATION CHART, 1966.

ENGINE CYL.	C.I.D.	TRANS.	TIRE SIZE	DRIVE GEAR	DRIVEN GEAR
			2.80 REAR AXLE RATIO		
6	170,200	C4	6.95x14	$	CODD-17271-A
8	289				
8	289	3/M/T	6.95x14	C8OZ-17285-B	CODD-17271-B
8	289	F/4/S	6.95x14	C4DZ-17285-A	C2DZ-17271-J
8	289	W/4/S	6.95x14	C6ZZ-17285-A	C2DZ-17271-J
			2.83 REAR AXLE RATIO		
6	170,200	C4	6.95x14	$	CODD-17271-A
8	289				
6	170,200	3/M/T	6.95x14	C2OZ-17285-B	CODD-17271-A
			3.00 REAR AXLE RATIO		
6	170,200	C4	6.95x14	$	C3DZ-17271-B
8	289				
8	289	3/M/T	6.95x14	C8OZ-17285-B	CODD-17271-C
8	289	W/4/S	6.95x14	C6ZZ-17285-A	C2OZ-17271-G
8	289	F/4/S	6.95x14	C4DZ-17285-A	C2DZ-17271-G
			3.20 REAR AXLE RATIO		
6	170,200	C4	6.95x14	$	C3DZ-17271-B
8	289				
6	170,200	3/M/T	6.95x14	C2OZ-17285-B	CODD-17271-B
6	170,200	D/4/S	6.95x14	C5ZZ-17285-B	C2DZ-17271-F
			3.25 REAR AXLE RATIO		
8	289	C4	6.95x14	$	C3DZ-17271-B
6	170,200	C4	6.95x14	$	CODD-17271-B
8	289				
6	170,200	3/M/T	6.95x14	C2OZ-17285-B	CODD-17271-B
8	289	3/M/T	6.95x14	C8OZ-17285-B	C4OZ-17271-A
8	289	F/4/S	6.95x14	C4DZ-17285-A	C2DZ-17271-F
			3.50 REAR AXLE RATIO		
6	170,200	C4	6.95x14	$	C1DD-17271-A
8	289				
6	170,200	3/M/T	6.95x14	C2OZ-17285-B	C1DD-17271-A
			3.89 REAR AXLE RATIO		
6	170,200	C4	6.95x14	$	C4OZ-17271-A
8	289				
8	289	F/4/S	6.95x14	C5ZZ-17285-A	C4DZ-17271-A

$ Part of transmission output shaft

MUSTANG SPEEDOMETER DRIVE GEAR APPLICATION CHART, 1967, Part One.

ENGINE CYL.	C.I.D.	TRANS.	TIRE SIZE	DRIVE GEAR	DRIVEN GEAR
			2.75 REAR AXLE RATIO		
8	390	C6	F70x14	$	C3DZ-17271-B
			2.79 REAR AXLE RATIO		
8	289	C4	6.95x14,7.35x14	$	C7SZ-17271-B
8	390	C4	F70x14	$	C7SZ-17271-A
8	289	3/M/T	6.95x14,7.35x14	C8OZ-17285-B	CODD-17271-B
8	289	3/M/T	F70x14	C8OZ-17285-B	C3DZ-17271-B
8	289	4/M/T	6.95x14,7.35x14	C4DZ-17285-A	C2DZ-17271-J
			2.80 REAR AXLE RATIO		
6	200	C4	7.35x14	$	CODD-17271-A
8	289	C4	6.95x14	$	C7SZ-17271-B
8	289	C4	F70x14	$	CODD-17271-A
8	289	3/M/T	6.95x14,7.35x14	C8OZ-17285-B	CODD-17271-B
8	289	3/M/T	F70x14	C8OZ-17285-B	CODD-17271-B
8	289	4/M/T	6.95x14,7.35x14	C4DZ-17285-A	C2DZ-17271-J
			2.83 REAR AXLE RATIO		
6	170,200	C4	6.95x14,7.35x14	$	CODD-17271-A
6	170,200	3/M/T(RED)	6.95x14,7.35x14	C2OZ-17285-B	CODD-17271-A
6	170,200	3/M/T(RAN)	6.95x14,7.35x14	C3OZ-17285-C	CODD-17271-A
			3.00 REAR AXLE RATIO		
6	200	C4	7.35x14	$	C3DZ-12721-B
8	289	C4	6.95x14	$	C7VY-17271-A
8	289	C4	7.35x14,F70x14	$	C7VY-17271-A
8	289,390	C6	7.35x14,F70x14	$	CODD-17271-C
8	289,390	3/M/T(RAN)	7.35x14,F70x14	C8OZ-17285-B	CODD-17271-C
8	289,390	3/M/T(RAN)	6.95x14	C8OZ-17285-B	CODD-17271-B
8	289,390	3/M/T(RAT) 4/M/T	6.95x14, 7.35x14	C8OZ-17285-B	CODD-17271-C
8	289,390	3/M/T(RAT) 4/M/T	F70x14	C4DZ-17285-A	C2DZ-17271-J
			3.20 REAR AXLE RATIO		
6	170,200	C4	6.95x14,7.35x14	C2OZ-17285-B	CODD-17271-B
6	170,200	3/M/T(RED)	6.95x14,7.35x14	C3OZ-17285-C	CODD-17271-B
			3.20 REAR AXLE RATIO		
6	170,200	3/M/T(RAN)	6.95x14,7.35x14	C3OZ-17285-C	CODD-17271-B
			3.25 REAR AXLE RATIO		
6	170,200	C4	7.35x14	$	CODD-17271-B
8	390	C6	F70x14	$	C1DD-17271-A
6	170,200	3/M/T(RAN)	7.35x14	C3OZ-17285-C	CODD-17271-B
8	289,390	3/M/T(RAT) 4/M/T	7.35x14	C8OZ-17285-B	C4OZ-17271-A
8	289,390	3/M/T(RAN)	F70x14	C4DZ-17285-A	C2DZ-17271-G

MUSTANG SPEEDOMETER DRIVE GEAR APPLICATION CHART, 1967, Part Two.

ENGINE CYL.	C.I.D.	TRANS.	TIRE SIZE	DRIVE GEAR	DRIVEN GEAR
			3.50 REAR AXLE RATIO		
6	170,200	C4	7.35x14	$	CODD-17271-C
8	289	C4	F70x14	$	CODD-17271-C
6	200	3/M/T(RAN)	7.35x14	C3OZ-17285-C	CODD-17271-C
8	289	4/M/T	F70x14	C5ZZ-17285-A	C2DZ-17271-J
			3.89 REAR AXLE RATIO		
8	289	C4	F70x14	$	C4OZ-17271-A
8	289	4/M/T	F70x14	C5ZZ-17285-A	C2DZ-17271-F

$ Part of transmission output shaft

MUSTANG SPEEDOMETER DRIVE GEAR APPLICATION CHART, 1968, Part One.

64)					
CYL.	C.I.D.	TRANS.	TIRE SIZE	DRIVE GEAR	DRIVEN GEAR
2.75 REAR AXLE RATIO					
8	390	C6	F70x14	$ - 8 teeth	C3DZ-17271-B
8	390	C6	7.35x14,E70x14, FR70x14	$ - 8 teeth	C3DZ-17271-B
8	390	C6	185Rx14	$ - 8 teeth	CODD-17271-B
2.79 REAR AXLE RATIO					
8	289,302	C4	6.95x14,185Rx14,E70x14	$	C7SZ-17271-B
8	289,302	C4	F70x14	$	C7SZ-17271-A
8	289,302	C4	7.35x14,FR70x14	$	C7SZ-17271-A
8	289,302	3/M/T	6.95x14,185Rx14,E70x14	C8OZ-17285-B	CODD-17271-B
8	289,302	3/M/T	F70x14	C8OZ-17285-B	C3DZ-17271-B
8	289,302	3/M/T	FR70x14	C8OZ-17285-B	C3DZ-17271-B
8	289,302	4/M/T	6.95x14,185Rx14,7.35x14, E70x14,F70x14,FR70x14	C4DZ-17285-A	C2DZ-17271-J
2.83 REAR AXLE RATIO					
6	170,200	C4	6.95x14,7.35x14	$	C7SZ-17271-B
6	170,200	3/M/T	6.95x14	C3OZ-17285-C	CODD-17271-A
3.00 REAR AXLE RATIO					
6	200	C4	6.95x14	$	C7VY-17171-A
8	289,302	C4	7.35x14,E70x14, F70x14,FR70x14,185Rx14	$ - 8 teeth	CODD-17271-C
6	200	3/M/T	7.35x14	C3OZ-17285-C	C3DZ-17271-B
8	289,302	3/M/T	6.95x14	C8OZ-17285-B	CODD-17271-C
8	289,302	3/M/T	7.35x14,E70x14, F70x14,FR70x14,185Rx14	C8OZ-17285-B	CODD-17271-C
8	390	3/M/T	7.35x14,185Rx14,E70x14	C4DZ-17285-A	C2DZ-17271-G
3.00 REAR AXLE RATIO					
8	289,302 390	4/M/T	6.95x14,E70x14,185Rx14	C4DZ-17285-A	C2DZ-17271-G
8	390	3/M/T	F70x14	C4DZ-17285-A	C2DZ-17271-J
8	289,302 390	4/M/T	F70x14	C4DZ-17285-A	C2DZ-17271-J
8	390	3/M/T	FR70x14	C4DZ-17285-A	C2DZ-17271-J
8	289,302 390	4/M/T	FR70x14,7.35x14	C4DZ-17285-A	C2DZ-17271-J
3.20 REAR AXLE RATIO					
6	170,200	C4	6.95x14	$	C8SZ-17271-B
6	170,200	3/M/T	6.95x14	C3OZ-17285-C	CODD-17271-B
6	200	3/M/T	7.35x14	C3OZ-17285-C	C3DZ-17271-B

MUSTANG SPEEDOMETER DRIVE GEAR APPLICATION CHART, 1968, Part Two.

3.25 REAR AXLE RATIO					
8	302	C4	F70x14,FR70x14	$	C8SZ-17271-B
8	170,200	C4	7.35x14	$	DOOZ-17271-A
8	289,302	C4	7.35x14	$	C8SZ-17271-B
8	289,302	C4	C70 x 14	$	DOOZ-17271-A
8	390	C6	7.35x14	$ - 8 teeth	C1DD-17271-A
8	390	C6	F70x14,FR70x14	$ - 8 teeth	C1DD-17271-A
8	390	C6	E70x14,185Rx14	$ - 8 teeth	C4OZ-17271-A
6	170,200	3/M/T	7.35x14,E70x14	C3OZ-17285-C	CODD-17271-B
8	289	3/M/T	7.35x14	C8OZ-17285-B	C1DD-17271-A
8	289	3/M/T	F70x14	C8OZ-17285-B	C1DD-17271-A
8	289	3/M/T	6.95x14,185Rx14, E70x14,FR70x14	C8OZ-17285-B	C4OZ-17271-A
8	390	3/M/T	7.35x14,F70x14, FR70x14,E70x14,185Rx14	C4DZ-12785-A	C2DZ-17271-F
8	289,302 390	4/M/T	7.35x14,F70x14, FR70x14,E70x14,185Rx14	C4DZ-12785-A	C2DZ-17271-F
8	289,302	4/M/T	6.95x14	C4DZ-12785-A	C2DZ-17271-F

$ Part of transmission output shaft

MUSTANG SPEEDOMETER DRIVE GEAR APPLICATION CHART, 1969, Part One.

ENGINE					
CYL.	C.I.D.	TRANS.	TIRE SIZE	DRIVE GEAR	DRIVEN GEAR
2.33 REAR AXLE RATIO					
6	250	C4	E78x14	$	CODD-17271-A
2.79 REAR AXLE RATIO					
6 8	250 302	C4	C78x14	$	C7SZ-17271-B
8	302	C4	F70x14	$	C7SZ-17271-A
6 8	250 302	C4	E78x14	$	C7SZ-17271-A
6 8	250 302	3/M/T	C78x14	C8OZ-17285-B	CODD-17271-B
6 8	250 302	3/M/T	E78x14	C8OZ-17285-B	C3DZ-17271-B
8	302	4/M/T	F70x14,C78x14,E78x14	C4DZ-17285-A	C2DZ-17271-J
2.83 REAR AXLE RATIO					
6	200	C4	E78x14	$	C7SZ-17271-B
6	200	C4	C78x14	$	C7SZ-17271-B
3.00 REAR AXLE RATIO					
6 8	200,250 302,351 390,428	C4,C6	F70x14 FR70x14 E78x14	$ - 8 teeth	C7VY-17271-A
6 8	200,250 302	C4	C78x14	$ - 8 teeth	C8SZ-17271-B
6 8	250 302,351	3/M/T	F70x14,FR70x14,E78x14	C8OZ-17285-B	CODD-17271-C
6 8	250 302,351	3/M/T	C78x14	C8OZ-17285-B	C1DD-17271-A
8	390	3/M/T	FR70x14,E78x 14	C4DZ-17285-A	C2DZ-17271-J
8	390	3/M/T	F70x14	C4DZ-17285-A	C2DZ-17271-J
8	302,351 390	4/M/T	E78x14 FR70x14	C4DZ-17285-A	C2DZ-17271-J
8	302,351	4/M/T	C78x14	C4DZ-17285-A	C2DZ-17271-G
3.08 REAR AXLE RATIO					
6	200	C4	C78x14	$	C8SZ-17271-B
6	200	C4	E78x14	$	C7VY-17271-A
6	200	3/M/T	C78x14	C8OZ-17285-B	C1DD-17271-A
6	200	3/M/T	E78x14	C8OZ-17285-B	CODD-17271-C
3.25 REAR AXLE RATIO					
6 8	250 302,351	C4	E78x14	$ - 8 teeth	C8SZ-17271-B
6 8	250 302,351 390,429	C4	F70x14 FR70x14	$ - 8 teeth	C8SZ-17271-B
6 8	250 302,351	C4	E70x14 C78x14	$ - 8 teeth	DOOZ-17271-A

MUSTANG SPEEDOMETER DRIVE GEAR APPLICATION CHART, 1969, Part Two.

ENGINE					
CYL	C.I.D.	TRANS.	TIRE SIZE	DRIVE GEAR	DRIVEN GEAR
3.25 REAR AXLE RATIO (Continued)					
6 8	250 302,351	3/M/T	E78x14	$ - 8 teeth	C1DD-17271-A
8	302,351	3/M/T	F70x14,FR70x14	C8OZ-17285-B	C1DD-17271-A
6 8	250 302,351	3/M/T	C78x14	C8OZ-17285-B	C4OZ-17271-A
8	390	3/M/T	F70x14,FR70x14,E78x14	C4DZ-17285-A	C2DZ-17271-F
8	428	4/M/T	F70x14,FR70x14	C5ZZ-17285-A	C2DZ-17271-J
3.50 REAR AXLE RATIO					
8	390,428	C6	F70x14,FR70x14,E78x14	$ - 8 teeth	DOOZ-17271-A
8	351,428	4/M/T	FR70x14	C5ZZ-17285-A	C2DZ-17271-J
8	351,428	4/M/T	F70x14	C5ZZ-17285-A	C2DZ-17271-J
8	302,351 390	4/M/T	F70x14,FR70x14,E78x14	C4DZ-17285-A	C4DZ-17271-A
3.91 REAR AXLE RATIO					
8	390,428	C6	F70x14,FR70x14	$ - 8 teeth	C7SZ-17271-A
8	302,390	4/M/T	F70x14,FR70x14	C4DZ-17285-A	C2DZ-17271-J
8	351,428	4/M/T	F70x14,FR70x14	C5ZZ-17285-A	C2DZ-17271-F
4.30 REAR AXLE RATIO					
8	428	C6	F70x14,FR70x14	$ - 8 teeth	C7SZ-17271-B
8	302,390	4/M/T	F70x14,FR70x14	C4DZ-17285-A	C2DZ-17271-G
8	351,428	4/M/T	F70x14,FR70x14	C5ZZ-17285-A	C2DZ-17271-H

$ Part of transmission output shaft

MUSTANG SPEEDOMETER DRIVE GEAR APPLICATION CHART, 1970.

ENGINE CYL	ENGINE C.I.D.	TRANS.	TIRE SIZE	DRIVE GEAR	DRIVEN GEAR
2.75 REAR AXLE RATIO					
8	351	C4	E78x14	$	C7SZ-17271-A
8	351	C4	F70x14	$	C7SZ-17271-A
8	351	3/M/T	E78x14	C8OZ-17285-B	C3DZ-17271-B
8	351	3/M/T	F70x14	C8OZ-17285-B	C3DZ-17271-B
2.79 REAR AXLE RATIO					
6 8	250 302	C4	E78x14	$	C7SZ-17271-A
6 8	250 302	3/M/T	E78x14	C8OZ-17285-B	C3DZ-17271-B
8	302	3/M/T	F70x14	C8OZ-17285-B	C3DZ-17271-B
3.00 REAR AXLE RATIO					
6 8	200,250 302,351	C4	F70x14,E78x14	$ - 8 teeth	C7VY-17271-A
6 8	250 302,351	3/M/T	F70x14,E78x14	C8OZ-17285-B	CODD-17271-C
8	302,351	4/M/T	E78x14	C4DZ-17285-A	C2DZ-17271-G
8	302,351	4/M/T	F70x14	C4DZ-17285-A	C2DZ-17271-J
3.08 REAR AXLE RATIO					
6	170,200	3/M/T	E78x14	C8OZ-17285-B	CODD-17271-C
3.25 REAR AXLE RATIO					
6 8	250 302,351	C4	E78x14	$ - 8 teeth	C8SZ-17271-B
6 8	250 302,351	3/M/T	E78x14	C8OZ-17285-B	C1DD-17271-A
6 8	200,250 302,351	3/M/T	F70x14	C8OZ-17285-B	C1DD-17271-A
8	302,351 428	4/M/T	F70x14	C4DZ-17285-A	C2DZ-17271-F
3.50 REAR AXLE RATIO					
8	351,428	C6	F70x14	$ - 8 teeth	DOOZ-17271-A
8	302,351 428,429	4/M/T	F70x14	C4DZ-17285-A	C2DZ-17271-H
3.91 REAR AXLE RATIO					
8	302,429	4/M/T	F60x15	C4DZ-17285-A	C2DZ-17271-J

$ Part of transmission output shaft

Mustang Rear Axle Tag Code Identification Chart, Part One.

Axle Code	Model Year	Axle Ratio	Ring Gear Diameter	Carrier Type	Axle Splines
WCY-E	65-70	3.20:1	7-1/4	Integral	24 Spline
WCY-F	65	3.50:1	7-1/4	Integral	24 Spline
WCY-L	65-66	3.20:1	7-1/4	Integral	24 Spline
WCY-R	65-70	2.83:1	7-1/4	Integral	24 Spline
WCY-AA	65-66	2.83:1	7-1/4	Integral	24 Spline
WCY-AJ	66-68	3.20:1	7-1/4	Integral	24 Spline
WCY-AJ1	66-68	3.20:1	7-1/4	Integral	24 Spline
WCY-AJ2	69-70	3.08:1	7-1/4	Integral	24 Spline
WCZ-E	65-67	2.80:1	8	Removable	28 Spline
WCZ-F	65-70	3.00:1	8	Removable	28 Spline
WCZ-F1	65-70	3.00:1	8	Removable	28 Spline
WCZ-G	65	3.50:1	8	Removable	28 Spline
WCZ-H	65	3.89:1	9	Removable	28 Spline
WCZ-J	65	4.11:1	9	Removable	28 Spline
WCZ-P	65	3.50:1	9	Removable	28 Spline
WCZ-R	66	3.89:1	9	Removable	28 Spline
WCZ-S	66	3.50:1	9	Removable	28 Spline
WCZ-T	66	3.50:1	9	Removable	28 Spline
WCZ-V1	67-70	2.79:1	8	Removable	28 Spline
WCZ-W	68	3.25:1	8	Removable	28 Spline
WES-F	67-70	3.00:1	9	Removable	28 Spline
WES-G	67	3.25:1	9	Removable	28 Spline
WES-H	67-68	3.50:1	9	Removable	28 Spline
WES-J	67	3.89:1	9	Removable	28 Spline
WES-K	67	3.50:1	9	Removable	28 Spline
WES-M	68-70	3.25:1	9	Removable	28 Spline

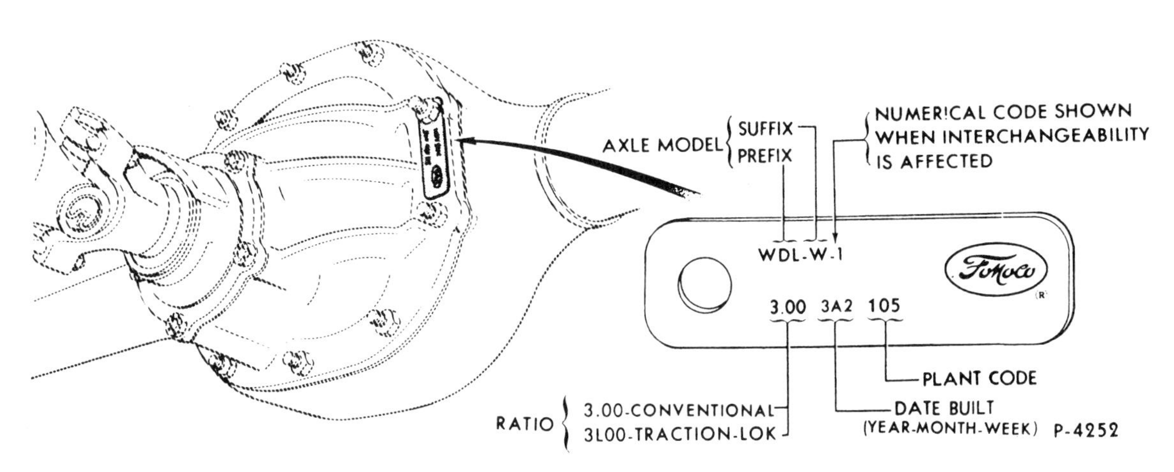

A typical rear axle identification tag contains numerous bits of information that are essential for cataloging and locating repair parts.

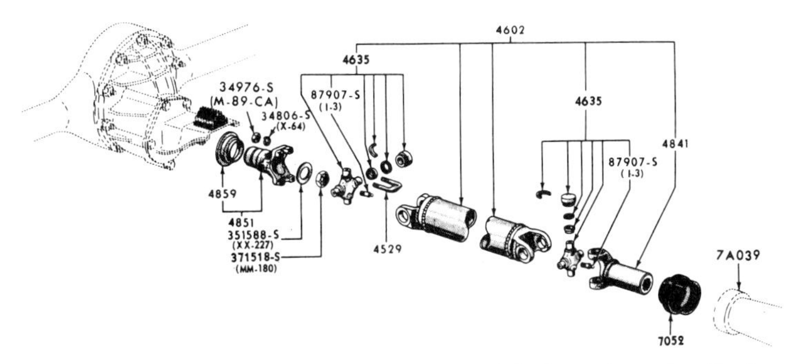

Driveshaft assembly for 1965-1972 Mustangs. The length, diameter, type of slip yoke (4841), and type of rear U-joint flange (4851) as well as the size of the universal joints (4635) depend on the displacement of the engine and the type of transmission used.

Mustang Rear Axle Tag Code Identification Chart, Part Two.

Axle Code	Model Year	Axle Ratio	Ring Gear Diameter	Carrier Type	Axle Splines
WES-N	68	3.00:1	9	Removable	28 Spline
WES-P	68	3.25:1	9	Removable	28 Spline
WES-R	68	3.25:1	9	Removable	28 Spline
WES-T	69-70	2.75:1	9	Removable	28 Spline
WES-T1	70-72	2.75:1	9	Removable	28 Spline
WES-U	68	3.50:1	9	Removable	31 Spline
WES-V	68	3.00:1	9	Removavle	28 Spline
WES-Y	68	3.50:1	9	Removable	31 Spline
WES-Z	68	3.00:1	9	Removable	28 Spline
WES-AA	69-70	3.00:1	9	Removable	28 Spline
WES-AB	69-70	3.25:1	9	Removable	28 Spline
WES-AC	69-70	3.00:1	9	Removable	28 Spline
WES-AD	69-70	3.25:1	9	Removable	28 Spline
WES-AE	69-70	3.50:1	9	Removable	31 Spline
WES-AG	69-70	2.75:1	9	Removable	28 Spline
WES-AH	69-70	3.00:1	9	Removable	31 Spline
WES-AJ	69-70	3.25:1	9	Removable	31 Spline
WFB-A	69-70	3.25:1	9	Removable	28 Spline
WFB-C	69-70	3.25:1	9	Removable	28 Spline
WFB-D	69-70	3.00:1	9	Removable	28 Spline
WFD-A	69-70	3.50:1	9	Removable	31 Spline
WFD-B	68-70	3.91:1	9	Removable	31 Spline
WFD-C	68-70	4.30:1	9	Removable	31 Spline
WFD-D	68-70	3.91:1	9	Removable	31 Spline
WFD-E	69-70	4.30:1	9	Removable	31 Spline
WFD-F	69-70	3.50:1	9	Removable	31 Spline
WFD-J	69	3.25:1	9	Removable	31 Spline
WFD-K	69-70	3.00:1	9	Removable	31 Spline
WFD-L	69-70	3.00:1	9	Removable	31 Spline
WFD-M	70	3.25:1	9	Removable	31 Spline
WFL-A	70	3.00:1	8	Removable	28 Spline
WFU-E	70	4.30:1	9	Removable	31 Spline
WDJ-B	65-66	2.80:1	8	Removable	28 Spline
WDJ-C	65-69	3.00:1	8	Removable	28 Spline
WDJ-C1	65-69	3.00:1	8	Removable	28 Spline
WDJ-C2	69	3.00:1	8	Removable	28 Spline

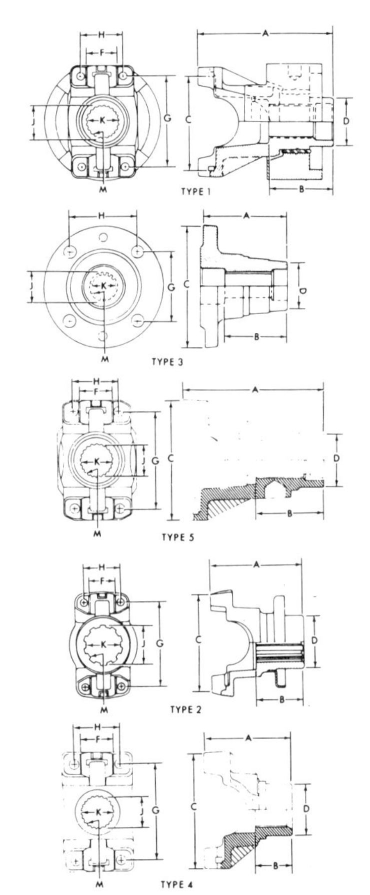

Rear axle universal joint flange assembly.

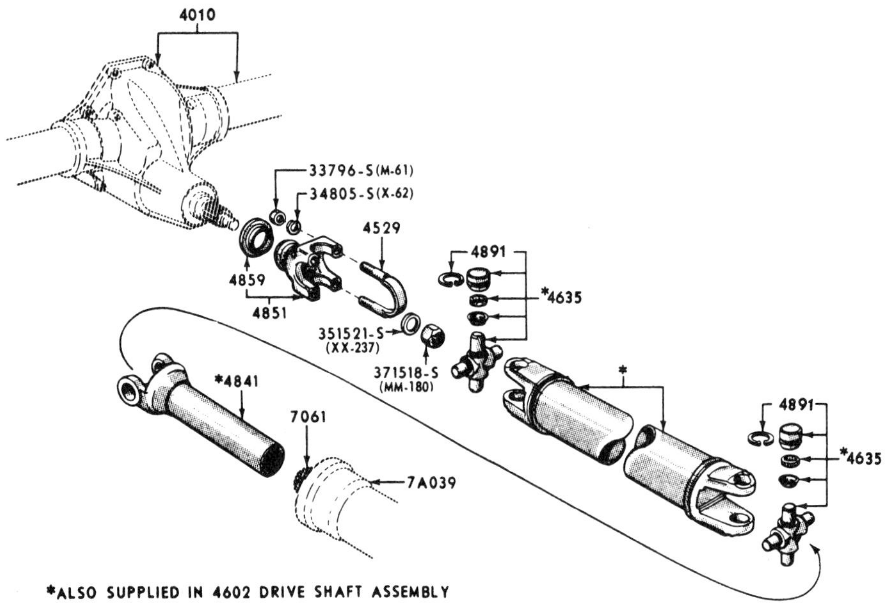

Power is transmitted from the transmission to the rear axle via a cylindrical steel propeller shaft or driveshaft. Because the rear axle is part of the suspension system, a universal joint at each end of the driveshaft allows for misalignment between the two units as the tires follow the contour of the road. The six-cylinder-equipped Mustang had a smaller-diameter driveshaft compared to a V-8-powered car. The front slip yoke is keyed to the transmission family. The rear yoke or pinion flange is keyed to the rear axle assembly.

MUSTANG DRIVESHAFT/REAR AXLE UNIVERSAL JOINT FLANGE DIMENSION CHART

Part Number	A	B	C	D	F	G	H	J	K	Splines
C2DZ-4851-A	2-15/16"	1-15/64"		1-21/32"	1"	2-7/32"	1-23/64"	1-1/64"	1-5/16"	23
D2ZF-4851-A	2-15/16"	1-15/64"		1-21/32"	1"	2-7/32"	1-23/64"	1-1/64"	1-5/16"	23
C3OZ-4851-A	4-13/32"	2-5/8"	3-7/32"	1-13/16"	1-1/8"	2-31/32"	1-37/64"	1-7/64"	1-1/64"	25
C3AZ-4851-L	5-1/16"	2-27/64"	3-7/32"	1-13/16"	1-1/8"	2-31/32"	1-37/64"	1-15/64"	1-9/64"	28
D2SZ-4851-A	5-1/16"			3-5/8"	1-13/16"	3-5/16"	1-37/64"	1-15/64"	1-9/64"	28
D2OZ-4851-A	4-25/32"	2-9/64"	3-5/8"	1-23/64"	1-1/8"	3-5/16"	1-37/64"	1-15/64"	1-9/64"	28

Note: Use with illustration labled: Flange Assy.- Rear Axle Universal Joint.

MUSTANG DRIVESHAFT/REAR AXLE UNIVERSAL JOINT FLANGE ASSEMBLY APPLICATION CHART

Year	Ring Gear	Identification	Part Number	Type
65/68	7-1/4	WCY	C2DZ-4851-A	2
		(exc. R,AA,AJ)		
65/70	7-1/4	WCY-R,AA,AJ	D2ZF-4851-A	2
65/70	8	WCZ,WDJ	C3OZ-4851-A	2
65/70	9		C3AZ-4851-L	2
67/71	9	28 spline axles	D2SZ-4851-A	5
68/71	9	31 spline axles	D2OZ-4851-A	4

Note: Use with illustration labled: Flange Assy.- Rear Axle Universal Joint.

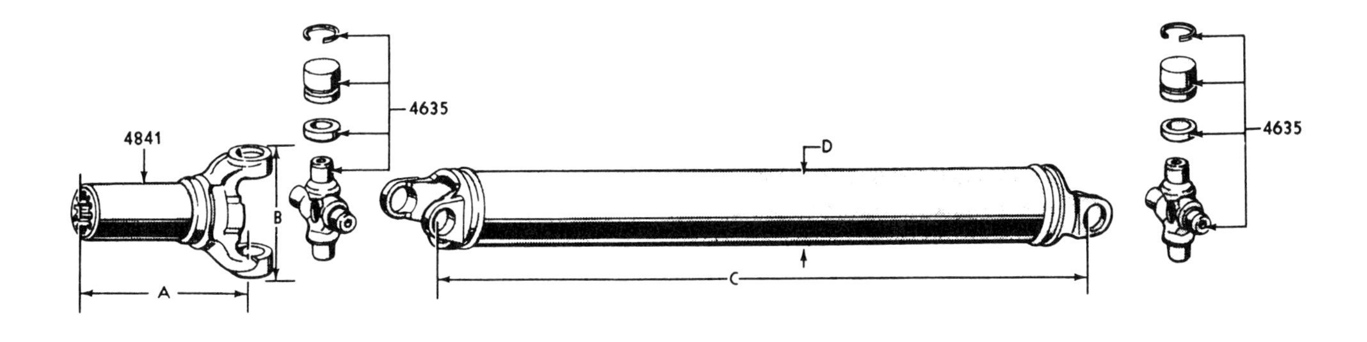

Typical 1965-1970 Mustang driveshaft, slip yoke, and universal joints.

MUSTANG DRIVESHAFT ASSEMBLY APPLICATION CHART, PART ONE, 1965-1968

Year	Engine	Transmission Type	Yoke Code	Yoke Type	Part Number	Notes	Driveshaft Diam.	Length	U-Joint Front	U-Joint Rear
65	170,200	3-S, 4-S	5	2	C1UZ-4841-A		2-1/2	51	J	J
65	170,200	C4	14	1	C5ZZ-4841-B		2-1/2	51	J	J
65	200	C4	6	2	C5DZ-4841-A		2-3/4	50-25/32	M	L
65	289	3-S, 4-S	6	2	C5DZ-4841-A	1	2-3/4	50-25/32	M	L
65	289	3-S, 4-S	7	2	C5DZ-4841-B	2	2-3/4	50-25/32	M	L
65	260,289	C4	6	2	C5DZ-4841-A	3	2-3/4	51-1/16	M	L
65	289 HiPo	4-S	7	2	C5DZ-4841-B	4	2-3/4	49-23/64	M	L
65	289 HiPo	4-S	3	2	C3AZ-4841-D	5	2-3/4	49-23/64	C	D
66	200	3-S	5	2	C1UZ-4841-A		2-1/2	49-63/64	J	J
66	200	4-S	5	2	C1UZ-4841-A		2-1/2	51	J	J
66	200	C4	6	2	C5DZ-4841-A		2-3/4	50-25/32	M	L
66	289	3-S, 4-S	6	2	C5DZ-4841-A	6	2-3/4	50-25/32	M	L
66	289	C4	6	2	C5DZ-4841-A	6	2-3/4	51-1/16	M	L
66	289	3-S, 4-S	6	2	C5DZ-4841-A	7	2-3/4	49-23/32	M	L
67	200	3-S	14	2	C5ZZ-4841-B		2-1/2	51	J	J
67	200	C4	3	4	C3AZ-4841-D		2-3/4	50-31/32	C	D
67	289	C4, 3-S, 4-S	3	4	C3AZ-4841-D	6	3	50-31/32	E	F
67	289	C4	4	2	C3AZ-4841-G	7	3	50-1/64	G	H
67	390	3-S, 4-S	4	2	C3AZ-4841-G	8	3	50-1/64	G	H
67	390	C6	16	2	C7SZ-4841-A	8	3-1/2	46-19/32	G	H
67	289,428	4-S	16	2	C7SZ-4841-A	9	3	50-1/64	G	H
68	289	C4, 4-S	4	2	C3AZ-4841-G	8	3	50-1/64	G	H
68	390	3-S, 4-S	4	2	C3AZ-4841-G	8	3	50-1/64	G	H
68	390,428CJ	C6	16	2	C7SZ-4841-A		3-1/2	46-19/32	G	H
68	200	3-S	14	2	C5ZZ-4841-B		2-1/2	51	J	J
68	200	C4	3	4	C3AZ-4841-D		3	50-31/32	E	F
68	289-2V,302	C4, 3-S, 4-S	3	4	C3AZ-4841-D	10	3	50-31/32	E	F
68	289-2V,302	C4, 3-S, 4-S	3	4	C3AZ-4841-D	11	3	50-7/32	E	F
68	428 CJ	4-S	3	2	C3AZ-4841-D		3	50	G	H

Notes:

1 From 12/15/64. Except HiPo
2 Before 12/15/64. Except HiPo
3 Mechanics Type. Except HiPo
4 From 10/5/64
5 Before 10/5/64
6 2V & 4V Except HiPo
7 HiPo
8 4V
9 Shelby GT-350, GT-500
10 Except GT
11 GT

Note: The drive shaft and slip yoke illustrations must be included with this chart

MUSTANG DRIVESHAFT ASSEMBLY APPLICATION CHART, PART TWO, 1969-1970

Year	Engine	Transmission Type	Yoke Code	Yoke Type	Part Number	Notes	Driveshaft Diam.	Length	U-Joint Front	U-Joint Rear
69	200	3-S	14	2	C5ZZ-4841-B		2-1/2	51	J	J
69	200	C4	3	4	C3AZ-4841-D	1	3	50-31/32	E	F
69	200,302	C4, 3-S, 4-S	3	4	C3AZ-4841-D	2	3	50-31/32	E	F
69	250	C4, 3-S	3	4	C3AZ-4841-D		3	50-11/16	E	F
69	302	C4, 3-S, 4-S	3	4	C3AZ-4841-D	3	3	50-7/32	E	F
69	250	C4	3	4	C3AZ-4841-D		3	50-7/32	E	F
69	351	3-S, 4-S	3	4	C3AZ-4841-D		3	50-7/32	E	F
69	351	FMX	2	4	C3AZ-4841-A		3	50-5/8	E	F
69	390	3-S, 4-S	4	2	C3AZ-4841-G	4	3	50-1/64	G	H
69	Boss 302	3-S, 4-S	4	2	C3AZ-4841-G	4	3	50-1/64	G	H
69	390,428	C6	16	2	C7SZ-4841-A		3-1/2	46-1932	G	H
69	428	4-S	16	2	C7SZ-4841-A		3	50	G	H
69	Boss 429	4-S	16	2	C7SZ-4841-A		3	50-7/8	G	H
70	200,250	C4, 3-S	3	4	C3AZ-4841-D		3	50-11/16	E	F
70	200,302	C4, 3-S, 4-S	3	4	C3AZ-4841-D	1	3	50-31/32	E	F
70	351	C4, 3-S, 4-S	3	4	C3AZ-4841-D	5	3	50-7/32	E	F
70	351	FMX	2	4	C3AZ-4841-A	5	3	50-5/8	E	F
70	Boss 302	4-S	4	2	C3AZ-4841-G	6	3	50-1/64	G	H
70	351-4V	4-S	4	2	C3AZ-4841-G	6	3	50-1/64	G	H
70	351	FMX	16	2	C7SZ-4841-A	6	3	50-5/8	G	H
70	428	C6	16	2	C7SZ-4841-A		3-1/2	46-19/32	G	H
70	428	4-S	16	2	C7SZ-4841-A		3	50	G	H
70	Boss 429	4-S	16	2	C7SZ-4841-A		3	50-7/8	G	H

Notes:

1 7-1/4" Ring Gear
2 7-1/4" Ring Gear, Except Boss
3 8-3/4" Ring Gear, Except Boss
4 Boss 4-Speed only
5 w/ 28 spline axles
6 w/ 31 spline axles

Note: The drive shaft and slip yoke illustrations must be included with this chart

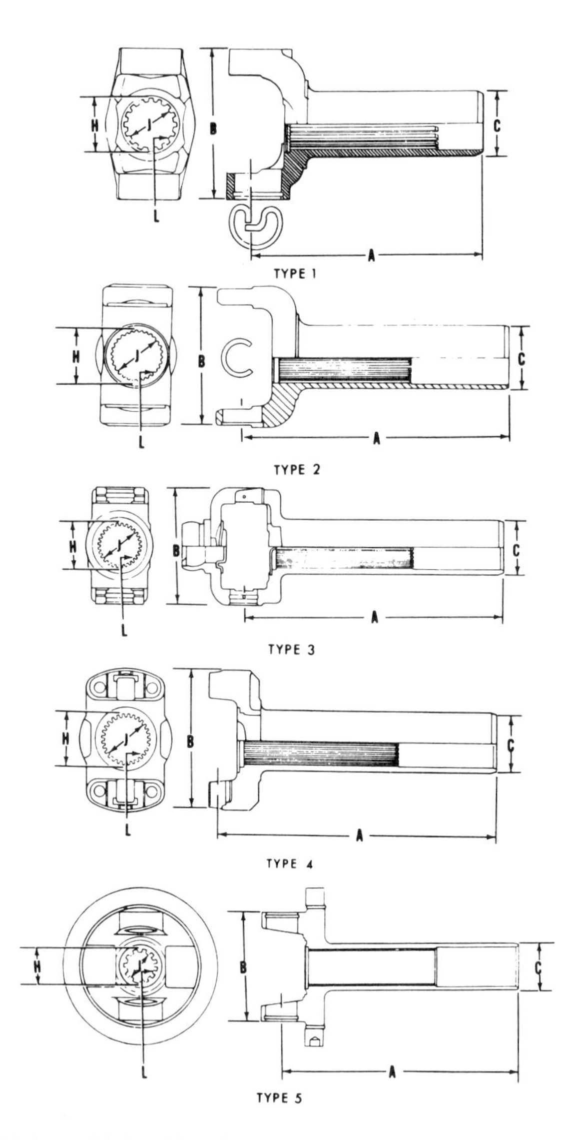

Universal joint slip yoke identification chart.

MUSTANG UNIVERSAL JOINT CODES AND DESCRIPTIONS CHART

Part Number	U-Joint Code	Type
C3AZ-4635-E	C	Dana (Spicer) or Ford-1260, 3-15/16" span
C3AZ-4635-F	D	Dana (Spicer) or Ford-1260, 3-15/16" span
C3AZ-4635-E	E	Cleveland Steel-1310, 3-1/4" span
C3AZ-4635-F	F	Cleveland Steel-1310, 3-1/4" span
C3AZ-4635-G	G	Cleveland Steel-1330, 3-5/8" span
C3AZ-4635-H	H	Cleveland Steel-1330, 3-5/8" span
C3UZ-4635-B	J	Dana (Spicer)-1100, 2-21/32" span
C5DZ-4635-A	L	Mechanics-1260, 3-15/32" span
C5DZ-4635-B	M	Mechanics-1260, 3-15/32" span

This chart correlates with the Universal Joint Slip Yoke Identification Chart, the Universal Joint Identification Illustration, and the Mustang Driveshaft Assembly Application Chart.

MUSTANG UNIVERSAL JOINT IDENTIFICATION CHART

Part Number	U-Joint Code	Industry Designation	A	B	C	D	E	F	G	H	Type
C3AZ-4635-E	C	1260	3-15/32"	3-15/32"	2-31/32"	2-31/32"	1-1/16"	1-1/16"	19/32"	9/16"	D
C3AZ-4635-F	D	1260	3-15/32"	3-15/32"	2-31/32"	2-31/32"	1-1/16"	1-1/8"	19/32"	9/16"	D
C3AZ-4635-E	E	1310	3-1/4"	3-1/4"	3"	3"	1-1/16"	1-1/16"	11/16"	17/32"	C
C3AZ-4635-F	F	1310	3-1/4"	3-1/4"	3"	3"	1-1/16"	1-1/8"	11/16"	17/32"	C
C3AZ-4635-G	G	1330	3-5/8"	3-5/8"	3-13/32"	3-13/32"	1-1/16"	1-1/16"	11/16"	19/32"	C
C3AZ-4635-H	H	1330	3-5/8"	3-5/8"	3-13/32"	3-13/32"	1-1/16"	1-1/8"	11/16"	19/32"	C
C3UZ-4635-B	J	1100	2-21/32"	2-21/32"	2-13/64"	2-1/2"	1"	1"	35/64"	17/32"	D
C5DZ-4635-A	L	1260	3-15/32"	3-15/32"	2-31/32"	2-31/32"	1"	1-1/8"	19/32"	29/64"	M
C5DZ-4635-B	M	1260	3-15/32"	3-15/32"	2-31/32"	2-31/32"	1"	1"	19/32"	29/64"	M

Types:

- D — Dana (Spicer) or Ford
- C — Cleveland
- M — Mechanics

This chart correlates with the Universal Joint Slip Yoke Identification Chart, the Universal Joint Identification Illustration, and the Mustang Driveshaft Assembly Application Chart.

MUSTANG DRIVESHAFT SLIP YOKE SPECIFICATIONS CHART

Part Number	Slip Yoke Code	Dimensions: A	B	C	H	J	Splines	Type
C3AZ-4841-A	2	8-7/32"	3-1/2"	1-11/16"	1-3/8"	1-17/64"	31	1
C3AZ-4841-D	3	8-3/32"	3-1/2"	1-1/2"	1-7/32"	1-3/32"	28	1
C3AZ-4841-G	4	8-7/32"	3-7/8"	1-1/2"	1-7/32"	1-3/32"	28	1
C1UZ-4841-A	5	5-29/32"	2-15/32"	1-3/8"	1-7/64"	63/64"	25	2
C5DZ-4841-A	6	8-3/32"	3-3/16"	1-1/2"	1-7/32"	1-3/32"	28	2
C5DZ-4841-B	7	8-1/4"	3-3/16"	1-1/2"	1-7/64"	63/64"	25	2
C5ZZ-4841-B	14	5-29/32"		1-1/2"	1-7/32"	1-3/32"	28	2
C7SZ-4841-A	16	8-5/32"	3-29/32"	1-11/16"	1-3/8"	1-17/64"	31	1

This chart correlates with the Universal Joint Slip Yoke Identification Chart, the Universal Joint Identification Illustration, and the Mustang Driveshaft Assembly Application Chart.

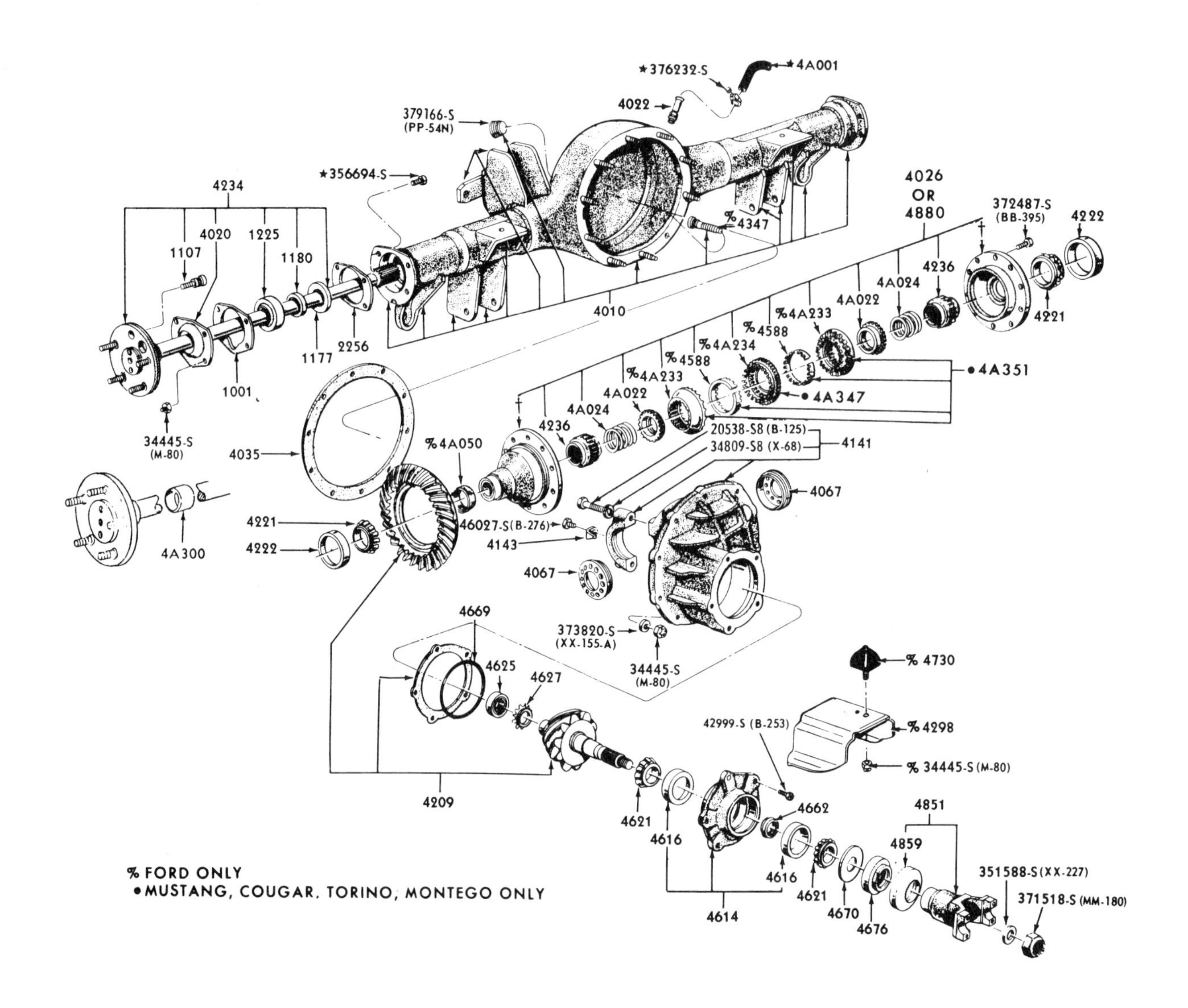

All V-8-powered Mustangs were equipped with axle housings with a removable gear carrier. That means the cast-iron center housing containing the ring and pinion gears can be removed from the rear axle as an assembly. Unlike the integral rear axle, the ring and pinion gears, the pinion carrier, and the associated bearings remain in adjustment or "set up" within the portable center housing. Besides having a removable or integral housing, all Ford rear axles are categorized by the diameter of the ring gear. The ring gear in an integral housing will measure 6¾, 7¼, or 7¾in. Only the 7¼in integral rear axle was installed under a Mustang. Rear axles having a removable carrier were manufactured in two sizes. The smallest was equipped with an 8in diameter ring gear and the larger a 9in gear. The 8 and 9in axles are similar in appearance, assembly, and fundamental operation, but share few interchangeable components. Virtually each piece of a 9in rear axle is larger and stronger compared to its little brother. The steel "banjo" housing, so called because of its shape, is the basic component of an 8 or 9in rear axle. The transfer of power through the differential is exactly the same as the integral rear axle housing.

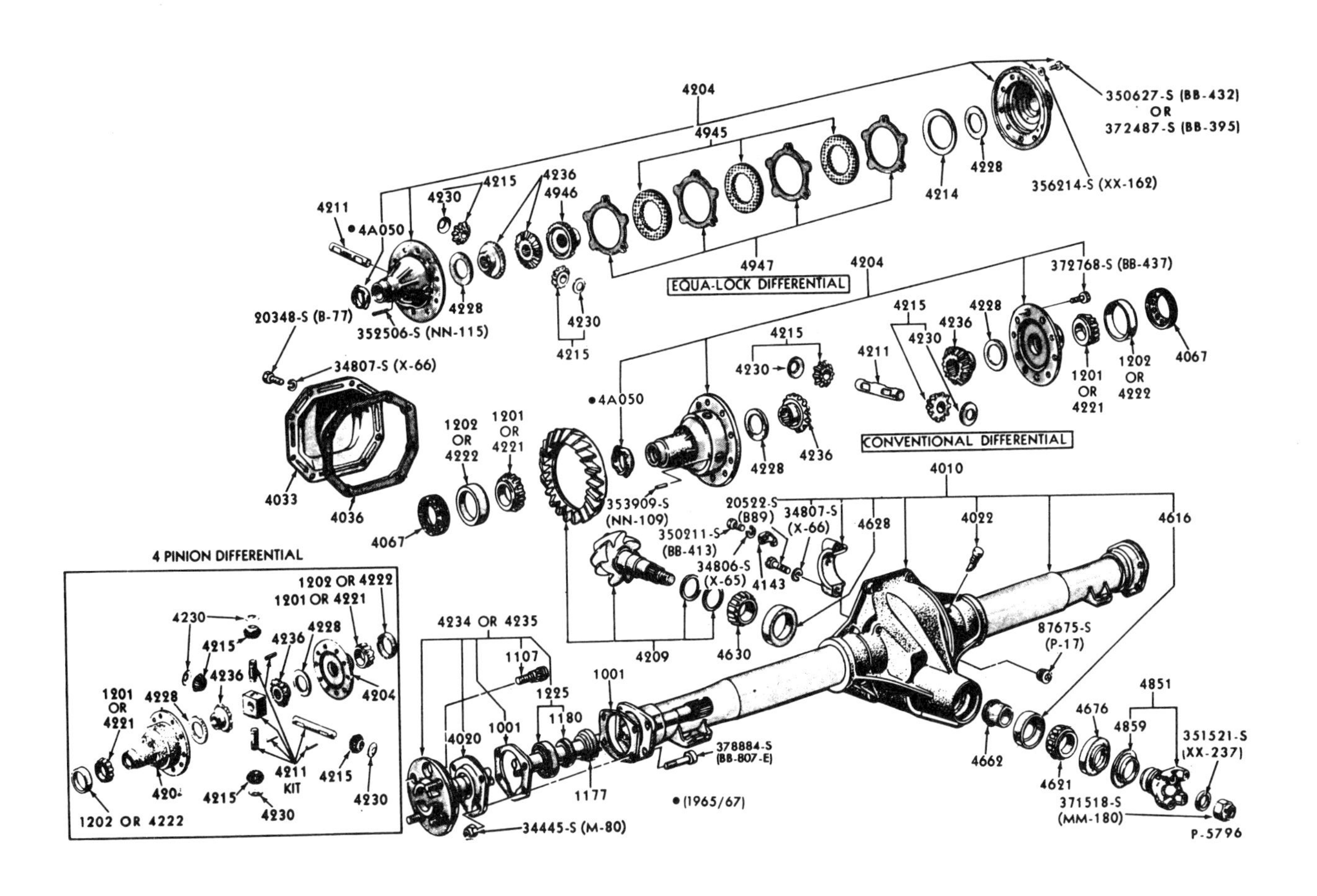

Ford manufactured two types of rear axle housings between 1965 and 1970. The type used under lightweight economy cars powered by six-cylinder engines like the Falcon, the Mustang, and the Fairlane housed the ring and pinion gears within the rear axle assembly, and was appropriately called an integral rear axle. It is a lightweight unit consisting of a one-piece cast-iron differential center housing outfitted with steel axle tubes that are pressed into place and tack welded. The center housing is machined to support the ring gear carrier and pinion gear. The outboard ends of the axle tubes are designed to accept a bearing, which supports the outer end of each axle. A four bolt flange, welded to the outer end of each tube, accommodates the brake backing plate and a bearing retainer. As with all modern-day differentials, power is transmitted from the pinion gear to the ring gear (4209), which rotates the carrier housing (4204). Each axle is attached to the carrier housing via an internally splined "side gear" (4236). The side gears mesh with the spider gears (4215), which are pinned to the carrier housing with a case-hardened steel rod (4211). The axles in a non-locking rear axle or conventional differential, rotate independently from each other. Power delivered by the driveshaft will be transmitted to the wheel with the least rolling resistance. An EquaLock differential incorporates a locking system that consists of three double-faced clutch discs, four friction discs, and a spring-steel bevel washer that, when assembled in the special ring gear carrier, locks both axles together as power is applied. Under light load conditions the clutches will slip, thus allowing each axle to rotate independently.

Mustang Differential Case Assembly Application Chart

Year	Axle Ident.	Notes	Part Number	Description
65	WCL & WCZ		D0OZ-4204-A	w/ 8" R/G, medium differential bearing, 2 pinion, non-Locking
65/69	WDJ		C4OZ-4204-B	w/ 8" R/G, medium differential bearing, 2 pinion Equalock
69/70	WFL		C9OZ-4204-F	w/ 8" R/G, medium differential bearing, 2 pinion Traction-Lok
65/70	WCZ & WES	1	D0OZ-4204-D	w/ 9" R/G, medium differential bearing, 2 pinion non-Locking
67/70	WES	1	C2AZ-4204-K	w/ 9" R/G, large differential bearing,
				2 or 4 pinion non-Locking, w/ 28 spline axles
69/70	WES	1	C9OZ-4204-B	w/ 9" R/G, large differential bearing,
				4 pinion non-Locking w/ 31 spline axles
67/68	WES		C4AZ-4204-F	w/ 9" R/G, large differential bearing,
				2 or 4 pinion Equalock w/ 28 spline axles
				7/16" or 1/2" diameter bolt holes optional.
69/70	WFB	1	C0OZ-4204-C	w/ 9" R/G, large differential bearing,
				4 pinion Traction-Lok w/ 28 spline axles
69/70	WFD	1	C0OZ-4204-C	w/ 9" R/G, large differential bearing,
				4 pinion Traction-Lok w/ 31 spline axles
70/72	All	2	D0OZ-2404-D	w/ 9" R/G, slim line differential bearing, 4 pinion non-Locking
				w/ 31 spline axles. Includes C9AZ-4221-A bearing,
				C9AZ-4222-A cup, (2) C9AZ-4228-A washers.
		2	D2AZ-4204-D	w/ 9" R/G, slim line differential bearing,
				2 pinion non-Locking w/ 28 spline axles.
		2	D0OZ-4204-C	w/ 9" R/G, slim line differential bearing,
				2 or 4 pinion Traction-Lok w/ 28 or 31 spline axles.
				Includes C9AZ-4221-A bearing, C9AZ-4222-A cup.
		2	DO0OZ-4204-E	w/ 9" R/G, slim line differential bearing,
				no-spin 2 pinion w/ 31 spline axles.
				Includes C9AZ-4221-A bearing, C9AZ-4222-A cup.
70	WFU-E	1	D0OZ-4204-B	w/ 9" R/G, large differential bearing,
				no-spin differential w/ 31 spline axles.
70	WFU-E	2	D0OZ-4204-E	w/ 9" R/G, slim line differential bearing,
				no-spin Locker w/ 31 spline axles.
65/68	WCY		C4DZ-4204-A	w/ 7-1/4" R/G, small differential bearing, 2 pinion Equalock
66	WCZ		C4DZ-4204-E	w/ 9" R/G, medium differential bearing,
				2 pinion Equalock w/ 28 spline axles.
68	WES-R		C4DZ-4204-E	w/ 9" R/G, medium differential bearing,
				2 pinion Equalock w/ 28 spline axles.
65/70	WCY		C1DW-4204-B	w/ 7-1/4" R/G, small differential bearing,
				2 or 4 pinion non-Locking.

Notes:

1 Used before 5/13/70

2 Used after 5/13/70

MUSTANG REAR AXLE DIFFERENTIAL SIDE GEAR KIT CHART

YEAR	MODEL	AXLE IDENT.	PART NUMBER	DESCRIPTION
65/68	8", 9" R/G		C9AZ-4236-B	28 splines- 4 omitted- with hub.
				Consists of (2) C2AW-4236-A gears
				(Use 1 piece for locking axle) and
				(2) C9OZZ-4228-A washers.
70	8 cylinder 427	WFU-D, E	C4AZ-4236-A	31 splines- 4 omitted- with hub.
	Exc. locking axle			Consists of (2) C2AW-4236-A gears and
				(2) C9OZZ-4228-A washers.
69/70	8", 9" R/G		C9AZ-4236-A	With locking differential- 28 spline incl.
				(1) C9OW-4236-A gear- 21/32" long,
				(1) C9OW-4236-B gear- 1 3/16" long,
				(2) C9OZ-4228-A washers.
			C9AZ-4236-B	With std. differential- 28 spline includes
				(2) C9OW-4236-B gears 1 3/16" long,
				(2) C9OZ-4228-A washers.
65/70	6-3/4", 7",		C1DZ-4236-C	24 splines- consists of:
	7-1/4" R/G			(2) DODZ 4236-A gears and
				(2) CODW 4228-A washers.

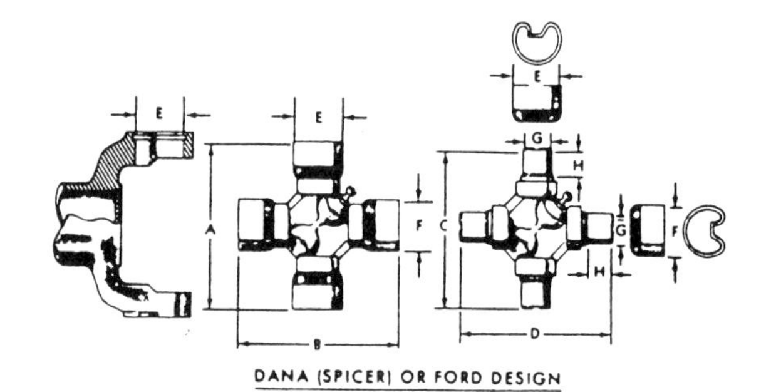

DANA (SPICER) OR FORD DESIGN

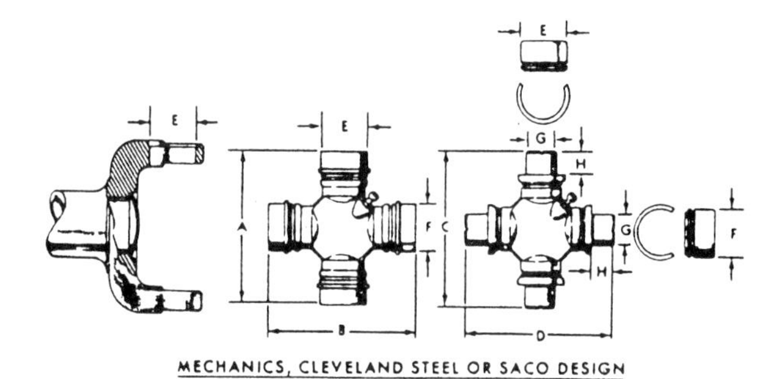

MECHANICS, CLEVELAND STEEL OR SACO DESIGN

Universal joint identification chart.

MUSTANG REAR AXLE DIFFERENTIAL ASSEMBLY APPLICATION CHART

YEAR	AXLE TAG I.D.	PART NUMBER	RATIO	GEAR SIZE	LOCKING	DESCRIPTION
67	WCZ-V,V2	C7OW-4200-F	2.79	8	NO	
67	WCZ-F,F2	C7OW-4200-G	3.00	8	NO	
68	WCZ-W	C7OW-4200-H	3.25	8	NO	
68	WFD-B,D,D2	C8OW-4200-C	3.91	9	YES	
69/70	WES-AC	C9OW-4200-D	3.00	9	NO	
69	WES-T,T1,T2	C9OW-4200-H	2.75	9	NO	
69	WES-AA	C9OW-4200-G	3.00	9	NO	
69	WES-AD,AD2	C9OW-4200-K	3.25	9		
69	WFB-C,C2	C9OW-4200-V	3.25	9	YES	TRACTION-LOK
70	WFL-A,A2	COOW-4200-A	3.00	8	YES	TRACTION-LOK

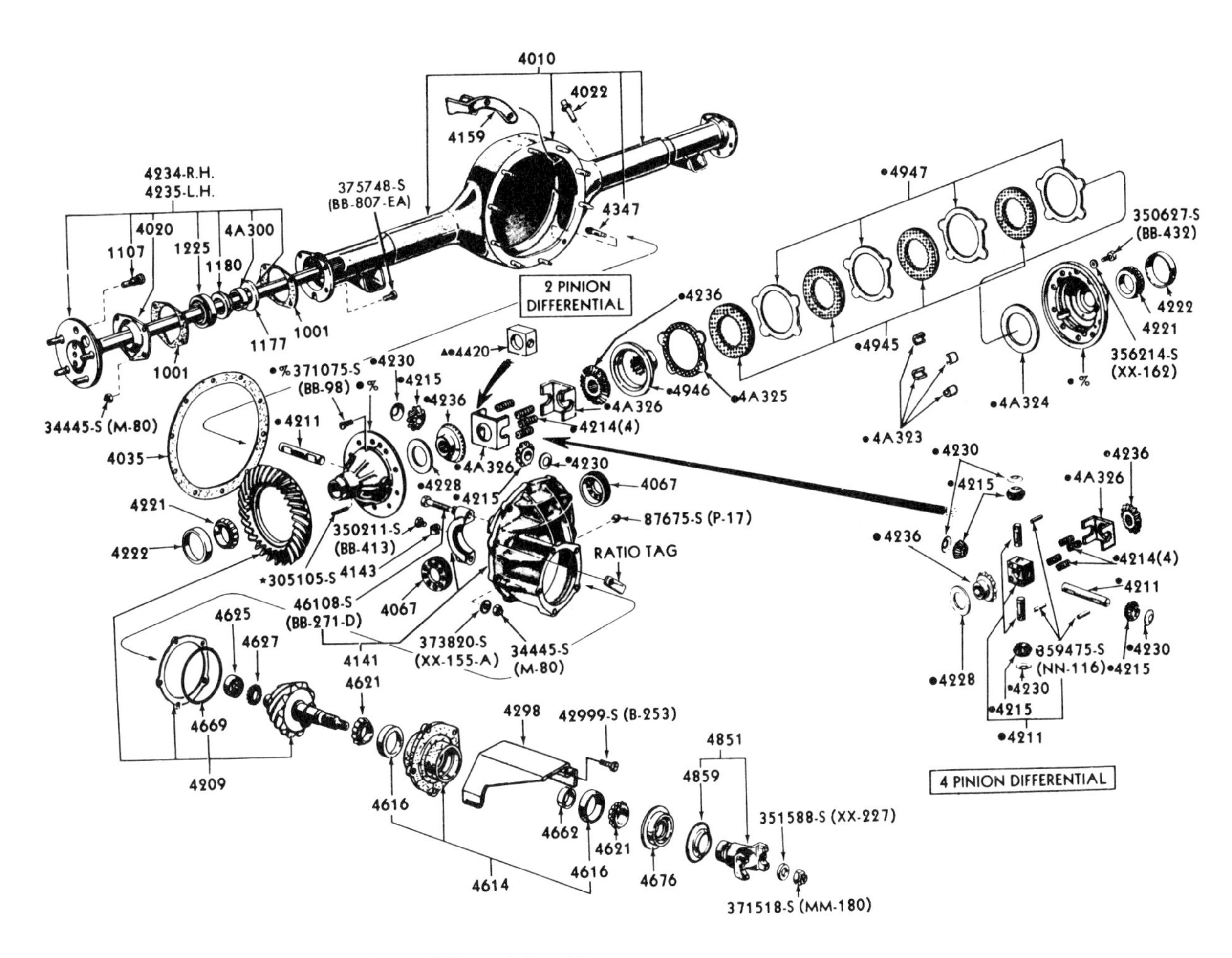

Typical Mustang 8in, 9in, and 9³/8in differential with Traction-Lok.

MUSTANG REAR AXLE SEAL ASSEMBLY CHART

YEAR	MODEL	AXLE IDENT.	PART NUMBER	DESCRIPTION
65/67	9", 9 3/8" R/G	Exc. WER	D1UZ-4676-A	1 13/16" I.D.- 3 5/16"" flange dia.,
				3" seal dia.- 1/2" thick-double lip seal
68/70	9", 9 3/8" R/G	Exc. WER	D1UZ-4676-A	1 13/16" I.D.- 3 5/16"" flange dia.,
				3" seal dia.- 1/2" thick
65/67	7 1/4", 8" R/G		C2OZ-4676-B	1 13/16" I.D.- 2 15/16" flange dia.,
				2 5/8" seal dia. - 29/64" thick
68	7", 8" R/G		C2OZ-4676-B	1 13/16" I.D.- 2 15/16" flange dia.,
				2 5/8" seal dia. - 29/64" thick
65/69	7 1/4" R/G		C1UZ-4676-C	1 21/32" I.D.- 2 5/8" O.D.- 29/64"
				thick- use with 1100 univ. joints only
68/70	9", 9 3/8" R/G	Exc. WER	C2AZ-4676-B	
	7", 8" R/G		C2OZ-4676-B	

MUSTANG REAR AXLE PINION BEARING RETAINER ASSEMBLY CHART

YEAR	MODEL APPLICATION	RETAINER AND CUP ASSEMBLY	CUPS	BEARINGS
65/70	7 1/4" R/G--	None required	CODW-4616-A	CODW-4621-A
	integral carrier		TBAA-4616-A	C1DZ-4630-A
65/67	8" R/G	C6OZ-4614-B	CODW-4616-A	CODW-4621-A
68/70	8" R/G	C6OZ-4614-B	CODW-4616-A	CODW-4621-A
65/70	9" R/G- 28 spline shaft	D2SZ-4614-A	B7A-4616-A	B7A-4621-A
67/70	9" R/G- 31 spline shaft	C3AZ-4614-B	B7A-4616-A	B7A-4621-A
			TBAA-4616-A	TBAA-4621-A

MUSTANG REAR AXLE SHAFT ASSEMBLY APPLICATION CHART

YEAR	MODEL	AXLE IDENTIFICATION	PART NUMBER	DESCRIPTION
65/66	7 1/4" R/G	WCY	#C4DZ-4234-A	24 splines-R.H. 29-15/16"
(U.S. only)			#C4DZ-4234-B	L.H. 26 7/16"
65/66	7 1/4" R/G	WCY	#C4DZ-4234-A	24 splines-R.H. 29-15/16"
(Can. only)			C4DW-4275-A	L.H. 26-7/16"
65/66	8" R/G	WCZ, WDJ	*C20Z-4234-A	28 splines-R.H. 30-1/8"
			*C20Z-4235-B	L.H. 26-5/64"
65/66	9" R/G	WCZ	*C50Z-4234-B	28 splines-R.H. 30-1/8"
			*C50Z-4234-A	L.H. 26-5/64"
67/69	8", 9" R/G	WCZ, WDJ, WES, WFB	*C60Z-4234-F	28 splines-R.H. 31-1/8"
	Before 3/15/69		*C90Z-4234-B	L.H. 27-1/16"
69/70	8", 9" R/G		*C90Z-4234-A	28 splines-R.H. 31 1/8"
	From 3/15/69		#C6DZ-4234-C	L.H. 27-1/16"
67/70	7 1/4" R/G	WCY	#C6DZ-4234-D	24 splines-R.H. 30-15/16"
			*C90Z-4234-C	L.H. 27-7/16"
67/70	9" R/G		*C90Z-4234-D	31 splines-R.H. 31-1/8"
				L.H. 27-1/16"

Includes CODZ-1225-C bearing (small).
* Includes DOAZ-1225-A bearing (medium).

MUSTANG REAR AXLE DIFFERENTIAL BEARING CONE CHART - ROLLER and CUP

YEAR	MODEL	CONE-ROLLER	CUP	DESCRIPTION
65/70	7 1/4" R/G	C1UW-4221-A	C1UW-4222-A	S - small size
	7 3/4", 8" R/G	B7A-4221-A	B7A-4222-A	M - medium size
	9" R/G Before 5/13/70	B7A-4221-A	B7A-4222-A	M - medium size
	From 5/13/70	B7A-4221-B	B7A-4222-B	L - large size
70	9" R/G From 5/13/70	C9AZ-4221-A	C9AZ-4222-A	SL - slim-line size

MUSTANG REAR AXLE DIFFERENTIAL PINION KIT CHART

YEAR	MODEL	AXLE IDENTIFICATION	PART NUMBER	DESCRIPTION
65/69	8", 9" R/G		C8OZ-4215-A	(2) C1AW-4215-C pinions
	with equalock axles			(2) B7A-4230-A washers
65/72	8", 9" R/G	WER-F, G, H	C9AZ-4215-A	(2) DOAW-4215-A pinions
	with conventional axles			(2) DOAZ-4230-A washers
68/70	8" and 9" R/G		C8OZ-4215-A	(4) C9OW-4215-A gears
	with traction lok axles			(4) C9OZ-4230-A washers
65/70	7 1/4" R/G		CODZ-4215-C	(2) CODW-4215-C pinions
				(2) CODW-4230-C washers

MUSTANG REAR AXLE HOUSING ASSEMBLY APPLICATION CHART

YEAR	MODEL	AXLE IDENTIFICATION	PART NUMBER	DESCRIPTION
65/66	7 1/4" R/G	WCY	C4DZ-4010-G	Integral carrier
67/69	7 1/4" R/G	WCY	C7ZZ-4010-A	Integral carrier
70	7 1/4" R/G	WCY	DOZZ-4010-B	Integral carrier
65/66	8" R/G	WCZ,WDJ	C4DZ-4010-F	Incl. (1) 378511-S clamp
67/69	8" R/G	WCZ,WDJ	C7ZZ-4010-E	
70	8" R/G	WCZ,WFL	C9ZZ-4010-B	
67/68	9" R/G Before 4/01/68	WES,WFD	DOZZ-4010-C	
68/70	9" R/G From 4/01/68	WES,WFB,WFD,WFU	DOZZ-4010-D	
	To 6/01/70			
70	9" R/G From 6/01/70	WES,WFB,WFD,WFU	DOZZ-4010-E	
	To 9/25/70			
70	9" R/G From 9/25/70	WES,WFB,WFD,WFU	D1ZZ-4010-B	

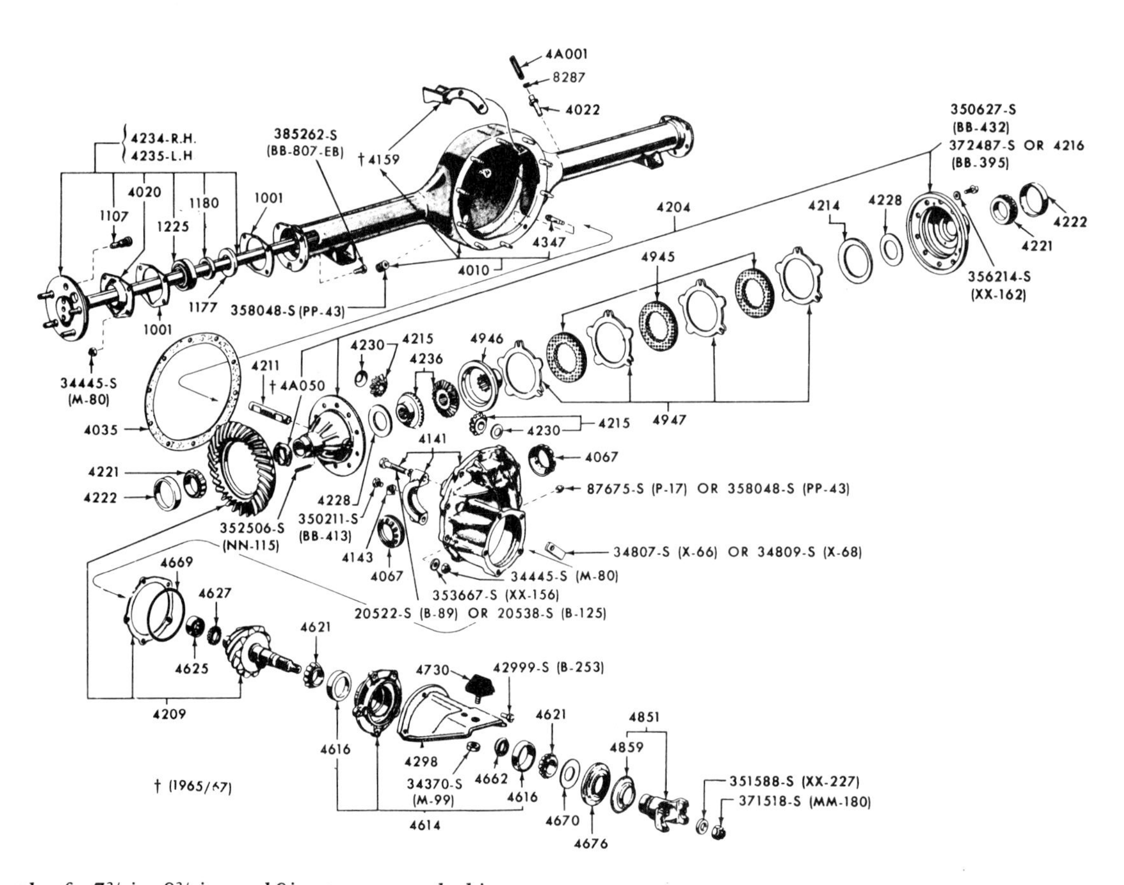

An example of a 7³/4in, 8³/4in, and 9in open or non-locking rear axle with removable carrier used in 1965-1970 Mustangs.

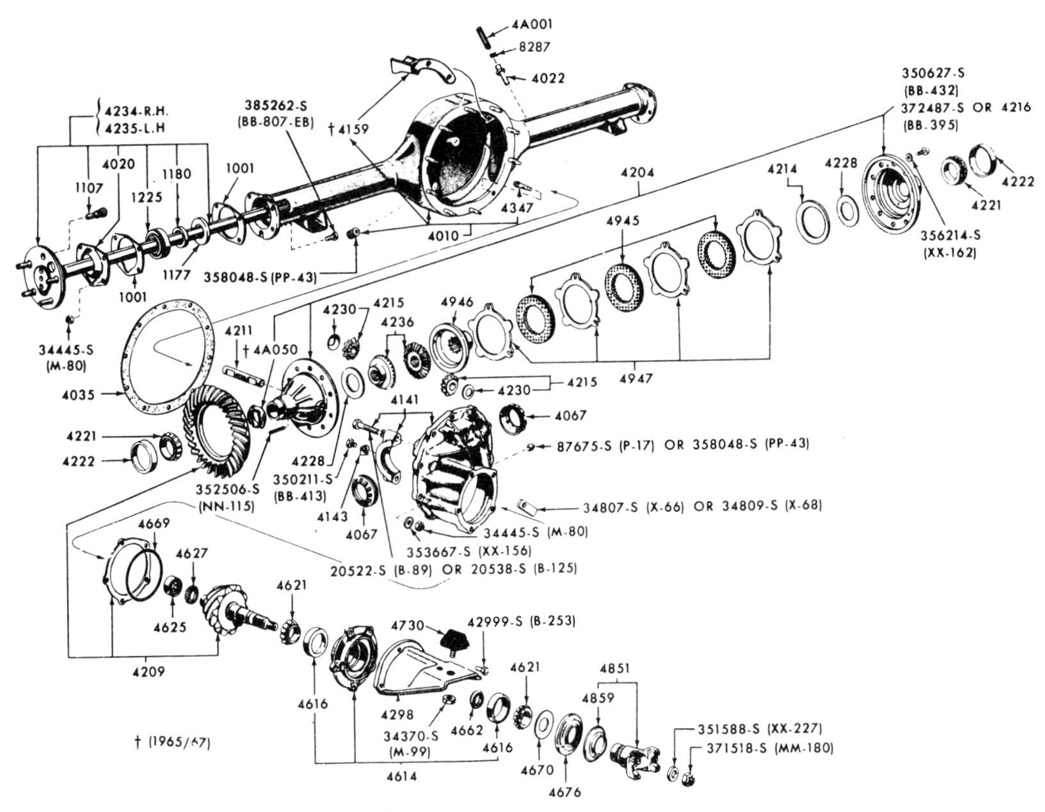

An example of a 7³/4in, 8³/4in, and 9in Equa Lock rear axle with removable carrier used in 1965-1970 Mustangs.

MUSTANG RING GEAR and PINION SETS APPLICATION CHART

Year	Ring Gear Diameter	Ratio	Part Number	Number of Teeth Gear	Pinion
65	7-1/4	2.83	C5DZ-4209-C	34	12
65	7-1/4	3.50	C1DZ-4209-Z	35	10
66/70	7-1/4	2.83	C5DZ-4209-C	34	12
69/70	7-1/4	3.08	C9ZZ-4209-A	37	12
65/70	8	2.80/2.79	C4OZ-4209-K	39	14
65/70	8	3.00	C4OZ-4209-H	39	13
65/70	8	3.50	C4OZ-4209-J	35	12
68	8	3.25	C4DZ-4209-D	39	12
65	9	4.11	B7AZ-4209-K	37	9
65/70	9	3.50	C9OZ-4209-A	35	10
65/67	9	3.89	B7AZ-4209-N	35	9
67/70	9	3.00	C0AZ-4209-E	39	13
67/70	9	3.25	B8AZ-4209-C	39	12
68/70	9	2.75	C7AZ-4209-J	44	16
68/70	9	3.91	C8OZ-4209-A	43	11
68/70	9	4.30	C8OZ-4209-B	43	10

Mustang Locking Differential Conversion and Replacment Kits.

1965/69 Part Number: C9OZ-4026-B

Items:	Part Number:	Description
1	B9A-4159-A	Lubricator
1	C4AZ-4204-F	Case
1	C2AW-4236-A	Side Gear
1	C4AZ-4946-A	Hub
1	352506-S	Pin
1	C4AZ-4214-A	Spring
3	B9AZ-4945-A	Clutch Plate
1	C2AZ-4236-B	Side Gear
4	B9AZ-4947-A	Clutch Plate
10	350627-S	Bolt-7/16x20
5.5 pints		Lubricant

1965/70 Part Number: C4DZ-4880-B

Notes: (7 & 7-3/4 inch ring gear) Use with integral carrier.

Items:	Part Number:	Description
1	B9A-4159-A	Lubricator
1	C4DZ-4204-A	Case
1	C4DZ-4236-A	Side Gear
1	C4DZ-4946-A	Hub
1	305086-S	Pin
1	C4DZ-4214-A	Spring
3	B9AZ-4945-A	Clutch Plate
1	C4DZ-4236-B	Side Gear
4	C4DZ-4947-A	Clutch Plate
8	379237-S	Bolt-3/8" 24 X 11/64"

Chapter 6

Suspension, Brakes, Steering, Wheels, and Exhaust

The groupings of topics throughout this book are primarily based on Ford's system of categorization within their original parts catalogs. For 1964 to 1970, an enormous variety of drivetrain assemblies and components were utilized in the quest of such myriad goals as ultra economy and rocket-like acceleration. The 1969 Mustang E, a plain six-cylinder SportsRoof, was manufactured with fuel mileage in mind; the 1969 Mach 1 428ci Cobra Jet with the Drag Pack option was intended to spin the shortest time imaginable in quarter-mile competition. Ironically, because weight always remained a strong factor in Ford's drivetrain development programs, the CJ and the E were not that far apart in concept and material.

In between the two extremes, millions of Mustangs were made that represented complete compromise between economy and power. Each of those could have had any number of different drivetrain component combinations depending on engine sizes, handling options, owner preferences, and ease of driveability.

The drivetrain is almost wholly out of sight of the average show judge or auto admirer. But many Mustangs, when restored, are forced to accommodate replacement assemblies, and compatibility of these assemblies or subassemblies is an absolute necessity to the car owner. While compatibility is obviously important in situations of engine swaps, for instance, the replacement of worn slip yokes and U-joints or damaged driveshafts requires exactitude even in "unmodified" cars. That is

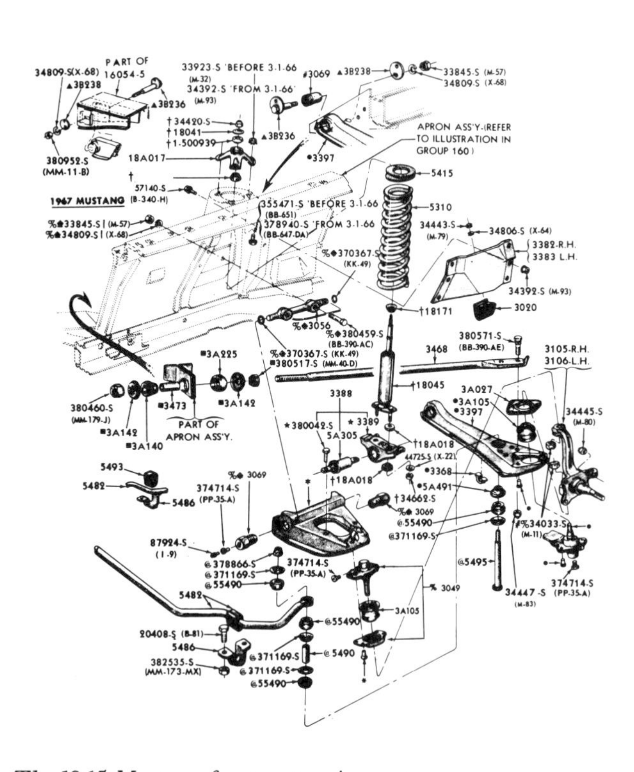

The 1965 Mustang front suspension system.

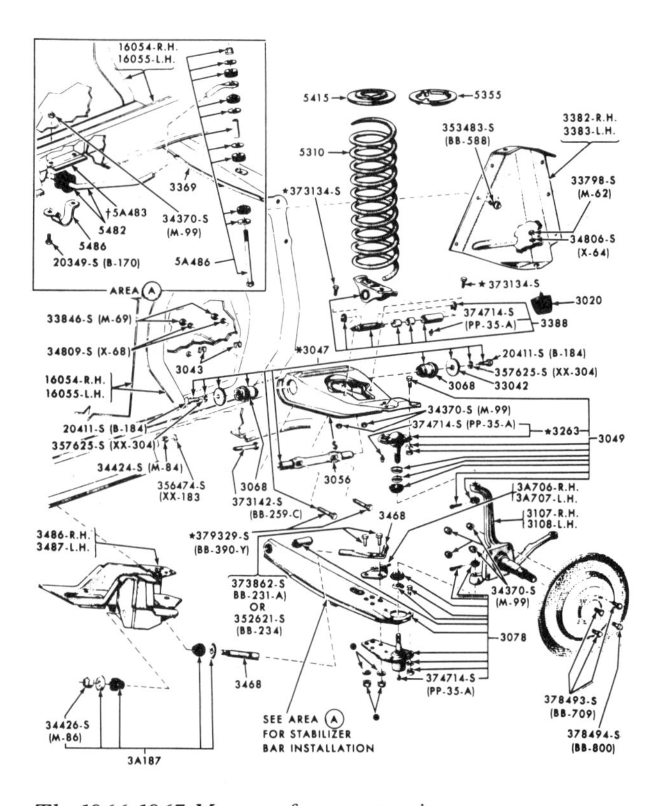

The 1966-1967 Mustang front suspension system.

where the charts and drawings of this chapter will benefit all.

In the realm of the drivetrain, Ford has managed over the years to maintain a surprisingly broad and complete stock of parts. Its decision to keep such parts in the supply system is based upon supply and demand and, obviously, over the years, owners have been interested in keeping a significant number of Mustangs on the road. Hence, the reader may find, in this chapter, part numbers that still apply and that will be, if not perfectly accurate, genuine guidelines for Ford service department personnel in locating needed items.

▲ SUPPLIED IN 18124 SHOCK ABSORBER KIT
*SUPPLIED IN 18198 KIT
®Before 3/1/66 #From 3/1/66

The 1965-1969 Mustang front shock absorber and mounting hardware.

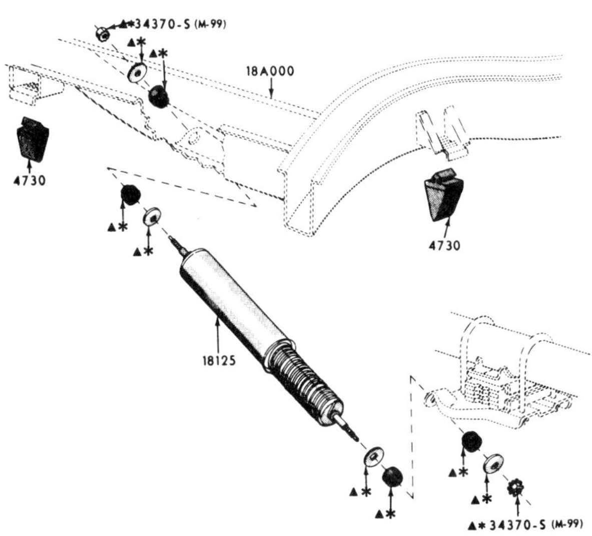

The 1965-1968 Mustang rear shock absorber and mounting hardware.

1964-1/2 Front Spring Application Chart, Hardtops before 8/20/64

CYL/CID	Load	Transmission	Steering	A/C	Marking Stripes	Part Number
6/170	1304	3/S, 4/S, C4	M	N	1 Pink	C5ZZ-5310-A
						r/b C5ZZ-5310-B
6/170	1304	3/S	P	N	1 Pink	C5ZZ-5310-A
						r/b C5ZZ-5310-B
6/170	1369	4/S, C4	P	N	1 Green	C5ZZ-5310-B
6/170	1369	3/S		A	1 Green	C5ZZ-5310-B
6/170	1438	3/S, 4/S, C4	P	A	1 Violet	C5ZZ-5310-D
						r/b C5ZZ-5310-C
6/170	1438	4/S, C4	M	A	1 Violet	C5ZZ-5310-D
						r/b C5ZZ-5310-C
6/200	1279	4/S	M	N	1 Pink	C5ZZ-5310-A
						r/b C5ZZ-5310-B
8/260	1438	3/S, C4	M	N	1 Violet	C5ZZ-5310-D
						r/b C5ZZ-5310-C
8/260	1438	3/S	P	N	1 Violet	C5ZZ-5310-D
						r/b C5ZZ-5310-C
8/260	1438	3/S, 4/S, C4	M	N	1 Red	C5ZZ-5310-F
						r/b C5ZZ-5310-E
8/260	1516	C4	P	N	1 Gray	C5ZZ-5310-C
8/260	1516	3/S, C4	M	A	1 Gray	C5ZZ-5310-C
8/260	1516	3/S, 4/S, C4	M/P	N	1 Brown	C5ZZ-5310-E
8/260	1516	3/S, 4/S, C4	M/P	A	1 Brown	C5ZZ-5310-E
8/260	1516	3/S, C4	P	A	1 Gray	C5ZZ-5310-C
8/289-4V	1438	4/S, C4	M	N	1 Violet	C5ZZ-5310-D
						r/b C5ZZ-5310-C
8/289-4V	1516	4/S, C4	P	N	1 Gray	C5ZZ-5310-C
8/289-4V	1516	4/S, C4	M	A	1 Gray	C5ZZ-5310-C
8/289-4V	1516	4/S, C4	P	A	1 Gray	C5ZZ-5310-C
8/289-4V	1438	4/S	M	N	1 Red	C5ZZ-5310-F
						r/b C5ZZ-5310-E
8/289-4V	1516	4/S	P	N	1 Brown	C5ZZ-5310-E
8/289-4V	1516	4/S	M	A	1 Brown	C5ZZ-5310-E
8/289-4V	1516	4/S	P	A	1 Brown	C5ZZ-5310-E

1964-1/2 Front Spring Application Chart, Convertibles before 8/20/64

CYL/CID	Load	Transmission Types	Steering	A/C	Marking Stripes	Part Number
6/170	1304	3/S, 4/S, C4	M	N	1 Pink	C5ZZ-5310-A
						r/b C5ZZ-5310-B
6/170	1369	3/S, 4/S	P	N	1 Green	C5ZZ-5310-B
6/170	1438	3/S	P	A	1 Violet	C5ZZ-5310-D
						r/b C5ZZ-5310-C
6/170	1438	C4	P	N	1 Violet	C5ZZ-5310-D
						r/b C5ZZ-5310-C
6/170	1438	3/S, 4/S, C4	M	A	1 Violet	C5ZZ-5310-D
						r/b C5ZZ-5310-C
6/170	1516	4/S, C4	P	A	1 Gray	C5ZZ-5310-C
8/260	1516	3/S, C4	M/P	N	1 Gray	C5ZZ-5310-C
8/260	1516	3/S, 4/S, C4	M/P	N	1 Brown	C5ZZ-5310-E
8/260	1516	3/S, 4/S, C4	M/P	A	1 Brown	C5ZZ-5310-E
8/260	1516	3/S, C4	M/P	A	1 Gray	C5ZZ-5310-C
8/289-4V	1516	3/S, C4	M/P	A	1 Gray	C5ZZ-5310-C
8/289-4V	1516	4/S, C4	M/P	N	1 Brown	C5ZZ-5310-E
8/289-4V	1516	3/S, 4/S	M/P	N	1 Gray	C5ZZ-5310-C
8/289-4V	1516	4/S, C4	M/P	A	1 Brown	C5ZZ-5310-E

1965 Mustang Front Spring Application Chart, 8/20/64-4/1/65
Hardtop and Fastback models.

CYL/CID	Model	Load	Transmission	Steering	A/C	Marking Stripes	Part Number
6/200	HT/FB	1279	3/S	M	N	1 Pink	C5ZZ-5310-A
							r/b C5ZZ-5310-B
6/200	HT	1279	4/S	M	N	1 Pink	C5ZZ-5310-A
							r/b C5ZZ-5310-B
6/200	HT/FB	1344	3/S, 4/S, C4	P	N	1 Green	C5ZZ-5310-B
6/200	HT/FB	1344	3/S	M	A	1 Green	C5ZZ-5310-B
6/200		1344	4/S, C4	M	N	1 Green	C5ZZ-5310-B
6/200	HT	1344	4/S	M	A	1 Green	C5ZZ-5310-B
6/200	HT	1344	C4	M	N	1 Green	C5ZZ-5310-B
6/200	HT/FB	1413	3/S, 4/S, C4	P	A	1 Violet	C5ZZ-5310-D
							r/b C5ZZ-5310-C
6/200	HT/FB	1413	C4	M	A	1 Violet	C5ZZ-5310-D
							r/b C5ZZ-5310-C
6/200	FB	1413	4/S	M	A	1 Violet	C5ZZ-5310-D
							r/b C5ZZ-5310-C
8/289	HT/FB	1413	3/S	M	N	1 Violet	C5ZZ-5310-D
							r/b C5ZZ-5310-C
8/289	HT/FB	1490	3/S	P	N	1 Gray	C5ZZ-5310-C
8/289	HT/FB	1490	3/S	M	A	1 Gray	C5ZZ-5310-C
8/289	HT/FB	1490	4/S, C4	M/P	N	1 Gray	C5ZZ-5310-C
8/289	HT/FB	1490	3/S, 4/S, C4	P	A	1 Gray	C5ZZ-5310-E
8/289	HT/FB	1490	4/S, C4	M	A	1 Gray	C5ZZ-5310-E
8/289-4V	HT/FB	1413	3/S, 4/S	M	N	1 Red	C5ZZ-5310-F
							r/b C5ZZ-5310-E
8/289-4V	HT/FB	1491	3/S, 4/S	P	N	1 Brown	C5ZZ-5310-E
8/289-4V	HT/FB	1491	3/S, 4/S	M	A	1 Brown	C5ZZ-5310-E
8/289-4V	HT/FB	1491	3/S, 4/S	P	A	1 Brown	C5ZZ-5310-E

1965 Mustang Front Spring Application Chart, 8/20/64-4/1/65
Convertibles.

CYL/CID	Model	Load	Transmission	Steering	A/C	Marking Stripes	Part Number
6/200	Conv	1344	3/S	M/P	N	1 Green	C5ZZ-5310-B
6/200	Conv	1344	4/S	M	N	1 Green	C5ZZ-5310-B
6/200	Conv	1344	C4	M	N	1 Green	C5ZZ-5310-B
6/200	Conv	1490	4/S	M	N	1 Gray	C5ZZ-5310-C
6/200	Conv	1490	C4	P	A	1 Gray	C5ZZ-5310-C
6/200	Conv	1413	3/S	M/P	A	1 Violet	C5ZZ-5310-D
							r/b C5ZZ-5310-C
6/200	Conv	1413	4/S	P	N	1 Violet	C5ZZ-5310-D
							r/b C5ZZ-5310-C
6/200	Conv	1413	4/S	M	A	1 Violet	C5ZZ-5310-D
							r/b C5ZZ-5310-C
6/200	Conv	1413	C4	P	N	1 Violet	C5ZZ-5310-D
							r/b C5ZZ-5310-C
6/200	Conv	1413	C4	M	A	1 Violet	C5ZZ-5310-D
							r/b C5ZZ-5310-C
8/289	Conv	1490	3/S, 4/S, C4	M	N	1 Gray	C5ZZ-5310-C
8/289	Conv	1490	3/S	P	N	1 Gray	C5ZZ-5310-C
8/289	Conv	1490	3/S, 4/S	M	A	1 Gray	C5ZZ-5310-C
8/289	Conv	1490	3/S, 4/S, C4	P	A	1 Gray	C5ZZ-5310-C
8/289	Conv	1490	4/S, C4	P	N	1 Gray	C5ZZ-5310-C
8/289-4V	Conv	1516	3/S, 4/S	M/P	A	1 Brown	C5ZZ-5310-E
8/289-4V	Conv	1491	3/S, 4/S	M/P	N	1 Brown	C5ZZ-5310-E

1965 Mustang Front Spring Application Chart, from 4/1/65

CYL/CID	Model	Load	Transmission	Steering	A/C	Marking Stripes	Part Number
6/200	HT/FB	1279	3/S, 4/S, C4	M	N	1 Pink	C5ZZ-5310-A
							r/b C5ZZ-5310-B
6/200	HT/FB	1344	3/S, 4/S, C4	M/P	A	1 Green	C5ZZ-5310-B
6/200	HT/FB	1344	3/S, 4/S, C4	P	N	1 Green	C5ZZ-5310-B
8/289	HT/FB	1413	3/S, 4/S, C4	M/P	N	1 Violet	C5ZZ-5310-D
							r/b C5ZZ-5310-C
8/289	HT/FB	1491	3/S, 4/S, C4	M/P	A	1 Gray	C5ZZ-5310-C
8/289 K	HT/FB	1413	3/S, 4/S	M	N	1 Red	C5ZZ-5310-F
							r/b C5ZZ-5310-E
8/289 GT	HT/FB	1413	C4	M/P	N	1 Red	C5ZZ-5310-F
							r/b C5ZZ-5310-E
8/289 GT	HT/FB	1491	3/S, 4/S, C4	M/P	A	1 Brown	C5ZZ-5310-E
6/200	Conv	1279	3/S	M	N	1 Pink	C5ZZ-5310-A
							r/b C5ZZ-5310-B
6/200	Conv	1344	4/S, C4	M/P	N	1 Green	C5ZZ-5310-B
6/200	Conv	1344	3/S	P	N	1 Green	C5ZZ-5310-B
6/200	Conv	1413	3/S, 4/S, C4	M/P	A	1 Violet	C5ZZ-5310-D
							r/b C5ZZ-5310-C
8/289	Conv	1491	3/S, 4/S, C4	M	N	1 Gray	C5ZZ-5310-C
8/289	Conv	1491	3/S, 4/S, C4	M	A	1 Gray	C5ZZ-5310-C
8/289	Conv	1491	4/S, C4	P	N	1 Gray	C5ZZ-5310-C
8/289	Conv	1491	3/S, 4/S, C4	P	A	1 Gray	C5ZZ-5310-C
8/289 K	Conv	1491	3/S, 4/S	M	N	1 Brown	C5ZZ-5310-E
8/289 GT	Conv	1491	3/S, 4/S, C4	M	N	1 Brown	C5ZZ-5310-E

1966 Mustang Front Spring Application Chart, from 2/14/66
(Hardtops and Fastbacks, Without Thermactor)

CYL/CID	Model	Load	Transmission	Steering	A/C	Marking Stripes	Part Number
6/200	HT/FB	1279	3/S, 4/S, C4	M/P	N	1 Pink	C5ZZ-5310-A
							r/b C5ZZ-5310-B
6/200	HT/FB	1279	3/S	M	A	1 Pink	C5ZZ-5310-A
							r/b C5ZZ-5310-B
6/200	HT/FB	1344	4/S, C4	M	A	1 Green	C5ZZ-5310-B
6/200	HT/FB	1344	3/S, 4/S, C4	P	A	1 Green	C5ZZ-5310-B
8/289	HT/FB	1413	3/S, 4/S, C4	M/P	N	1 Violet	C5ZZ-5310-D
							r/b C5ZZ-5310-C
8/289	HT/FB	1490	3/S, 4/S, C4	M/P	A	1 Gray	C5ZZ-5310-C
8/289-2V HD	HT/FB	1413	3/S, 4/S, C4	M/P	N	1 Red	C5ZZ-5310-F
							r/b C5ZZ-5310-E
8/289-2V HD	HT/FB	1413	3/S, 4/S, C4	M	A	1 Brown	C5ZZ-5310-E
8/289-4V	HT/FB	1413	3/S, 4/S, C4	M/P	N	1 Violet	C5ZZ-5310-D
							r/b C5ZZ-5310-C
8/289-4V	HT/FB	1490	3/S, 4/S, C4	M/P	A	1 Gray	C5ZZ-5310-C
8/289-4V HD	HT/FB	1413	3/S, 4/S, C4	M/P	N	1 Red	C5ZZ-5310-F
							r/b C5ZZ-5310-E
8/289-4V HD	HT/FB	1491	3/S, 4/S, C4	M/P	A	1 Brown	C5ZZ-5310-E
8/289-4V K	HT/FB	1413	3/S, 4/S	M/P	N	1 Red	C5ZZ-5310-F
							r/b C5ZZ-5310-E
8/289-4V K	HT/FB	1491	3/S, 4/S	M/P	A	1 Brown	C5ZZ-5310-E
8/289-4V GT	HT/FB	1413	3/S, 4/S, C4	M/P	N	1 Red	C5ZZ-5310-F
							r/b C5ZZ-5310-E
8/289-4V GT	HT/FB	1491	3/S, 4/S, C4	M/P	A	1 Brown	C5ZZ-5310-E

1966 Mustang Front Spring Application Chart, before 2/14/66
(All models, Without Thermactor)

CYL/CID	Model	Load	Transmission	Steering	A/C	Marking Stripes	Part Number
6/200	HT/FB	1279	3/S, 4/S, C4	M/P	N	1 Pink	C5ZZ-5310-A
							r/b C5ZZ-5310-B
6/200	HT/FB	1279	3/S	M	A	1 Pink	C5ZZ-5310-A
							r/b C5ZZ-5310-B
6/200	HT/FB	1344	4/S, C4	M	A	1 Green	C5ZZ-5310-B
6/200	HT/FB	1344	3/S, 4/S, C4	P	A	1 Green	C5ZZ-5310-B
8/289	HT/FB	1413	3/S, 4/S, C4	M/P	N	1 Violet	C5ZZ-5310-D
							r/b C5ZZ-5310-C
8/289	HT/FB	1491	3/S, 4/S, C4	M/P	A	1 Gray	C5ZZ-5310-C
8/289 HD	HT/FB	1413	3/S, 4/S, C4	M/P	N	1 Red	C5ZZ-5310-F
							r/b C5ZZ-5310-E
8/289 HD	HT/FB	1413	3/S, 4/S, C4	M/P	A	1 Brown	C5ZZ-5310-E
6/200	Conv	1279	3/S	M	N	1 Pink	C5ZZ-5310-A
							r/b C5ZZ-5310-B
6/200	Conv	1344	4/S, C4	M	N	1 Green	C5ZZ-5310-B
6/200	Conv	1344	3/S, 4/S, C4	P	N	1 Green	C5ZZ-5310-B
6/200	Conv	1344	3/S	M	A	1 Green	C5ZZ-5310-B
6/200	Conv	1413	4/S, C4	M	A	1 Violet	C5ZZ-5310-D
							r/b C5ZZ-5310-C
6/200	Conv	1413	3/S, 4/S, C4	P	A	1 Violet	C5ZZ-5310-D
							r/b C5ZZ-5310-C
8/289	Conv	1491	3/S, 4/S, C4	M/P	N	1 Gray	C5ZZ-5310-C
8/289	Conv	1491	3/S, 4/S, C4	M/P	A	1 Gray	C5ZZ-5310-C
8/289 HD	Conv	1491	3/S, 4/S, C4	M/P	N	1 Brown	C5ZZ-5310-E
8/289 HD	Conv	1491	3/S, 4/S, C4	M/P	A	1 Brown	C5ZZ-5310-E

1966 Mustang Front Spring Application Chart, from 2/14/66
(Convertibles, Without Thermactor)

CYL/CID	Load	Transmission	Steering	A/C	Marking Stripes	Part Number
6/200	1279	3/S	M	N	1 Pink	C5ZZ-5310-A
						r/b C5ZZ-5310-B
6/200	1344	4/S, C4	M	N	1 Green	C5ZZ-5310-B
6/200	1344	3/S, 4/S, C4	P	N	1 Green	C5ZZ-5310-B
6/200	1344	3/S	M	A	1 Green	C5ZZ-5310-B
6/200	1413	3/S	P	A	1 Violet	C5ZZ-5310-D
						r/b C5ZZ-5310-C
6/200	1413	4/S, C4	M/P	A	1 Violet	C5ZZ-5310-D
						r/b C5ZZ-5310-C
8/289-2V	1490	3/S, 4/S, C4	M/P	N	1 Gray	C5ZZ-5310-C
8/289-2V	1490	3/S, 4/S, C4	M/P	A	1 Gray	C5ZZ-5310-C
8/289-2V HD	1491	3/S, 4/S, C4	M/P	N	1 Brown	C5ZZ-5310-E
8/289-2V HD	1491	3/S, 4/S, C4	M/P	A	1 Brown	C5ZZ-5310-E
8/289-4V	1490	3/S, 4/S, C4	M/P	N	1 Gray	C5ZZ-5310-C
8/289-4V	1490	3/S, 4/S, C4	M/P	A	1 Gray	C5ZZ-5310-C
8/289-4V HD	1491	3/S, 4/S, C4	M/P	N	1 Brown	C5ZZ-5310-E
8/289-4V HD	1491	3/S, 4/S, C4	M/P	A	1 Brown	C5ZZ-5310-E
8/289-4V K	1491	3/S, 4/S	M/P	N	1 Brown	C5ZZ-5310-E
8/289-4V K	1491	3/S, 4/S	M/P	A	1 Brown	C5ZZ-5310-E
8/289-4V GT	1491	3/S, 4/S, C4	M/P	N	1 Brown	C5ZZ-5310-E
8/289-4V GT	1491	3/S, 4/S, C4	M/P	A	1 Brown	C5ZZ-5310-E

1966 Mustang Front Spring Application Chart, before 2/14/66
(Hardtops and Fastbacks, With Thermactor)

CYL/CID	Model	Load	Transmission	Steering	A/C	Marking Stripes	Part Number
6/200	HT/FB	1279	3/S, 4/S, C4	M/P	N	1 Pink	C5ZZ-5310-A
							r/b C5ZZ-5310-B
6/200	HT/FB	1344	3/S, 4/S, C4	M	A	1 Green	C5ZZ-5310-B
6/200	HT/FB	1344	3/S, 4/S, C4	P	A	1 Green	C5ZZ-5310-B
6/200	FB	1344	4/S, C4	P	A	1 Green	C5ZZ-5310-B
6/200	HT	1413	C4	P	A	1 Violet	C5ZZ-5310-D
							r/b C5ZZ-5310-C
8/289-2V	HT/FB	1413	3/S, 4/S, C4	M	N	1 Violet	C5ZZ-5310-D
							r/b C5ZZ-5310-C
8/289-2V	FB	1413	3/S, 4/S, C4	P	N	1 Violet	C5ZZ-5310-D
							r/b C5ZZ-5310-C
8/289-2V	HT	1413	3/S, C4	P	N	1 Violet	C5ZZ-5310-D
							r/b C5ZZ-5310-C
8/289-2V	HT	1491	4/S	P	N	1 Gray	C5ZZ-5310-C
8/289-2V	HT/FB	1491	3/S, 4/S, C4	M/P	A	1 Gray	C5ZZ-5310-C
8/289-2V HD	HT/FB	1413	3/S, 4/S, C4	M	N	1 Red	C5ZZ-5310-F
							r/b C5ZZ-5310-E
8/289-2V HD	FB	1413	3/S, 4/S, C4	P	N	1 Red	C5ZZ-5310-F
							r/b C5ZZ-5310-E
8/289-2V HD	HT	1413	3/S, C4	P	N	1 Red	C5ZZ-5310-F
							r/b C5ZZ-5310-E
8/289-2V HD	HT	1491	4/S	P	N	1 Brown	C5ZZ-5310-E
8/269-2V HD	HT/FB	1491	3/S, 4/S, C4	M/P	A	1 Brown	C5ZZ-5310-E
8/289-4V	HT/FB	1413	3/S, 4/S, C4	M	N	1 Violet	C5ZZ-5310-D
							r/b C5ZZ-5310-C
8/289-4V	HT/FB	1413	3/S	P	N	1 Violet	C5ZZ-5310-D
							r/b C5ZZ-5310-C
8/289-4V	HT/FB	1491	4/S, C4	P	N	1 Gray	C5ZZ-5310-C
8/289-4V	HT/FB	1491	3/S, 4/S, C4	M/P	A	1 Gray	C5ZZ-5310-C
8/289-4V HD	HT/FB	1413	3/S, 4/S, C4	M	N	1 Red	C5ZZ-5310-F
							r/b C5ZZ-5310-E
8/289-4V HD	HT/FB	1413	3/S	P	N	1 Red	C5ZZ-5310-F
							r/b C5ZZ-5310-E
8/289-4V HD	HT/FB	1491	4/S, C4	P	N	1 Brown	C5ZZ-5310-E
8/289-4V HD	HT/FB	1491	3/S, 4/S, C4	M/P	A	1 Brown	C5ZZ-5310-E
8/289-4V K	HT/FB	1413	3/S, 4/S	M	N	1 Red	C5ZZ-5310-F
							r/b C5ZZ-5310-E
8/289-4V K	FB	1413	3/S	P	N	1 Red	C5ZZ-5310-F
							r/b C5ZZ-5310-E
8/289-4V K	HT	1491	3/S	P	N	1 Brown	C5ZZ-5310-E
8/289-4V K	HT/FB	1491	4/S	P	N	1 Brown	C5ZZ-5310-E
8/289-4V K	HT/FB	1491	3/S	M/P	A	1 Brown	C5ZZ-5310-E
8/289-4V K	FB	1491	4/S	M/P	A	1 Brown	C5ZZ-5310-E
8/289-4V K	HT	1491	4/S	M	A	1 Brown	C5ZZ-5310-E
8/289-4V K	HT	1491	3/S, 4/S, C4	M/P	A	1 Brown	C5ZZ-5310-E

1966 Mustang Front Spring Application Chart, Convertibles, before 2/14/66
(With Thermactor)

CYL/CID	Model	Load	Transmission	Steering	A/C	Marking Stripes	Part Number
6/200	Conv	1344	3/S, 4/S, C4	M	N	1 Green	C5ZZ-5310-B
6/200	Conv	1344	4/S, C4	P	N	1 Green	C5ZZ-5310-B
6/200	Conv	1413	C4	P	N	1 Violet	C5ZZ-5310-D
							r/b C5ZZ-5310-C
6/200	Conv	1413	3/S, 4/S, C4	M/P	A	1 Violet	C5ZZ-5310-D
							r/b C5ZZ-5310-C
8/289	Conv	1491	3/S, 4/S, C4	M/P	N	1 Gray	C5ZZ-5310-C
8/289	Conv	1491	3/S, 4/S, C4	M/P	A	1 Gray	C5ZZ-5310-C
8/289	Conv	1491	3/S, 4/S, C4	M/P	N	1 Brown	C5ZZ-5310-E
8/289	Conv	1491	3/S, 4/S, C4	M/P	A	1 Brown	C5ZZ-5310-E
							r/b C5ZZ-5310-C

1966 Mustang Front Spring Application Chart, from 2/14/66 (With Thermactor)

CYL/CID	Model	Load	Transmission	Steering	A/C	Marking Stripes	Part Number
6/200	HT/FB	1279	3/S, 4/S, C4	M	N	1 Pink	C5ZZ-5310-A
							r/b C5ZZ-5310-B
6/200	HT/FB	1279	3/S, 4/S	P	N	1 Pink	C5ZZ-5310-A
							r/b C5ZZ-5310-B
6/200	FB	1279	C4	P	N	1 Pink	C5ZZ-5310-A
							r/b C5ZZ-5310-B
6/200	HT	1344	C4	P	N	1 Green	C5ZZ-5310-B
6/200	HT/FB	1344	3/S, 4/S	M/P	A	1 Green	C5ZZ-5310-B
6/200	HT/FB	1344	C4	M	A	1 Green	C5ZZ-5310-B
6/200	FB	1344	C4	P	A	1 Green	C5ZZ-5310-B
6/200	HT	1413	C4	P	A	1 Violet	C5ZZ-5310-D
							r/b C5ZZ-5310-C
8/289 2V	HT/FB	1413	3/S, 4/S, C4	M	N	1 Violet	C5ZZ-5310-D
							r/b C5ZZ-5310-C
8/289 2V	HT/FB	1413	3/S, C4	P	N	1 Violet	C5ZZ-5310-D
							r/b C5ZZ-5310-C
8/289 2V	FB	1413	4/S	P	N	1 Violet	C5ZZ-5310-D
							r/b C5ZZ-5310-C
8/289 2V	HT	1490	4/S	P	N	1 Gray	C5ZZ-5310-C
8/289 2V	HT/FB	1490	3/S, 4/S, C4	M/P	A	1 Gray	C5ZZ-5310-C
8/289 2V HD	HT/FB	1413	3/S, 4/S, C4	M	N	1 Red	C5ZZ-5310-F
							r/b C5ZZ-5310-E
8/289 2V HD	HT/FB	1413	3/S, C4	P	N	1 Red	C5ZZ-5310-F
							r/b C5ZZ-5310-E
8/289 2V HD	FB	1413	4/S	P	N	1 Red	C5ZZ-5310-F
							r/b C5ZZ-5310-E
8/289 2V HD	HT	1491	4/S	P	N	1 Brown	C5ZZ-5310-E
8/289 2V HD	HT/FB	1491	3/S, 4/S, C4	M/P	A	1 Brown	C5ZZ-5310-E
8/289 4V	HT/FB	1413	3/S, 4/S, C4	M	N	1 Violet	C5ZZ-5310-D
							r/b C5ZZ-5310-C
8/289 4V	HT/FB	1413	3/S	P	N	1 Violet	C5ZZ-5310-D
							r/b C5ZZ-5310-C
8/289 4V	HT/FB	1490	4/S, C4	P	N	1 Gray	C5ZZ-5310-C
8/289 4V	HT/FB	1490	3/S, 4/S, C4	M/P	A	1 Gray	C5ZZ-5310-C
8/289 4V HD	HT/FB	1413	3/S, 4/S, C4	M	N	1 Red	C5ZZ-5310-F
							r/b C5ZZ-5310-E
8/289 4V HD	HT/FB	1413	3/S	P	N	1 Red	C5ZZ-5310-F
							r/b C5ZZ-5310-E
8/289 4V HD	HT/FB	1491	4/S, C4	P	N	1 Brown	C5ZZ-5310-E
8/289 4V HD	HT/FB	1491	3/S, 4/S, C4	M/P	A	1 Brown	C5ZZ-5310-E
8/289 4V K	HT/FB	1413	3/S, 4/S	M	N	1 Red	C5ZZ-5310-F
							r/b C5ZZ-5310-E
8/289 4V K	FB	1413	3/S	P	N	1 Red	C5ZZ-5310-F
							r/b C5ZZ-5310-E
8/289 4V K	HT	1491	3/S	P	N	1 Brown	C5ZZ-5310-E
8/289 4V K	HT/FB	1491	4/S	P	N	1 Brown	C5ZZ-5310-E
8/289 4V K	HT/FB	1491	3/S, 4/S	M/P	A	1 Brown	C5ZZ-5310-E
8/289 4V GT	HT/FB	1413	3/S, 4/S, C4	M	N	1 Red	C5ZZ-5310-F
							r/b C5ZZ-5310-E
8/289 4V GT	HT/FB	1413	3/S	P	N	1 Red	C5ZZ-5310-F
							r/b C5ZZ-5310-E
8/289 4V GT	HT/FB	1491	4/S, C4	P	N	1 Brown	C5ZZ-5310-E
8/289 4V GT	HT/FB	1491	3/S, 4/S, C4	M/P	A	1 Brown	C5ZZ-5310-E
6/200	Conv	1344	3/S, 4/S, C4	M	N	1 Green	C5ZZ-5310-B
6/200	Conv	1344	3/S, 4/S	P	N	1 Green	C5ZZ-5310-B
6/200	Conv	1413	C4	P	N	1 Violet	C5ZZ-5310-D
							r/b C5ZZ-5310-C
6/200	Conv	1413	3/S, 4/S, C4	M/P	A	1 Violet	C5ZZ-5310-D
							r/b C5ZZ-5310-C
8/289 2V	Conv	1490	3/S, 4/S, C4	M/P	N	1 Gray	C5ZZ-5310-C
8/289 2V	Conv	1490	3/S, 4/S, C4	M/P	A	1 Gray	C5ZZ-5310-C
8/289 2V HD	Conv	1491	3/S, 4/S, C4	M/P	N	1 Brown	C5ZZ-5310-E
8/289 2V HD	Conv	1491	3/S, 4/S, C4	M/P	A	1 Brown	C5ZZ-5310-E
8/289 4V	Conv	1490	3/S, 4/S, C4	M/P	N	1 Gray	C5ZZ-5310-C
8/289 4V	Conv	1490	3/S, 4/S, C4	M/P	A	1 Gray	C5ZZ-5310-C
8/289 4V HD	Conv	1491	3/S, 4/S, C4	M/P	N	1 Brown	C5ZZ-5310-E
8/289 4V HD	Conv	1491	3/S, 4/S, C4	M/P	A	1 Brown	C5ZZ-5310-E
8/289 4V K	Conv	1491	3/S, 4/S	M/P	N	1 Brown	C5ZZ-5310-E
8/289 4V K	Conv	1491	3/S, 4/S	M/P	A	1 Brown	C5ZZ-5310-E
8/289 4V GT	Conv	1491	3/S, 4/S, C4	M/P	N	1 Brown	C5ZZ-5310-E
8/289 4V GT	Conv	1491	3/S, 4/S, C4	M/P	A	1 Brown	C5ZZ-5310-E

1967 Mustang Front Spring Application Chart

(Manual Transmissions, Without Thermactor)

Note: Refer to chart for spring code identification.

			Three-Speed Options:				Four-Speed Options:			
Engine	Model	Note	None	A/C	P/S	A/C & P/S	None	A/C	P/S	A/C & P/S
200	F/B		4T	4G	4T	4W	4T	4W	4G	4W
289-2V	F/B		4H	4J	4J	4V	4J	4V	4J	4V
289-4V	F/B		4H	4V	4J	4V	4J	4V	4J	4V
390	F/B		4F	4E	4E	4U	4F	4E	4E	4U
390	F/B	1	4S	4R	4R	4R	4S	4R	4R	4R
390	F/B	2	4K	4L	4L	4L	4K	4L	4L	4L
200	H/T		4T	4G	4G	4W	4T	4W	4G	4W
289-2V	H/T		4H	4V	4J	4V	4J	4V	4J	4V
289-4V	H/T		4J	4V	4J	4V	4J	4V	4J	25
390	H/T		4F	4E	4E	4U	4F	4E	4E	4U
390	H/T	1	4S	4R	4R	4P	4S	4R	4R	4P
390	H/T	2	4K	4L	4L	4L	4K	4L	4L	4L
200	Conv	3	4G	4W	4W	68	4W	68	4W	68
200	Conv	4	4G	4W	4W	4J	4W	4J	4W	4J
289-2V	Conv	3	4J	25	4V	25	4V	25	4V	25
289-2V	Conv	4	4J	4F	4T	4F	4T	4F	4T	4F
289-4V	Conv	3	4V	25	4V	25	4V	25	4V	25
289-4V	Conv	4	4T	4F	4T	4F	4T	4F	4T	4F
390	Conv		4E	4U	4U	4D	4E	4U	4U	4D
390	Conv	1	4R	4P	4R	4P	4R	4P	4P	4P
390	Conv	2	4L	4M	4L	4M	4L	4M	4L	4M

Notes:

1. Improved Handling Package
2. Racing only
3. Before 9/1/66
4. From 9/1/66

1967 Mustang Front Spring Application Chart

(Manual Transmissions, With Thermactor)

Note: Refer to chart for spring code identification.

			Three-Speed Options:				Four-Speed Options:			
Engine	Model	Note	None	A/C	P/S	A/C & P/S	None	A/C	P/S	A/C & P/S
200	F/B	3	4T	4G	4G	4W	4G	4W	4G	68
200	F/B	4	4T	4G	4G	4W	4G	4W	4G	4J
289-2V	F/B	3	4J	4V	4J	4V	4J	4V	4V	25
289-2V		4	4J	4T	4J	4T	4J	4T	4T	4F
289-4V	F/B	3	4J	4V	4J	25	4J	4V	4V	25
289-4V		4	4J	4T	4J	4F	4J	4T	4T	4F
390	F/B		4F	4E	4E	4U	4F	4E	4E	4U
390	F/B	1	4S	4R	4R	4P	4S	4R	4R	4P
390	F/B	2	4K	4L	4L	4M	4K	4L	4L	4M
200	H/T	3	4T	4W	4G	4W	4G	4W	4W	68
200	H/T	4	4T	4W	4G	4W	4G	4W	4W	4J
289-2V	H/T	3	4J	4V	4J	25	4J	4V	4V	25
289-2V	H/T	4	4J	4T	4J	4F	4J	4T	4T	4F
289-4V	H/T	3	4J	4V	4V	25	4J	4V	4V	25
289-4V	H/T	4	4J	4T	4T	4F	4J	4T	4T	4F
390	H/T		4E	4U	4E	4U	4E	4U	4E	4U
390	H/T	1	4R	4R	4R	4P	4R	4R	4R	4P
390	H/T	2	4L	4L	4L	4M	4L	4L	4L	4M
200	Conv	3	4G	68	4W	68	4W	68	68	4V
200	Conv	4	4G	4J	4W	4J	4R	4J	4J	4T
289-2V	Conv	3	4V	25	4V	4E	4V	25	25	4E
289-2V	Conv	4	4T	4F	4T	4E	4T	4F	4F	4E
289-4V	Conv	3	4V	25	25	4E	4V	25	25	4E
289-4V	Conv	4	4T	4F	4F	4E	4T	4F	4F	4E
390	Conv		4E	4U	4U	4D	4E	4U	4U	4D
390	Conv	1	4R	4P	4P	4N	4R	4P	4P	4N
390	Conv	2	4L	4M	4M	4M	4L	4M	4M	4M

Notes:

1. Improved Handling Package
2. Racing only
3. Before 9/1/66
4. From 9/1/66

1967 Mustang Front Spring Application Chart
(Automatic Transmissions, Without Thermactor)
Note: Refer to chart for spring code identification.

			C4 Options:			A/C &	C6 Options:			A/C &
Engine	Model	Note	None	A/C	P/S	P/S	None	A/C	P/S	P/S
200	F/B		4T	4G	4G	4W				
289-2V	F/B		4H	4J	4J	4V				
289-4V	F/B		4J	4V	4J	4V				
390	F/B						4F	4E	4E	4U
390	F/B	1					4S	4R	4R	4P
390	F/B	2					4K	4L	4L	4L
200	H/T		4T	4W	4G	4W				
289-2V	H/T		4J	4V	4J	4V				
289-4V	H/T		4J	4V	4J	4V				
390	H/T						4F	4E	4E	4U
390	H/T	1					4S	4R	4R	4P
390	H/T	2					4K	4L	4L	4M
200	Conv	3	4G	4W	4W	68				
200	Conv	4	4G	4W	4W	4J				
289-2V	Conv	3	4V	25	4V	25				
289-2V	Conv	4	4T	4F	4T	4F				
289-4V	Conv	3	4V	25	4V	25				
289-4V	Conv	4	4T	4F	4T	4F				
390	Conv						4E	4U	4U	4D
390	Conv	1					4R	4P	4P	4P
390	Conv	2					4L	4M	4L	4M

Notes:
1 Improved Handling Package
2 Racing only
3 Before 9/1/66
4 From 9/1/66

1967 Mustang Front Spring Application Chart
(Automatic Transmissions, With Thermactor)
Note: Refer to chart for spring code identification.

			C4 Options:			A/C &	C6 Options:			A/C &
Engine	Model	Note	None	A/C	P/S	P/S	None	A/C	P/S	P/S
200	F/B	3	4G	4W	4G	4W				
200	F/B	4	4G	4W	4G	4W				
289-2V	F/B	3	4J	4V	4J	4V				
289-2V		4	4J	4T	4J	4T				
289-4V	F/B	3	4J	4V	4V	25				
289-4V		4	4J	4T	4T	4F				
390	F/B						4E	4U	4E	4U
390	F/B	1					4R	4R	4R	4P
390	F/B	2					4L	4L	4L	4M
200	H/T	3	4G	4W	4G	68				
200	H/T	4	4G	4W	4G	4J				
289-2V	H/T	3	4J	4V	4V	25				
289-2V	H/T	4	4J	4T	4T	4F				
289-4V	H/T	3	4J	4V	4V	25				
289-4V	H/T	4	4J	4J	4T	4F				
390	H/T		4E	4U	4E	4U				
390	H/T	1	4R	4P	4R	4P				
390	H/T	2	4L	4L	4L	4M				
200	Conv	3	4W	68	4W	4V				
200	Conv	4	4W	4J	4W	4T				
289-2V	Conv	3	4V	25	25	4E				
289-2V	Conv	4	4T	4F	4F	4E				
289-4V	Conv	3	4V	25	25	4E				
289-4V	Conv	4	4T	4F	4F	4E				
390	Conv						4U	4U	4U	4D
390	Conv	1					4R	4P	4P	4N
390	Conv	2					4L	4M	4M	4M

Notes:
1 Improved Handling Package
2 Racing only
3 Before 9/1/66
4 From 9/1/66

1968 MUSTANG Front Spring Application Chart
(Automatic Transmissions, Standard Springs)
Note: Refer to chart for spring code identification.

			C4 Options:			A/C &	C6 Options:			A/C &
Engine	Model	Note	None	A/C	P/S	P/S	None	A/C	P/S	P/S
200	F/B	1	4G	4W	4W	4J				
200	F/B	2	4W	4J	4J	4V				
289-2V	F/B		4J	4F	4V	4F				
289-4V	F/B		4V	4F	4V	4E				
302	F/B		4V	4F	4V	4F				
390-2	F/B						4U	4D	4D	4Y
390-4	F/B	1,3					4L	4M	4M	4M
390-4	F/B	2,3					4L	4M	4M	4M
390-4	F/B						4U	4D	4U	4D
427	F/B	3					4M	4M	4M	1Q
200	H/T	1	4G	4J	4W	4J				
200	H/T	2	4W	4J	4J	4V				
289-2V	H/T		4V	4F	4V	4F				
289-4V	H/T		4V	4F	4F	4E				
302	H/T		4V	4F	4V	4F				
390-2	H/T						4U	4D	4D	4Y
390-4	H/T	1,3					4M	4M	4M	6D
390-4	H/T	2,3					4M	4M	4M	1Q
390-4	H/T						4U	4D	4U	4D
427	H/T	3					4M	1Q	4M	1Q
200	Conv	1	4W	4J	4J	4V				
200	Conv	2	4J	4V	4J	4V				
289-2V	Conv		4V	4F	4F	4E				
289-4V	Conv		4F	4E	4F	4E				
302	Conv		4F	4E	4F	4E				
390-2	Conv						4D	4Y	4D	4Y
390-4	Conv	1,3					4M	6D	4M	6D
390-4	Conv	2,3					4M	1Q	4M	1Q
390-4	Conv						4D	4Y	4D	4Y
427	Conv	3					4M	1Q	1Q	1Q

Notes:
1 Before 12/11/67
2 From 12/11/67
3 Improved Handling Package

1968 MUSTANG Front Spring Application Chart
(Manual Transmissions, Standard Springs)
Note: Refer to chart for spring code identification.

			Three-Speed Options:			A/C &	Four-Speed Options:			A/C &
Engine	Model	Note	None	A/C	P/S	P/S	None	A/C	P/S	P/S
200	F/B	1	4G	4W	4G	4J				
200	F/B	2	4W	4J	4W	4J				
289-2V	F/B		4V	4F	4V	4F	4V	4F	4V	4F
289-4V	F/B						4V	4F	4F	4E
302	F/B		4F	4F	4F	4E	4V	4F	4F	4E
390-2	F/B		4U	4D	4U	4D	4U	4D	4U	4D
390-4	F/B	1,3	4L	4M	4M	4M	4M	4M	4M	6D
390-4	F/B	2,3	4L	4M	4M	4M	4M	4M	4M	1Q
390-4	F/B		4U	4D	4U	4D	4U	4D	4D	4D
427	F/B	3	4M	4M	4M	1Q	4M	1Q	4M	1Q
200	H/T	1	4G	4W	4W	4J				
200	H/T	2	4W	4J	4W	4J				
289-2V	H/T		4V	4F	4V	4F	4V	4F	4F	4E
289-4V	H/T						4V	4F	4F	4E
302	H/T		4F	4F	4F	4E	4V	4F	4F	4E
390-2	H/T		4U	4D	4U	4D	4U	4D	4D	4D
390-4	H/T	1,3	4L	4M	4M	6D	4M	4M	4M	6D
390-4	H/T	2,3	4L	4M	4M	1Q	4M	4M	4M	1Q
390-4	H/T		4U	4D	4U	4D	4U	4D	4D	4Y
427	H/T	3	4M	1Q	4M	1Q	4M	1Q	4M	1Q
200	Conv	1	4W	4J	4W	4J				
200	Conv	2	4J	4V	4J	4V				
289-2V	Conv		4F	4E	4F	4E	4F	4E	4F	4E
289-4V	Conv						4F	4E	4F	4E
302	Conv		4F	4E	4F	4E	4F	4E	4E	4E
390-2	Conv		4U	4D	4D	4D	4D	4D	4D	4Y
390-4	Conv	1,3	4M	6D	4M	6D	4M	6D	6D	6D
390-4	Conv	2,3	4M	1Q	4M	1Q	4M	1Q	1Q	1Q
390-4	Conv		4D	4D	4D	4Y	4D	4Y	4D	4Y
427	Conv	3	4M	1Q	1Q	1Q	4M	1Q	1Q	1Q

Notes:
1 Before 12/11/67
2 From 12/11/67
3 Improved Handling Package

1968 MUSTANG Front Spring Application Chart
(Automatic Transmissions, Heavy Duty Suspension)
Note: Refer to chart for spring code identification.

		C4				C6			
		Options:			A/C &	Options:			A/C &
Engine	Model	None	A/C	P/S	P/S	None	A/C	P/S	P/S
289-2V	F/B	4X	4S	4X	4S				
289-4V	F/B	4X	4S	4S	4R				
302	F/B	4X	4S	4S	4S				
390-2V	F/B					4P	4N	4N	4N
390-4V	F/B					4P	4N	4P	4N
428	F/B					4P	4N	4N	4C
289-2V	H/T	4X	4S	4S	4S				
289-4V	H/T	4S	4S	4S	4R				
302	H/T	4X	4S	4S	4R				
390-2V	H/T					4P	4N	4N	4N
390-4V	H/T					4P	4N	4P	4N
428	H/T					4N	4C	4N	4C
289-2V	Conv	4S	4R	4S	4R				
289-4V	Conv	4S	4R	4S	4R				
302	Conv	4S	4R	4S	4R				
390-2V	Conv					4N	4N	4N	4C
390-4V	Conv					4N	4N	4N	4C
428	Conv					4N	4C	4C	4C

1968 MUSTANG Front Spring Application Chart
(Manual Transmissions, Heavy Duty Suspension)
Note: Refer to chart for spring code identification.

		Three-Speed				Four-Speed			
		Options:			A/C &	Options:			A/C &
Engine	Model	None	A/C	P/S	P/S	None	A/C	P/S	P/S
289-2V	F/B	4X	4S	4S	4S	4X	4S	4S	4R
289-4V	F/B					4S	4S	4S	4R
302	F/B	4S	4S	4S	4R	4S	4S	4S	4R
390-2V	F/B	4P	4N	4P	4N	4P	4N	4P	4N
390-4V	F/B	4P	4N	4P	4N	4P	4N	4P	4N
428	F/B	4P	4N	4N	4N	4P	4N	4N	4C
289-2V	H/T	4X	4S	4S	4R	4S	4S	4S	4R
289-4V	H/T					4S	4R	4S	4R
302	H/T	4S	4S	4S	4R	4S	4R	4S	4R
390-2V	H/T	4P	4N	4P	4N	4P	4N	4P	4N
390-4V	H/T	4P	4N	4P	4N	4P	4N	4P	4N
428	H/T	4P	4N	4N	4N	4N	4N	4N	4C
289-2V	Conv	4S	4R	4S	4R	4S	4R	4S	4R
289-4V	Conv					4S	4R	4R	4R
302	Conv	4S	4R	4R	4R	4S	4R	4R	4R
390-2V	Conv	4P	4N	4N	4C	4P	4N	4N	4C
390-4V	Conv	4P	4N	4N	4C	4N	4C	4N	4C
428	Conv	4N	4C	4N	4C	4N	4C	4N	4C

1969 MUSTANG Front Spring Application Chart
Hardtop models.
(Manual Transmissions, Standard Suspension.)
Note: Refer to chart for spring code identification.

		Three-Speed				Four-Speed			
		Options:			A/C &	Options:			A/C &
Engine	Notes.	None	A/C	P/S	P/S	None	A/C	P/S	P/S
200	1,2,5	4W	4W	4W	4J				
200	1,2,6	4G	4G	4G	4W				
250	1,2,5	4J	4J	4J	4V				
250	1,2,6	4G	4J	4W	4J				
302	1,2,5	4V	4F	4V	4F	4V	4F	4F	4E
302	1,2,6	4J	4V	4J	4V	4J	4V	4J	4F
351-2V	1,2,5	4F	4E	4F	4E	4F	4E	4F	4E
351-2V	1,2,6	4V	4F	4V	4E	4V	4F	4F	4E
351-4V	1,2,5	4F	4E	4F	4E	4F	4E	4E	4E
351-4V	1,2,6	4V	4F	4F	4E	4V	4F	4F	4E
390	1,2,7	4E	4U	4E	4U				
390	1,2,8	4F	4E	4E	4U	4F	4E	4E	4U
428	1,2,9	4M	4M	4M	1Q	4M	4M	4M	1Q
428	1,2,10	4M	4M	4M	1Q	4M	4M	4M	1Q
200	3,4,5	4W	4J	4W	4J				
200	3,4,6	4G	4G	4G	4W				
250	3,4,5	4J	4V	4J	4V				
250	3,4,6	4G	4J	4W	4J				
302	3,4,5	4V	4F	4V	4E	4V	4F	4F	4E
302	3,4,6	4J	4V	4J	4F	4J	4V	4J	4F
351-2V	3,4,5	4F	4E	4F	4E	4F	4E	4E	4E
351-2V	3,4,6	4V	4F	4V	4E	4V	4F	4F	4E
351-4V	3,4,5	4F	4E	4E	4E	4F	4E	4E	4U
351-4V	3,4,6	4V	4F	4F	4E	4V	4F	4F	4E
390	3,4,7	4E	4U	4E	4U	4E	4U	4U	4U
390	3,4,8	4F	4E	4E	4U	4F	4E	4E	4U
428	3,4,9	4M	4M	4M	1Q	4M	4M	4M	1Q
428	3,4,10	4M	4M	4M	1Q	4M	4M	4M	1Q

Notes:

1. Std. Interior, Bucket Seats
2. Deluxe Interior, Bucket Seats
3. Std. Interior, Bench Seat
4. Deluxe Interior, Bench Seat
5. Before 9/11/68, with .69" Dia. Stabilizer Bar
6. From 9/11/68, with .69" Dia. Stabilizer Bar
7. Before 9/11/68, with .72" Dia. Stabilizer Bar
8. From 9/11/68, with .72" Dia. Stabilizer Bar
9. Before 9/11/68, with .95" Dia. Stabilizer Bar, Competition Suspension
10. From 9/11/68, with .95" Dia. Stabilizer Bar, Competition Suspension

1969 MUSTANG Front Spring Application Chart
SportsRoof and Mach 1 models.
(Manual Transmissions, Standard Suspension.)
Note: Refer to chart for spring code identification.

Engine	Model	Three-Speed Options: None	A/C	P/S	A/C & P/S	Four-Speed Options: None	A/C	P/S	A/C & P/S
200	1,2,4	4W	4W	4W	4J				
200	1,2,5	4G	4G	4G	4W				
250	1,2,4	4W	4J	4J	4V				
250	1,2,5	4G	4W	4W	4J				
302	1,2,4	4J	4F	4V	4F	4V	4F	4V	4E
302	1,2,5	4W	4J	4J	4V	4J	4V	4J	4V
351-2V	1,2,4	4V	4E	4F	4E	4F	4E	4F	4E
351-2V	1,2,5	4J	4F	4V	4F	4J	4F	4V	4E
351-4V	1,2,4	4F	4E	4F	4E	4F	4E	4F	4E
351-4V	1,2,5	4J	4F	4V	4E	4J	4F	4V	4E
390	1,2,6	4E	4U	4E	4U	4E	4U	4E	4U
390	1,2,7	4F	4E	4E	4U	4F	4E	4E	4U
428	1,2,8	4L	4M	4M	1Q	4M	4M	4M	1Q
428	1,2,9	4L	4M	4M	1Q	4M	4M	4M	1Q
429	1,2							6L	
351-2V	3,4	4F	4E	4F	4E	4F	4E	4E	4E
351-2V	3,5	4J	4F	4J	4E	4V	4F	4F	4E
351-4V	3,4	4F	4E	4E	4E	4F	4E	4E	4U
351-4V	3,5	4V	4F	4F	4E	4V	4F	4F	4E
390	3,6	4E	4U	4E	4U	4E	4U	4U	4U
390	3,7	4F	4E	4E	4U	4F	4E	4E	4U
428	3,8	4M	4M	4M	1Q	4M	4M	4M	1Q
428	3,9	4M	4M	4M	1Q	4M	4M	4M	1Q
429	3							6L	

Notes:

1 Std. Interior, Bucket Seats
2 Deluxe Interior, Bucket Seats
3 Mach 1
4 Before 9/11/68, with .69" Dia. Stabilizer Bar
5 From 9/11/68, with .69" Dia. Stabilizer Bar
6 Before 9/11/68, with .72" Dia. Stabilizer Bar
7 From 9/11/68, with .72" Dia. Stabilizer Bar
8 Before 9/11/68, with .95" Dia. Stabilizer Bar, Competition Suspension
9 From 9/11/68, with .95" Dia. Stabilizer Bar, Competition Suspension

1969 MUSTANG Front Spring Application Chart
Grande models.
(Manual Transmissions, Standard Suspension.)
Note: Refer to chart for spring code identification.

Engine	Notes.	Three-Speed Options: None	A/C	P/S	A/C & P/S	Four-Speed Options: None	A/C	P/S	A/C & P/S
200	1	4W	4J	4W	4J				
200	2	4G	4W	4G	4W				
250	1	4J	4V	4J	4F				
250	2	4W	4J	4W	4J				
302	1	4V	4F	4F	4E	4V	4F	4F	4E
302	2	4J	4V	4J	4F	4J	4V	4V	4F
351-2V	1	4F	4E	4E	4U	4F	4E	4E	4U
351-2V	2	4V	4F	4F	4E	4V	4F	4F	4E
351-4V	1	4F	4E	4E	4U	4F	4E	4E	4U
351-4V	2	4V	4F	4F	4E	4V	4E	4F	4E
390	3	4E	4U	4U	4D	4E	4U	4U	4D
390	4	4F	4U	4E	4U	4E	4U	4E	4U
428	5	4M	1Q	4M	1Q	4M	1Q	4M	1Q
428	6	4M	1Q	4M	1Q	4M	1Q	4M	1Q

Notes:

1 Before 9/11/68, with .69" Dia. Stabilizer Bar
2 From 9/11/68, with .69" Dia. Stabilizer Bar
3 Before 9/11/68, with .72" Dia. Stabilizer Bar
4 From 9/11/68, with .72" Dia. Stabilizer Bar
5 Before 9/11/68, with .95" Dia. Stabilizer Bar, Competition Suspension
6 From 9/11/68, with .95" Dia. Stabilizer Bar, Competition Suspension

1969 MUSTANG Front Spring Application Chart
Convertible models.
(Manual Transmissions, Standard Suspension.)
Note: Refer to chart for spring code identification.

Engine	Notes	Three-Speed Options: None	A/C	P/S	A/C & P/S	Four-Speed Options: None	A/C	P/S	A/C & P/S
200	1	4W	4J	4J	4V				
200	2	4G	4W	4G	4J				
250	1	4J	4V	4V	4F				
250	2	4W	4J	4J	4V				
302	1	4V	4F	4F	4E	4V	4F	4F	4E
302	2	4J	4V	4V	4F	4J	4V	4V	4F
351-2V	1	4F	4E	4E	4U	4F	4E	4E	4U
351-2V	2	4V	4E	4F	4E	4V	4E	4F	4E
351-4V	1	4F	4E	4E	4U	4F	4E	4E	4U
351-4V	2	4V	4E	4F	4E	4V	4E	4F	4E
390	3	4E	4U	4U	4D	4E	4U	4U	4D
390	4	4E	4U	4E	4D	4E	4U	4E	4D
428	5	4M	1Q	4M	1Q	4M	1Q	4M	1Q
428	6	4M	1Q	4M	1Q	4M	1Q	4M	1Q

Notes:

1 Before 9/11/68, with .69" Dia. Stabilizer Bar
2 From 9/11/68, with .69" Dia. Stabilizer Bar
3 Before 9/11/68, with .72" Dia. Stabilizer Bar
4 From 9/11/68, with .72" Dia. Stabilizer Bar
5 Before 9/11/68, with .95" Dia. Stabilizer Bar, Competition Suspension
6 From 9/11/68, with .95" Dia. Stabilizer Bar, Competition Suspension

1969 MUSTANG Front Spring Application Chart
SportsRoof and Mach 1 models.
(Automatic Transmissions, Standard Suspension.)
Note: Refer to chart for spring code identification.

Engine	Note	FMX Options: None	A/C	P/S	A/C & P/S	C4 Options: None	A/C	P/S	A/C & P/S	C6 Options: None	A/C	P/S	A/C & P/S
200	1,2,4					4W	4W	4W	4J				
200	1,2,5					4G	4G	4G	4G				
250	1,2,4					4W	4J	4J	4V				
250	1,2,5					4G	4W	4W	4J				
302	1,2,4					4J	4V	4V	4F				
302	1,2,5					4W	4J	4J	4V				
351-2V	1,2,4	4F	4E	4F	4E								
351-2V	1,2,5	4J	4F	4V	4E								
351-4V	1,2,4	4F	4E	4F	4E								
351-4V	1,2,5	4J	4F	4V	4E								
390	1,2,6									4E	4U	4E	4U
390	1,2,7									4F	4E	4E	4U
428	1,2,8									4M	4M	4M	1Q
428	1,2,9									4M	4M	4M	1Q
351-2V	3,4	4F	4E	4E	4E								
351-2V	3,5	4V	4F	4F	4E								
351-4V	3,4	4F	4E	4E	4U								
351-4V	3,5	4V	4F	4F	4E								
390	3,6									4E	4U	4U	4D
390	3,7									4F	4U	4E	4U
428	3,8									4M	1Q	4M	1Q
428	3,9									4M	1Q	4M	1Q

Notes:

1 Std. Interior, Bucket Seats
2 Deluxe Interior, Bucket Seats
3 Mach 1
4 Before 9/11/68, with .69" Dia. Stabilizer Bar
5 From 9/11/68, with .69" Dia. Stabilizer Bar
6 Before 9/11/68, with .72" Dia. Stabilizer Bar
7 From 9/11/68, with .72" Dia. Stabilizer Bar
8 Before 9/11/68, with .95" Dia. Stabilizer Bar, Competition Suspension
9 From 9/11/68, with .95" Dia. Stabilizer Bar, Competition Suspension

1969 MUSTANG Front Spring Application Chart
Hardtop models.
(Automatic Transmissions, Standard Suspension.)
Note: Refer to chart for spring code identification.

		FMX				C4				C6			
		Options:			A/C &	Options:			A/C &	Options:			A/C &
Engine	Notes.	None	A/C	P/S	P/S	None	A/C	P/S	P/S	None	A/C	P/S	P/S
200	1,2,5					4W	4W	4W	4J				
200	1,2,6					4G	4G	4G	4W				
250	1,2,5					4W	4J	4J	4V				
250	1,2,6					4G	4J	4W	4W				
302	1,2,5					4V	4F	4V	4F				
302	1,2,6					4J	4J	4J	4V				
351-2V	1,2,5	4F	4E	4F	4E								
351-2V	1,2,6	4V	4F	4F	4E								
351-4V	1,2,5	4F	4E	4E	4E								
351-4V	1,2,6	4V	4F	4F	4E								
390	1,2,7									4E	4U	4U	4D
390	1,2,8									4F	4U	4E	4U
428	1,2,9									4M	1Q	4M	1Q
428	1,2,10									4M	1Q	4M	1Q
200	3,4,5					4W	4W	4W	4J				
200	3,4,6					4G	4G	4G	4W				
250	3,4,5					4J	4J	4J	4V				
250	3,4,6					4G	4W	4W	4J				
302	3,4,5					4V	4F	4V	4F				
302	3,4,6					4J	4V	4J	4V				
351-2V	3,4,5	4F	4E	4E	4E								
351-2V	3,4,6	4V	4F	4F	4E								
351-4V	3,4,5	4F	4E	4E	4U								
351-4V	3,4,6	4V	4F	4F	4E								
390	3,4,7									4E	4U	4U	4D
390	3,4,8									4F	4U	4E	4U
428	3,4,9									4M	1Q	4M	1Q
428	3,4,10									4M	1Q	4M	1Q

Notes:

1. Std. Interior, Bucket Seats
2. Deluxe Interior, Bucket Seats
3. Std. Interior, Bench Seat
4. Deluxe Interior, Bench Seat
5. Before 9/11/68, with .69" Dia. Stabilizer Bar
6. From 9/11/68, with .69" Dia. Stabilizer Bar
7. Before 9/11/68, with .72" Dia. Stabilizer Bar
8. From 9/11/68, with .72" Dia. Stabilizer Bar
9. Before 9/11/68, with .95" Dia. Stabilizer Bar, Competition Suspension
10. From 9/11/68, with .95" Dia. Stabilizer Bar, Competition Suspension

1969 MUSTANG Front Spring Application Chart
Grande models.
(Automatic Transmissions, Standard Suspension.)
Note: Refer to chart for spring code identification.

		FMX				C4				C6			
		Options:			A/C &	Options:			A/C &	Options:			A/C &
Engine	Notes.	None	A/C	P/S	P/S	None	A/C	P/S	P/S	None	A/C	P/S	P/S
200	1					4W	4W	4W	4J				
200	2					4G	4G	4G	4W				
250	1					4J	4V	4J	4V				
250	2					4G	4J	4W	4J				
302	1					4V	4F	4F	4F				
302	2					4J	4V	4J	4F				
351-2V	1	4F	4E	4E	4U								
351-2V	2	4V	4F	4F	4E								
351-4V	1	4F	4E	4E	4U								
351-4V	2	4V	4E	4F	4E								
390	3									4E	4U	4U	4D
390	4									4E	4U	4E	4D
428	5									4M	1Q	1Q	1Q
428	6									4M	1Q	1Q	1Q

Notes:

1. Before 9/11/68, with .69" Dia. Stabilizer Bar
2. From 9/11/68, with .69" Dia. Stabilizer Bar
3. Before 9/11/68, with .72" Dia. Stabilizer Bar
4. From 9/11/68, with .72" Dia. Stabilizer Bar
5. Before 9/11/68, with .95" Dia. Stabilizer Bar, Competition Suspension
6. From 9/11/68, with .95" Dia. Stabilizer Bar, Competition Suspension

1969 MUSTANG Front Spring Application Chart
Convertible models.
(Automatic Transmissions, Standard Suspension.)
Note: Refer to chart for spring code identification.

		FMX				C4				C6			
		Options:			A/C &	Options:			A/C &	Options:			A/C &
Engine	Notes	None	A/C	P/S	P/S	None	A/C	P/S	P/S	None	A/C	P/S	P/S
200	1					4W	4J	4W	4J				
200	2					4G	4W	4G	4W				
250	1					4J	4V	4J	4F				
250	2					4W	4J	4W	4J				
302	1					4V	4F	4F	4E				
302	2					4J	4V	4V	4F				
351-2V	1	4F	4E	4E	4U								
351-2V	2	4V	4E	4F	4E								
351-4V	1	4F	4E	4E	4U								
351-4V	2	4V	4E	4F	4E								
390	3									4U	4U	4U	4D
390	4									4E	4U	4U	4D
428	5									4M	1Q	1Q	1Q
428	6									4M	1Q	1Q	1Q

Notes:

1. Before 9/11/68, with .69" Dia. Stabilizer Bar
2. From 9/11/68, with .69" Dia. Stabilizer Bar
3. Before 9/11/68, with .72" Dia. Stabilizer Bar
4. From 9/11/68, with .72" Dia. Stabilizer Bar
5. Before 9/11/68, with .95" Dia. Stabilizer Bar, Competition Suspension
6. From 9/11/68, with .95" Dia. Stabilizer Bar, Competition Suspension

1969 MUSTANG Front Spring Application Chart
SportsRoof and Mach 1 models.
(Manual Transmissions, Heavy Duty Suspension.)
Note: Refer to chart for spring code identification.

		Three-Speed				Four-Speed			
		Options:			A/C &	Options:			A/C &
Engine	Model	None	A/C	P/S	P/S	None	A/C	P/S	P/S
302	1,2,4	4V	4S	4V	4S	4V	4S	4S	4S
302	1,2,5	4X	4S	4X	4S	4X	4S	4S	4S
351	1,2,4	4S	4R	4S	4R	4S	4R	4S	4R
351-2V	1,2,5	4S	4R	4S	4R	4S	4R	4S	4R
351-4V	1,2,5	4S	4R	4S	4R	4S	4R	4S	4R
390	1,2,4	4R	4R	4R	4P	4R	4R	4R	4P
390	1,2,5	4R	4R	4R	4P	4R	4R	4R	4P
302	3,4	4V	4S	4S	4R	4V	4S	4S	4R
302	3,5	4X	4S	4S	4R	4X	4S	4S	4R
351-2V	3,4	4S	4R	4S	4R	4S	4R	4R	4R
351-2V	3,5	4S	4R	4S	4R	4S	4R	4R	4R
351-4V	3,4	4S	4R	4R	4R	4S	4R	4R	4R
351-4V	3,5	4S	4R	4R	4R	4S	4R	4R	4R
390	3,4	4R	4P	4R	4P	4R	4P	4R	4P
390	3,5	4R	4P	4R	4P	4R	4P	4R	4P

Notes:

1. Std. Interior, Bucket Seats
2. Deluxe Interior, Bucket Seats
3. Mach 1
4. Before 9/11/68, with .85" Dia. Stabilizer Bar With GT or Handling Suspension Option
5. From 9/11/68, with .85" Dia. Stabilizer Bar With GT or Handling Suspension Option

1969 MUSTANG Front Spring Application Chart
Hardtop and Grande models.
(Manual Transmissions, Heavy Duty Suspension.)
Note: Refer to chart for spring code identification.

Engine	Model	Three-Speed Options: None	A/C	P/S	A/C & P/S	Four-Speed Options: None	A/C	P/S	A/C & P/S
302	1,2,6	4V	4S	4S	4S	4S	4V	4S	4R
302	1,2,7	4X	4S	4S	4S	4X	4S	4S	4R
351-2V	1,2,6	4S	4R	4S	4R	4S	4R	4S	4R
351-2V	1,2,7	4S	4R	4S	4R	4S	4R	4S	4R
351-4V	1,2,6	4S	4R	4S	4R	4S	4R	4S	4R
351-4V	1,2,7	4S	4R	4S	4R	4S	4R	4R	4R
390	1,2,6	4R	4P	4R	4P	4R	4P	4R	4P
390	1,2,7	4R	4P	4R	4P	4R	4P	4R	4P
302	3,4,6	4V	4S	4S	4R	4V	4S	4S	4R
302	3,4,7	4X	4S	4S	4R	4X	4S	4S	4R
351-2V	3,4,6	4S	4R	4S	4R	4S	4R	4R	4R
351-2V	3,4,7	4S	4R	4S	4R	4S	4R	4R	4R
351-4V	3,4,6	4S	4R	4R	4R	4S	4R	4R	4R
351-4V	3,4,7	4S	4R	4R	4R	4S	4R	4R	4R
390	3,4,6	4R	4P	4R	4P	4R	4P	4R	4P
390	3,4,7	4R	4P	4R	4P	4R	4P	4R	4P
302	5,6	4V	4S	4S	4R	4V	4S	4S	4R
302	5,7	4X	4S	4S	4R	4X	4S	4S	4R
351-2V	5,6	4S	4R	4R	4R	4S	4R	4R	4P
351-2V	5,7	4S	4R	4R	4R	4S	4R	4R	4R
351-4V	5,6	4S	4R	4R	4P	4S	4R	4R	4P
351-4V	5,7	4S	4R	4R	4P	4S	4R	4R	4P
390	5,6	4R	4P	4R	4P	4R	4P	4P	4P
390	5,7	4R	4P	4R	4P	4R	4P	4P	4P

Notes:

1 Std. Interior, Bucket Seats
2 Deluxe Interior, Bucket Seats
3 Std. Interior, Bench Seats
4 Deluxe Interior, Bench Seats
5 Grande model.
6 Before 9/11/68, with .85" Dia. Stabilizer Bar With GT or Handling Suspension Option
7 From 9/11/68, with .85" Dia. Stabilizer Bar With GT or Handling Suspension Option

1969 MUSTANG Front Spring Application Chart
Convertible models.
(Manual Transmissions, Heavy Duty Suspension.)
Note: Refer to chart for spring code identification.

Engine	Model	Three-Speed Options: None	A/C	P/S	A/C & P/S	Four-Speed Options: None	A/C	P/S	A/C & P/S
302	1	4S	4S	4S	4R	4S	4S	4S	4R
302	2	4S	4S	4S	4R	4S	4S	4S	4R
351	1	4S	4R	4R	4P	4S	4R	4R	4P
351-2V	2	4S	4R	4R	4P	4S	4R	4R	4P
351-4V	2	4S	4R	4R	4P	4S	4R	4R	4P
390	1	4R	4P	4P	4P	4R	4P	4P	4N
390	2	4R	4P	4P	4P	4R	4P	4P	4N

Notes:

1 Before 9/11/68, with .85" Dia. Stabilizer Bar With GT or Handling Suspension Option
2 From 9/11/68, with .85" Dia. Stabilizer Bar With GT or Handling Suspension Option

1969 MUSTANG Front Spring Application Chart
SportsRoof and Mach 1 models.
(Automatic Transmissions, Heavy Duty Suspension.)
Note: Refer to chart for spring code identification.

Engine	Model	FMX Options: None	A/C	P/S	A/C & P/S	C4 Options: None	A/C	P/S	A/C & P/S	C6 Options: None	A/C	P/S	A/C & P/S
302	1,2,4					4V	4S	4V	4S				
302	1,2,5					4X	4S	4X	4S				
351	1,2,4	4S	4R	4S	4R								
351-2V	1,2,5	4S	4R	4S	4R								
351-4V	1,2,5	4S	4R	4S	4R								
390	1,2,4									4R	4P	4R	4P
390	1,2,5									4R	4P	4R	4P
390	3,4					4V	4S	4S	4S				
390	3,5					4X	4S	4S	4S				
351-2V	3,4	4S	4R	4R	4R								
351-2V	3,5	4S	4R	4R	4R								
351-4V	3,4	4S	4R	4R	4R								
351-4V	3,5	4S	4R	4R	4R								
390	3,4									4R	4P	4P	4P
390	3,5									4R	4P	4P	4P

Notes:

1 Std. Interior, Bucket Seats
2 Deluxe Interior, Bucket Seats
3 Mach 1
4 Before 9/11/68, with .85" Dia. Stabilizer Bar With GT or Handling Suspension Option.
5 From 9/11/68, with .85" Dia. Stabilizer Bar With GT or Handling Suspension Option.

1969 MUSTANG Front Spring Application Chart
Hardtop and Grande models.
(Automatic Transmissions, Heavy Duty Suspension.)
Note: Refer to chart for spring code identification.

Engine	Notes.	FMX Options: None	A/C	P/S	A/C & P/S	C4 Options: None	A/C	P/S	A/C & P/S	C6 Options: None	A/C	P/S	A/C & P/S
302	1,2,6					4V	4S	4S	4S				
302	1,2,7					4X	4S	4S	4S				
351-2V	1,2,6	4S	4R	4S	4R								
351-2V	1,2,7	4S	4R	4S	4R								
351-4V	1,2,6	4S	4R	4R	4R								
351-4V	1,2,7	4S	4R	4R	4R								
390	1,2,6									4R	4P	4P	4P
390	1,2,7									4R	4P	4P	4P
302	3,4,6					4V	4S	4S	4S				
302	3,4,7					4X	4S	4S	4S				
351-2V	3,4,6	4S	4R	4R	4R								
351-2V	3,4,7	4S	4R	4R	4R								
351-4V	3,4,6	4S	4R	4R	4R								
351-4V	3,4,7	4S	4R	4R	4R								
390	3,4,6									4R	4P	4P	4P
390	3,4,7									4R	4P	4P	4P
302	5,6					4V	4S	4S	4R				
302	5,7					4X	4S	4S	4R				
351-2V	5,6	4S	4R	4R	4R								
351-2V	5,7	4S	4R	4R	4R								
351-4V	5,6	4S	4R	4R	4P								
351-4V	5,7	4S	4R	4R	4P								
390	5,6									4R	4P	4P	4N
390	5,7									4R	4P	4P	4N

Notes:

1 Std. Interior, Bucket Seats
2 Deluxe Interior, Bucket Seats
3 Std. Interior, Bench Seat
4 Deluxe Interior, Bench Seat
5 Grande model.
6 Before 9/11/68, with .85" Dia. Stabilizer Bar With GT or Handling Suspension Option.
7 From 9/11/68, with .85" Dia. Stabilizer Bar

1969 MUSTANG Front Spring Application Chart Convertible models.

(Automatic Transmissions, Heavy Duty Suspension.)

Note: Refer to chart for spring code identification.

Engine	Notes	FMX None	FMX A/C	FMX P/S	FMX A/C & P/S	C4 None	C4 A/C	C4 P/S	C4 A/C & P/S	C6 None	C6 A/C	C6 P/S	C6 A/C & P/S
302	1					4V	4S	4S	4R				
302	2					4X	4S	4S	4R				
351	1	4S	4R	4R	4P								
351-2V	2	4S	4R	4R	4P								
351-4V	2	4S	4R	4R	4P								
390	1									4R	4P	4P	4N
390	2									4R	4P	4P	4N

Notes:

1. Before 9/11/68, with .85" Dia. Stabilizer Bar With GT or Handling Suspension Option.
2. From 9/11/68, with .85" Dia. Stabilizer Bar With GT or Handling Suspension Option.

1970 MUSTANG Front Spring Application Chart SportsRoof and Mach 1 models.

(Manual Transmissions, Standard Suspension.)

Note: Refer to chart for spring code identification.

Engine	Model	Three-Speed None	Three-Speed A/C	Three-Speed P/S	Three-Speed A/C & P/S	Four-Speed None	Four-Speed A/C	Four-Speed P/S	Four-Speed A/C & P/S
200	1,2,4	4G	4J	4W	4J				
200	1,2,5	4G	4G	4G	4W				
250	1,2,4	4W	4V	4J	4V				
250	1,2,5	4G	4J	4W	4J				
302-2V	1,2,4	4J	4F	4V	4F	4V	4F	4V	4E
302-2V	1,2,5	4W	4V	4J	4V	4J	4V	4J	4F
302-4V	1,2					6Y	6Y	6Y	4K
351-2V	1,2,4	4V	4E	4F	4E	4F	4E	4F	4E
351-2V	1,2,5	4J	4F	4V	4E	4J	4F	4V	4E
351-4V	1,2,4	4F	4E	4F	4E	4F	4E	4F	4E
351-4V	1,2,5	4J	4F	4V	4E	4J	4F	4V	4E
428	1,2					4L	4M	4M	1Q
429	1,2							6L	
428	3					4M	1Q	4M	1Q
429	3							6L	

Notes:

1. Std. Interior, Bucket Seats
2. Deluxe Interior, Bucket Seats
3. Mach 1
4. Before 8/28/69
5. From 8/28/69

1970 MUSTANG Front Spring Application Chart Hardtop and Grande models.

(Manual Transmissions, Standard Suspension.)

Note: Refer to chart for spring code identification.

Engine	Model	Three-Speed None	Three-Speed A/C	Three-Speed P/S	Three-Speed A/C & P/S	Four-Speed None	Four-Speed A/C	Four-Speed P/S	Four-Speed A/C & P/S
200	1,2,4	4G	4J	4W	4J				
200	1,2,5	4G	4W	4G	4W				
250	1,2,4	4W	4V	4J	4V				
250	1,2,5	4G	4J	4W	4J				
302	1,2,4	4V	4F	4V	4E	4V	4F	4V	4E
302	1,2,5	4J	4V	4J	4F	4J	4V	4J	4F
351-2V	1,2,4	4F	4E	4F	4E	4F	4E	4F	4U
351-2V	1,2,5	4J	4F	4V	4E	4V	4F	4V	4E
351-4V	1,2,4	4F	4E	4F	4U	4F	4E	4F	4U
351-4V	1,2,5	4V	4F	4V	4E	4V	4E	4V	4E
428	1,2					4M	1Q	4M	1Q
200	3,4	4W	4J	4W	4V				
200	3,5	4G	4W	4G	4W				
250	3,4	4J	4V	4J	4F				
250	3,5	4W	4J	4W	4J				
302	3,4	4V	4F	4V	4E	4V	4E	4F	4E
302	3,5	4J	4V	4J	4F	4J	4F	4J	4F
351-2V	3,4	4F	4E	4F	4U	4F	4E	4E	4U
351-2V	3,5	4V	4E	4F	4E	4V	4E	4F	4E
351-4V	3,4	4F	4E	4E	4U	4F	4E	4E	4U
351-4V	3,5	4V	4E	4F	4E	4V	4E	4F	4E
428	3					4M	1Q	4M	1Q

Notes:

1. Std. Interior, Bucket Seats
2. Deluxe Interior, Bucket Seats
3. Grande
4. Before 8/28/69
5. From 8/28/69

1970 MUSTANG Front Spring Application Chart SportsRoof and Mach 1 models.

(Automatic Transmissions, Standard Suspension.)

Note: Refer to chart for spring code identification.

Engine	Model	FMX (or C6; see note.) None	FMX A/C	FMX P/S	FMX A/C & P/S	C4 None	C4 A/C	C4 P/S	C4 A/C & P/S
200	1,2,4					4G	4W	4G	4J
200	1,2,5					4G	4G	4G	4W
250	1,2,4					4W	4J	4J	4V
250	1,2,5					4G	4W	4G	4J
302-2V	1,2,4					4J	4F	4V	4F
302-2V	1,2,5					4W	4V	4J	4V
351-2V	1,2,4	4F	4E	4F	4E				
351-2V	1,2,5	4J	4F	4V	4E				
351-4V	1,2,4	4F	4E	4F	4E				
351-4V	1,2,5	4J	4F	4V	4E				
428	1,2,6	4M	1Q	4M	1Q				
428	3,6	4M	1Q	4M	1Q				

Notes:

1. Std. Interior, Bucket Seats
2. Deluxe Interior, Bucket Seats
3. Mach 1
4. Before 8/28/69
5. From 8/28/69
6. C6 Transmission; not FMX.

1970 MUSTANG Front Spring Application Chart
Convertible models.
(Manual Transmissions, Standard Suspension.)
Note: Refer to chart for spring code identification.

		Three-Speed Options:				Four-Speed Options:			
Engine	Model	None	A/C	P/S	A/C & P/S	None	A/C	P/S	A/C & P/S
200	1,2,3	4W	4J	4W	4V				
200	1,2,4	4G	4W	4G	4J				
250	1,2,3	4J		4J	4F				
250	1,2,4	4W	4J	4W	4V				
302	1,2,3	4V	4E	4F	4E	4V	4E	4F	4E
302	1,2,4	4J	4F	4J	4F	4J	4F	4V	4E
351-2V	1,2,3	4F	4E	4E	4U	4F	4U	4E	4U
351-2V	1,2,4	4V	4E	4F	4E	4V	4E	4F	4U
351-4V	1,2,3	4F	4U	4E	4U	4F	4U	4E	4U
351-4V	1,2,4	4V	4E	4F	4U	4V	4E	4F	4U
428	1,2					4M	1Q	4M	1Q

Notes:
1. Std. Interior, Bucket Seats
2. Deluxe Interior, Bucket Seats
3. Before 8/28/69
4. From 8/28/69

1970 MUSTANG Front Spring Application Chart
Convertible models.
(Automatic Transmissions, Standard Suspension.)
Note: Refer to chart for spring code identification.

		FMX (or C6; see note.) Options:				C4 Options:			
Engine	Model	None	A/C	P/S	A/C & P/S	None	A/C	P/S	A/C & P/S
200	1,2,3					4G	4J	4W	4J
200	1,2,4					4G	4W	4G	4W
250	1,2,3					4J	4V	4J	4F
250	1,2,4					4W	4J	4W	4J
302	1,2,3					4V	4F	4F	4E
302	1,2,4					4J	4F	4J	4F
351-2V	1,2,3	4F	4U	4E	4U				
351-2V	1,2,4	4V	4E	4F	4U				
351-4V	1,2,3	4F	4U	4E	4U				
351-4V	1,2,4	4V	4E	4F	4U				
428	1,2,5	4M	1Q	4M	1Q				

Notes:
1. Std. Interior, Bucket Seats
2. Deluxe Interior, Bucket Seats
3. Before 8/28/69
4. From 8/28/69
5. C6 Transmission; not FMX.

1970 MUSTANG Front Spring Application Chart
Hardtop and Grande models.
(Automatic Transmissions, Standard Suspension.)
Note: Refer to chart for spring code identification.

		FMX (or C6; see note.) Options:				C4 Options:			
Engine	Model	None	A/C	P/S	A/C & P/S	None	A/C	P/S	A/C & P/S
200	1,2,4					4G	4W	4G	4J
200	1,2,5					4G	4G	4G	4W
250	1,2,4					4W	4V	4J	4V
250	1,2,5					4G	4J	4W	4J
302	1,2,4					4V	4F	4V	4F
302	1,2,5					4J	4V	4J	4F
351-2V	1,2,4	4F	4E	4F	4U				
351-2V	1,2,5	4V	4F	4V	4E				
351-4V	1,2,4	4F	4E	4F	4U				
351-4V	1,2,5	4V	4E	4V	4E				
428	1,2,6	4M	1Q	4M	1Q				
200	3,4					4G	4J	4W	4J
200	3,5					4G	4W	4G	4W
250	3,4					4J	4V	4J	4V
250	3,5					4G	4J	4W	4J
302	3,4					4V	4F	4V	4E
302	3,5					4J	4V	4J	4F
351-2V	3,4	4F	4E	4F	4U				
351-2V	3,5	4V	4E	4F	4E				
351-4V	3,4	4F	4E	4E	4U				
351-4V	3,5	4V	4E	4F	4E				
428	3,6	4M	1Q	4M	1Q				

Notes:
1. Std. Interior, Bucket Seats
2. Deluxe Interior, Bucket Seats
3. Grande
4. Before 8/28/69
5. From 8/28/69
6. C6 Transmission; not FMX.

1970 MUSTANG Front Spring Application Chart
Heavy Duty Suspension, All Models with Manual Transmissions.
Note: Refer to chart for spring code identification.

		Three-Speed Options:				Four-Speed Options:			
Engine	Notes	None	A/C	P/S	A/C & P/S	None	A/C	P/S	A/C & P/S
302-2V	1	4X	4S	4X	4S	4X	4S	4X	4S
351	1	4K	4L	4K	4L	4K	4L	4K	4L
351	2	4K	4L	4K	4L	4K	4L	4K	4L
302	3	4X	4S	4X	4R	4X	4S	4X	4R
351	3	4K	4L	4K	4L	4K	4L	4K	4L
302	4	4X	4S	4S	4R	4X	4S	4S	4R
351	4	4K	4L	4K	4L	4K	4L	4K	4L
302	5	4X	4R	4S	4R	4S	4R	4S	4R
351	5	4K	4L	4L	4M	4K	4L	4L	4M

Notes:
1. SportsRoof, Standard or Deluxe Interior
2. Mach 1
3. Hardtop, Standard or Deluxe Interior
4. Hardtop, Grande model
5. Convertible, Standard or Deluxe Interior

1970 MUSTANG Front Spring Application Chart
Heavy Duty Suspension, All Models with Automatic Transmissions.
Note: Refer to chart for spring code identification.

		FMX Options:				C4 Options:			
Engine	Notes	None	A/C	P/S	A/C & P/S	None	A/C	P/S	A/C & P/S
302-2V	1					4X	4S	4X	4S
351	1	4K	4L	4K	4L				
351	2	4K	4L	4K	4L				
302	3					4X	4S	4X	4S
351	3	4K	4L	4K	4L				
302	4					4X	4S	4S	4R
351	4	4K	4L	4K	4L				
302	5					4S	4R	4S	4R
351	5	4K	4L	4L	4M				

Notes:
1. SportsRoof, Standard or Deluxe Interior
2. Mach 1
3. Hardtop, Standard or Deluxe Interior
4. Hardtop, Grande model
5. Convertible, Standard or Deluxe Interior

1965-1970 Mustang Front Spring Identification and Cross Reference Chart

Spring Code	Part Number	Color Stripe	Normal Load Lbs.	No. of Coils	Wire Dia. Inches	Free Length
25	C7ZZ-5310-AJ	1 Orange	1610	9-3/4	.675	18-1/2
	r/b C7ZZ-5310-C (4F)					
68	C7ZZ-5310-AL	1 Brown	1454	9-3/16	.575	16-5/8
	r/b C7ZZ-5310-Z (4J)					
82	C5ZZ-5310-A	1 Pink	1279	8-1/2	.570	15-1/8
	r/b C5ZZ-5310-B					
84	C5ZZ-5310-B	1 Green	1344	8-1/2	.570	15-3/8
86	C5ZZ-5310-C	1 Gray	1490	8-1/2	.585	15-1/2
88	C5ZZ-5310-D	1 Violet	1413	8-1/2	.585	15-1/4
	r/b C5ZZ-5310-C (86)					
89	C5ZZ-5310-E	1 Brown	1491	8	.600	14-1/8
90	C5ZZ-5310-F	1 Red	1413	8	.600	13-7/8
	r/b C5ZZ-5310-E (89)					
1Q	C7ZZ-5310-AR	Gold/Gray	1850	8-3/4	.650	15-1/8
4A	C7ZZ-5310-H	Gray/Green	2130	8-1/2	.760	15 1/8
	r/b C9AZ-5310-Z (4J)					
4C	C7ZZ-5310-BB	Gray/Yellow	1900	9	.635	16
4D	C7ZZ-5310-A	Pink/Violet	1790	9-1/2	.610	16-7/8
4E	C7ZZ-5310-B	Tan/Violet	1650	9	.600	16-3/8
4F	C7ZZ-5310-C	Orange/Violet	1580	9	.600	16-1/8
4G	C7ZZ-5310-F	1 Green	1342	9-1/4	.575	16-1/8
4H	C7ZZ-5310-Y	Orange/Red	1400	8-3/4	.585	15-3/4
	r/b C7ZZ-5310-Z (4J)					
4J	C7ZZ-5310-Z	Gold/Red	1460	8-3/4	.585	16
4K	C7ZZ-5310-AA	Gold/Orange	1580	8	.635	14-3/8
4L	C7ZZ-5310-AB	Gold/Tan	1670	8-3/4	.650	14-5/8
4M	C7ZZ-5310-AC	Gold/Pink	1760	8-3/4	.650	14-7/8
4N	C7ZZ-5310-AD	Green/Yellow	1820	9	.635	15-3/4
4P	C7ZZ-5310-AD	Gold/Yellow	1740	9	.635	15-1/2
4R	C7ZZ-5310-AF	Tan/Yellow	1660	8-3/4	.625	15-1/4
4S	C7ZZ-5310-AG	Orange/Yellow	1580	8-3/4	.625	15
4T	C7ZZ-5310-AH	1 Pink	1286	9-1/4	.575	15-7/8
	r/b C7ZZ-5310-F (4G)					
4U	C7ZZ-5310-AK	Gold/Violet	1720	9-1/2	.610	16-5/8
4V	C7ZZ-5310-AN	Red/Yellow	1520	9 1/2	.600	16-1/4
4W	C7ZZ-5310-AP	Brown/Gray	1398	9-1/4	.575	16-3/8
	r/b C7ZZ-5310-Z (4J)					
4X	C7ZZ-5310-AZ	Violet/Yellow	1500	8-3/4	.625	14-3/4
4Y	C7ZZ-5310-BA	Gray/Violet	1860	9-1/2	.610	17-1/8
6D	C8ZZ-5310-A	Gray/Gold	1850	8-3/4	.650	15-1/8
6L	C9ZZ-5310-B	Data not provided in Master Parts Chart				
6Y	D0ZZ-5310-A	2 Gold/Violet	1490	8-1/8	.629	14-5/16
	r/b C7ZZ-5310-AA (4K)					

MUSTANG FRONT STABILIZER BAR ASSEMBLY CHART #1, 1965-1968

Year	Cyl/CID	Model	Type	Dia.	Identification	Part Number
65	6/170		S			C5ZZ-5482-B
65	6/200		S			C5ZZ-5482-B
65	8/260		S			C5ZZ-5482-B
65	8/289	EXC. GT350 K	S			C5ZZ-5482-B
65	8/289	GT350	S			S1MS-5482-A
65	8/289	K	S	13/16	NONE	C5ZZ-5482-A
66	6/200		S			C5ZZ-5482-B
66	8/289	EXC. GT350 K	S			C5ZZ-5482-B
66	8/289	GT350	S			S1MS-5482-A
66	8/289	K	S	13/16	NONE	C5ZZ-5482-A
67	6/200		S	'3/4		C7WY-5482-A
67	8/289	EXC. GT350, GT K	S	'3/4		C7WY-5482-A
67	8/289	GT350	S			C9ZZ-5482-E
67	8/289	GT w/ IMPROVED	S	'7/8	1G	C7ZZ-5482-C
67		HANDLING PKG				
67	8/289	w/ MAXIMUM	S			C9ZZ-5482-E
		HANDLING PKG				
67	8/289	K	S			C9ZZ-5482-E
67	8/390		S	'3/4	1Y-1B	C7ZZ-5482-B
67	8/390	GT w/ IMPROVED	S	'7/8	1G	C7ZZ-5482-C
		HANDLING PKG				
67	8/390	GT500	S			C9ZZ-5482-E
68	6/200		S	'3/4		C7WY-5482-A
68	8/289	EXC. GT 350, GT	S	'3/4		C7WY-5482-A
68	8/289	GT350				C9ZZ-5482-E
68	8/289	GT w/ IMPROVED	S	'7/8	1G	C7ZZ-5482-C
		HANDLING PKG				
68	8/289	GT w/ MAXIMUM	S			C9ZZ-5482-E
		HANDLING PKG				
68	8/302	EXC. GT350,GT	S	'3/4		C7WY-5482-A
68	8/302	GT350	S			C9ZZ-5482-E
68	8/302	GT w/ IMPROVED	S	'7/8	1G	C7ZZ-5482-C
		HANDLING PKG				
68	8/302	GT w/ MAXIMUM	S			C9ZZ-5482-E
		HANDLING PKG				
68	8/390	EXC. GT	S	'3/4	1Y-1B	C7ZZ-5482-B
68	8/390	GT w/ IMPROVED	S	'7/8	1G`	C7ZZ-5482-C
		HANDLING PKG.				
68	8/390	GT w/ MAXIMUM	S			C9ZZ-5482-E
		HANDLING PKG				
68	8/428	GT w/ MAXIMUM	S			C9ZZ-5482-E
		HANDLING PKG				
68	8/428	GT350/500	S			C9ZZ-5482-E
68	8/428	GT350/500	S			C9ZZ-5482-E

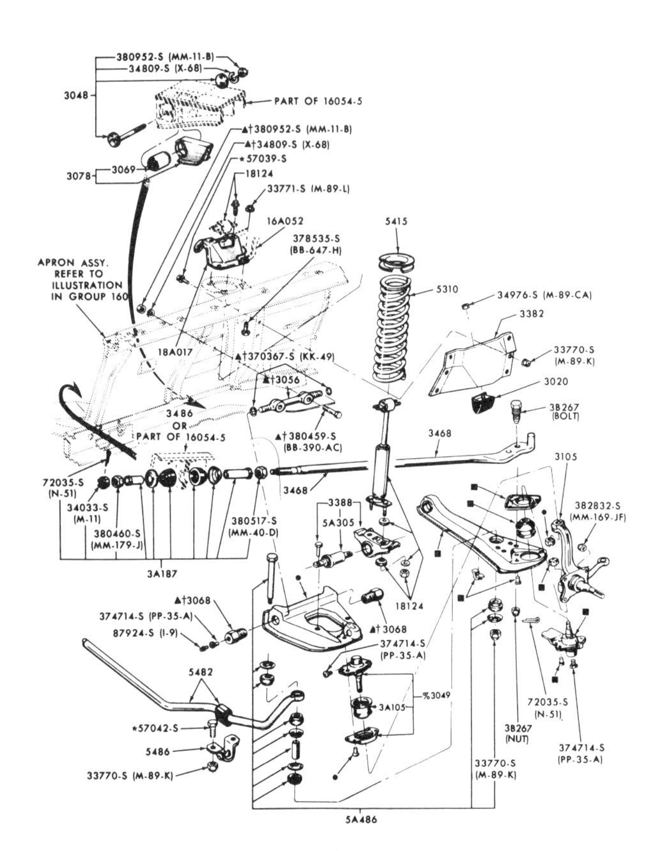

Typical 1968-1970 Mustang front suspension system.

MUSTANG FRONT STABILIZER BAR ASSEMBLY CHART #2, 1969 and 1970

Year	Cyl/CID	Model	Type	Dia.	Ident.	Part Number
69	6/200		S	11/16	1P-1GO	C9ZZ-5482-B
69	6/250		S	11/16	1P-1GO	C9ZZ-5482-B
69	8/302-2B		S	11/16	1P-1GO	C9ZZ-5482-B
69	8/302-2B	GT w/ IMPROVED	S	7/8	1G	C7ZZ-5482-C
		HANDLING PKG				
69	8/302-4B	BOSS (Note 1)	S	7/8	1G	C9ZZ-5482-D
69	8/302-4B	BOSS (Note 2)	S	3/4	1Y-1B	C9ZZ-5482-C
69	8/302-4B	GT w/ IMPROVED	S	7/8	1G	C7ZZ-5482-C
		HANDLING PKG				
69	8/351	All, non-Mach 1	S	11/16	1P-2GO	C9ZZ-5482-B
69	8/351	MACH 1	S	7/8	1G	C9ZZ-5482-D
69	8/390	All, non-Mach 1	S	3/4	1Y-1B	C9ZZ-5482-C
		w/o SPECIAL				
		HANDLING PKG				
69	8/390	w/ SPECIAL	S	7/8	1G	C9ZZ-5482-D
		HANDLING PKG				
69	8/390	GT w/ IMPROVED	S	7/8	1G	C7ZZ-5482-C
		HANDLING PKG				
69	8/428		S	15/16	1G-1Y	C9ZZ-5482-E
69	8/429	BOSS	S	15/16	1G-1Y	C9ZZ-5482-E
70	6/200		S			D0ZZ-5482-A
70	6/250		S			D0ZZ-5482-A
70	8/302-2/B		S			D0ZZ-5482-A
70	8/302-2/B		H/D	11/16	1P-2GO	C9ZZ-5482-B
70	8/302-2/B	W/ COMPETITION	S	7/8	1G	C9ZZ-5482-D
		HANDLING PKG				
70	8/302-4/B	BOSS	S	15/16	1G-1Y	C9ZZ-5482-E
70	8/351		S			D0ZZ-5482-A
70	8/351		H/D	15/16	1G-1Y	C9ZZ-5482-E
70	8/428			15/16	1G-1Y	C9ZZ-5482-E

Notes:

1 Before 4/14/69

2 From 4/14/69

MUSTANG STABILIZER BAR BRACKET APPLICATION CHART

Year	Model	Description	Part Number
65/66	Exc. GT-350		C3DZ-5486-A
65/66	GT-350		C0DD-5486-A
67	6 & 8 Cyl		C6OZ-5486-B
	Exc. 289-4V		
67	8 Cyl 289 4V		C7ZZ-5486-A
68	6 & 8 Cyl.		C6OZ-5486-B
	Exc. 390 w/ Competition		
	Handling Package		
	or GT Equipment Group		
68	8 Cyl 390 w/ Competition		C7ZZ-5486-A
	Handling Package		
	or GT Equipment Group		
69	6/8 Cyl. Exc. 390 w/ Comp.		C6OZ-5486-B
	Handling Pkg, 428, Boss		
69	8 Cyl 390 w/ Competition		C7ZZ-5486-A
	Handling Package		
	w/ GT Equipment Group		
69	8 Cyl 428 W/GT & Comp.		C7ZZ-5486-A
	Handling Pkg.		
69	Boss 302	Before 4/14/69	C6OZ-5486-B
69	Boss 302	From 4/14/69	C7ZZ-5486-A
69	Boss 429	R.H., for Front Stab. Bar	C9ZZ-5486-A
69	Boss 429	L.H., for Front Stab. Bar	C9ZZ-5486-B
69	Boss 429	for Rear Stab. Bar	C9ZZ-5486-C
70	6 & 8 Cyl 302	for Front Stab. Bar	C6OZ-5486-B
70	8 Cyl 351 Exc. Mach 1 &	for Front Stab. Bar	C6OZ-5486-B
	all Comp. Handling Pkg		
70	Boss 302 8 Cyl, 428	for Rear Stab. Bar	D0OZ-5486-A
	Competition Handling Pkg		
70	8 Cyl 351, Exc. Mach 1,	for Front Stab. Bar	C7ZZ-5486-A
	w/ GT & Comp. Hand. Pkgs		
70	Mach 1, 8 Cyl 351	for Front Stab. Bar	C7ZZ-5486-A
70	8 Cyl. 428	for Front Stab. Bar	C7ZZ-5486-A
70	Boss 429	R.H., for Front Stab. Bar	C9ZZ-5486-A
70	Boss 429	L.H., for Front Stab. Bar	C9ZZ-5486-B
70	Boss 429	for Rear Stab. Bar	D0OZ-5486-A

Mustang Rear Spring Application Chart, Part One, 1965-1967

Year	Model	Description	Part Number
65/66	F/B & Conv., 6 cyl. & 8 cyl. 260,	4 leaf-650 lb. load rate	C5ZZ-5560-D
	289-2V, 289-4V Premium Fuel	C4ZA-5556-B,D,F,N,U	
65/66	Hardtop, 6 Cyl., & 8 cyl., 260,	4 leaf-610 lb. load rate	C5ZZ-5560-C
	289-2V, 289-4V Premium Fuel	C4ZA-5556-C,E,M,T	
65/66	F/B & Conv., 8 cyl. 289-4V Hi-Po	4 leaf-650 lb. load rate	C5ZZ-5560-E
		C4ZA-5556-K,L,S,Y	
		r/b C7ZZ-5560-M	
65/66	Hardtop, 8 cyl. 289-4V Hi-Po	4 leaf-610 lb. load rate	C5ZZ-5560-F
		C4ZA-5556-H,J,R,V	
67	Hardtop, 6 Cyl. & 8 cyl. 289	4 leaf-650 lb. load rate	C4DZ-5560-J
	Before 9/20/66	C4DA-5556-U	
67	F/B, 6 cyl. & 8 cyl., 289-2V &	4 leaf-650 lb. load rate	C4DZ-5560-J
	289-4V (P.F.) from 9/20/66	C4DA-5556-U	
67	Hardtop, 6 cyl., 8 cyl. 289-2V,	4 leaf-650 lb. load rate	C4DZ-5560-J
	289-4V & 390 from 9/20/66	C4DA-5556-U	
67	F/B, 6 cyl. & 8 cyl.,	4 leaf-650 lb. load rate	C7ZZ-5560-A
	Before 9/20/66	C7ZA-5556-J,AB	
		r/b C4DZ-5560-J	
67	F/B, 8 Cyl. 390, from 9/20/66	4 leaf-650 lb. load rate	C7ZZ-5560-A
		C7ZA-5556-J,AB	
		r/b C4DZ-5560-J	
67	Conv. 6 cyl., 8 cyl. 289-2V,	4 leaf-650 lb. load rate	C7ZZ-5560-A
	289-4V (P.F.), 390 from 9/20/66	C7ZA-5556-J,AB	
		r/b C4DZ-5560-J	
67	F/B, Conv. 8 cyl. 390	4 leaf-665 lb. load rate	C7ZZ-5560-C
	Before 10/26/66	C7ZA-5556-N,AD	
		r/b C7ZZ-5560-M	
67	Hardtop 8 cyl. 390	4 leaf-625 lb. load rate	C7ZZ-5560-D
	Before 10/26/66	C7ZA-5556-S,AF	
		r/b C7ZZ-5560-M	
67	F/B, 8 cyl. 289, 390, from	4 leaf-665 lb. load rate	C7ZZ-5560-M
	10/26/66, Handling Package	C7ZA-5556-AS,AT,AU,AV	
67	H/T, Conv. 8 cyl., from	4 leaf-665 lb. load rate	C7ZZ-5560-M
	10/26/66, Handling Package	C7ZA-5556-AS,AT,AU,AV	
67	F/B, Conv. 8 cyl. 289, 390, before	4 leaf-665 lb. load rate	C7ZZ-5560-J
	11/15/66, Comp. Handling GT only	C7ZA-5556-AG	
		r/b C7ZZ-5560-M	
67	Hardtop, 8 cyl. 289, 390, before	4 leaf-625 lb. load rate	C7ZZ-5560-K
	11/15/66, Comp. Handling GT only	C7ZA-5556-AJ	
		r/b C7ZZ-5560-T	

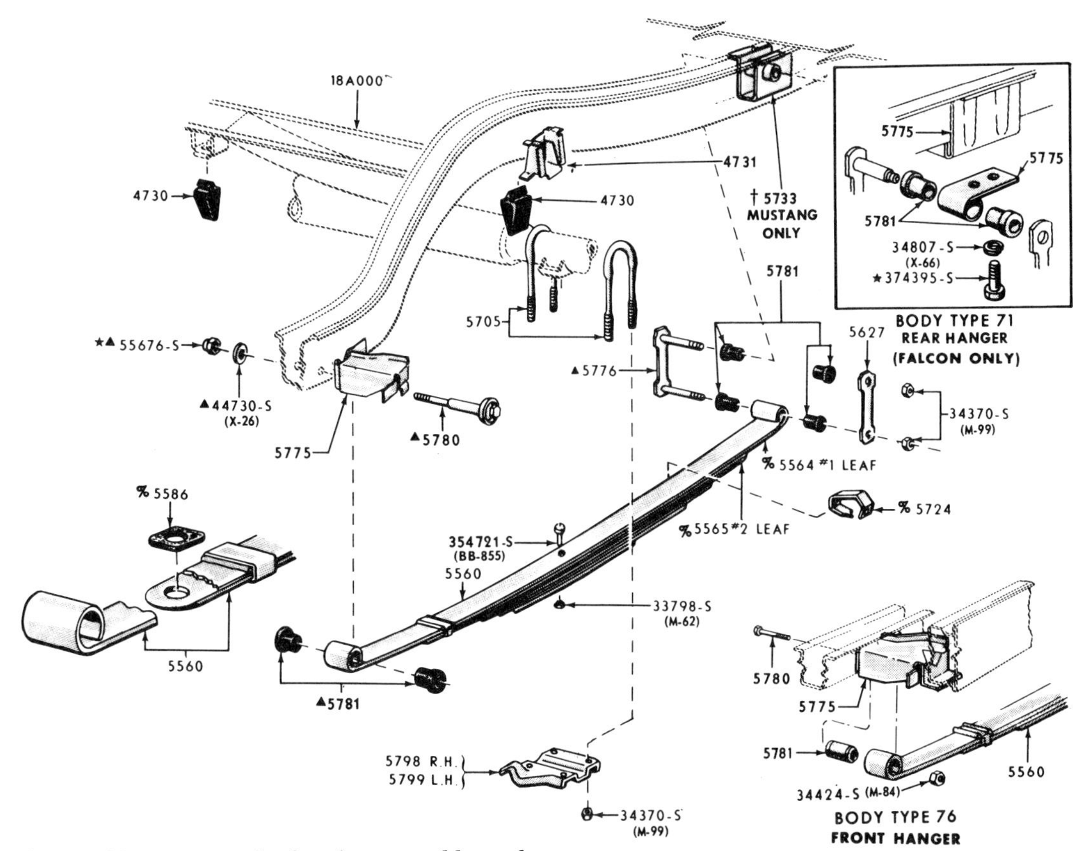

The 1965-1970 Mustang rear leaf spring assembly and attaching hardware.

Mustang Rear Spring Application Chart, Part Two, 1967-1968

Year	Model	Description	Part Number
67/68	F/B 8 cyl. 289 for competition	4 leaf-625 lb. load rate	C7ZZ-5560-U
	handling GT only, before 11/15/66	C7ZA-5556-AN,AR	
		r/b C7ZZ-5560-T	
67/68	H/T 8 cyl. 289, 390 for comp.	4 leaf-625 lb. load rate	C7ZZ-5560-U
	handling GT only, before 11/15/66	C7ZA-5556-AN,AR	
		r/b C7ZZ-5560-T	
67/68	GT-350; GT-500	4 leaf-625 lb. load rate	C7ZZ-5560-U
		C7ZA-5556-AN,AR	
		r/b C7ZZ-5560-T	
67/68	F/B 8 cyl. 390 for competition	4 leaf-665 lb. load rate	C7ZZ-5560-T
	handling GT only, before 11/15/66	C7ZA-5556-AL,AM	
67/68	Conv. 8 cyl. 289, 390 for comp	4 leaf-665 lb. load rate	C7ZZ-5560-T
	handling GT only, before 11/15/66	C7ZA-5556-AL,AM	
67/68	GT-350; GT-500	4 leaf-665 lb. load rate	C7ZZ-5560-T
		C7ZA-5556-AL,AM	
68	F/B, 6 cyl., 8 cyl., 289-2V	4 leaf-650 lb. load rate	C4DZ-5560-J
	& 302-4V except GT-350	C4DA-5556-U	
68	Hardtop	4 leaf-650 lb. load rate	C4DZ-5560-J
		C4DA-5556-U	
68	F/B 8 cyl. 289,302-4V	4 leaf-625 lb. load rate	C7ZZ-5560-D
	Except GT-350	C7ZA-5556-S, AF	
		r/b C7ZZ-5560-M	
68	H/T 8 cyl. 289,302-4V	4 leaf-625 lb. load rate	C7ZZ-5560-D
	Except GT-350	C7ZA-5556-S, AF	
		r/b C7ZZ-5560-M	
68	F/B 8 cyl. 289,302,390-4V	4 leaf-625 lb. load rate	C7ZZ-5560-M
	Except GT-350	C7ZA-5556-AS,AT,AU,AV	
68	H/T, Conv. 8 cyl. 289, 302-4V	4 leaf-625 lb. load rate	C7ZZ-5560-M
	& 390-4V, except GT-350	C7ZA-5556-AS,AT,AU,AV	
68	F/B, Conv. 8 cyl. 390-2VPF & 4V	4 leaf-650 lb. load rate	C7ZZ-5560-A
		C7ZA-5556-J,AB	
		r/b C7ZZ-5560-J	
68	Conv. 6 cyl., 8 cyl. 289-2V,	4 leaf-650 lb. load rate	C7ZZ-5560-A
	& 302-4V, except GT-350	C7ZA-5556-J,AB	
		r/b C7ZZ-5560-J	
68	F/B, Conv. 8 cyl. 390-4V	4 leaf-665 lb. load rate	C7ZZ-5560-C
		C7ZA-5556-N,AD	
		r/b C7ZZ-5560-M	
68	Conv. 8 cyl. 289-2V,	4 leaf-665 lb. load rate	C7ZZ-5560-C
	& 302-4V, except GT-350	C7ZA-5556-N,AD	
		r/b C7ZZ-5560-M	
68	F/B,H/T,Conv. 8 cyl. 390-2V(P.F.),	4 leaf-665 lb. load rate	C7ZZ-5560-C
	428CJ-4V w/ impr. handling	C7ZA-5556-N,AD	
	package, except GT-500	r/b C7ZZ-5560-M	

Mustang Rear Spring Application Chart, Part Three, 1969

Year	Model	Description	Part Number
69	F/B 8 Cyl. 302, Before 9/10/68	4 leaf-700 lb. load rate	C9ZZ-5560-A
		C9ZA-5556-B	
69	F/B 351, 390-4V; exc. GT-350;	4 leaf-700 lb. load rate	C9ZZ-5560-A
	Exc. Mach 1; Before 9/10/68	C9ZA-5556-B	
69	Conv. 6 Cyl.; 8 Cyl. 302, 351,	4 leaf-700 lb. load rate	C9ZZ-5560-A
	390-4V; exc. GT-350	C9ZA-5556-B	
69	F/B 8 Cyl. 302 w/ Improved	4 leaf-665 lb. load rate	C7ZZ-5560-M
	Handling Package	C7ZA-5556-AS,AT,AU,AV	
69	Hardtop, 8 Cyl. 302, 351, 390-4V	4 leaf-665 lb. load rate	C7ZZ-5560-M
	w/ Improved Handling Package	C7ZA-5556-AS,AT,AU,AV	
69	F/B 8 Cyl. 390 for Competition	4 leaf-665 lb. load rate	C7ZZ-5560-T
	Handling GT only.	C7ZA-5556-AL,AM	
69	Conv. 8 Cyl. 289, 390 for	4 leaf-665 lb. load rate	C7ZZ-5560-T
	Competition Handling GT only.	C7ZA-5556-AL,AM	
69	Hardtop, exc. Grandé, 6 Cyl.,	4 leaf-650 lb. load rate	C4DZ-5560-J
	8 Cyl. 302,351,390	C4DA-5556-U	
69	Hardtop, Grandé from 9/10/68	4 leaf-650 lb. load rate	C4DZ-5560-J
		C4DA-5556-U	
69	F/B 8 Cyl. 289 for Competition	4 leaf-625 lb. load rate	C7ZZ-5560-U
	Handling GT only.	C7ZA-5556-AN,AR	
		r/b C7ZZ-5560-T	
69	H/T 8 Cyl. 289, 390 for Comp.	4 leaf-625 lb. load rate	C7ZZ-5560-U
	Handling GT only.	C7ZA-5556-AN,AR	
		C7ZA-5556-AN,AR	
69	Hardtop, Grandé , all 6 Cyl., 8 Cyl.	4 leaf-665 lb. load rate	C9ZZ-5560-E
	302, 351,390-4V Before 9/10/68	C9ZA-5556-K,M	
		r/b C4DZ-5560-J	
69	Conv. 6 Cyl., 8 Cyl. 302,351,390	4 leaf-650 lb. load rate	C7ZZ-5560-A
		C7ZA-5556-J,AB	
		r/b C4DZ-5560-J	
69	F/B 6 Cyl., 8 Cyl. 351,390, Exc.	4 leaf-595 lb. load rate	C9ZZ-5560-H
	GT-350, Mach 1, From 9/10/68	C9ZA-5556-R	
69	F/B, 8 cyl. 302, Exc. Boss 302	4 leaf-595 lb. load rate	C9ZZ-5560-H
	From 9/10/68	C9ZA-5556-R	
69	F/B, H/T, 428 Exc. GT-500 &	4 leaf-690 lb. load rate	C9ZZ-5560-D
	Grandé; w/o trunk battery	C9ZA-5556-H	
		2 Gold, 1 Violet Stripes	
69	F/B, H/T, 428 Exc. GT-500 &	4 leaf-730 lb. load rate	C9ZZ-5560-C
	Grandé; with trunk battery	C9ZA-5556-F	
		2 Yellow, 2 Brown Stripes	
69	Conv., 428 Exc. GT-500;	4 leaf-730 lb. load rate	C9ZZ-5560-C
	with or without trunk battery	C9ZA-5556-F	
		2 Yellow, 2 Brown Stripes	

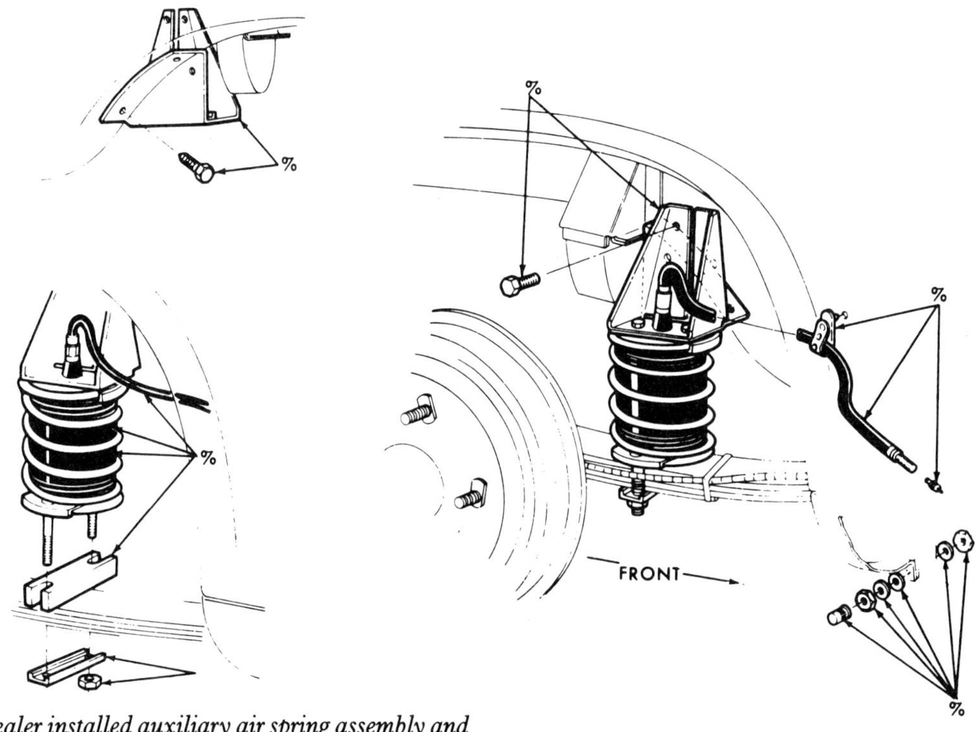

Optional dealer installed auxiliary air spring assembly and mounting hardware for all 1965-1970 Mustangs.

MUSTANG REAR SPRING PLATE APPLICATION CHART

Year	Model	Right Hand	Left Hand	Notes
'65'66		C5DZ-5796-A	C5DZ-5796-B	
'67	Exc. 428-CJ	C6DZ-5796-A	C6DZ-5796-B	
'68	Exc. 428-CJ	C6DZ-5796-A	C6DZ-5796-B	
'68	8 Cyl. 428-CJ	C8ZZ-5796-A	C8ZZ-5796-B	
'69	Exc. Grandé, Boss	C6DZ-5796-A	C6DZ-5796-B	
	302, 428-CJ			
'69	Grandé	C9OZ-5796-C	C9OZ-5796-B	Before 9/10/68
'69	Grandé	C6DZ-5796-A	C6DZ-5796-B	From 9/10/68
'69	8 Cyl 428-CJ	C8ZZ-5796-A	C8ZZ-5796-B	
'69	Boss 302	C8ZZ-5796-A	C8ZZ-5796-B	
'70	6 Cyl., 8 Cyl.	C6DZ-5796-A	C6DZ-5796-B	
	302-2V, 351			
'70	8 Cyl. 428-CJ,	C6DZ-5796-A	C6DZ-5796-B	
	429-CJ A/T			
'70	8 Cyl. 428-CJ,	C8ZZ-5796-A	C8ZZ-5796-B	
	429-CJ M/T			
'70	Boss 302	C8ZZ-5796-A	C8ZZ-5796-B	

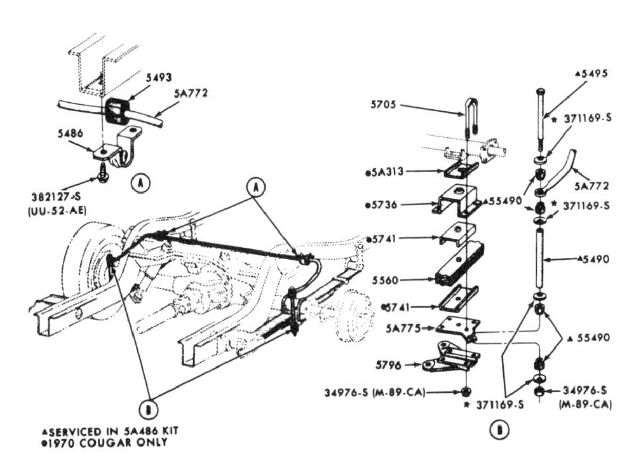

Rear sway bar and attaching hardware for 1970 Mustangs equipped with optional handling package.

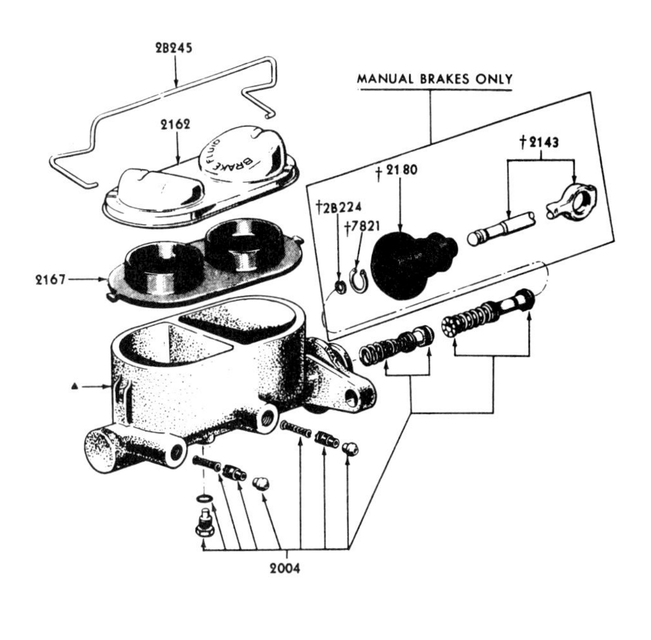

The 1967-1970 Mustang brake master cylinder with dual reservoir.

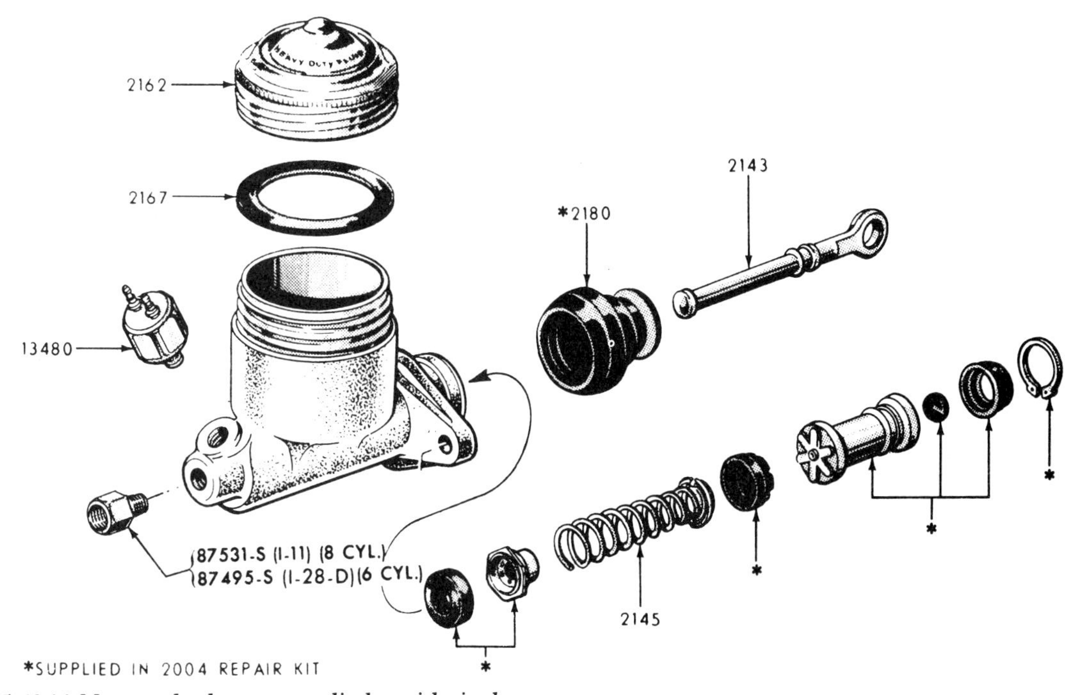

The 1965-1966 Mustang brake master cylinder with single reservoir.

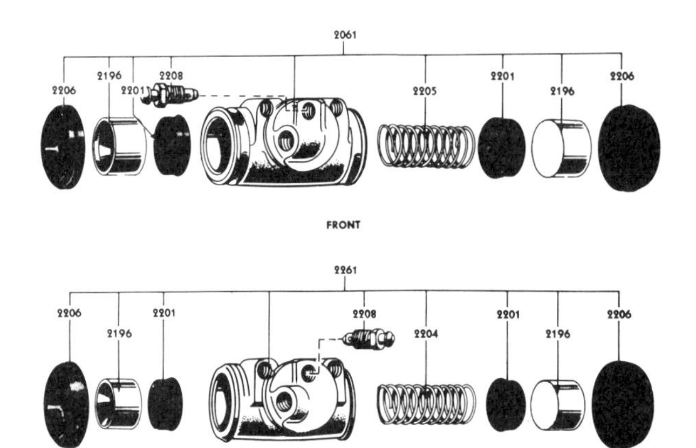

The 1966-1970 Mustang front and rear wheel cylinder assemblies.

MUSTANG FRONT BRAKE CYLINDER ASSEMBLY APPLICATION CHART

YEAR	MODEL	LOCATION	DIAMETER	PART NUMBER	NOTES
65/69	9" x 2-1/4" brakes	R.H.	1-1/16"	C5DZ-2061-A	
		L.H.		C5DZ-2061-A	
65/69	10" x 2-1/4" brakes	R.H.	1-1/8"	C3OZ-2061-B	
		L.H.		C3OZ-2062-B	
66/69	10" x 2-1/2" brakes	R.H.	1-3/32"	C6OZ-2061-A	1
				C9TZ-2062-B	2
		L.H.	7/8"	C6OZ-2062-A	1
				C9TZ-2062-B	2
70	9" x 2-1/4" brakes	R.H.	1-1/16"	C5DZ-2061-A	
		L.H.		C5DZ-2062-A	
70	10" x 2-1/4", or	R.H.	1-1/8"	C3OZ-2061-B	
	10" x 2-1/2" brakes	L.H.		C3OZ-2062-B	

NOTES:
1 Original
2 Service Replacement

MUSTANG REAR BRAKE CYLINDER ASSEMBLY APPLICATION CHART

YEAR	MODEL	LOCATION	DIAMETER	PART NUMBER	NOTES
65		R.H.	13/16"	C3DZ-2261-A	
		L.H.		C3DZ-2261-A	
65	6 cylinder	R.H.	29/32"	C5DZ-2261-B	
		L.H.		C5DZ-2261-B	
65/66	6 cylinder	R.H.	27/32"	C5DZ-2261-A	
		L.H.		C5DZ-2261-A	
66		R.H.	7/8"	C6OZ-2261-A	1
				D2ZZ-2261-A	2
		L.H.		C6OZ-2262-A	1
				D2ZZ-2261-A	2
65/66	8 cylinder	R.H.	29/32"	C3OZ-2261-B	
		L.H.		C3OZ-2262-B	1
				D2ZZ-2262-B	2
67/68	8 cylinder 390-4/B	R.H.	13/16"	C7ZZ-2262-A	
		L.H.		C7ZZ-2262-A	
67/69	6 cylinder	R.H.	27/32"	C5DZ-2261-A	
		L.H.		C5DZ-2261-A	
67/69	8 cyl. 289-4/B-GT	R.H.	7/8"	C6OZ-2262-A	1
				D2ZZ-2262-A	2
		L.H.		C6OZ-2262-A	1
				D2ZZ-2262-A	2
67/69		R.H.	7/8"	C6OZ-2261-A	1
				D2ZZ-2261-A	2
		L.H.		C6OZ-2262-A	1
				D2ZZ-2262-A	2
70	6 cylinder 200	R.H.	27/32"	C5DZ-2261-A	
		L.H.		C5DZ-2261-A	
70	6 cyl. 250, 8 cyl. 302	R.H.	7/8"	C6OZ-2261-A	1
				D2ZZ-2261-A	2
		L.H.		C6OZ-2262-A	1
				D2ZZ-2262-A	2
70	8 cylinder 351, 428	R.H.	29/32"	C3OZ-2262-B	1
				D2ZZ-2262-B	2
		L.H.		C3OZ-2262-B	1
				D2ZZ-2262-B	2

NOTES:
1 * Original
2 ^ Service replacement

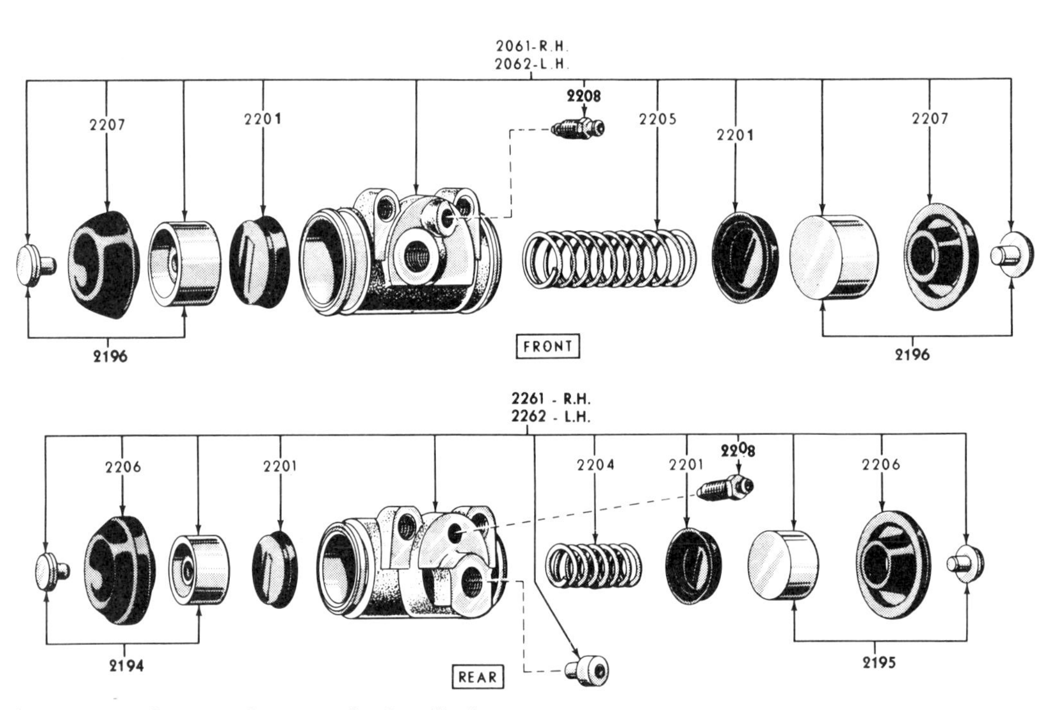

The 1965 Mustang front and rear wheel cylinder assemblies.

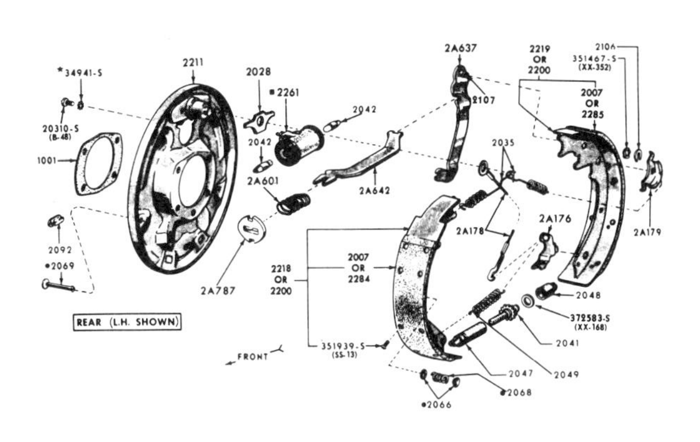

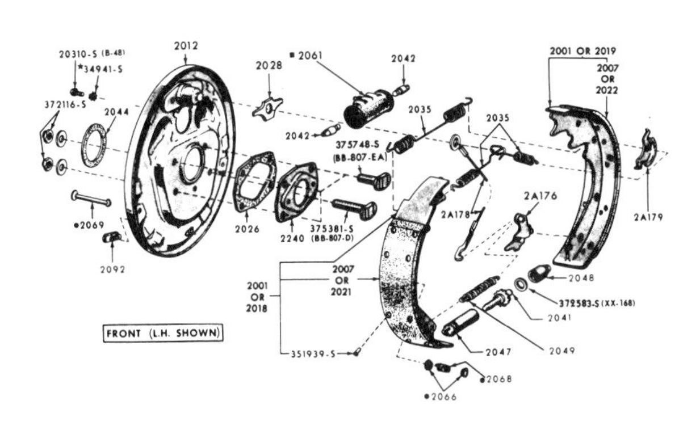

The 1965-1970 Mustang front and rear drum brake assemblies.

By 1967 the chassis had been redesigned and the new body was designed to accept a common brake booster and new dual reservoir master cylinder. The brake system was split to separate the front brakes from the rear brakes hence the reason for the two-chamber master cylinder. The split system provided two wheel brakes if either the front or the rear circuits failed.

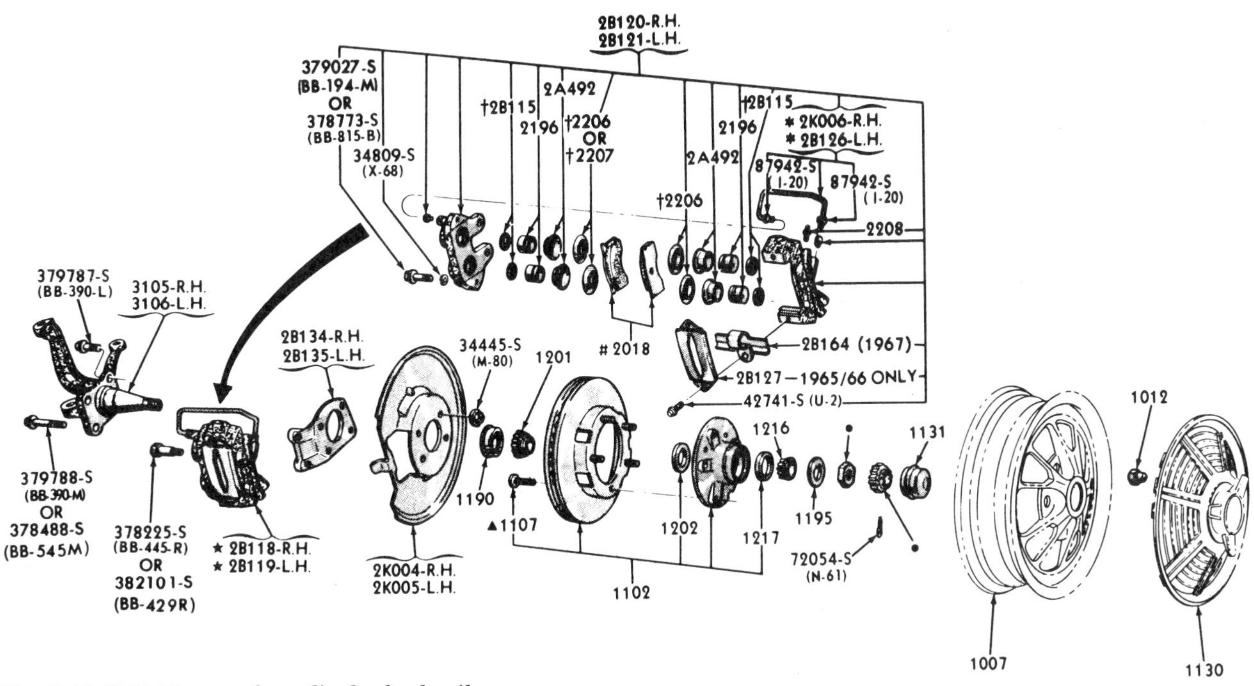

The 1965-1967 Mustang front disc brake details.

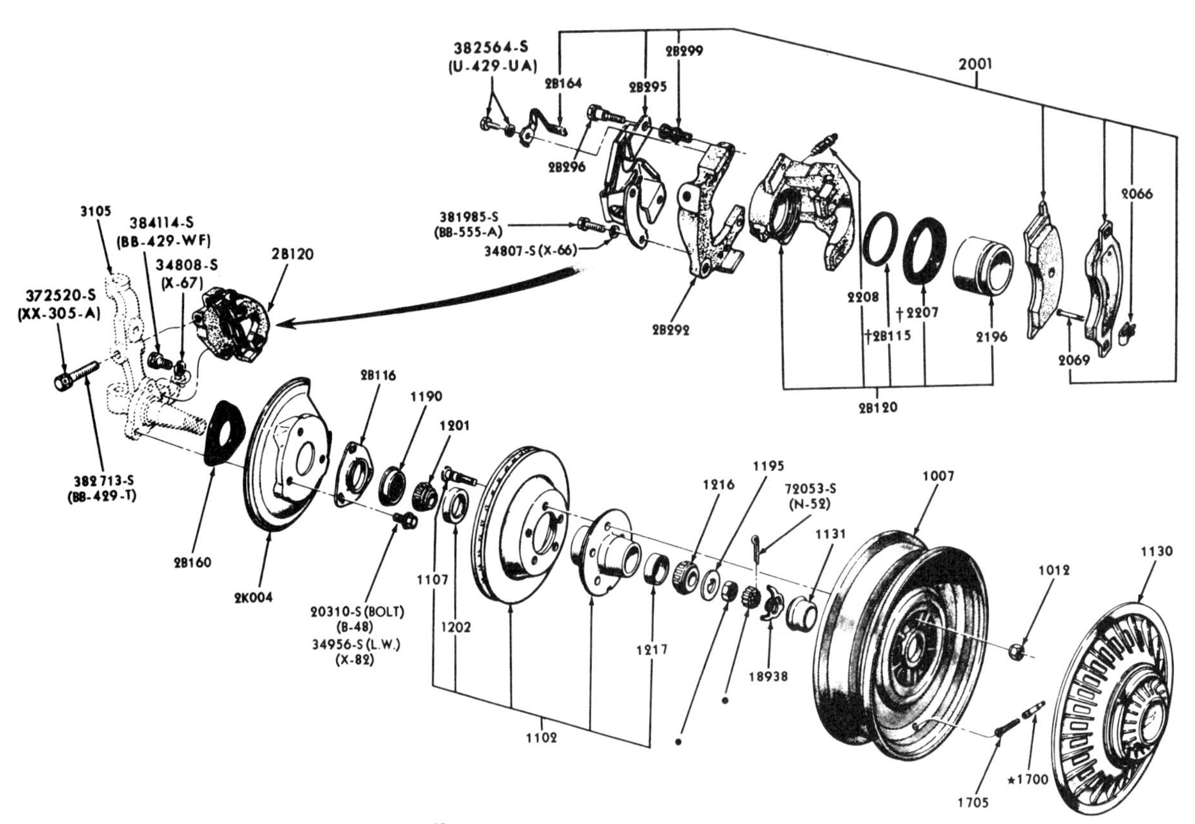

The 1968-1970 Mustang front disc brake details.

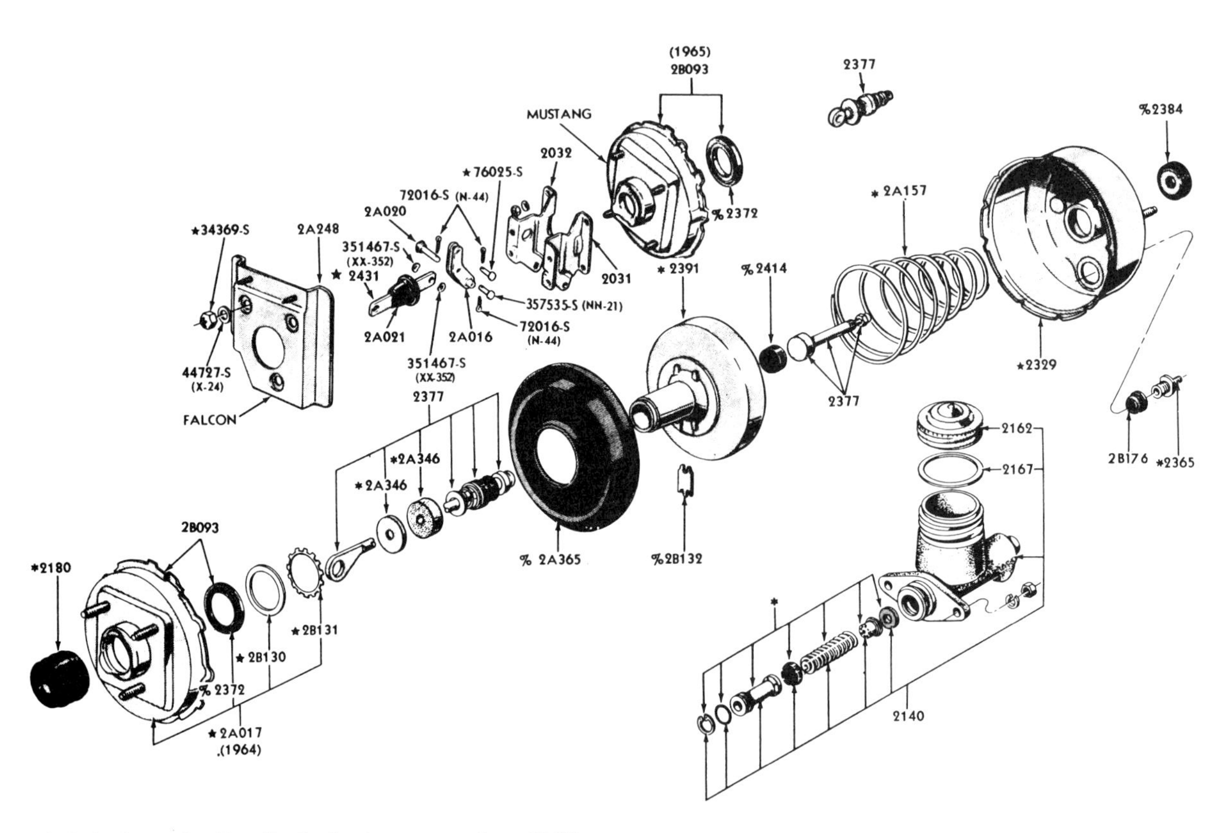

Exploded view of a Bendix brake booster used on 1965-1966 Mustangs.

As we mentioned earlier, the first Mustang was adapted from the Falcon. Neither car was engineered for power brakes because there was not enough distance provided between the left shock tower and the firewall for a brake booster and master cylinder. So, in an effort to retrofit the old chassis, an elaborate bracket and lever assembly was used to adapt a compact Bendix booster and single piston master cylinder. To clear the shock tower, the bracket positioned the assembly higher than stock and angled it slightly upward. The linkage connected the remote booster to the brake pedal. At $42.29, power brakes was not a popular option in 1966.

POWER BRAKE BOOSTER INSTALLATION KIT
Retrofit for Mustangs

YEAR	MODEL					
65/66	**Drum brakes**					**C6ZZ-2A091-A**
	Consists of:					
	C6ZA-2A040-C	Tube	C6ZA-2420-A	Tube	(2) 376428-S8	Clamp
	B9SS-2A047-A	Hose	C6AZ-9A474-A	Fitting	(2) 376977-S2	Screw, washer
	C6ZA-2B195-A	Booster, M. cyl.	C6AZ-9A474-B	Fitting	380613-S	Cap
	C6OA-2420-D	Tube			380614-S	Cap
67	**Drum brakes without collapsible steering column**					**C7ZZ-2A091-A**
	Consists of:					
	C7ZA-2A040-F	Tube	C7ZZ-2455-B	Pedal (M/T)	(4) 55748-S	Locknut
	C1AA-2A047-D	Hose	C7ZZ-2455-K	Pedal (A/T)	351053-S	Strap
	C7ZA-2A188-A	Spacer	(2) B7AZ-2471-A	Bushing	353574-S7	Washer
	C7WA-2B195-E	Booster, M. cyl.	C7OA-7B086-A	Connector	(2) 376428-S8	Clamp
	C7ZA-2B253-E	Tube	C6AZ-9A474-A	Fitting	376977-S2	Screw,washer
	C6OA-2420-D	Tube	C6AZ-9A474-B	Fitting	380613-S	Cap
	C70A-2420-A	Tube	34419 - S2	Locknut	380614-S	Cap
	C7ZA-2420-A	Tube	(4) 44728-S	Washer		Installation sketch
	C7ZA-2420-B	Tube	45534-S2	Bolt		
68	**Drum brakes**					**C8ZZ-2AO91-A**
	Consists of:					
	C8WY-2OO5-A	Booster	C8AA-2420-G	Tube	45534-S2	Bolt
	C8ZA-2A040-E	Tube	C8ZZ-2455-A	Pedal (A/T)	(4) 55748-S	Locknut
	C8AA-2A047-A	Hose	C8ZZ-2455-C	Pedal (M/T)	351053-S	Strap
	C8ZA-2169-A	Piston	(2) 33799-S2	Nut	353574-S7	Washer
	C7ZA-2A188-A	Spacer	34419-S2	Locknut	353777-S	Connector
	C8ZA-2B253-E	Tube	(2) 34807-S2	Lockwasher	(2) 376428-S8	Clamp
			(4) 44728-S	Washer	376977-S2	Screw,washer
69	**Drum brakes**			With manual transmission use C9ZZ-2455-D in place of kit pedal.		**C9ZZ-2A091-A**
	Consists of:					
	C9ZJ-2005-B	Booster	C8ZA-2B253-E	Tube	353777-S	Connector
	C9ZZ-2B022-A	Gasket	C8AA-2420-G	Tube	(2) 376428-S8	Clamp
	C8ZA-2A040-E	Tube	C9ZZ-2455-E	Pedal (A/T)	376977-S2	Screw,washer
	C8ZZ-2A047-A	Hose	56349-S4	Bolt	(6) 377706-S	Locknut
	C9AZ-2169-F	Piston	351053-S	Strap	380699-S100	Pin
	C7ZA-2A188-A	Spacer	35374-S7	Washer	382802-S100	Locknut
70	**Drum brakes**			With man. trans., use D1ZZ-2455-A in place of kit pedal. Not replaced.		**DOZZ-2A091-A**
	Consists of:					
	C9ZJ-2005-B	Booster	C7ZA-2A188-A	Spacer	353574-S7	Washer
	C9ZZ-2B022-A	Gasket	DOZJ-2B253-A	Tube	(2) 376428-S8	Clamp
	DOZJ-2A040-A	Tube	C9ZZ-2455-A	Pedal (A/T)	376977-S2	Screw,washer
	B9SS-2A047-A	Hose	DOZA-3A762-A	Clamp	(6) 377706-S	Locknut
	C9AZ-2169-F	Piston	56349-S4	Bolt	380699-S100	Pin
					382802-S100	Locknut

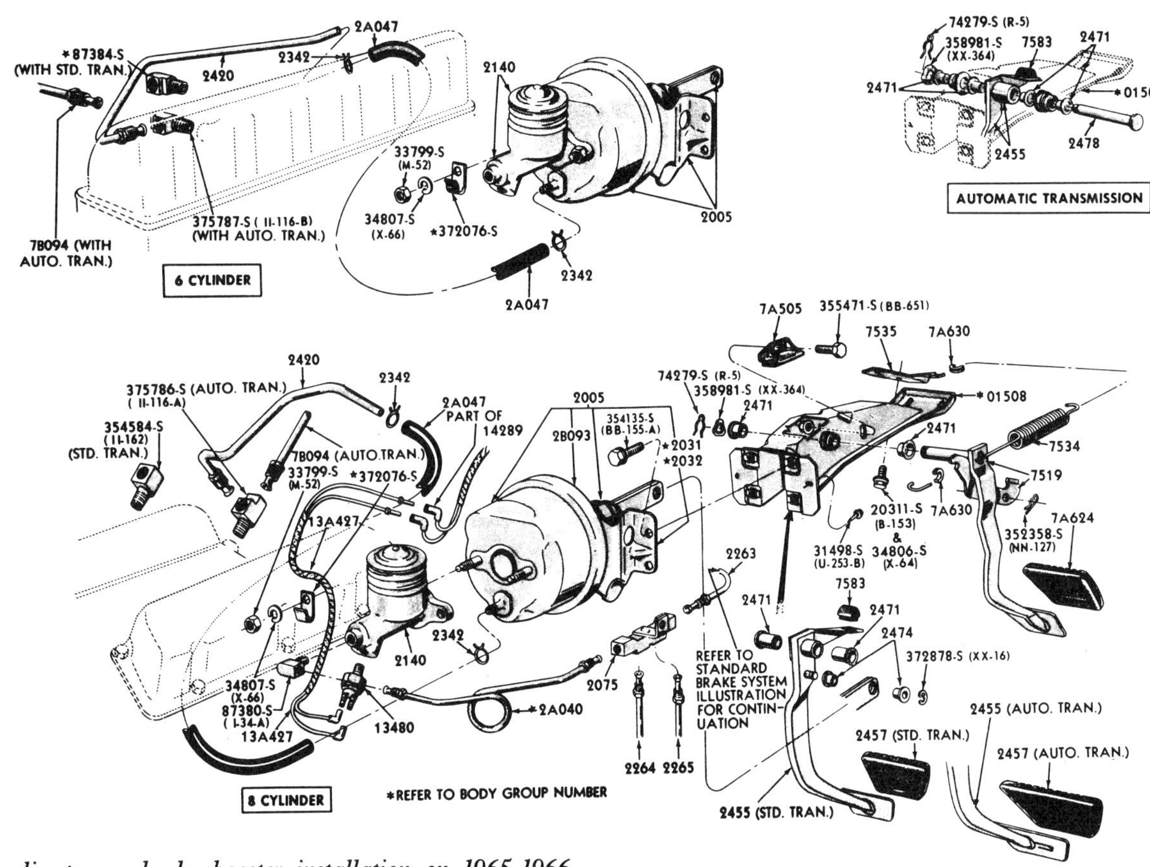

Bendix power brake booster installation on 1965-1966 Mustangs.

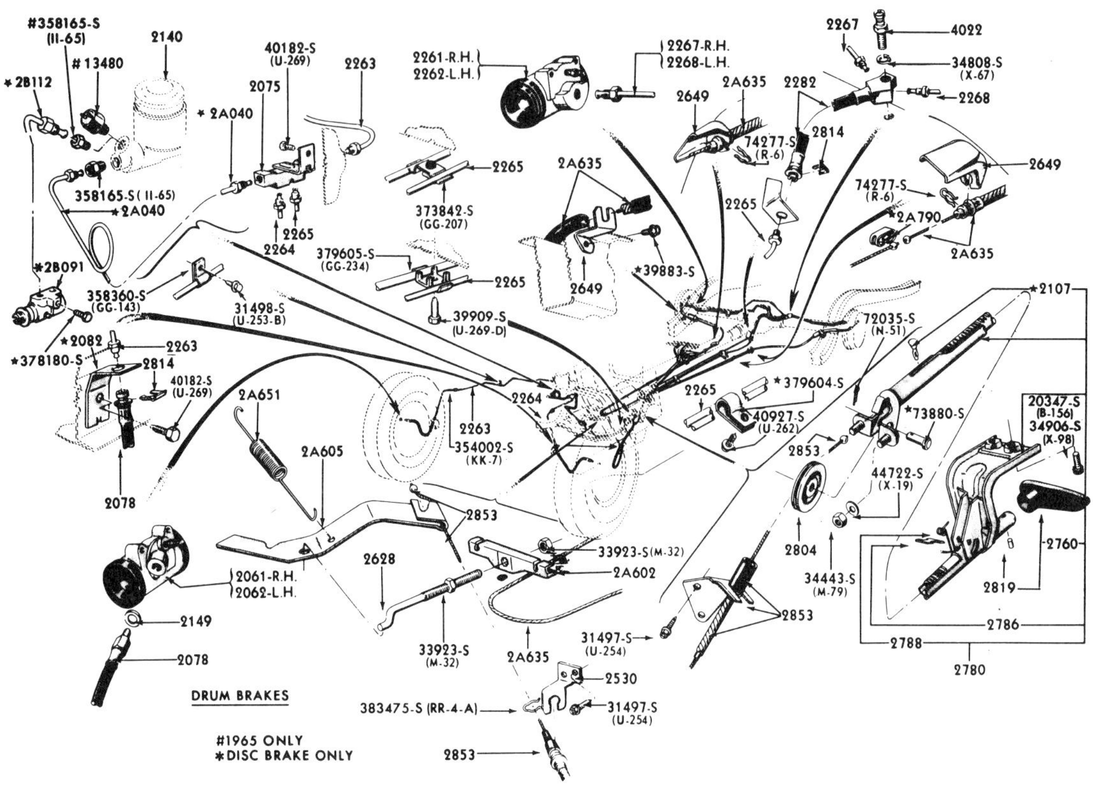

Exploded view of 1965-1966 Mustang disc and drum brake systems.

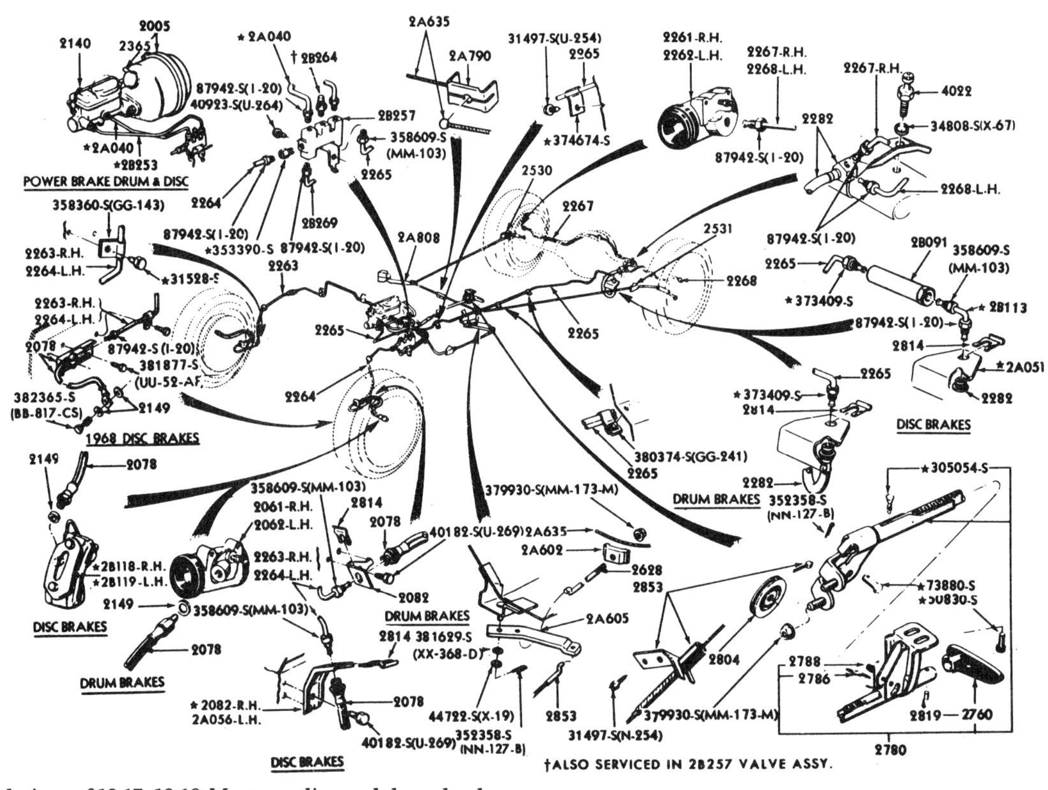

Exploded view of 1967-1968 Mustang disc and drum brake systems.

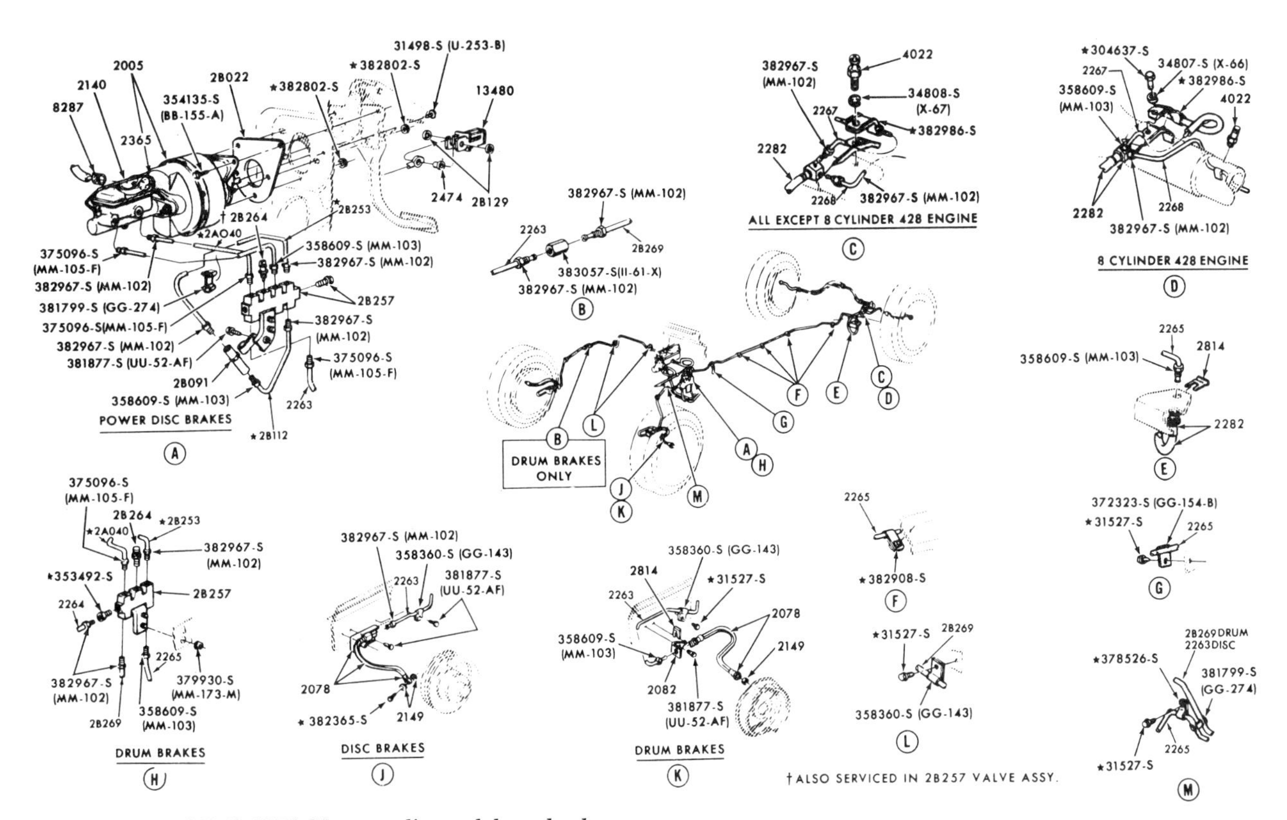

Exploded view of 1969-1970 Mustang disc and drum brake systems.

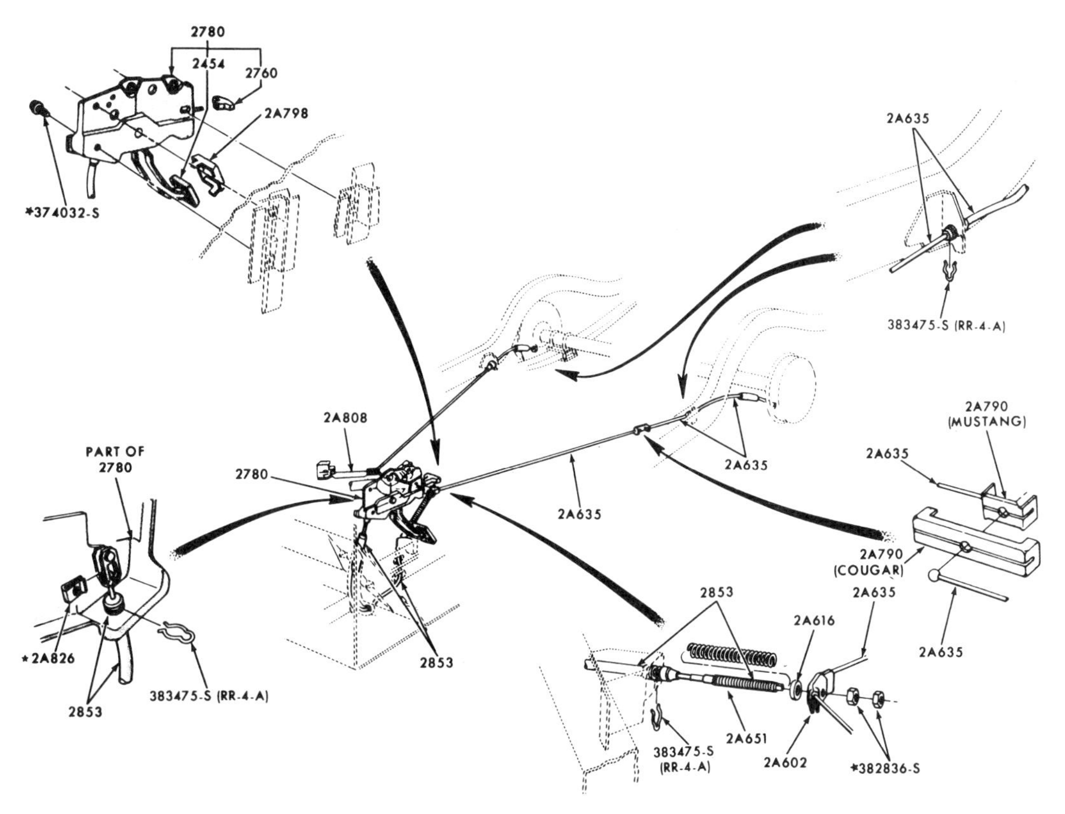

The 1969-1970 Mustang parking brake pedal and cable assembly.

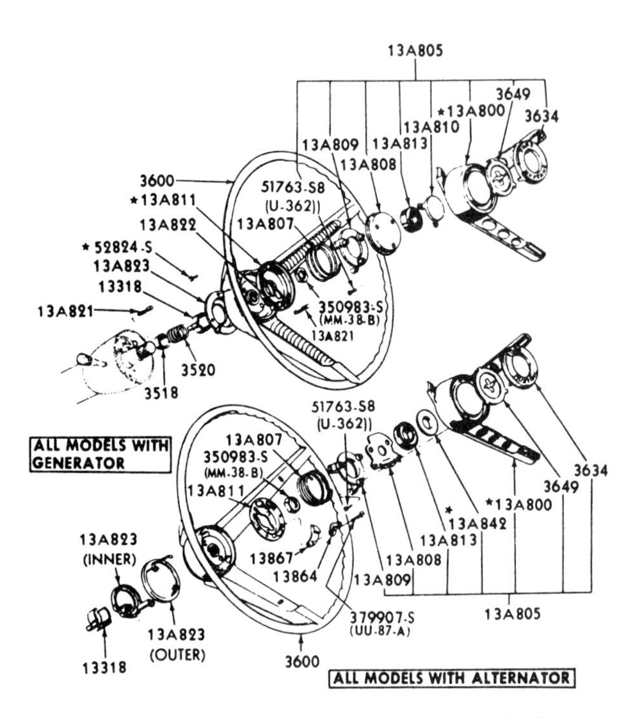

The 1965-1966 Mustang standard steering wheel, horn button, and related parts.

A bulbous crash pad, part of the new government-imposed safety standards, adorns the steering wheel horn ring in 1967.

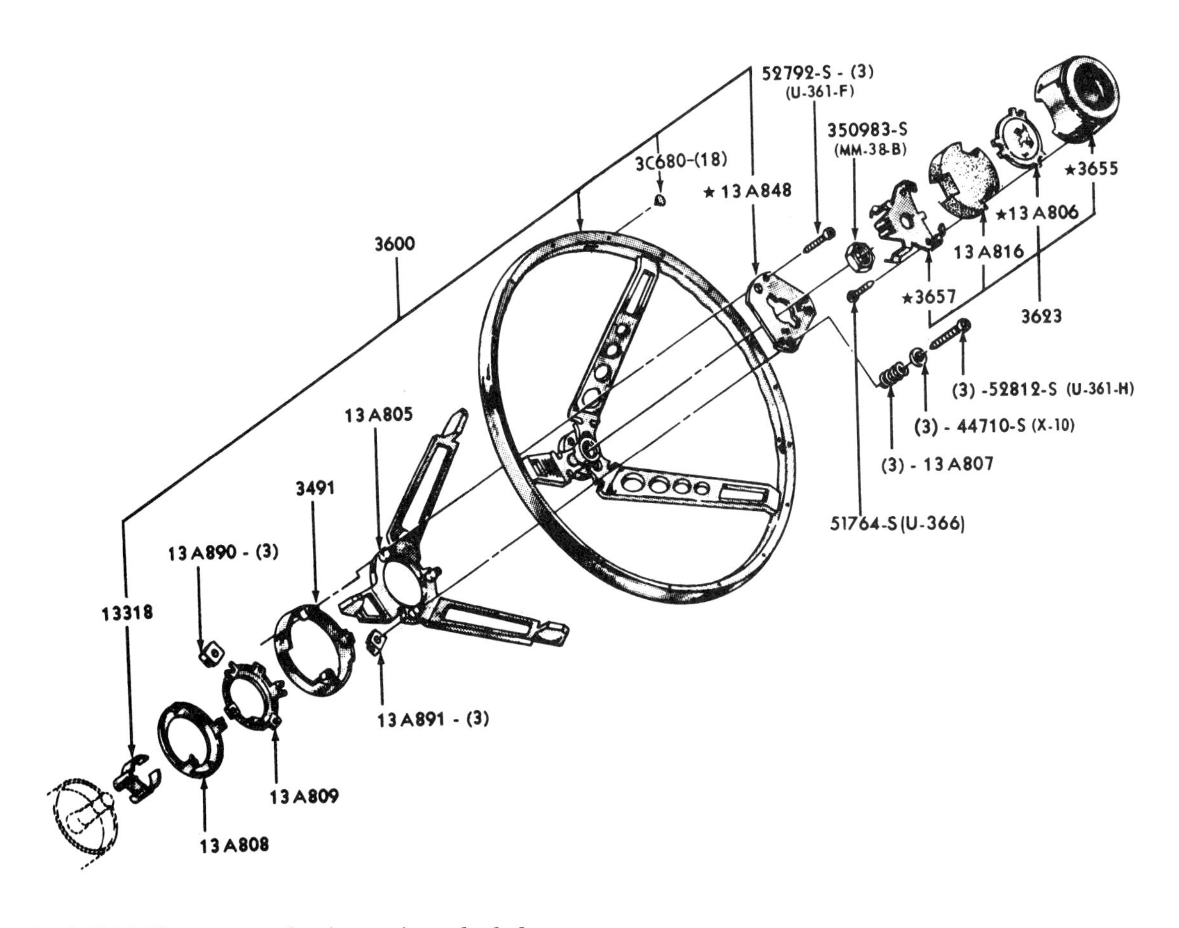

The 1965-1966 Mustang woodgrain steering wheel, horn button, and related parts.

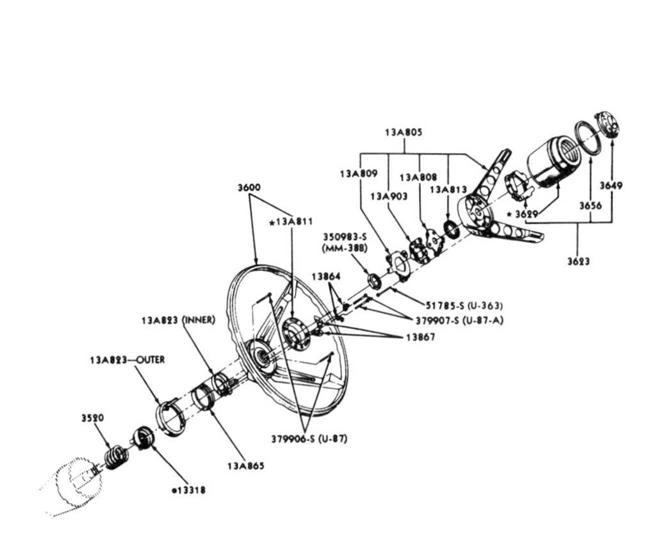

The 1967 Mustang standard three-spoke steering wheel, horn button, and related parts.

A Rim Blow steering wheel was a Mustang option in 1969 and 1970. The horn would sound when a rubber strip, encircling the inside diameter of the wheel, was pressed. This three-spoke center pad is part of the standard interior package.

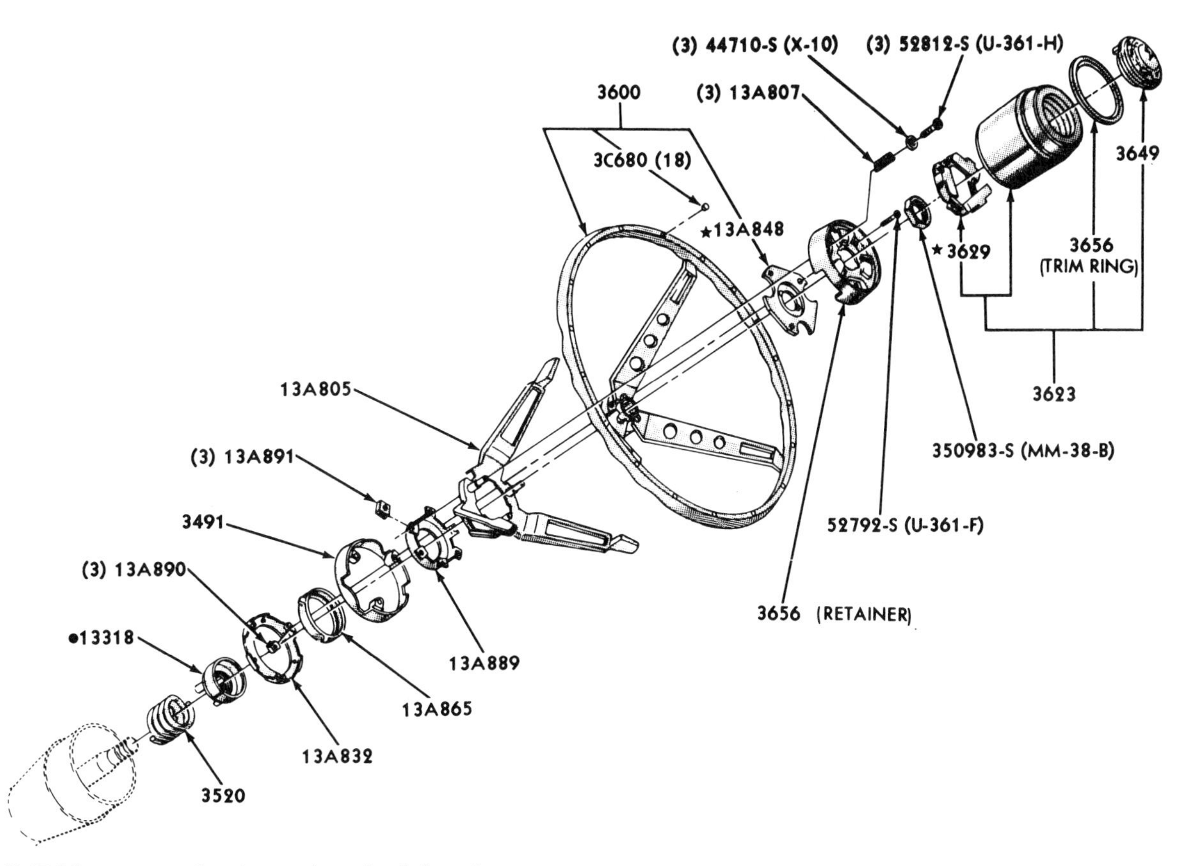

The 1967 Mustang woodgrain steering wheel, horn button, and related parts.

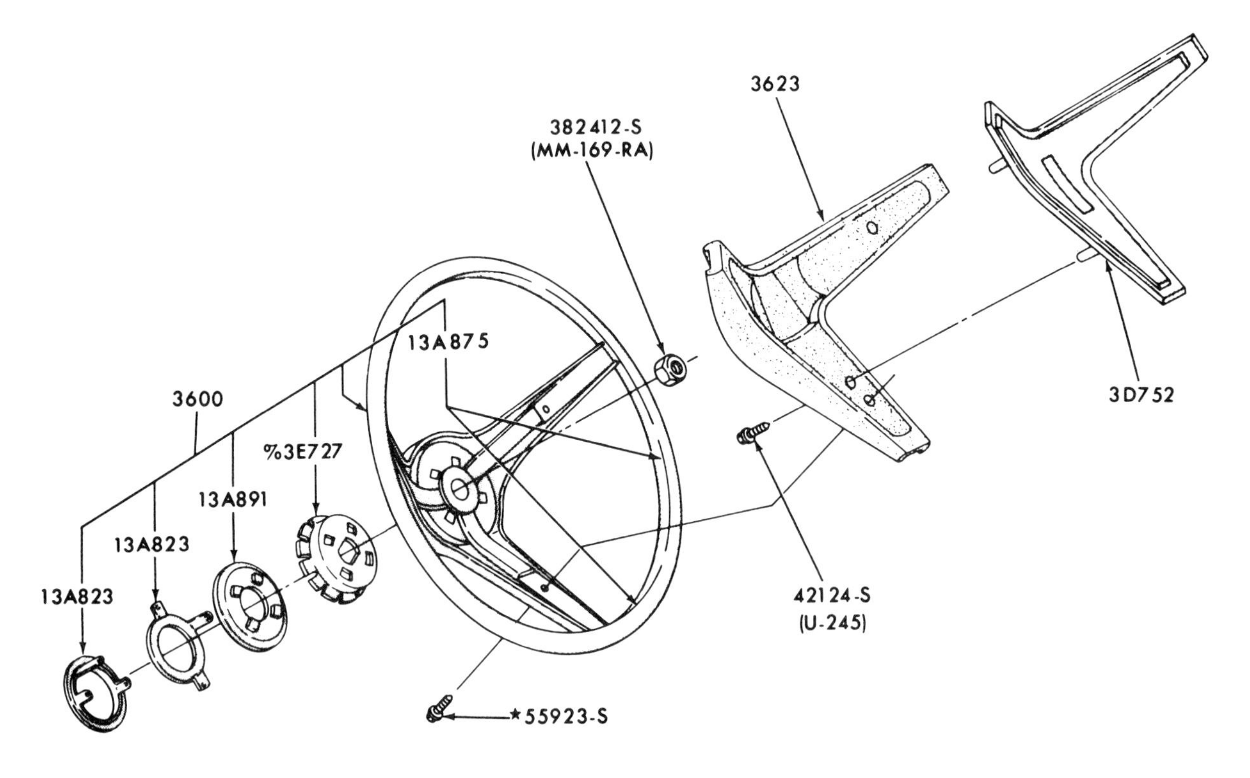

The 1969-1970 Shelby Mustang three-spoke steering wheel with Rim Blow horn.

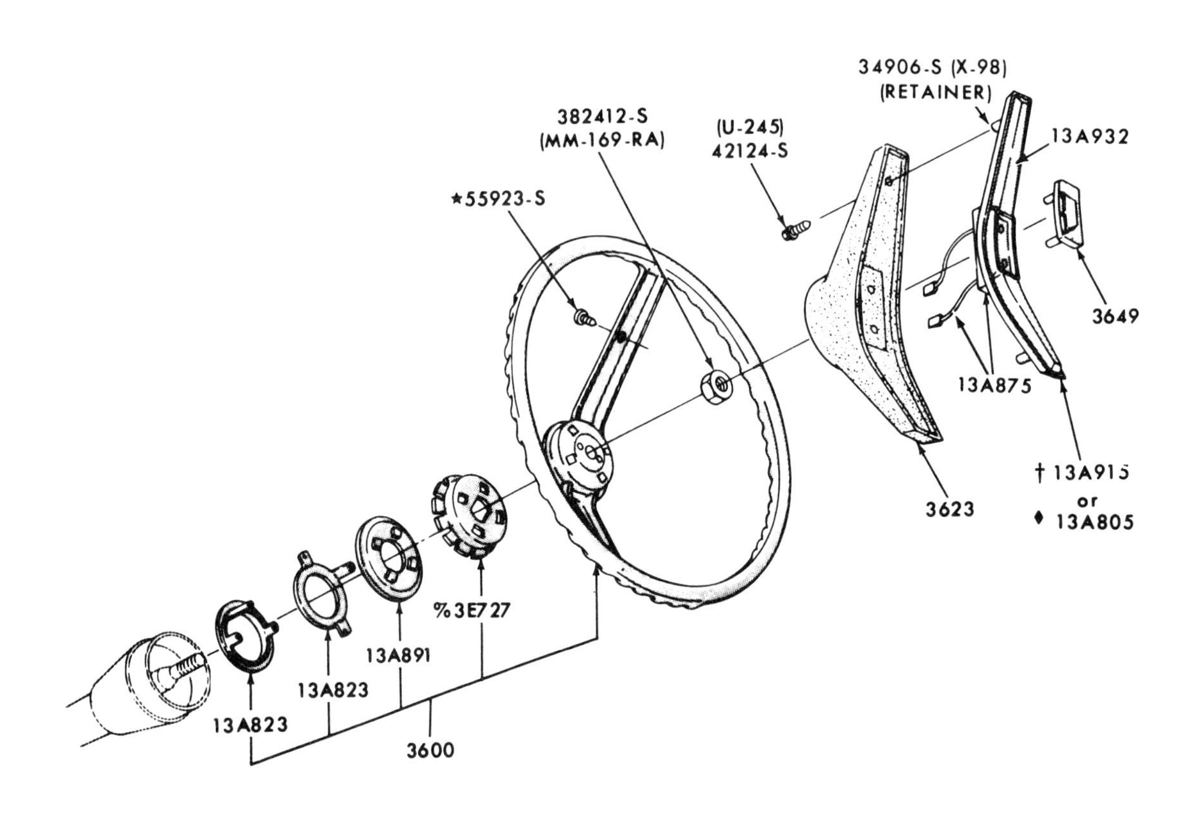

The 1969-1970 Mustang standard two-spoke steering wheel.

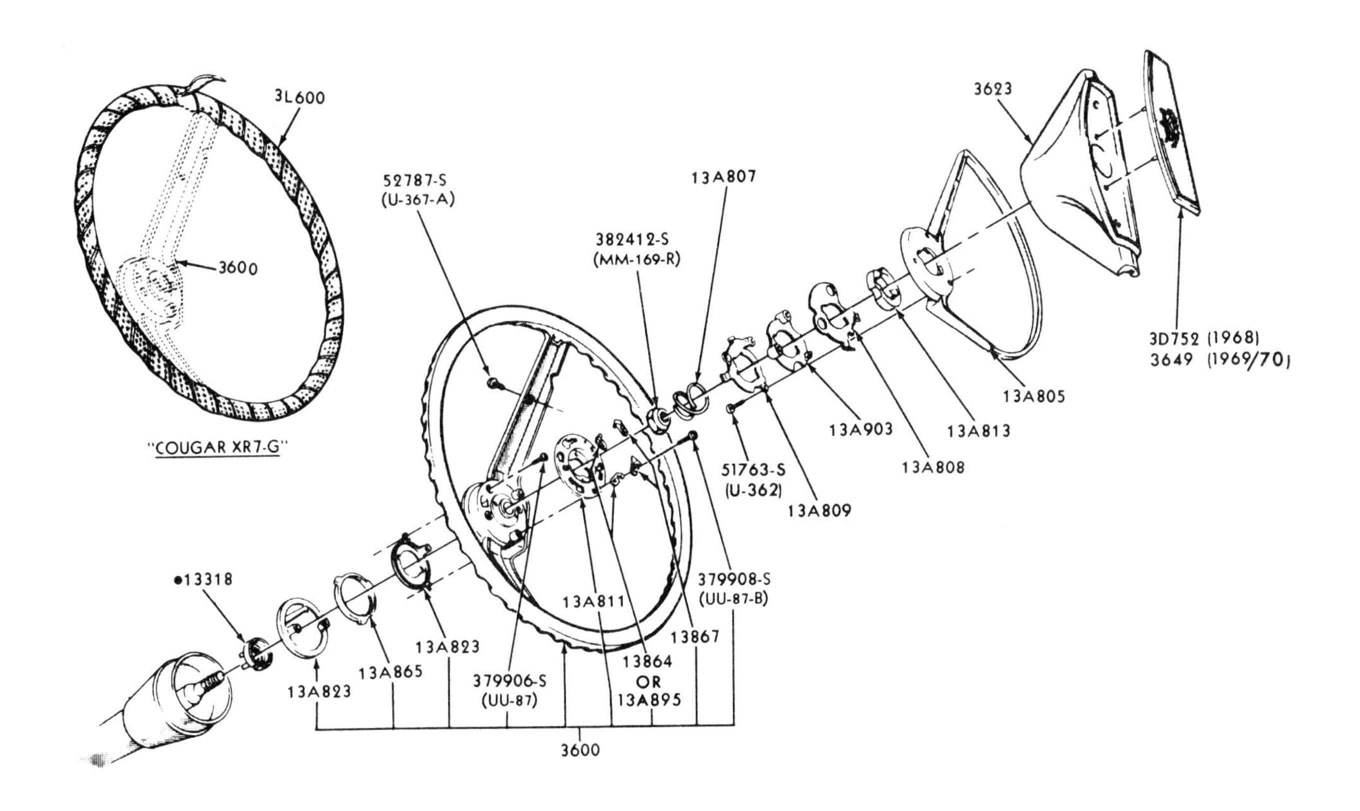

The 1968 Mustang two-spoke steering wheel, horn ring, and related parts.

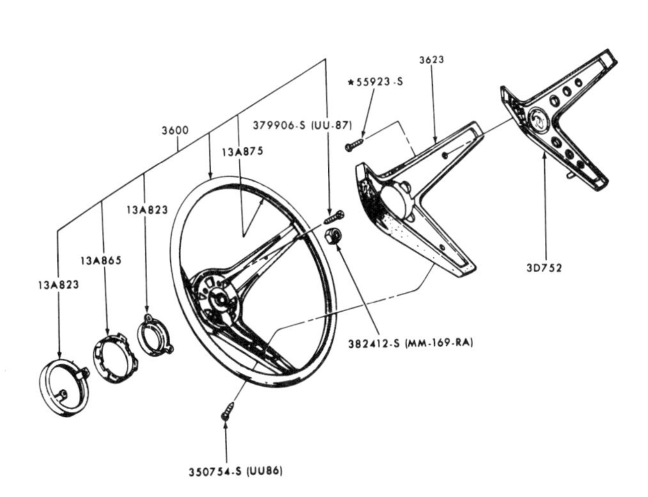

The 1969 Mustang three-spoke steering wheel with Rim Blow horn.

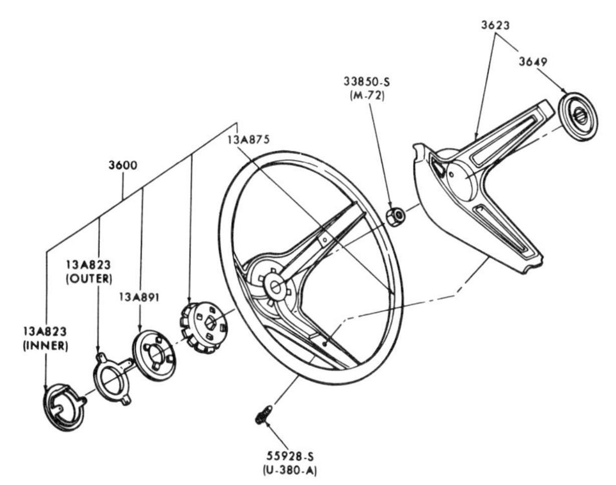

The 1969-1970 Mustang three-spoke steering wheel with Rim Blow horn.

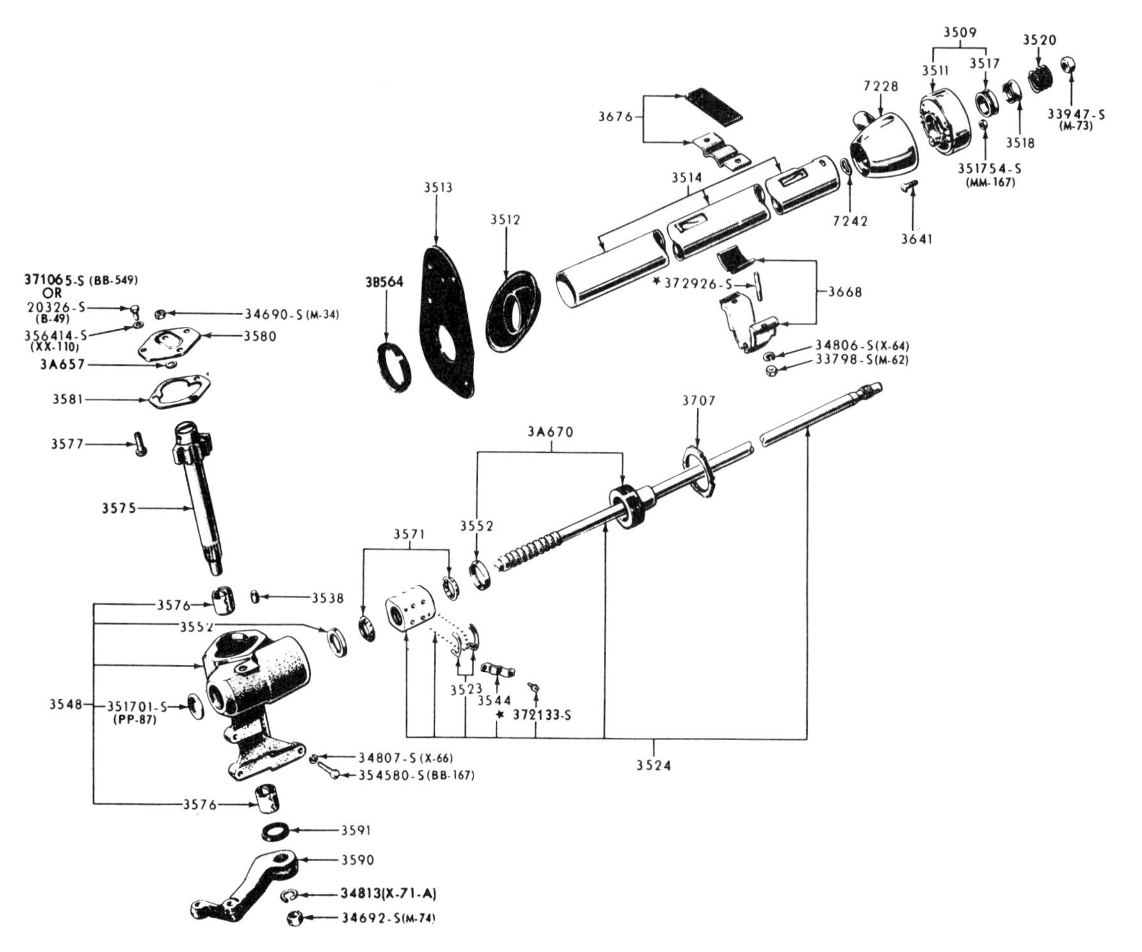

An exploded view of a typical 1965 Mustang steering column and gearbox. Note the steering shaft (3524) extends the length of the steering column. The steering wheel attaches to the splined upper end. When installed, the wheel, column, and gearbox become a complete assembly.

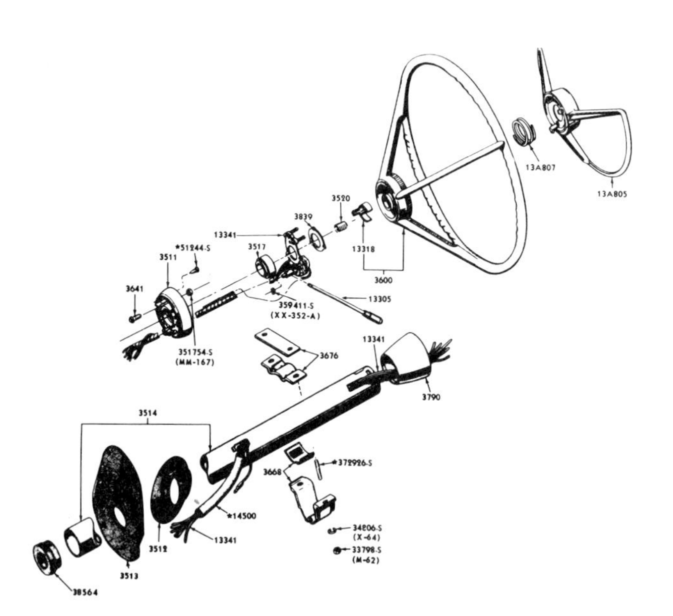

Steering column and wheel for 1965 and 1966 Mustangs with standard interior.

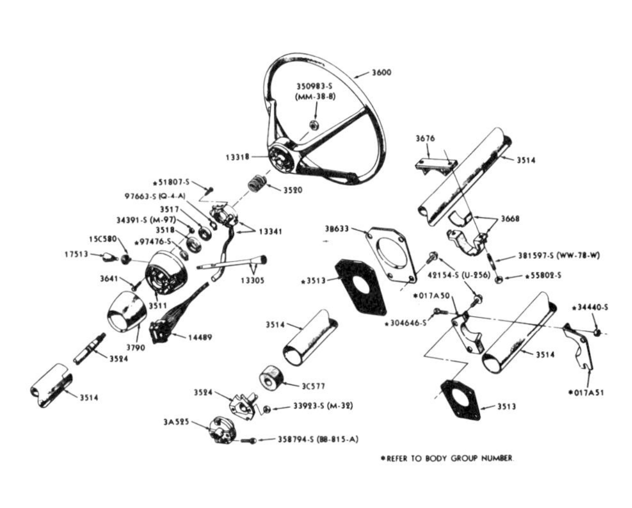

Steering column assembly and related parts for 1967 Mustangs with fixed wheel (as opposed to tilt wheel).

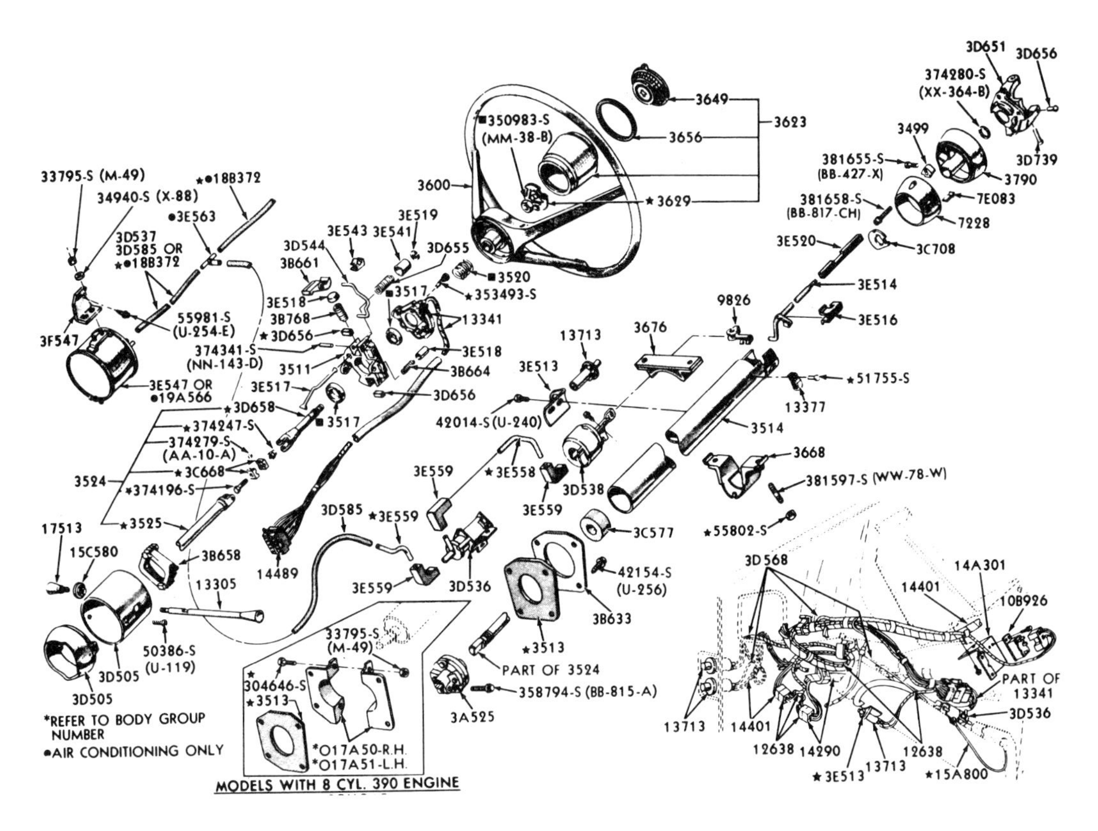

Steering column assembly and related parts for 1967 Mustangs with tilt wheel.

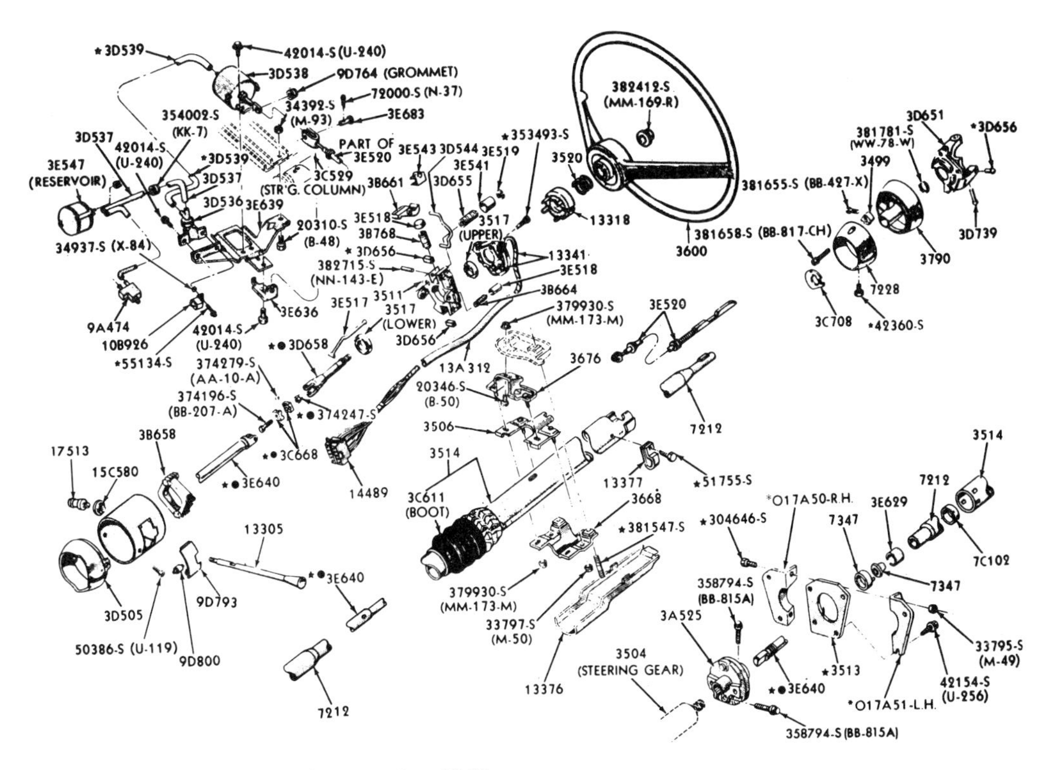

Steering column assembly and related parts for 1968 Mustangs with tilt wheel.

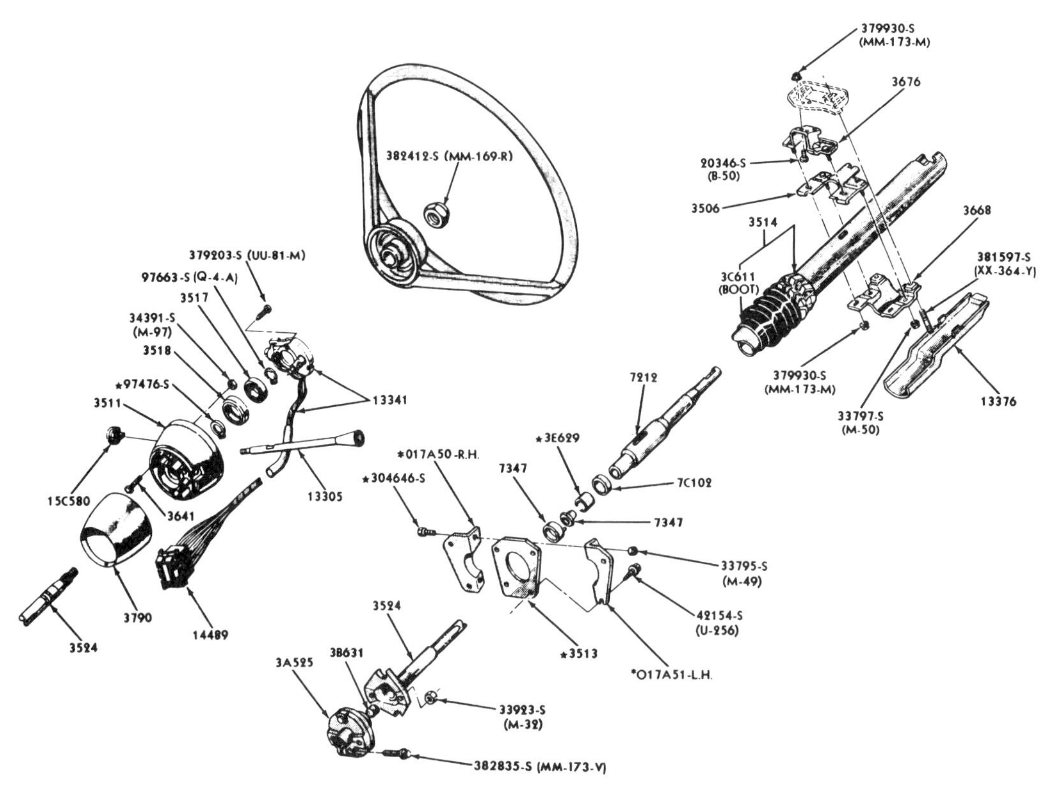

Steering column assembly and related parts for 1968 Mustangs with fixed wheel (as opposed to tilt wheel).

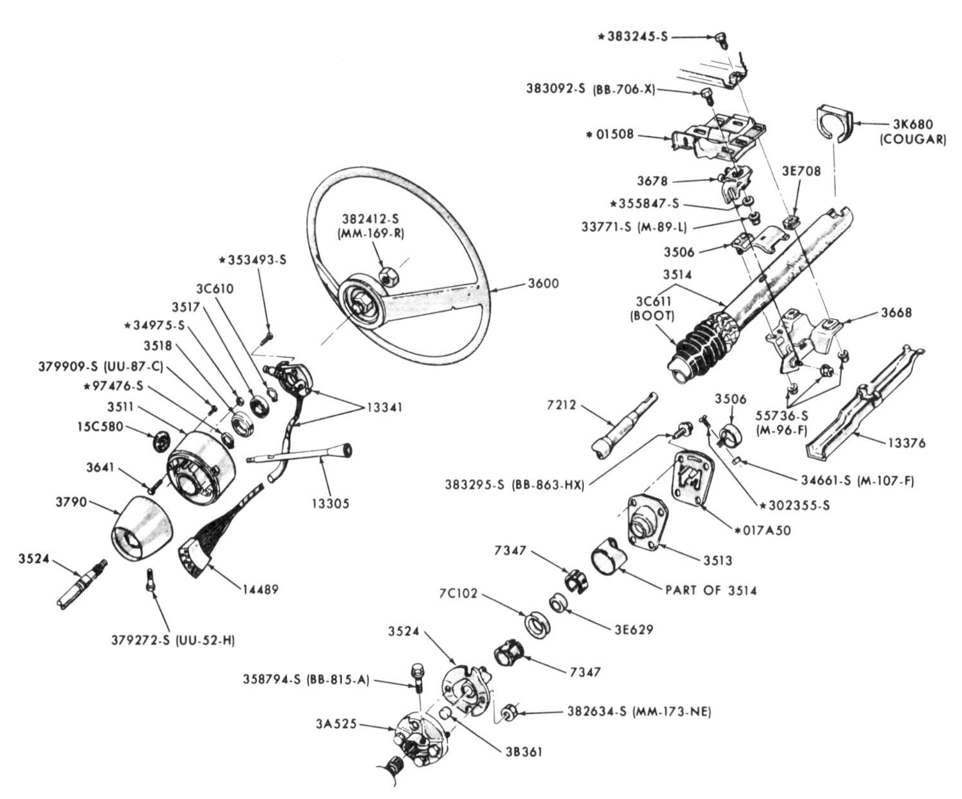

Steering column assembly and related parts for 1969 Mustangs with fixed wheel (as opposed to tilt wheel).

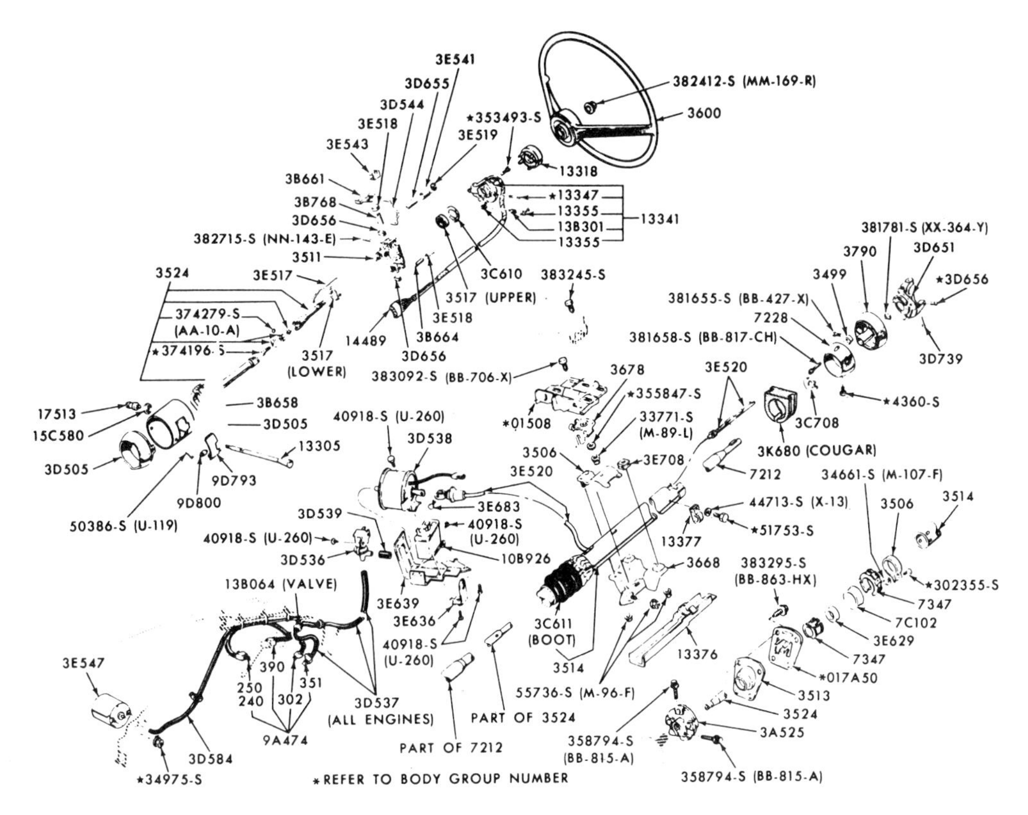

Steering column assembly and related parts for 1969 Mustangs with tilt wheel.

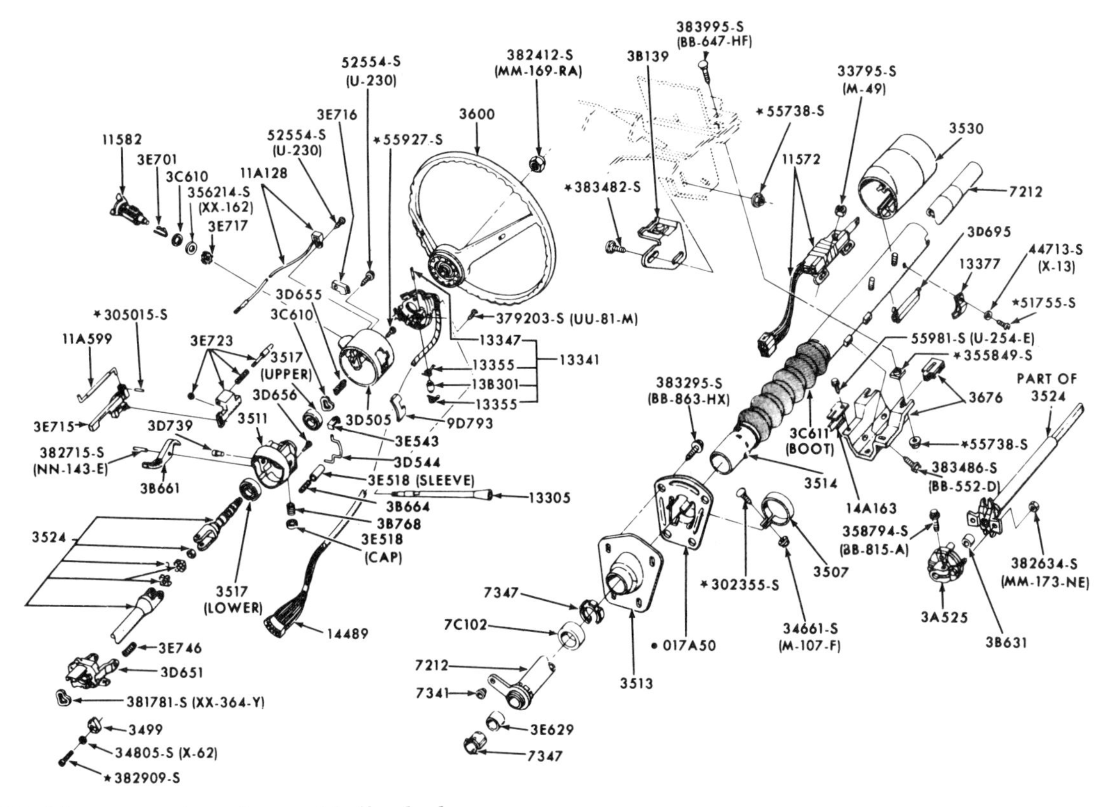

The 1970 Mustang steering column with tilt wheel.

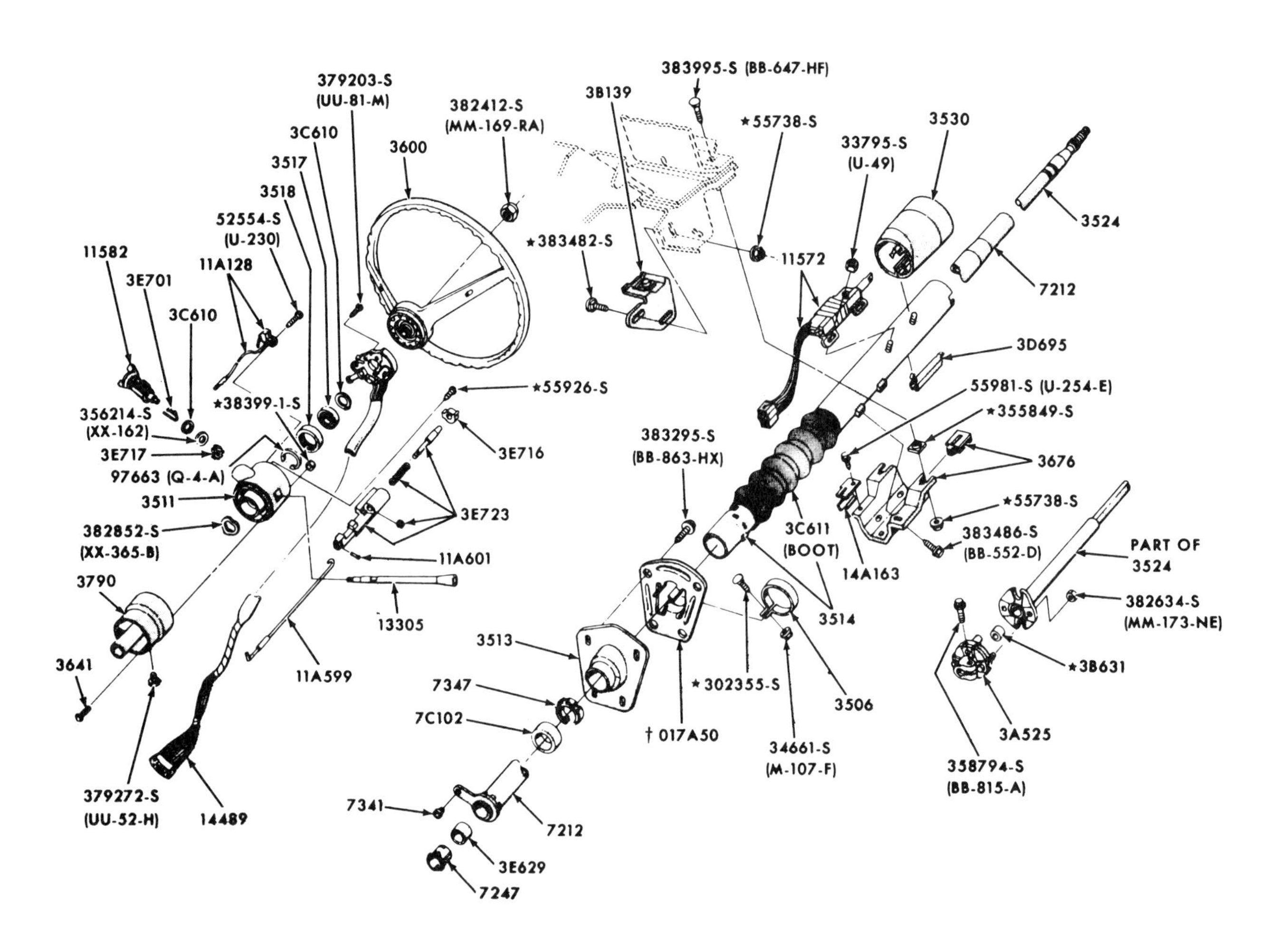

Steering column assembly and related parts for 1970 Mustangs with fixed wheel (as opposed to tilt wheel).

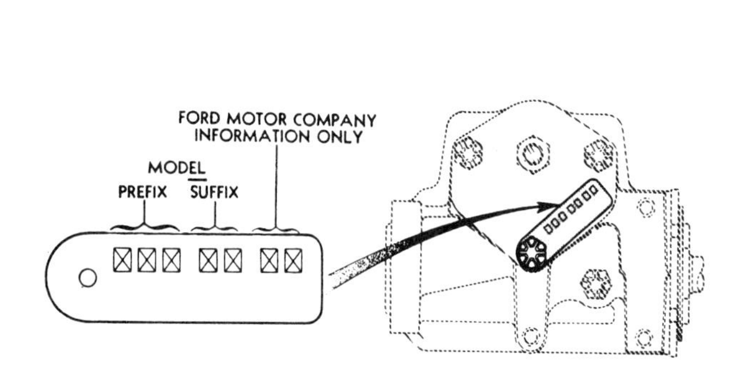

The model prefix and suffix are listed on the stamped identification tag attached to the steering gearbox. The information can be used to validate a unit's correct application and date of manufacture.

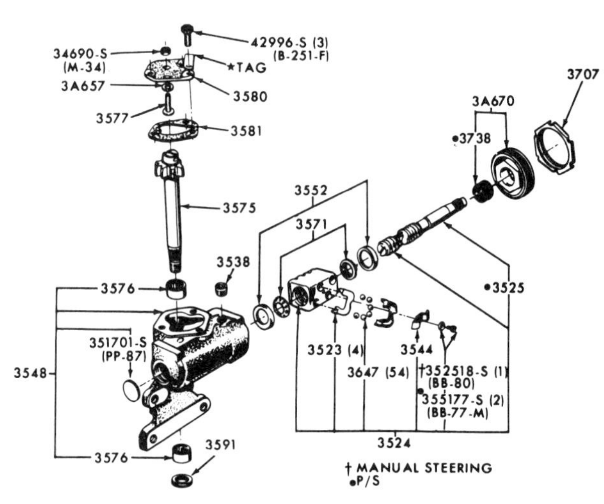

An exploded view of a 1965-1970 steering gearbox. The only difference betwen a power and non-power steering box is the gear ratio.

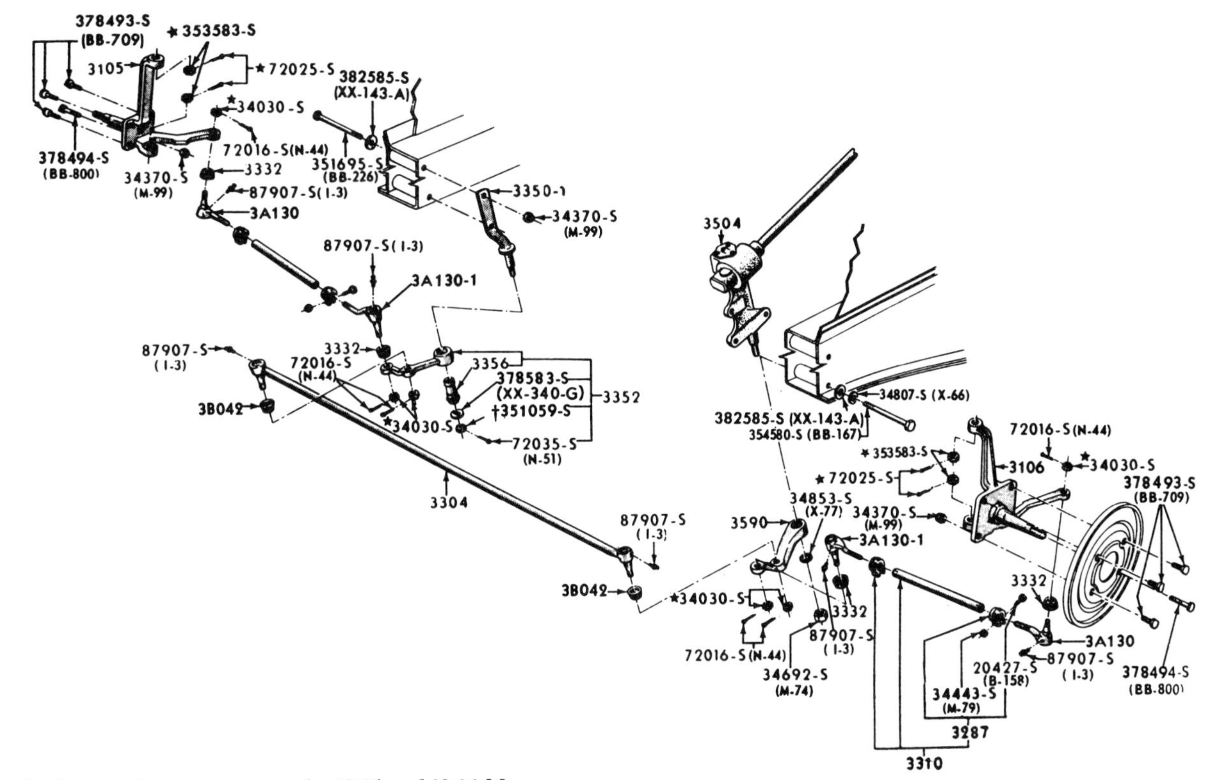

Typical manual steering system for 1965 and 1966 Mustangs equipped with six-cylinder engines.

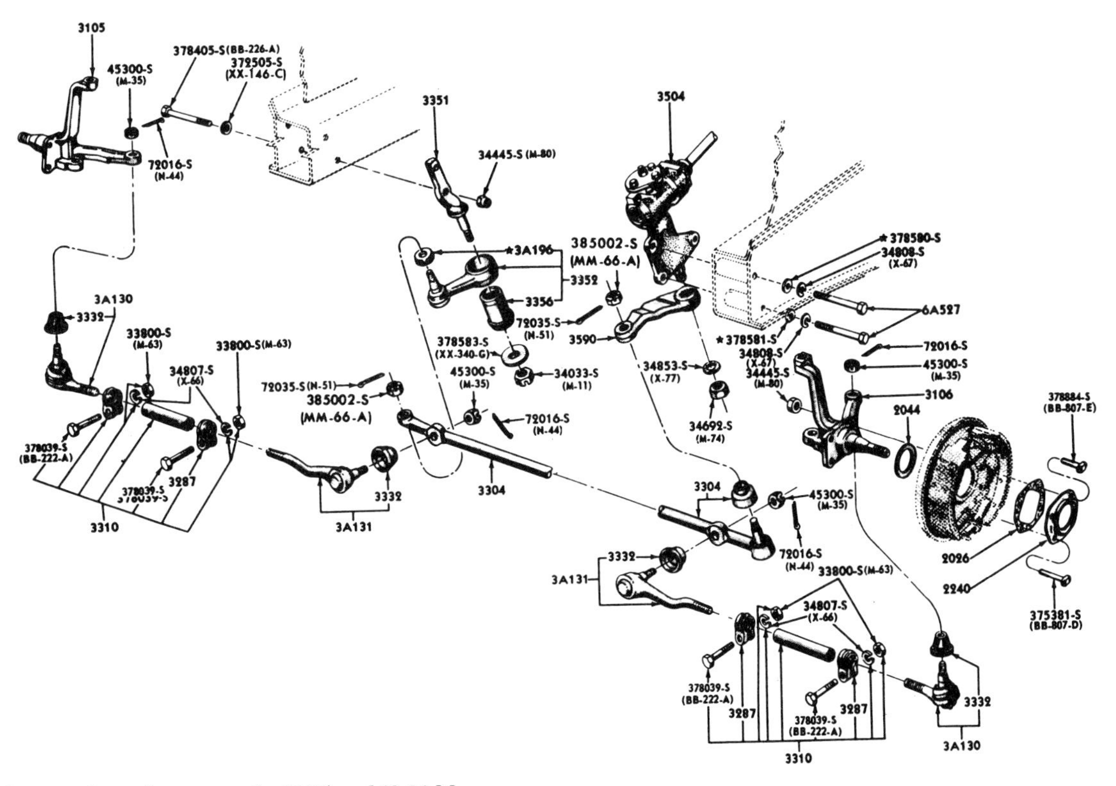

Typical manual steering system for 1965 and 1966 Mustangs equipped with V-8 engines.

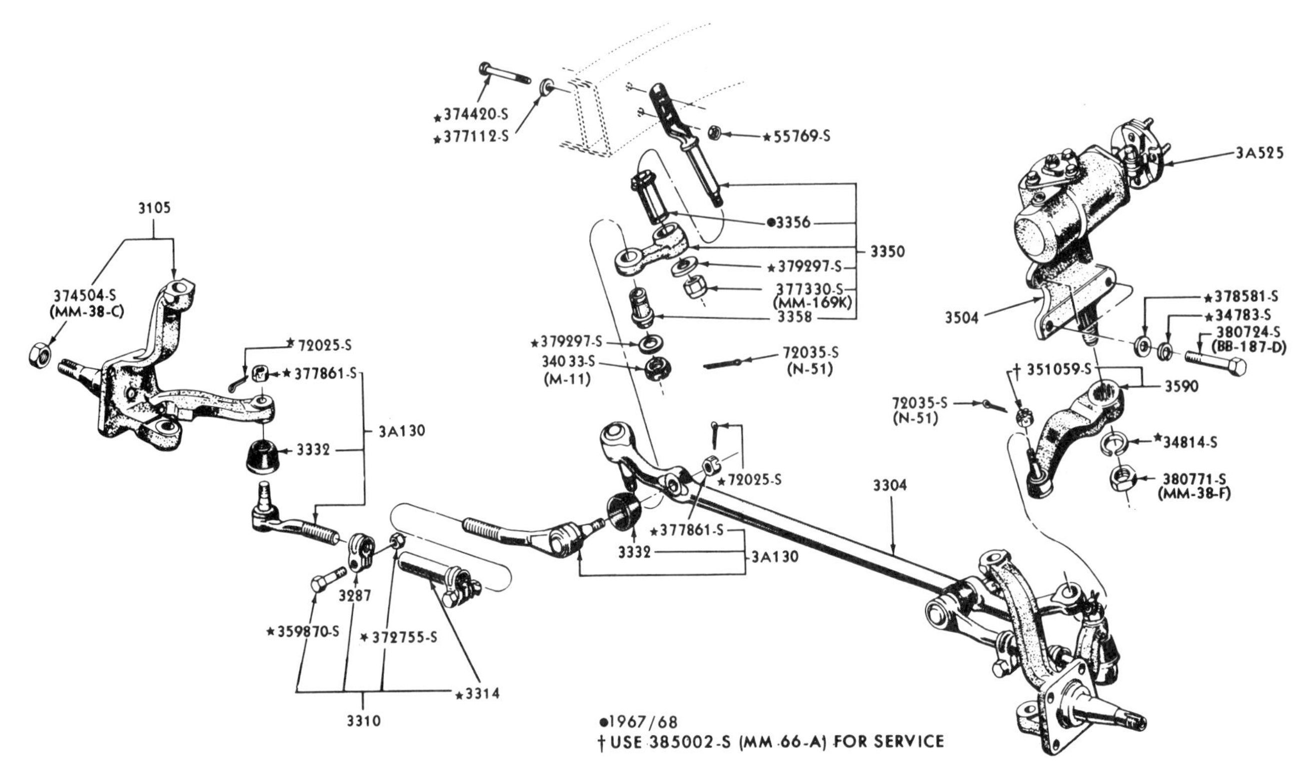

Typical manual steering system for 1966-1970 Mustangs.

MUSTANG POWER STEERING INSTALLATION KIT APPLICATION CHART

Year	Model	Notes	Cyl/CID	Linkage Kit	Pump Kit	Air Conditioner Adapter Kit
65	w/ alternator	1,3,5	8/289	C5ZZ-3A634-D	C5ZZ-3A635-F	C5ZZ-3A635-H
	w/ alternator, Before 4/22/65	4,6	8/289	C5ZZ-3A634-D	C5ZZ-3A635-K	C5PZ-3A635-G
	w/ alternator, From 4/22/65	2,4	8/289	C5ZZ-3A634-D	C5ZZ-3A635-K	C5ZZ-3A635-L
66			6/200	C5ZZ-3A634-F	C6ZZ-3A635-C	C5DZ-3A635-C
		2,4	8/289	C5ZZ-3A634-G	C5ZZ-3A635-K	C5ZZ-3A635-L
67			6/200	C5ZZ-3A634-A	C7OZ-3A635-B	C7DZ-3A635-B
	w/ alternator, Before 5/01/67	8,9	8/289	C5ZZ-3A634-A	C7OZ-3A635-C	C7ZZ-3A635-E
	w/ alternator, From 5/01/67	7,9	8/289	C7ZZ-3A634-B	C7OZ-3A635-C	C7ZZ-3A635-E
	w/ alternator, Before 5/01/67	10	8/390	C7ZZ-3A634-A	C7ZZ-3A635-C	C7OZ-3A635-E
	w/ alternator, From 5/01/67	10	8/390	C7ZZ-3A634-B	C7ZZ-3A635-C	C7OZ-3A635-E
68			6/200	C8ZZ-3A634-A	C8OZ-3A635-A	None required
			289,302	C8ZZ-3A634-A	C8OZ-3A635-C	C8OZ-3A635-E
			8/390	C8ZZ-3A634-A	C8ZZ-3A635-A	C8OZ-3A635-E
69			6/200	C8ZZ-3A634-B	C9ZZ-3A635-A	None required
			6/250	C8ZZ-3A634-B	C9OZ-3A635-D	None required
			8/302	C8ZZ-3A634-B	C9ZZ-3A635-B	C9ZZ-3A635-E
			8/351	C8ZZ-3A634-B	C9ZZ-3A635-C	C9ZZ-3A635-F
			8/390	C8ZZ-3A634-B	C9ZZ-3A635-D	C9ZZ-3A635-E
			8/428	Not available	Not available	Not available
70		11	6/200	DOZZ-3A634-A	DOZZ-3A635-G	None required
			6/250	DOZZ-3A634-A	DOZZ-3A635-F	None required
		11	8/302	DOZZ-3A634-A	DOZZ-3A635-C	None required
		11	351(W)	DOZZ-3A634-A	DOZZ-3A635-D	None required
		11	351(C)	DOZZ-3A634-A	DOZZ-3A635-E	None required
			8/428	Not available	Not available	Not available

Notes:

1. For units with Eaton P.S. pump. Not for units with cast iron water pump.
2. Use with pumps ident. w/ tag number HBA-AC2, HBA-AC3, HBA-AC4, HBA-AD2, HBA-AD3.
3. Includes Eaton Pump.
4. Includes Ford Pump.
5. Use with pumps ident. w/ tag number HBA-A32, HBA-AE3, HBA-AE4, HBA-AF2, HBA-AF3.
6. Use with pumps ident. w/ tag number HBA-A3, HBA-AE1, HBA-AF, HBA-AAF1, HBA-AR.
7. Use C5ZZ-3590-A with this kit for models w/o special handling package (Before 5/01/67).
8. Use C7ZZ-3590-C with this kit for models with special handling package.
9. Not for use on High Performance engines.
10. Includes Ford gear and Ford pump.
11. Not to be used with C78 x 14 tires.

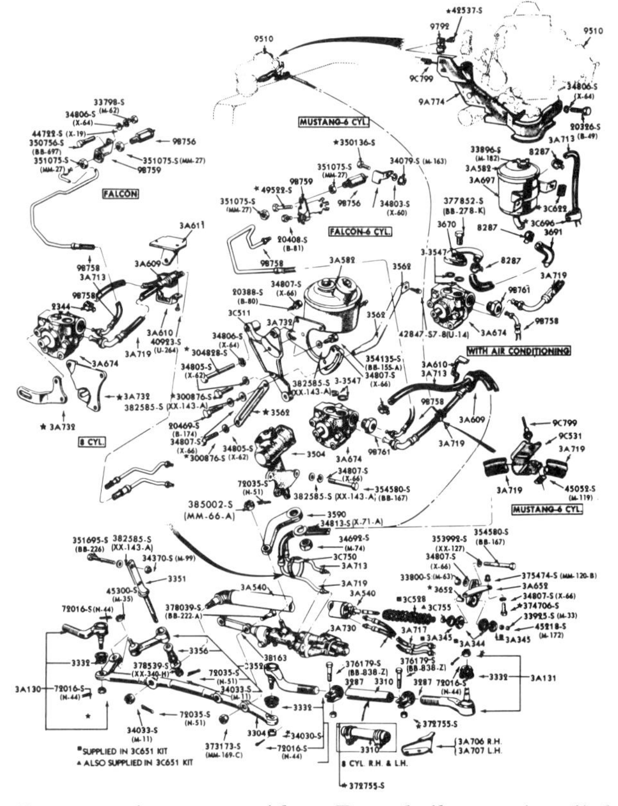

Power steering system with an Eaton-built pump installed on early 1965 Mustangs.

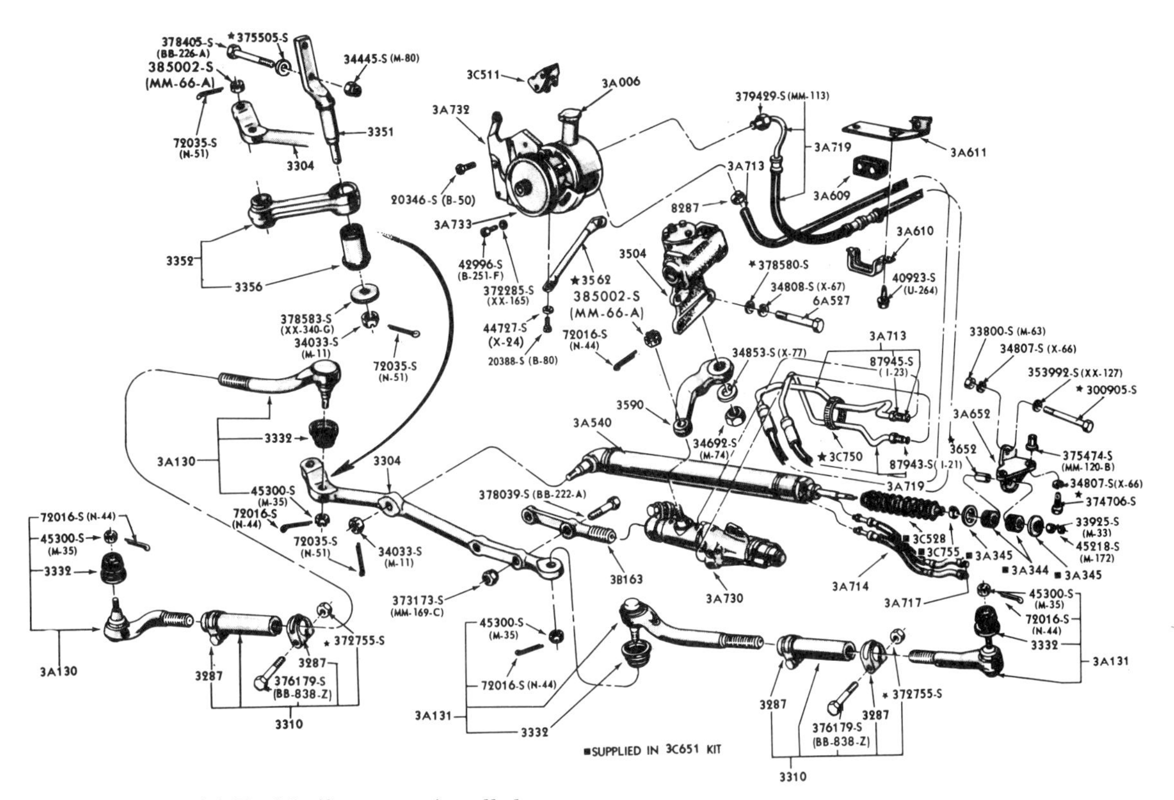

Power steering system with Ford-built pump as installed on 1965 and 1966 Mustangs equipped with a six-cylinder engine.

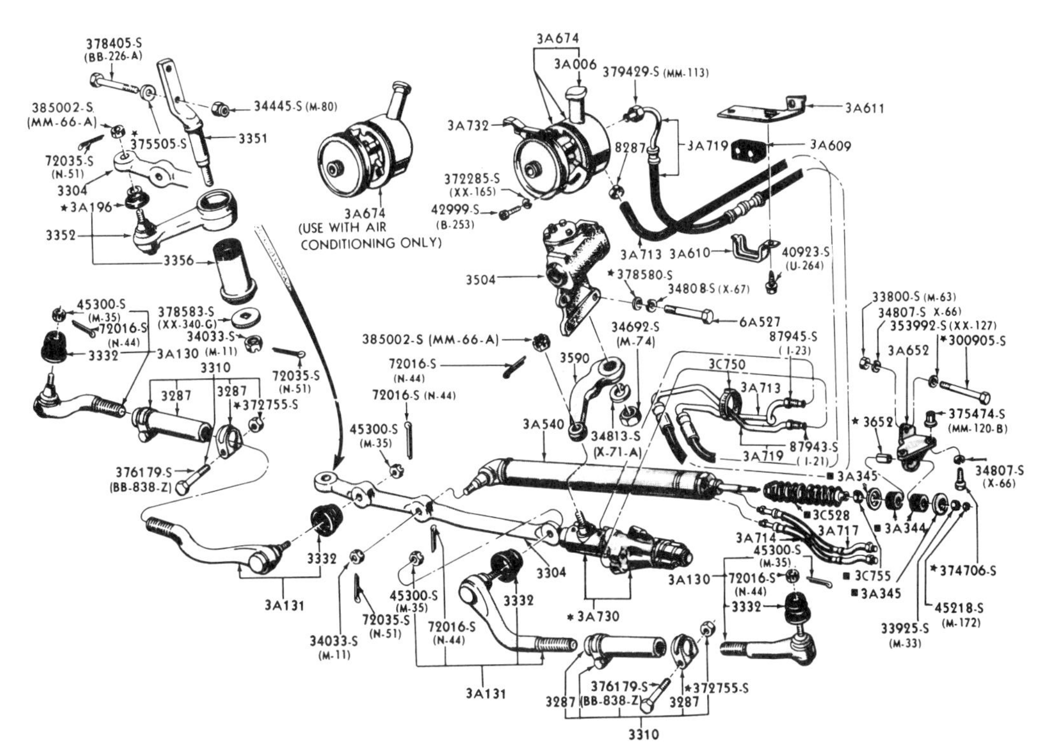

Power steering system with Ford-built pump installed on 1965 and 1966 Mustangs equipped with V-8 engines. Note the pump (3A674) with its slanted filler neck needed for cars equipped with optional air conditioning.

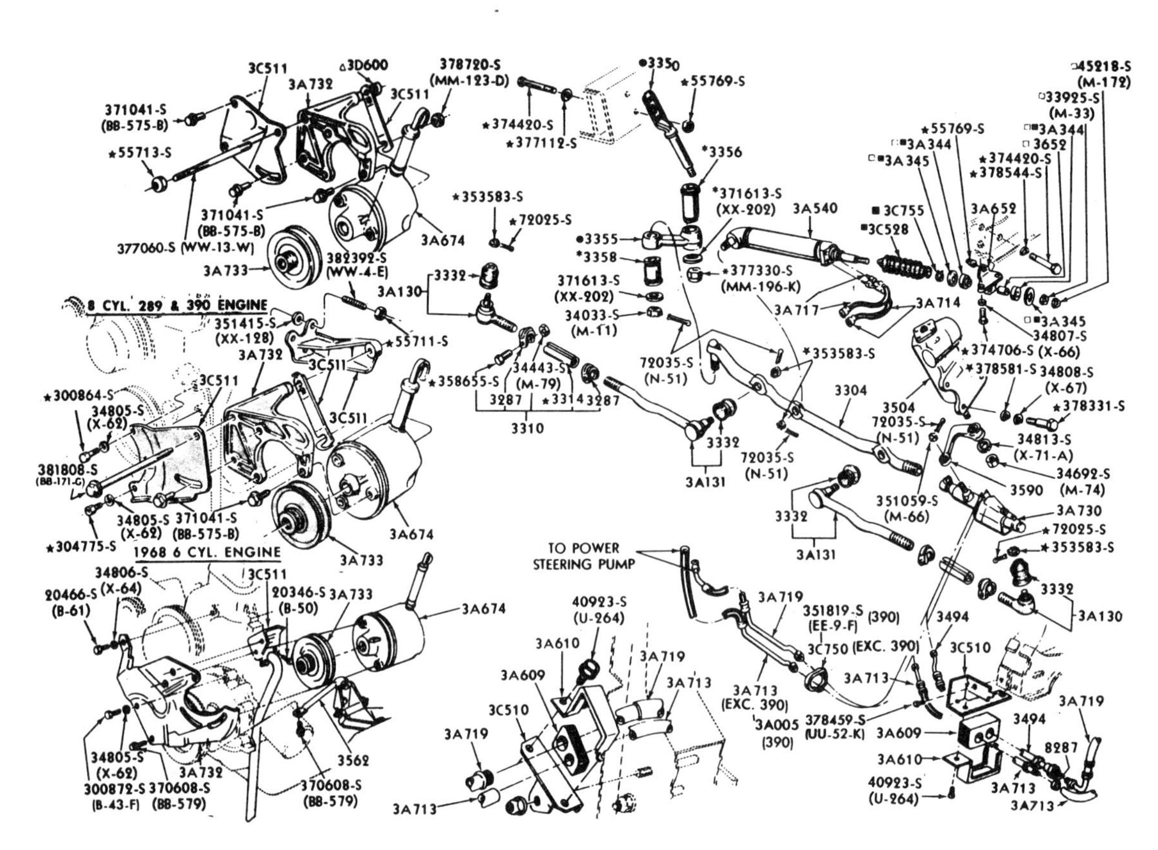

Variations of the power steering system installed on 1967 and 1968 Mustangs based on engine type and size.

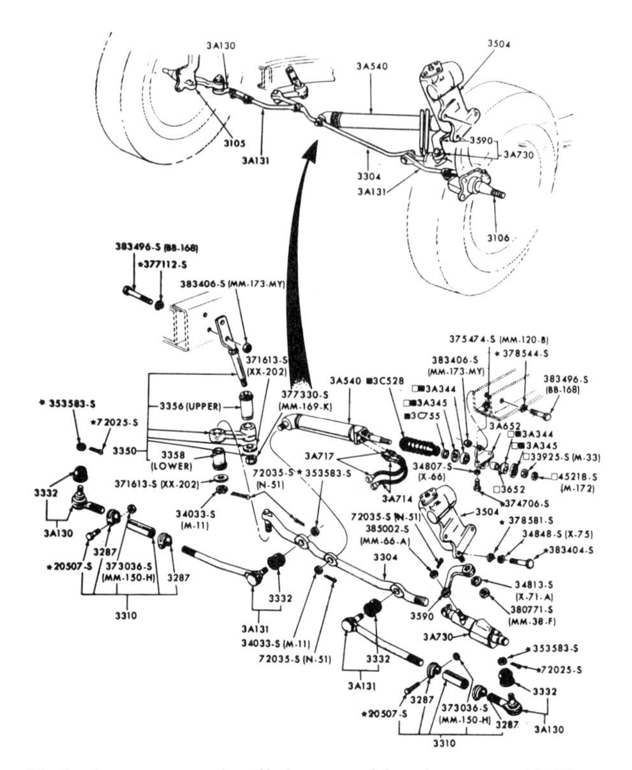

Typical power steering linkage and hardware available as an option on all 1965-1970 Mustangs.

MUSTANG POWER STEERING HOSE ASSEMBLY APPLICATION CHART, 1965-1966

YEAR	MODEL	DESCRIPTION	APPROX. DEV. LENGTH (INCHES)	PART NUMBER
65	170, 200, w/ Eaton pump	Control valve to pump-return	43-1/2	C5DZ-3A713-C
	Before 3/2/64	Control valve to cylinder	9	C3DZ-4A714-A
	From 3/2/64	Control valve to cylinder	9-1/4	C5DZ-3A714-A
	Before 3/2/64	Cylinder to valve	9-1/2	C3DZ-3A717-A
	From 3/2/64	Cylinder to valve	10	C5DZ-3A717-A
		Pump to control valve-pressure	34	C5DZ-3A719-D
	6 cyl. 170 with Ford pump	Control valve to pump-return	32	C5ZZ-3A713-E
		Control valve to cylinder	9	C3DZ-3A714-A
		Cylinder to valve	9-1/2	C3DZ-3A717-A
		Pump to control valve-pressure	31	C5ZZ-3A719-C
	6 cyl. 200 with Ford pump	Control valve to pump-return	45	C5DZ-3A713-A
		Control valve to cylinder	9-1/2	C5DZ-3A714-A
		Cylinder to valve	10	C5DZ-3A717-A
		Pump to control valve-pressure	70	C5DZ-3A719-A
	260, 289 with Eaton pump	Control valve to pump-return	32	C5ZZ-3A713-E
		Control valve to cylinder	9	C3DZ-3A714-A
		Cylinder to valve	9-1/2	C3DZ-3A717-A
		Pump to control valve-pressure	41	C5ZZ-3A719-E
	8 cyl. 260 with Ford pump	Con. valve to pump-return (w/o A/C)	29-3/4	C5ZZ-3A713-A
		Con. valve to pump-return (with A/C)	34-1/2	C5ZZ-3A713-B
		Control valve to cylinder	9	C3DZ-3A714-A
		Cylinder to valve	9-1/2	C3DZ-3A717-A
		Pump to control valve-pressure	29-3/4	C5ZZ-3A719-A
	8 cyl. 289 with Ford pump	Control valve to pump-return	35	C5ZZ-3A713-D
		Control valve to cylinder	9	C3DZ-3A714-A
		Cylinder to valve	9-1/2	C3DZ-3A717-A
		Pump to control valve-pressure	42-1/2	C5ZZ-3A719-D
66	6 cylinder 200	Control valve to pump-return	45	C5DZ-3A713-A
		Control valve to cylinder	9-1/2	C5DZ-3A714-A
		Cylinder to valve	10	C5DZ-3A717-A
		Pump to control valve-pressure	70	C5DZ-3A719-A
	8 cylinder 289	Control valve to pump-return	35	C5ZZ-3A713-D
		Control valve to cylinder	9	C3DZ-3A714-A
		Cylinder to valve	9-1/2	C3DZ-3A717-A
		Pump to control valve-pressure	42-1/2	C5ZZ-3A719-D

MUSTANG POWER STEERING HOSE ASSEMBLY APPLICATION CHART, 1967-1968

YEAR	MODEL	DESCRIPTION	APPROX. DEV. LENGTH (INCHES)	PART NUMBER
67	6 cylinder	Control valve to pump-return	32-1/2	C7OZ-3A713-A
		Control valve to cylinder	9-1/2	C6OZ-34714-A
		Cylinder to valve	9-3/4	C6OZ-3A717-A
		Pump to control valve-pressure	31	C7OZ-3A719-A
	8 cylinder 289	Control valve to pump-return	30	C70Z-3A713-D
		Control valve to cylinder	9-1/2	C60Z-3A714-A
		Cylinder to valve	9-3/4	C60Z-3A717-A
	(exc. GT350)	Pump to control valve-pressure	41	C7OZ-3A719-C
	(GT350)	Pump to control valve-pressure		C7ZX-3A719-A
	8 cylinder 390	Pump outlet tube to insulator-return	15-1/2	C7ZZ-3A713-C
		From insulator to valve-return		C7ZZ-3A713-B
		Control valve to cylinder	9-1/2	C6OZ-3A714-A
		Cylinder to valve	9-3/4	C6OZ-3A717-A
		Pump to valve outlet tube-pressure	17-1/2	C7ZZ-3A719-B
		Valve outlet tube to pressure hose	17-3/4	C7ZZ-3494-A
68	6 cylinder 200	Control valve to pump-return	32-1/2	C7OZ-3A713-A
		Control valve to cylinder	9-1/2	C6OZ-3A714-A
		Cylinder to valve	9-3/4	C6OZ-3A717-A
		Pump to control valve-pressure	30	C8OZ-3A719-B
	8 cyl. 289, 302	Control valve to pump-return	30	C7OZ-317A3-C
		Control valve to cylinder	9-1/2	C6OZ-3A714-A
		Cylinder to valve	9-3/4	C6OZ-3A717-A
		Pump to control valve-pressure	51	C8OZ-3A719-C
	390, 428CJ (w/o A/C)	Pump outlet tube to insulator-return	15-1/2	C7ZZ-3A713-C
		From insulator to valve-return	21	C8ZZ-3A713-A
		Control valve to cylinder	9-1/2	C60Z-3A714-A
		Cylinder to valve	9-3/4	C6OZ-3A717-A
	(exc. GT 350/500)	From pump to valve outlet tube-pressure	16	C9ZZ-3A719-B
	(GT 350/500)	From pump to valve outlet tube-pressure		C7ZX-3A719-A
		Valve outlet tube to pressure hose	17-1/2	C7ZZ-3494-B
	8 cyl. 428CJ with A/C	From valve to insulator-return	21	C8ZZ-3A713-A
		From insulator to cooler-return	15-1/2	C7ZZ-3A713-C
		From cooler to pump-return	AR	@ 3A005
		Control valve to cylinder	9-1/2	C60Z-3A714-A
		Cylinder to valve	9-3/4	C60Z-3A717-A
	(exc. GT 350/500)	From pump to valve outlet tube-pressure	16	C9ZZ-3A717-B
	(GT 350/500)	From pump to valve outlet tube-pressure		C7ZX-3A719-A
		Valve outlet tube to pressure hose	17-1/2	C7ZZ-3494-B

@ Cut to appropriate length from C7AZ-3A005-A (3/8" I.D. - 5/8" O.D.) bulk hose, and use (2) B5A-2344-A clamps

MUSTANG POWER STEERING HOSE ASSEMBLY APPLICATION CHART, 1969

MODEL	DESCRIPTION	APPROX. DEV. LENGTH (INCHES)	PART NUMBER
6 cylinder 200	Control valve to pump-return	32-1/2	C7OZ-3A713-A
	Control valve to cylinder		C6OZ-3A714-A
	Cylinder to valve	9-3/4	C6OZ-3A717-A
	Pump to control valve-pressure	30	C8PZ-31719-B
6 cylinder 250	Control valve to pump-return		C7OZ-3A713-A
	Control valve to cylinder	9-1/2	C6OZ-3A714-A
	Cylinder to valve	9-3/4	C6OZ-3A717-A
	Pump to control valve-pressure	32-1/2	C9OZ-3A719-B
302 (w/o oil cooler), 351	Control valve to pump-return		C8ZZ-3A713-A
	Control valve to cylinder	9-1/2	C6OZ-3A714-A
	Cylinder to valve	9-3/4	C6OZ-3A717-A
	Pump to valve outlet tube-pressure	27	C9ZZ-3A719-A
	Valve outlet tube to pressure hose	17-1/2	C7ZZ-3494-B
302 (with oil cooler)	Control valve to cooler-return		C8ZZ-3A713-A
	From cooler to pump-return		@ 3A005
	Insulator for return line		* 3A609
	Control valve to cylinder-inner	9-1/2	C6OZ-3A714-A
	Cylinder to valve-outer	9-3/4	C6OZ-3A717-A
	Pump to valve outlet tube-pressure		C9ZZ-3A719-C
	Valve outlet tube to pressure hose	17-1/2	C7ZZ-3494-B
390, 428 CJ, 429 Boss	From valve to insulator-return	21	C8ZZ-3A713-A
	From insulator to cooler-return	15-1/2	C7ZZ-3A713-C
	From cooler to pump-return		@ 3A005
	Insulator for return line	AR	* 3A609
	Control valve to cylinder	9-1/2	C6OZ-3A714-A
	Cylinder to valve	9-3/4	C6OZ-3A717-A
(exc. Boss)	Pump to valve outlet tube-pressure		C9ZZ-3A719-B
Boss 429	From pump to valve outlet-pressure		C9ZZ-3A719-D
	Valve outlet tube to pressure hose	17-1/2	C7ZZ-3494-B

@ Cut to appropriate length from C7AZ-3A005-A (3/8" I.D. - 5/8" O.D.) bulk hose, and use (2) B5A-2344-A clamps

* Cut to appropriate length from C7AZ-3A609-A bulk stock.

MUSTANG POWER STEERING HOSE ASSEMBLY APPLICATION CHART, 1970

MODEL	DESCRIPTION	APPROX. DEV. LENGTH (INCHES)	PART NUMBER
6 cylinder 200	Control valve to pump-return	29	DOZZ-3A713-A
	Control valve to cylinder	9-1/2	C6OZ-3A714-A
	Cylinder to valve	9-3/4	C60Z-3A717-A
	Pump to control valve-pressure	30	DOZZ-3A719-F
6 cylinder 250	Control valve to pump-return	29	DOZZ-3A713-A
	Control valve to cylinder	9-1/2	D6OZ-3A714-A
	Cylinder to valve	9-3/4	C6OZ-3A717-A
	Pump to control valve-pressure	32-1/2	DOZZ-3A719-F
302 (w.o oil cooler), 351	Control valve to pump-return	21	C8ZZ-3A713-A
	Control valve to cylinder	9-1/2	C6OZ-3A7114-A
	Cylinder to valve	9-3/4	C6OZ-3A717-A
Before 1/26/70	From pump to valve outlet tube-pressure	22-1/2	DOZZ-3A719-H
From 1/26/70	From pump to valve outlet tube-pressure	25	DOZZ-3A719-E
	Valve outlet tube to pressure hose	17-1/2	C7ZZ-3494-B
	From cooler to pump-return		@ 3A005
	Insulator for above return line		* 3A609
302 (with oil cooler)	Control valve to cooler-return	21	C8ZZ-3A7113-A
	From cooler to pump-return		@ 3A005
	Insulator for above return line		* 3A609
	Control valve to cylinder-inner	9-1/2	C6OZ-3A714-A
	Cylinder to valve-outer	9-3/4	C60Z-3A717-A
	From pump to valve outlet tube-pressure	24-1/2	DOZZ-3A719-G
	Valve outlet tube to pressure hose	17-1/2	C7ZZ-3494-B
8 cyl. 428CJ, 429 Boss	From valve to insulator-return	21	C8ZZ-3A713-A
	From insulator to cooler-return	AR	@ 3A005
	From cooler to pump-return	AR	@ 3A005
	Insulator for return line	AR	* 3A609
	Control valve to cylinder	9-1/2	C6OZ-3A714-A
	Cylinder to valve	9-3/4	C6OZ-31717-A
428 CJ	From pump to valve outlet tube-pressure		C9ZZ-3A719-B
429 Boss	From pump to valve outlet tube-pressure		C9ZZ-3A719-D

@ Cut to appropriate length from C7AZ-3A005-A (3/8" I.D. - 5/8" O.D.) bulk hose, and use (2) B5A-2344-A clamps

* Cut to appropriate length from C7AZ-3A609-A bulk stock.

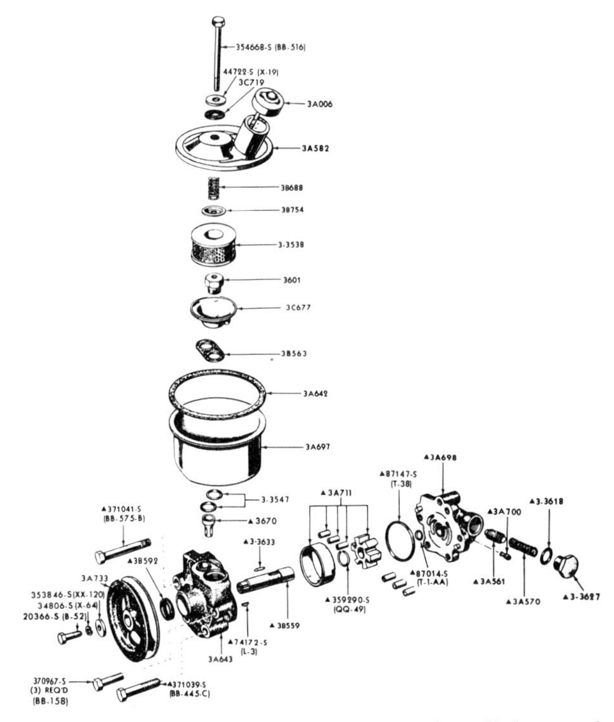

Eaton-built power steering pumps were installed on early 1965 Mustangs. The oil reservoir (3A697) is mounted on the left fender apron to make room for the compressor on air-conditioned cars.

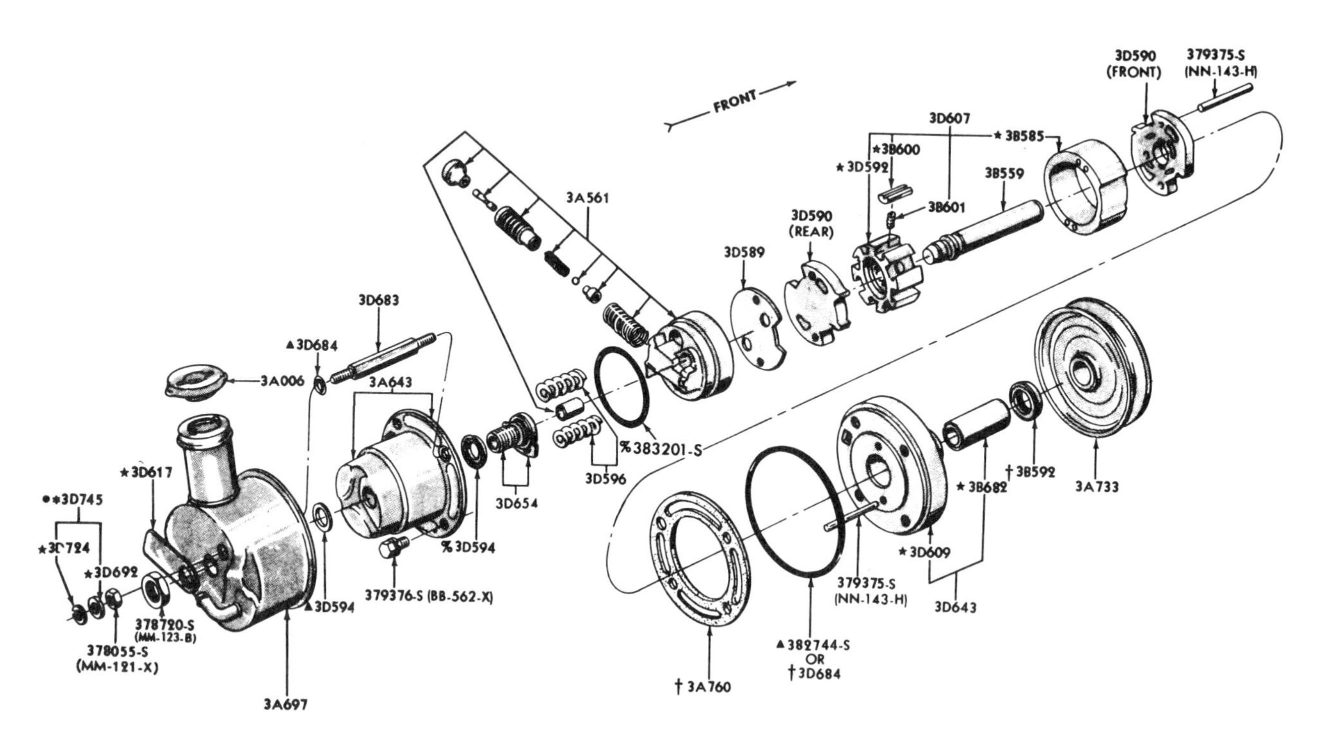

A Ford-built power steering pump with large diameter filler cap (3A006) was installed on all late 1965 and all 1966 Mustangs. The filler neck was canted rearward to make room for the compressor on air-conditioned cars.

The power steering filler neck changed places many times over the years. Pre-1969 power steering pumps (shown) had a large-diameter filler neck. If installed on a six-cylinder engine, the neck entered straight up from the center of the cannister. The air-conditioning compressor, if equipped, did not interfere with neck because it was located on the opposite side of the engine. V-8 engines used the same pump although the bracketry slanted the neck slightly to the driver's side. The filler neck was located to the back of the cannister and angled rearward (shown) in order to clear the compressor if the same V-8 engine was equipped with air conditioning.

Early 1965 Mustangs were equipped with Eaton-manufactured power steering pumps. Eaton pumps are easily identified by their separate reservoirs. This pump is on a non-air-conditioned car, however, when air conditioning is added, the reservoir is located on the left inner fender apron and connected to the pump by hoses.

MUSTANG POWER STEERING PUMP ASSEMBLY APPLICATION CHART

YEAR	MODEL	DESCRIPTION/IDENTIFICATION MARKING	PART NUMBER
65	6 cyl. 170 without A/C	Eaton pump	C2AZ-3A674-B
	8 cyl. 260 without A/C	Eaton pump	C5OZ-3A764-B
	8 cyl. 260 with A/C	Eaton pump	C5AZ-3A674-D
65/66	6 cyl. 200	HBA-AN,AN1,AN2,AN3,AN4 (Ford pump)	C5DZ-3A674-BRM
	8 cyl. 289 without A/C	HBA-AC,AC1,AC2,AC3,AC4 (Ford pump)	C5DZ-3A674-ARM
	8 cyl. 289 with A/C	HBA-AD,AD1,AD2,AD3,AD4 (Ford pump)	C5DZ-3A674-CRM
67/68	Exc. GT 350/500	HBA-BF,BF1	C7OZ-3A674-A
	GT350/500	HBA-BH,BH1	C7TZ-3A674-A
69	Exc. Boss 429	HBA-BF,BF1	C7OZ-3A674-A
	Boss 429	HBA-BZ,CD	DOZZ-3A674-A
70	8 cyl. 302,351	HBA-CC	DOZZ-3A674-B
	6 cyl. & 8 cyl. 428	HBA-BR	DOZZ-3A674-B
	Boss 429	HBA-BZ,CD	DOZZ-3A674-C

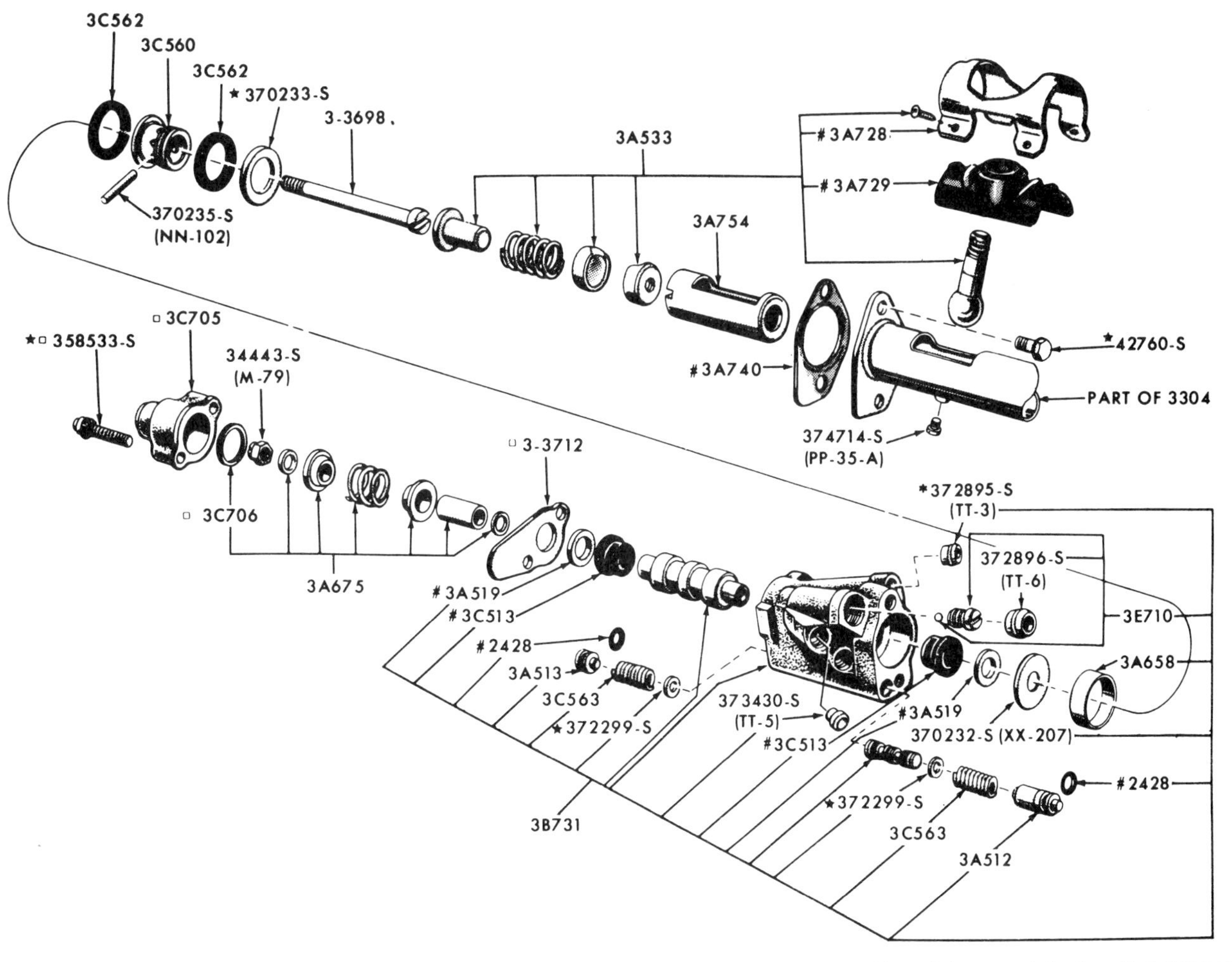

The power steering control valve used on 1965 and early 1966 Mustangs. Unlike the removable valve used on late 1966 and later cars, this unit is part of the drag link (3304). (See view of steering linkage.)

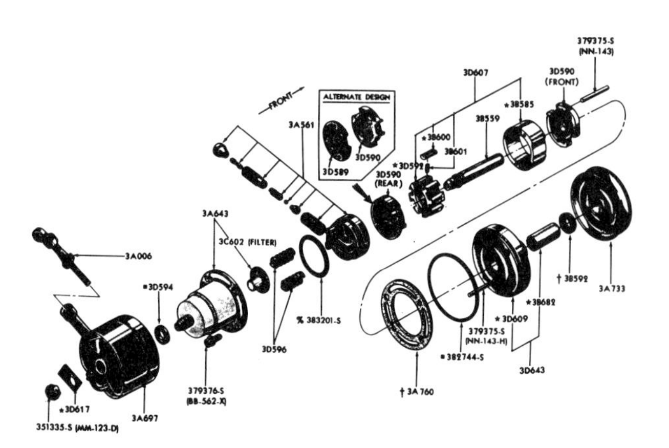

Typical Ford-built power steering pump used on all 1967-1970 Mustangs. Note the small-diameter dipstick (3A006) and tube.

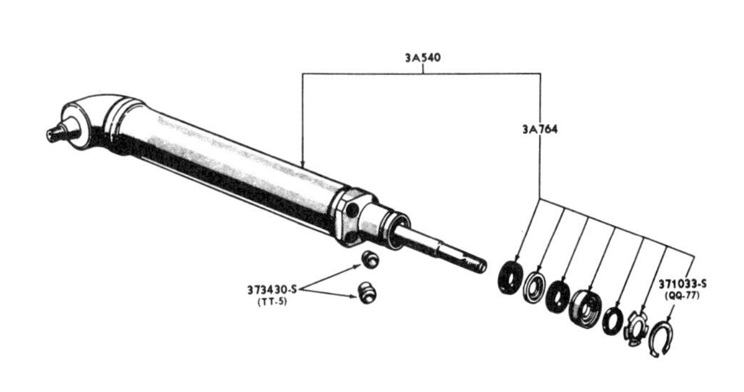

The power steering cylinder (3A540) is a welded unit and cannot be rebuilt. The rod packing (3A764) and the hose seats (373430-S) are common service items.

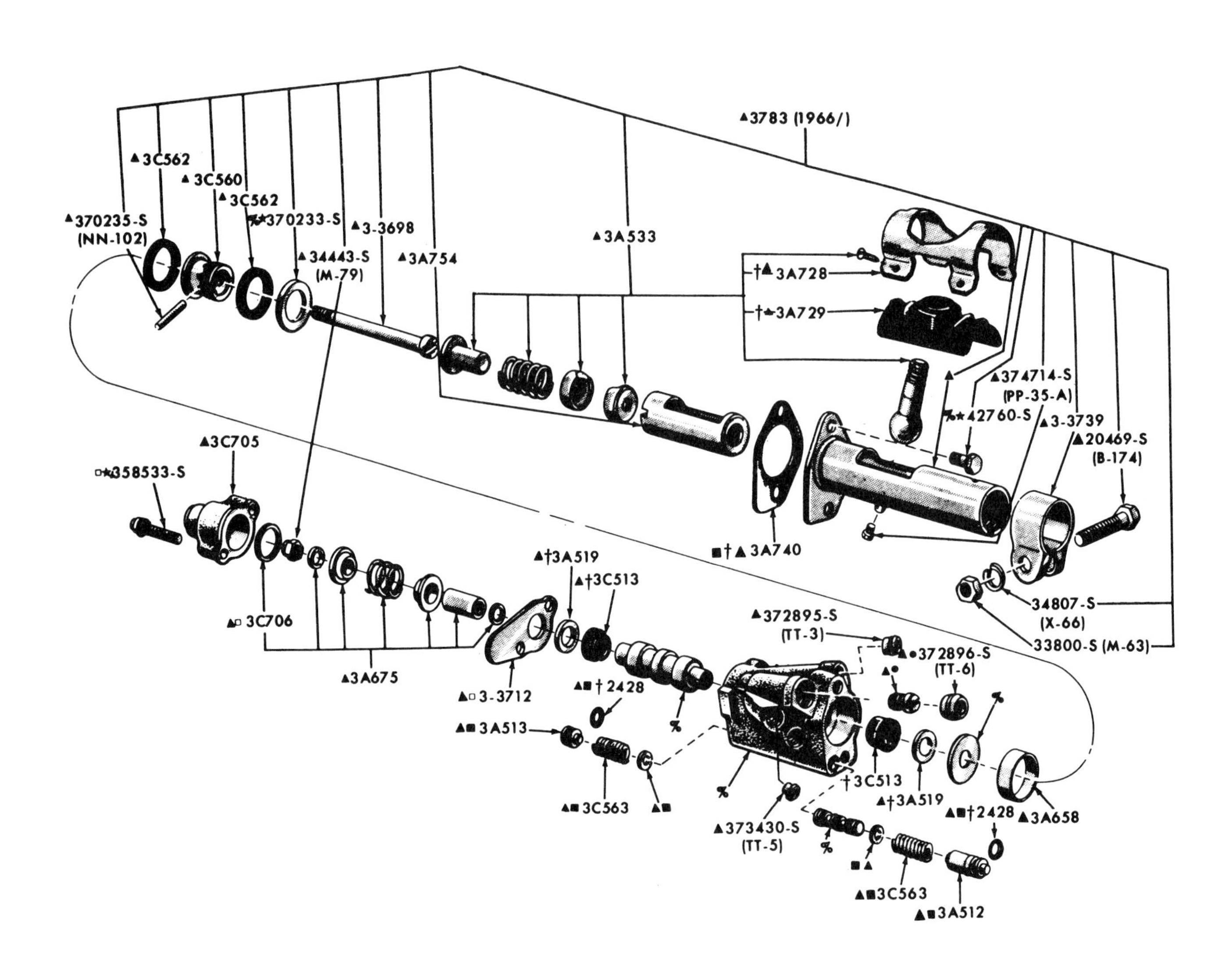

An exploded view of the power steering control valve used on all 1966-1970 Mustangs.

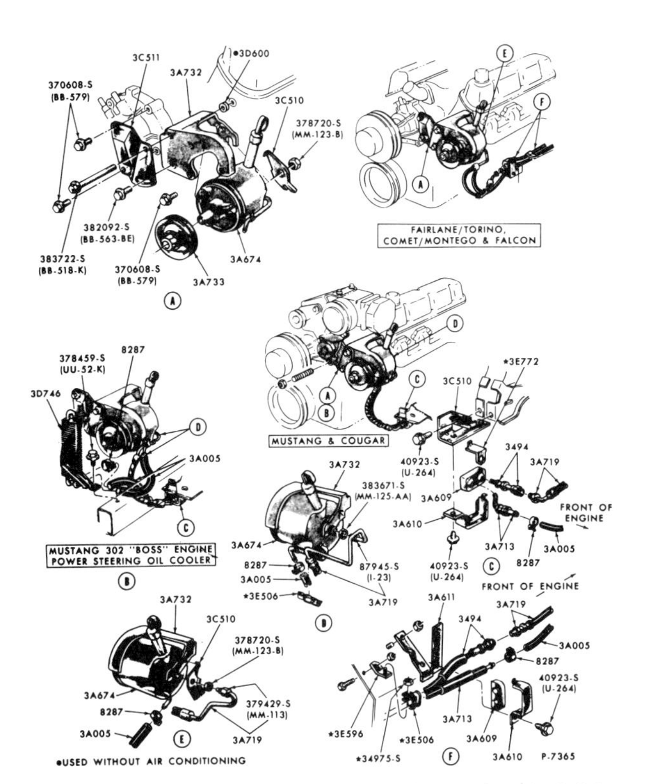

Power steering pump mounting hardware for 1970 Mustangs equipped with 302 or 351ci engines.

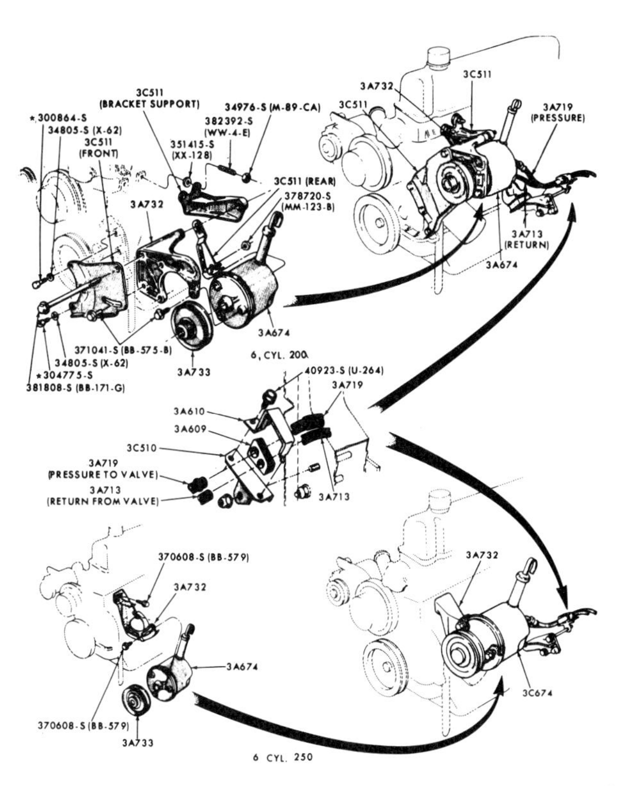

Power steering pump mounting hardware for 1969 and 1970 Mustangs equipped with a 250ci six-cylinder engine.

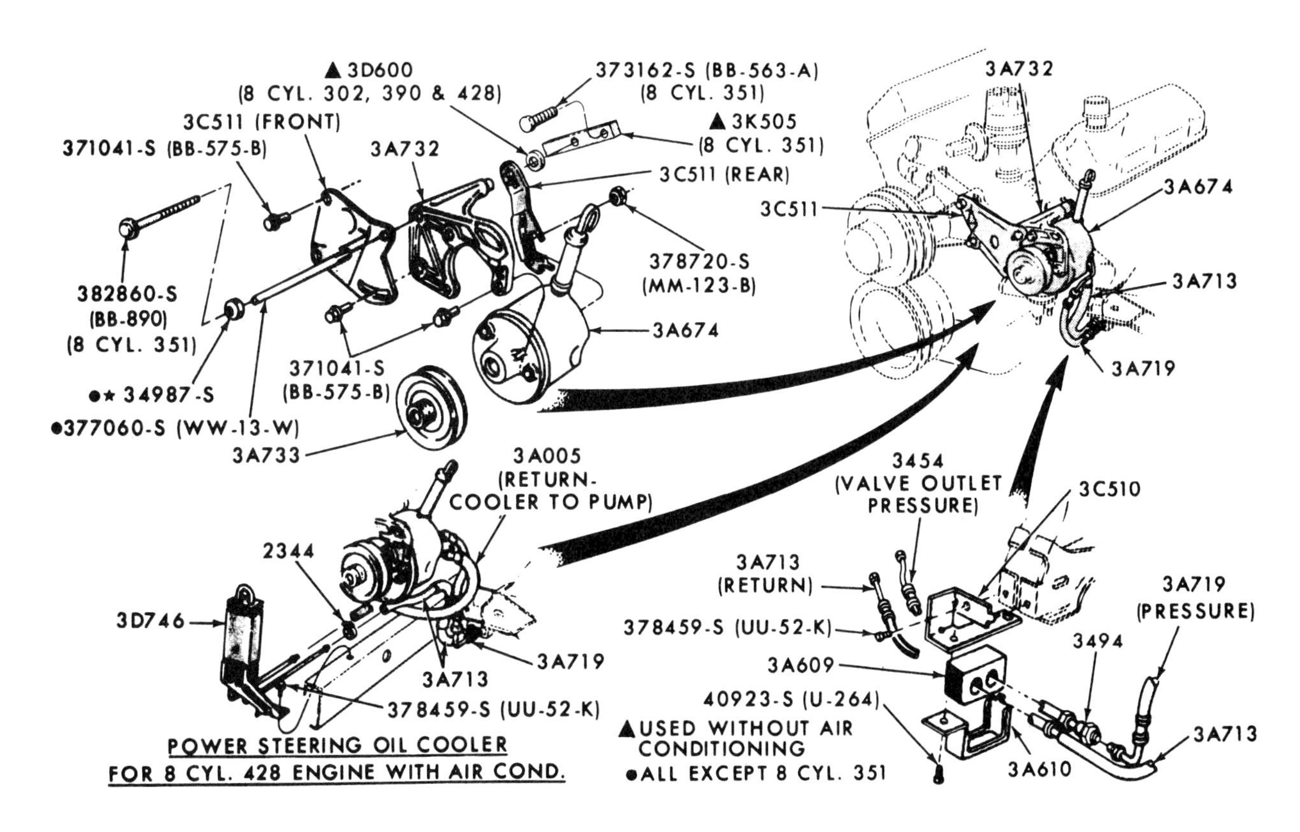

Power steering pump mounting hardware for 1970 Mustangs equipped with 428ci engines.

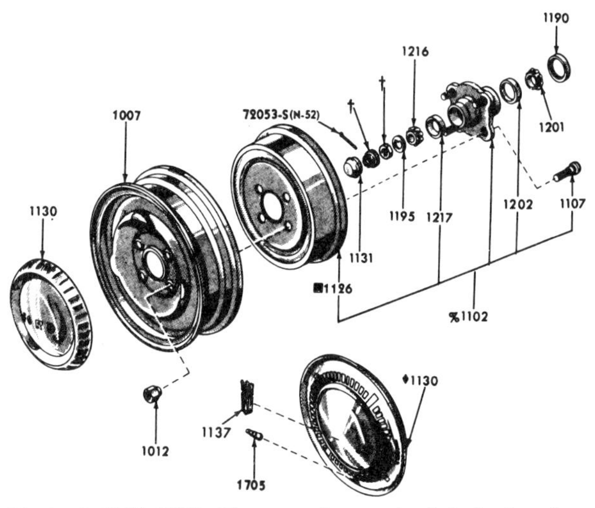

Typical 1965-1970 Mustang front wheel hub, bearings, rim, and cover assembly.

MUSTANG HUB CAP OR WHEEL COVER APPLICATION CHART, 1965 and 1966

1965

PART NUMBER	IDENTIFICATION	DIAM.	DESCRIPTION
C5ZZ-1130-W	C5ZA-1130-J	2-1/2"	Imprinted "Ford Mustang" - Incl. attaching parts (exc. GT350)
C5ZZ-1130-A	C4ZA-1130-F	13"	Imprinted "Mustang" - Black center (exc. GT350)
C5ZZ-1130-A	C4ZA-1130-H	13"	Imprinted "Mustang" - Black center - Incl. spinner (exc. GT350)
C5ZZ-1130-K	C4ZA-1130-K	13"	Imprinted "Mustang" - Black center (exc. GT350)
C5ZZ-1130-P	C5ZA-1130-D	13"	Imprinted "Mustang" - Black center (exc. GT350)
C5ZZ-1130-J	C4ZA-1130-M	13"	Imprinted "Mustang" - Black center - Incl. spinner (exc. GT350)
C5ZZ-1130-S	C5ZA-1130-F	13"	Imprinted "Mustang" - Black center - Incl. spinner (exc. GT350)
C4OZ-1130-H		13"	Wire (except GT350)
C5ZZ-1130-C	C4ZA-1130-E	14"	Imprinted "Mustang" - Black center (exc. GT350)
C5ZZ-1130-J	C4ZA-1130-J	14"	Imprinted "Mustang" - Black center (exc. GT350)
C5ZZ-1130-N	C5ZA-1130-C	14"	Imprinted "Mustang" - Black center (exc. GT350)
C5ZZ-1130-C	C4ZA-1130-L	14"	Imprinted "Mustang" - Black center - Incl. spinner (exc. GT350)
C5ZZ-1130-R	C5ZA-1130-E	14"	Imprinted "Mustang" - Black center - Incl. spinner (exc. GT350)
C4AZ-1130-H	C4AA-1130-F	14"	Wire (except GT350)
C4AZ-1130-L	C4AA-1130-R	14"	Simulated wire, Incl. spinner, R/W/B center (exc. GT350)
C3AZ-1130-Z	C3AA-1130-Z,AF	15"	Includes spinner - Stainless steel (exc. GT350)
S1MS-1130-A		3"	Cap assembly - Use with Cragar 15" wheel (GT350)
S2MS-1130-B	Consists of:	2-3/4"	Cap assembly - Use with S2MS-1007-B & C (GT350)
	S2MS-1131-B		Bracket
	S2MS-1132-A		Emblem (GT350)
	S2MS-1133-A		Plastic cap
	(2) S2MS-1135-A		Nut
S2MS-1130-C	Consists of:	2-1/2"	Cap assembly - Use with S2MS-1007-A (GT350)
	S2MS-1130-A		Cap - Die cast imprinted "CS"
	S7MS-1131-B		Bracket
			Hex bolt 1/4"-20 x 5/8"

1966

PART NUMBER	IDENTIFICATION	DIAM.	DESCRIPTION
C5ZZ-1130-W	C5ZA-1130-J	2-1/2"	Imprinted "Ford Mustang" - Includes attaching parts
C6DZ-1130-A	C6DA-1130-A	9-1/2"	Imprinted "Ford" - 36 depressions
C6ZZ-1130-A	C6ZA-1130-A	14"	Black center
C6ZZ-1130-A	C6ZA-1130-B	14"	Black center - Includes spinner
C6OZ-1130-N	C8OA-1130-M	14"	Simulated mag - Plain chrome center
C6OZ-1130-K	C6OA-1130-E	14"	Simulated mag, Chrome center, Also serviced in C6OZ-1130-J kit
S2MS-1130-B	Consists of:	2-3/4"	Cap assembly - Use with S2MS-1007-B & C kit
	S2MS-1131-B		Bracket
	S2MS-1132-A		Emblem (GT350)
	S2MS-1133-A		Plastic cap
	S2MS-1135-A		Nut
S2MS-1130-C	Consists of:	2-1/2"	Cap assembly - Use with S2MS-1007-A
	S2MS-1130-A		Cap - Die cast imprinted "CS"
	S7MS-1131-B		Bracket
			Hex bolt 1/4"-20 x 5/8"

MUSTANG HUB CAP OR WHEEL COVER APPLICATION CHART, 1967 and 1968

1967

PART NUMBER	IDENTIFICATION	DIAM.	DESCRIPTION
C5ZZ-1130-W	C5ZA-1130-J	2-1/2"	Imprinted "Ford Mustang" - Includes attaching parts (exc. GT350)
C7OZ-1130-C	C7OA-1130-C	2-1/2"	Push-in type - Includes crest - Red center (exc. GT350)
C7ZZ-1130-A	C7ZA-1130-E	3-3/4"	Push-in type (exc. GT350)
C7DZ-1130-A	C7DA-1130-D	9-5/8"	Black center - Ford emblem (exc. GT350)
C7AZ-1130-A	C7AA-1130-A	10-1/2"	Black center - Ford emblem (exc. GT350)
C6OZ-1130-K	C6OA-1130-E	14"	Simul. mag, Chrome center (exc. GT350)
C7OZ-1130-F	C7OA-1130-D	14"	Wire - Includes spinner - Blue center (exc. GT350)
C6OZ-1130-N	C8OA-1130-M	14"	Simulated mag - Chrome center (exc. GT350)
C7ZZ-1130-G	C7ZA-1130-H	14"	Imprinted "Mustang" - 21 openings - Red center (exc. GT350)
C7ZZ-1130-E	C7ZA-1130-G	14"	Simulated wire - Red center (exc. GT350)
C7ZZ-1130-B	C7ZA-1130-B	14-1/8"	Imprinted "Mustang" - 21 openings - Red center (exc. GT350)
C6AZ-1130-C	C6AA-1130-E	16-1/4"	Simul. steel wheel, Incl. crest - Red center (exc. GT350)
C7AZ-1130-H	C7AA-1130-E	15"	Wire - Includes spinner - Blue center (exc. GT350)
S7MS-1130-B	Consists of:	2-1/2"	Cap assembly - Use with S7MS-1007-D (GT350)
	S7MS-1131-B		Bracket
	S7MS-1132-A		Emblem (Shelby Cobra)
	S7MS-1133-B		Cap
			Hex bolt 1/4"-20 x 5/8"
S7MS-1130-C	Consists of:	2-3/4"	Cap assembly - Use with S7MS-1007-C (GT350)
	S7MS-1131-A		Bracket
	S2MS-1132-B		Emblem (Shelby Cobra)
	S7MS-1133-A		Cap
	(2) 42127-S		Screws
S7MS-1130-A		16-1/4"	Cover-styled (GT350)

1968

PART NUMBER	IDENTIFICATION	DIAM.	DESCRIPTION
C8OZ-1130-C	C8OA-1130-C,D	7-1/2"	Includes GT emblem
C8OZ-1130-G	C8OA-1130-H,J	7-1/2"	Plain center
C8ZZ-1130-F	C8ZA-1130-D	10-1/2"	Black center - Includes Mustang emblem
C6OZ-1130-K	C6OA-1130-E	14"	Simulated mag - Chrome center - Also svcd in C6OZ-1130-J kit
C6OZ-1130-N	C8OA-1130-M	14"	Simulated mag - Chrome center
C8ZZ-1130-A	C8ZA-1130-A	14"	Imprinted "Mustang" - 16 openings
C7ZZ-1130-E	C7ZA-1130-G	14"	Simulated wire - Red center
C8ZZ-1130-E	C8ZA-1130-E	14"	Includes Mustang emblem - 16 openings - Red center
C6AZ-1130-K	C6AA-1130-J C8AA-1130-H,L	15"	Simulated mag wheel
C8OZ-1130-J	C8OA-1130-N	14"	Simulated wire - Plain center
C8OZ-1130-C	C6AA-1130-E	16-1/4"	Simul. steel wheel - Bolt-on type crest - Red center
S8MS-1130-A		16-1/4"	Die-cast cover

MUSTANG HUB CAP OR WHEEL COVER APPLICATION CHART, 1969 and 1970

1969

PART NUMBER	IDENTIFICATION	DIAM.	DESCRIPTION
C9ZZ-1130-F		2-3/16"	Use with C9ZZ-1007-H wheel - Mustang emblem in center
C9ZZ-1130-E		2-3/4"	Black center - Use with styled steel wheels - Mustang emblem
C9ZZ-1130-D	S9MS-1130-B	2-3/4"	Black center - Use with snap-on type styled steel wheels
C8OZ-1130-C	C8OA-1130-C,D	7-1/2"	Includes GT emblem
C8OZ-1130-G	C8OA-1130-J	7-1/2"	Plain center
C8AZ-1130-D	C8AA-1130-D	10-1/2"	Black center - Includes Ford crest
C9ZZ-1130-A	C9ZA-1130-B	14"	Imprinted "Mustang," Plain center, Also svcd in C9ZZ-1130-B kit
C7ZZ-1130-E	C7ZA-1130-G	14"	Simulated wire - Red center
C7AZ-1130-C		16-1/4"	Simul. bolt-on type steel wheel - Includes crest - Red center

1970

PART NUMBER	IDENTIFICATION	DIAM.	DESCRIPTION
C9ZZ-1130-E	NONE	2-3/4"	Use w/ styled steel wheels, Mustang emblem, Black ctr (Boss)
C9ZZ-1130-D	S9MS-1130-B	2-3/4"	Use w/ snap-on type style steel wheels - Black ctr (GT350/500)
C8OZ-1130-G	C8OA-1130-H,J	7-1/2"	Use w/ styled steel wheels - Plain ctr (except Boss, GT350/500)
DOAZ-1130-E	DOZA-1130-H	10-1/4"	Imprint "Ford Motor Company," Chrome finish, Circ. black stripes (except Boss, GT350/500) - Before 12/24/69
DOAZ-1130-H	DOAA-1130-G D2AA-1130-AA	10-1/4"	Imprint "Ford Motor Company," Chrome finish, Circ. black stripes (except Boss, GT350/500) - From 12/24/69
D5DZ-1130-B	DOOA-1130-K DOZA-1130-J	10-1/4"	Imprint "Ford Motor Company" - Brushed chrome finish - Circular black stripes, Use w/ 1210 trim ring (exc. Boss, GT350/500)
C6OZ-1130-N	D8OA-1130-M	14"	Simulated mag - Plain chrome center (exc. Boss, GT350/500)
DOZZ-1130-A	DOZA-1130-K	14-1/8"	Imprint "Mustang,"Chrome bkgrd, 16 depressions Exc. Boss, GT350, GT500.
DOZZ-1130-D	DOZA-1130-E,L	15-1/4"	Simul. styled steel, Mustang embl., Black bkgrnd. - Use w/14x7 wheels (except Boss, GT350/500)
DOZZ-1130-E	DOZA-1130-F,M	15-1/4"	Simul. styled steel, Mustang emblem, Black bkgrnd. - Use w/14x6 wheels (except Boss, GT350/500)
C8OZ-1130-J	C8OA-1130-N	14"	Simulated wire - Plain center hub (exc. Boss, GT350/500)
DOZZ-1130-F	DOZA-1130-G,N	16-1/4"	Simul. styled steel, Mustang emblem, Black bkgrnd., Use w/15x7 wheels. (except Boss, GT350/500)

This wire cover with spinner with the vintage Ford crest in the center was used on early 1964½ Mustangs.

Mustangs built between April and September 1964 were outfitted with 13in wheels as standard equipment. Cars optioned with a V-8 engine and Special Handling Package received 14in wheels. This standard full disc wheelcover was manufactured in both 13in and 14in sizes through September 1965.

The vintage Ford crest center emblem in the wire wheelcover spinner was replaced by what would be later known as the emblem signifying Ford's new luxury full-size car, the LTD. This wire cover with spinner was introduced in the fall of 1964.

The standard wheelcover became optional when a three-prong spinner was added. It too was available in 13 and 14in sizes.

Styled Steel wheels were available only in 14in size. In 1965 Styled Steels featured a one-piece stamped center and chrome rim.

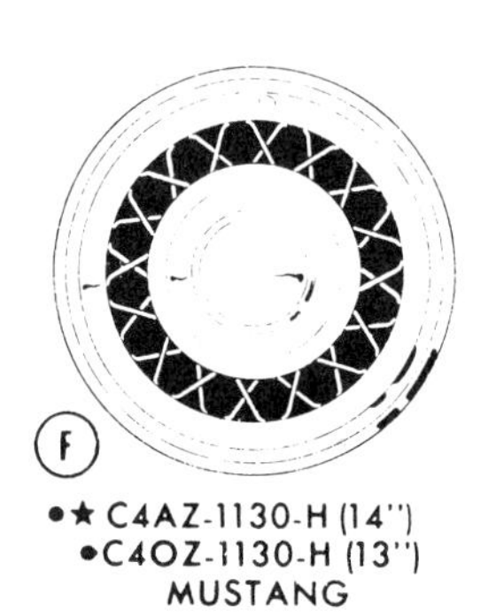

●★ C4AZ-1130-H (14")
●C4OZ-1130-H (13")
MUSTANG

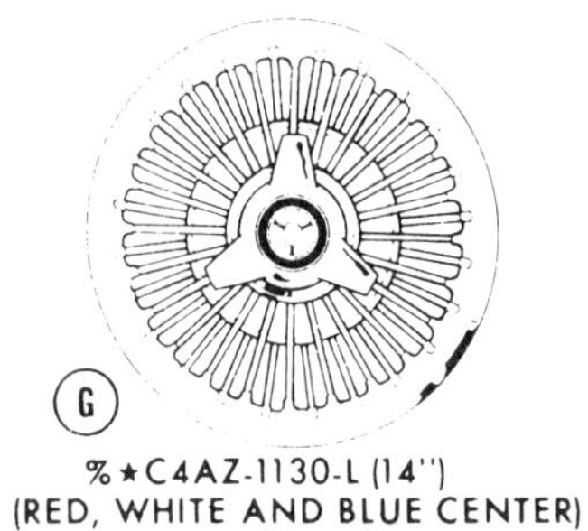

%★C4AZ-1130-L (14")
(RED, WHITE AND BLUE CENTER)
MUSTANG

★ S1MS-1130-A (3")
(GT-350)
MUSTANG-HUB CAP

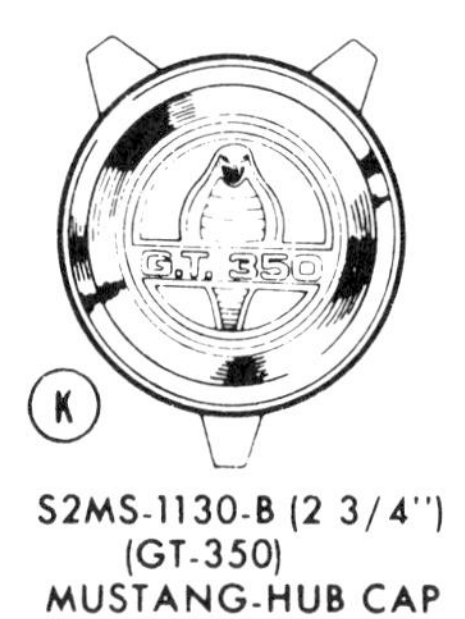

S2MS-1130-B (2 3/4")
(GT-350)
MUSTANG-HUB CAP

★S2MS-1130-C (2 1/2")
(GT-350)
MUSTANG-HUB CAP

Standard and optional 1965 Mustang wheelcovers.

In 1966, the Styled Steel wheel rim was painted, then fitted with a polished stainless-steel trim ring.

Only 14in tires and wheels were offered in 1966 and spinner wheelcovers were the in thing. This design, less the spinner, was the standard wheel cover for 1966. It too became an option when the spinner was added.

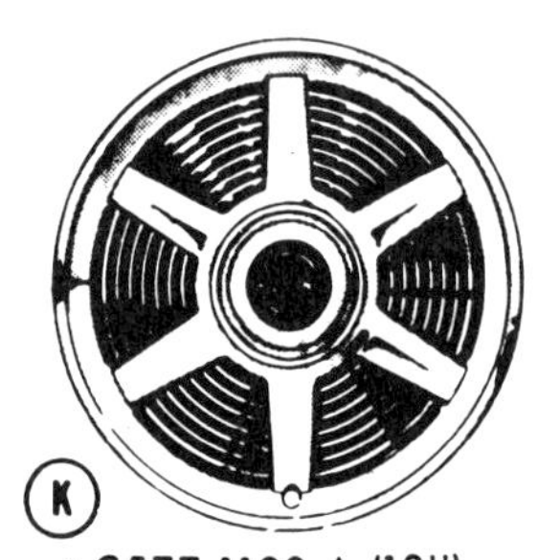

▲C5ZZ-1130-A (13")
▲C5ZZ-1130-K (13")
●C5ZZ-1130-P (13")
▲C5ZZ-1130-C (14")
▲C5ZZ-1130-J (14")
●C5ZZ-1130-N (14")
(BLACK CENTER)
MUSTANG

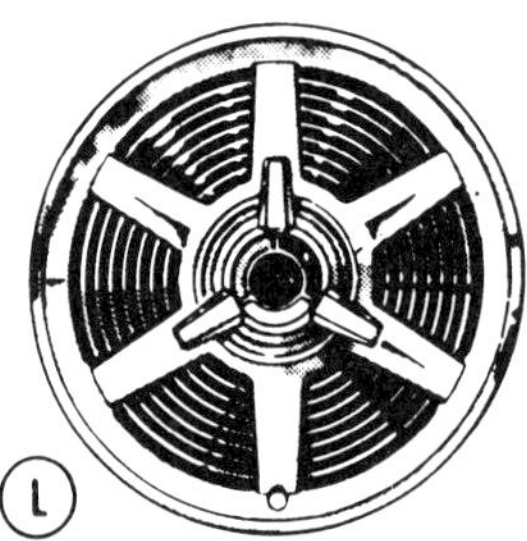

#▲★ C5ZZ-1130-B (13")
#▲★C5ZZ-1130-M (13")
#●C5ZZ-1130-S (13")
#▲★C5ZZ-1130-L (14")
#★C5ZZ-1130-R (14")
(BLACK CENTER)
MUSTANG

C5ZZ-1130-W (2 1/2")
(HUB CAP)
MUSTANG

●CHROME
▲RUSTLESS STEEL
#SPINNER IN CENTER
%SIMULATED WIRE
WITH SPINNER

Standard and optional 1965 Mustang wheelcovers (continued).

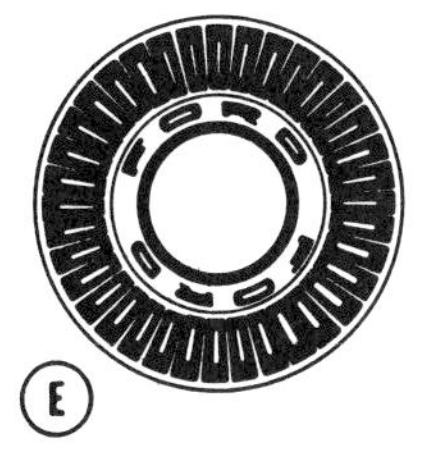

C5AZ-1130-A (10 1/2")
C6DZ-1130-A (9 1/2)
(HUB CAP)
MUSTANG

C5ZZ-1130-W (2 1/2")
("MUSTANG" EMBLEM)
MUSTANG-HUB CAP

S2MS-1130-B (2 3/4")
(GT-350)
MUSTANG-HUB CAP

S2MS-1130-C (2 1/2")
(GT-350)
MUSTANG-HUB CAP

Standard and optional 1966 Mustang wheelcovers.

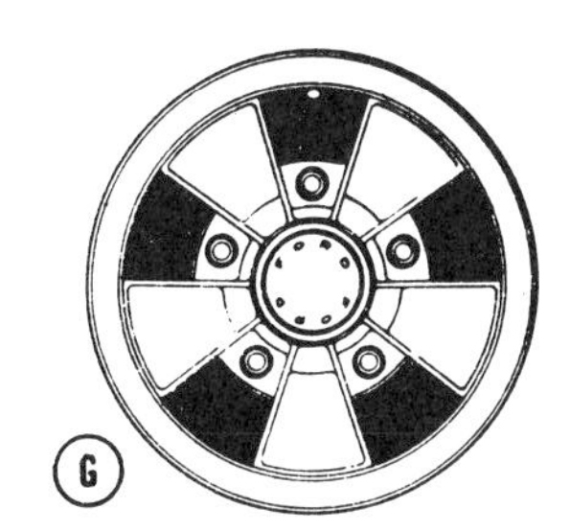

*C6OZ-1130-K (14")
#C6OZ-1130-N (14")
FAIRLANE, FALCON, MUSTANG

Standard and optional 1966 Mustang wheelcovers (continued).

By 1967 the spinner had been labeled unsafe by our overly protective government and was therefore banned from all factory production wheels and wheelcovers. This smooth centered wire cover was offered as a Mustang option from 1967 through 1970.

1967 Styled Steel wheels took on a slightly different shape which became unique to the year. The center cap was scalloped and the polished trim ring was wider than in 1966.

C5ZZ-1130-W (2 1/2")
("MUSTANG" EMBLEM)
MUSTANG-HUB CAP

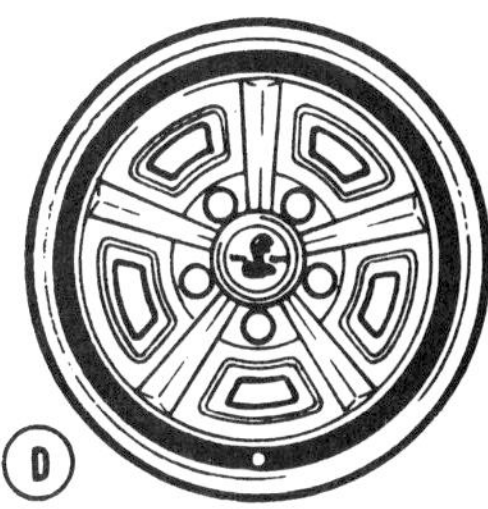

S7MS-1130-A (16 1/4")
(GT-350/500)
MUSTANG

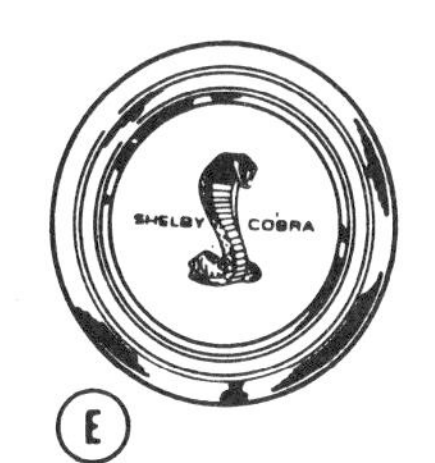

S7MS-1130-B (2 1/2")
GT-350/500
MUSTANG

S7MS-1130-C (2 3/4")
GT-350/500
MUSTANG

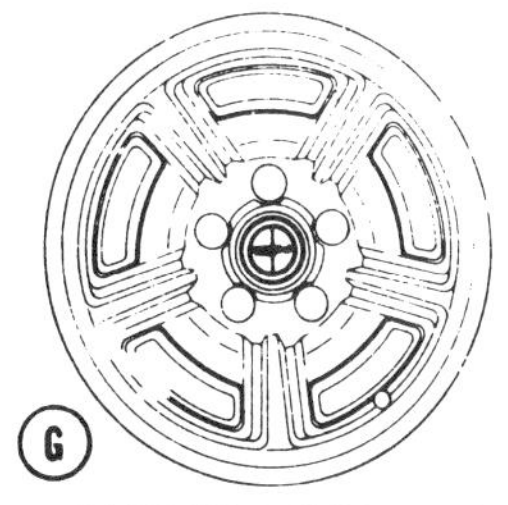

C6AZ-1130-C (16 1/4")
(RED CENTER)
FORD, MUSTANG

Standard and optional 1967 Mustang wheelcovers.

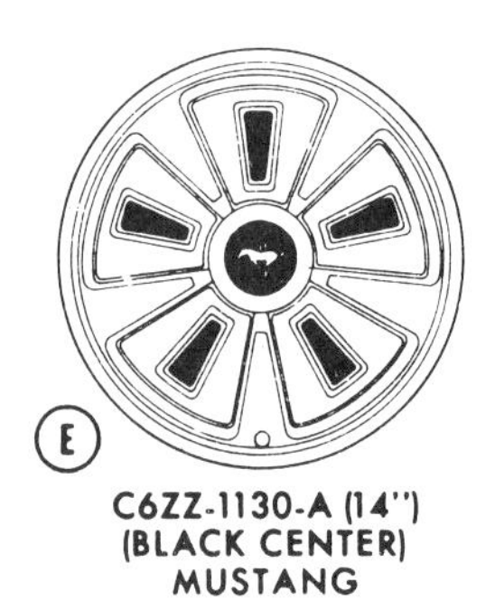

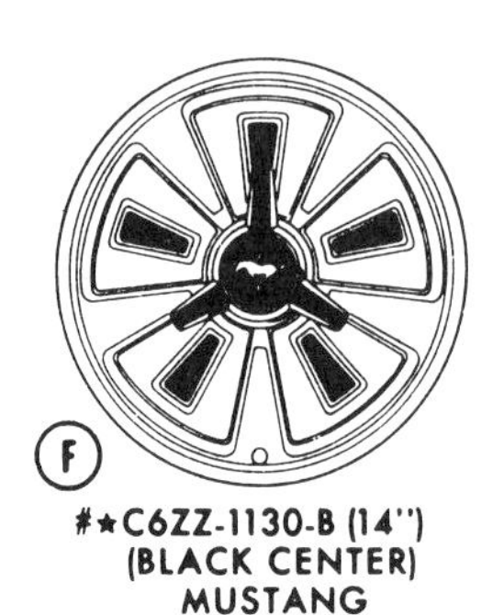

Standard and optional 1966 Mustang wheelcovers (continued).

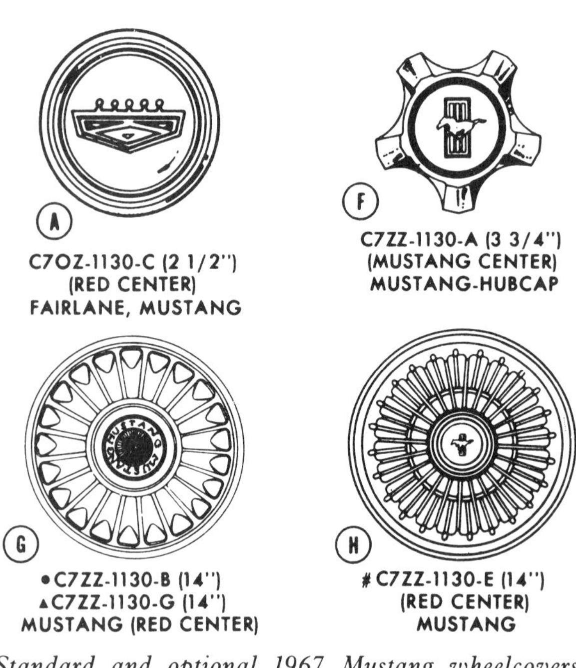

Standard and optional 1967 Mustang wheelcovers (continued).

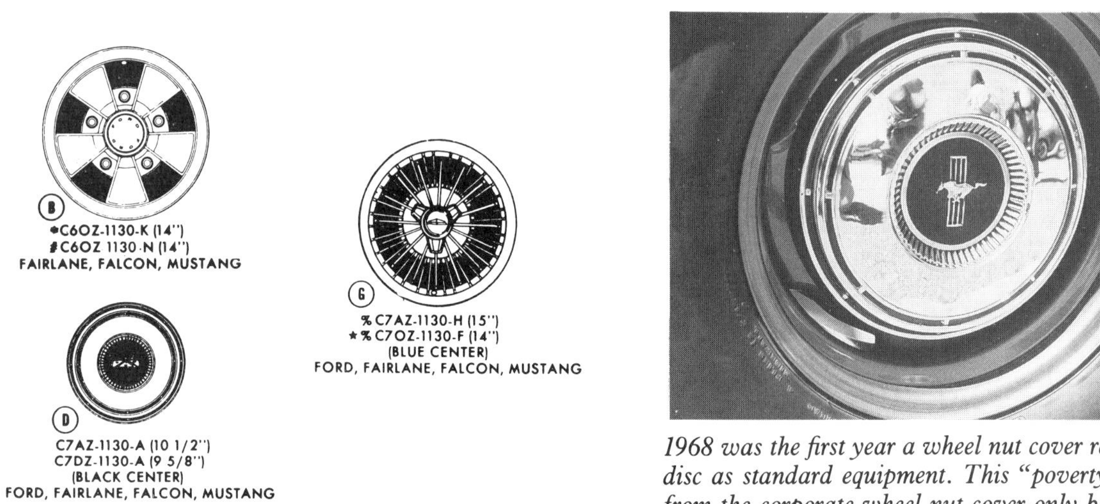

Standard and optional 1967 Mustang wheelcovers (continued).

Styled Steel wheels changed completely in 1968 and remained a separate option for all V-8-equipped body styles. The wheel was argent in color (pictured) or chromed. The polished stainless-steel trim ring and round center cap were the same in either case.

The letters GT graced the center of the cap when the Sport Handling Package was included. These wheels were available through the 1969 model year.

1968 was the first year a wheel nut cover replaced the full disc as standard equipment. This "poverty cap" differed from the corporate wheel nut cover only because it had a Mustang running horse emblem in its center instead of a Ford crest.

The base full disc cover for 1968 had the Mustang name stamped in block letters twice around its center.

Deluxe wheelcovers for 1968 are similar to the base full disc cover except they have a bright plastic center disc encasing a running horse emblem.

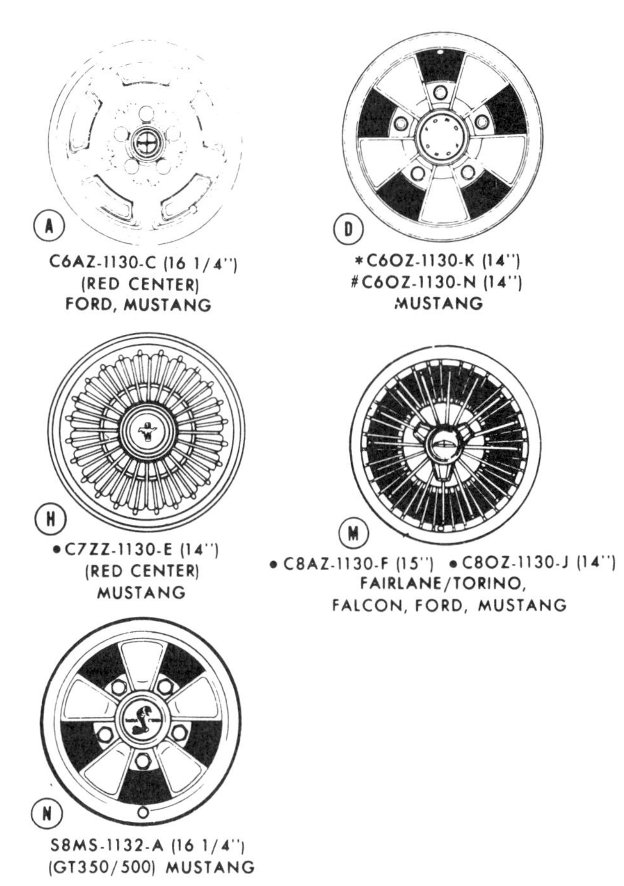

Standard and optional 1968 Mustang wheelcovers.

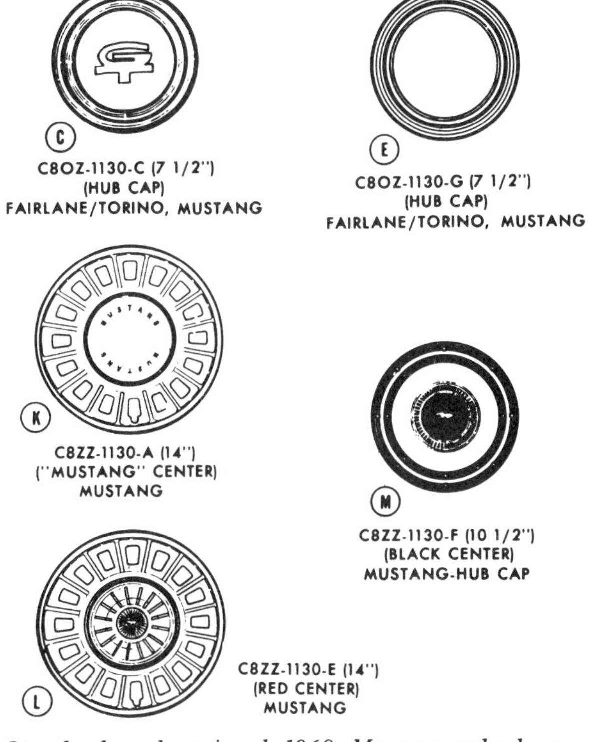

Standard and optional 1968 Mustang wheelcovers (continued).

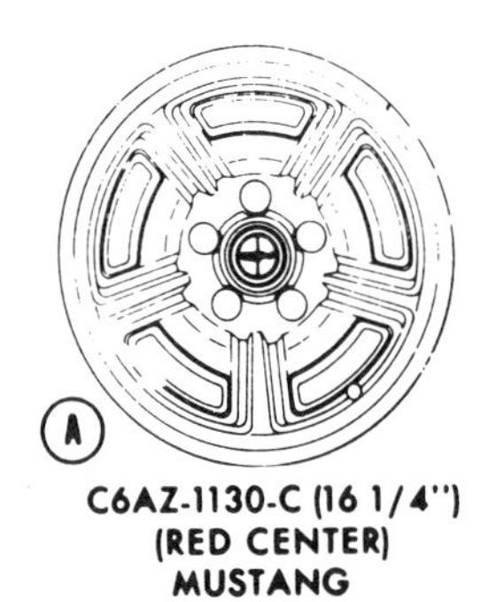

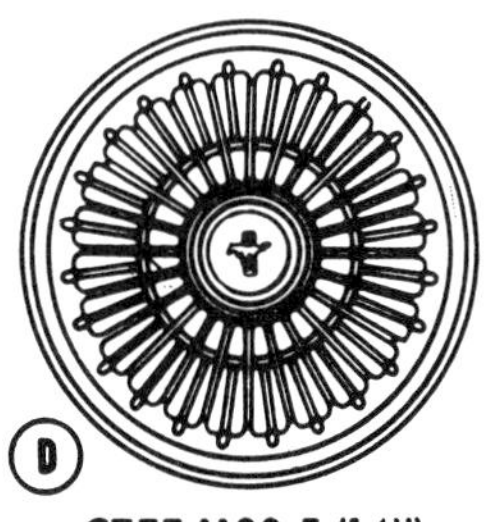

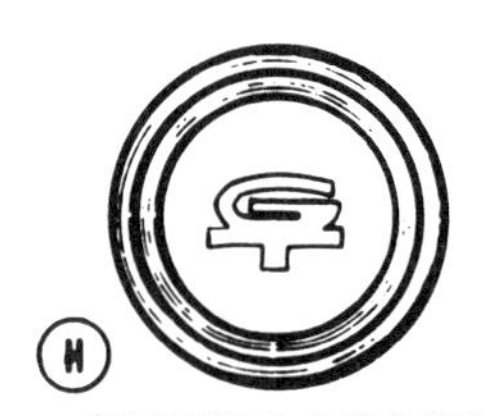

Standard and optional 1969 Mustang wheelcovers.

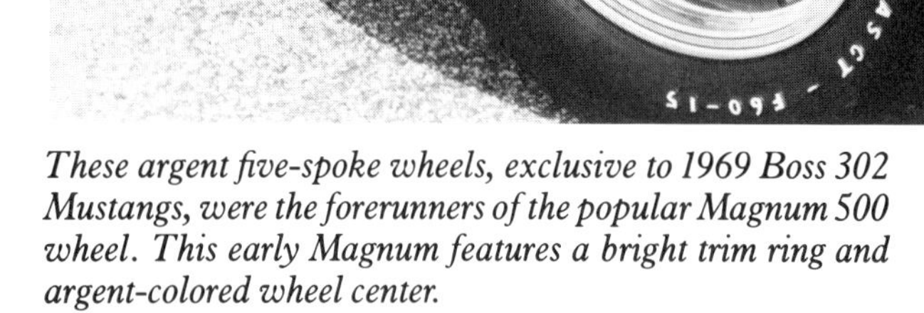

These argent five-spoke wheels, exclusive to 1969 Boss 302 Mustangs, were the forerunners of the popular Magnum 500 wheel. This early Magnum features a bright trim ring and argent-colored wheel center.

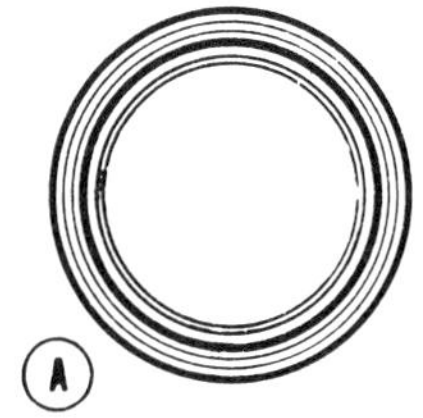

Standard and optional 1969 Mustang wheelcovers (continued).

The base full disc wheelcover for 1969 was this heavy-looking five-spoke version.

The chrome 1968 Styled Steel wheel with smooth center cap was the standard wheel for all 1969 Mach 1 Mustangs. Unfortunately, this wheel was not offered in 15in sizes.

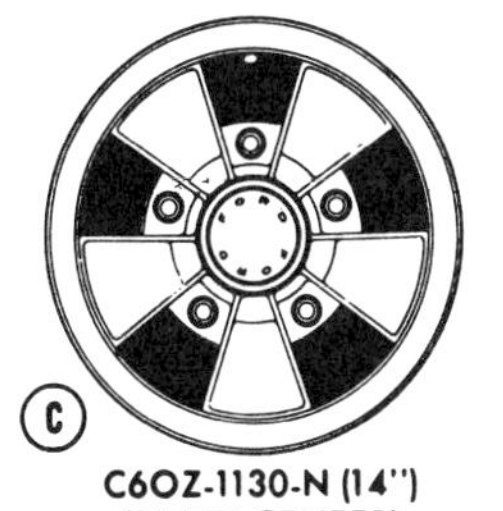

C6OZ-1130-N (14")
(PLAIN CENTER)
FAIRLANE/TORINO, MUSTANG

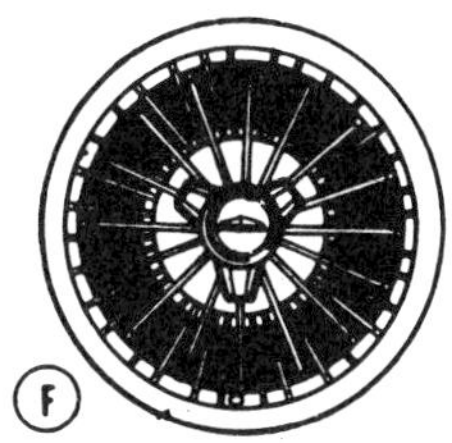

#C8AZ-1130-F (15")
#C8OZ-1130-J (14")
MUSTANG

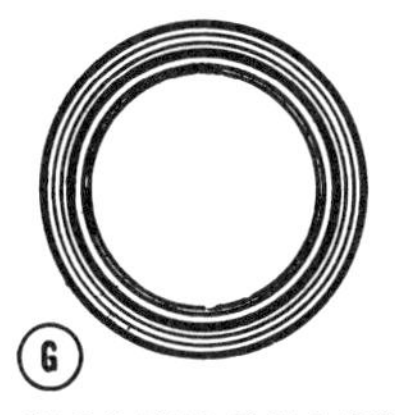

C8OZ-1130-G (7 1/2")
(HUB CAP)
FAIRLANE/TORINO, MUSTANG

C9ZZ-1130-D (2 3/4")
GT-350/500 (BLACK CENTER)
MUSTANG

C9ZZ-1130-E 2 3/4"
(BLACK CENTER)
MUSTANG

Standard and optional 1970 Mustang wheelcovers.

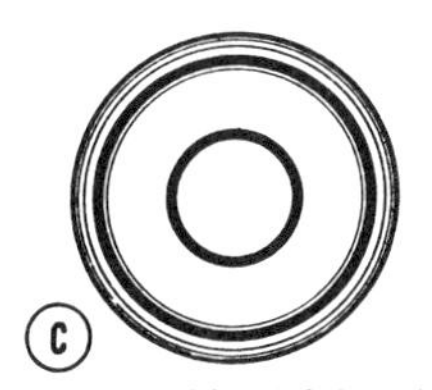

*DOAZ-1130-E (10 1/4")
*DOAZ-1130-H (10 1/4")
(CHROME FINISH-BLACK LETTERS)
#★ DOAZ-1130-G (10 1/4")
*D5DZ-1130-B(10 1/4")
FAIRLANE/TORINO, FORD,
MAVERICK, MUSTANG-HUB CAP

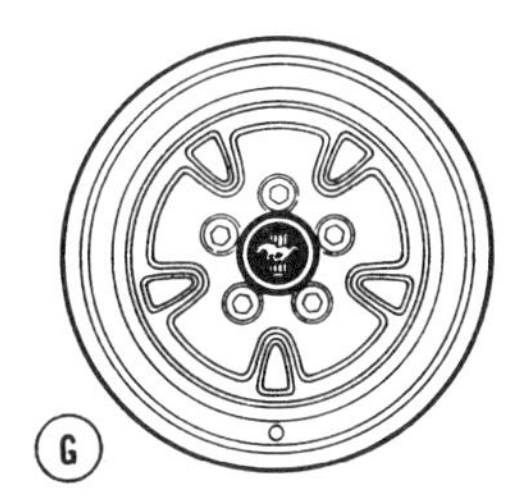

▼%DOOZ-1130-G (16 1/4")
▼□DOOZ-1130-D (15 1/4")
(RED CENTER)
FAIRLANE/TORINO
%DOZZ-1130-F (16 1/4")
□DOZZ-1130-D (15 1/4")
■DOZZ-1130-E (15 1/4")
(BLACK CENTER)
MUSTANG

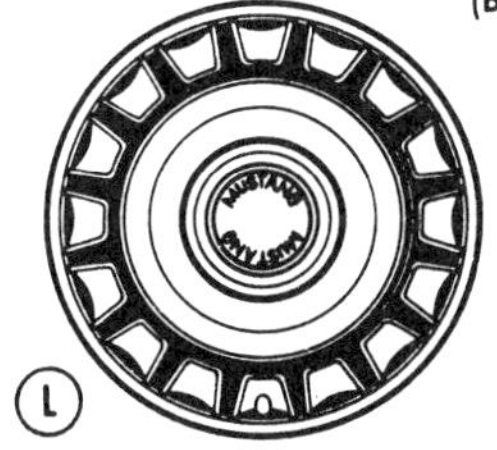

DOZZ-1130-A (14 1/8")
("MUSTANG" CENTER)
MUSTANG

Standard and optional 1970 Mustang wheelcovers (continued).

The Styled Steel wheel changed again in 1970. The rim was painted dark argent and the bright trim ring changed slightly.

More 1970 Mustangs were outfitted with this base full disc cover than the standard 10½in lug nut cover introduced in 1968. This attractive cover has sixteen radial spokes and a large brightmetal center disc.

A new simulated alloy wheelcover with five lug bolts and bright trim ring became standard equipment for the Mach 1 in 1970.

The black-centered Mangum 500 was the standard wheel for all Boss 429 Mustangs and an optional wheel for the 1970 Boss 302.

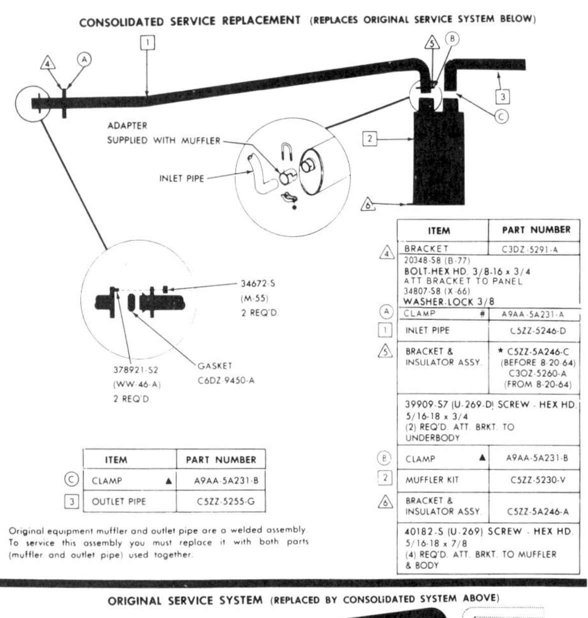

	ITEM	PART NUMBER
4	BRACKET	C3DZ-5291-A
	20348-S8 (B-77) BOLT-HEX HD. 3/8-16 x 3/4 ATT BRACKET TO PANEL 34807-S8 (X-66) WASHER-LOCK 3/8	
A	CLAMP #	A9AA-5A231-A
1	INLET PIPE	C5ZZ-5246-D
5	BRACKET & INSULATOR ASSY.	★ C5ZZ-5A246-C (BEFORE 8-20-64) C3OZ-5260-A (FROM 8-20-64)
	39909-S7 (U-269-D) SCREW - HEX HD. 5/16-18 x 3/4 (2) REQ'D. ATT. BRKT. TO UNDERBODY	
B	CLAMP ▲	A9AA-5A231-B
2	MUFFLER KIT	C5ZZ-5230-V
6	BRACKET & INSULATOR ASSY.	C5ZZ-5A246-A
	40182-S (U-269) SCREW - HEX HD. 5/16-18 x 7/8 (4) REQ'D. ATT. BRKT. TO MUFFLER & BODY	

	ITEM	PART NUMBER
C	CLAMP ▲	A9AA-5A231-B
3	OUTLET PIPE	C5ZZ-5255-G

Original equipment muffler and outlet pipe are a welded assembly. To service this assembly you must replace it with both parts (muffler and outlet pipe) used together.

ORIGINAL SERVICE SYSTEM (REPLACED BY CONSOLIDATED SYSTEM ABOVE)

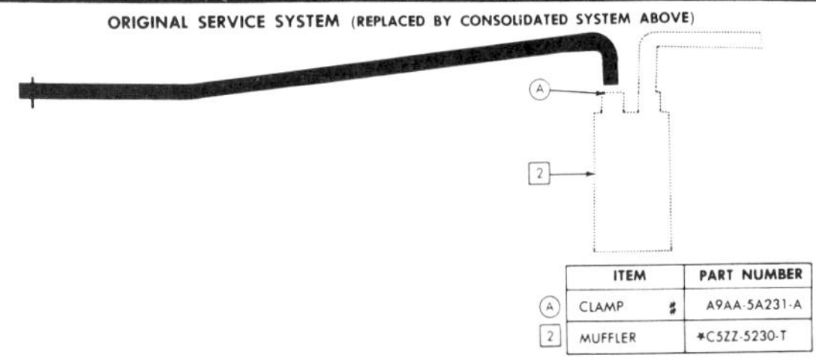

	ITEM	PART NUMBER
A	CLAMP #	A9AA-5A231-A
2	MUFFLER	★C5ZZ-5230-T

Complete exhaust system and related hardware for 1965 Mustangs equipped with 170 and 200ci engines.

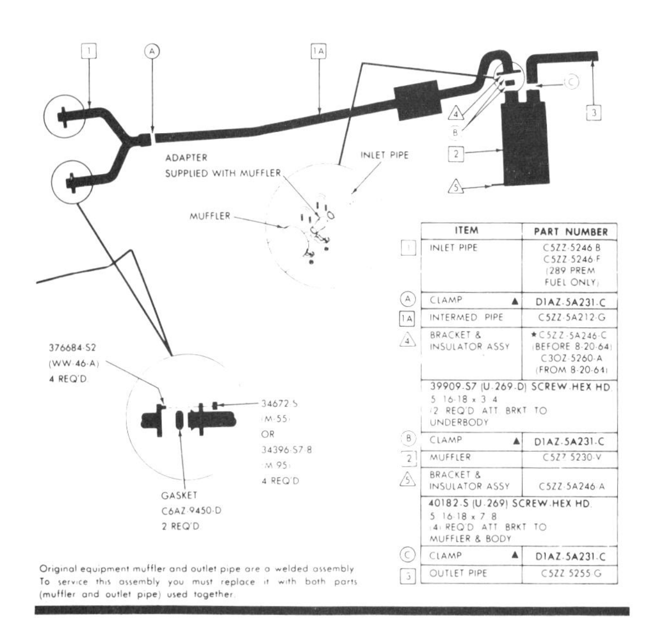

	ITEM	PART NUMBER
1	INLET PIPE	C5ZZ-5246-B C5ZZ-5246-F (289 PREM FUEL ONLY)
A	CLAMP ▲	D1AZ-5A231-C
1A	INTERMED. PIPE	C5ZZ-5A212-G
4	BRACKET & INSULATOR ASSY	★C5ZZ-5A246-C (BEFORE 8-20-64) C3OZ-5260-A (FROM 8-20-64)
	39909-S7 (U-269-D) SCREW-HEX HD. 5/16-18 x 3/4 (2) REQ'D ATT BRKT TO UNDERBODY	
B	CLAMP ▲	D1AZ-5A231-C
2	MUFFLER	C5ZZ-5230-V
5	BRACKET & INSULATOR ASSY	C5ZZ-5A246-A
	40182-S (U-269) SCREW-HEX HD. 5/16-18 x 7/8 (4) REQ'D ATT BRKT TO MUFFLER & BODY	
C	CLAMP ▲	D1AZ-5A231-C
3	OUTLET PIPE	C5ZZ-5255-G

Original equipment muffler and outlet pipe are a welded assembly. To service this assembly you must replace it with both parts (muffler and outlet pipe) used together.

ORIGINAL SERVICE SYSTEM (REPLACED BY CONSOLIDATED SYSTEM ABOVE)

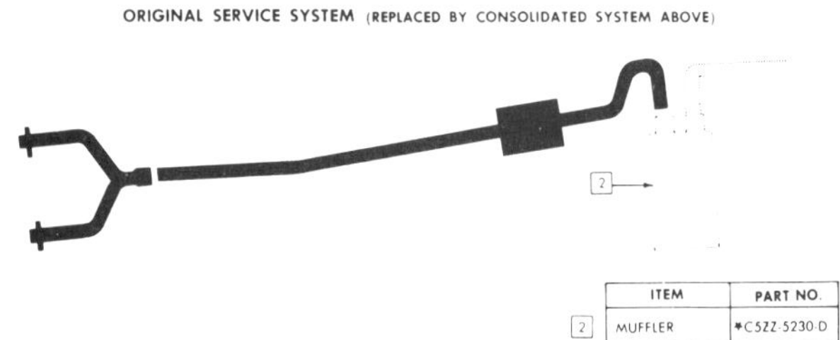

	ITEM	PART NO.
2	MUFFLER	★C5ZZ-5230-D

Complete exhaust system and related hardware for 1965 Mustangs equipped with 289 4-V premium-fuel engines and 260 and 289 2-V regular-fuel engines.

1965 MUSTANG—8 CYL. 289-4B SPECIAL (BEFORE 10-15-64)—INLINE MUFFLER

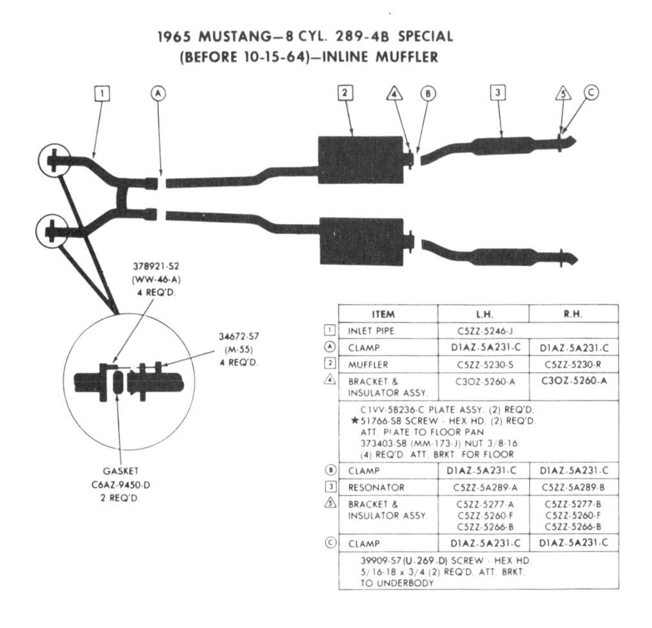

	ITEM	L.H.	R.H.
1	INLET PIPE	C5ZZ-5246-J	
A	CLAMP	D1AZ-5A231-C	D1AZ-5A231-C
2	MUFFLER	C5ZZ-5230-S	C5ZZ-5230-R
4	BRACKET & INSULATOR ASSY.	C3OZ-5260-A	C3OZ-5260-A
	C1VV-5B236-C PLATE ASSY. (2) REQ'D. ★51766-S8 SCREW - HEX HD. (2) REQ'D. ATT. PLATE TO FLOOR PAN 373403-S8 (MM-173-J) NUT 3/8-16 (4) REQ'D. ATT. BRKT. FOR FLOOR		
B	CLAMP	D1AZ-5A231-C	D1AZ-5A231-C
3	RESONATOR	C5ZZ-5A289-A	C5ZZ-5A289-B
5	BRACKET & INSULATOR ASSY.	C5ZZ-5277-A C5ZZ-5260-F C5ZZ-5266-B	C5ZZ-5277-B C5ZZ-5260-F C5ZZ-5266-B
C	CLAMP	D1AZ-5A231-C	D1AZ-5A231-C
	39909-S7 (U-269-D) SCREW - HEX HD. 5/16-18 x 3/4 (2) REQ'D. ATT. BRKT. TO UNDERBODY		

Complete exhaust system and related hardware for 1965 Mustangs equipped with 289 High Performance engines built before 10/15/64.

Exhaust tips or trumpets were more cosmetic than functional, and were a standard feature on all GT and Mach 1 Mustangs. The 1965-1966 GT exhaust trumpets protrude through a chrome-trimmed hole in the rear valance. Backup or reverse lights were optional in 1965 and standard equipment in 1966 on GT models.

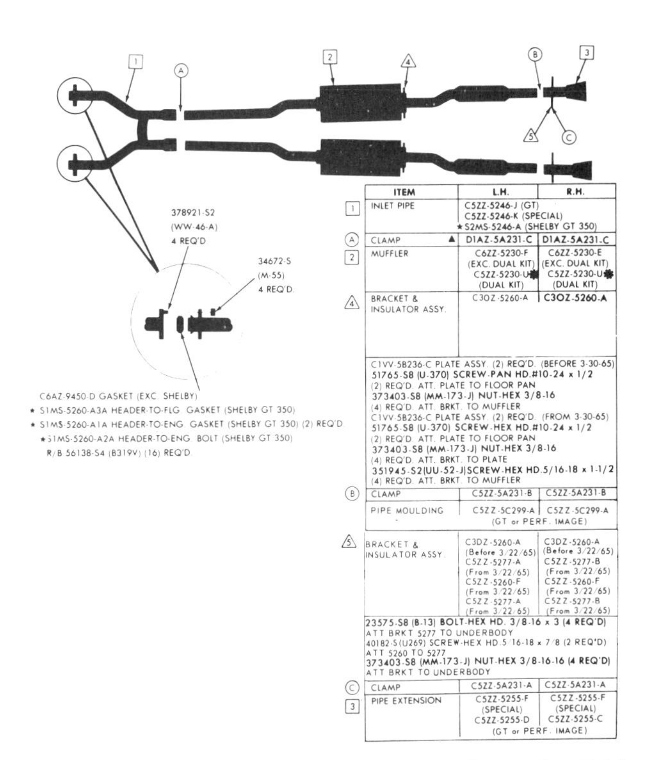

	ITEM	L.H.	R.H.
1	INLET PIPE	C5ZZ-5246-J (GT) C5ZZ-5246-K (SPECIAL) ★S2MS-5246-A (SHELBY GT 350)	
A	CLAMP ▲	D1AZ-5A231-C	D1AZ-5A231-C
2	MUFFLER	C6ZZ-5230-F (EXC. DUAL KIT) C5ZZ-5230-U✱ (DUAL KIT)	C6ZZ-5230-E (EXC. DUAL KIT) C5ZZ-5230-U✱ (DUAL KIT)
4	BRACKET & INSULATOR ASSY.	C3OZ-5260-A	C3OZ-5260-A
	C1VV-5B236-C PLATE ASSY. (2) REQ'D. (BEFORE 3-30-65) 51765-S8 (U-370) SCREW-PAN HD.#10-24 x 1/2 (2) REQ'D. ATT. PLATE TO FLOOR PAN 373403-S8 (MM-173-J) NUT-HEX 3/8-16 (4) REQ'D. ATT. BRKT. TO MUFFLER C1VV-5B236-C PLATE ASSY. (2) REQ'D. (FROM 3-30-65) 51765-S8 (U-370) SCREW-HEX HD.#10-24 x 1/2 (2) REQ'D. ATT. PLATE TO FLOOR PAN 373403-S8 (MM-173-J) NUT-HEX 3/8-16 (4) REQ'D. ATT. BRKT. TO PLATE 351945-S2(UU-52-J)SCREW-HEX HD.5/16-18 x 1-1/2 (4) REQ'D. ATT. BRKT. TO MUFFLER		
B	CLAMP	C5ZZ-5A231-B	C5ZZ-5A231-B
	PIPE MOULDING	C5ZZ-5C299-A (GT or PERF. IMAGE)	C5ZZ-5C299-A
5	BRACKET & INSULATOR ASSY.	C3DZ-5260-A (Before 3/22/65) C5ZZ-5277-A (From 3/22/65) C5ZZ-5260-F (From 3/22/65) C5ZZ-5277-A (From 3/22/65)	C3DZ-5260-A (Before 3/22/65) C5ZZ-5277-B (From 3/22/65) C5ZZ-5260-F (From 3/22/65) C5ZZ-5277-B (From 3/22/65)
	23575-S8 (B-13) BOLT-HEX HD. 3/8-16 x 3 (4 REQ'D) ATT BRKT 5277 TO UNDERBODY 40182-S(U269) SCREW-HEX HD.5/16-18 x 7/8 (2 REQ'D) ATT 5260 TO 5277 373403-S8 (MM-173-J) NUT-HEX 3/8-16-16 (4 REQ'D) ATT BRKT TO UNDERBODY		
C	CLAMP	C5ZZ-5A231-A	C5ZZ-5A231-A
3	PIPE EXTENSION	C5ZZ-5255-F (SPECIAL) C5ZZ-5255-D (GT or PERF. IMAGE)	C5ZZ-5255-F (SPECIAL) C5ZZ-5255-C

Complete exhaust system and related hardware for 1965 GT Mustangs equipped with 289 4-V engines, non-GT Mustangs equipped with 289 High Performance engines, and GT-350 Shelby Mustangs (all built from 10/15/64).

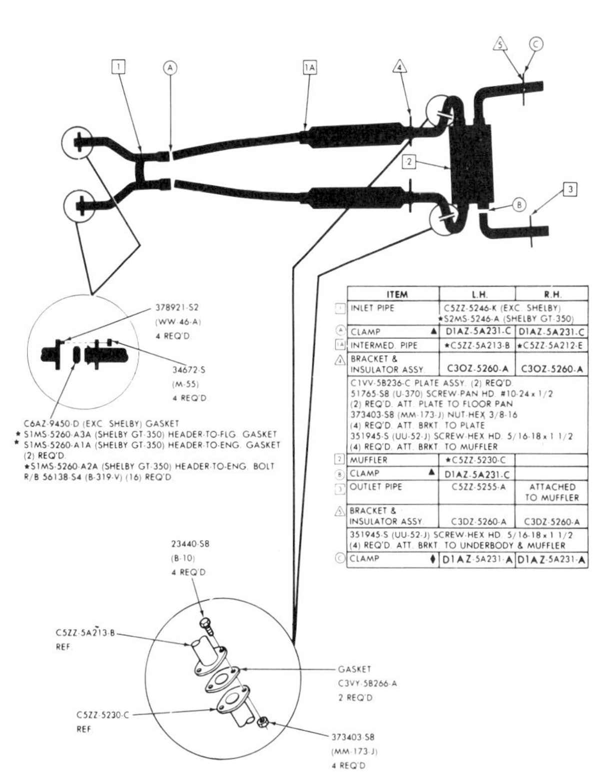

	ITEM	L.H.	R.H.
1	INLET PIPE	C5ZZ-5246-K (EXC. SHELBY) ★S2MS-5246-A (SHELBY GT-350)	
A	CLAMP ▲	D1AZ-5A231-C	D1AZ-5A231-C
1A	INTERMED. PIPE	★C5ZZ-5A213-B	★C5ZZ-5A212-E
4	BRACKET & INSULATOR ASSY.	C3OZ-5260-A	C3OZ-5260-A
	C1VV-5B236-C PLATE ASSY. (2) REQ'D. 51765-S8 (U-370) SCREW-PAN HD. #10-24 x 1/2 (2) REQ'D. ATT. PLATE TO FLOOR PAN 373403-S8 (MM-173-J) NUT-HEX 3/8-16 (4) REQ'D. ATT. BRKT. TO PLATE 351945-S (UU-52-J) SCREW-HEX HD. 5/16-18 x 1 1/2 (4) REQ'D. ATT. BRKT. TO MUFFLER		
2	MUFFLER	★C5ZZ-5230-C	
B	CLAMP ▲	D1AZ-5A231-C	
3	OUTLET PIPE	C5ZZ-5255-A	ATTACHED TO MUFFLER
5	BRACKET & INSULATOR ASSY.	C3DZ-5260-A	C3DZ-5260-A
	351945-S (UU-52-J) SCREW-HEX HD. 5/16-18 x 1 1/2 (4) REQ'D. ATT. BRKT. TO UNDERBODY & MUFFLER		
C	CLAMP ♦	D1AZ-5A231-A	D1AZ-5A231-A

Complete exhaust system and related hardware for 1965 Mustangs and GT-350 Shelby Mustangs (built before 10/15/64) equipped with 289 4-V engines.

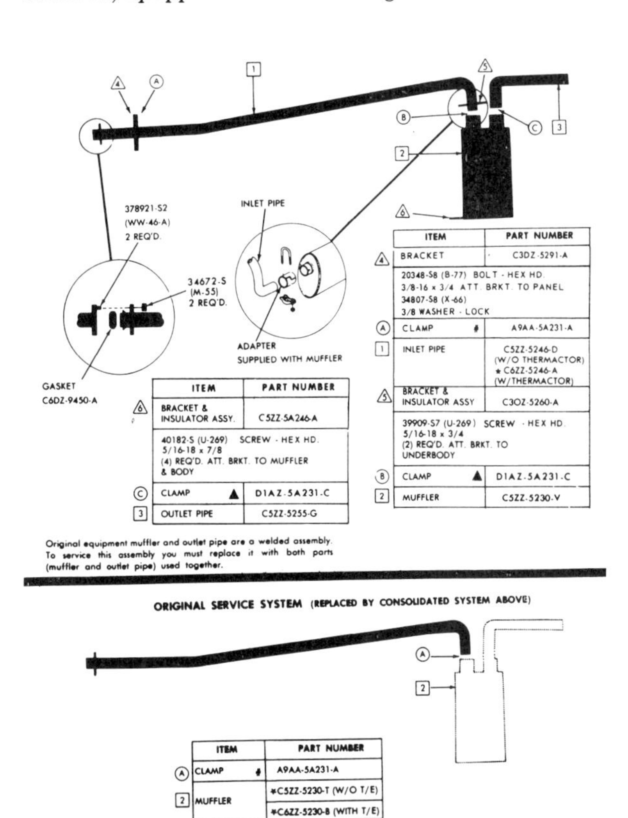

	ITEM	PART NUMBER
6	BRACKET & INSULATOR ASSY.	C5ZZ-5A246-A
	40182-S (U-269) SCREW - HEX HD. 5/16-18 x 7/8 (4) REQ'D. ATT. BRKT. TO MUFFLER & BODY	
C	CLAMP ▲	D1AZ-5A231-C
3	OUTLET PIPE	C5ZZ-5255-G

	ITEM	PART NUMBER
4	BRACKET	C3DZ-5291-A
	20348-S8 (B-77) BOLT - HEX HD. 3/8-16 x 3/4 ATT. BRKT. TO PANEL 34807-S8 (X-66) 3/8 WASHER - LOCK	
A	CLAMP #	A9AA-5A231-A
1	INLET PIPE	C5ZZ-5246-D (W/O THERMACTOR) ★ C6ZZ-5246-A (W/THERMACTOR)
5	BRACKET & INSULATOR ASSY	C3OZ-5260-A
	39909-S7 (U-269) SCREW - HEX HD. 5/16-18 x 3/4 (2) REQ'D. ATT. BRKT. TO UNDERBODY	
B	CLAMP ▲	D1AZ-5A231-C
2	MUFFLER	C5ZZ-5230-V

Original equipment muffler and outlet pipe are a welded assembly. To service this assembly you must replace it with both parts (muffler and outlet pipe) used together.

ORIGINAL SERVICE SYSTEM (REPLACED BY CONSOLIDATED SYSTEM ABOVE)

	ITEM	PART NUMBER
A	CLAMP ♦	A9AA-5A231-A
2	MUFFLER	★C5ZZ-5230-T (W/O T/E) ★C6ZZ-5230-B (WITH T/E)

Complete exhaust system and related hardware for 1966 Mustangs equipped with 200ci engines.

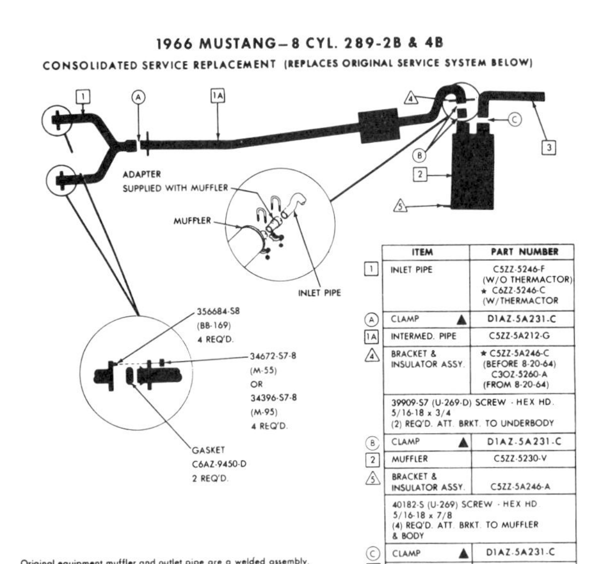

	ITEM	PART NUMBER
1	INLET PIPE	C5ZZ-5246-F (W/O THERMACTOR) ★ C6ZZ-5246-C (W/THERMACTOR
A	CLAMP ▲	D1AZ-5A231-C
1A	INTERMED. PIPE	C5ZZ-5A212-G
4	BRACKET & INSULATOR ASSY.	★ C5ZZ-5A246-C (BEFORE 8-20-64) C3OZ-5260-A (FROM 8-20-64)
	39909-S7 (U-269-D) SCREW - HEX HD. 5/16-18 x 3/4 (2) REQ'D. ATT. BRKT. TO UNDERBODY	
B	CLAMP ▲	D1AZ-5A231-C
2	MUFFLER	C5ZZ-5230-V
5	BRACKET & INSULATOR ASSY.	C5ZZ-5A246-A
	40182-S (U-269) SCREW - HEX HD 5/16-18 x 7/8 (4) REQ'D. ATT. BRKT. TO MUFFLER & BODY	
C	CLAMP ▲	D1AZ-5A231-C
3	OUTLET PIPE	C5ZZ-5255-G

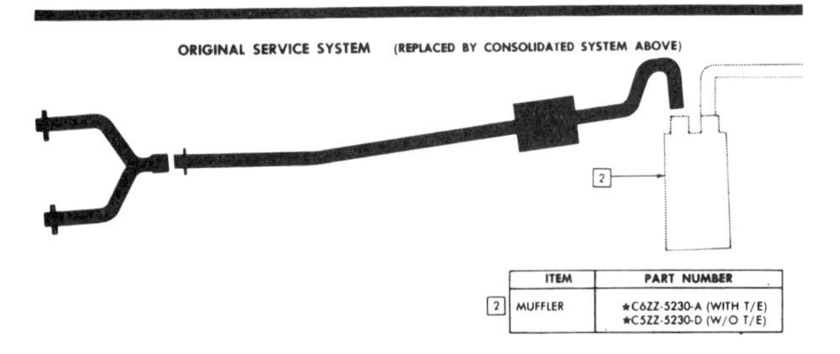

	ITEM	PART NUMBER
2	MUFFLER	★C6ZZ-5230-A (WITH T/E) ★C5ZZ-5230-D (W/O T/E)

Complete exhaust system and related hardware for 1966 Mustangs equipped with 289 2-V and 4-V engines.

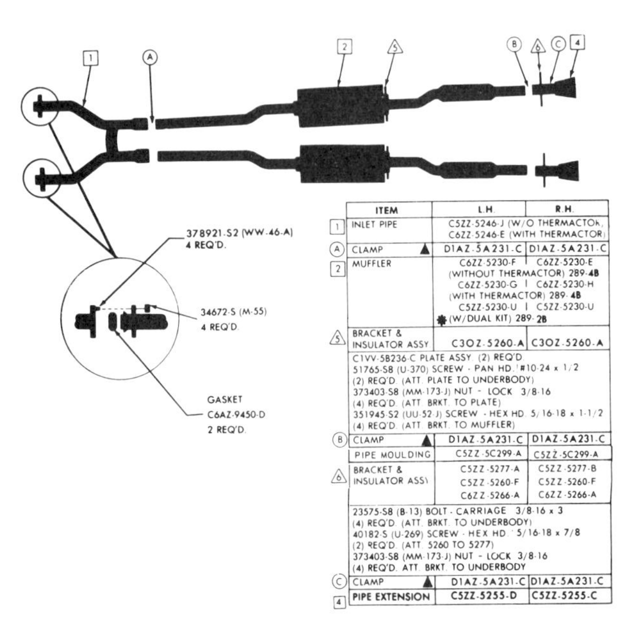

	ITEM	L.H.	R.H.
1	INLET PIPE	C5ZZ-5246-J (W/O THERMACTOR, C6ZZ-5246-E (WITH THERMACTOR)	
A	CLAMP ▲	D1AZ-5A231-C	D1AZ-5A231-C
2	MUFFLER	C6ZZ-5230-F (WITHOUT THERMACTOR) 289-4B C6ZZ-5230-G (WITH THERMACTOR) 289-4B C5ZZ-5230-U ✱ (W/DUAL KIT) 289-2B	C6ZZ-5230-E C6ZZ-5230-H C5ZZ-5230-U
5	BRACKET & INSULATOR ASSY	C3OZ-5260-A	C3OZ-5260-A
	C1VV-5B236-C PLATE ASSY. (2) REQ'D. 51765-S8 (U-370) SCREW - PAN HD. #10-24 x 1/2 (2) REQ'D. (ATT. PLATE TO UNDERBODY) 373403-S8 (MM-173-J) NUT - LOCK 3/8-16 (4) REQ'D. (ATT. BRKT. TO PLATE) 351945-S2 (UU-52-J) SCREW - HEX HD. 5/16-18 x 1-1/2 (4) REQ'D. (ATT. BRKT. TO MUFFLER)		
B	CLAMP ▲	D1AZ-5A231-C	D1AZ-5A231-C
	PIPE MOULDING	C5ZZ-5C299-A	C5ZZ-5C299-A
6	BRACKET & INSULATOR ASSY	C5ZZ-5277-A C5ZZ-5260-F C6ZZ-5266-A	C5ZZ-5277-B C5ZZ-5260-F C6ZZ-5266-A
	23575-S8 (B-13) BOLT - CARRIAGE 3/8-16 x 3 (4) REQ'D. (ATT. BRKT. TO UNDERBODY) 40182-S (U-269) SCREW - HEX HD. 5/16-18 x 7/8 (2) REQ'D. (ATT. 5260 TO 5277) 373403-S8 (MM-173-J) NUT - LOCK 3/8-16 (4) REQ'D. ATT. BRKT. TO UNDERBODY		
C	CLAMP ▲	D1AZ-5A231-C	D1AZ-5A231-C
4	PIPE EXTENSION	C5ZZ-5255-D	C5ZZ-5255-C

Complete exhaust system and related hardware for 1966 GT Mustangs equipped with 289 4-V engines, Mustangs equipped with optional performance image dual exhaust, and 298 V-2 engines with dual exhaust kit.

1966 MUSTANG—8 CYL. 289-4B PREM. FUEL - WITH STD. PIPE EXTENSIONS (INCLUDES MODELS W/289-2B & DUAL EXHAUST KIT - ✱)

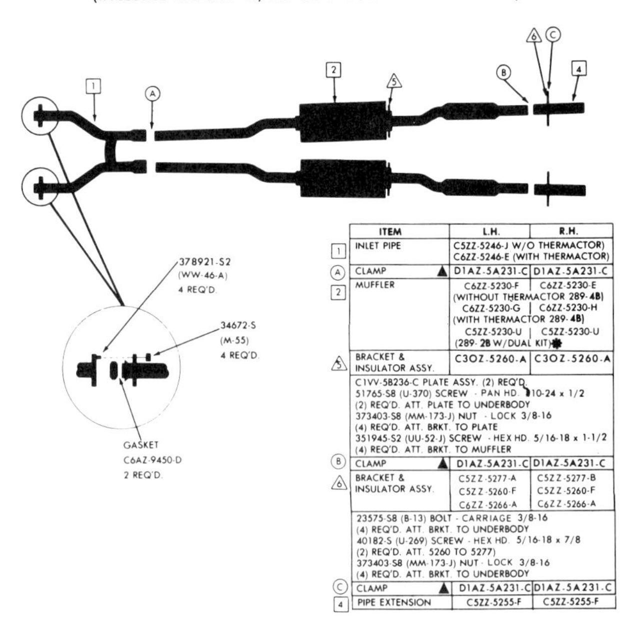

	ITEM	L.H.	R.H.
1	INLET PIPE	C5ZZ-5246-J W/O THERMACTOR) C6ZZ-5246-E (WITH THERMACTOR)	
A	CLAMP ▲	D1AZ-5A231-C	D1AZ-5A231-C
2	MUFFLER	C6ZZ-5230-F (WITHOUT THERMACTOR 289-4B) C6ZZ-5230-G (WITH THERMACTOR 289-4B) C5ZZ-5230-U (289-2B W/DUAL KIT)✱	C6ZZ-5230-E C6ZZ-5230-H C5ZZ-5230-U
5	BRACKET & INSULATOR ASSY.	C3OZ-5260-A	C3OZ-5260-A
	C1VV-5B236-C PLATE ASSY. (2) REQ'D. 51765-S8 (U-370) SCREW - PAN HD. #10-24 x 1/2 (2) REQ'D. ATT. PLATE TO UNDERBODY 373403-S8 (MM-173-J) NUT - LOCK 3/8-16 (4) REQ'D. ATT. BRKT. TO PLATE 351945-S2 (UU-52-J) SCREW - HEX HD. 5/16-18 x 1-1/2 (4) REQ'D. ATT. BRKT. TO MUFFLER		
B	CLAMP ▲	D1AZ-5A231-C	D1AZ-5A231-C
6	BRACKET & INSULATOR ASSY.	C5ZZ-5277-A C5ZZ-5260-F C6ZZ-5266-A	C5ZZ-5277-B C5ZZ-5260-F C6ZZ-5266-A
	23575-S8 (B-13) BOLT - CARRIAGE 3/8-16 (4) REQ'D. ATT. BRKT. TO UNDERBODY 40182-S (U-269) SCREW - HEX HD. 5/16-18 x 7/8 (2) REQ'D. ATT. 5260 TO 5277) 373403-S8 (MM-173-J) NUT - LOCK 3/8-16 (4) REQ'D. ATT. BRKT. TO UNDERBODY		
C	CLAMP ▲	D1AZ-5A231-C	D1AZ-5A231-C
4	PIPE EXTENSION	C5ZZ-5255-F	C5ZZ-5255-F

Complete exhaust system and related hardware for 1966 Mustangs equipped with 289 4-V premium-fuel engines with standard pipe extensions and 298 V-2 engines with dual exhaust kit.

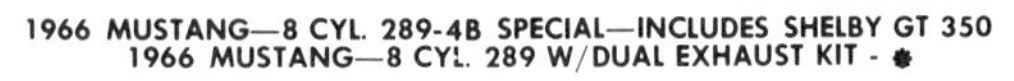

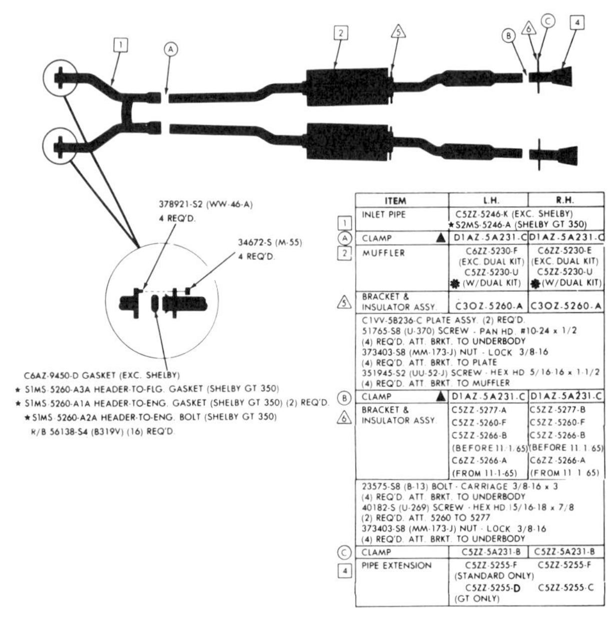

	ITEM	L.H.	R.H.
1	INLET PIPE	C5ZZ-5246-K (EXC. SHELBY) ★S2MS-5246-A (SHELBY GT 350)	
A	CLAMP ▲	D1AZ-5A231-C	D1AZ-5A231-C
2	MUFFLER	C6ZZ-5230-F (EXC. DUAL KIT) C5ZZ-5230-U ✱ (W/DUAL KIT)	C6ZZ-5230-E (EXC. DUAL KIT) C5ZZ-5230-U ✱(W/DUAL KIT)
5	BRACKET & INSULATOR ASSY.	C3OZ-5260-A	C3OZ-5260-A
	C1VV-5B236-C PLATE ASSY. (2) REQ'D. 51765-S8 (U-370) SCREW - PAN HD. #10-24 x 1/2 (4) REQ'D. ATT. BRKT. TO UNDERBODY 373403-S8 (MM-173-J) NUT - LOCK 3/8-16 (4) REQ'D. ATT. BRKT. TO PLATE 351945-S2 (UU-52-J) SCREW - HEX HD 5/16-16 x 1-1/2 (4) REQ'D. ATT. BRKT. TO MUFFLER		
B	CLAMP ▲	D1AZ-5A231-C	D1AZ-5A231-C
6	BRACKET & INSULATOR ASSY.	C5ZZ-5277-A C5ZZ-5260-F C5ZZ-5266-B (BEFORE 11-1-65) C6ZZ-5266-A (FROM 11-1-65)	C5ZZ-5277-B C5ZZ-5260-F C5ZZ-5266-B (BEFORE 11-1-65) C6ZZ-5266-A (FROM 11-1-65)
	23575-S8 (B-13) BOLT - CARRIAGE 3/8-16 x 3 (4) REQ'D. ATT. BRKT. TO UNDERBODY 40182-S (U-269) SCREW - HEX HD 5/16-18 x 7/8 (2) REQ'D. ATT. 5260 TO 5277 373403-S8 (MM-173-J) NUT - LOCK 3/8-16 (4) REQ'D. ATT. BRKT. TO UNDERBODY		
C	CLAMP	C5ZZ-5A231-B	C5ZZ-5A231-B
4	PIPE EXTENSION	C5ZZ-5255-F (STANDARD ONLY) C5ZZ-5255-D (GT ONLY)	C5ZZ-5255-F C5ZZ-5255-C

Complete exhaust system and related hardware for 1966 Mustangs equipped with 289 High Performance engines, Shelby GT-350 Mustangs, and 289 engines equipped with dual exhaust kit.

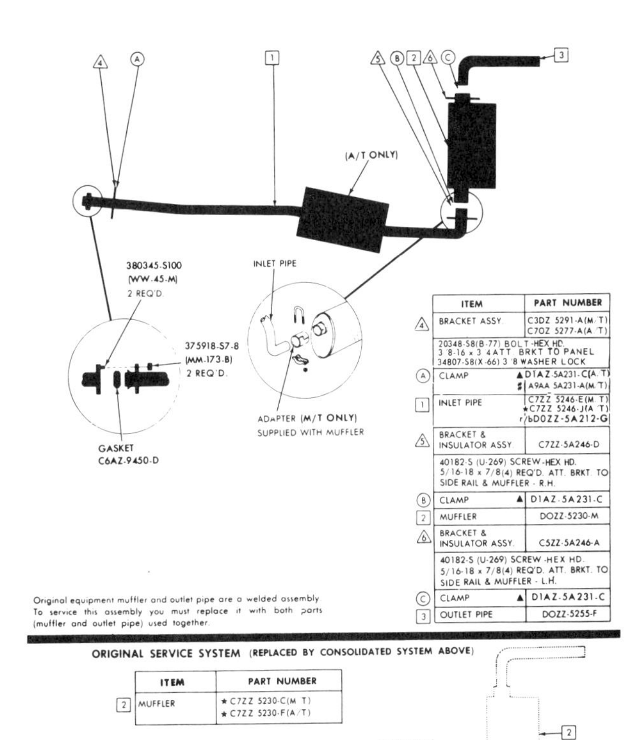

	ITEM	PART NUMBER
4	BRACKET ASSY.	C3DZ 5291-A(M/T) C7OZ 5277-A(A/T)
	20348-S8(B-77) BOLT-HEX HD. 3/8-16 x 3/4 ATT. BRKT TO PANEL 34807-S8(X-66) 3/8 WASHER LOCK	
A	CLAMP	▲ D1AZ-5A231-C(A/T) ‡ A9AA 5A231-A(M/T)
1	INLET PIPE	C7ZZ 5246-E(M/T) ★C7ZZ 5246-J(A/T) r/b D0ZZ-5A212-G
5	BRACKET & INSULATOR ASSY.	C7ZZ-5A246-D
	40182-S (U-269) SCREW-HEX HD. 5/16-18 x 7/8(4) REQ'D. ATT. BRKT. TO SIDE RAIL & MUFFLER - R.H.	
B	CLAMP	▲ D1AZ-5A231-C
2	MUFFLER	D0ZZ-5230-M
6	BRACKET & INSULATOR ASSY.	C5ZZ-5A246-A
	40182-S (U-269) SCREW-HEX HD. 5/16-18 x 7/8(4) REQ'D. ATT. BRKT. TO SIDE RAIL & MUFFLER - L.H.	
C	CLAMP	▲ D1AZ-5A231-C
3	OUTLET PIPE	D0ZZ-5255-F

	ITEM	PART NUMBER
2	MUFFLER	★C7ZZ 5230-C(M/T) ★C7ZZ 5230-F(A/T)

Complete exhaust system and related hardware for 1967 Mustangs equipped with 200ci engines.

These quad tips split each tailpipe into two smaller-diameter chrome tubes with angle-cut openings. This style exhaust outlet was standard equipment on 1967-1969 Mustang GTs and the 1969 Mach 1.

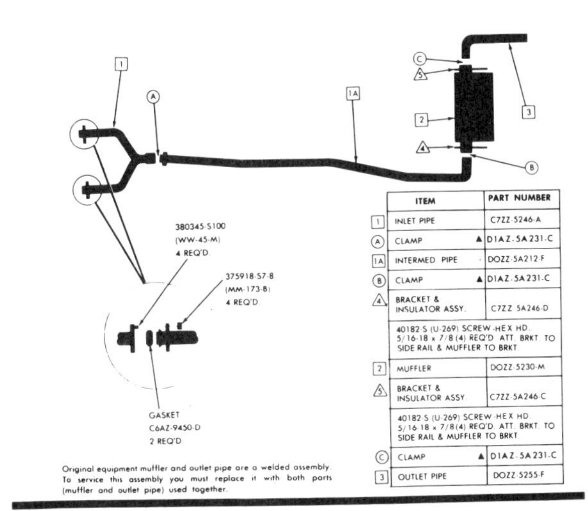

	ITEM	PART NUMBER
1	INLET PIPE	C7ZZ-5246-A
A	CLAMP	▲ D1AZ-5A231-C
1A	INTERMED. PIPE	D0ZZ-5A212-F
B	CLAMP	▲ D1AZ-5A231-C
4	BRACKET & INSULATOR ASSY.	C7ZZ 5A246-D
	40182-S (U-269) SCREW-HEX HD. 5/16-18 x 7/8 (4) REQ'D. ATT. BRKT. TO SIDE RAIL & MUFFLER TO BRKT.	
2	MUFFLER	D0ZZ-5230-M
5	BRACKET & INSULATOR ASSY.	C7ZZ-5A246-C
	40182-S (U-269) SCREW-HEX HD. 5/16-18 x 7/8(4) REQ'D. ATT. BRKT. TO SIDE RAIL & MUFFLER TO BRKT.	
C	CLAMP	▲ D1AZ-5A231-C
3	OUTLET PIPE	D0ZZ-5255-F

ORIGINAL SERVICE SYSTEM (REPLACED BY CONSOLIDATED SYSTEM ABOVE)

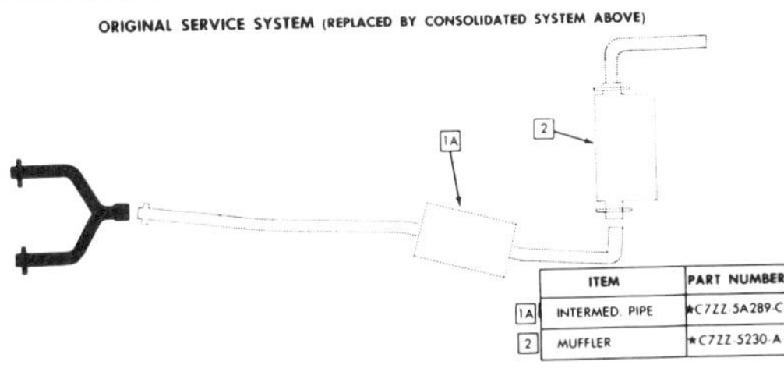

	ITEM	PART NUMBER
1A	INTERMED. PIPE	★C7ZZ-5A289-C
2	MUFFLER	★C7ZZ 5230-A

Complete exhaust system and related hardware for 1967 Mustangs equipped with 289 2-V and 4-V premium-fuel engines.

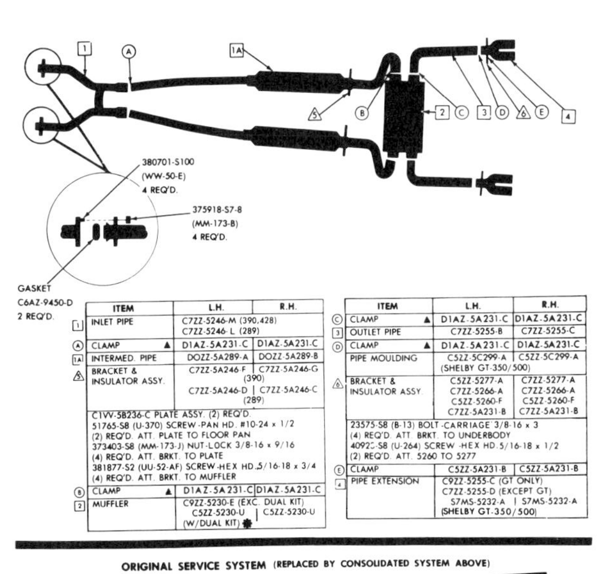

	ITEM	L.H.	R.H.
1	INLET PIPE	C7ZZ-5246-M (390,428) C7ZZ-5246-L (289)	
A	CLAMP ▲	D1AZ-5A231-C	D1AZ-5A231-C
1A	INTERMED. PIPE	D0ZZ-5A289-A	D0ZZ-5A289-B
5	BRACKET & INSULATOR ASSY.	C7ZZ-5A246-F (390) C7ZZ-5A246-D (289)	C7ZZ-5A246-G (390) C7ZZ-5A246-C (289)
	C1VV-5B236-C PLATE ASSY. (2) REQ'D. 51765-S8 (U-370) SCREW-PAN HD. #10-24 x 1/2 (2) REQ'D. ATT. PLATE TO FLOOR PAN 373403-S8 (MM-173-J) NUT-LOCK 3/8-16 x 9/16 (4) REQ'D. ATT. BRKT. TO PLATE 381877-S2 (UU-52-AF) SCREW-HEX HD. 5/16-18 x 3/4 (4) REQ'D. ATT. BRKT. TO MUFFLER		
B	CLAMP ▲	D1AZ-5A231-C	D1AZ-5A231-C
2	MUFFLER	C9ZZ-5230-E (EXC. DUAL KIT) C5ZZ-5230-U (W/DUAL KIT) ✱	C5ZZ-5230-U

	ITEM	L.H.	R.H.
C	CLAMP ▲	D1AZ-5A231-C	D1AZ-5A231-C
3	OUTLET PIPE	C7ZZ-5255-B	C7ZZ-5255-C
D	CLAMP ▲	D1AZ-5A231-C	D1AZ-5A231-C
	PIPE MOULDING	C5ZZ-5C299-A (SHELBY GT-350/500)	C5ZZ-5C299-A
6	BRACKET & INSULATOR ASSY.	C5ZZ-5277-A C7ZZ-5266-A C5ZZ-5260-F C7ZZ-5A231-B	C7ZZ-5277-A C7ZZ-5266-A C5ZZ-5260-F C7ZZ-5A231-B
	23575-S8 (B-13) BOLT-CARRIAGE 3/8-16 x 3 (4) REQ'D. ATT. BRKT. TO UNDERBODY 40920-S8 (U-264) SCREW-HEX HD. 5/16-18 x 1/2 (2) REQ'D. ATT. 5260 TO 5277		
E	CLAMP	C5ZZ-5A231-B	C5ZZ-5A231-B
4	PIPE EXTENSION	C9ZZ-5255-C (GT ONLY) C7ZZ-5255-D (EXCEPT GT) S7MS-5232-A (SHELBY GT-350/500)	S7MS-5232-A

ORIGINAL SERVICE SYSTEM (REPLACED BY CONSOLIDATED SYSTEM ABOVE)

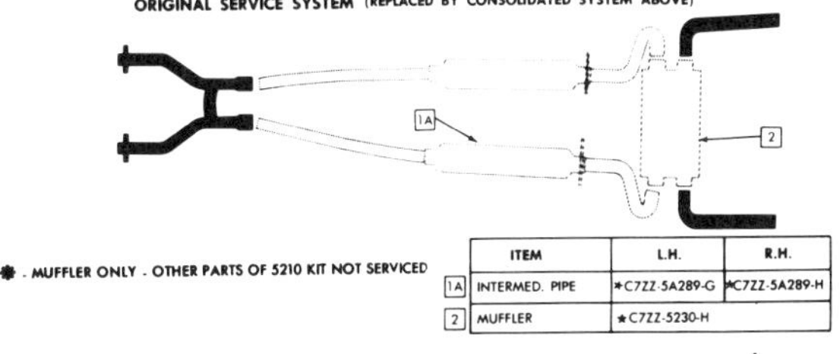

✱ - MUFFLER ONLY - OTHER PARTS OF 5210 KIT NOT SERVICED

	ITEM	L.H.	R.H.
1A	INTERMED. PIPE	★C7ZZ-5A289-G	★C7ZZ-5A289-H
2	MUFFLER	★C7ZZ-5230-H	

Complete exhaust system and related hardware for 1967 GT Mustangs equipped with 289 and 390 4-V engines, standard Mustangs equipped with 390 4-V engines, and GT-350 and GT-500 Shelby Mustangs.

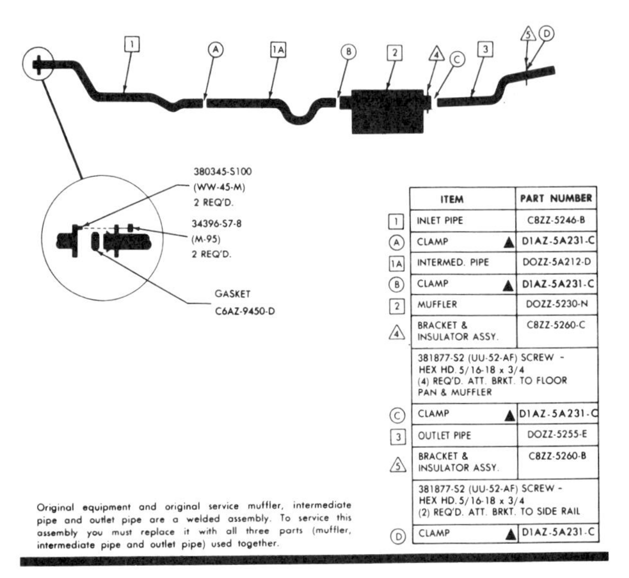

	ITEM	PART NUMBER
1	INLET PIPE	C8ZZ-5246-B
A	CLAMP ▲	D1AZ-5A231-C
1A	INTERMED. PIPE	DOZZ-5A212-D
B	CLAMP ▲	D1AZ-5A231-C
2	MUFFLER	DOZZ-5230-N
△4	BRACKET & INSULATOR ASSY.	C8ZZ-5260-C
	381877-S2 (UU-52-AF) SCREW - HEX HD. 5/16-18 x 3/4 (4) REQ'D. ATT. BRKT. TO FLOOR PAN & MUFFLER	
C	CLAMP ▲	D1AZ-5A231-C
3	OUTLET PIPE	DOZZ-5255-E
△5	BRACKET & INSULATOR ASSY.	C8ZZ-5260-B
	381877-S2 (UU-52-AF) SCREW - HEX HD. 5/16-18 x 3/4 (2) REQ'D. ATT. BRKT. TO SIDE RAIL	
D	CLAMP ▲	D1AZ-5A231-C

Original equipment and original service muffler, intermediate pipe and outlet pipe are a welded assembly. To service this assembly you must replace it with all three parts (muffler, intermediate pipe and outlet pipe) used together.

ORIGINAL SERVICE SYSTEM (REPLACED BY CONSOLIDATED SYSTEM ABOVE)

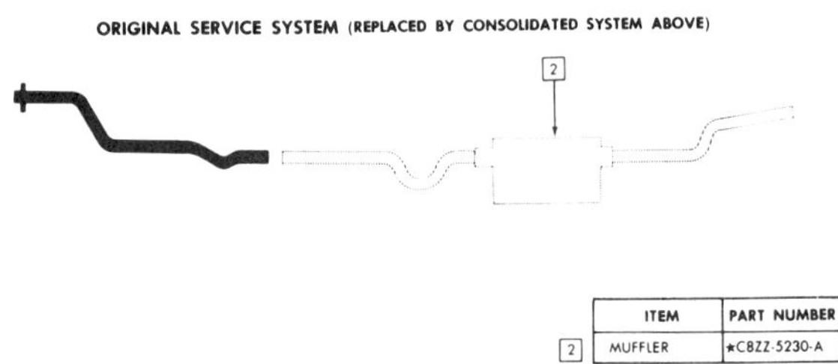

	ITEM	PART NUMBER
2	MUFFLER	★C8ZZ-5230-A

Complete exhaust system and related hardware for a 1968 Mustang equipped with a 200ci engine.

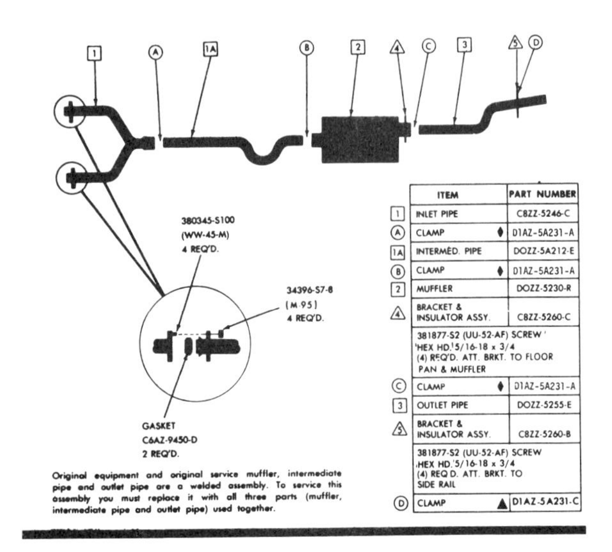

	ITEM	PART NUMBER
1	INLET PIPE	C8ZZ-5246-C
A	CLAMP ♦	D1AZ-5A231-A
1A	INTERMED. PIPE	DOZZ-5A212-E
B	CLAMP ♦	D1AZ-5A231-A
2	MUFFLER	DOZZ-5230-R
△4	BRACKET & INSULATOR ASSY.	C8ZZ-5260-C
	381877-S2 (UU-52-AF) SCREW HEX HD. 5/16-18 x 3/4 (4) REQ'D. ATT. BRKT. TO FLOOR PAN & MUFFLER	
C	CLAMP ♦	D1AZ-5A231-A
3	OUTLET PIPE	DOZZ-5255-E
△5	BRACKET & INSULATOR ASSY.	C8ZZ-5260-B
	381877-S2 (UU-52-AF) SCREW HEX HD. 5/16-18 x 3/4 (4) REQ D. ATT. BRKT. TO SIDE RAIL	
D	CLAMP ▲	D1AZ-5A231-C

Original equipment and original service muffler, intermediate pipe and outlet pipe are a welded assembly. To service this assembly you must replace it with all three parts (muffler, intermediate pipe and outlet pipe) used together.

ORIGINAL SERVICE SYSTEM (REPLACED BY CONSOLIDATED SYSTEM ABOVE)

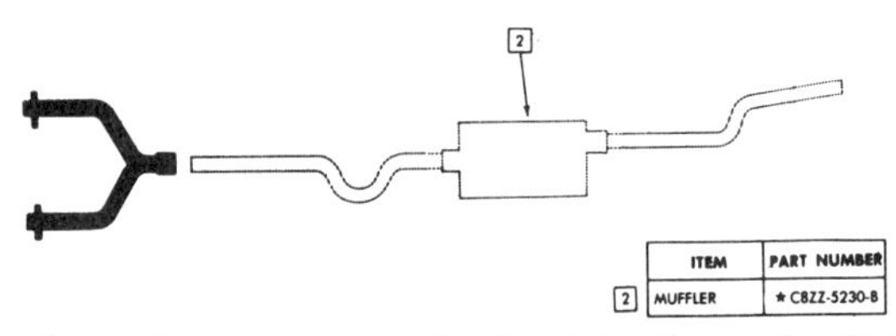

	ITEM	PART NUMBER
2	MUFFLER	★C8ZZ-5230-B

Complete exhaust system and related hardware for 1968 Mustangs equipped with 289 2-V and 302 4-V engines.

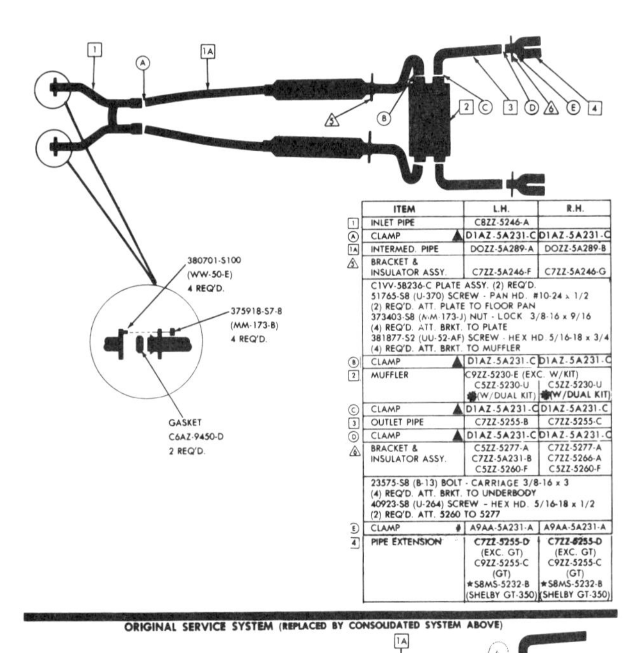

	ITEM	L.H.	R.H.
1	INLET PIPE	C8ZZ-5246-A	
A	CLAMP ▲	D1AZ-5A231-C	D1AZ-5A231-C
1A	INTERMED. PIPE	DOZZ-5A289-A	DOZZ-5A289-B
△5	BRACKET & INSULATOR ASSY.	C7ZZ-5A246-F	C7ZZ-5A246-G
	C1VV-5B236-C PLATE ASSY. (2) REQ'D. 51765-S8 (U-370) SCREW - PAN HD. #10-24 x 1/2 (2) REQ'D. ATT. PLATE TO FLOOR PAN 373403-S8 (MM-173-J) NUT - LOCK 3/8-16 x 9/16 (4) REQ'D. ATT. BRKT. TO PLATE 381877-S2 (UU-52-AF) SCREW - HEX HD. 5/16-18 x 3/4 (4) REQ'D. ATT. BRKT. TO MUFFLER		
B	CLAMP ▲	D1AZ-5A231-C	D1AZ-5A231-C
2	MUFFLER	C9ZZ-5230-E (EXC. W/KIT) C5ZZ-5230-U ✱(W/DUAL KIT)	C5ZZ-5230-U ✱(W/DUAL KIT)
C	CLAMP ▲	D1AZ-5A231-C	D1AZ-5A231-C
3	OUTLET PIPE	C7ZZ-5255-B	C7ZZ-5255-C
D	CLAMP ▲	D1AZ-5A231-C	D1AZ-5A231-C
△6	BRACKET & INSULATOR ASSY.	C5ZZ-5277-A C7ZZ-5A231-B C5ZZ-5260-F	C7ZZ-5277-A C7ZZ-5266-A C5ZZ-5260-F
	23575-S8 (B-13) BOLT - CARRIAGE 3/8-16 x 3 (4) REQ'D. ATT. BRKT. TO UNDERBODY 40923-S8 (U-264) SCREW - HEX HD. 5/16-18 x 1/2 (2) REQ'D. ATT. 5260 TO 5277		
E	CLAMP #	A9AA-5A231-A	A9AA-5A231-A
4	PIPE EXTENSION	C7ZZ-5255-D (EXC. GT) C9ZZ-5255-C (GT) ★S8MS-5232-B (SHELBY GT-350)	C7ZZ-5255-D (EXC. GT) C9ZZ-5255-C (GT) ★S8MS-5232-B (SHELBY GT-350)

ORIGINAL SERVICE SYSTEM (REPLACED BY CONSOLIDATED SYSTEM ABOVE)

	ITEM	L.H.	R.H.
1A	INTERMED. PIPE	★C7ZZ-5A289-G	★C7ZZ-5A289-H
2	MUFFLER	★C7ZZ-5230-H	

Complete exhaust system and related hardware for 1968 GT Mustangs equipped with 289 2-V and 302 4-V engines and Shelby GT-350 Mustangs.

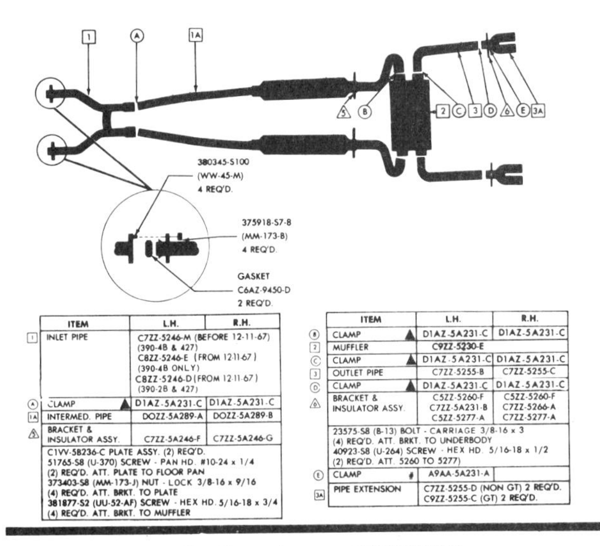

	ITEM	L.H.	R.H.
1	INLET PIPE	C7ZZ-5246-M (BEFORE 12-11-67) (390-4B & 427) C8ZZ-5246-E (FROM 12-11-67) (390-4B ONLY) C8ZZ-5246-D (FROM 12-11-67) (390-2B & 427)	
A	CLAMP ▲	D1AZ-5A231-C	D1AZ-5A231-C
1A	INTERMED. PIPE	DOZZ-5A289-A	DOZZ-5A289-B
△5	BRACKET & INSULATOR ASSY.	C7ZZ-5A246-F	C7ZZ-5A246-G
	C1VV-5B236-C PLATE ASSY. (2) REQ'D. 51765-S8 (U-370) SCREW - PAN HD. #10-24 x 1/4 (2) REQ'D. ATT. PLATE TO FLOOR PAN 373403-S8 (MM-173-J) NUT - LOCK 3/8-16 x 9/16 (4) REQ'D. ATT. BRKT. TO PLATE 381877-S2 (UU-52-AF) SCREW - HEX HD. 5/16-18 x 3/4 (4) REQ'D. ATT. BRKT. TO MUFFLER		

	ITEM	L.H.	R.H.
B	CLAMP ▲	D1AZ-5A231-C	D1AZ-5A231-C
2	MUFFLER	C9ZZ-5230-E	
C	CLAMP ▲	D1AZ-5A231-C	D1AZ-5A231-C
3	OUTLET PIPE	C7ZZ-5255-B	C7ZZ-5255-C
D	CLAMP ▲	D1AZ-5A231-C	D1AZ-5A231-C
△6	BRACKET & INSULATOR ASSY.	C5ZZ-5260-F C7ZZ-5A231-B C5ZZ-5277-A	C5ZZ-5260-F C7ZZ-5266-A C7ZZ-5277-A
	23575-S8 (B-13) BOLT - CARRIAGE 3/8-16 x 3 (4) REQ'D. ATT. BRKT. TO UNDERBODY 40923-S8 (U-264) SCREW - HEX HD. 5/16-18 x 1/2 (2) REQ'D. ATT. 5260 TO 5277)		
E	CLAMP #	A9AA-5A231-A	
3A	PIPE EXTENSION	C7ZZ-5255-D (NON GT) 2 REQ'D. C9ZZ-5255-C (GT) 2 REQ'D.	

ORIGINAL SERVICE SYSTEM (REPLACED BY CONSOLIDATED SYSTEM ABOVE)

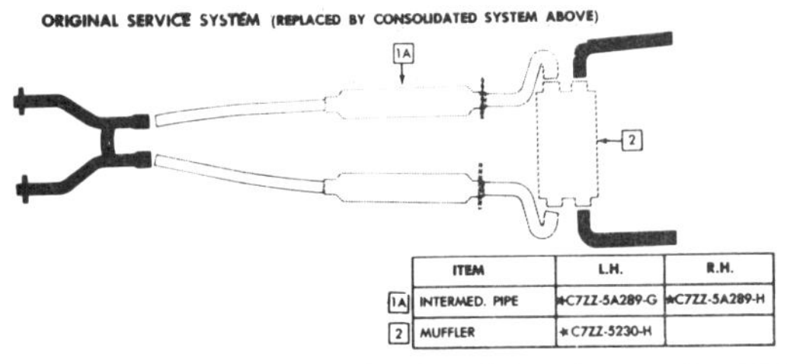

	ITEM	L.H.	R.H.
1A	INTERMED. PIPE	★C7ZZ-5A289-G	★C7ZZ-5A289-H
2	MUFFLER	★C7ZZ-5230-H	

Complete exhaust system and related hardware for 1968 Mustangs equipped with 390 2-V and 4-V engines.

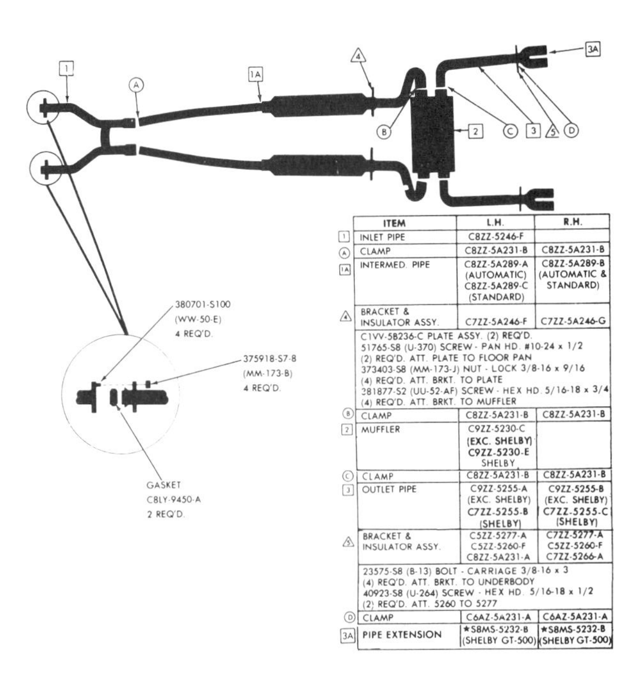

	ITEM	L.H.	R.H.
1	INLET PIPE	C8ZZ-5246-F	
A	CLAMP	C8ZZ-5A231-B	C8ZZ-5A231-B
1A	INTERMED. PIPE	C8ZZ-5A289-A (AUTOMATIC) C8ZZ-5A289-C (STANDARD)	C8ZZ-5A289-B (AUTOMATIC & STANDARD)
△4	BRACKET & INSULATOR ASSY.	C7ZZ-5A246-F	C7ZZ-5A246-G
	C1VV-5B236-C PLATE ASSY. (2) REQ'D. 51765-S8 (U-370) SCREW - PAN HD. #10-24 x 1/2 (2) REQ'D. ATT. PLATE TO FLOOR PAN 373403-S8 (MM-173-J) NUT - LOCK 3/8-16 x 9/16 (4) REQ'D. ATT. BRKT. TO PLATE 381877-S2 (UU-52-AF) SCREW - HEX HD. 5/16-18 x 3/4 (4) REQ'D. ATT. BRKT. TO MUFFLER		
B	CLAMP	C8ZZ-5A231-B	C8ZZ-5A231-B
2	MUFFLER	C9ZZ-5230-C (EXC. SHELBY) C9ZZ-5230-E SHELBY	
C	CLAMP	C8ZZ-5A231-B	C8ZZ-5A231-B
3	OUTLET PIPE	C9ZZ-5255-A (EXC. SHELBY) C7ZZ-5255-B (SHELBY)	C9ZZ-5255-B (EXC. SHELBY) C7ZZ-5255-C (SHELBY)
△5	BRACKET & INSULATOR ASSY.	C5ZZ-5277-A C5ZZ-5260-F C8ZZ-5A231-A	C7ZZ-5277-A C5ZZ-5260-F C7ZZ-5266-A
	23575-S8 (B-13) BOLT - CARRIAGE 3/8-16 x 3 (4) REQ'D. ATT. BRKT. TO UNDERBODY 40923-S8 (U-264) SCREW - HEX HD. 5/16-18 x 1/2 (2) REQ'D. ATT. 5260 TO 5277		
D	CLAMP	C6AZ-5A231-A	C6AZ-5A231-A
3A	PIPE EXTENSION	★S8MS-5232-B (SHELBY GT-500)	★S8MS-5232-B (SHELBY GT-500)

Complete exhaust system and related hardware for 1968 GT Mustangs equipped with 428 4-V engines and Shelby GT-500 Mustangs.

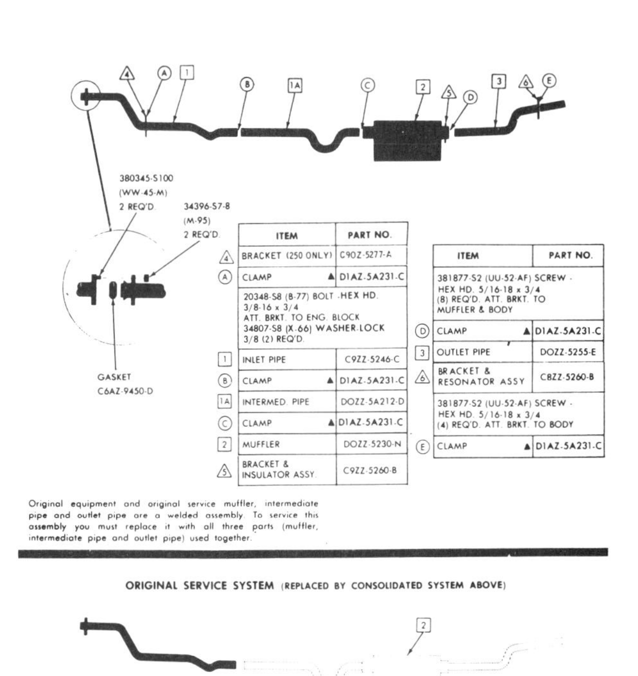

	ITEM	PART NO.
△4	BRACKET (250 ONLY)	C9OZ-5277-A
A	CLAMP ▲	D1AZ-5A231-C
	20348-S8 (B-77) BOLT -HEX HD. 3/8-16 x 3/4 ATT. BRKT. TO ENG. BLOCK 34807-S8 (X-66) WASHER-LOCK 3/8 (2) REQ'D.	
1	INLET PIPE	C9ZZ-5246-C
B	CLAMP ▲	D1AZ-5A231-C
1A	INTERMED. PIPE	DOZZ-5A212-D
C	CLAMP ▲	D1AZ-5A231-C
2	MUFFLER	DOZZ-5230-N
△5	BRACKET & INSULATOR ASSY.	C9ZZ-5260-B

	ITEM	PART NO.
	381877-S2 (UU-52-AF) SCREW - HEX HD. 5/16-18 x 3/4 (8) REQ'D. ATT. BRKT. TO MUFFLER & BODY	
D	CLAMP ▲	D1AZ-5A231-C
3	OUTLET PIPE	DOZZ-5255-E
△6	BRACKET & RESONATOR ASSY	C8ZZ-5260-B
	381877-S2 (UU-52-AF) SCREW - HEX HD. 5/16-18 x 3/4 (4) REQ'D. ATT. BRKT. TO BODY	
E	CLAMP ▲	D1AZ-5A231-C

Original equipment and original service muffler, intermediate pipe and outlet pipe are a welded assembly. To service this assembly you must replace it with all three parts (muffler, intermediate pipe and outlet pipe) used together.

ORIGINAL SERVICE SYSTEM (REPLACED BY CONSOLIDATED SYSTEM ABOVE)

	ITEM	PART NO.
2	MUFFLER	★C9ZZ-5230-D

Complete exhaust system and related hardware for a 1969 Mustang equipped with a 200 or 250ci engine.

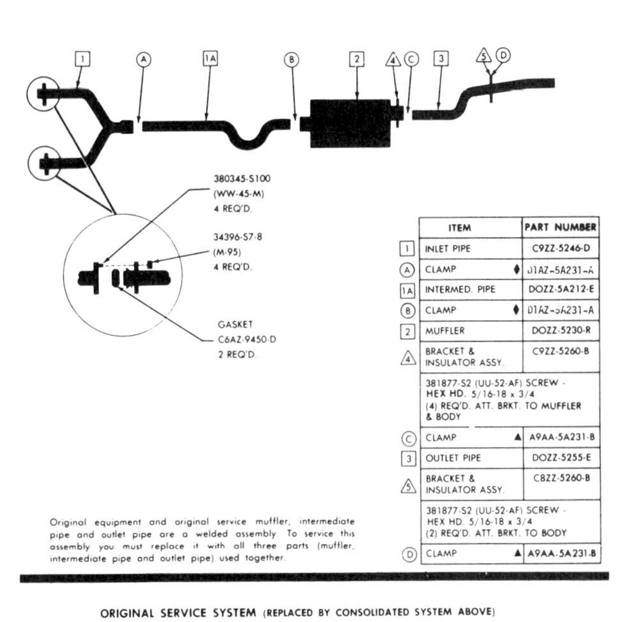

	ITEM	PART NUMBER
1	INLET PIPE	C9ZZ-5246-D
A	CLAMP ♦	D1AZ-5A231-A
1A	INTERMED. PIPE	DOZZ-5A212-E
B	CLAMP ♦	D1AZ-5A231-A
2	MUFFLER	DOZZ-5230-R
△4	BRACKET & INSULATOR ASSY.	C9ZZ-5260-B
	381877-S2 (UU-52-AF) SCREW - HEX HD. 5/16-18 x 3/4 (4) REQ'D. ATT. BRKT. TO MUFFLER & BODY	
C	CLAMP ▲	A9AA-5A231-B
3	OUTLET PIPE	DOZZ-5255-E
△5	BRACKET & INSULATOR ASSY.	C8ZZ-5260-B
	381877-S2 (UU-52-AF) SCREW - HEX HD. 5/16-18 x 3/4 (2) REQ'D. ATT. BRKT. TO BODY	
D	CLAMP ▲	A9AA-5A231-B

Original equipment and original service muffler, intermediate pipe and outlet pipe are a welded assembly. To service this assembly you must replace it with all three parts (muffler, intermediate pipe and outlet pipe) used together.

ORIGINAL SERVICE SYSTEM (REPLACED BY CONSOLIDATED SYSTEM ABOVE)

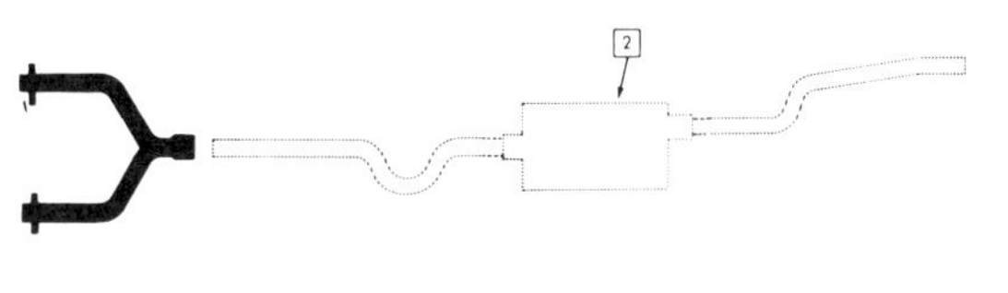

	ITEM	PART NUMBER
2	MUFFLER	★C9ZZ-5230-F

Complete exhaust system and related hardware for a 1969 Mustang equipped with a 302 2-V engine.

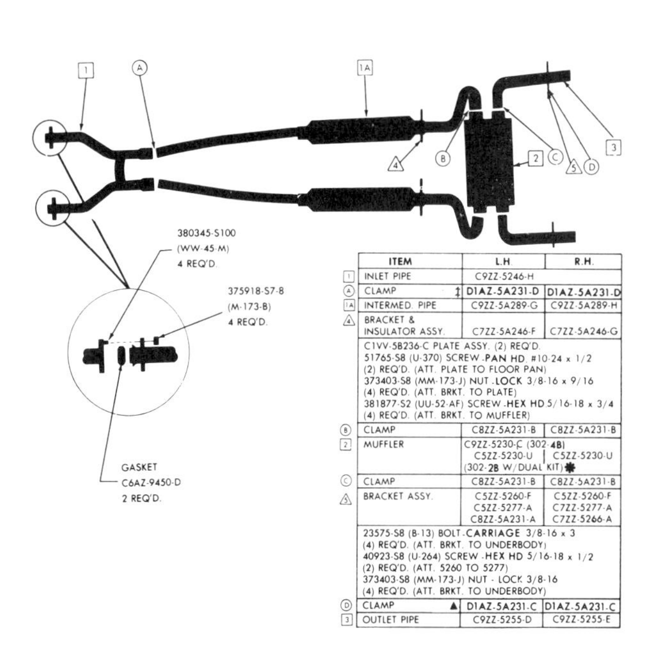

	ITEM	L.H.	R.H.
1	INLET PIPE	C9ZZ-5246-H	
A	CLAMP ‡	D1AZ-5A231-D	D1AZ-5A231-D
1A	INTERMED. PIPE	C9ZZ-5A289-G	C9ZZ-5A289-H
△4	BRACKET & INSULATOR ASSY.	C7ZZ-5A246-F	C7ZZ-5A246-G
	C1VV-5B236-C PLATE ASSY. (2) REQ'D. 51765-S8 (U-370) SCREW -PAN HD. #10-24 x 1/2 (2) REQ'D. (ATT. PLATE TO FLOOR PAN) 373403-S8 (MM-173-J) NUT -LOCK 3/8-16 x 9/16 (4) REQ'D. (ATT. BRKT. TO PLATE) 381877-S2 (UU-52-AF) SCREW -HEX HD 5/16-18 x 3/4 (4) REQ'D. (ATT. BRKT. TO MUFFLER)		
B	CLAMP	C8ZZ-5A231-B	C8ZZ-5A231-B
2	MUFFLER	C9ZZ-5230-Ꞔ (302-4B) C5ZZ-5230-U (302-2B W/DUAL KIT) ✱	C5ZZ-5230-U
C	CLAMP	C8ZZ-5A231-B	C8ZZ-5A231-B
△5	BRACKET ASSY.	C5ZZ-5260-F C5ZZ-5277-A C8ZZ-5A231-A	C5ZZ-5260-F C7ZZ-5277-A C7ZZ-5266-A
	23575-S8 (B-13) BOLT -CARRIAGE 3/8-16 x 3 (4) REQ'D. (ATT. BRKT. TO UNDERBODY) 40923-S8 (U-264) SCREW -HEX HD 5/16-18 x 1/2 (2) REQ'D. (ATT. 5260 TO 5277) 373403-S8 (MM-173-J) NUT - LOCK 3/8-16 (4) REQ'D. (ATT. BRKT. TO UNDERBODY)		
D	CLAMP ▲	D1AZ-5A231-C	D1AZ-5A231-C
3	OUTLET PIPE	C9ZZ-5255-D	C9ZZ-5255-E

Complete exhaust system and related hardware for a 1969 Boss 302 Mustang and models equipped with a 302 2-V engine and dual exhausts.

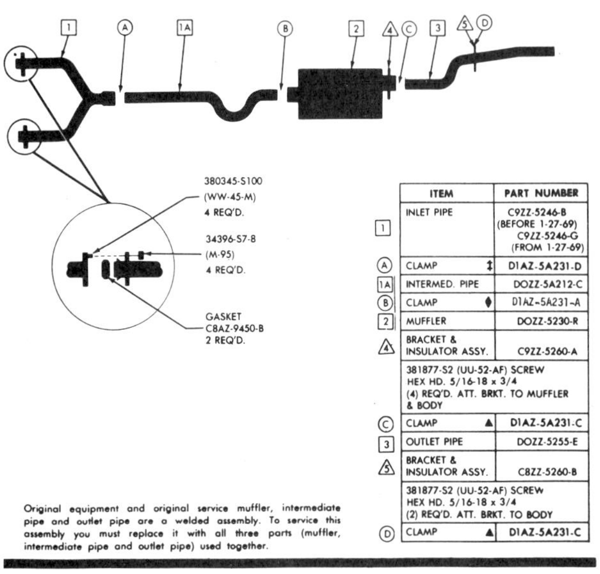

	ITEM	PART NUMBER
1	INLET PIPE	C9ZZ-5246-B (BEFORE 1-27-69) C9ZZ-5246-G (FROM 1-27-69)
A	CLAMP ↕	D1AZ-5A231-D
1A	INTERMED. PIPE	DOZZ-5A212-C
B	CLAMP ◆	D1AZ-5A231-A
2	MUFFLER	DOZZ-5230-R
4	BRACKET & INSULATOR ASSY.	C9ZZ-5260-A
	381877-S2 (UU-52-AF) SCREW HEX HD. 5/16-18 x 3/4 (4) REQ'D. ATT. BRKT. TO MUFFLER & BODY	
C	CLAMP ▲	D1AZ-5A231-C
3	OUTLET PIPE	DOZZ-5255-E
5	BRACKET & INSULATOR ASSY.	C8ZZ-5260-B
	381877-S2 (UU-52-AF) SCREW HEX HD. 5/16-18 x 3/4 (2) REQ'D. ATT. BRKT. TO BODY	
D	CLAMP ▲	D1AZ-5A231-C

Original equipment and original service muffler, intermediate pipe and outlet pipe are a welded assembly. To service this assembly you must replace it with all three parts (muffler, intermediate pipe and outlet pipe) used together.

ORIGINAL SERVICE SYSTEM (REPLACED BY CONSOLIDATED SYSTEM ABOVE)

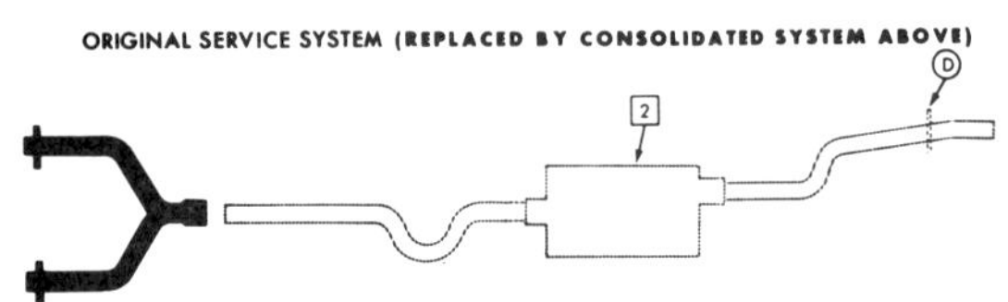

	ITEM	PART NUMBER
2	MUFFLER	★C9ZZ-5230-A
D	CLAMP ◆	D1AZ-5A231-A

Complete exhaust system and related hardware for a 1969 Mustang equipped with a 351 2-V engine.

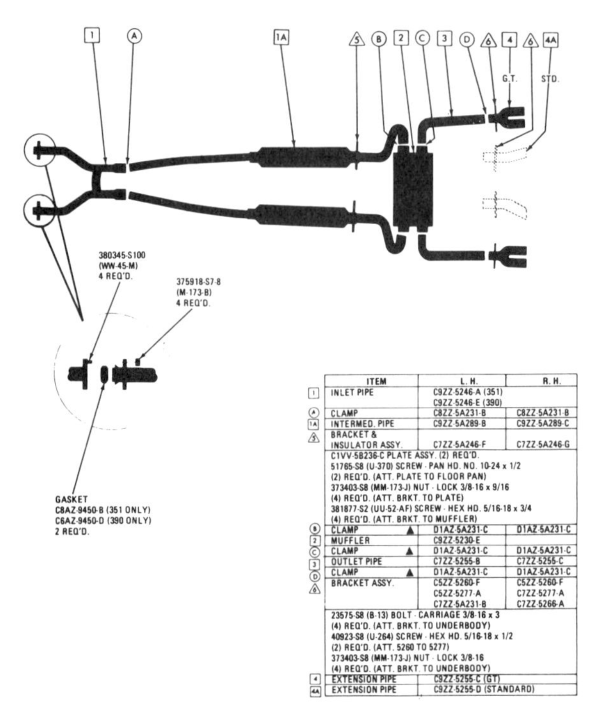

	ITEM	L. H.	R. H.
1	INLET PIPE	C9ZZ-5246-A (351) C9ZZ-5246-E (390)	
A	CLAMP	C8ZZ-5A231-B	C8ZZ-5A231-B
1A	INTERMED. PIPE	C9ZZ-5A289-B	C9ZZ-5A289-C
5	BRACKET & INSULATOR ASSY.	C7ZZ-5A246-F	C7ZZ-5A246-G
	C1VV-5B236-C PLATE ASSY. (2) REQ'D. 51765-S8 (U-370) SCREW - PAN HD. NO. 10-24 x 1/2 (2) REQ'D. (ATT. PLATE TO FLOOR PAN) 373403-S8 (MM-173-J) NUT - LOCK 3/8-16 x 9/16 (4) REQ'D. (ATT. BRKT. TO PLATE) 381877-S2 (UU-52-AF) SCREW - HEX HD. 5/16-18 x 3/4 (4) REQ'D. (ATT. BRKT. TO MUFFLER)		
B	CLAMP ▲	D1AZ-5A231-C	D1AZ-5A231-C
2	MUFFLER	C9ZZ-5230-E	
C	CLAMP ▲	D1AZ-5A231-C	D1AZ-5A231-C
3	OUTLET PIPE	C7ZZ-5255-B	C7ZZ-5255-C
D	CLAMP ▲	D1AZ-5A231-C	D1AZ-5A231-C
6	BRACKET ASSY.	C5ZZ-5260-F C5ZZ-5277-A C7ZZ-5A231-B	C5ZZ-5260-F C7ZZ-5277-A C7ZZ-5266-A
	23575-S8 (B-13) BOLT - CARRIAGE 3/8-16 x 3 (4) REQ'D. (ATT. BRKT. TO UNDERBODY) 40923-S8 (U-264) SCREW - HEX HD. 5/16-18 x 1/2 (2) REQ'D. (ATT. 5260 TO 5277) 373403-S8 (MM-173-J) NUT - LOCK 3/8-16 (4) REQ'D. (ATT. BRKT. TO UNDERBODY)		
4	EXTENSION PIPE	C9ZZ-5255-C (GT)	
4A	EXTENSION PIPE	C9ZZ-5255-D (STANDARD)	

Complete exhaust system and related hardware for 1969 Standard and GT Mustangs equipped with either 351 4-V or 390 4-V engines.

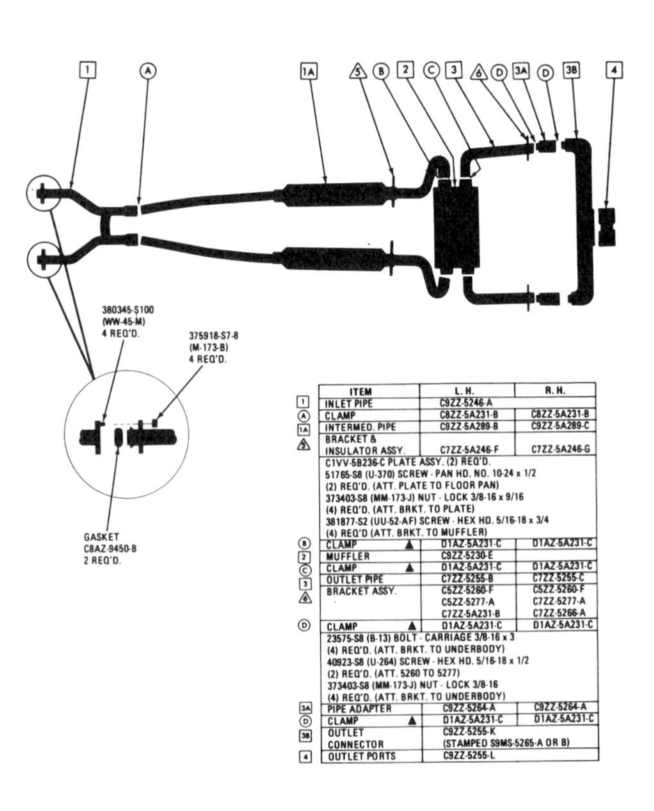

	ITEM	L. H.	R. H.
1	INLET PIPE	C9ZZ-5246-A	
A	CLAMP	C8ZZ-5A231-B	C8ZZ-5A231-B
1A	INTERMED. PIPE	C9ZZ-5A289-B	C9ZZ-5A289-C
5	BRACKET & INSULATOR ASSY.	C7ZZ-5A246-F	C7ZZ-5A246-G
	C1VV-5B236-C PLATE ASSY. (2) REQ'D. 51765-S8 (U-370) SCREW - PAN HD. NO. 10-24 x 1/2 (2) REQ'D. (ATT. PLATE TO FLOOR PAN) 373403-S8 (MM-173-J) NUT - LOCK 3/8-16 x 9/16 (4) REQ'D. (ATT. BRKT. TO PLATE) 381877-S2 (UU-52-AF) SCREW - HEX HD. 5/16-18 x 3/4 (4) REQ'D (ATT. BRKT. TO MUFFLER)		
B	CLAMP ▲	D1AZ-5A231-C	D1AZ-5A231-C
2	MUFFLER	C9ZZ-5230-E	
C	CLAMP ▲	D1AZ-5A231-C	D1AZ-5A231-C
3	OUTLET PIPE	C7ZZ-5255-B	C7ZZ-5255-C
6	BRACKET ASSY.	C5ZZ-5260-F C5ZZ-5277-A C7ZZ-5A231-B	C5ZZ-5260-F C7ZZ-5277-A C7ZZ-5266-A
D	CLAMP ▲	D1AZ-5A231-C	D1AZ-5A231-C
	23575-S8 (B-13) BOLT - CARRIAGE 3/8-16 x 3 (4) REQ'D. (ATT. BRKT. TO UNDERBODY) 40923-S8 (U-264) SCREW - HEX HD. 5/16-18 x 1/2 (2) REQ'D. (ATT. 5260 TO 5277) 373403-S8 (MM-173-J) NUT - LOCK 3/8-16 (4) REQ'D. (ATT. BRKT. TO UNDERBODY)		
3A	PIPE ADAPTER	C9ZZ-5264-A	C9ZZ-5264-A
D	CLAMP ▲	D1AZ-5A231-C	D1AZ-5A231-C
3B	OUTLET CONNECTOR	C9ZZ-5255-K (STAMPED S9MS-5265-A OR B)	
4	OUTLET PORTS	C9ZZ-5255-L	

Complete exhaust system and related hardware for a 1969 Shelby GT-350 Mustang.

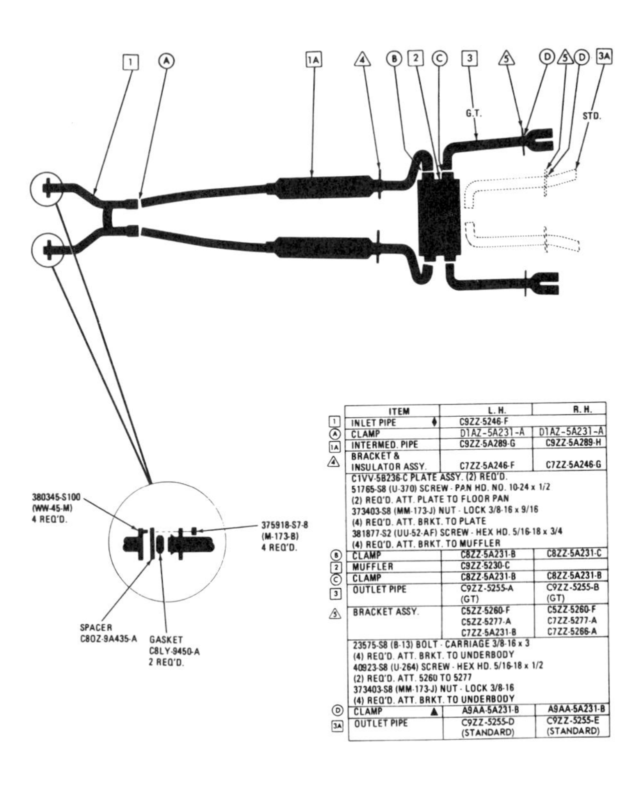

	ITEM	L. H.	R. H.
1	INLET PIPE ◆	C9ZZ-5246-F	
A	CLAMP	D1AZ-5A231-A	D1AZ-5A231-A
1A	INTERMED. PIPE	C9ZZ-5A289-G	C9ZZ-5A289-H
4	BRACKET & INSULATOR ASSY.	C7ZZ-5A246-F	C7ZZ-5A246-G
	C1VV-5B236-C PLATE ASSY. (2) REQ'D. 51765-S8 (U-370) SCREW - PAN HD. NO. 10-24 x 1/2 (2) REQ'D. ATT. PLATE TO FLOOR PAN 373403-S8 (MM-173-J) NUT - LOCK 3/8-16 x 9/16 (4) REQ'D. ATT. BRKT. TO PLATE 381877-S2 (UU-52-AF) SCREW - HEX HD. 5/16-18 x 3/4 (4) REQ'D. ATT. BRKT. TO MUFFLER		
B	CLAMP	C8ZZ-5A231-B	C8ZZ-5A231-C
2	MUFFLER	C9ZZ-5230-C	
C	CLAMP	C8ZZ-5A231-B	C8ZZ-5A231-B
3	OUTLET PIPE	C9ZZ-5255-A (GT)	C9ZZ-5255-B (GT)
5	BRACKET ASSY.	C5ZZ-5260-F C5ZZ-5277-A C7ZZ-5A231-B	C5ZZ-5260-F C7ZZ-5277-A C7ZZ-5266-A
	23575-S8 (B-13) BOLT - CARRIAGE 3/8-16 x 3 (4) REQ'D. ATT. BRKT. TO UNDERBODY 40923-S8 (U-264) SCREW - HEX HD. 5/16-18 x 1/2 (2) REQ'D. ATT. 5260 TO 5277 373403-S8 (MM-173-J) NUT - LOCK 3/8-16 (4) REQ'D. ATT. BRKT. TO UNDERBODY		
D	CLAMP ▲	A9AA-5A231-B	A9AA-5A231-B
3A	OUTLET PIPE	C9ZZ-5255-D (STANDARD)	C9ZZ-5255-E (STANDARD)

Complete exhaust system and related hardware for 1969 standard and GT Mustangs equipped with a 428 4-V engine.

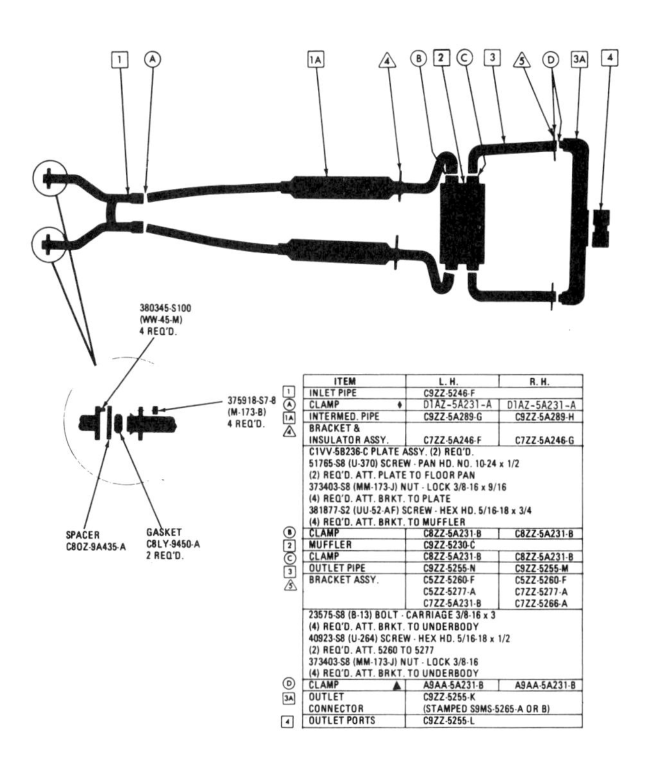

	ITEM	L. H.	R. H.
1	INLET PIPE	C9ZZ-5246-F	
A	CLAMP ♦	D1AZ-5A231-A	D1AZ-5A231-A
1A	INTERMED. PIPE	C9ZZ-5A289-G	C9ZZ-5A289-H
4	BRACKET & INSULATOR ASSY.	C7ZZ-5A246-F	C7ZZ-5A246-G
	C1VV-5B236-C PLATE ASSY. (2) REQ'D. 51765-S8 (U-370) SCREW - PAN HD. NO. 10-24 x 1/2 (2) REQ'D. ATT. PLATE TO FLOOR PAN 373403-S8 (MM-173-J) NUT - LOCK 3/8-16 x 9/16 (4) REQ'D. ATT. BRKT. TO PLATE 381877-S2 (UU-52-AF) SCREW - HEX HD. 5/16-18 x 3/4 (4) REQ'D. ATT. BRKT. TO MUFFLER		
B	CLAMP	C8ZZ-5A231-B	C8ZZ-5A231-B
2	MUFFLER	C9ZZ-5230-C	
C	CLAMP	C8ZZ-5A231-B	C8ZZ-5A231-B
3	OUTLET PIPE	C9ZZ-5255-N	C9ZZ-5255-M
5	BRACKET ASSY.	C5ZZ-5260-F C5ZZ-5277-A C7ZZ-5A231-B	C5ZZ-5260-F C7ZZ-5277-A C7ZZ-5266-A
	23575-S8 (B-13) BOLT - CARRIAGE 3/8-16 x 3 (4) REQ'D. ATT. BRKT. TO UNDERBODY 40923-S8 (U-264) SCREW - HEX HD. 5/16-18 x 1/2 (2) REQ'D. ATT. 5260 TO 5277 373403-S8 (MM-173-J) NUT - LOCK 3/8-16 (4) REQ'D. ATT. BRKT. TO UNDERBODY		
D	CLAMP ▲	A9AA-5A231-B	A9AA-5A231-B
3A	OUTLET CONNECTOR	C9ZZ-5255-K (STAMPED S9MS-5265-A OR B)	
4	OUTLET PORTS	C9ZZ-5255-L	

Complete exhaust system and related hardware for a 1969 Shelby GT-500 Mustang.

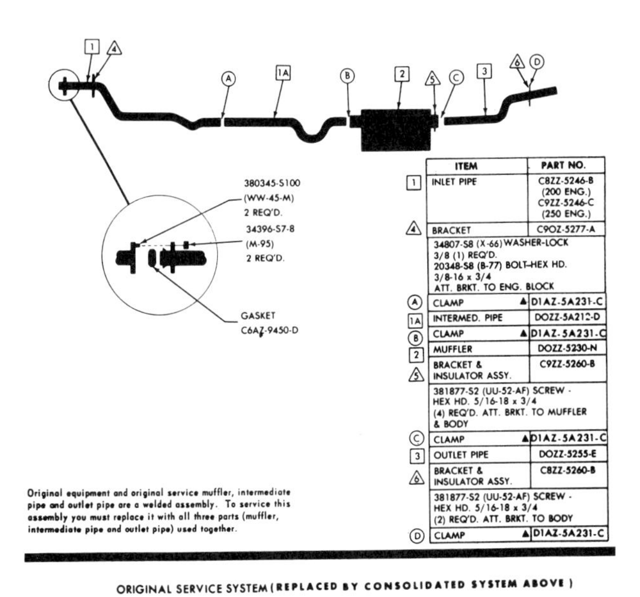

	ITEM	PART NO.
1	INLET PIPE	C8ZZ-5246-B (200 ENG.) C9ZZ-5246-C (250 ENG.)
4	BRACKET	C9OZ-5277-A
	34807-S8 (X-66) WASHER-LOCK 3/8 (1) REQ'D. 20348-S8 (B-77) BOLT-HEX HD. 3/8-16 x 3/4 ATT. BRKT. TO ENG. BLOCK	
A	CLAMP ▲	D1AZ-5A231-C
1A	INTERMED. PIPE	DOZZ-5A212-D
B	CLAMP ▲	D1AZ-5A231-C
2	MUFFLER	DOZZ-5230-N
5	BRACKET & INSULATOR ASSY.	C9ZZ-5260-B
	381877-S2 (UU-52-AF) SCREW - HEX HD. 5/16-18 x 3/4 (4) REQ'D. ATT. BRKT. TO MUFFLER & BODY	
C	CLAMP ▲	D1AZ-5A231-C
3	OUTLET PIPE	DOZZ-5255-E
6	BRACKET & INSULATOR ASSY.	C8ZZ-5260-B
	381877-S2 (UU-52-AF) SCREW - HEX HD. 5/16-18 x 3/4 (2) REQ'D. ATT. BRKT. TO BODY	
D	CLAMP ▲	D1AZ-5A231-C

ORIGINAL SERVICE SYSTEM (REPLACED BY CONSOLIDATED SYSTEM ABOVE)

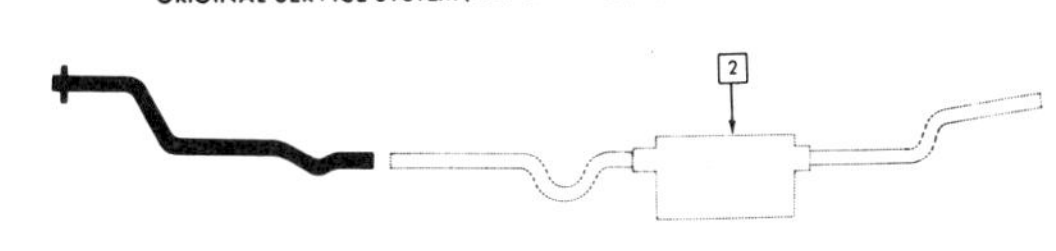

	ITEM	PART NO.
2	MUFFLER	★DOZZ-5230-L

Complete exhaust system and related hardware for a 1970 Mustang equipped with a 200 or 250ci engine.

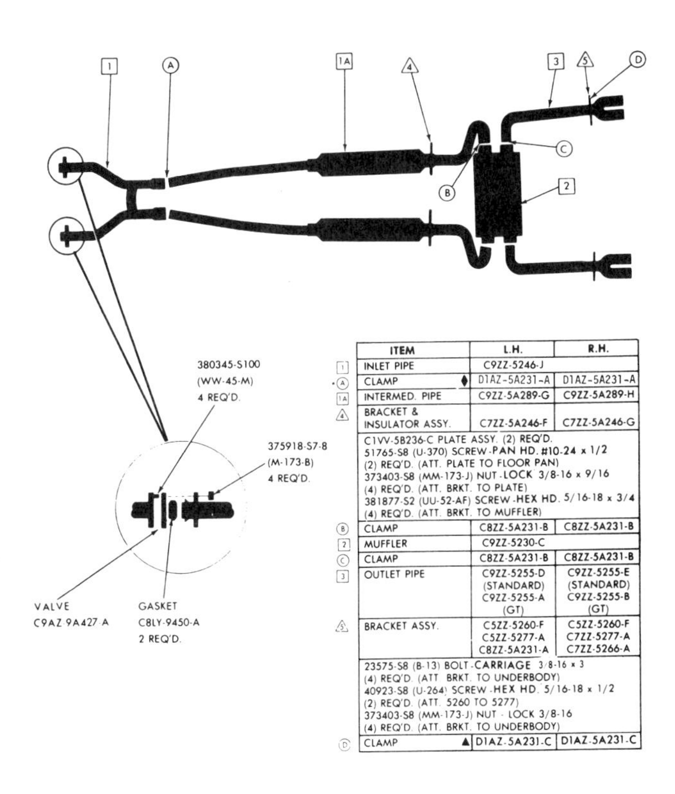

	ITEM	L.H.	R.H.
1	INLET PIPE	C9ZZ-5246-J	
A	CLAMP ♦	D1AZ-5A231-A	D1AZ-5A231-A
1A	INTERMED. PIPE	C9ZZ-5A289-G	C9ZZ-5A289-H
4	BRACKET & INSULATOR ASSY.	C7ZZ-5A246-F	C7ZZ-5A246-G
	C1VV-5B236-C PLATE ASSY. (2) REQ'D. 51765-S8 (U-370) SCREW-PAN HD. #10-24 x 1/2 (2) REQ'D. (ATT. PLATE TO FLOOR PAN) 373403-S8 (MM-173-J) NUT-LOCK 3/8-16 x 9/16 (4) REQ'D. (ATT. BRKT. TO PLATE) 381877-S2 (UU-52-AF) SCREW-HEX HD. 5/16-18 x 3/4 (4) REQ'D. (ATT. BRKT. TO MUFFLER)		
B	CLAMP	C8ZZ-5A231-B	C8ZZ-5A231-B
2	MUFFLER	C9ZZ-5230-C	
C	CLAMP	C8ZZ-5A231-B	C8ZZ-5A231-B
3	OUTLET PIPE	C9ZZ-5255-D (STANDARD) C9ZZ-5255-A (GT)	C9ZZ-5255-E (STANDARD) C9ZZ-5255-B (GT)
5	BRACKET ASSY.	C5ZZ-5260-F C5ZZ-5277-A C8ZZ-5A231-A	C5ZZ-5260-F C7ZZ-5277-A C7ZZ-5266-A
	23575-S8 (B-13) BOLT-CARRIAGE 3/8-16 x 3 (4) REQ'D. (ATT. BRKT. TO UNDERBODY) 40923-S8 (U-264) SCREW-HEX HD. 5/16-18 x 1/2 (2) REQ'D. (ATT. 5260 TO 5277) 373403-S8 (MM-173-J) NUT - LOCK 3/8-16 (4) REQ'D. (ATT. BRKT. TO UNDERBODY)		
D	CLAMP ▲	D1AZ-5A231-C	D1AZ-5A231-C

Complete exhaust system and related hardware for a 1969 Boss 429 Mustang.

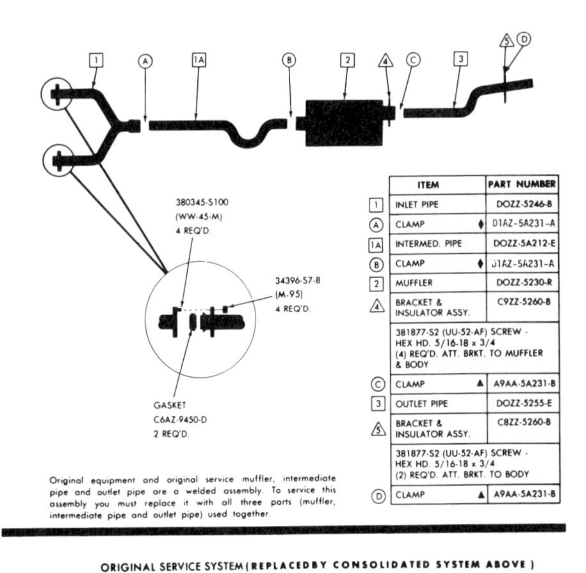

	ITEM	PART NUMBER
1	INLET PIPE	DOZZ-5246-B
A	CLAMP ♦	D1AZ-5A231-A
1A	INTERMED. PIPE	DOZZ-5A212-E
B	CLAMP ♦	D1AZ-5A231-A
2	MUFFLER	DOZZ-5230-R
4	BRACKET & INSULATOR ASSY.	C9ZZ-5260-B
	381877-S2 (UU-52-AF) SCREW - HEX HD. 5/16-18 x 3/4 (4) REQ'D. ATT. BRKT. TO MUFFLER & BODY	
C	CLAMP ▲	A9AA-5A231-B
3	OUTLET PIPE	DOZZ-5255-E
5	BRACKET & INSULATOR ASSY.	C8ZZ-5260-B
	381877-S2 (UU-52-AF) SCREW - HEX HD. 5/16-18 x 3/4 (2) REQ'D. ATT. BRKT. TO BODY	
D	CLAMP ▲	A9AA-5A231-B

ORIGINAL SERVICE SYSTEM (REPLACED BY CONSOLIDATED SYSTEM ABOVE)

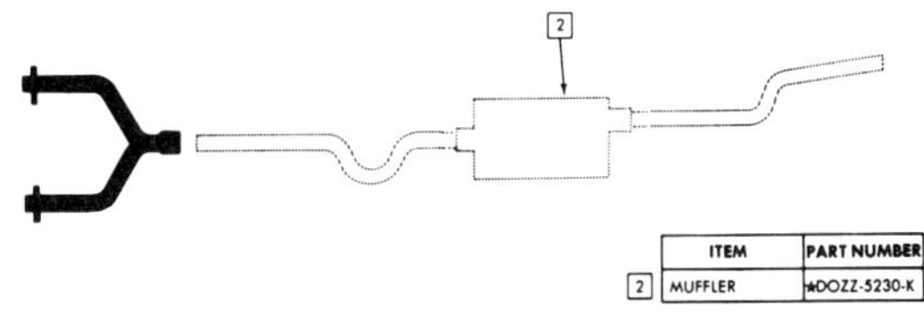

	ITEM	PART NUMBER
2	MUFFLER	★DOZZ-5230-K

Complete exhaust system and related hardware for a 1970 Mustang equipped with a 302ci engine.

Mustang's leaner, meaner high-performance breeds, the Boss 302 and Boss 429, were designed to be more functional than visual. Both were outfitted with the base Mustang rear valance and ordinary turn downs at the end of their dual tailpipes.

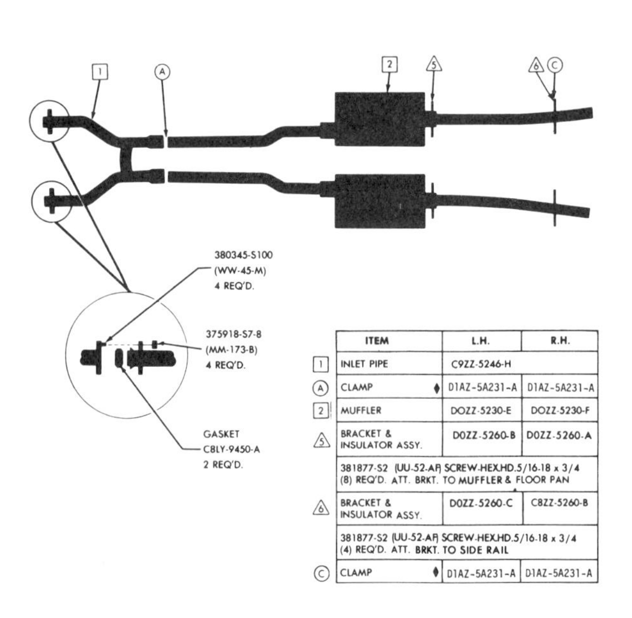

	ITEM	L.H.	R.H.
1	INLET PIPE	C9ZZ-5246-H	
A	CLAMP ♦	D1AZ-5A231-A	D1AZ-5A231-A
2	MUFFLER	DOZZ-5230-E	DOZZ-5230-F
5	BRACKET & INSULATOR ASSY.	D0ZZ-5260-B	D0ZZ-5260-A
	381877-S2 (UU-52-AF) SCREW-HEX.HD.5/16-18 x 3/4 (8) REQ'D. ATT. BRKT. TO MUFFLER & FLOOR PAN		
6	BRACKET & INSULATOR ASSY.	D0ZZ-5260-C	C8ZZ-5260-B
	381877-S2 (UU-52-AF) SCREW-HEX.HD.5/16-18 x 3/4 (4) REQ'D. ATT. BRKT. TO SIDE RAIL		
C	CLAMP ♦	D1AZ-5A231-A	D1AZ-5A231-A

Complete exhaust system and related hardware for a 1970 Boss 302 Mustang.

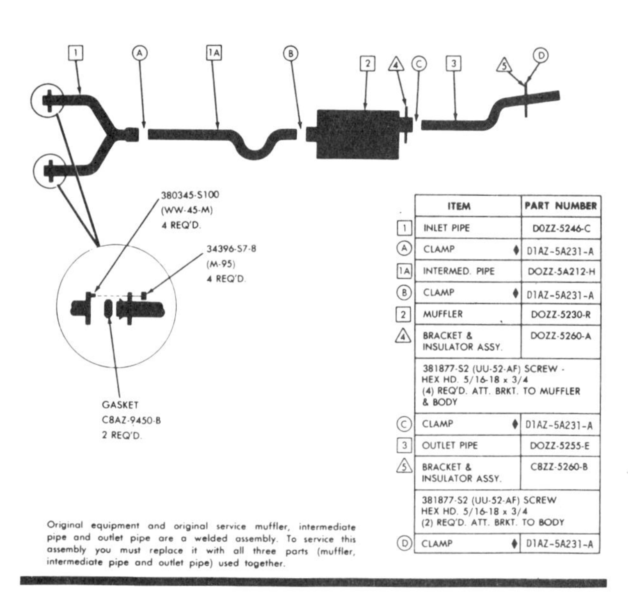

	ITEM	PART NUMBER
1	INLET PIPE	D0ZZ-5246-C
A	CLAMP ♦	D1AZ-5A231-A
1A	INTERMED. PIPE	DOZZ-5A212-H
B	CLAMP ♦	D1AZ-5A231-A
2	MUFFLER	DOZZ-5230-R
4	BRACKET & INSULATOR ASSY.	DOZZ-5260-A
	381877-S2 (UU-52-AF) SCREW - HEX HD. 5/16-18 x 3/4 (4) REQ'D. ATT. BRKT. TO MUFFLER & BODY	
C	CLAMP ♦	D1AZ-5A231-A
3	OUTLET PIPE	DOZZ-5255-E
5	BRACKET & INSULATOR ASSY.	C8ZZ-5260-B
	381877-S2 (UU-52-AF) SCREW HEX HD. 5/16-18 x 3/4 (2) REQ'D. ATT. BRKT. TO BODY	
D	CLAMP ♦	D1AZ-5A231-A

Original equipment and original service muffler, intermediate pipe and outlet pipe are a welded assembly. To service this assembly you must replace it with all three parts (muffler, intermediate pipe and outlet pipe) used together.

ORIGINAL SERVICE SYSTEM (REPLACED BY CONSOLIDATED SYSTEM ABOVE)

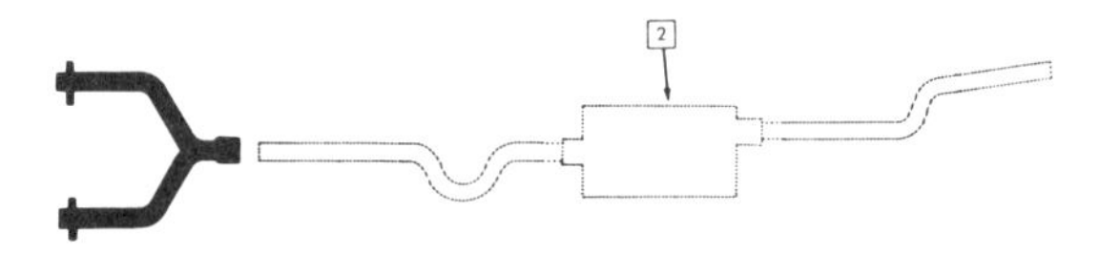

Complete exhaust system and related hardware for a 1970 Mustang equipped with a 351 2-V Cleveland engine.

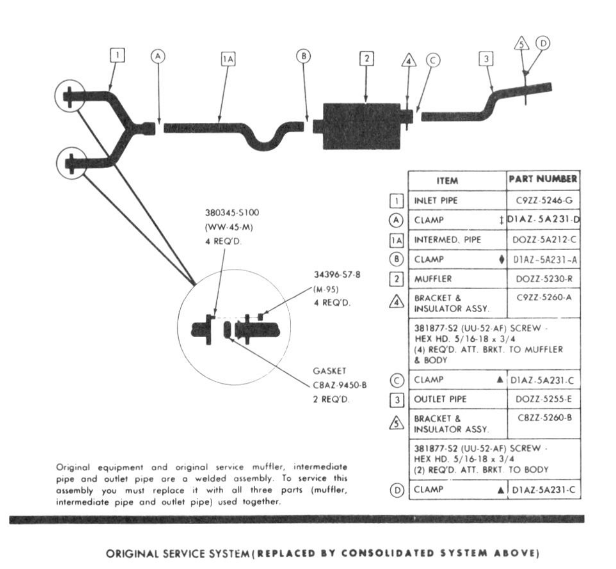

	ITEM	PART NUMBER
1	INLET PIPE	C9ZZ-5246-G
A	CLAMP ‡	D1AZ-5A231-D
1A	INTERMED. PIPE	DOZZ-5A212-C
B	CLAMP ♦	D1AZ-5A231-A
2	MUFFLER	DOZZ-5230-R
4	BRACKET & INSULATOR ASSY.	C9ZZ-5260-A
	381877-S2 (UU-52-AF) SCREW - HEX HD. 5/16-18 x 3/4 (4) REQ'D. ATT. BRKT. TO MUFFLER & BODY	
C	CLAMP ▲	D1AZ-5A231-C
3	OUTLET PIPE	DOZZ-5255-E
5	BRACKET & INSULATOR ASSY.	C8ZZ-5260-B
	381877-S2 (UU-52-AF) SCREW - HEX HD. 5/16-18 x 3/4 (2) REQ'D. ATT. BRKT. TO BODY	
D	CLAMP ▲	D1AZ-5A231-C

Original equipment and original service muffler, intermediate pipe and outlet pipe are a welded assembly. To service this assembly you must replace it with all three parts (muffler, intermediate pipe and outlet pipe) used together.

ORIGINAL SERVICE SYSTEM (REPLACED BY CONSOLIDATED SYSTEM ABOVE)

2
D

	ITEM	PART NUMBER
2	MUFFLER	★C9ZZ-5230-A
D	CLAMP ♦	D1AZ-5A231-A

Complete exhaust system and related hardware for a 1970 Mustang equipped with a 351 2-V Windsor engine.

The split exhaust tips were replaced with this flat trumpet with an oval, angular-cut opening for 1970.

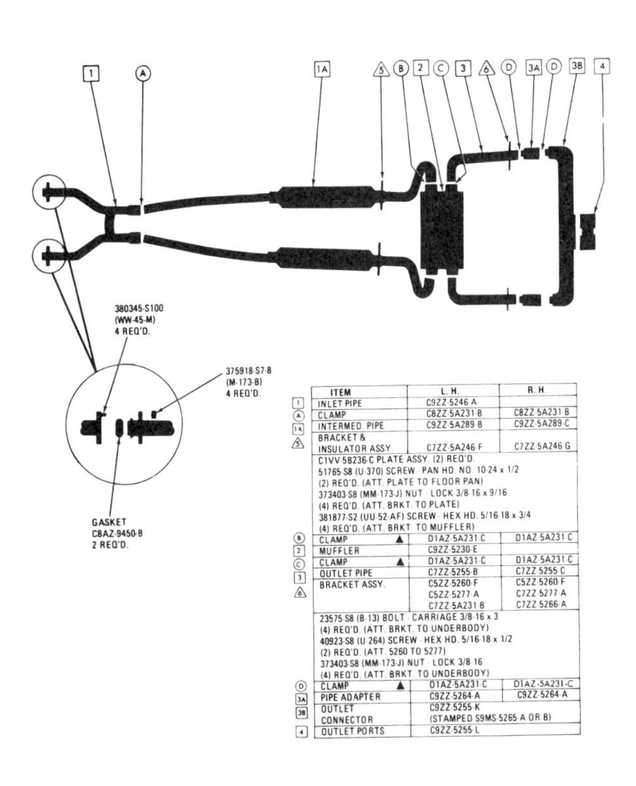

	ITEM	L. H.	R. H.
1	INLET PIPE	C9ZZ-5246-A	
A	CLAMP	C8ZZ-5A231-B	C8ZZ-5A231-B
1A	INTERMED. PIPE	C9ZZ-5A289-B	C9ZZ-5A289-C
5	BRACKET & INSULATOR ASSY.	C7ZZ-5A246-F	C7ZZ-5A246-G
	C1VV-5B236-C PLATE ASSY. (2) REQ'D. 51765-S8 (U-370) SCREW - PAN HD. NO. 10-24 x 1/2 (2) REQ'D. (ATT. PLATE TO FLOOR PAN) 373403-S8 (MM-173-J) NUT - LOCK 3/8-16 x 9/16 (4) REQ'D. (ATT. BRKT. TO PLATE) 381877-S2 (UU-52-AF) SCREW - HEX HD. 5/16-18 x 3/4 (4) REQ'D. (ATT. BRKT. TO MUFFLER)		
B	CLAMP ▲	D1AZ-5A231-C	D1AZ-5A231-C
2	MUFFLER	C9ZZ-5230-E	
C	CLAMP ▲	D1AZ-5A231-C	D1AZ-5A231-C
3	OUTLET PIPE	C7ZZ-5255-B	C7ZZ-5255-C
6	BRACKET ASSY.	C5ZZ-5260-F C5ZZ-5277-A C7ZZ-5A231-B	C5ZZ-5260-F C7ZZ-5277-A C7ZZ-5266-A
	23575-S8 (B-13) BOLT - CARRIAGE 3/8-16 x 3 (4) REQ'D. (ATT. BRKT. TO UNDERBODY) 40923-S8 (U-264) SCREW - HEX HD. 5/16-18 x 1/2 (2) REQ'D. (ATT. 5260 TO 5277) 373403-S8 (MM-173-J) NUT - LOCK 3/8-16 (4) REQ'D. (ATT. BRKT. TO UNDERBODY)		
D	CLAMP ▲	D1AZ-5A231-C	D1AZ-5A231-C
3A	PIPE ADAPTER	C9ZZ-5264-A	C9ZZ-5264-A
3B	OUTLET CONNECTOR	C9ZZ-5255-K (STAMPED S9MS-5265-A OR B)	
4	OUTLET PORTS	C9ZZ-5255-L	

Complete exhaust system and related hardware for a 1970 Shelby GT-350 Mustang equipped with a 351 4-V engine.

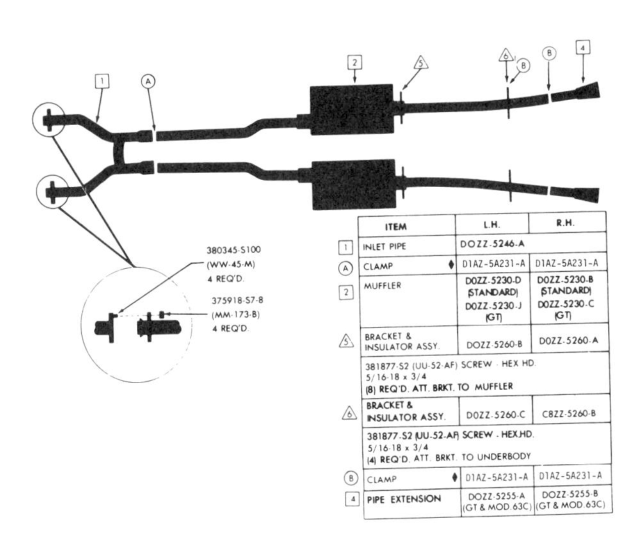

	ITEM	L.H.	R.H.
1	INLET PIPE	DOZZ-5246-A	
A	CLAMP ♦	D1AZ-5A231-A	D1AZ-5A231-A
2	MUFFLER	DOZZ-5230-D (STANDARD) DOZZ-5230-J (GT)	DOZZ-5230-B (STANDARD) DOZZ-5230-C (GT)
5	BRACKET & INSULATOR ASSY.	DOZZ-5260-B	DOZZ-5260-A
	381877-S2 (UU-52-AF) SCREW - HEX HD. 5/16-18 x 3/4 (8) REQ'D. ATT. BRKT. TO MUFFLER		
6	BRACKET & INSULATOR ASSY.	DOZZ-5260-C	C8ZZ-5260-B
	381877-S2 (UU-52-AF) SCREW - HEX.HD. 5/16-18 x 3/4 (4) REQ'D. ATT. BRKT. TO UNDERBODY		
B	CLAMP ♦	D1AZ-5A231-A	D1AZ-5A231-A
4	PIPE EXTENSION	DOZZ-5255-A (GT & MOD.63C)	DOZZ-5255-B (GT & MOD.63C)

Complete exhaust system and related hardware for a 1970 standard and GT Mustang equipped with a 351 4-V engine.

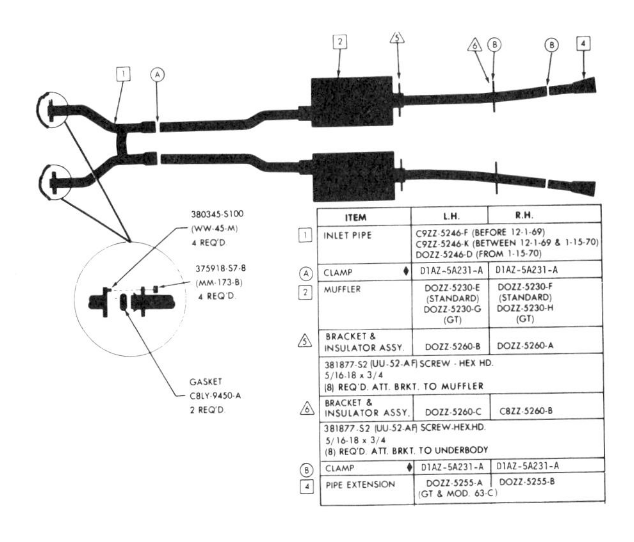

	ITEM	L.H.	R.H.
1	INLET PIPE	C9ZZ-5246-F (BEFORE 12-1-69) C9ZZ-5246-K (BETWEEN 12-1-69 & 1-15-70) DOZZ-5246-D (FROM 1-15-70)	
A	CLAMP ♦	D1AZ-5A231-A	D1AZ-5A231-A
2	MUFFLER	DOZZ-5230-E (STANDARD) DOZZ-5230-G (GT)	DOZZ-5230-F (STANDARD) DOZZ-5230-H (GT)
5	BRACKET & INSULATOR ASSY.	DOZZ-5260-B	DOZZ-5260-A
	381877-S2 (UU-52-AF) SCREW - HEX HD. 5/16-18 x 3/4 (8) REQ'D. ATT. BRKT. TO MUFFLER		
6	BRACKET & INSULATOR ASSY.	DOZZ-5260-C	C8ZZ-5260-B
	381877-S2 (UU-52-AF) SCREW-HEX.HD. 5/16-18 x 3/4 (8) REQ'D. ATT. BRKT. TO UNDERBODY		
B	CLAMP ♦	D1AZ-5A231-A	D1AZ-5A231-A
4	PIPE EXTENSION	DOZZ-5255-A (GT & MOD. 63-C)	DOZZ-5255-B

Complete exhaust system and related hardware for a 1970 standard and GT Mustang equipped with a 428 4-V engine.

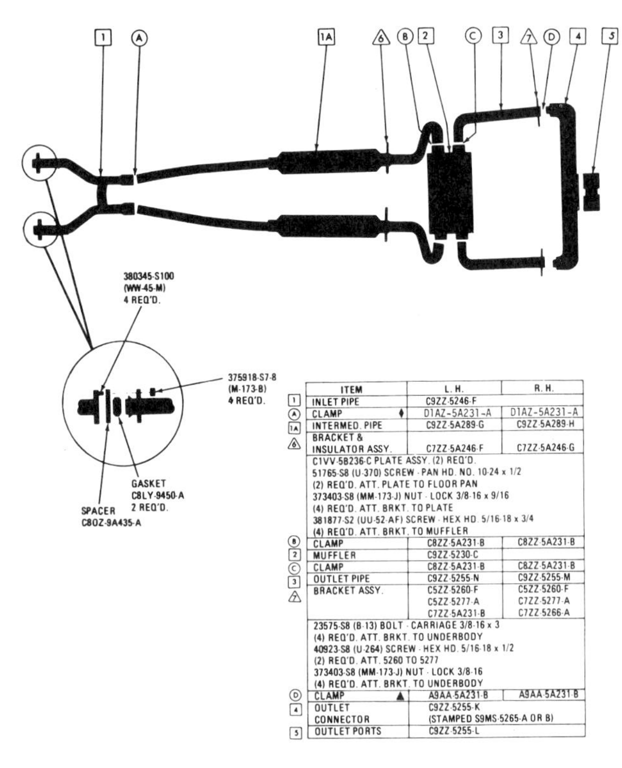

	ITEM	L. H.	R. H.
1	INLET PIPE	C9ZZ-5246-F	
A	CLAMP ♦	D1AZ-5A231-A	D1AZ-5A231-A
1A	INTERMED. PIPE	C9ZZ-5A289-G	C9ZZ-5A289-H
6 (triangle)	BRACKET & INSULATOR ASSY.	C7ZZ-5A246-F	C7ZZ-5A246-G
	C1VV-5B236-C PLATE ASSY. (2) REQ'D. 51765-S8 (U-370) SCREW - PAN HD. NO. 10-24 x 1/2 (2) REQ'D. ATT. PLATE TO FLOOR PAN 373403-S8 (MM-173-J) NUT - LOCK 3/8-16 x 9/16 (4) REQ'D. ATT. BRKT. TO PLATE 381877-S2 (UU-52-AF) SCREW - HEX HD. 5/16-18 x 3/4 (4) REQ'D. ATT. BRKT. TO MUFFLER		
B	CLAMP	C8ZZ-5A231-B	C8ZZ-5A231-B
2	MUFFLER	C9ZZ-5230-C	
C	CLAMP	C8ZZ-5A231-B	C8ZZ-5A231-B
3	OUTLET PIPE	C9ZZ-5255-N	C9ZZ-5255-M
7 (triangle)	BRACKET ASSY.	C5ZZ-5260-F C5ZZ-5277-A C7ZZ-5A231-B	C5ZZ-5260-F C7ZZ-5277-A C7ZZ-5266-A
	23575-S8 (B-13) BOLT - CARRIAGE 3/8-16 x 3 (4) REQ'D. ATT. BRKT. TO UNDERBODY 40923-S8 (U-264) SCREW - HEX HD. 5/16-18 x 1/2 (2) REQ'D. ATT. 5260 TO 5277 373403-S8 (MM-173-J) NUT - LOCK 3/8-16 (4) REQ'D. ATT. BRKT. TO UNDERBODY		
D	CLAMP ▲	A9AA-5A231-B	A9AA-5A231-B
4	OUTLET CONNECTOR	C9ZZ-5255-K (STAMPED S9MS-5265-A OR B)	
5	OUTLET PORTS	C9ZZ-5255-L	

Complete exhaust system and related hardware for a 1970 Shelby GT-500 Mustang equipped with a 428 4-V engine.

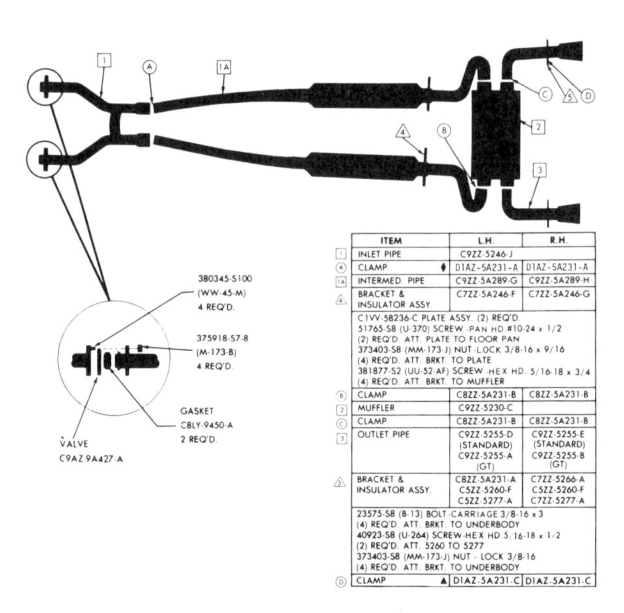

	ITEM	L.H.	R.H.
1	INLET PIPE	C9ZZ-5246-J	
A	CLAMP ♦	D1AZ-5A231-A	D1AZ-5A231-A
1A	INTERMED. PIPE	C9ZZ-5A289-G	C9ZZ-5A289-H
4 (triangle)	BRACKET & INSULATOR ASSY.	C7ZZ-5A246-F	C7ZZ-5A246-G
	C1VV-5B236-C PLATE ASSY. (2) REQ'D. 51765-S8 (U-370) SCREW -PAN HD #10-24 x 1/2 (2) REQ'D. ATT. PLATE TO FLOOR PAN 373403-S8 (MM-173-J) NUT -LOCK 3/8-16 x 9/16 (4) REQ'D. ATT. BRKT. TO PLATE 381877-S2 (UU-52-AF) SCREW -HEX HD. 5/16-18 x 3/4 (4) REQ'D. ATT. BRKT. TO MUFFLER		
B	CLAMP	C8ZZ-5A231-B	C8ZZ-5A231-B
2	MUFFLER	C9ZZ-5230-C	
C	CLAMP	C8ZZ-5A231-B	C8ZZ-5A231-B
3	OUTLET PIPE	C9ZZ-5255-D (STANDARD) C9ZZ-5255-A (GT)	C9ZZ-5255-E (STANDARD) C9ZZ-5255-B (GT)
5 (triangle)	BRACKET & INSULATOR ASSY.	C8ZZ-5A231-A C5ZZ-5260-F C5ZZ-5277-A	C7ZZ-5266-A C5ZZ-5260-F C7ZZ-5277-A
	23575-S8 (B-13) BOLT -CARRIAGE 3/8-16 x 3 (4) REQ'D. ATT. BRKT. TO UNDERBODY 40923-S8 (U-264) SCREW-HEX HD.5/16-18 x 1/2 (2) REQ'D. ATT. 5260 TO 5277 373403-S8 (MM-173-J) NUT - LOCK 3/8-16 (4) REQ'D. ATT. BRKT. TO UNDERBODY		
D	CLAMP ▲	D1AZ-5A231-C	D1AZ-5A231-C

Complete exhaust system and related hardware for a 1970 Boss 429 Mustang.

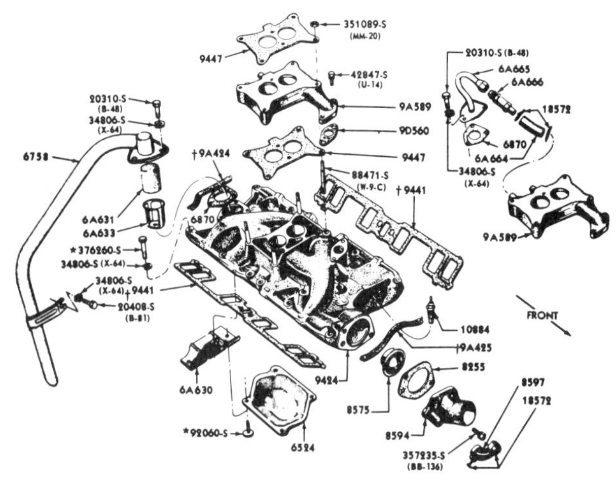

Intake manifold, gaskets, and related parts for 260, 289, 302, and 351W engines.

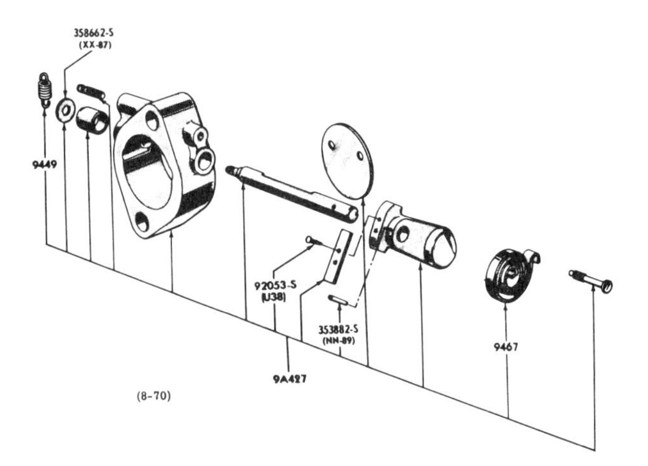

Typical thermostatically controlled exhaust valve (heat riser).

Chapter 7

Wiring and Electrical Systems

Major wiring harnesses were basically similar for all body styles. But pigtails and supplementary harnesses, especially in the realm of factory-installed air conditioning, were added where needed to adapt accessories. Early four-way flashers, Rally-Pacs, convertible tops, and fog-lamps all had their own pigtail harnesses. An owner upgrading to the use of an in-dash tachometer in a 1969 or 1970 Mach 1, for instance, must replace the entire underdash harness to rewire the gauges and lights as well as the added instrument.

One of the great frustrations in restoring Mustangs is an owner's discovery after a car is assembled that incorrect harness routings or incomplete harnesses are costing either safety or concours points. Add to those problems the possibilities of fire or poor engine performance, and one realizes the importance of getting it right the first time.

This chapter provides cutaway drawings to demonstrate proper placement of the several harnesses needed to wire any Mustang or Shelby. As a supplement, the reader may wish to obtain model-specific wiring diagrams to assist in a complete rewiring. A good source for such reproduction diagrams is Jim Osborn Reproductions in Lawrenceville, Georgia (404-962-7556).

Also included in this area are drawings of correct light buckets, guides to proper lenses, bulbs, and fuses, and a basis for the restorer to plan and establish priorities in budgeting for rebuilding a Mustang.

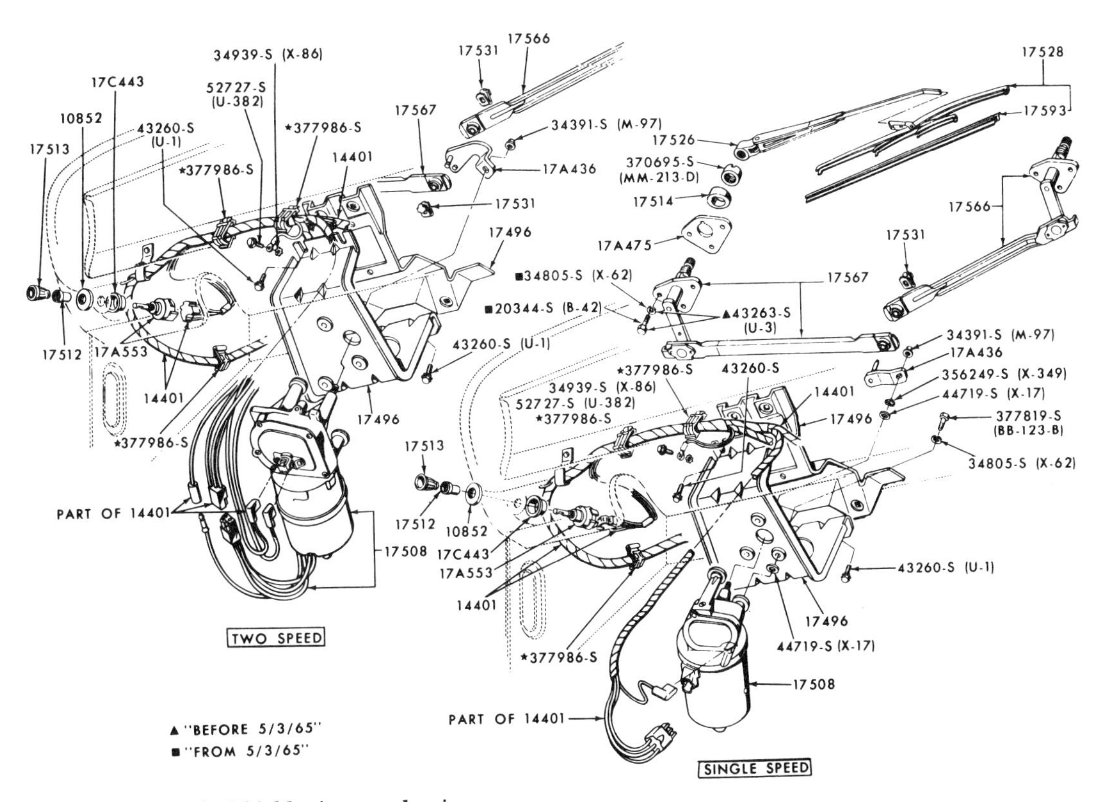

The 1965-1968 Mustang windshield wiper mechanism.

MUSTANG WINDSHIELD WIPER MOTOR ASSEMBLY APPLICATION CHART

YEAR	PART NUMBER	DESCRIPTION
65/66	C3UZ-17508-C	Single speed electric - less drive
65	C3DZ-17508-A	2-speed electric - less drive
66	C6ZZ-17508-A	Used with 2-speed wipers - Includes bracket & drive
67/68	C6OZ-17508-D	Used with 2-speed wipers
69/70	DOZZ-17508-A	Used with 2-speed or intermittent wipers

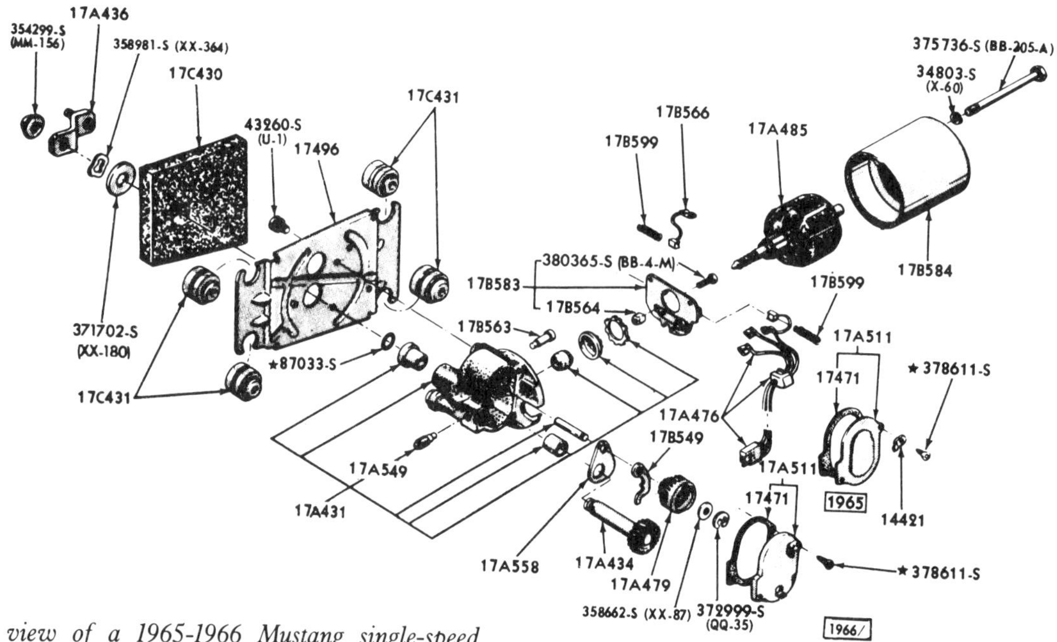

Exploded view of a 1965-1966 Mustang single-speed windshield wiper motor.

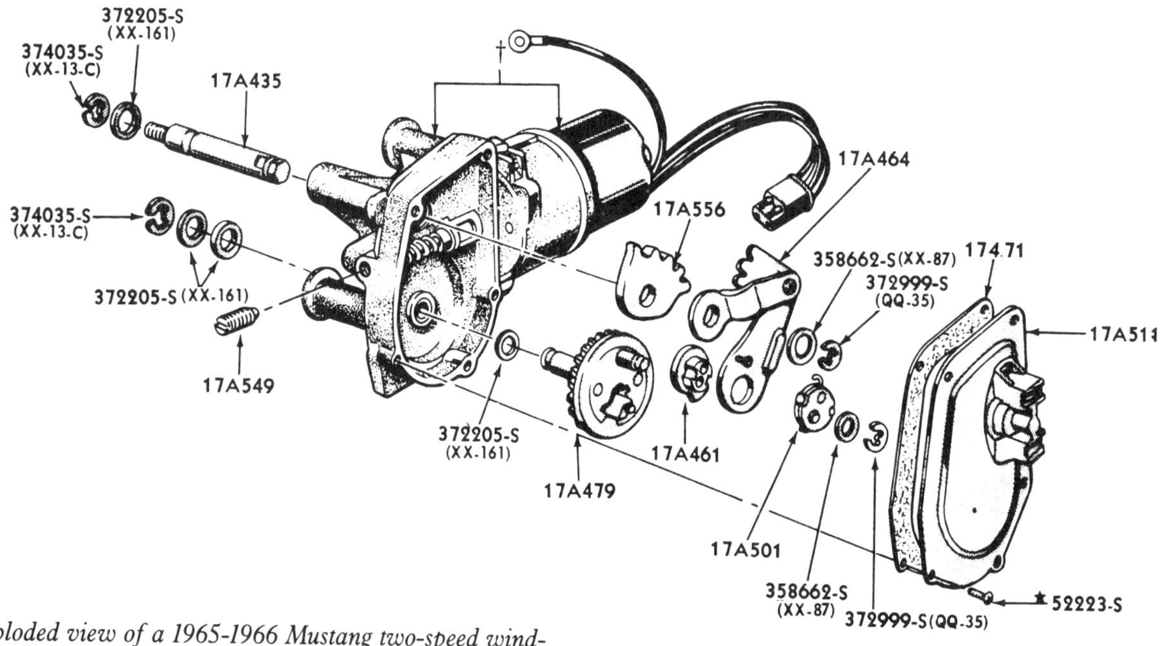

Exploded view of a 1965-1966 Mustang two-speed windshield wiper motor.

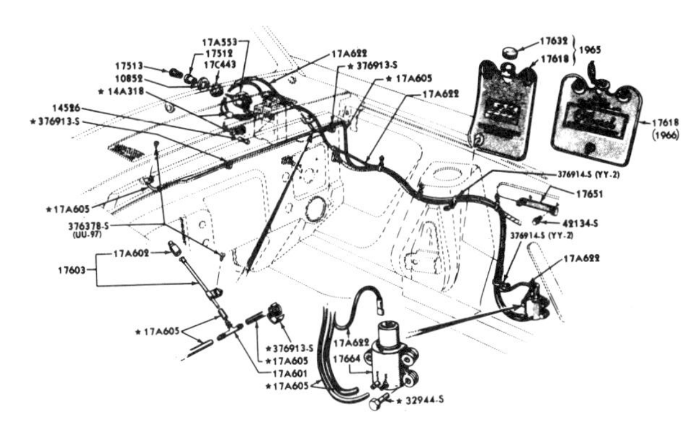

The 1965-1966 Mustang windshield washer system.

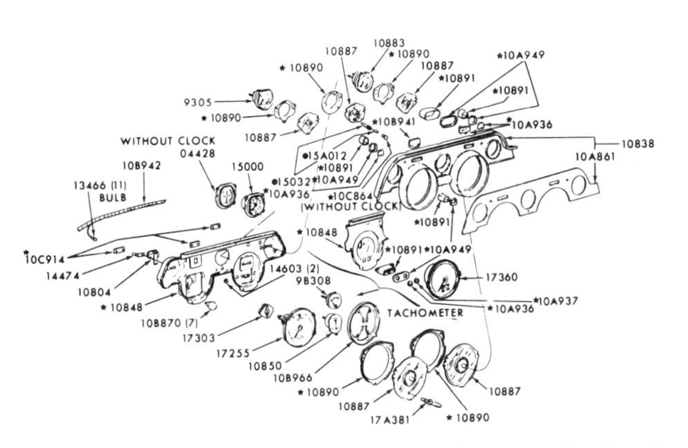

The 1967-1968 Mustang instrument cluster and related parts.

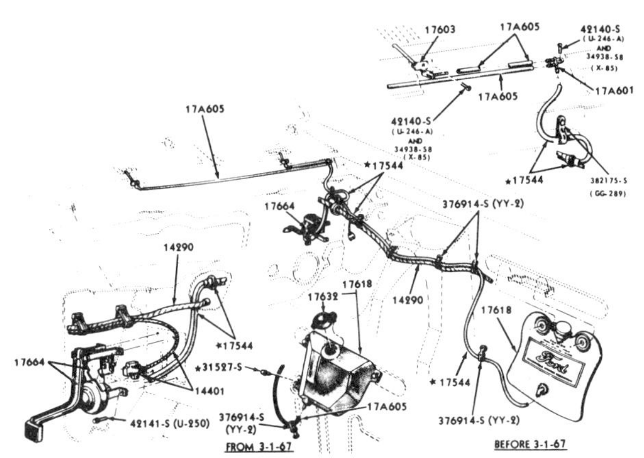

The 1967-1968 Mustang windshield washer system.

MUSTANG SPEEDOMETER ASSEMBLY APPLICATION CHART

Year	Model	Description	Part Number
65		w/o Camera Case Cluster	C5DZ-17255-A
		Rectangular, 120 mph	
65/66		Woodgrain & Camera Case	C5DZ-17255-A
		Cluster, Round, 140 mph	
67/68	Exc. Shelby	Miles-w/o Tripodometer	C7ZZ-17255-E
67/68	Exc. Shelby	Miles-w/ Tripodometer	C7ZZ-17255-F
67/68	All Shelby	140 mph w/ Tripodometer	C8AF-17282-E
69	Exc. Mach 1 & Shelby	Miles-w/ Tripodometer	C9ZZ-17255-B
69/70	All Shelby	140 mph w/ Tripodometer	C9ZZ-17255-N
69/70	Exc. Grandé & Shelby	w/o Tripodometer, 10-mile	D0ZZ-17255-A
		Graduations-Black Dial	
69/70	Grandé	w/o Tripodometer, 5-mile	D0ZZ-17255-B
		Graduations-Gray Blue Dial	
70	Exc. Mach 1	Miles w/o Tripodometer	D0ZZ-17255-C
69/70	Mach 1	Miles-w/ Tripodometer	D0ZZ-17255-D

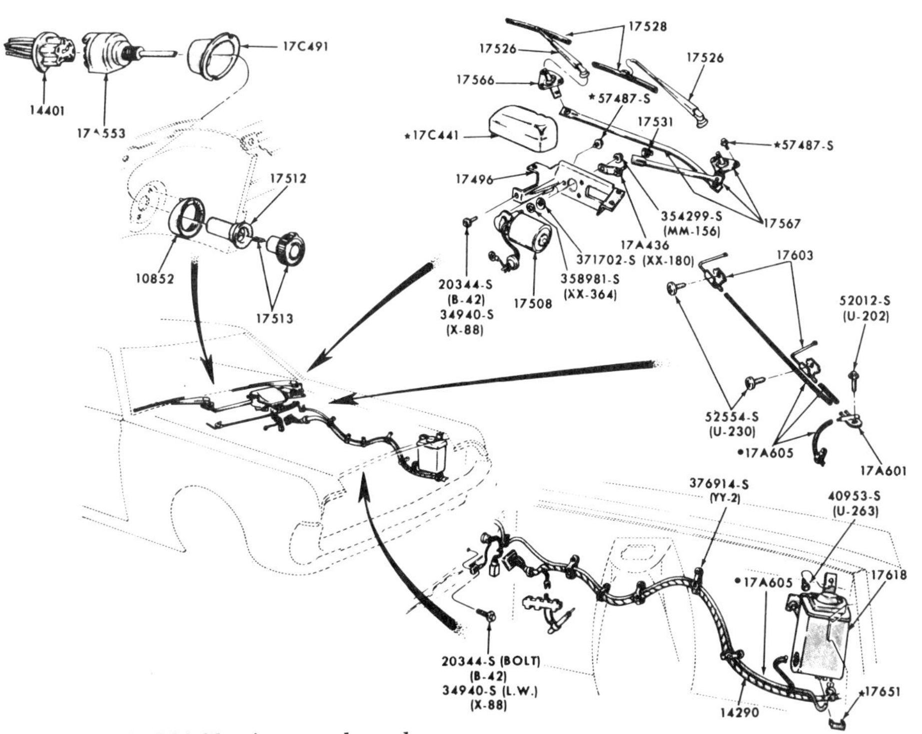

The 1969-1970 Mustang windshield wipers and washer system.

The 1965 Mustang instrument cluster, like this example in a convertible with standard black interior, positions the round fuel and temperature gauges at opposite ends of the horizontal speedometer. Generator charge and oil-pressure warning lights are on either side of the odometer.

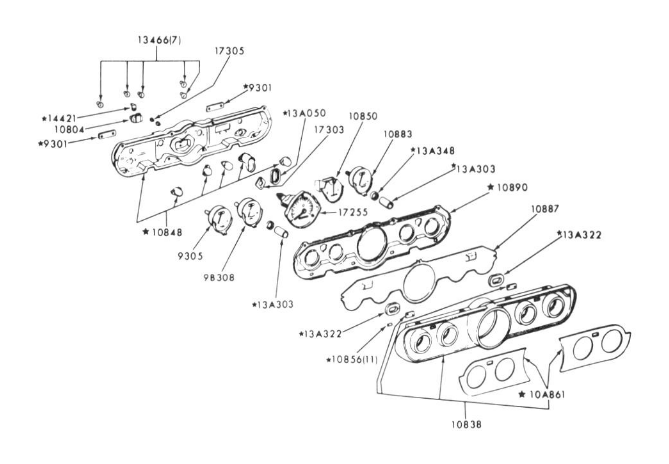

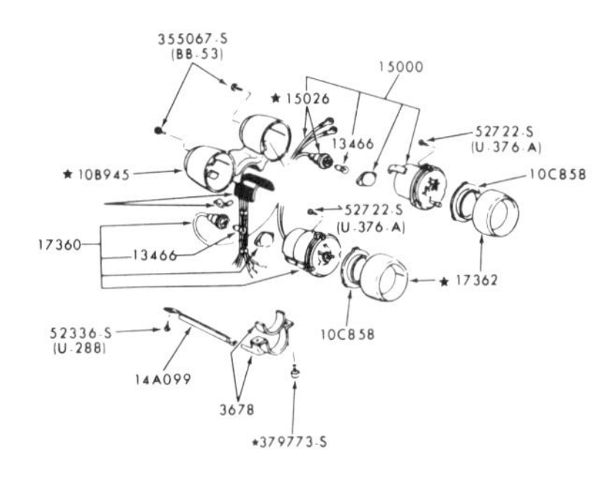

The 1965-1966 Mustang instrument cluster (with gauges) and Rally-Pac.

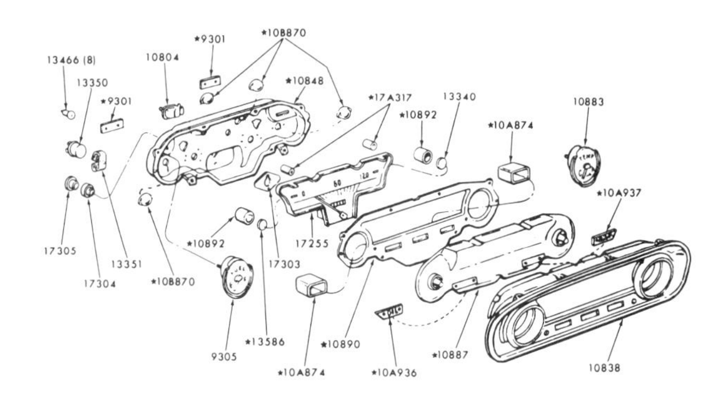

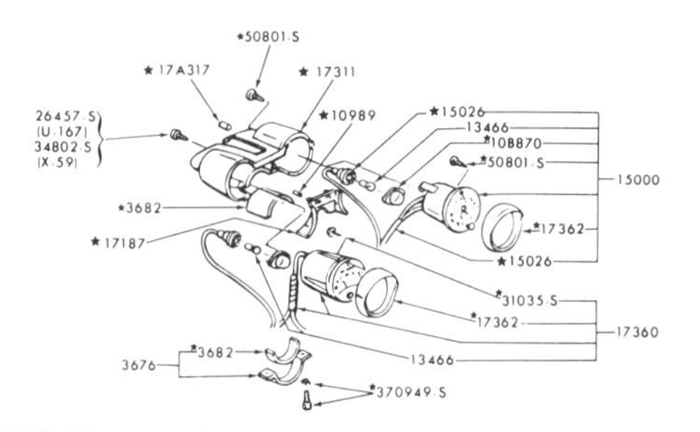

The 1965 Mustang instrument cluster (with warning lights) and Rally-Pac.

A low-profile Rally-Pac was introduced for 1966. Its case could either be black or color-keyed to the steering column.

A tachometer was part of the Rally-Pac option. The standard tachometer stopped at 6000rpm. An 8000rpm tach was mated with the 271hp 289ci HiPo engine.

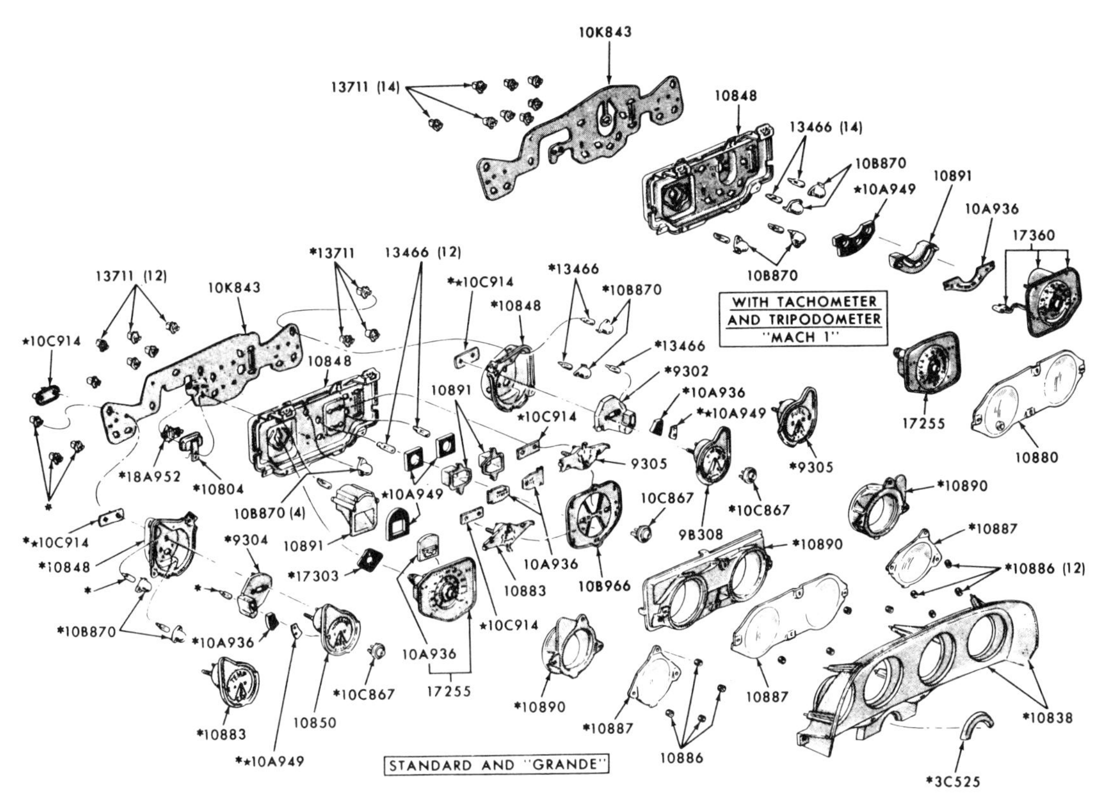

The 1969-1970 Mustang instrument cluster and related parts.

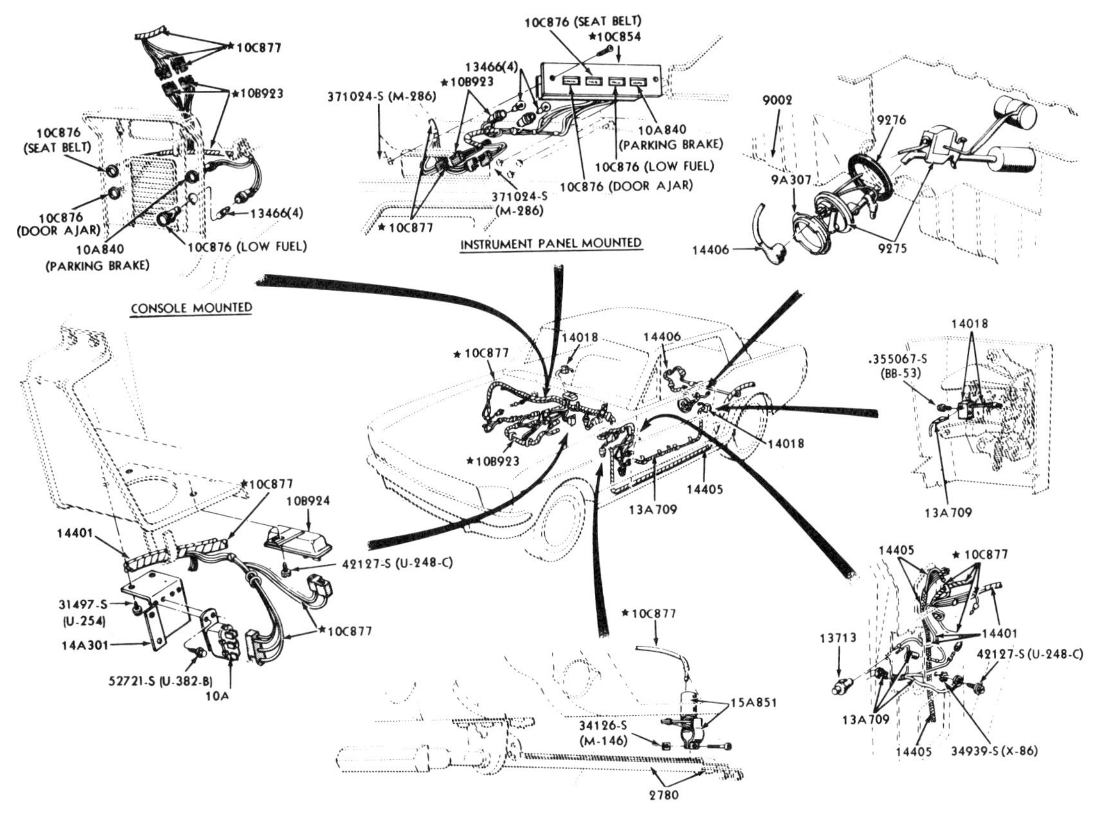

The 1967-1968 Mustang optional warning indicator panel assembly, wiring, and related hardware.

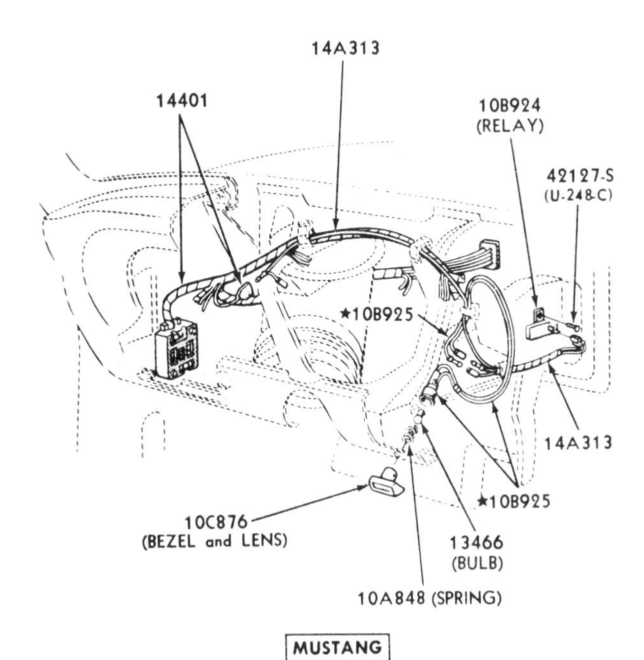

The 1966-1969 Mustang seat belt warning lamp, wiring, and related hardware.

MUSTANG ELECTRICAL CIRCUIT BREAKER ASSEMBLY APPLICATION CHART

Year	Model	Amp	Length	Width	Ident.	Diam.	Part Number
65/67	Fog Lamp	10	1.25"	.78"	White	9/32"	C5ZZ-14526-A
'68	with Fog Lamps	10	1.25"	.79"	White	13/64"	C8ZZ-14526-B
69/70	GT-350/500 Stop Lamp, Fog Lamp	15					C9ZZ-14526-A
'70	w/ automatic seat back hatch	20	1.07"	.40"	Red		C4DZ-14526-C
'65	Conv. Top Control	20	1.07"	.40"	Red		C4DZ-14526-C
'69	GT-350/500 Fog Lamp	20	1.25"	.75"			C9ZZ-14526-B
'67	with A/C	25	1.07"	.40"	Red, Yellow		C5AZ-14526-C
'67	with A/C	25	1.26"	.79"	Red, Yellow		C7AZ-14526-A
'68	6 Cyl. 200, 8 Cyl. 289 with A/C	25	1.25	.79"	Red, Green	13/64"	C8ZZ-14526-A
'69	GT-350/500 w/ Fog Lamp	25	1.25	.79"	Red, Green	13/64"	C8ZZ-14526-A

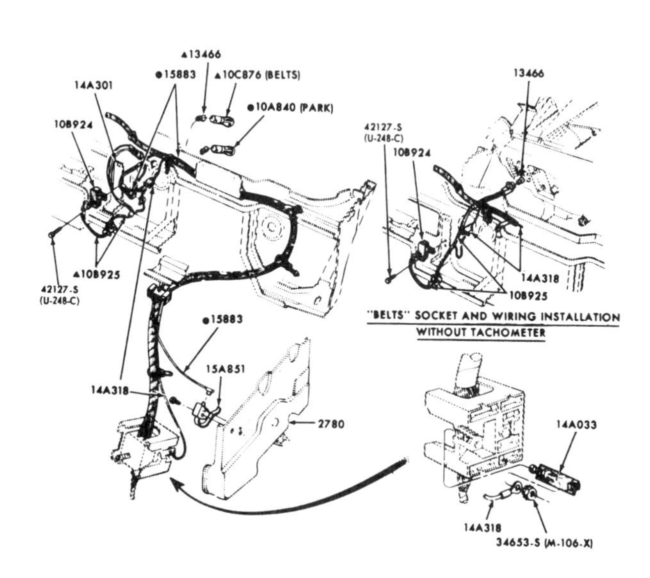

The 1969-1970 Mustang seat belt and parking lamp warning lamp wiring assembly.

MUSTANG GLASS TUBE FUSE BREAKER REFERENCE CHART

Amperage	Length	Industry Number	Part Number
1	.62"	AGA-1	B6AZ-14526-D
2	.62"	1AG-2	D4AZ-14526-C
2.5	.62"	1AG-2 1/2	C4GY-14526-A
3	.62"	AGA-3	A6AZ-14526-A
3	1.00"	8AG-3	D4AZ-14526-E
4	.62"	SFE-4	C9SZ-14526-C
4	.84"	AGW-4	COLY-14526-B
5	.62"	AGA-5	B6AZ-14526-B
6	.75"	SFE-6	C1VY-14526-A
7.5	.88"	SFE-7 1/2	B6AZ-14526-A
9	.88"	SFE-9	A6AZ-14526-B
10	1.25"	AGC-10	B8TZ-14526-A
14	1.06"	SFE-14	A0AZ-14526-A
15	1.25	AGC-15	A2AZ-14526-A
15	.88"	AGW-15	C4SZ-14526-B
20	.62"	AGA-20	8HC-14526
20	1.62"	AGX-20	C5AZ-14526-B
20	1.25"	SGE-20	A0AZ-14526-B
25	1.25"	3AG-25	C6TZ-14526-AB
30	1.25"	AGC-30	A9AZ-14526-A
30	1"	AGX-30	C8OZ-14526-A

FUSIBLE LINK WIRE TYPE FUSE BREAKER CHART

Year	Model	Description	Part Number
'68	Conv. top control	6.00" long - #C8ZB-14A094-A	C8ZZ-14526-C
		16-Gauge Wire	
'69	w/55, 61 OR 65 amp	9" long - 14 Gauge wire	C9AZ-14526-D
	Ford Alternator	r/b D3AZ-14526-D (4-74)	
'69	w/ 38, 42 OR 45 amp	9" long, 16 Gauge Wire Fuse Link	C9AZ-14526-A
	Ford Alternator	(Alternator Protection)	

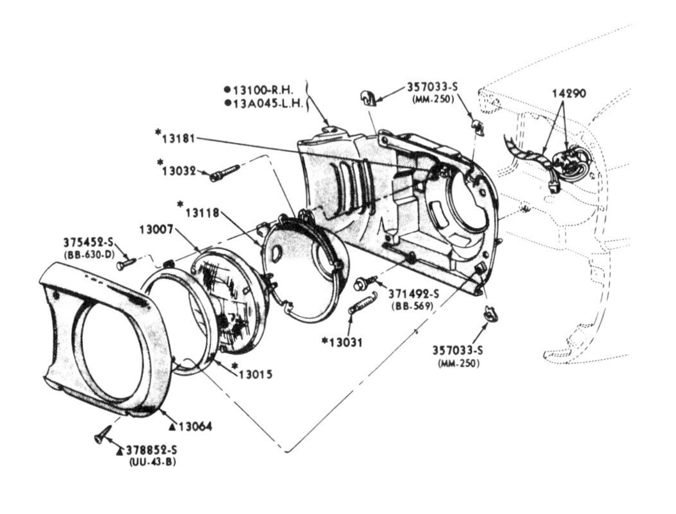

The 1965-1966 Mustang headlamp assembly and attaching hardware.

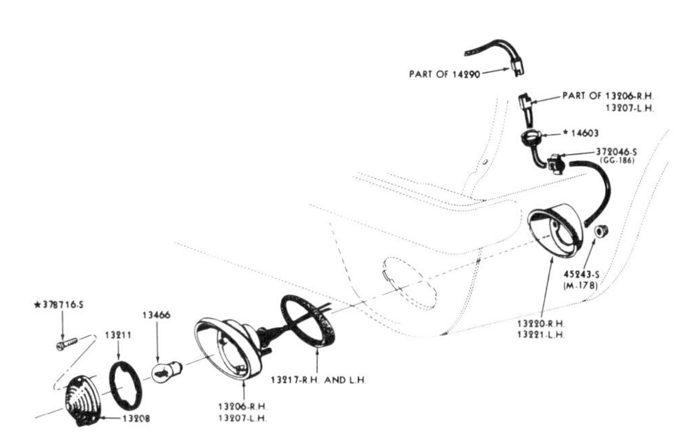

The 1965-1966 Mustang parking lamp, wiring, and attaching hardware.

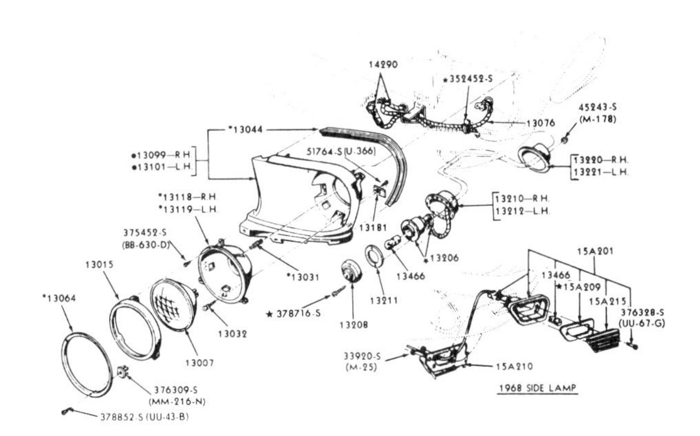

The 1967-1968 Mustang front lighting, wiring, and attaching hardware.

INDUSTRY NUMBER TO SERVICE PART NUMBER GUIDE
Mustang Related Light Bulbs

Industry Number	Contact Base	Candle Power	Part Number
53	Single	1	C4SZ-13466-C
90	Double	6	C6SZ-13466-A
97	Single	4	C3AZ-13466-G
105	Single	12	D0AZ-13466-A
161	Wedge	1	C3AZ-13466-D
194	Wedge	2	C2AZ-13466-C
194A	Wedge	2	C9DZ-13466-A
212-1	Double End	6	C6VY-13466-A
256	Single	1.6	C4SZ-13466-A
257	Single	2	C0AZ-13466-A
562	Single	6	D0ZZ-13466-A
631	Single	6	C3VY-13466-A
1003	Single	15	B6AZ-13466-A
1004	Double	15	C3SZ-13466-A
1142	Double	21	C5ZZ-13466-A
1156	Single	32	C3DZ-13466-A
1157	Double	32.3	C8TZ-13466-A
1157A	Double	32.3	C9AZ-13466-A
1157NA	Double	32.3	C9MY-13466-A
1178A	Double	4	C8ZZ-13466-A
1232	Double	4	C5TZ-13466-A
1445	Single	1	B6A-13466-B
1891	Single	1.9	B9SZ-13466-A
1892	Single	1.3	B9MY-13466-A
1895 (57)	Single	2	C3AZ-13466-B
1895	Single	2	C4SZ-13466-B

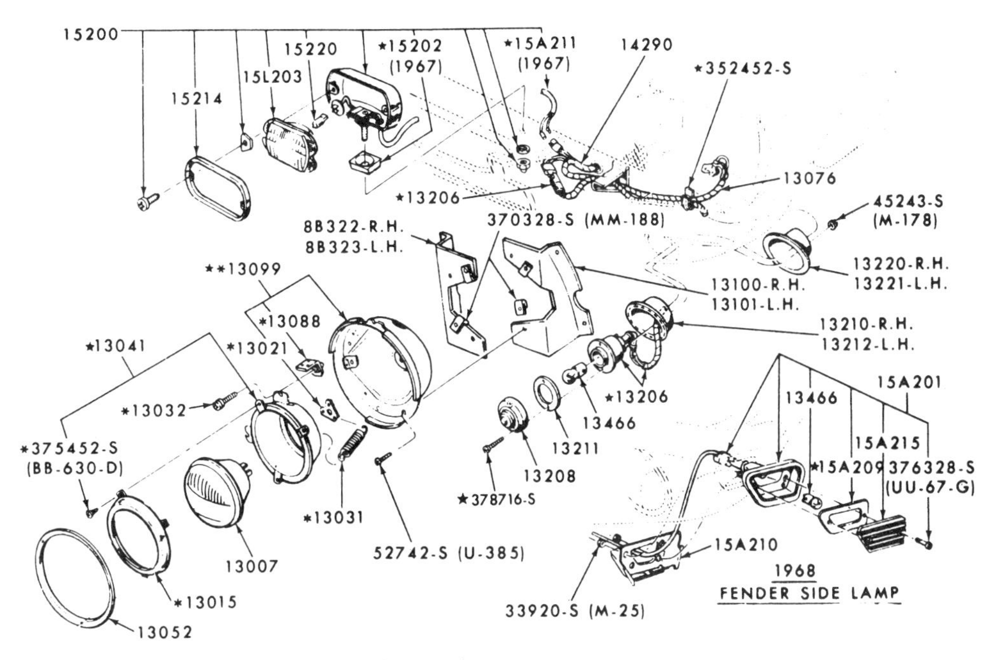

The 1967-1968 Shelby Mustang front lighting, wiring, and attaching hardware.

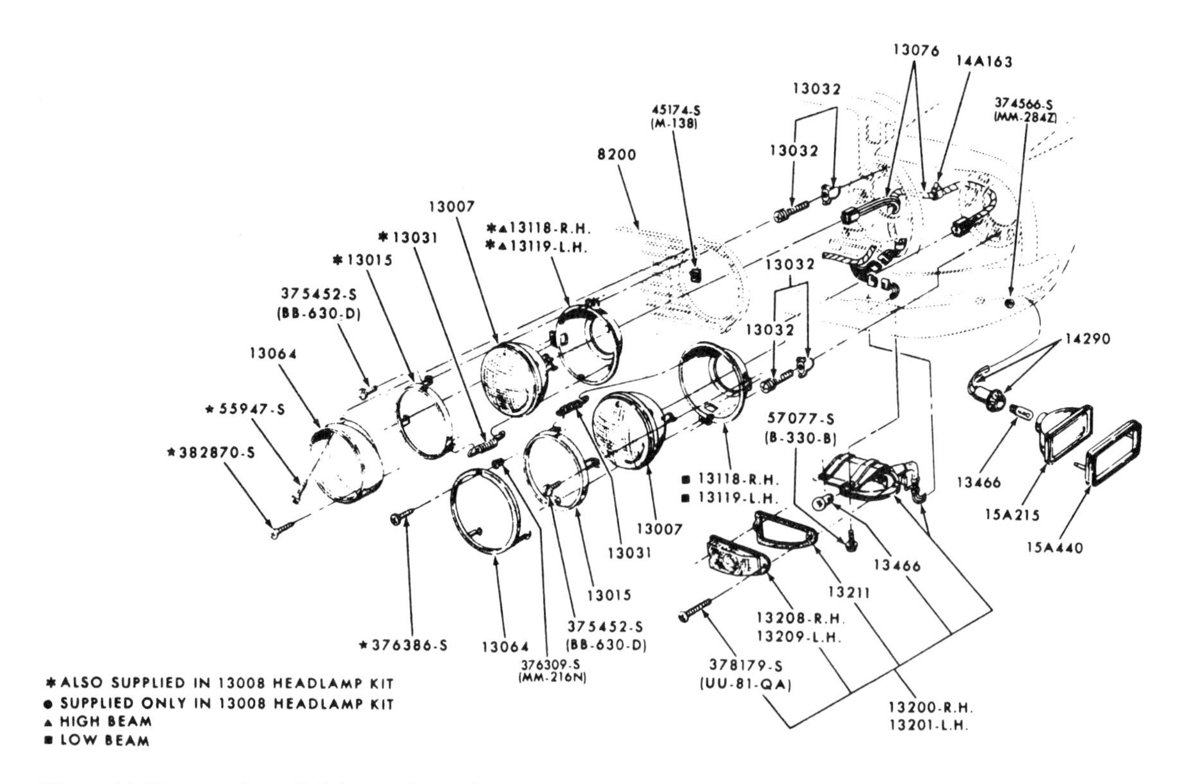

The 1969 Mustang front lighting and attaching hardware.

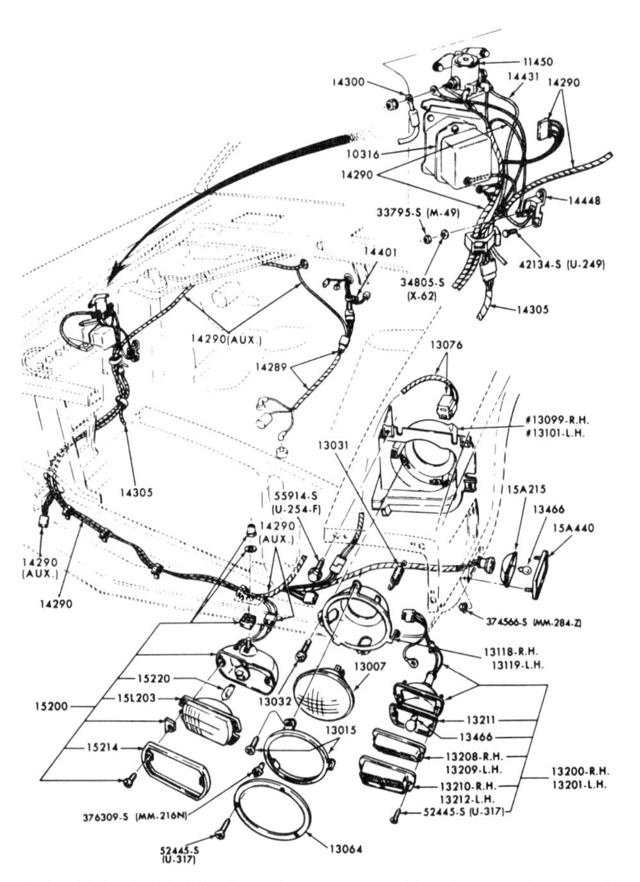

The 1969-1970 Shelby Mustang front lighting, wiring, and attaching hardware.

Mustang Dash Panel to Headlamp Junction Wire Assembly Application Chart

Year	Model	Description	Part Number
65	With generator		C5ZZ-14290-B
65	With Standard Interior;		C5ZZ-14290-C
	alternator & charge or oil indicator		
	warning lamps		
65	With Deluxe Interior		C5ZZ-14290-D
65	Standard Interior 8 cyl. w/		C5ZZ-14290-D
	ammeter & oil pressure gauges		
66	Before 11/15/65		C6ZZ-14290-A
66	From 11/15/65		C6ZZ-14290-B
67	Exc. GT and Exc. w/tachometer		C7ZZ-14290-AG
67	Exc. GT, w/tachometer		C7ZZ-14290-AD
67	(GT-350/500)	Not replaced	S7MS-14290-A
67	GT w/o tachometer	Not replaced	C7ZZ-14290-AJ
67	GT w/ tachometer		C7ZZ-14290-AF
68	Exc. GT, w/o tachometer		C8ZZ-14290-A
68	Exc. GT, with tachometer	Not replaced	C8ZZ-14290-B
68	GT & GT/CS w/o tachometer		C8ZZ-14290-C
68	GT & GT/CS w/ tachometer		C8ZZ-14290-D
69	Without tachometer		C9ZZ-14290-S
69	With tachometer		C9ZZ-14290-R
69	(GT-350/500)	Instrument panel	C9ZZ-14290-E
		to fog lamp	
70	Exc. Mach 1		D0ZZ-14290-A
70	Mach 1, Before 10/15/69		D0ZZ-14290-K
70	Mach 1, From 10/15/69		D0ZZ-14290-N

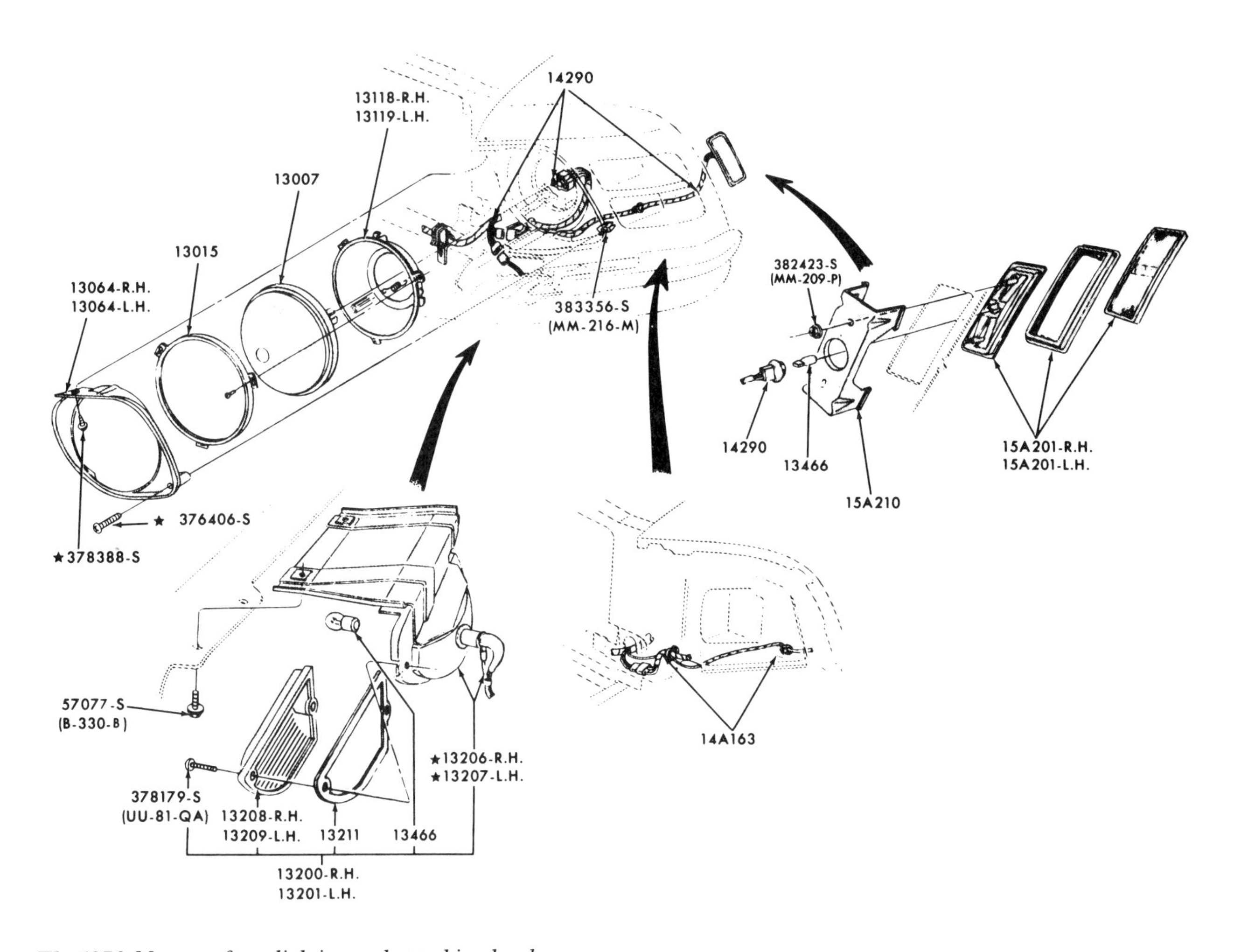

The 1970 Mustang front lighting and attaching hardware.

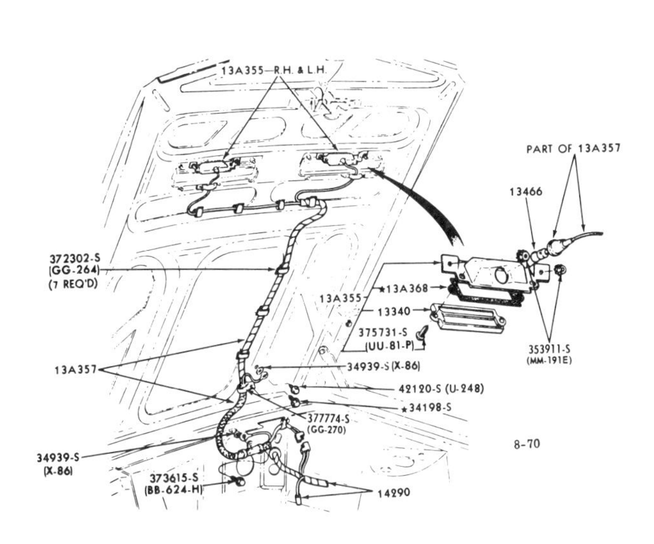

The 1967-1968 Mustang hood-mounted turn indicator lamps, wiring, and attaching hardware.

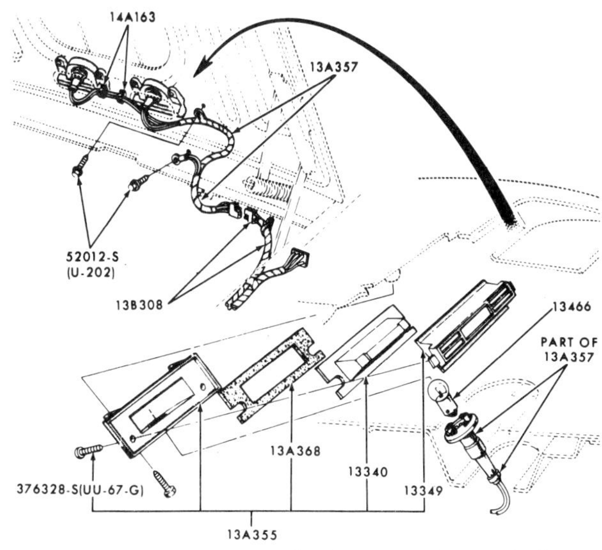

The 1969-1970 Mustang hood-mounted turn indicator lamps, wiring, and attaching hardware.

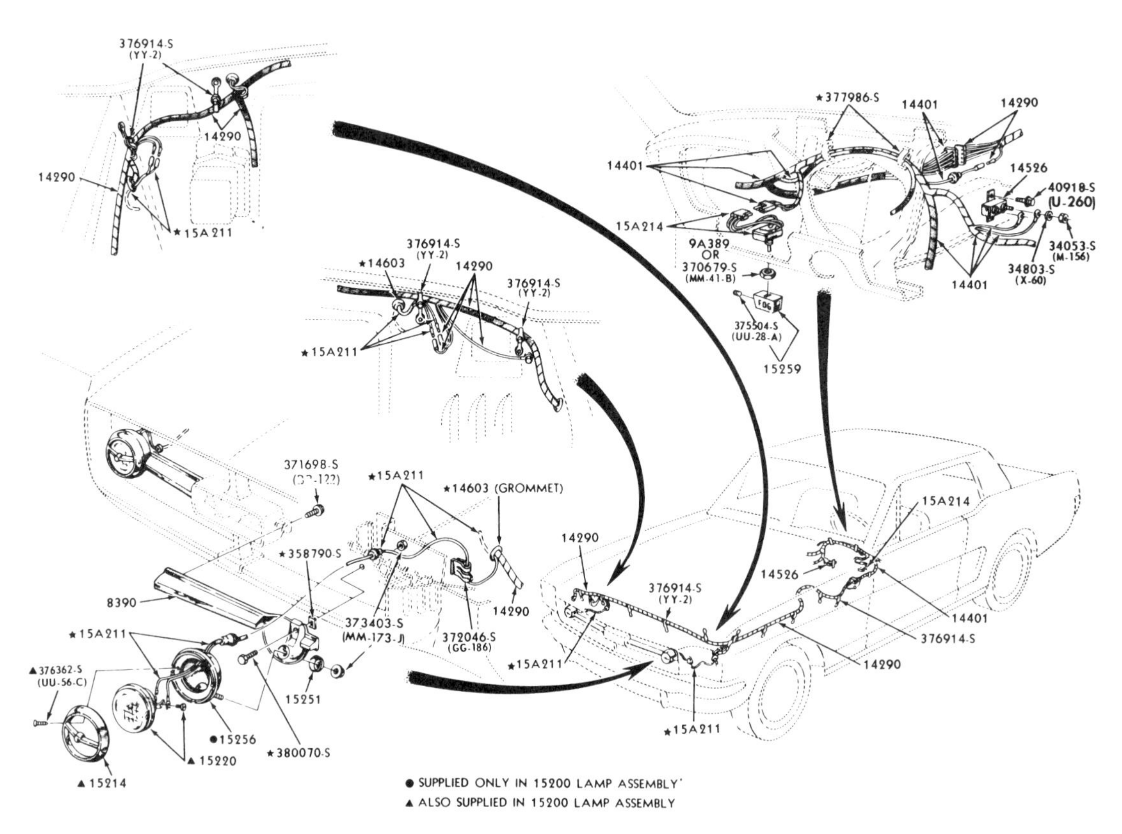

The 1965 Mustang foglamp, wiring, and related parts.

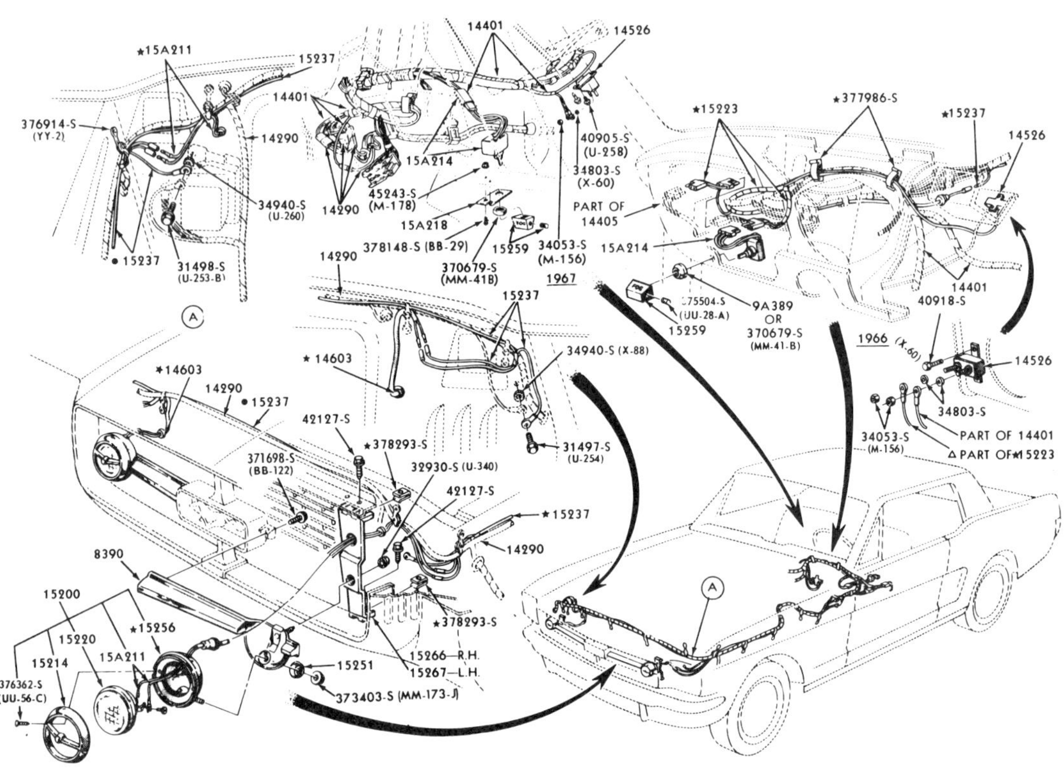

The 1966-1968 Mustang foglamp, wiring, and related parts.

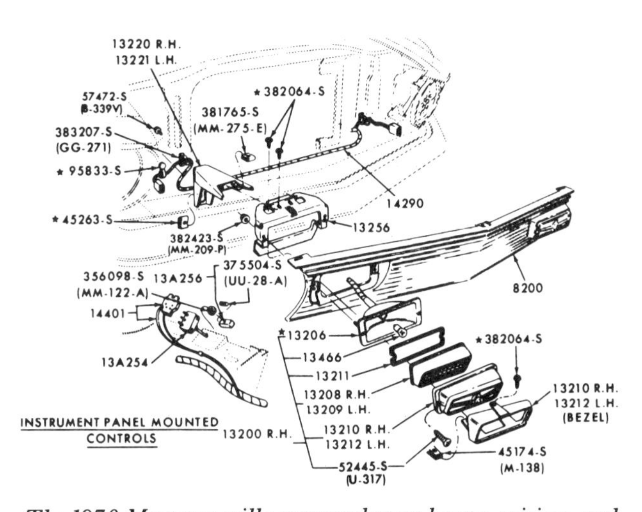

The 1970 Mustang grille-mounted sport lamps, wiring, and mounting hardware.

The 1965-1966 Mustang taillamp.

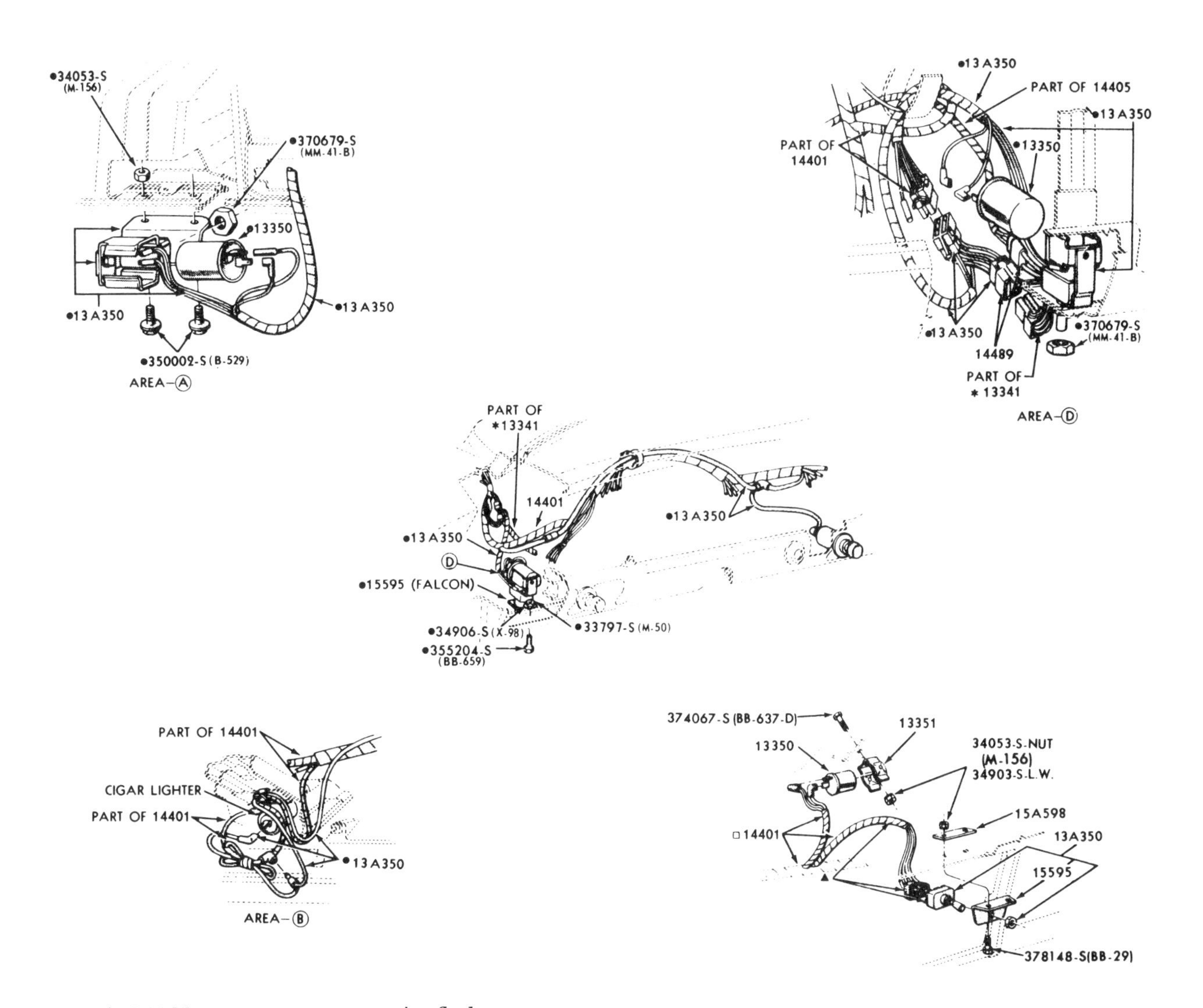

The 1965-1966 Mustang emergency warning flasher system, related wiring, and mounting hardware.

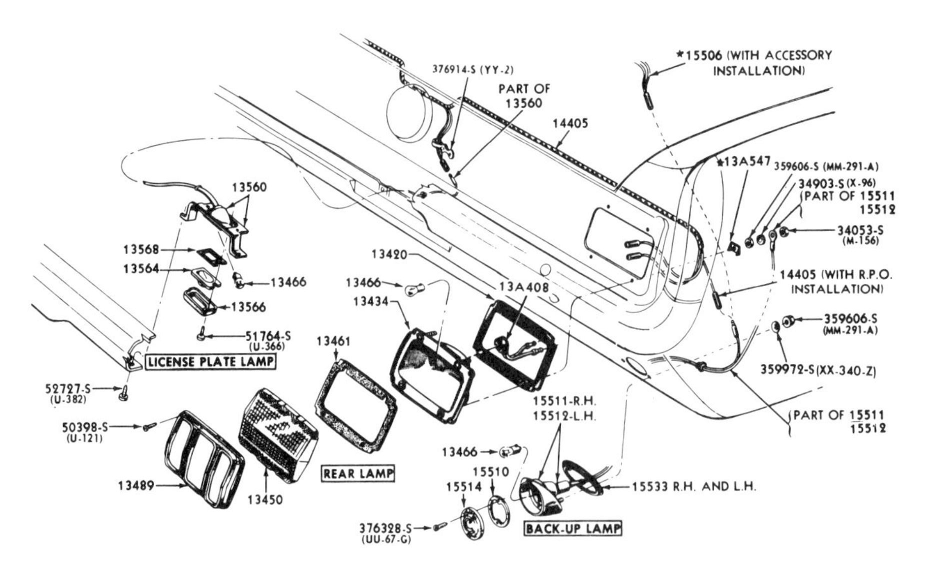

The 1965-1966 Mustang rear light fixtures and related hardware.

The 1967 taillamps on a fastback. The same taillamps were used in 1968, but the surrounding inner groove of the individual chrome bezels was painted black.

The 1967 Mustang tail panel with rear grille option.

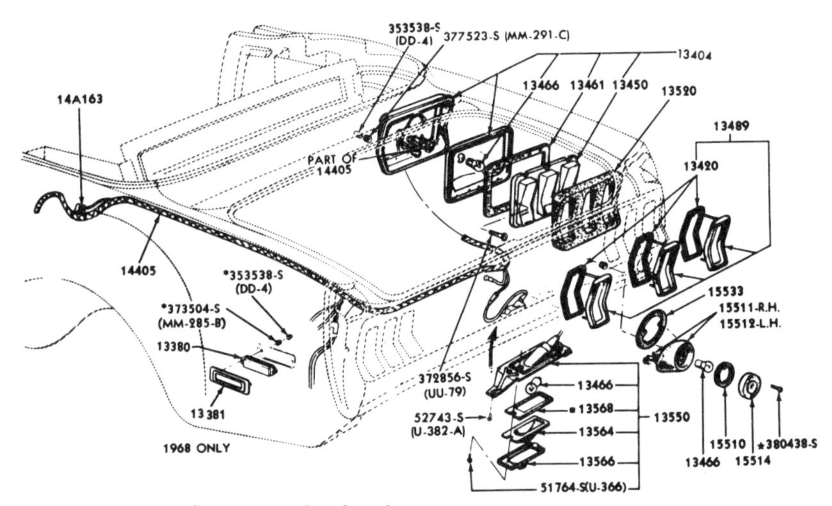

The 1967-1968 Mustang rear light fixtures and related hardware.

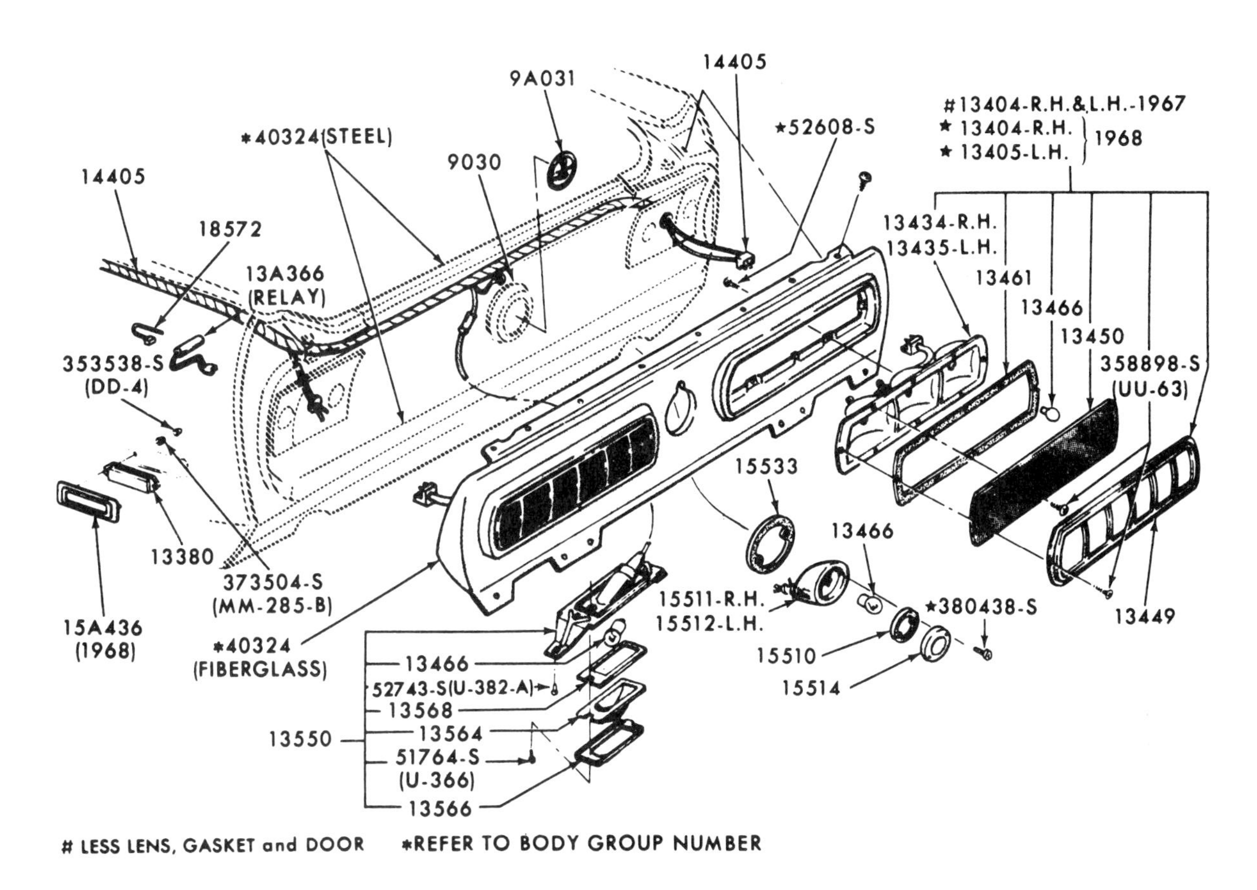

The 1967-1968 Shelby Mustang rear light fixtures and related hardware.

Side marker lights were mandatory on all cars built after 1967, so 1968 Mustangs (shown) are easily distinguished from their third-year cousins. The Mustang's marker lights were unique to each year through 1970.

While the front marker light remained the same throughout the entire 1968 model year, the rear lights changed in January. Early 1968 rear-quarter markers were flush mounted and bordered within a body-colored frame.

Late 1968 rear-quarter marker lights were surface mounted and bordered with chrome trim.

The 1969 Mustang's rectangular front marker lights were installed in the front valance below the front bumper.

Mustang rear markers were the same size and shape as the front markers in 1969. Front marker lights have a clear lens and an amber tinted bulb. The rear lights have a red lens and clear bulb.

Mustang rear marker lights were recessed using a chrome frame in 1970. The lens is perpendicular in contrast to the contour of the sheet metal.

The 1969 Mustang taillamps.

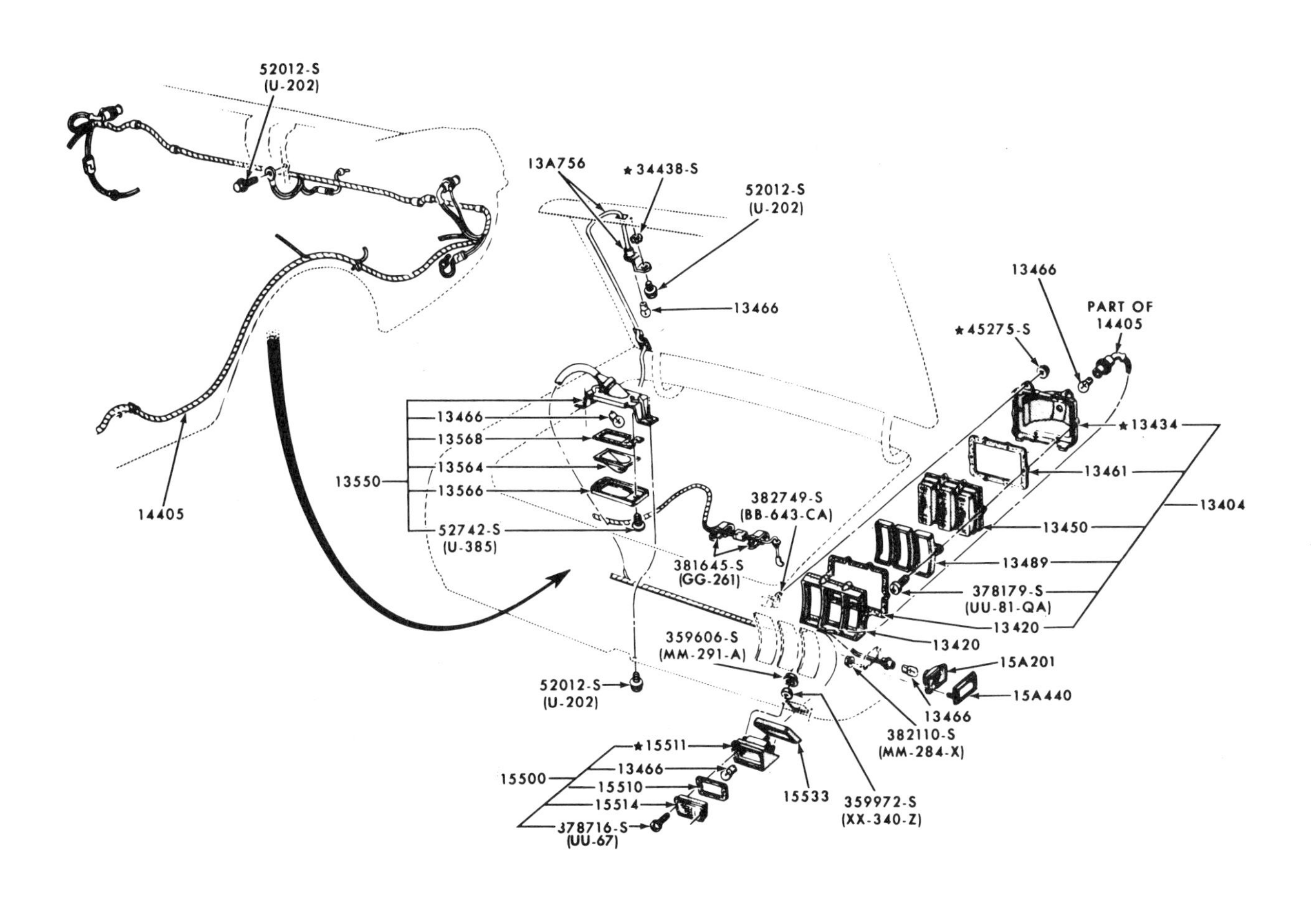

The 1969 Mustang rear light fixtures and related hardware.

The 1970 Mustang taillamp.

The 1970 Boss 302 tail panel.

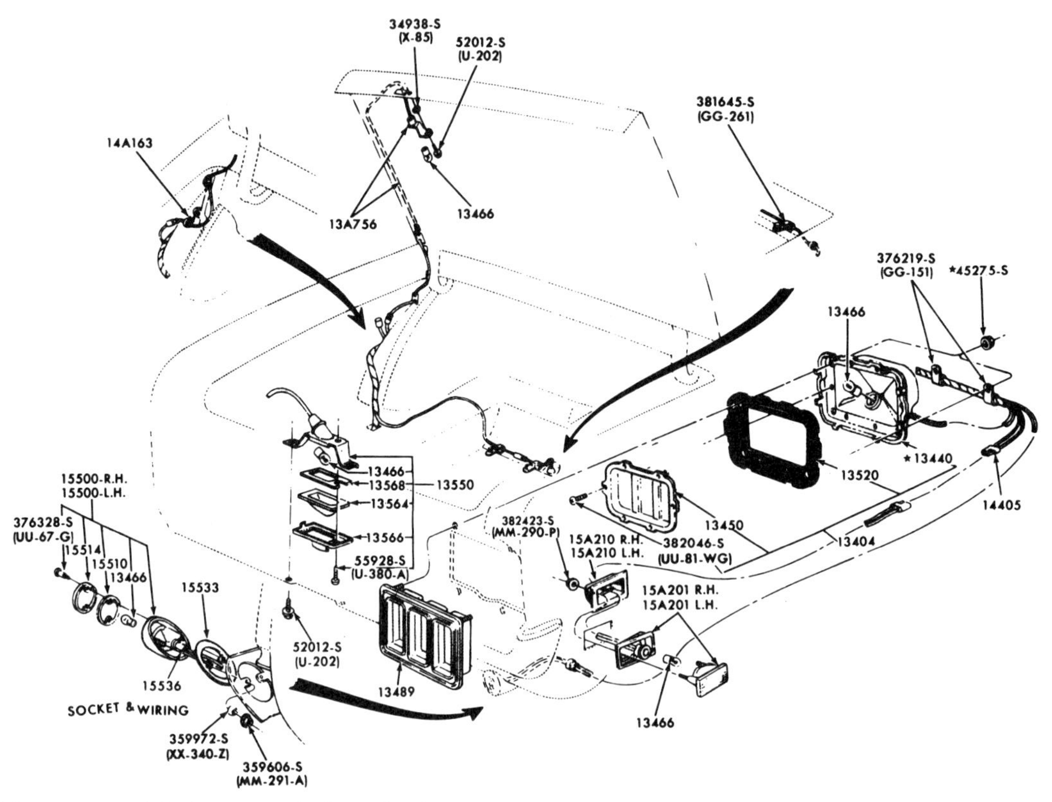

The 1970 Mustang rear light fixtures and related hardware.

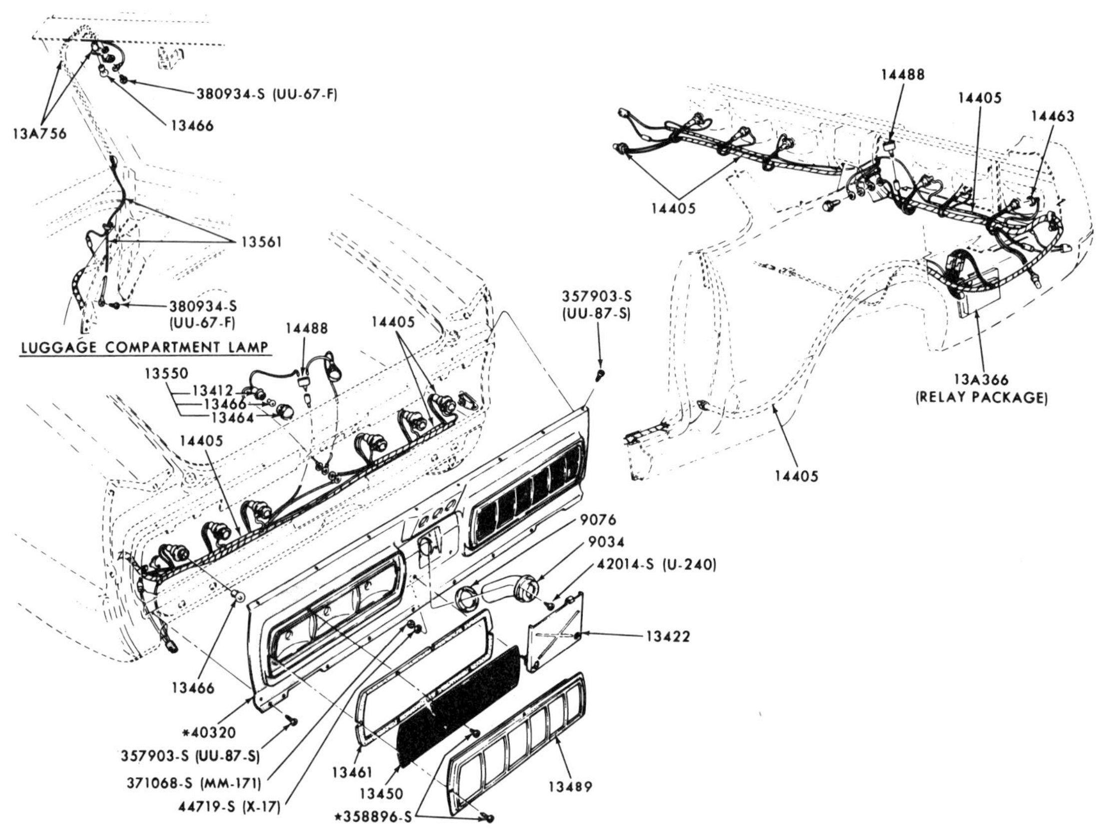

The 1969-1970 Shelby Mustang rear light fixtures and related hardware.

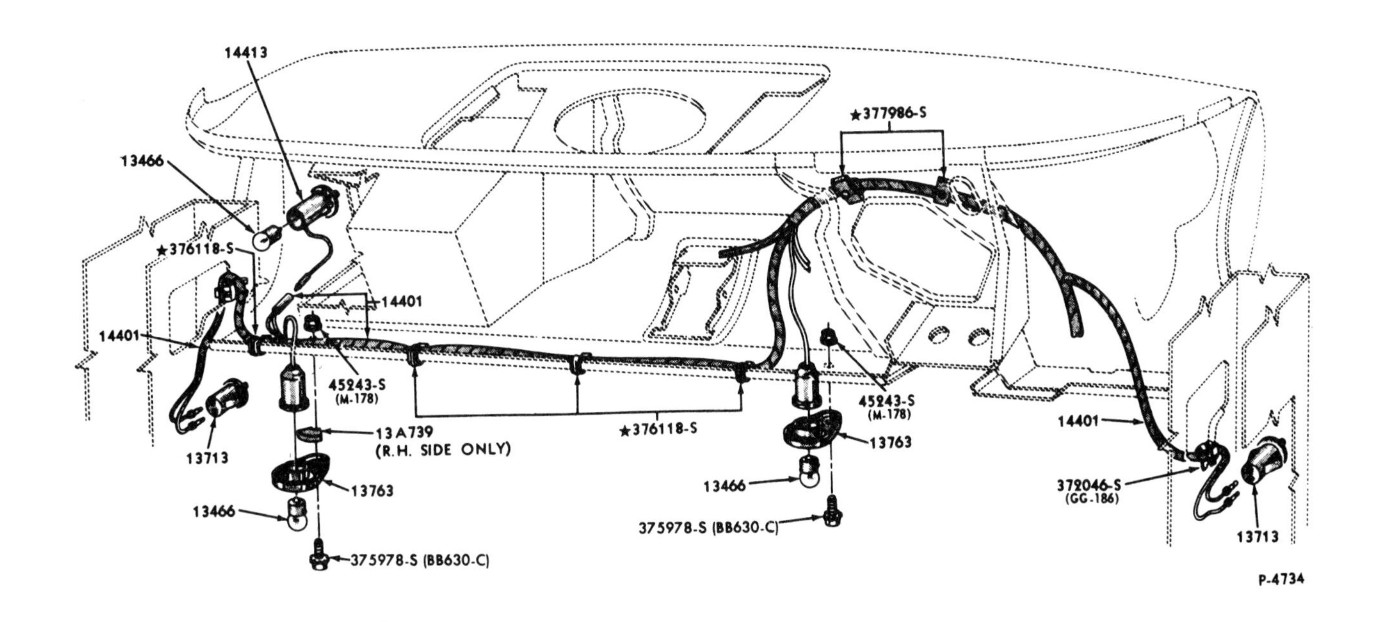

The 1965-1966 Mustang instrument panel mounted courtesy lamps and wiring.

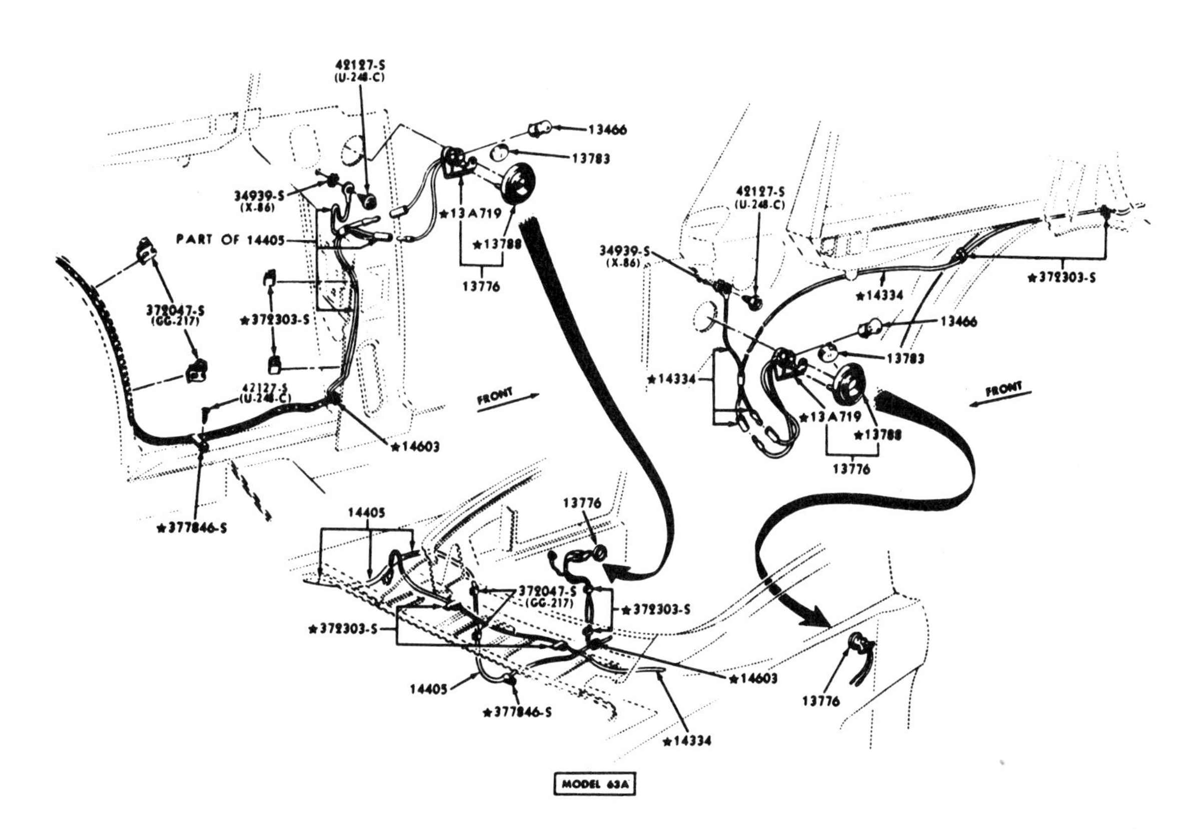

The 1965-1970 Mustang fastback or SportsRoof interior quarter panel courtesy lamps and wiring.

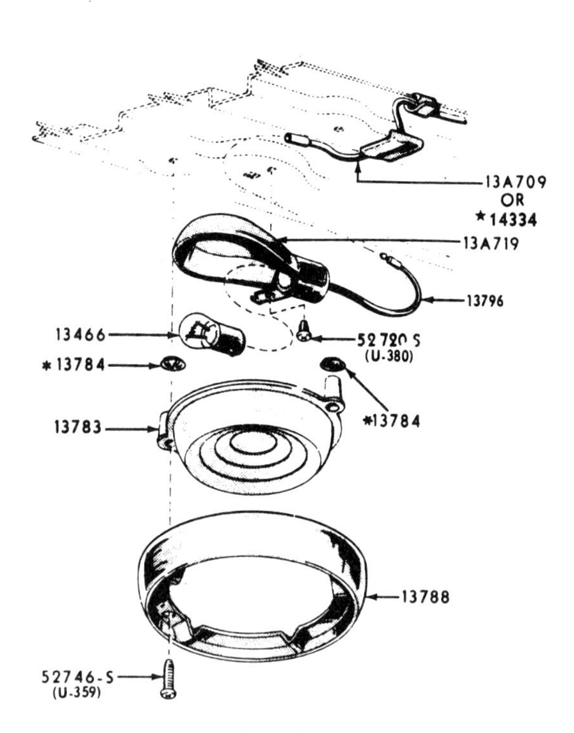

The 1967-1969 Mustang dome lamp.

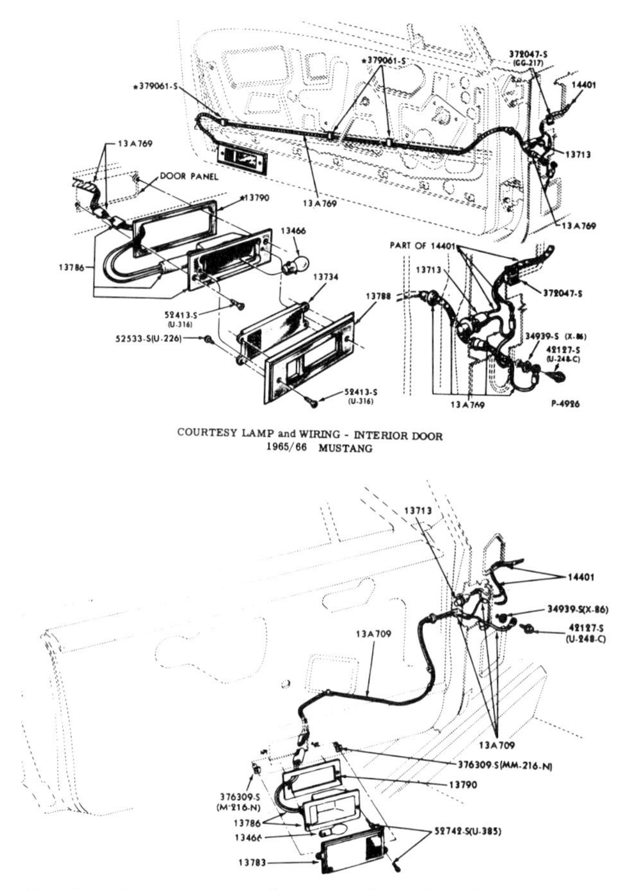

Interior door courtesy lamp and wiring for 1965-1966 Mustang (top) and 1967-1968 Mustang (bottom).

Mustang Rear Lamp Wiring Harness Application Chart

Year	Model	Description	Part Number
68	Except overhead console	Courtesy lamp to switch	C8ZZ-14405-A
	Before 11/1/67	fuel gauge to sender	
68	Except overhead console or safety	Courtesy lamp to switch	C8ZZ-14405-E
	convenience panel from 11/1/67	fuel gauge to sender	
63	With overhead console and/or	Courtest lamp to switch	C8ZZ-14405-F
	safety convenience panel	to fuse panel-fuel gauge	
		sender to gauge to warning	
		relay-defogger	
		switch to defogger motor	
68	(63A, C, 65A, C) Rear window	Courtest lamp to switch	C8ZZ-14405-F
	defogger, and/or safety convenience	to fuse panel-fuel gauge	
		relay-defogger	
		switch to defogger motor	
68	(65) GT - CS		C8ZZ-14405-C
69	Except safety convenience package		C9ZZ-14405-A
69	With safety convenience package		C9ZZ-14405-B
67	(GT350/500)	R.H. & L.H.	S7MS-14405-A
69	(GT350/500) ground		C9ZZ-14405-J
69	(GT350/500)		C9WY-14405-B
70	'6 & 8 cyl. 302-4V Boss 351-390-4V		D0ZZ-14405-B
	'428 CJ with visibility light group		
	Before 12/15/69		
70	From 12/15/69		D0ZZ-14405-B

BULB CHART, Part One, for '65-'70 MUSTANG

Year	Function	Industry Number	Part Number	Notes
65/70	Alternator Charge Indicator	1894	C3AZ-13466-B	
70	Ash Tray Console	1892	B7MY-13466-A	
70	Ash Tray Inst. Panel	1445	B6A-13466-B	
65/68	Back Up Lamp	1142	C5ZZ-13466-A	
69	Back Up Lamp	1156	C3DZ-13466-A	
69	Body Side Lamp - Rear	194A	C9DZ-13466-A	
70	Body Side Lamp - Rear	161	C3AZ-13466-D	
66	Cigar Lighter	1895	C3AZ-13466-B	
65/70	Clock Lamp	1895	C3AZ-13466-B	
66/70	Clock Lamp	1816	B8AZ-13466-A	Rally Pak
65/69	Courtesy Lamp (Instr. Panel)	631	C3VY-13466-A	
65/70	(Console Panel)	1816	B8AZ-13466-A	
65/70	(Quarter Panel)	1003	B6AZ-13466-A	
65/66	(Door Mount)	1004	C3SZ-13466-A	
67/68	(Door Mount)	1004	C3SZ-13466-A	Exc. California
70	(Door Mount)	212	C6VY-13466-A	
70	(Instrument Panel	562	D0ZZ-13466-A	
67/68	Courtesy Lamp (Door Mount)	90	C6SZ-13466-A	California
70	Courtesy Lamp (Pillar)	105	D0AZ-13466-A	
67/69	Dome Lamp	1003	B6AZ-13466-A	Before 6/1/71
70	Dome Lamp	105	D0AZ-13466-A	Before 6/1/71
67/70	Door Ajar	256	C4SZ-13466-A	
68	Front Fender Side Lamp	1178A	C8ZZ-13466-A	
69/70	Front Fender Side Lamp	194A	C9DZ-13466-A	
65	Fuel Indicator	1895	C3AZ-13466-B	
65	Generator Warning Lamp	1895	C3AZ-13466-B	
65	Glove Compt. Instru. Panel	1895	C3AZ-13466-B	
65	Glove Compt. Console Panel	1445	B6A-13466-B	
65/70	Glove Compt. Console Panel	1895	C3AZ-13466-B	
65/70	High Beam Indicator	1895	C3AZ-13466-B	
66	Illuminated Emblem	1003	B6AZ-13466-A	
67/68	Illuminated Emblem	631	C3VY-13466-A	
65/68	Instrument Panel	1895	C3AZ-13466-B	
69/70	Instrument Panel	194	C2AZ-13466-C	
70	Instrument Panel	212-1	C6VY-13466-A	
66/68	Interior Lamp (Mounts on Lower	97	C3AZ-13466-G	
	Quarter Trim Panel 65 Models			

BULB CHART, Part Two, for '65-'70 MUSTANG

Year	Function	Industry Number	Part Number	Notes
'70	Lights on Lamp	1895	C3AZ-13466-B	
'67	Low Fuel Warning	1891	B9SZ-13466-A	Before 9/6/66
'67/69	Low Fuel Warning	1445	B6A-13466-B	From 9/6/66
'70	Low Fuel Warning	1895	C3AZ-13466-B	
'69	Map Lamp Instrument	1004	C3SZ-13466-A	
	Panel			
'67/70	Map Lamp Roof Console	631	C3VY-13466-A	
'69/70	Marker Lamp-Front	194A	C9DZ-13466-A	
'65/70	Oil Pressure Warning	1895	C3AZ-13466-B	
'65/69	Parking Brake Warning	257	C0AZ-13466-A	
	on Safety Convenience Pkg.			
'67	Parking Brake Warning	256	C4SZ-13466-A	
	on Safety Convenience Pkg.			
'65/66	Parking Lamp	1157	C8TZ-13466-A	
'67/70	Parking Lamp	1157NA	C9MY-13466-A	
		Amber		
'67	Portable Trunk Lamp	1003	B6AZ-13466-A	
'65/66	Radio Dial Lamp	1891	B9SZ-13466-A	
'67/70	Radio Dial Lamp	1893	C1VY-13466-A	
'65	Rear Lamp	1157	C8TZ-13466-A	
'65	Rear License Plate	97	C3AZ-13466-A	
'69	GT350/500	1232	C5TZ-13466-A	
'67/70	Safety Package Lamps	1895	C3AZ-13466-B	
'66/68	Seat Belt Warning (On Safety	1895	C3AZ-13466-B	Before 9/6/66
	Convenience Panel)			
'67	Seat Belt Warning (On Safety	1891	B9SZ-13466-A	Before 9/6/66
	Convenience Panel)			
'67	Seat Belt Warning (On Safety	1445	B6A-13466-B	From 9/6/66
	Convenience Panel)			
'65/70	Speedometer	1895	C3AZ-13466-B	
'70	Sport Lamp	1157A	C9AZ-13466-A	
'70	Stereo Tape Deck	1893	C1VY-13466-A	
'65	Tachometer	1895	C3AZ-13466-B	
'65/70	Temperature Indicator	1895	C3AZ-13466-B	
'65	Transmission Control	1445	B6A-13466-B	
	Selector Dial			
'65/69	Transmission Control	1893	C1VY-13466-A	
	Selector Dial			
'65/71	Turn Signal Indicator	1895	C3AZ-13466-B	
'67/68	Turn Signal Indicator			
	(Hood Mount)	53	C4SZ-13466-C	
'69	Turn Signal Indicator	1895	C3AZ-13466-B	
	(Hood Mount)			
'70	Transmission Control	1445	B6A-13466-B	

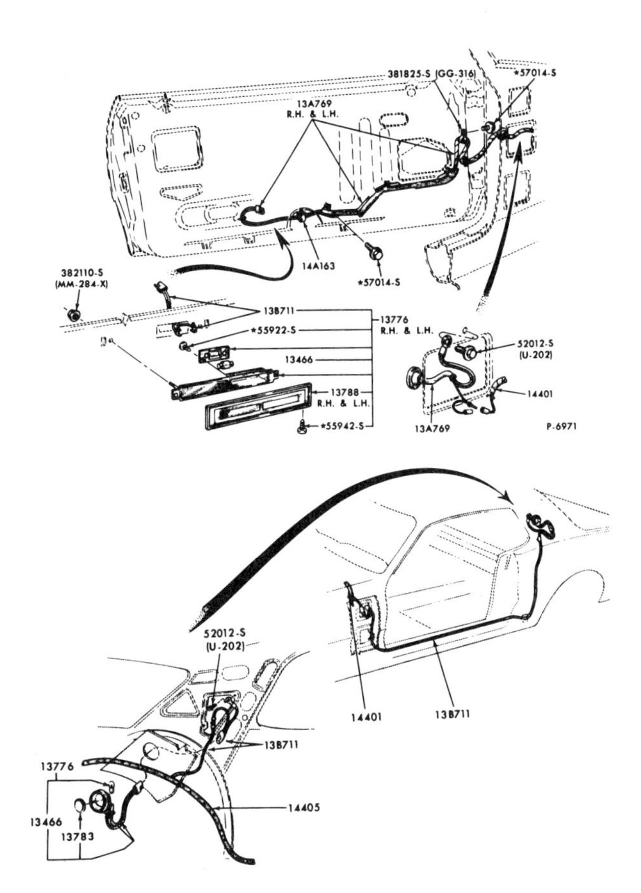

The 1969-1970 Mustang interior door courtesy lamp and wiring (top) and rear pillar-mounted courtesy lamp and wiring (bottom).

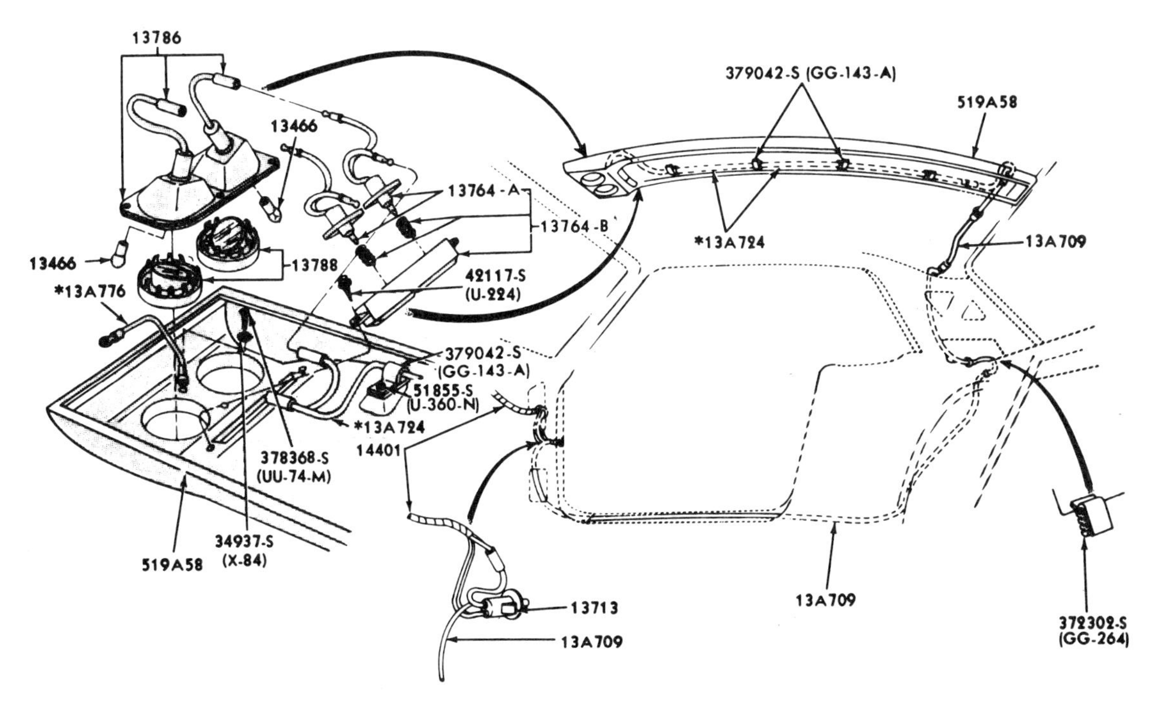

The 1967-1968 Mustang roof console map lamps and wiring.

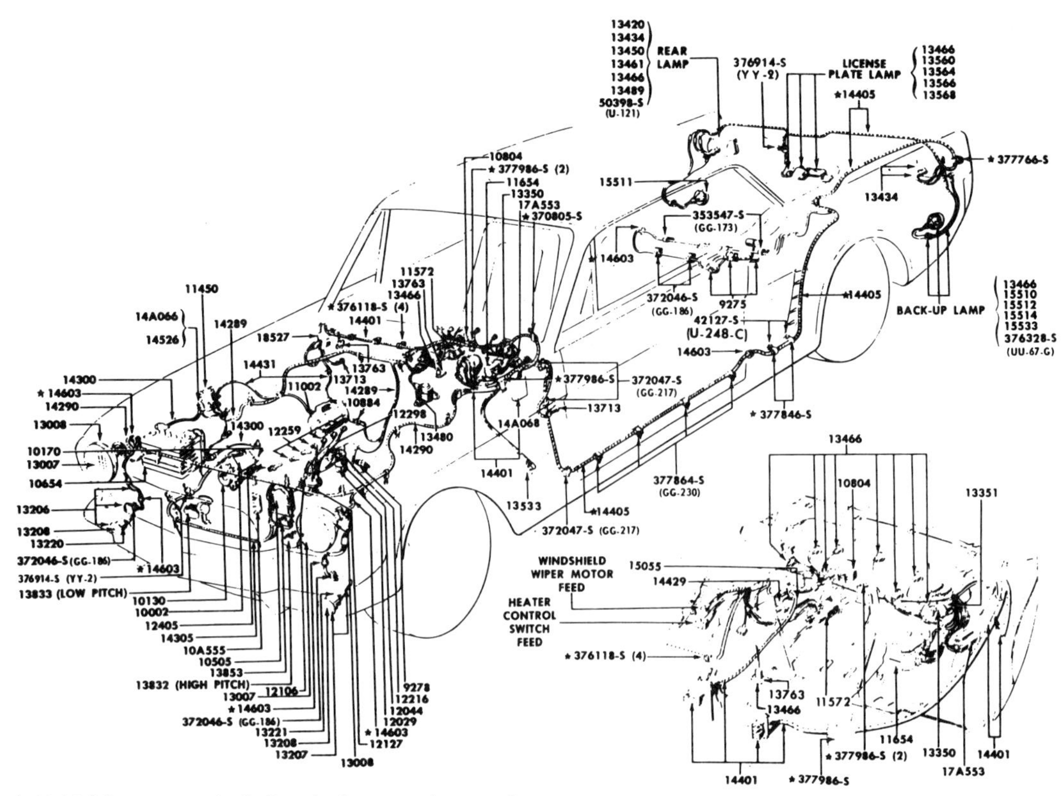

The 1965-1966 Mustang typical electrical connections and wiring.

Mustang Generator/Alternator to Voltage Regulator Wiring Ass'y Application Chart

Year	Model	Description	Part Number
65	6 cyl., with generator		C5ZZ-14305-A
65	8 cyl., with generator		C5ZZ-14305-B
65	6 cyl., with alternator - except		C5DZ-14305-A
	luxury trim		
65	8 cyl., with alternator - except		C5DZ-14305-B
	luxury trim		
65	6 & 8 cyl., with luxury trim		C5GY-14305-B
66	6 cylinder		C6ZZ-14305-A
66	8 cylinder	Heavy duty wiring	C6ZZ-14305-B
67/68	6 cylinder		C7ZZ-14305-A
67	8 cyl. 289, with tachometer		C7ZZ-14305-B
68	289, 302, without tachometer		C7ZZ-14305-B
67	8 cyl. 289, with tachometer		C7ZZ-14305-E
68	289, 302, with tachometer		C7ZZ-14305-E
67/68	8 cyl., 390, 428		C7ZZ-14305-H
	without tachometer		
67/68	8 cyl. 390, 428		C7ZZ-14305-J
	with tachometer		
69	8 cyl. 302, 351, 390, 428		C9WY-14305-G
	with tachometer		
69	6 cyl., or Boss 429		C9WY-14305-G
	with tachometer		
69	8 cyl. 302, 351, 390, 428		C9ZZ-14305-S
	without tachometer		
69	Boss 429 without tachometer		C9ZZ-14305-S
70	8 cyl. 302-2V, 351-2V, 4V,	With tachometer	D0WY-14305-A
	428CJ Before 5/15/70		
70	8 cyl. 302-2V, 351-2V, 4V,	With tachometer	D0WY-14305-C
	428CJ From 5/15/70		
70	6 & 8 cylinder	Without tachometer	D0ZZ-14305-C

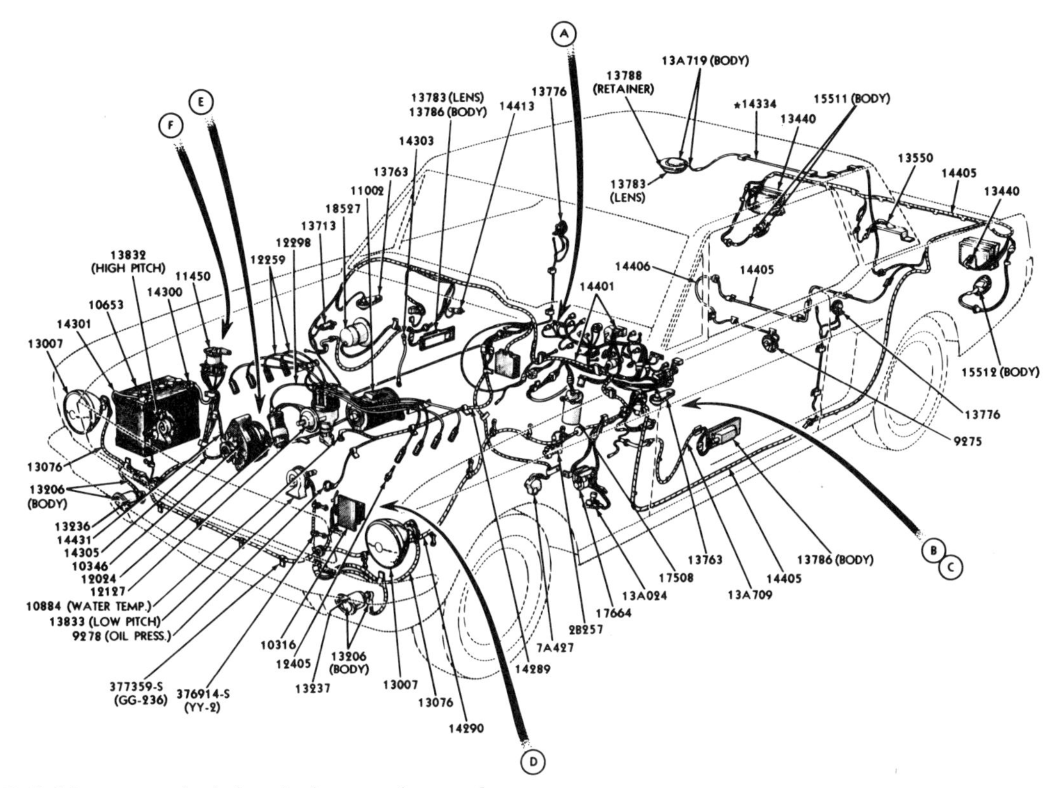

The 1967-1968 Mustang typical electrical connections and wiring.

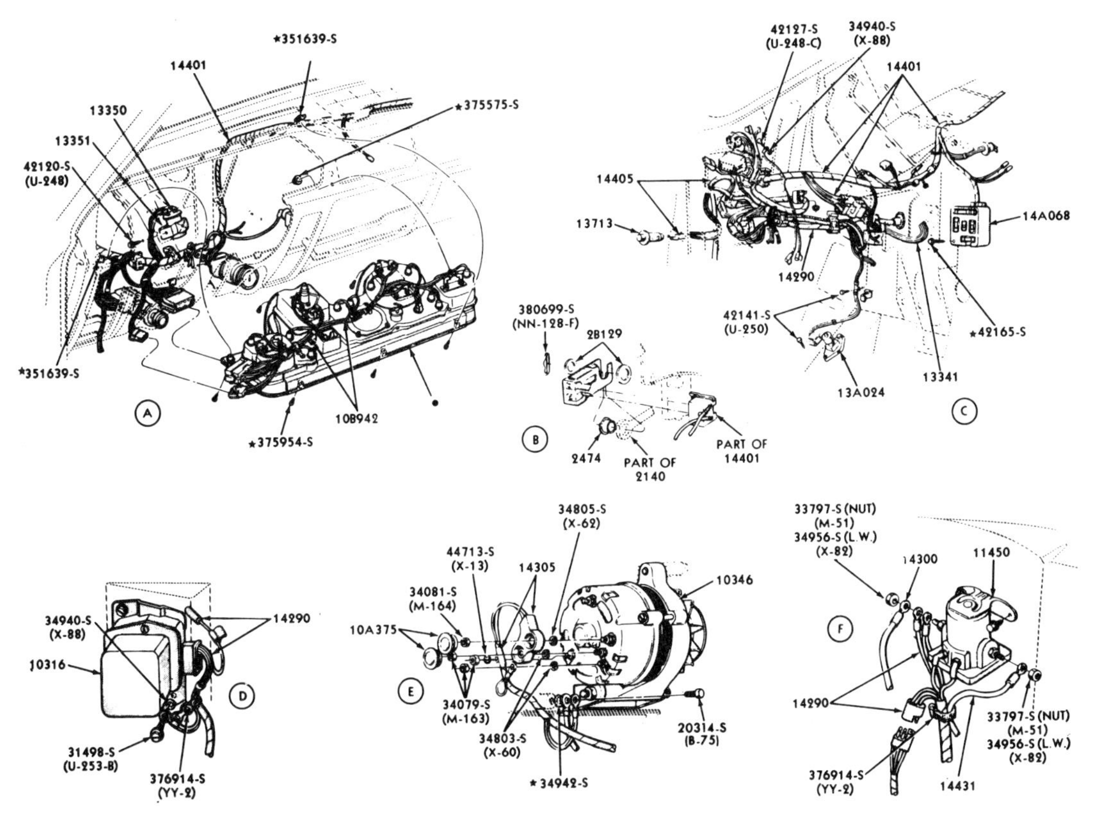

The 1967-1968 Mustang typical electrical connections and wiring (continued).

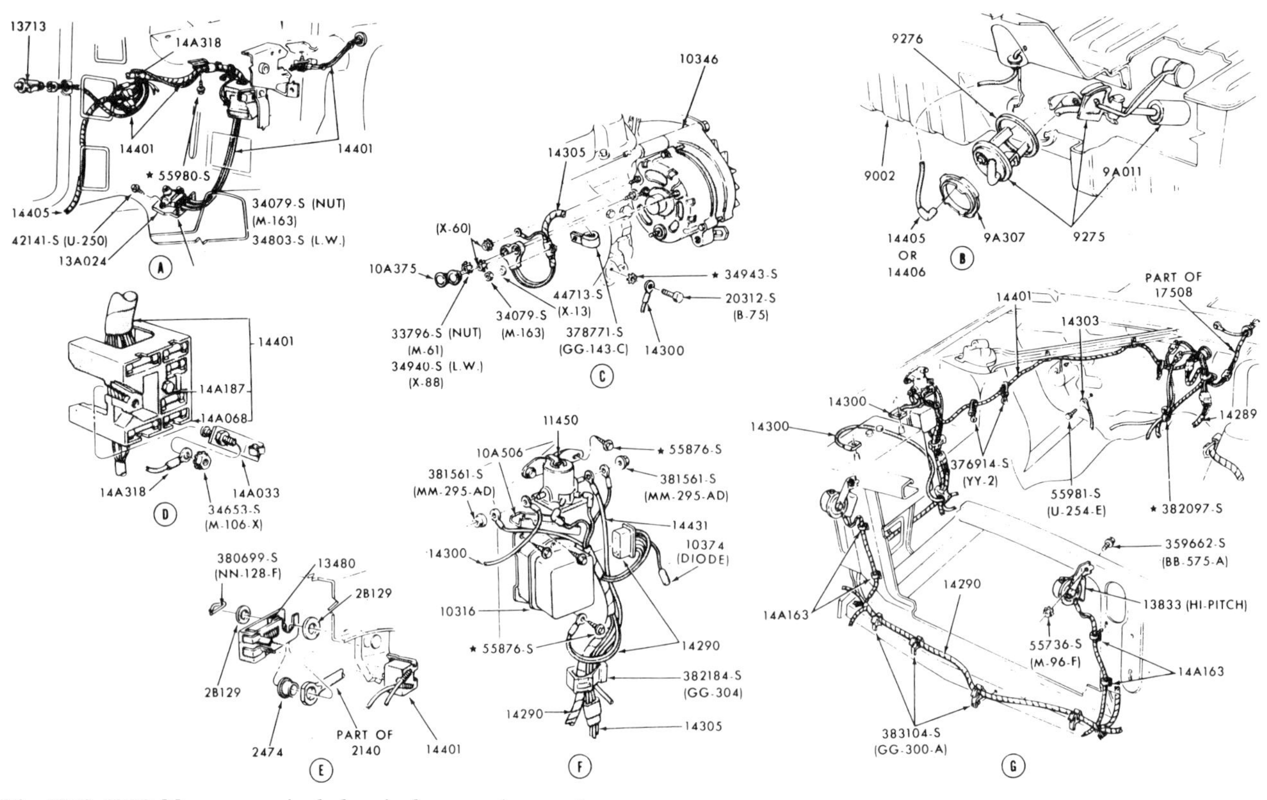

The 1969-1970 Mustang typical electrical connections and wiring.

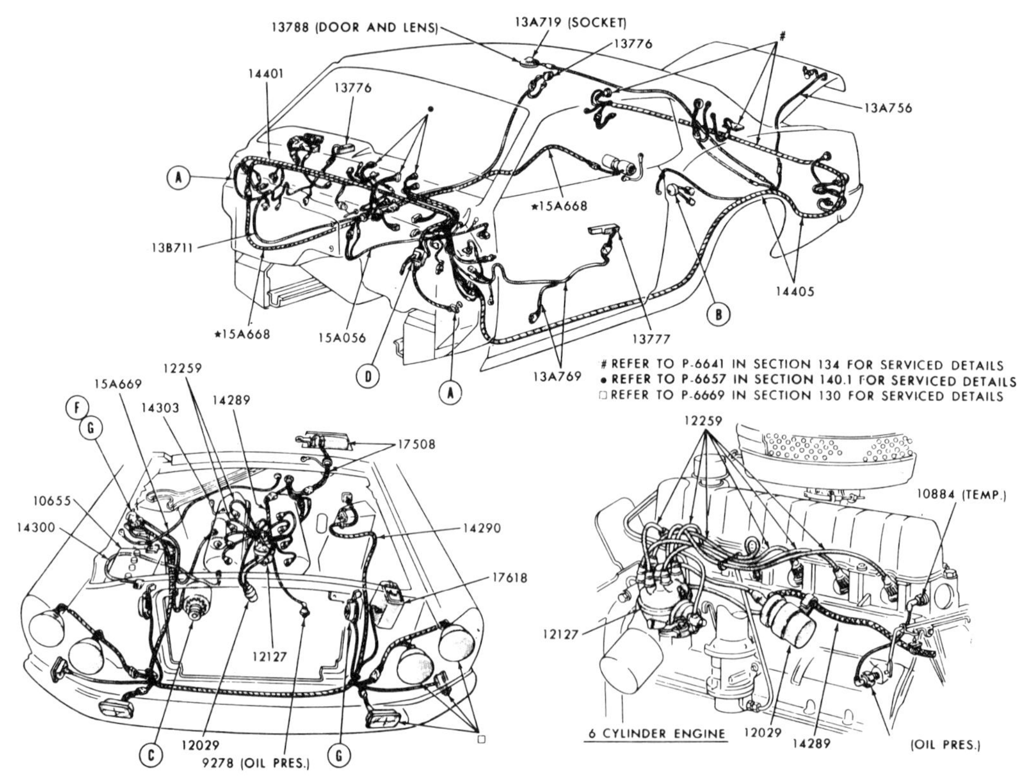

The 1969-1970 Mustang typical electrical connections and wiring (continued).

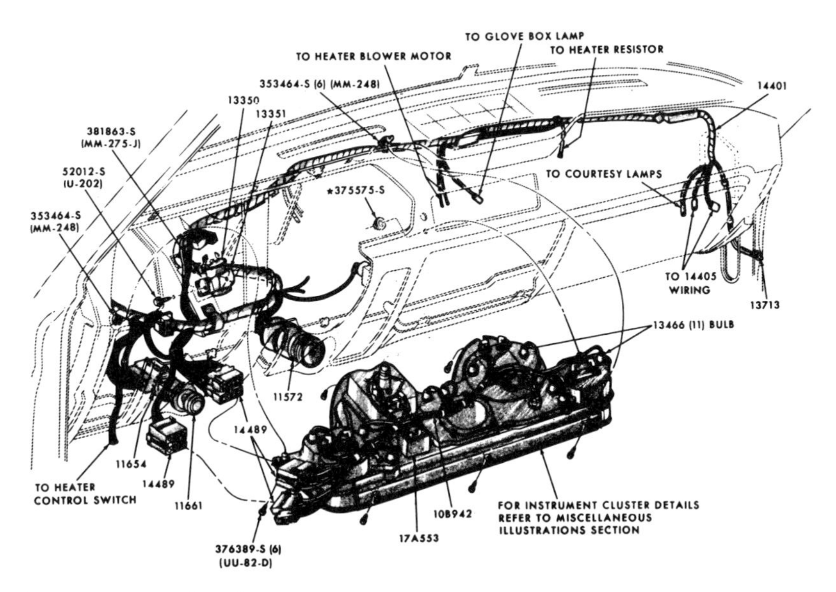

The 1967-1968 Mustang instrument panel wiring.

Mustang Dash to Engine Gauge Feed Wire Assembly Application Chart

Year	Model	Part Number
65	Hardtop, Conv., Std. Interior, 6 cyl.	C5ZZ-14289-J
65	8 cyl. with generator	C5ZZ-14289-J
65	Standard interior, 8 cyl.	C5ZZ-14289-A
	with alternator	
65	Deluxe interior, 8 cyl.	C5DZ-14289-B
65	Standard interior, 8 cyl. with	C5DZ-14289-B
	ammeter & oil pressure gauges	
66	6 cyl.	C6ZZ-14289-A
66	8 cyl. with heavy duty wiring	C6ZZ-14289-B
67	8 cyl. 289	C6ZZ-14289-B
68	8 cyl. 289, 302 before 4/8/68	C6DZ-14289-B
68	8 cyl. 289, 302 from 4/8/68	C8DZ-14289-A
67 / 68	6 cyl.	C7DZ-14289-A
67 / 68	8 cyl. 390, 428	C7ZZ-14289-A
69	8 cyl. 390, 428	D0AZ-14289-A
70	8 cyl. 428CJ	D0AZ-14289-A
69	8 cyl. 429	D0AZ-14289-A
69	8 cyl. 302 Boss	D0ZZ-14289-E
70	Hardtop, 8 cyl. 302-4V w/tachometer	D0ZZ-14289-E
70	SportsRoof, 8 cyl. 302-4V Boss	D0ZZ-14289-E
69	6 cyl.	C9ZZ-14289-A
70/71	6 cyl.	D0DZ-14289-C
69/70	8 cyl. 351-2V (Windsor)	D0OZ-14289-A
69/70	8 cyl. 302-2V	D0OZ-14289-A
70	8 cyl. 351-4V	D0ZZ-14289-A
70	8 cyl. 351-2V (Cleveland)	D0ZZ-14289-A

Mustang Instrument Panel to Dash Panel Wiring Assembly Application Chart

Year	Model	Part Number
69	(GT-350/500)	C5ZZ-5560-D
70	non-Mach 1 w/o A/C or tachometer	C5ZZ-5560-C
70	non-Mach 1 w/o A/C with tachometer	C5ZZ-5560-E
	Before 11/1/69	
70	non-Mach 1 w/o A/C w/tachometer	C5ZZ-5560-F
	From 11/1/69	
70	non-Mach 1 w/o tachometer	C4DZ-5560-J
	Before 11/1/69	
70	non-Mach 1 w/ A/C, w/o tachometer	C4DZ-5560-J
	From 11/1/69	
70	Mach 1 w/o tachometer	C4DZ-5560-J
	Before 11/1/69	
70	Mach 1, w/o tachometer, From 11/1/69	C7ZZ-5560-A
70	Mach 1 w/ tachometer, Before 11/1/69	C7ZZ-5560-A
70	Mach 1, w/ tachometer, From 11/1/69	C7ZZ-5560-A

Mustang Instrument Panel to Dash Panel Wiring Assembly Application Chart

Year	Model	Description	Part Number
65	With single speed wipers	Not replaced	C5ZZ-14401-E
	with generator		
65	With 2 speed wipers w/generator	Not replaced	C5ZZ-14401-F
65	Ecx. Deluxe interior, with single	Must be used w/basic	C5ZZ-14401-M
	or 2 speed wipers Before 4/1/65	17B587 wiring-not used	
		w/fog lamp	
65	Exc. Deluxe interior, with single	Must be used w/basic	C5ZZ-14401-N
	or 2 speed wipers From 4/1/65	17B587 wiring-not used	
		w/fog lamp	
65	Deluxe interior	When used on units w/	C5ZZ-14401-G
		2 speed heater blower	
		motor, basic 18A336	
		is required	
66			C6ZZ-14401-F
67	Exc. GT, Exc. w/tachometer		C7ZZ-14401-AK
67	Exc. GT w/tachometer		C7ZZ-14401-AL
67	GT, Exc. w/ tach., Before 1/3/67	Not replaced	C7ZZ-14401-G
67	GT, Exc. w/ tachometer	Not replaced	C7ZZ-14401-AC
	From 1/3/67 to 4/15/67		
67	GT, Exc. w/ tach., From 4/15/67	Not replaced	C7ZZ-14401-AM
67	GT w/ tachometer Before 1/3/67	Replaced 2/74 by	C7ZZ-14401-H
		C7ZZ-14401-AD	
67	GT, w/ tachometer, From 1/3/67	Not replaced	C7ZZ-14401-AD
	to 4/15/67		
67	GT, w/ tachometer, From 4/15/67	Not replaced	C7ZZ-14401-AN
68	GT/CS, w/ tachometer		C7ZZ-14401-AN
68	With or without fog lamps		C8ZZ-14401-C
	less tachometer		
68	GT/CS, w/o tachometer		C8ZZ-14401-C
68	With or without fog lamps		C8ZZ-14401-D
	with tachometer		
69	Exc. A/C or tach. Before 4/15/69	Except 302-4V Boss	C9ZZ-14401-A
69	Exc. A/C or tach. From 4/15/69	Except 302-4V Boss	C9ZZ-14401-AL
69	Less tachometer, From 4/15/69	With 302-4V Boss	C9ZZ-14401-BC
	to 5/19/69		
69	Less tachometer From 5/19/69	With 302-4V Boss	C9ZZ-14401-BE
69	With tachometer, w/ or w/o A/C	Except 302-4V Boss	C9ZZ-14401-B
	Before 4/15/69		
69	With tachometer, w/ or w/o A/C	Except 302-4V Boss	C9ZZ-14401-AM
	From 4/15/69	not replaced	
69	W/ A/C, w/o tach. Before 4/15/69	Except 302-4V Boss	C9ZZ-14401-C
69	W/ A/C, w/o tach. From 4/15/69	Except 302-4V Boss	C9ZZ-14401-AN
69	W/ Tach. From 4/15/69 to 5/19/69	With 302-4V Boss	C9ZZ-14401-BB
69	W/ tachometer, From 5/19/69	With 302-4V Boss	C9ZZ-14401-BD

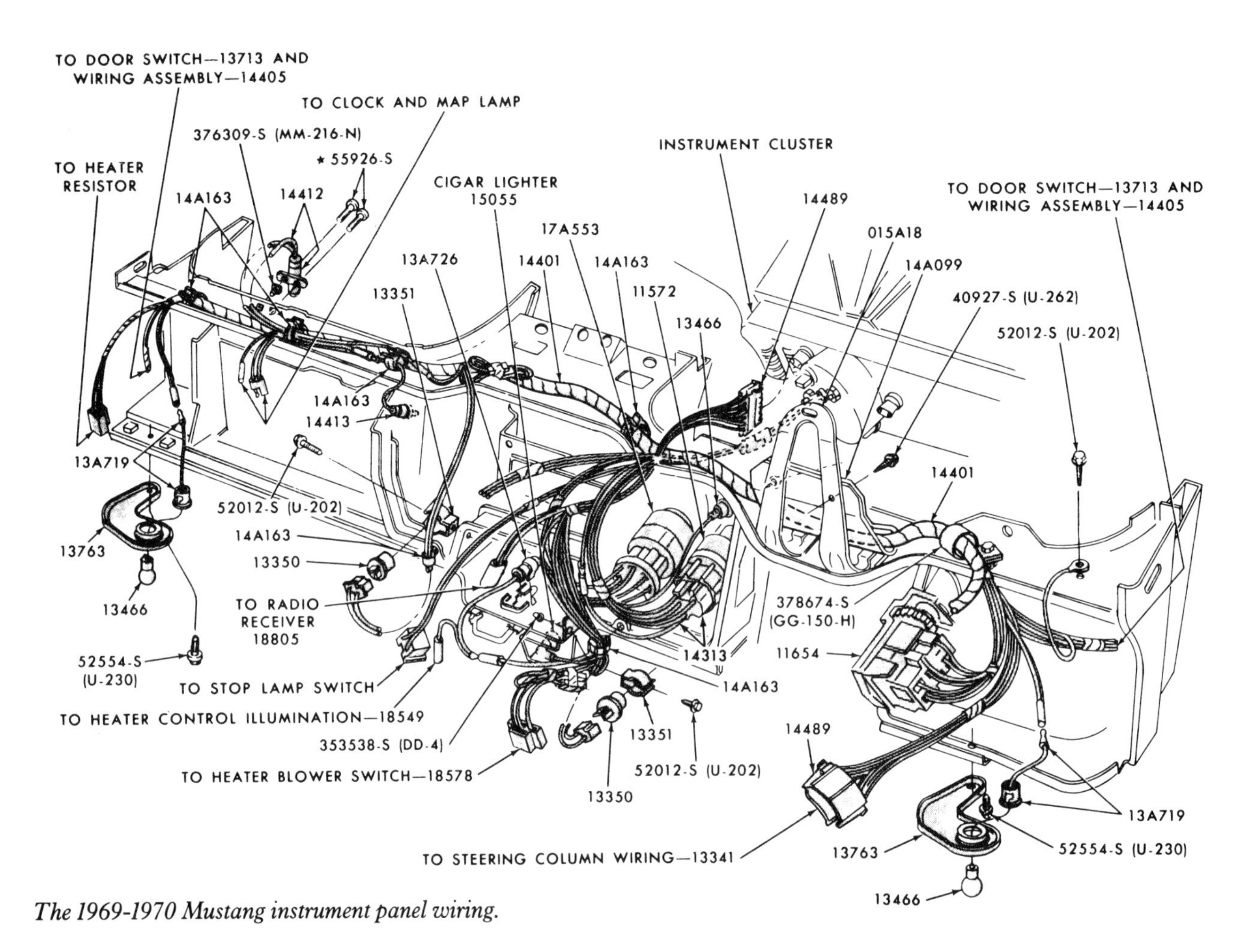

The 1969-1970 Mustang instrument panel wiring.

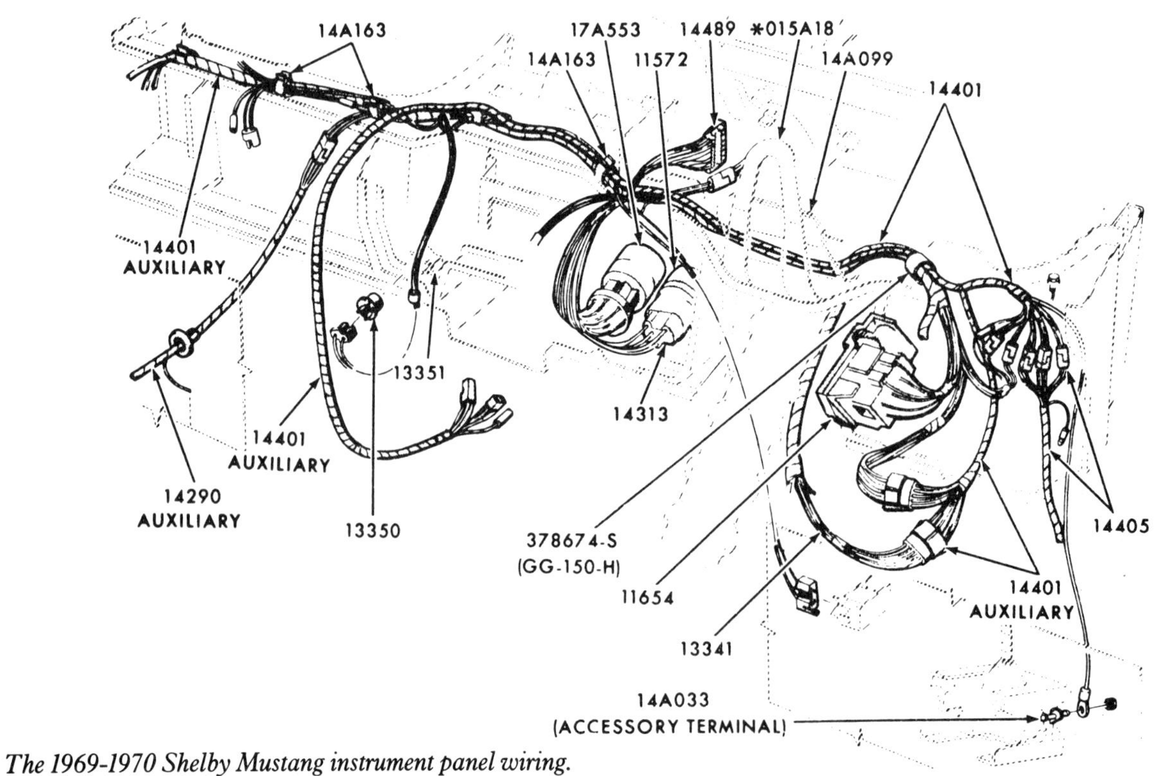

The 1969-1970 Shelby Mustang instrument panel wiring.

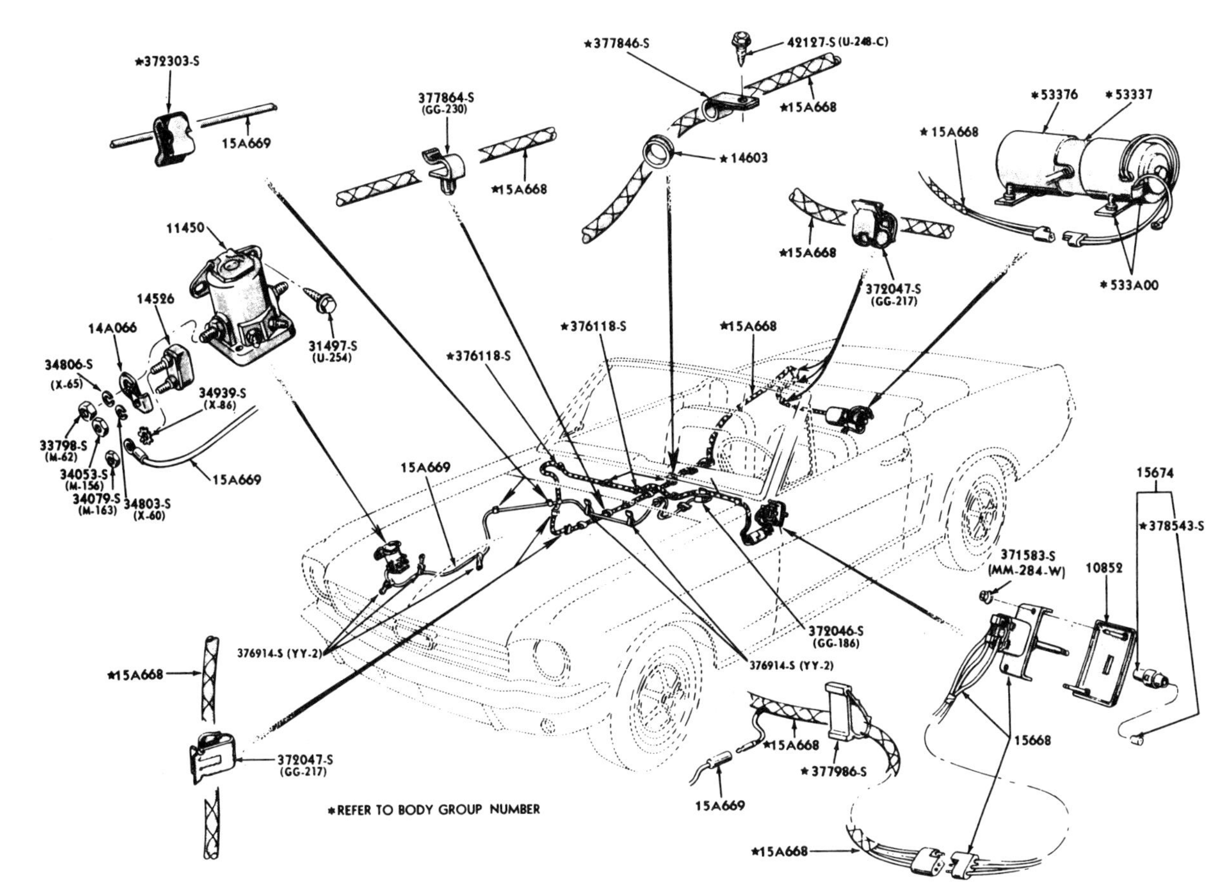

The 1965-1966 Mustang convertible top electrical system.

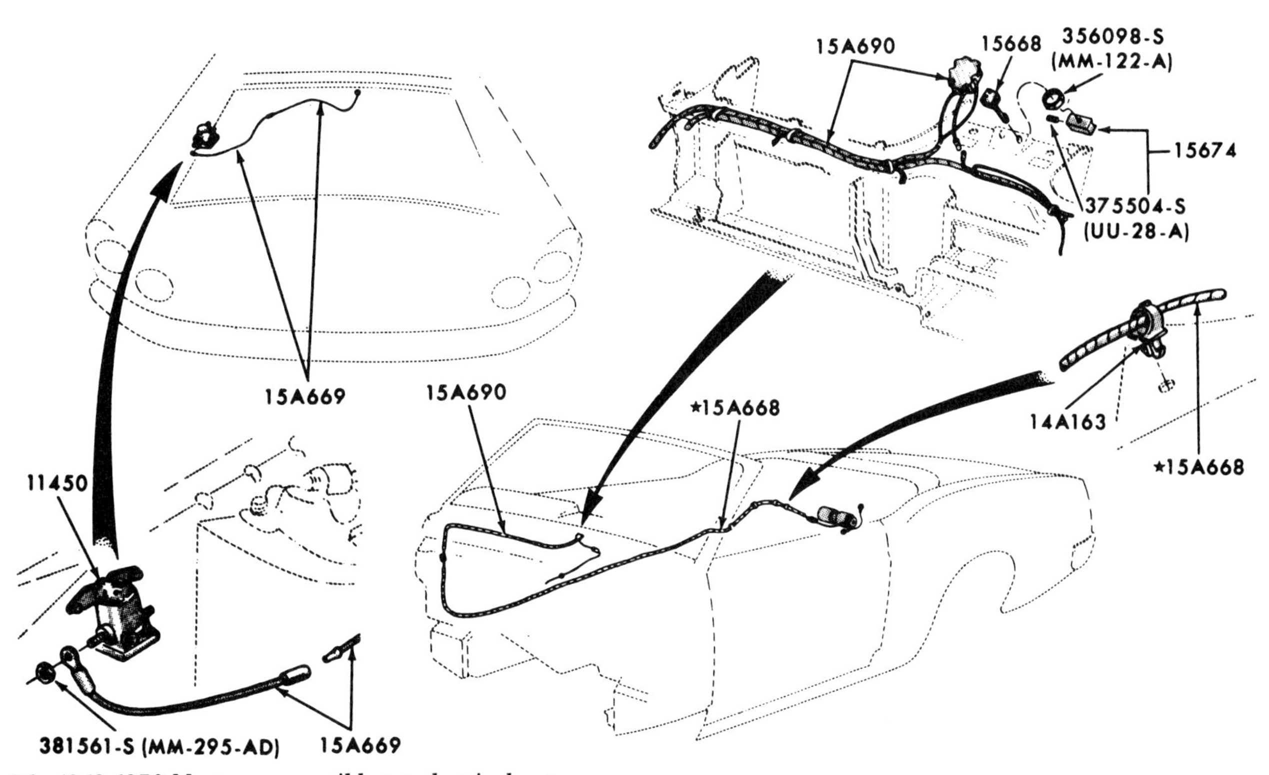

The 1969-1970 Mustang convertible-top electrical system.

MUSTANG GENERATOR I.D. and APPLICATION CHART, 1965.

Engine	Amp.	Notes	Identification	Service Generator	Voltage Regulator	Pulley
170	25	Exc. A/C	C3DF-10000-A	C1DZ-10002-A	C3DZ-10505-A	C2OZ-10130-C
170	30	w/ A/C	C2OF-10000-J C3DF-10000-D	C1DZ-10002-A	C3TZ-10505-B	C2OZ-10130-C
260,289	30		C2OF-10000-G,H C3OF-10000-B C4OF-10000-A,B,C	C1TZ-10002-A	C3TZ-10505-B	C2OF-10130-A

MUSTANG GENERATOR BRACKET APPLICATION CHART, 1965.

Engine	Amp.	Notes	Adjustment Arm	Side Bracket
170	25	Exc. A/C	C2DZ-10145-B	C1DE-10039-A
170	30	w/ A/C	C4DZ-2882-A	C1DE-10039-A
260,289	30	Exc. A/C	C4OZ-10145-A	C2OZ-10151-A
260,289	30	w/ A/C	C3OZ-10145-C	C2OZ-10151-A

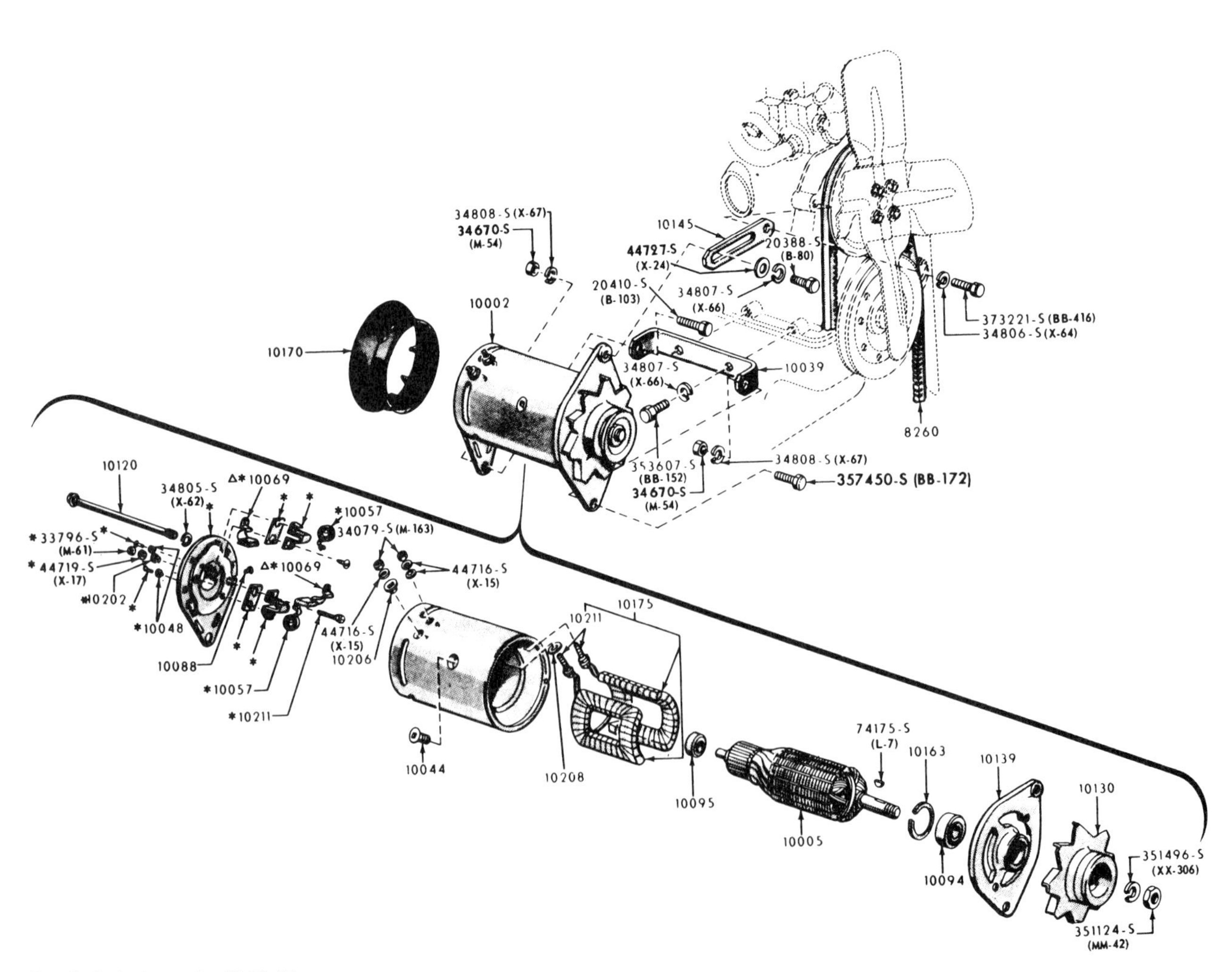

Exploded view of a 1965 Mustang generator.

MUSTANG GENERATOR PULLEY APPLICATION CHART, 1965.

Engine	Amp.	Notes	Identification	Type	Dimensions: A	B	C	D	E	F	Part Number
170	25		C20F-10130-G	1	2.67	.670	.38	5.57	1.62		C2OZ-10130-C
260,289	30	1	C20F-10130-A	1	2.67	.669	.38	5.57	1.92		C2OF-10130-A
260,289	30	2	C20F-10130-B	2	2.67	.670	.38	5.57	2.22	.64	C2OZ-10130-B
260,289	30	3	C30F-10130-B	1	4.32	.670	.38	5.57	1.94		C3OZ-10130-B

Notes

1 Except A/C & Premium Fuel

2 A/C except Premium Fuel

3 Premium Fuel

Note: This chart to accompany Illustrations of Pulley Type 1 and Pulley Type 2, Section 100, Page 3.

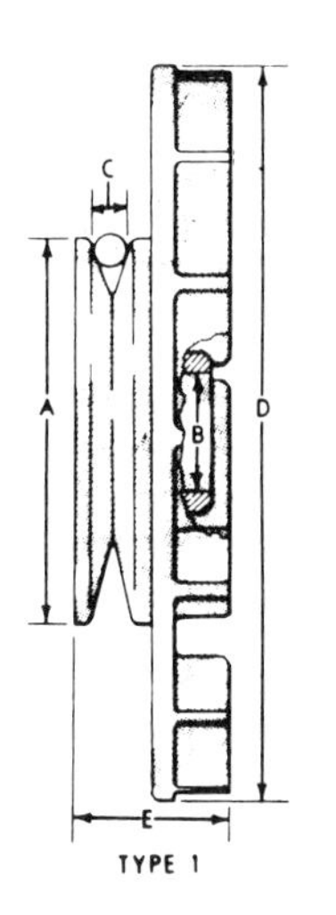

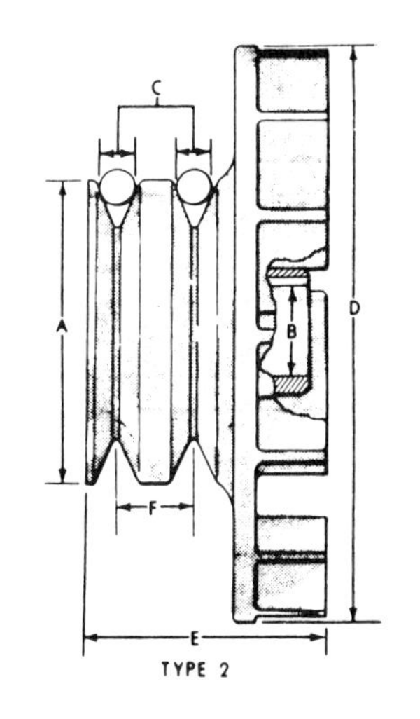

Generator mounting parts.

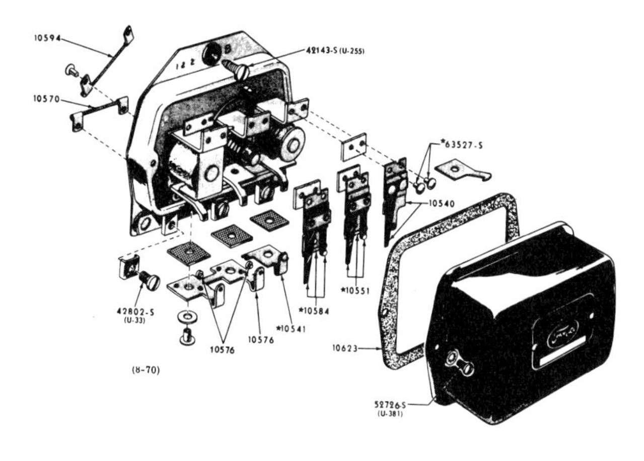

Exploded view of a 1965 Mustang voltage regulator used with a generator.

MUSTANG GENERATOR REGULATOR APPLICATION CHART

Year	Model	Description	Make	Part Number
65	6-cyl. 170 cid	15 volt, 25 amp., 3 terminals Identified C2DF-10505-A Use with generator C1DZ-10002-A	Ford	C3DZ-10505-A
65	All 6-cyl.,8-cyl.	15 volts, 30 amp., 3 terminals Identified C2AF-10505-A; C2TF-10505-B Use with generator C1DZ-10002-A; or C1TZ-10002-A	Ford	C9ZZ-10505-B

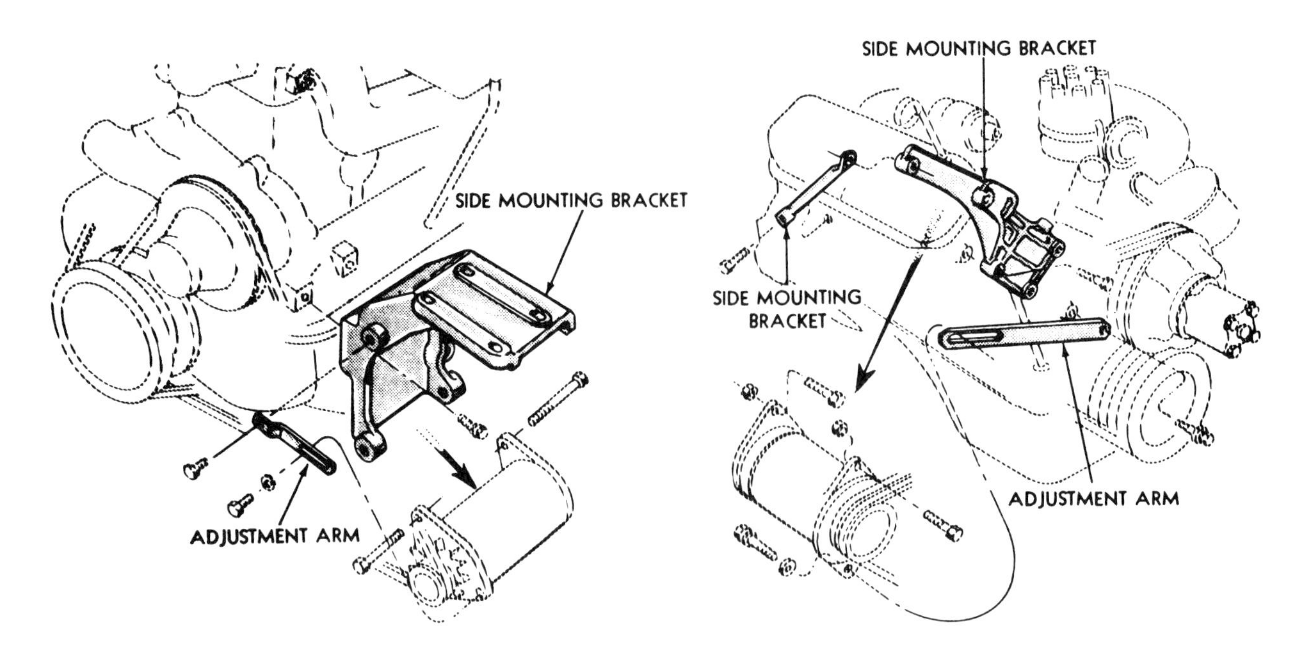

The generator mounting brackets for V-8 and six-cylinder engines of the 1964 1/2 model year.

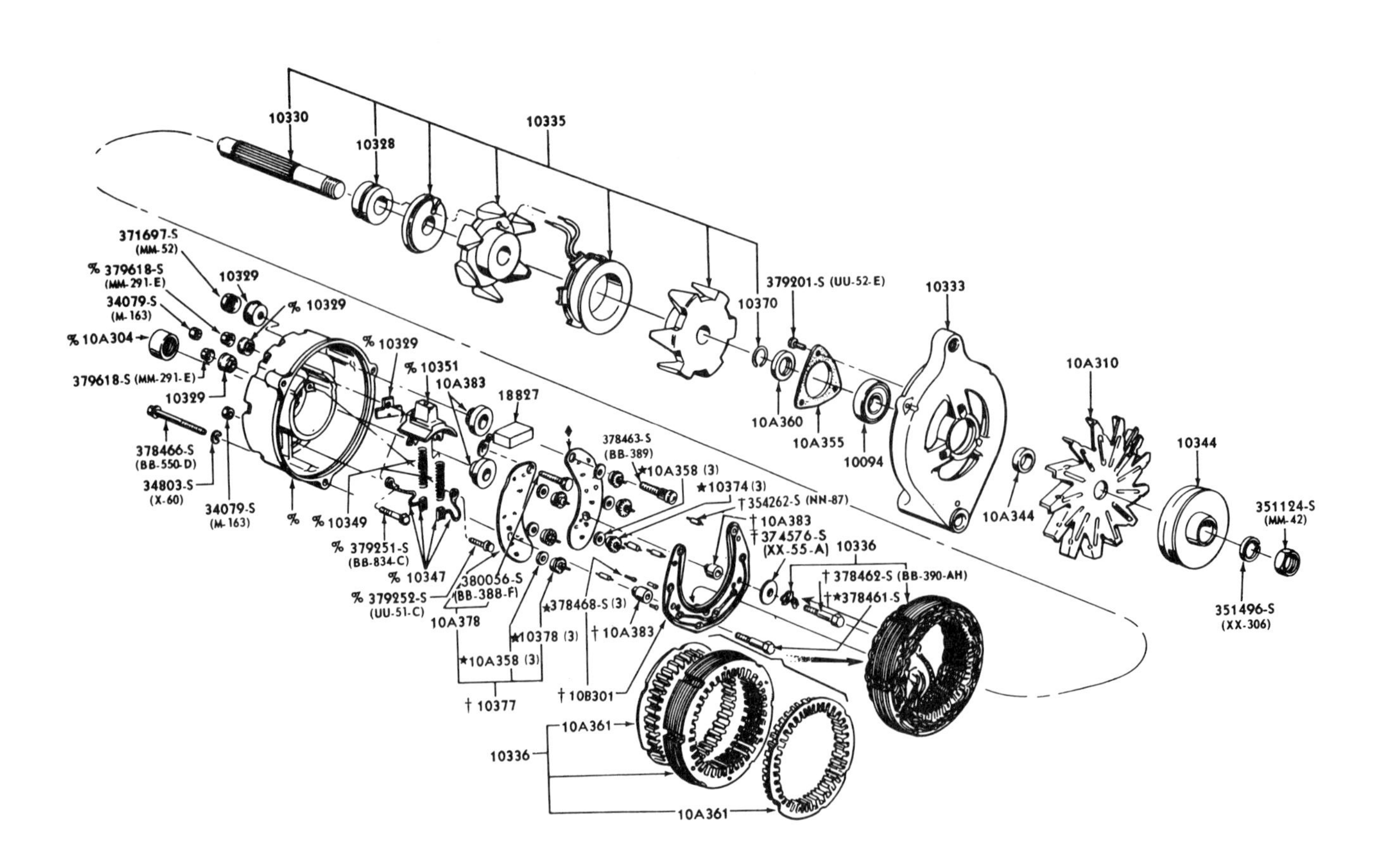

Exploded view of a 1965-1970 Mustang alternator.

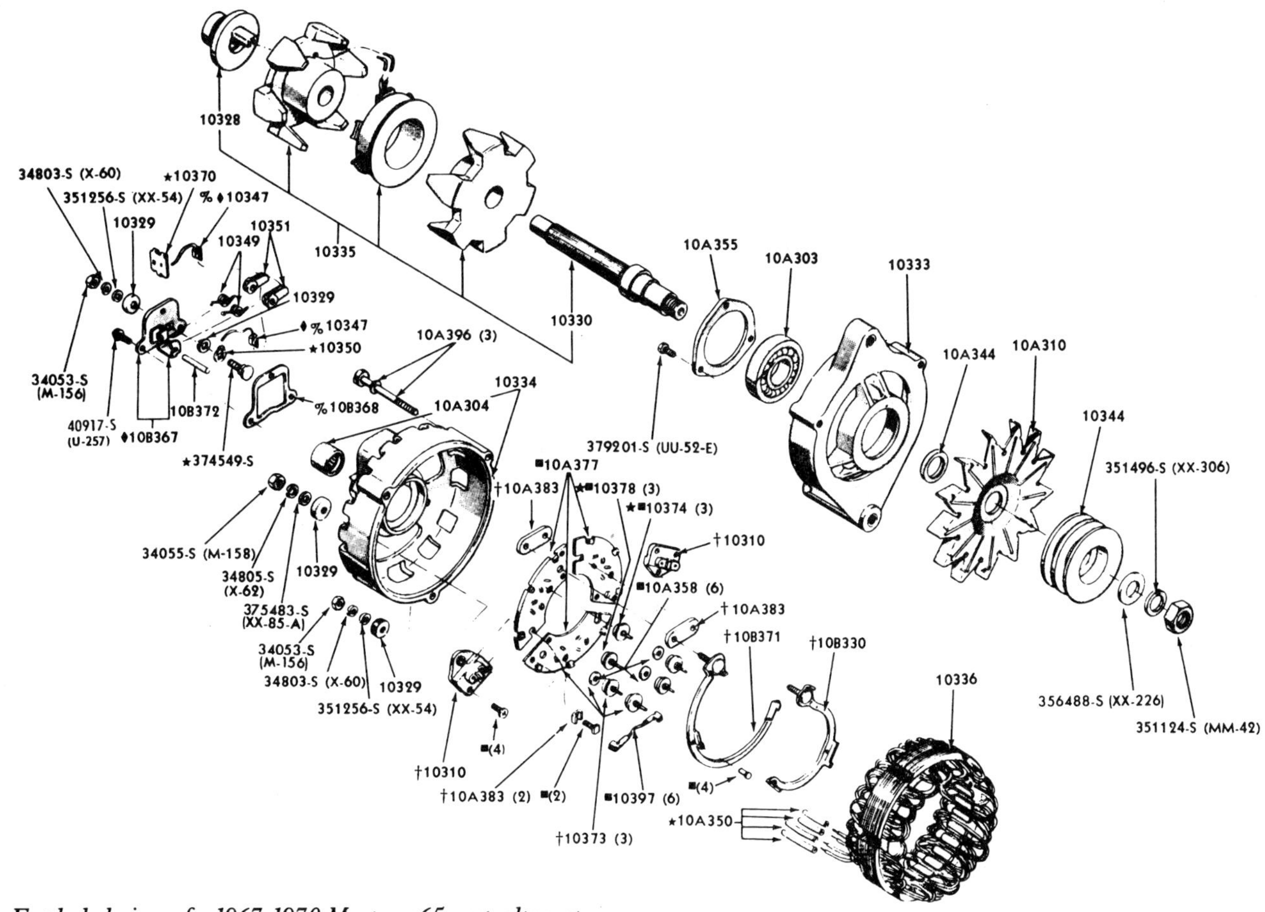

Exploded view of a 1967-1970 Mustang 65 amp alternator.

MUSTANG ALTERNATOR APPLICATION CHART

Year	Engine	Amp. Hrs.	Make	Service Alternator	Voltage Regulator
65	All 6&8	38 42 45	Motorcraft	D2AZ-10346-C	C3SZ-10316-B r/b D4TZ-10316-A
66	All 6&8	38 42	Motorcraft	D2AZ-10346-C	C3SZ-10316-B r/b D4TZ-10316-A
67	All 6&8	38 42 45	Motorcraft	D2AZ-10346-C	C3SZ-10316-B r/b D4TZ-10316-A
67	200,289, 390	55	Motorcraft	D0AZ-10346-F	C3SZ-10316-B r/b D4TZ-10316-A
68	All 6&8	38 42	Motorcraft	D2AZ-10346-C	C3SZ-10316-B r/b D4TZ-10316-A
68	All 6&8	55	Motorcraft	D0AZ-10346-F	C3SZ-10316-B r/b D4TZ-10316-A
69	All 6&8	38 42	Motorcraft	D2AZ-10346-C	C3SZ-10316-B r/b D4TZ-10316-A
69	All 8	55	Motorcraft	D0AZ-10346-F	C3SZ-10316-B r/b D4TZ-10316-A
69	428CJ Boss 429	55	Motorcraft	D0ZZ-10346-B	C3SZ-10316-B r/b D4TZ-10316-A
69	All 8	70	Motorcraft	D1AZ-10346-F	C3SZ-10316-B r/b D4TZ-10316-A
70	All 6&8	38 42	Motorcraft	D2AZ-10346-C	C3SZ-10316-B r/b D4TZ-10316-A
70	428CJ Boss 302	55	Motorcraft	D0ZZ-10346-B	C3SZ-10316-B r/b D4TZ-10316-A
70	6 w/ A/C All 6&8	55	Motorcraft	D0AZ-10346-F	C3SZ-10316-B r/b D4TZ-10316-A

MUSTANG ALTERNATOR PULLEY APPLICATION CHART, Part One, 1965-1967

Year	Engine	Identification	Type	Dimensions: A	B	C	E	F	Amp.	Part Number
65	170,200	C5AF-10A352-B,G,K C9AF-10A352-A,B	3	2.84	.670	.38	1.015		38,42,45	C5AZ-10344-K
65	289, w/o A/C	C5AF-10A352-B,G,K C9AF-10A352-A,B	3	2.84	.670	.38	1.015		38,42,45	C5AZ-10344-K
65	289, w/ A/C	C5AF-10A352-D,J C5TF-10A352-B,K C9AF-10A352-C	6	2.84	.670	.38	1.40	.64	All	C5AZ-10344-L
66	200	C5AF-10A352-B,F, G,K,L C9AF-10A352-A,B	3	2.84	.670	.38	1.015		38,42,55	C5AZ-10344-K
66	289	C5AF-10A352-B,F, G,K,L C9AF-10A352-A,B	3	2.84	.670	.38	1.015		38,42,55	C5AZ-10344-K
66	289	C5AF-10A352-D,J C5TF-10A352-B,K C9AF-10A352-C	6	2.84	.670	.38	1.40	.64	38,42,45	C5AZ-10344-L
66	289	C5AF-10A352-H D1ZF-10A352-A	5	3.05	670	.38	.91		38,42,55	D1ZZ-10344-A
67	200	C5AF-10A352-B,F, G,K,L C9AF-10A352-A,B	3	2.84	.670	.38	1.02		38,42,55	C5AZ-10344-K
67	289	C5AF-10A352-D,J C5TF-10A352-B,K C9AF-10A352-C	6	2.84	.670	.38	1.40	.64	38,45	C5AZ-10344-L
67	289	C5AF-10A352-B,G,K C9AF-10A352-A,B	3	2.84	.670	.38	1.015		38,45	C5AZ-10344-K
67	390	C5AF-10A352-B,F, G,K,L C9AF-10A352-A,B	3	2.84	.670	.38	1.015		38,55	C5AZ-10344-K
67	390 w/ T/E	C5AF-10A352-D,J C5TF-10A352-B,K C9AF-10A352-C	6	2.84	.670	.38	1.40	.64	38,55	C5AZ-10344-L

MUSTANG ALTERNATOR PULLEY APPLICATION CHART, Part Two, 1968-1969

Year	Engine	Identification	Type	Dimensions: A	B	C	E	F	Amp.	Part Number
68	200	C5AF-10A352-D,J C5TF-10A352-B,K C9AF-10A352-C	6	2.84	.670	.38	1.40	.64	55	C5AZ-10344-L
68	200	C5AF-10A352-B,F, G,K,L C9AF-10A352-A,B	3	2.84	.670	.38	1.015		38	C5AZ-10344-K
68	289	C5AF-10A352-B,J,K C9AF-10A352-A,B	3	2.84	.670	.38	1.015		38 55 w/ P/S	C5AZ-10344-K
68	289	D1ZF-10A352-A	5	3.05	.670	.38	.91		42,55	D1ZZ-10344-A
68	289	C5AF-10A352-D,J C5TF-10A352-B,K C9AF-10A352-C	6	2.84	.670	.38	1.40	.64	55	C5AZ-10344-L
68	302 A/C exc. P/S	C5AF-10A352-D,J C5TF-10A352-B,K C9AF-10A352-C	6	2.84	.670	.38	1.40	.64	55	C5AZ-10344-L
68	302	C5AF-10A352-B,G,K C9AF-10A352-A,B	3	2.84	.670	.38	1.015		38 55 w/ P/S	C5AZ-10344-K
68	390 GT, A/C w/ P/S	C5AF-10A352-B,F, G,K,L C9AF-10A352-A,B	5	2.84	.670	.38	1.015		55	C5AZ-10344-K
68	390 GT	C5AF-10A352-D,J C5TF-10A352-B,K C9AF-10A352-C	6	2.84	.670	.38	1.40	.64	38	C5AZ-10344-L
68	390 A/C exc. P/S	C5AF-10A352-D,J C5TF-10A352-B,K C9AF-10A352-C	6	2.84	.670	.38	1.40	.64	55 w/ P/S	C5AZ-10344-L
68	428CJ	C5AF-10A352-D,J C5TF-10A352-B,K C9AF-10A352-C	6	2.84	.670	.38	1.40	.64	55	C5AZ-10344-L
69	200	C5AF-10A352-B,F, G,K,L C9AF-10A352-A,B	3	2.84	.670	.38	1.015		38,42	C5AZ-10344-K
69	250	C5AF-10A352-B,G,K C9AF-10A352-A,B	3	2.84	.670	.38	1.015		38	C5AZ-10344-K
69	302	C5AF-10A352-B,G,K C9AF-10A352-A,B	3	2.84	.670	.38	1.015		38 42 w/ P/S	C5AZ-10344-K
69	351	C5AF-10A352-B,G,K C9AF-10A352-A,B	3	2.84	.670	.38	1.015		42	C5AZ-10344-K
69	390GT	C5AF-10A352-B,G,K C9AF-10A352-A,B	3	2.84	.670	.38	1.015		38,55	C5AZ-10344-K
69	390GT	C5TF-10A352-B,K C9AF-10A352-C	6	2.84	.670	.38	1.40	.64	55	C5AZ-10344-L
69	428CJ	C5AF-10A352-D,J C5TF-10A352-B,K C9AF-10A352-C	6	2.84	.670	.38	1.40	.64	55	C5AZ-10344-L
69	Boss 429	C5AF-10A352-B,F, G,K,L C9AF-10A352-A,B	3	2.84	.670	.38	1.015		55	C5AZ-10344-K

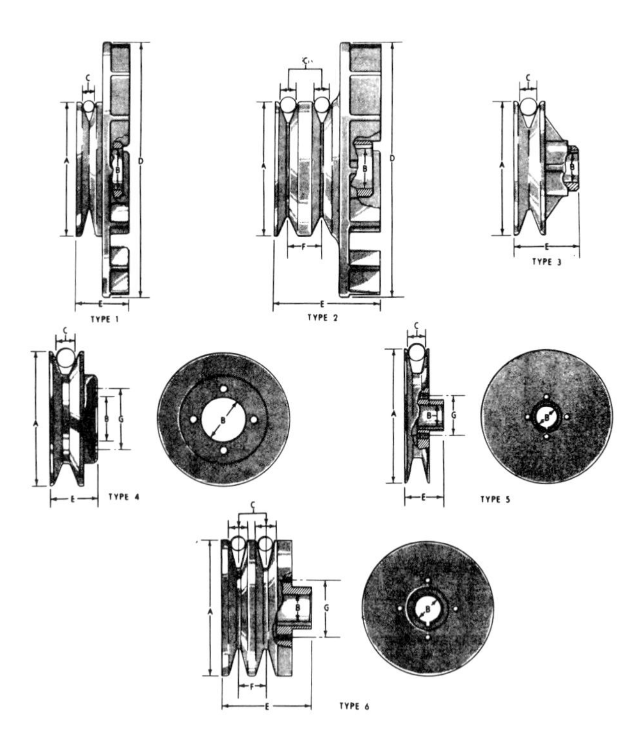

Exploded view of a 1967-1970 Mustang 65 amp alternator.

MUSTANG ALTERNATOR PULLEY APPLICATION CHART, Part Three, 1970

Year	Engine	Identification	Type	Dimensions: A	B	C	E	F	Amp.	Part Number
70	200	C5AF-10A352-B,F, G,K,L C9AF-10A352-A,B	3	2.84	.670	.38	1.015		42	C5AZ-10344-K
70	250	C5AF-10A352-B,G,K C9AF-10A352-A,B	3	2.84	.670	.38	1.015		38	C5AZ-10344-K
70	250 w/ A/C	C5AF-10A352-D,J C5TF-10A352-B,K C9AF-10A352-C	6	2.84	.670	.38	1.40	.64	55	C5AZ-10344-L
70	302	C5AF-10A352-B,F, G,K,L C9AF-10A352-A,B	3	2.84	.670	.38	1.015		38,42 55 w/ A/C	C5AZ-10344-K
70	Boss 302	D1ZF-10A352-A	5	3.05	.670	.38	.91		55	D1ZZ-10344-A
70	351	C5AF-10A352-B,G,K C9AF-10A352-A,B	3	2.84	.670	.38	1.015		38,42,55	C5AZ-10344-K
70	428CJ	C5AF-10A352-D,J C5TF-10A352-B,K C9AF-10A352-C	6	2.84	.670	.38	1.40	.64	55	C5AZ-10344-L
70	429	D1ZF-10A352-A	5	3.05	.670	.38	.91		55	D1ZZ-10344-A

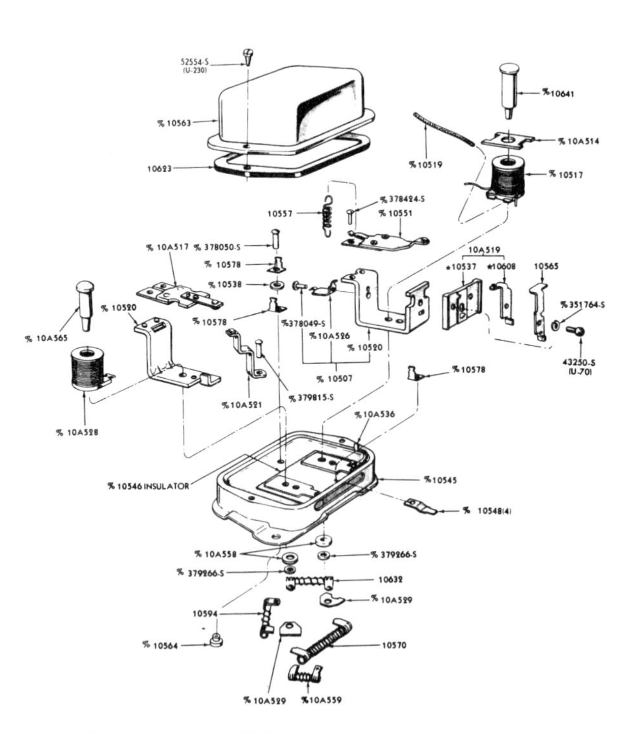

Exploded view of a 1965-1970 Mustang voltage regulator used with an alternator.

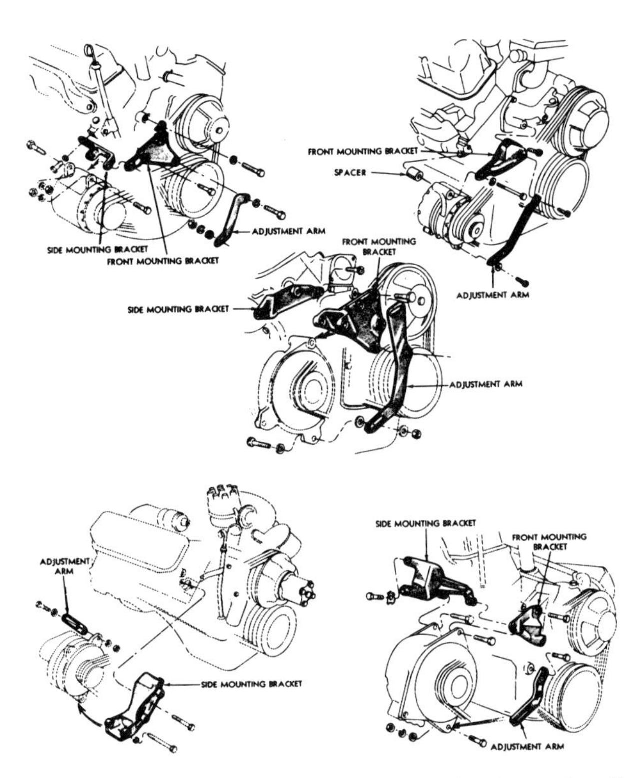

Typical alternator mounting brackets and hardware for all 1965-1970 Ford V-8 engines.

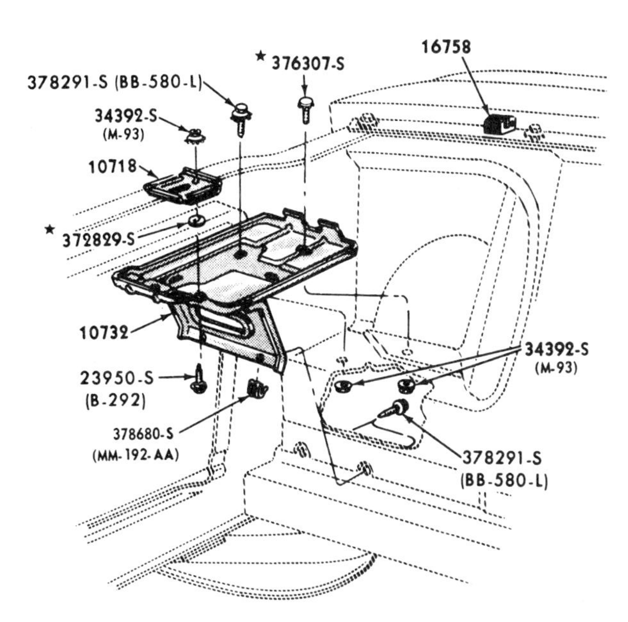

The 1965-1968 Mustang battery tray and attaching hardware.

MUSTANG BATTERY APPLICATION CHART
All engines, 1965-1970.

Motorcraft Group No.	Case Width	Case Length	Case Height	Type	Amperes	Plates
22FC	6-13/16"	9-27/64"	8-5/16"	Premium, dry	49	54
				Standard, dry	42	54
22HF	6-3/4"	9"	8	Premium, dry	45	54
				Premium, dry	54	66
24F	6-13/16"	10-3/4"	8-61/64"	Premium, H/D, dry	81	78
				Premium, dry	55	66
				Standard, dry	41	42
27F	6-13/16"	12-1/2"	9-9/64"	Premium, dry	70	78
				Premium, H/D, dry	94	90
				Premium, dry	77	78

MUSTANG BATTERY MOUNTING PARTS CHART

Year	Amp. Hrs.	Hold Down Clamp Part Number	Carrier or Tray Part Number	Clamp Bolt Part Number	Brace or Bracket Part Number
65	40,45	C2DZ-10718-A	C5ZZ-10732-C		
65	55,65	C5DZ-10718-A	C5ZZ-10732-C		
66	45	C2DZ-10718-A	C5ZZ-10732-C		
66	55	C5DZ-10718-A	C5ZZ-10732-C		
67	45	C5AZ-10718-A	C7ZZ-10732-C	D0AZ-10756-A	C7ZZ-10753-A
67	55	C5AZ-10718-A	C7ZZ-10732-D	D0AZ-10756-A	C7ZZ-10753-A
68	45	C5AZ-10718-A	C7ZZ-10732-C	D0AZ-10756-A	C7ZZ-10753-A
68	55	C5AZ-10718-A	C7ZZ-10732-D	D0AZ-10756-A	C7ZZ-10753-A
69	45	C5AZ-10718-A	C7ZZ-10732-C	D0AZ-10756-A	C7ZZ-10753-A
69	55	C5AZ-10718-A	C7ZZ-10732-D	D0AZ-10756-A	C7ZZ-10753-A
69	85 (note 1)	C5AZ-10718-A	C9ZZ-10732-B	D0AZ-10756-A	C9ZZ-10679-B
70	45 (note 2)	D0OZ-10718-A	C7ZZ-10732-C	D0AZ-10756-A	C9ZZ-10A705-A
70	55 (note 2)	D0OZ-10718-A	C7ZZ-10732-D	D0AZ-10756-A	C9ZZ-10A705-A
70	70 (note 2)	D0OZ-10718-A	C9ZZ-10732-A	D0AZ-10756-A	C9ZZ-10A705-A
70	All (note 3)	D0OZ-10718-A	C9ZZ-10732-B	D0AZ-10756-A	C9ZZ-10679-B
70	All (note 3)	D0OZ-10718-A	C9ZZ-10732-B	D0AZ-10756-A	C9ZZ-10A674-B
70	All (note 3)	D0OZ-10718-A	C9ZZ-10732-B	D0AZ-10756-A	C9ZZ-10A710-B

Notes:
1 Trunk-mounted battery
2 Except Boss
3 Boss

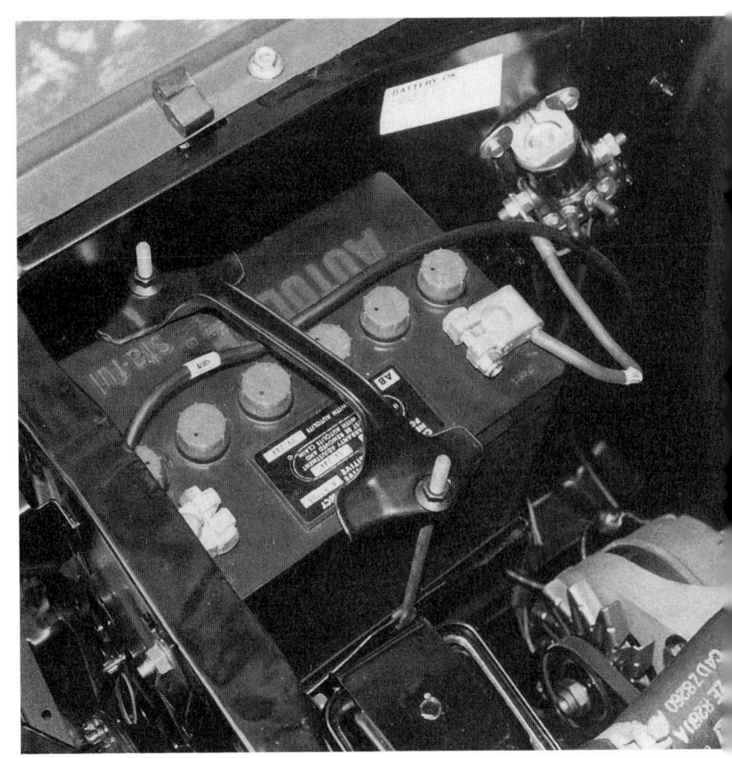

A heavy-duty battery and charging system was an option for every Mustang. Additional optional equipment was required in some cases. The maximum output varied from 55 to 70 amps depending on the year and whether the car was equipped with factory-installed air conditioning. Mustangs built through the 1971 model year were equipped with Autolite batteries. With the 1972 model year, Ford dropped the Autolite name and began equipping Mustangs with Motorcraft batteries.

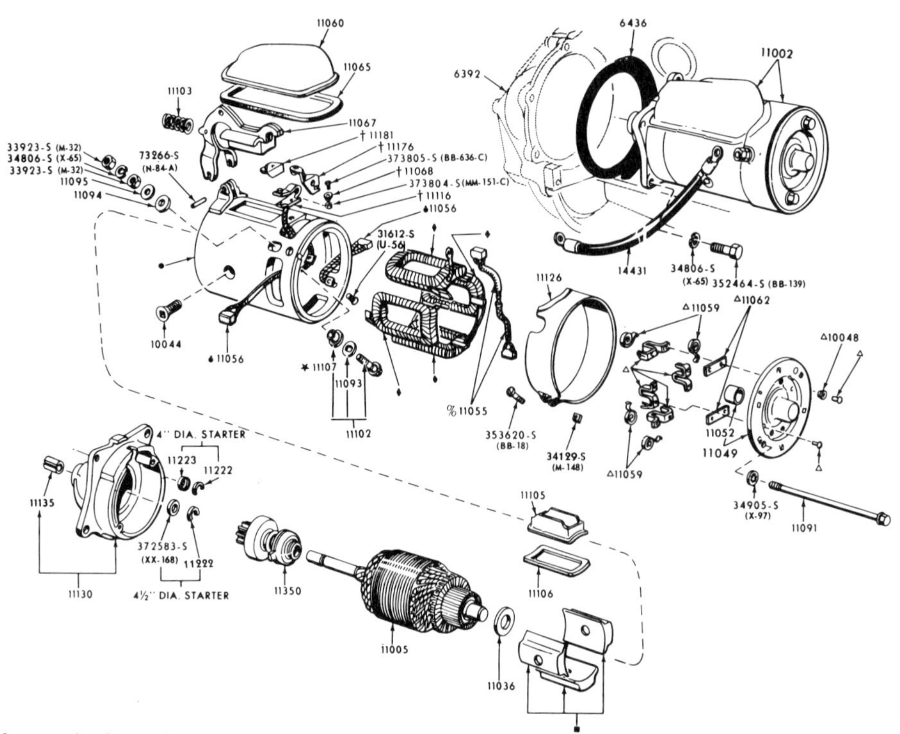

Exploded view of a typical 1965-1970 Mustang starter assembly.

MUSTANG STARTER APPLICATION CHART, 1965-1968

Year	Cyls.	CID	Ident. No.	Serv. Starter Ass'y
1965	6 & 8	260/289	C4OF-11001-A	C2OZ-11002-B
1965	6	200	C5AF-11001-A	C3OZ-11002-C
1965	6	170	C4DF-11001-A,B	C2DZ-11002-A
1966	8	289	C5TF-11001-A	C5TZ-11002-D
1966	8	289	C4ZF-11001-A	C2OZ-11002-B
1966	6	200	C5AF-11001-A	C3OZ-11002-C
1966	6	200	C6VF-11001-A	C6VY-11002-A
1966	6	200	C6OF-11001-A	C6OZ011002-A
1967	6	200	C6OF-11001-A	C6OZ-11002-A
1967	6	200	C7ZF-11001-A	C6OZ-11002-A
1967	6	200	C5AF-11001-A	C3OZ-11002-C
1967	8	289	C4OF-11001-A	C2OZ-11002-B
1967	8	289	C7AF-11001-B	C2OZ-11002-B
1967	8	390	C7AF-11001-A,C	C3OZ-11002-C
1968	6	200	C7ZF-11001-A,B	C6OZ-11002-A
1968	8	289	C7AF-11001-B	C2OZ-11002-B
1968	8	302	C7AF-11001-B	C2OZ-11002-B
1968	8	289	C7AF-11001-D	C5TZ-11002-A
1968	8	302	C7AF-11001-F	C5TZ-11002-D
1968	8	390	C7AF-11001-C	C3OZ-11002-C
1968	8	427	C7AF-11001-C	C3OZ-11002-C
1968	6	200	C7OF-11001-A	C3OZ-11002-C
1968	8	390	C7OF-11001-A	C3OZ-11002-C
1968	8	390	C7AF-1001-E	C6AZ-11002-A
1968	8	427/428	C8AF-11001-A	C8AZ-11002-A

MUSTANG STARTER APPLICATION CHART, 1969-1970

Year	Cyls.	CID	Ident. No.	Serv. Starter Ass'y
1969	6	200	C7ZF-11001-A	C6OZ-11002-A
1969	6	200	C7OF-11001-A	C3OZ-11002-C
1969	6	250	C9ZF-11001-A	C2OZ-11002-B
				r/b D4OZ-11002-A
1969	8	302/351	C9ZF-11001-A	C2OZ-11002-B
				r/b D4OZ-11002-A
1969	8 (s/t)	302/351	C7AF-11001-F	C5TZ-11002-D
1969	8	390	C7AF-11001-C	C3OZ-11002-C
1969	8	390	C7OF-11001-A	C3OZ-11002-C
1969	8	390	C9AF-11001-B	C3OZ-11002-C
1969	8	428/428CJ	C8AF-11001-A	C8AZ-11002-A
				r/b C4TZ-11002-B
1969	8	Boss 429	C9AF-11001-A	C8VY-11002-C
1970	6	200	C7ZF-11001-A	C6OZ-11002-A
1970	6	200	C7DF-11001-A	C2DZ-11002-A
				r/b C3OZ-11002-C
1970	6	200	C7OF-11001-A	C3OZ-11002-C
1970	6	200	C7AF-11001-B	C2OZ-11002-B
				r/b D4OZ-11002-A
1970	6	200	DOOF-11001-A	C3OZ-11002-C
1970	6	200	DOZF-11001-B	C6OZ-11002-A
1970	6	250	DOZF-11001-A	C2OZ-11002-B
				r/b D4OZ-11002-A
1970	6	250	C9ZF-11001-A	C2OZ-11002-B
				r/b D4OZ-11002-A
1970	8	302 (a/t)	C7AF-11001-B	C2OZ-11002-B
				r/b D4OZ-11002-A
1970	8	351 (a/t)	C7AF-11001-B	C2OZ-11002-B
				r/b D4OZ-11002-A
1970	8	302 (s/t)	C7AF-11001-F	C5TZ-11002-D
1970	8	351 (s/t)	C7AF-11001-F	C5TZ-11002-D
1970	8	302351	DOAF-11001-B	C2OZ-11002-B
				r/b D4OZ-11002-A
1970	8	302/351	DOAF-11001-C	C5TZ-11002-D
1970	8	428CJ	C8AF-11001-A	C8AZ-11002-A
				r/b C4TZ-11002-B
1970	8	428CJ	DOTF-11001-A	C4TZ-11002-B

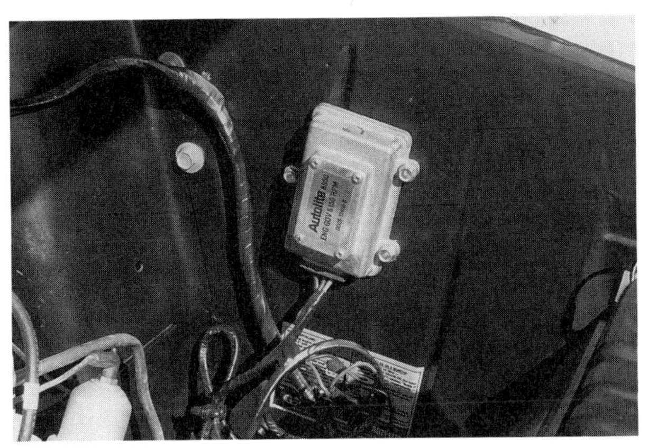

In an attempt to prevent premature destruction, Boss engines were equipped with an electronic rpm-limiting device. The mandatory accessory prevented the engine from exceeding a safe rpm range. This 1970 model interrupts the ignition system at a predetermined 6150rpm as noted on the cover.

At a preset rpm, the Boss 302 rev limiter intercepts the signal to and from the stock ignition coil.

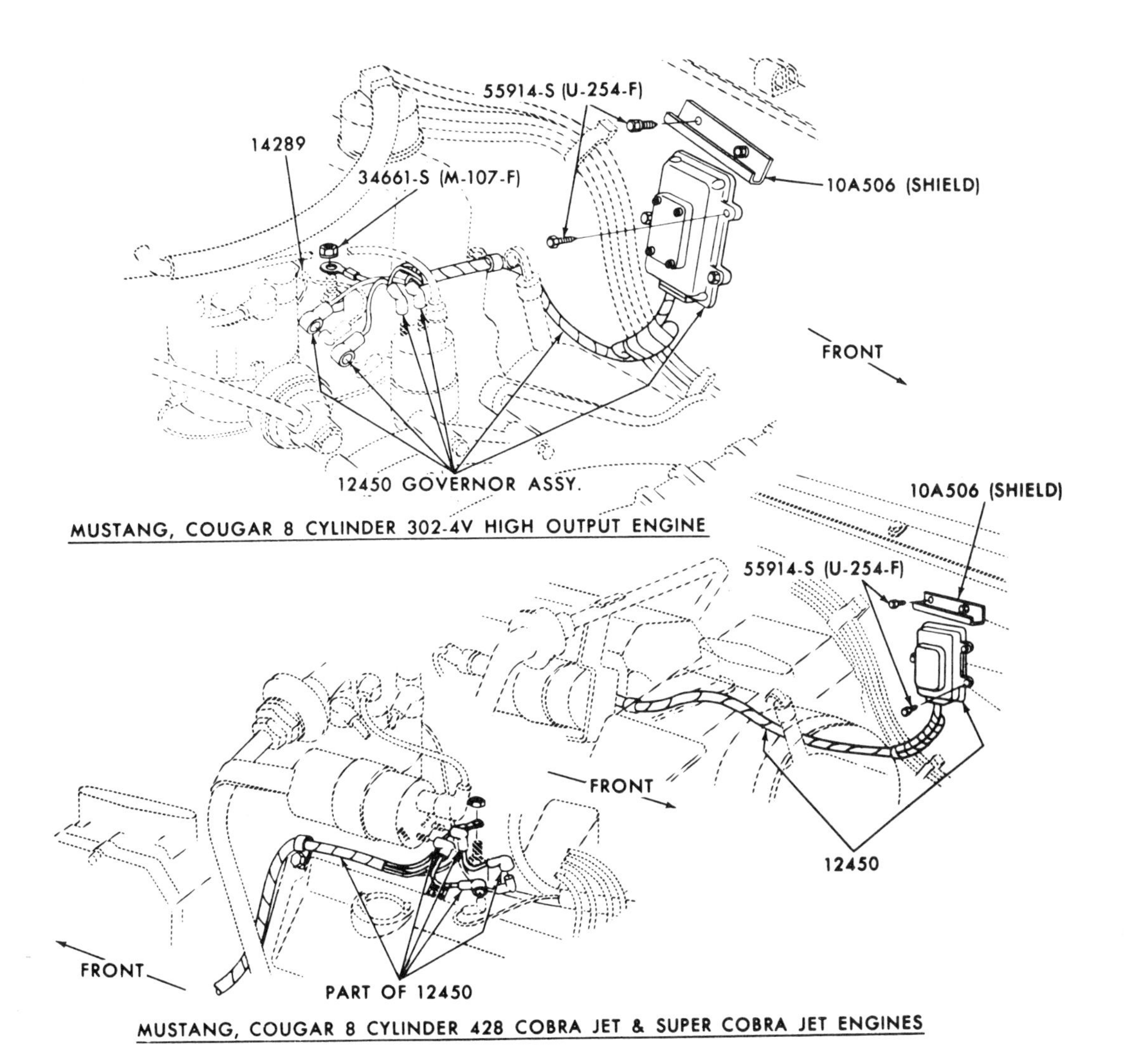

Transistorized engine governor for 1969-1970 Mustang equipped with a Boss 302, 428, or 429 engine.

Chapter 8

Options

The Mustang and Shelby restorer is constantly taunted by the possibilities of adding options to the cars. From the introduction of the Mustang in April 1964, Ford attempted to offer as many options as possible in order that the owner or prospective owner might "customize" his or her purchase. For that reason, today's owners are tempted to locate and install certain extras in search of added luxury, convenience, or, as before, personality. It has been said that the early Mustang offered more individual options than any other vehicle in the Ford product line. The reader should note that this chapter does not cover many options that are covered in previous chapters; radios, brakes, upholstery, exterior trim, air conditioning, and steering are good examples of this.

Of course, to the concours camp, there is the ever-present danger of "over restoration," the addition of so many after-the-fact options that a restored Mustang or Shelby bears no resemblance to any car that ever left the Metuchen, Dearborn, or San Jose assembly plants. In the 1960s, the norm was simplicity, and few Mustangs were heavily optioned. In fact, there existed a school of thought that promoted the "sleeper," a car optioned for performance with no regard for chrome and convenience extras.

The majority of the components depicted and described here are no longer available through standard Ford parts channels, and must be located through specialized vendors, reproduction sources, or swap meets. Many vendors advertise original and reproduction parts in periodicals, and many also travel the Mustang show circuit from late spring to early fall. Often, finding an exotic option can be a matter of luck; just as often, the purchase of that item is a matter of expense. A good example would be the discovery of an NOS

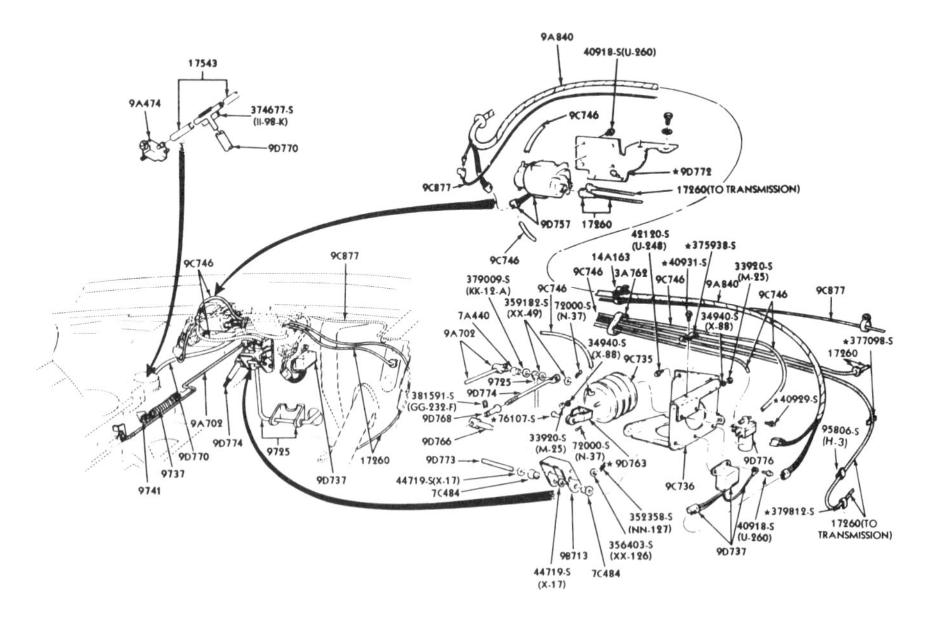

Optional cruise control system (engine compartment) for 1967 Mustang.

hang-on air-conditioning system, complete with both engine compartment and interior brackets, bolts, hoses, belts, and additional hardware. They exist, but it takes a clever researcher to find and buy them.

The drawings in this chapter should allow the restorer to comprehend the choices available, the accuracy needed, and the tasks necessary to upgrade a plain Mustang or to replace damaged original option pieces.

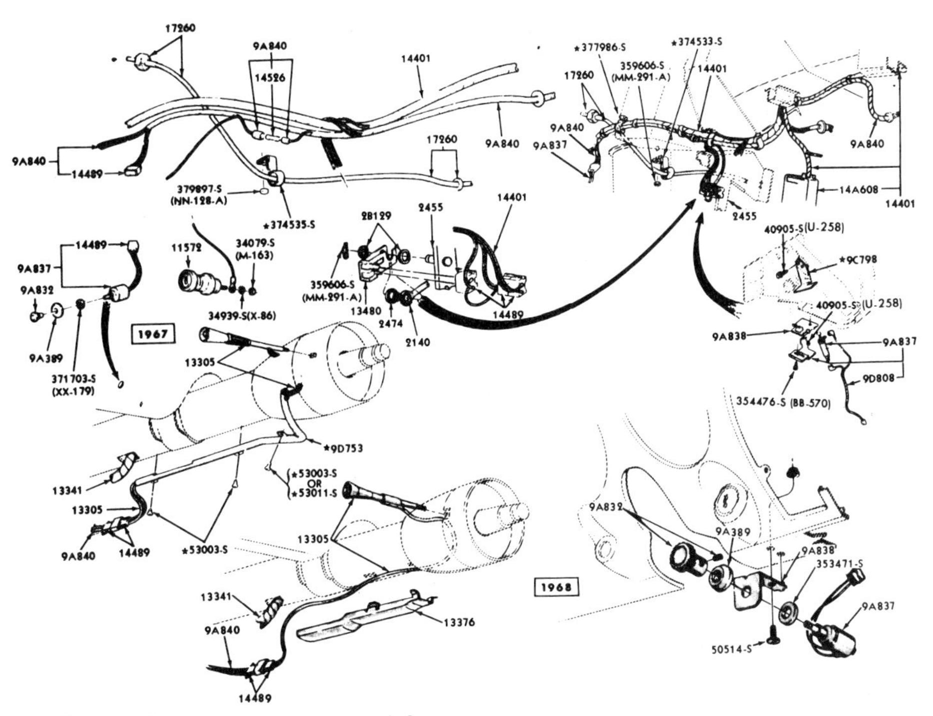

Optional cruise control system (passenger compartment) for 1967-1968 Mustang.

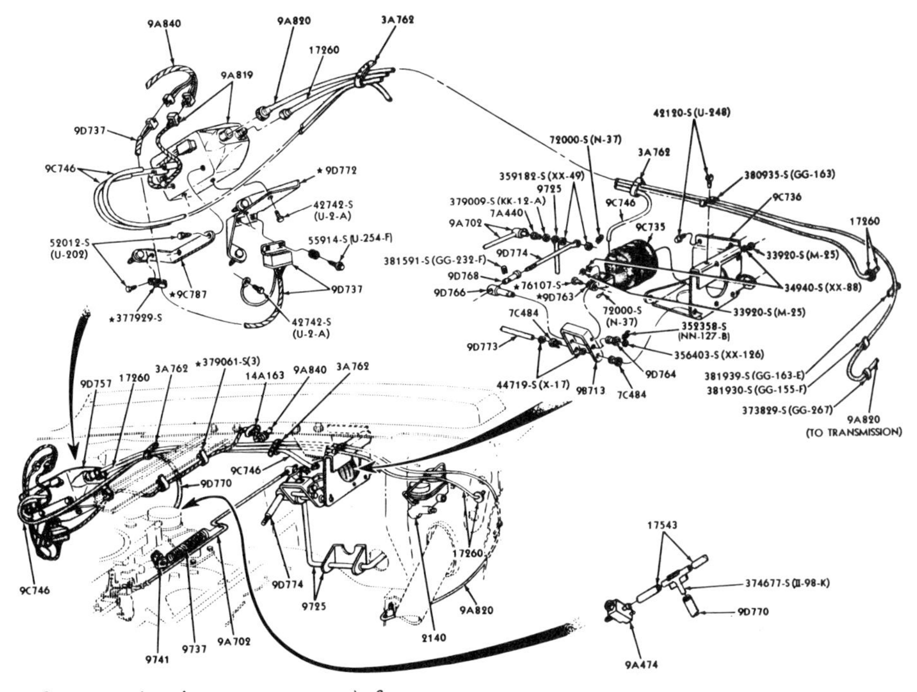

Optional cruise control system (engine compartment) for 1968 Mustang.

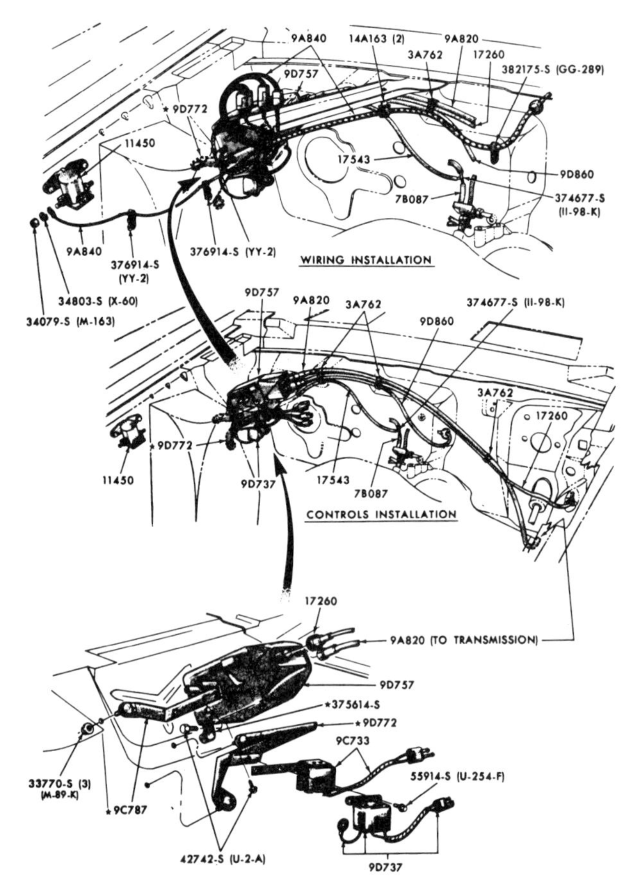

Optional cruise control system (engine compartment) for 1969 Mustang.

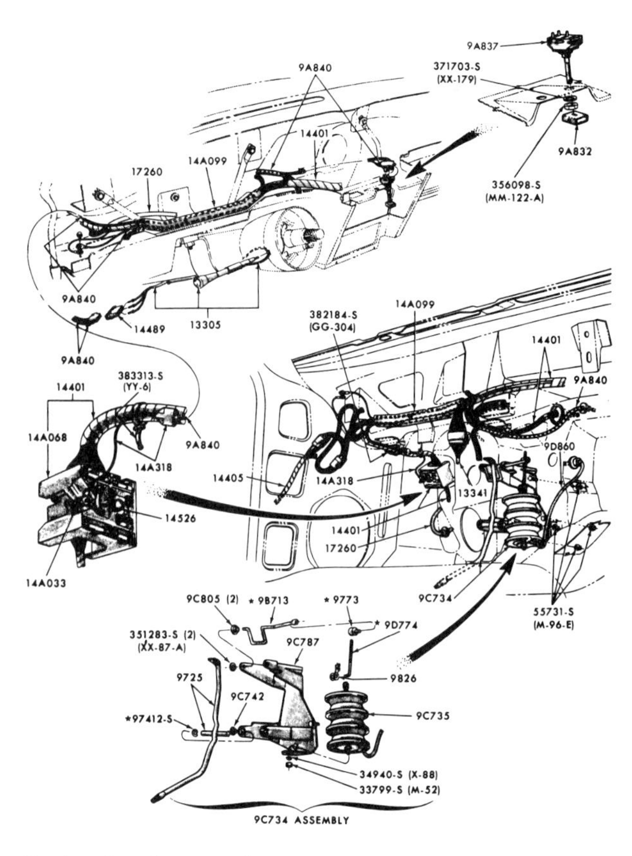

Optional cruise control system (passenger compartment) for 1969 Mustang.

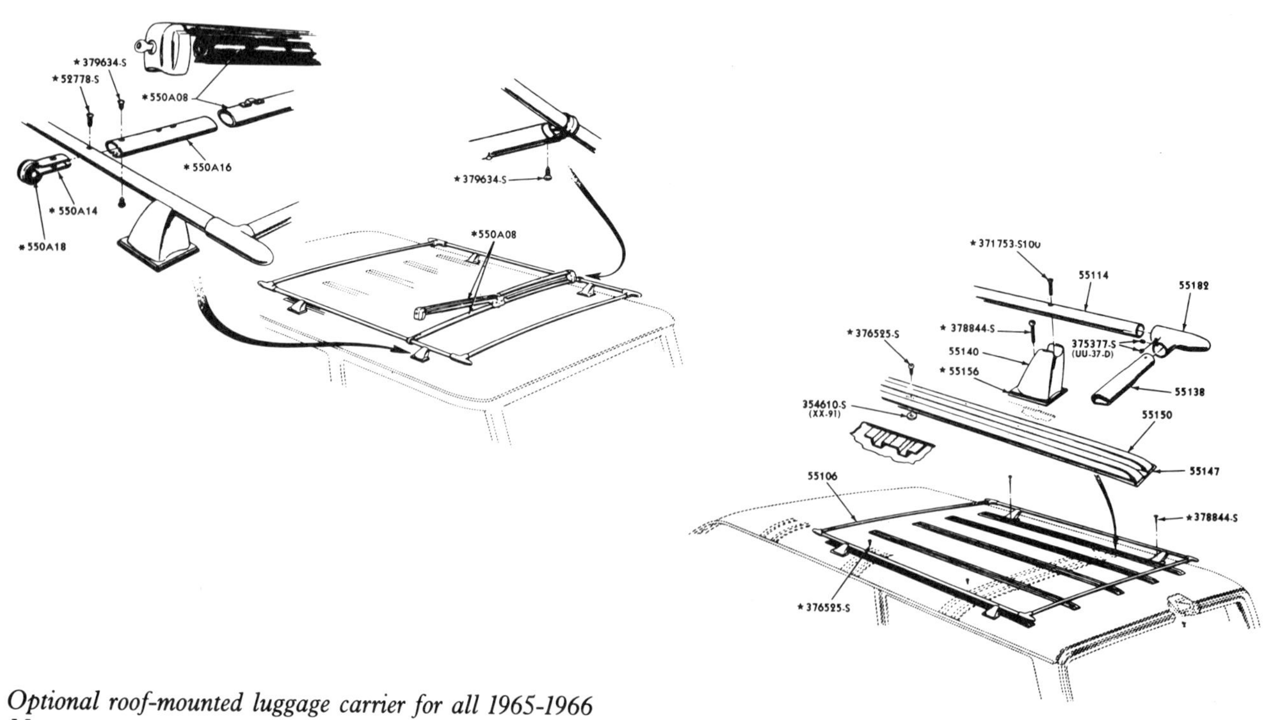

Optional roof-mounted luggage carrier for all 1965-1966 Mustangs.

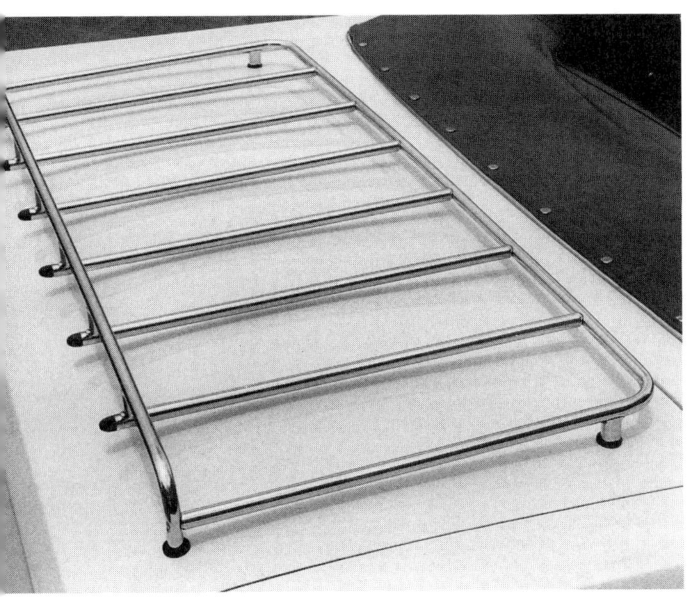

Chrome trunk-mounted luggage carriers were options for all years, but only for hardtops and convertibles. Roof-mounted carriers were available for Mustang fastbacks.

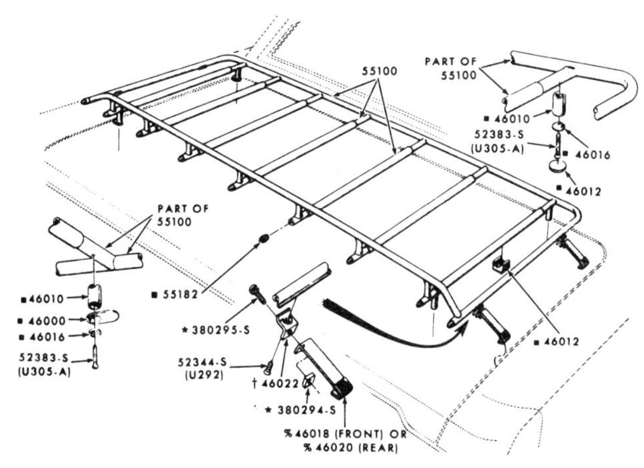

Optional removable rear-deck-mounted luggage carrier for 1969-1970 Mustang.

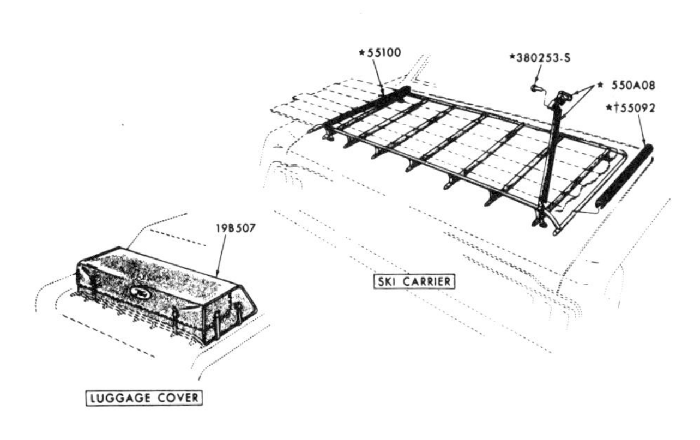

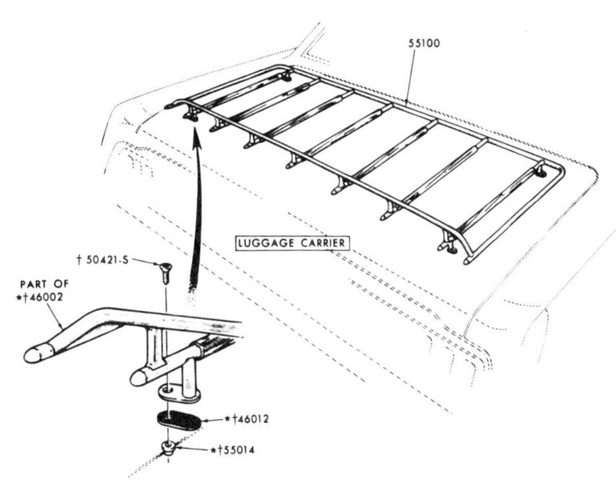

Optional fixed (non-removable) rear-deck-mounted luggage carrier, ski rack, and luggage cover for 1970 Mustang.

Factory-installed rear window louvers were a popular option on 1969-1970 Mustang SportsRoofs.

High-quality OEM louvers featured heavy chrome hinges and spring-loaded latches.

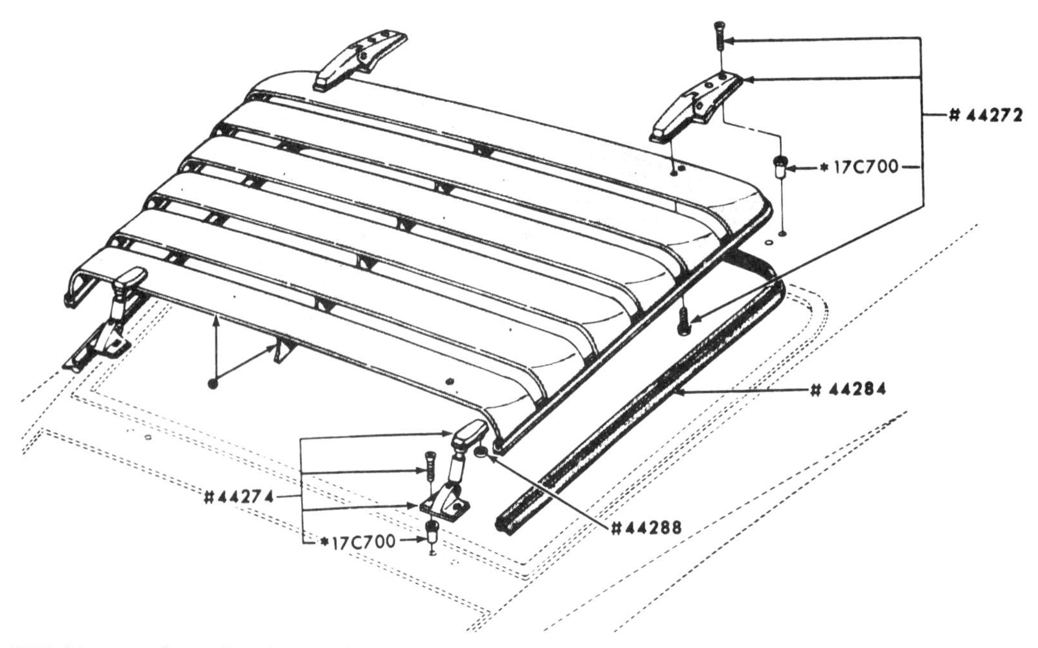

The 1969-1970 Mustang SportsRoof rear window louvers and attaching hardware.

MUSTANG RADIO ANTENNA KIT APPLICATION CHART

YEAR	PART NUMBER	DESCRIPTION
65/68	C5ZZ-18813-B	Manual - front
67	C1SZ-17696-B	F/B, H/T - Univ. electric type - rear qtr. mounted
69/70	C2OZ-17696-A	R.H. - manual - front
69/70	D3AZ-18813-C	(Shelby GT350/500) - Rear antenna

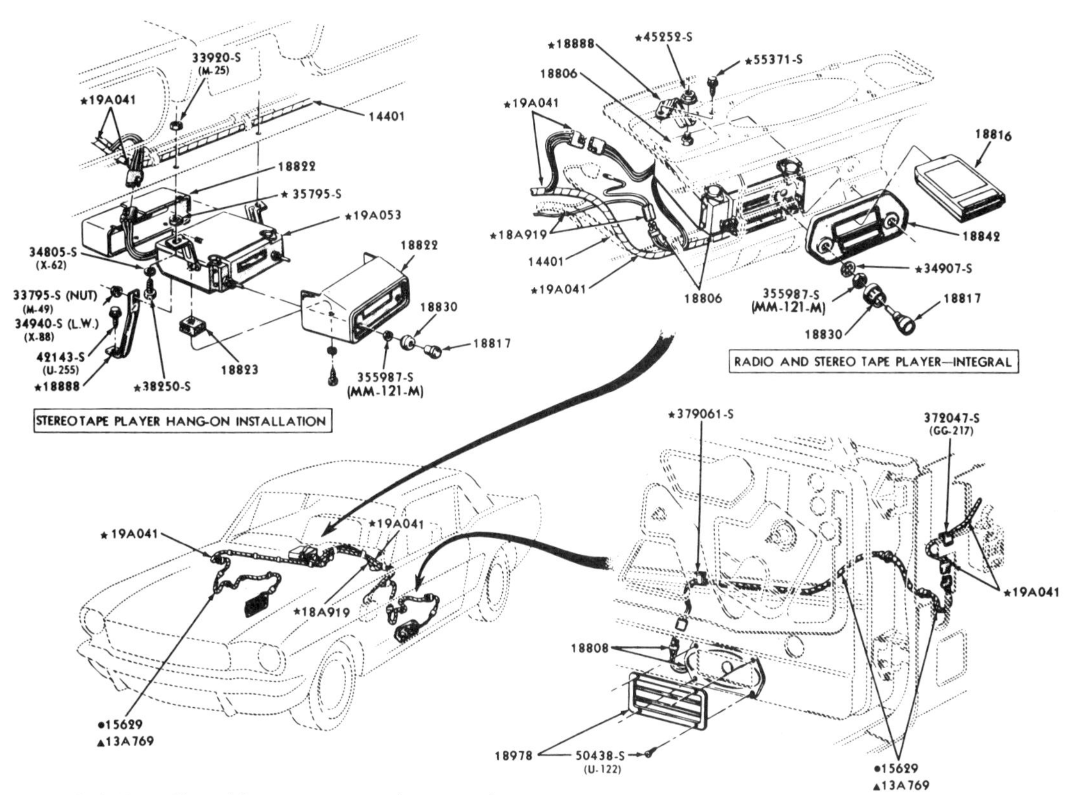

Optional integral AM radio with stereo tape player and hang-on stereo tape deck for 1965-1966 Mustang.

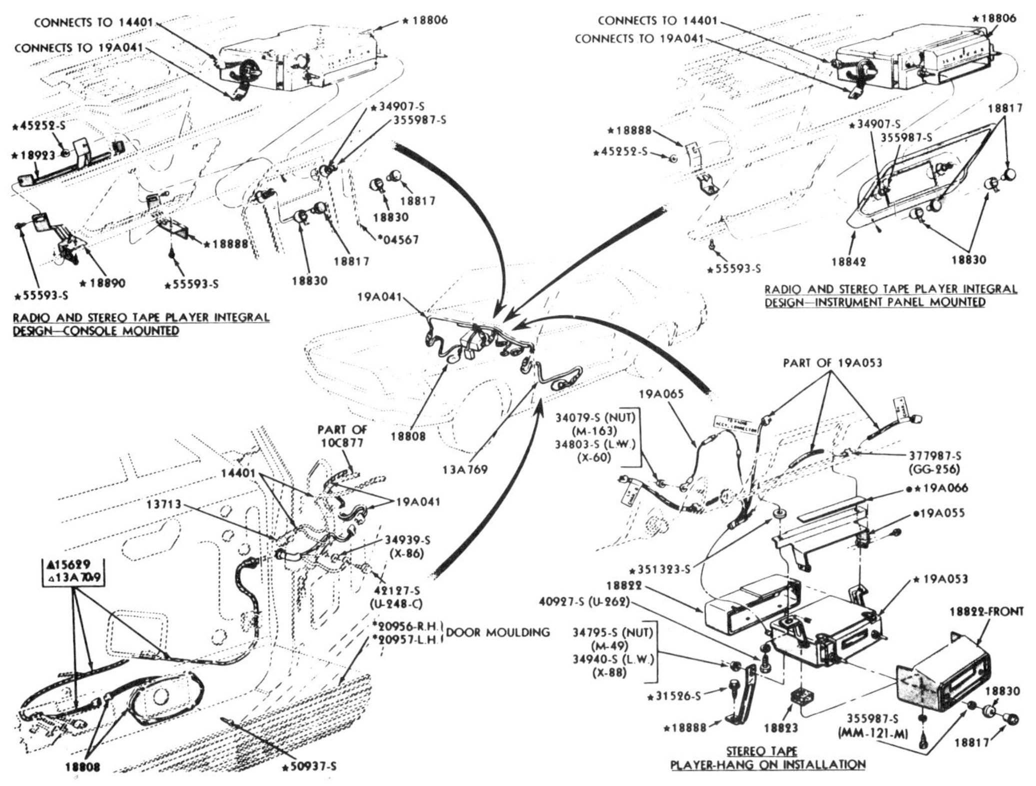

Optional integral AM radio with stereo tape player and hang-on tape deck for 1967 Mustang.

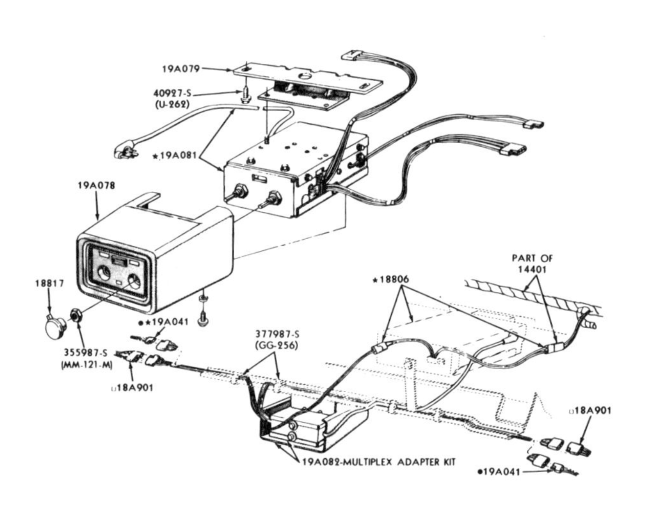

Optional AM/FM radio multiplex adapter for 1967 Mustang.

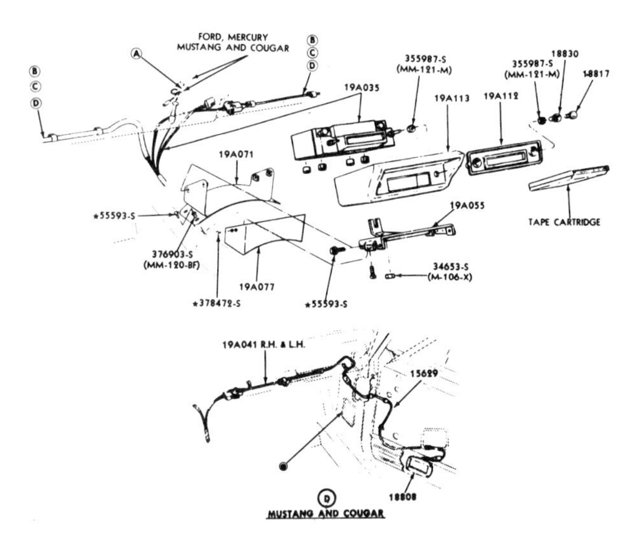

Optional hang-on stereo tape deck, speakers, and wiring for 1968-1970 Mustangs.

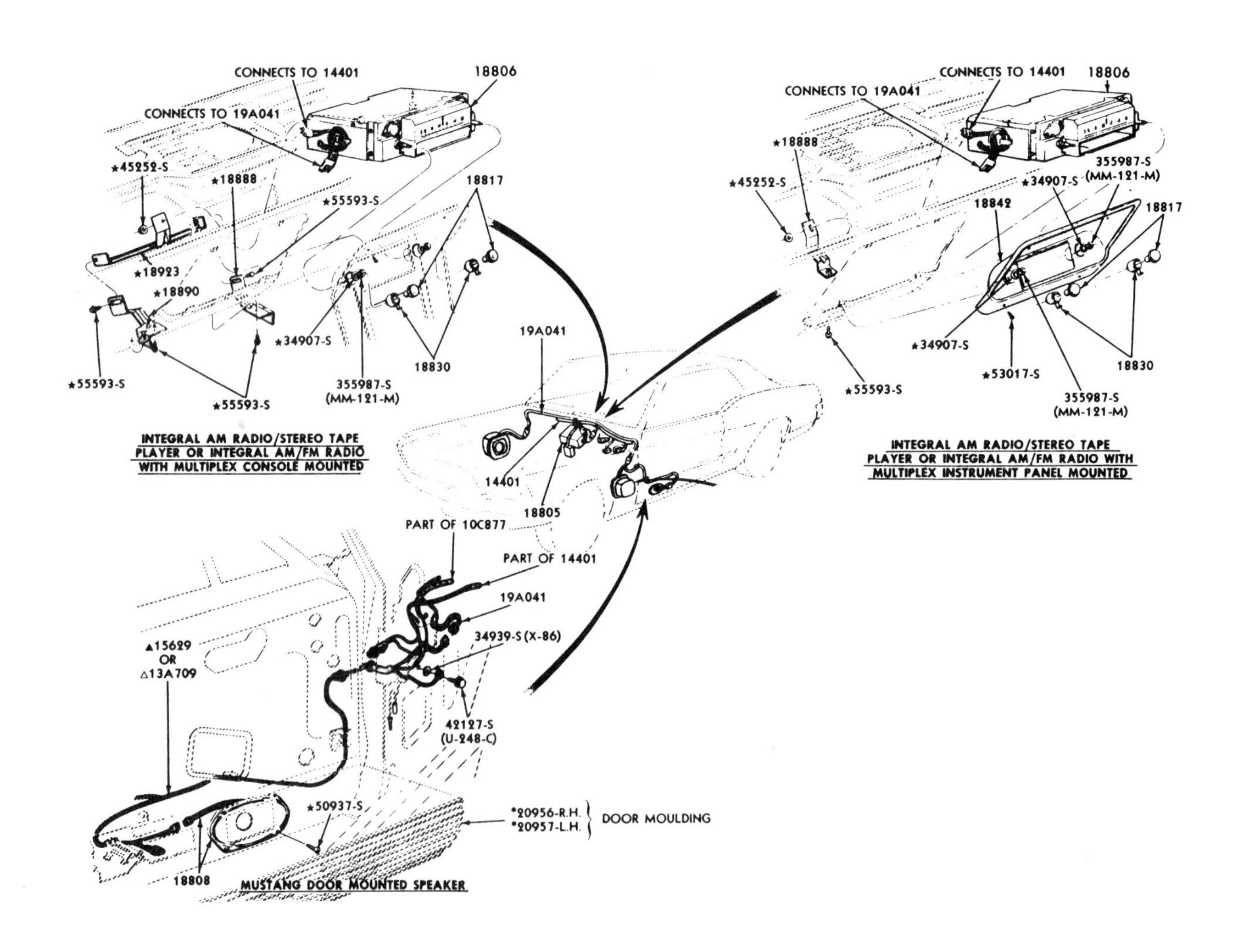

Optional integral AM stereo tape player and AM/FM multiplex radio for 1968 Mustang.

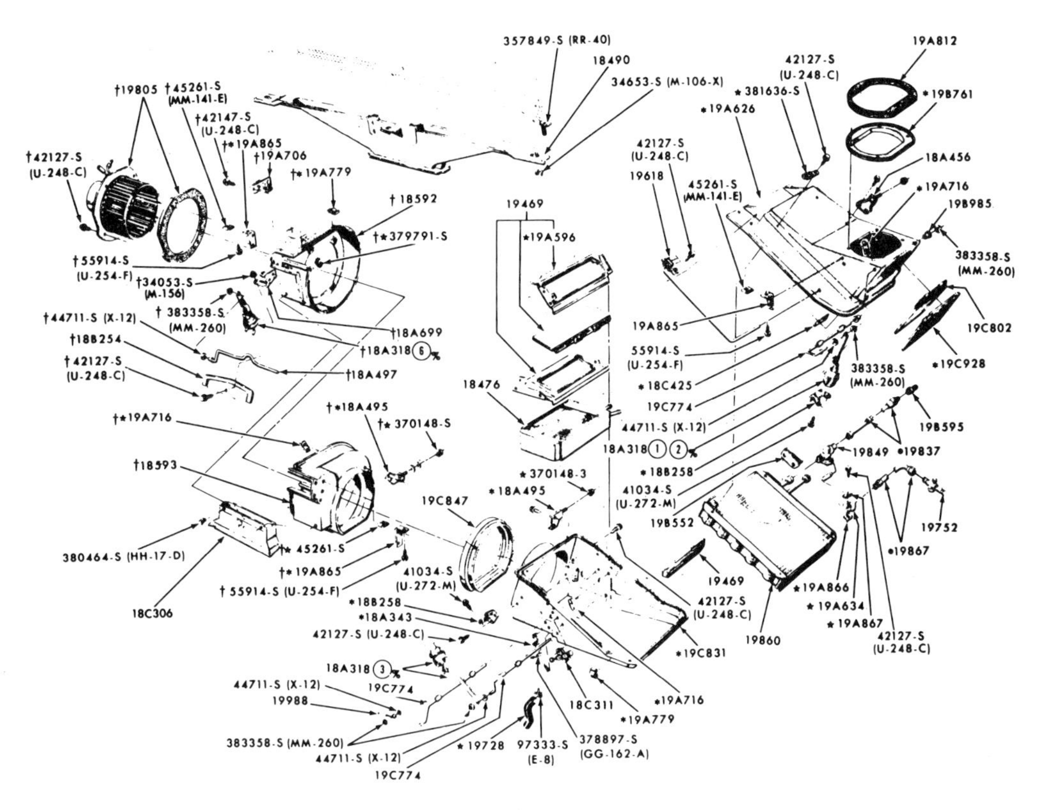

Factory-installed integral air-conditioning evaporator, plenum, blower, and related parts for 1969-1970 Mustangs.

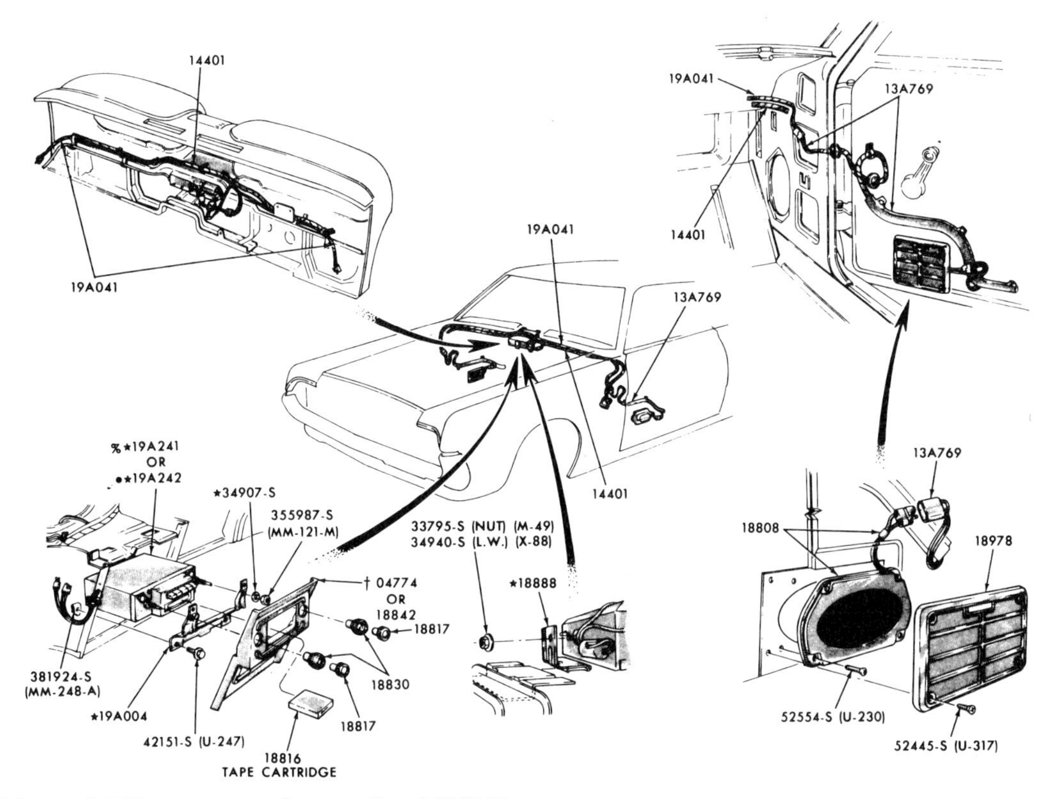

Optional integral AM stereo tape player and/or AM/FM multiplex radio for 1969-1970 Mustangs.

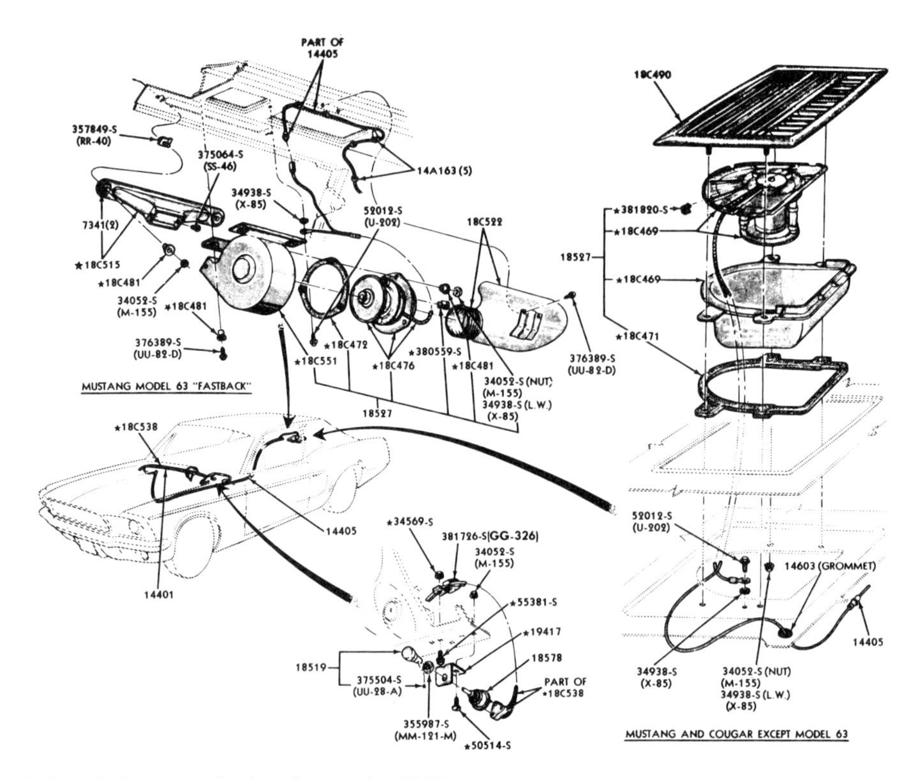

Optional rear window defogger and related parts for 1967 Mustang hardtops and fastbacks.

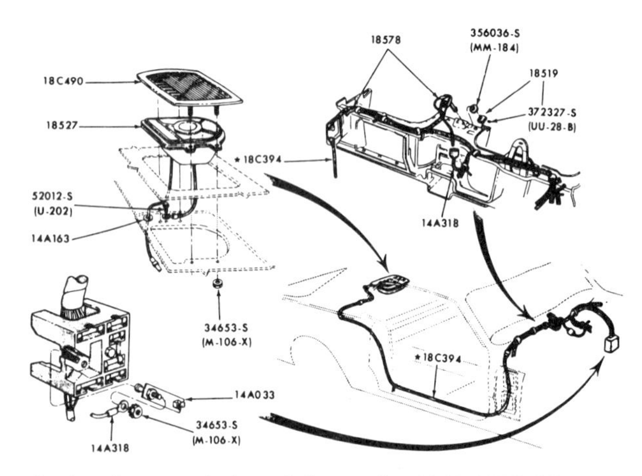

Optional rear window defogger for 1969-1970 Mustang hardtops.

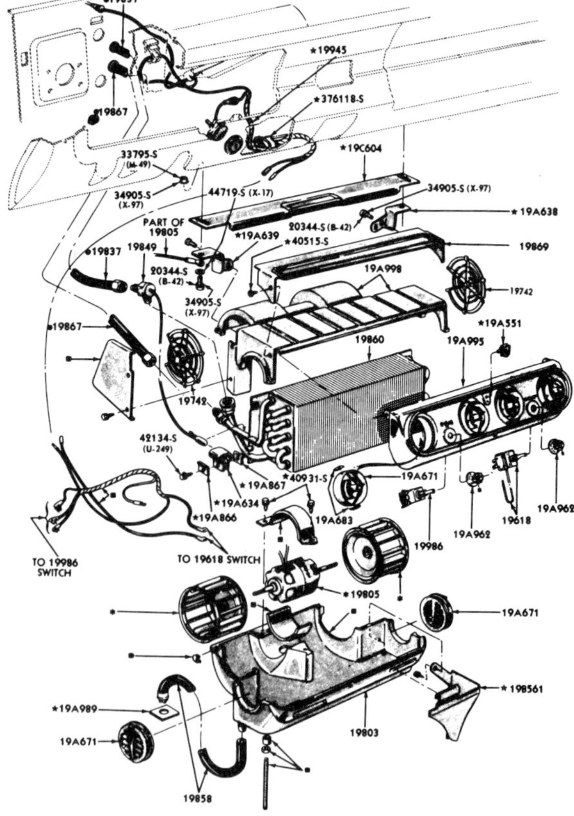

The 1965-1966 Mustang air-conditioner evaporator, blower housing, wiring, and related components.

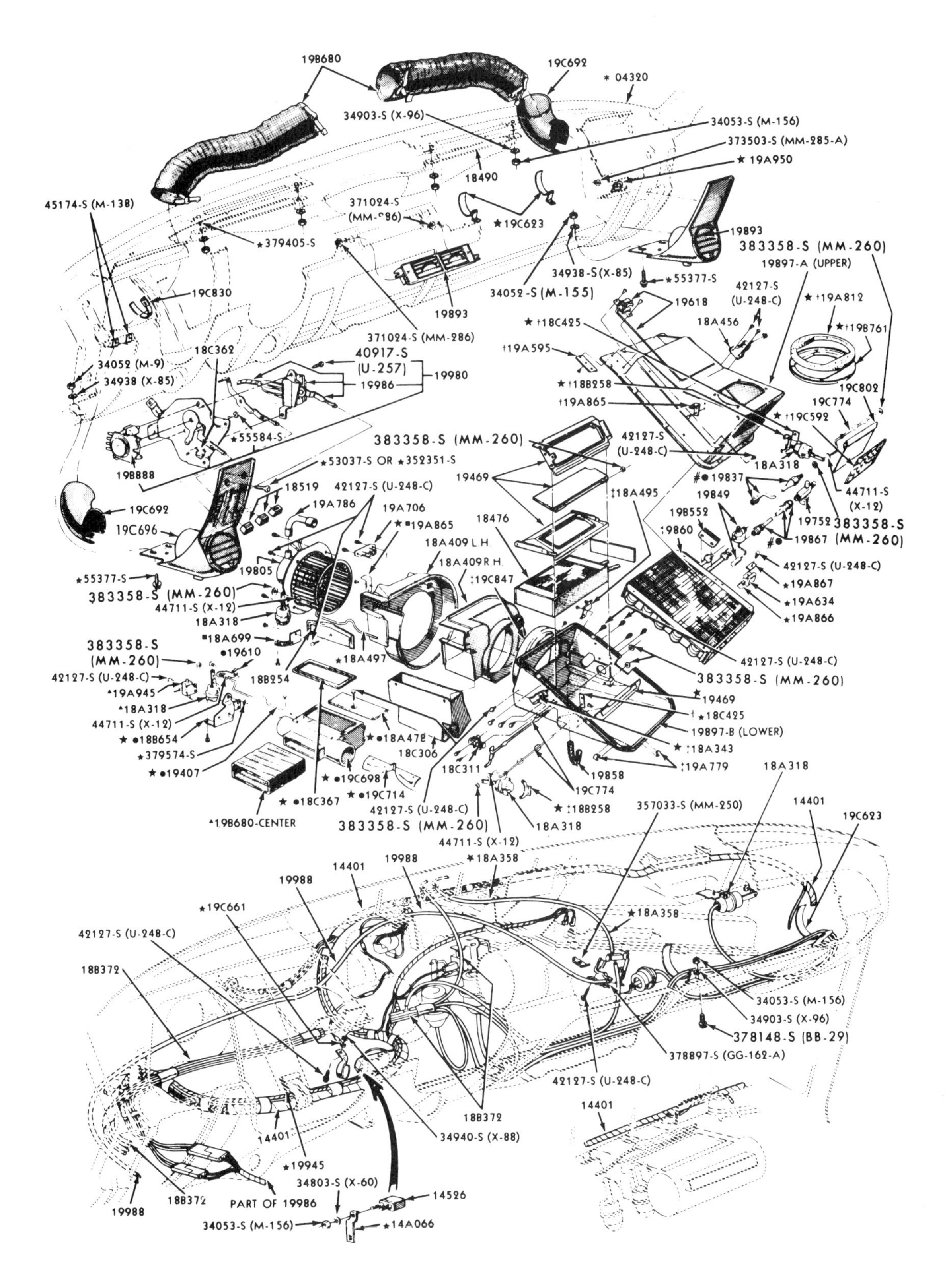

Factory-installed integral air conditioner for 1967-1968 Mustangs.

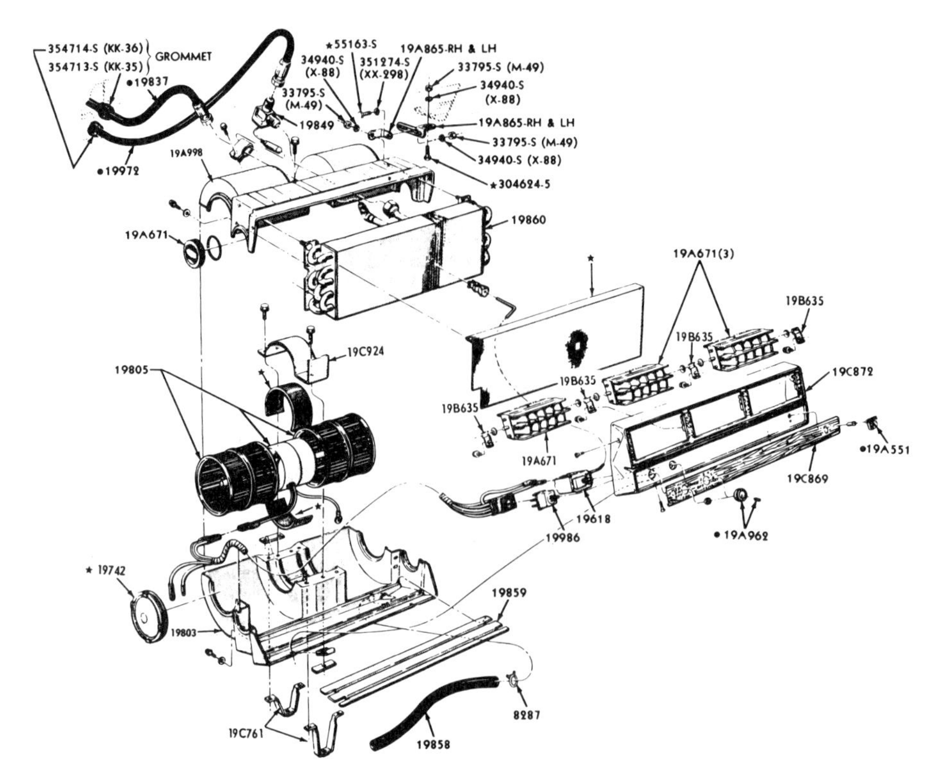

Optional economy hang-on air conditioner for 1967-1968 Mustangs.

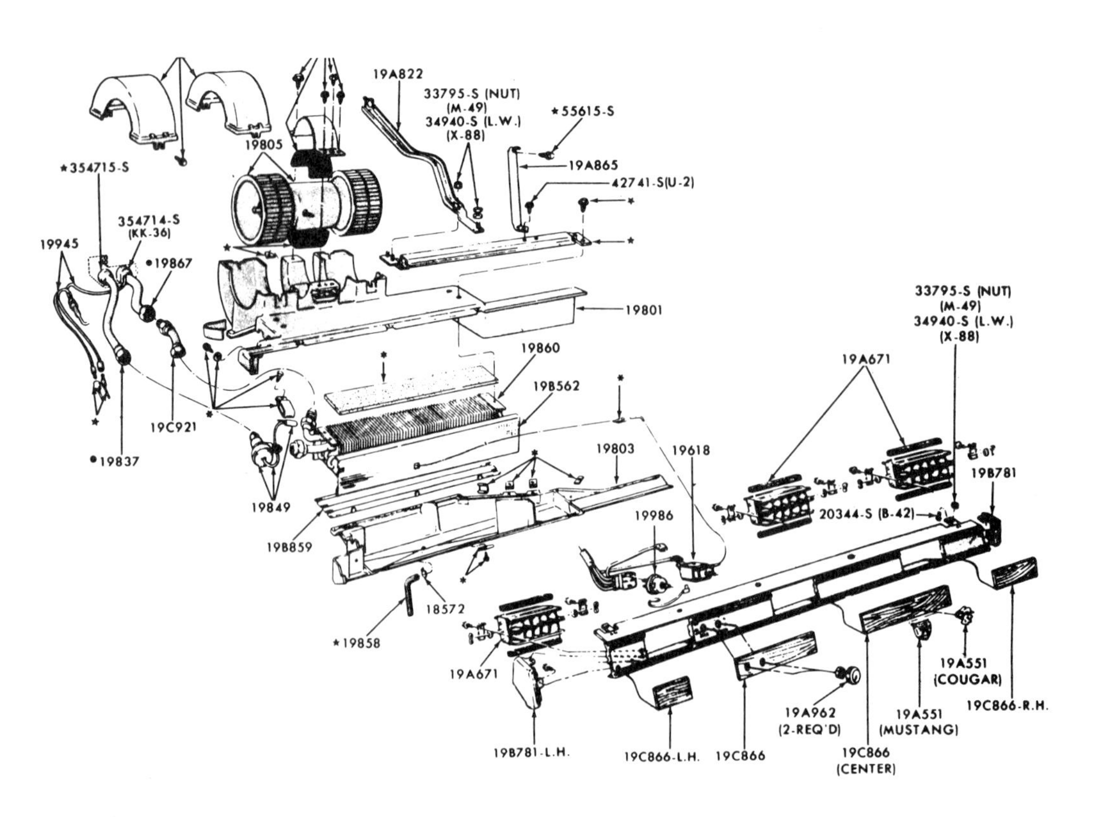

Optional Deluxe hang-on air conditioner for 1967-1968 Mustangs.

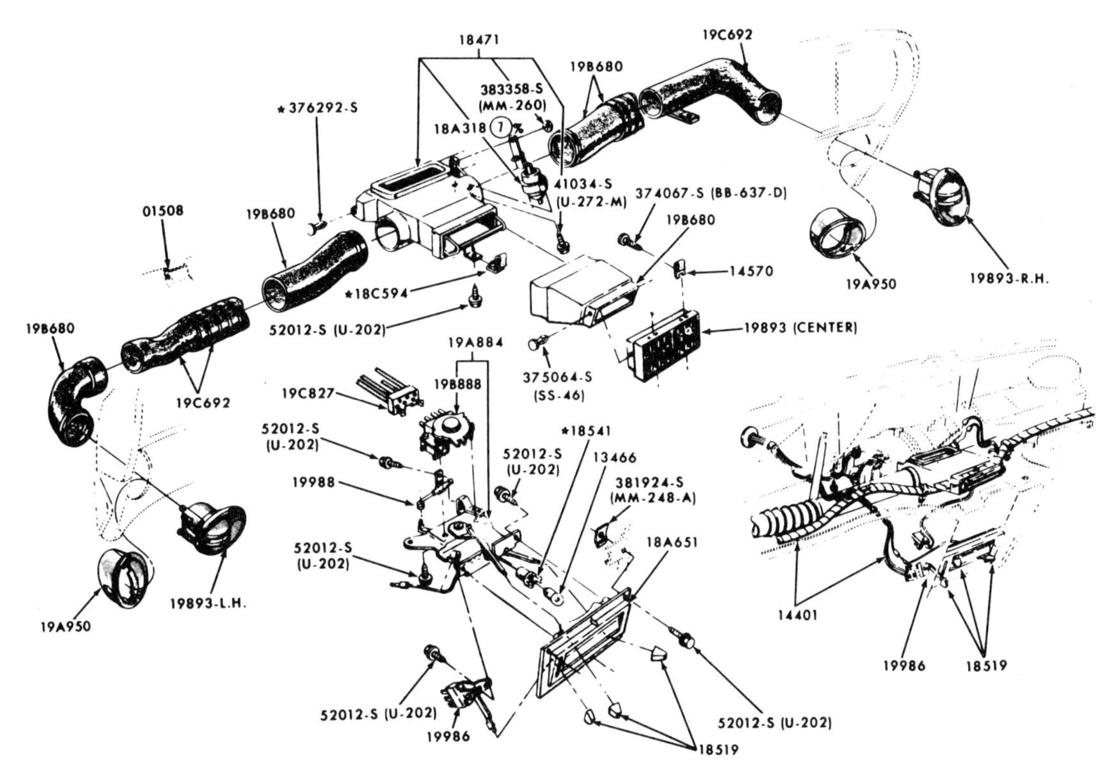

Factory-installed integral air conditioning controls, ducting, wiring, and related parts for 1969-1970 Mustangs.

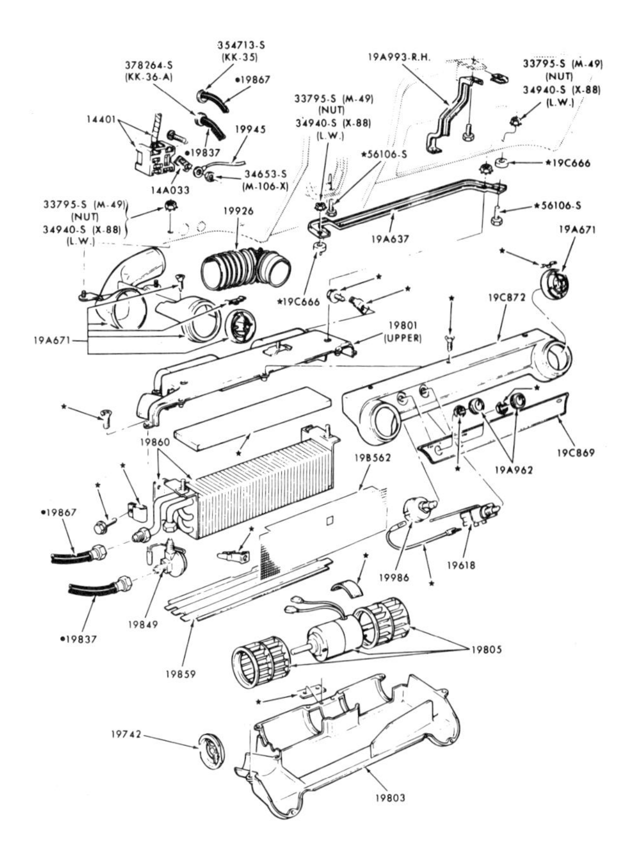

Optional Slimline hang-on air conditioner for 1969-1970 Mustangs.

Index

Air cleaners, 178-186
Air conditioning, 123, 168-176, 348-351
Alternators, 334-337
Axle codes, 8

Batteries, 334
Body
  buck tags, 11
  components, 28-35
  style codes, 7
Brake system components, 263-269
Bumpers
  front, 35-44
  rear, 70-71

Carburetors and components, 176-186, 198-206
  identification tags, 196
Clutch and pressure plates, 222
Console panel and components, 122-126
Convertible tops and components, 61-63
Cooling systems, 161-167
Cruise control components, 340-342

Data plates
  engine codes, 132-133
  transmission codes, 8-9
Decals, 12-23, 74-87
Distributors, 147-150
District Sales Office codes, 7-8
Doorhandles and components, 58
Driveshafts, 237-241

Emblems and insignia, 27, 74-87
Emissions equipment, 186-195
Engines
  170, 133-146
  200, 133-146
  250, 133-146
  260, 133-146
  280 HiPo, 130, 133-146
  289 HiPo, 129-130, 133-146
  302 Boss, 133-136, 142, 145
  351 Cleveland, 132, 133-146
  351 Windsor, 132, 133-146
  390, 133-146
  427, 133-146
  428 Cobra Jet, 131, 133-146
  428 Drag Pack, 132
  428 Super Cobra Jet, 131-132, 133-146
  428, 131, 133-146
  429 Boss, 133-136, 143
  components, 151-159
  identification tags, 133-135
  mounts, 160-161
Exhaust systems, 296-306

Fenders and components, 45-49
Fuel pumps, 206-211
Fuel tanks, 209-211

Gas caps, 26, 68-70
Generators, 332-334
Grilles, 27, 35-44

Heaters and components, 89-90
Hoods and components, 49-51

Instrument panel and components, 88, 91-94
Interior trim and components, 95-102
  scheme codes, 103-109

Lights
  head, 312-317
  interior, 323-325
  side marker, 319-320
  tail, 318-322
Luggage racks, 342-343

Mirrors
  rearview, 52, 71-74
  side, 71-74
Molding, 74-87

Paint codes, 25-26

Radios and components, 344-347
Rear axles, 237-247
  tag codes, 237

Seatbelts, 116-119
Seats and components
  front, 110-115
  rear, 120-121
Spark plugs, 147
Speedometer, cables and gears, 233-237
Springs
  front, 252-260
  rear, 262-263
Starter motors, 338-339
Steering components, 270-287
Steering wheels and components, 270-278
Suspension components
  front, 248-261
  rear, 261-263

Tags, 12
Transmissions and linkages, 212-233
  tag codes, 218
Trunk and components, 126-128
Trunk lids and components, 66-68

Underbody components, 28-35

Vehicle Identification Numbers, 6
  plates, 11
Vents, quarter, 59-60

Warranty plates, 9-10
  date codes, 7
Wheel covers and hubcaps, 288-296
Windows
  defrosters, 348
  front, 28-35
  louvers, 343-344
  quarter, 54-55
  rear, 52-53, 64-66
  side, 56-57
  wind-up mechanisms, 56-57
Wiring harnesses
  engine compartment, 312-317, 329-339
  interior, 307-312
  trunk, 318-322

OCT '05 SOUTH HILL